D1170829

THE MIND'S EYE

COGNITIVE AND APPLIED ASPECTS OF EYE MOVEMENT RESEARCH

Historical note on the origins of precise eye movement monitoring

The picture on the front cover of our book illustrates a historic breakthrough in the methodology of eye movement research. It depicts the first camera-based device for eye movement registration as described by Dodge and Cline (1901), often referred to as the "falling plate camera" or "photochronograph".

The right part of the figure shows a bellows camera mounted on a perimeter together with a head rest. The perimeter carried knitting needles with bits of white paper serving as adjustable fixation points and a holder for printed matter. Using the camera, the corneal reflection of a white piece of cardboard illuminated by sunlight was photographed. A continuous time record was achieved via oscillations of a pendulum inside the plate holder resulting in a periodical darkening at the edge of the moving photographic plate. The pendulum was set in motion by interrupting a circuit to an electric magnet fixating it as long as the plate did not move.

The left part of the figure depicts a magnification of the plate holder attached to the back of the camera. This wooden box, seen in cross section, contains in its upper part the photographic plate and in its lower part a conventional bicycle pump. This simple device, together with a second pump not visible in this figure, provided for a smooth and continuous descent of the plate during the course of registration.

In their original work, Dodge and Cline (1901) used the falling plate camera to provide the first precise analyses of the duration and velocity of saccades. Dodge (1903) published a report on recordings that were aimed at defining "Five Types of Eye-Movement in the Horizontal Meridian Plane of the Field of Regard". An improved version of the apparatus served in the classic study by Dearborn (1906) on the metrics of eye movements in reading. Subsequently the photochronograph was professionally manufactured by Spindler & Hoyer, Göttingen and sold for the price of 375 Mark.

It is interesting to note that many fundamental observations on basic properties of eye movements were made in the context of research on reading. Thus, the tradition of using the eyes in the study of cognitive processes is as old as research on the oculomotor system as such. It appears that being interested in both sides of the coin was the norm rather than the exception for the early 20th century researcher. Seen in this perspective, the idea of integrating the two main streams of current work, research *with* and *about* eye movements as it is advocated in the present volume essentially constitutes a return to the origins of our field.

The editors are grateful to Dieter Heller (Technical University of Aachen) and Nicholas Wade (University of Dundee) for suggesting this motif for the cover of the present volume. For a detailed account of the role played by Dodge and his contemporaries in the history of oculomotor research we refer to the recent work by Wade, Tatler and Heller (under review).

References

Dodge, R., & Cline, T. S. (1901). The angle velocity of eye-movements. *Psychological Review, 8,* 145–157.

Dearborn, W. F. (1906). *The Psychology of Reading.* Columbia University Contributions to Philosophy, Psychology and Education. Vol. XIV, No. 1. New York: Columbia University Press.

Dodge, R. (1903). Five types of eye-movement in the horizontal meridian plane of the field of regard. *American Journal of Physiology, VIII*, 307–329.

Wade, N., Tatler, B. W., & Heller, D. Dodge-ing the issue: Dodge, Javal, and the discovery of "saccades" in eye movement research. *Perception*, under review.

THE MIND'S EYE

COGNITIVE AND APPLIED ASPECTS OF EYE MOVEMENT RESEARCH

EDITED BY

J. HYÖNÄ

Department of Psychology, University of Turku, Finland

R. RADACH

Institute of Psychology, Technical University of Aachen, Germany

H. DEUBEL

Department of Psychology, Ludwig-Maximilians-University, Munich, Germany

2003

Elsevier

Amsterdam – Boston – London – New York – Oxford – Paris
San Diego – San Francisco – Singapore – Sydney – Tokyo

ELSEVIER SCIENCE BV
Sara Burgerhartstraat 25
1055 KV Amsterdam, The Netherlands

First edition 2003

Library of Congress Cataloging in Publication Data
A catalog record from the Library of Congress has been applied for.

British Library Cataloguing in Publication Data
A catalogue record from the British Library has been applied for.

ISBN: 0-444-51020-6

♾ The paper used in this publication meets the requirements of ANSI/NISO Z39.48-1992 (Permanence of Paper).
Printed in The Netherlands.

Contents

Contributors

Christian Balkenius	Department of Cognitive Science, Lund University, Lund, Sweden
Maria Barthelson	Department of Cognitive Science, Lund University, Lund, Sweden
Rob Brooks	Department of Psychology, University of Dundee, Dundee, UK
Reinier Cozijn	Department of Communication and Cognition, University of Tilburg, The Netherlands
Martha E. Crosby	Department of ICS, University of Hawaii, Honolulu, HI, USA
Géry d'Ydewalle	Department of Psychology, University of Leuven, Leuven, Belgium
Wim De Bruycker	Department of Psychology, University of Leuven, Leuven, Belgium
L. Dempere-Marco	Imperial College of Science, Technology and Medicine, London, UK
Claudio de'Sperati	Fac. Psicologia, Universitá S. Raffaele, Milano, Italy
Heiner Deubel	Department of Psychology, Ludwig-Maximilians-University, Munich, Germany
John Dingliana	Department of Computer Science, Trinity College Dublin, Dublin, Ireland
Christian Dobel	Max Plank Institute for Psycholinguistics, Nijmegen, The Netherlands
Melanie Doyle	Department of Psychology, Royal Holloway, University of London, Egham, Surrey, UK
Ralf Engbert	Department of Psychology, University of Potsdam, Potsdam, Germany
John M. Findlay	Department of Psychology, University of Durham, Durham, UK

Lesley-Anne Flynn	Department of Psychology, University of Dundee, Dundee, UK
Alastair G. Gale	Institute of Behavioural Sciences, University of Derby, Derby, UK
Elizabeth Gilman	Department of Psychology, University of Sheffield, Sheffield, UK
Richard Godijn	Department of Cognitive Psychology, Vrije Universiteit, Amsterdam, The Netherlands
Joseph H. Goldberg	Advanced User Interfaces, Oracle Corporation, Redwood Shores, CA, USA
Charlotte Golding	Division of Neuroscience, Imperial College, London, UK
Jonathan Grainger	Centre de Recherche en Psychologie Cognitive, Université de Provence, Aix-en-Provence, France
Marianne Gullberg	Max Planck Institute for Psycholinguistics, Nijmegen, The Netherlands
Dieter Heller	Institute of Psychology, Technical University of Aachen, Aachen, Germany
Timothy L. Hodgson	School of Psychology, Washington Singer Laboratories, Exeter, UK
Kenneth Holmqvist	Department of Cognitive Science, Lund University, Lund, Sweden
Jana Holsanova	Department of Cognitive Science, Lund University, Lund, Sweden
K. Horii	Department of Industrial Engineering, Kansai University, Osaka, Japan
Sarah Howlett	Department of Computer Science, Trinity College Dublin, Dublin, Ireland
X. P. Hu	Imperial College of Science, Technology and Medicine, London, UK
Jukka Hyönä	Department of Psychology, University of Turku, Turku, Finland
Marko Illi	Department of Computer and Information Sciences, University of Tampere, Tampere, Finland
Albrecht W. Inhoff	Department of Psychology, State University of New York at Binghamton, Binghamton, NY, USA
Poika Isokoski	Department of Computer and Information Sciences, University of Tampere, Tampere, Finland

Robert J. K. Jacob	Department of Computer Science, Tufts University, versity of South Carolina, Columbia, SC, USA
Gretchen Kambe	Department of Psychology, University of Nevada, Las Vegas, NV, USA
Keith S. Karn	Xerox Corporation, Industrial Design/Human Interface Department, Rochester, NY, USA
Matthew Kean	Research Centre for Cognitive Neuroscience, Department of Psychology, University of Auckland, Auckland, New Zealand
Alan Kennedy	Department of Psychology, University of Dundee, Dundee, UK
Y. Kitamura	Institute of Foreign Language Education and Research, Kansai University, Takatsuki, Osaka, Japan
Reinhold Kliegl	Department of Psychology, University of Potsdam, Potsdam, Germany
K. Kotani	Department of Industrial Engineering, Kansai University, Osaka, Japan
Anthony Lambert	Research Centre for Cognitive Neuroscience, Department of Psychology, University of Auckland, Auckland, New Zealand
Stefanie Lemmer	Institute of Psychology, Technical University of Aachen, Aachen, Germany
Simon P. Liversedge	Department of Psychology, University of Durham, Durham, UK and Department of Psychology, University of Massachusetts, Amherst, MA, USA
Robert F. Lorch Jr	Department of Psychology, University of Kentucky, Lexington, KY, USA
Daniel Lundqvist	Department of Clinical Neuroscience, Karolinska Institutet, Stockholm, Sweden
George W. McConkie	Beckham Institute, University of Illinois, Urbana, IL, USA
Antje S. Meyer	Behavioural Brain Sciences Centre, University of Birmingham, Edgbaston, Birmingham, UK
Brett Miller	Department of Psychology, University of Massachusetts, Amherst, MA, USA
Jan Morén	Department of Cognitive Science, Lund University, Lund, Sweden
Robin K. Morris	Department of Psychology, University of South Carolina, Columbia, SC, USA

Carol O'Sullivan	Department of Computer Science, Trinity College Dublin, Dublin, Ireland
Alexander Pollatsek	Department of Psychology, University of Massachusetts, Amherst, MA, USA
Marc Pomplun	Department of Psychology, University of Toronto, Toronto, Canada
Carrie Prophet	Department of Psychology, University of Dundee, Dundee, UK
Karina Radach	Power and Radach Advertising & Web Design Inc., Aachen, Germany
Ralph Radach	Institute of Psychology, Technical University of Aachen, Aachen, Germany
Keith Rayner	Department of Psychology, University of Massachusetts, Amherst, MA, USA
Erik Reichle	Department of Psychology, University of Pittsburgh, Pittsburgh, PA, USA
Ronan G. Reilly	Department of Computer Science, University of Dublin, Dublin, Ireland
Eyal M. Reingold	Department of Psychology, University of Toronto, Toronto, Canada
Mike Rinck	Department of Psychology, Dresden University of Technology, Dresden, Germany
Jiye Shen	Department of Psychology, University of Toronto, Toronto, Canada
Catherine Sophian	Department of ICS, University of Hawaii, Honolulu, HI, USA
Dave M. Stampe	Department of Psychology, University of Toronto, Toronto, Canada
Iréne Stenfors	Area 17 AB, Lund, Sweden
Sascha Stowasser	Institute for Human and Industrial Engineering, University of Karlsruhe, Karlsruhe, Germany
Veikko Surakka	Department of Computer and Information Sciences, University of Tampere, Tampere, Finland
O. Takeuchi	Institute of Foreign Language Education and Research, Kansai University, Takatsuki, Osaka, Japan
J. Tchalenko	Camberwell College of Arts, London, UK

Jan Theeuwes	Department of Cognitive Psychology, Vrije Universiteit, Amsterdam, The Netherlands
Jie-Li Tsai	Laboratory for Cognitive Neuropsychology, National Yang-Ming University, Pei-Tou, Taipei, Taiwan
Geoffrey Underwood	Department of Psychology, University of Nottingham, Nottingham, UK
Wietske Vonk	Max Planck Institute for Psycholinguistics, Nijmegen, The Netherlands
Christian Vorstius	Institute of Psychology, Technical University of Aachen, Aachen, Germany
Robin Walker	Department of Psychology, Royal Holloway, University of London, Egham, Surrey, UK
Ulrich Weger	Department of Psychology, State University of New York at Binghamton, Binghamton, NY, USA
Sarah J. White	Department of Psychology, University of Durham, Durham, UK
Anna M. Wichansky	Advanced User Interfaces, Oracle Corporation, Redwood Shores, CA, USA
Diane E. Williams	Department of Psychology, University of Toronto, Toronto, Canada
Rihana S. Williams	Department of Psychology, University of South Carolina, Columbia, SC, USA
G. Z. Yang	Imperial College of Science, Technology and Medicine, London, UK
Shun-nan Yang	Beckham Institute, University of Illinois, Urbana, IL, USA
Gert Zülch	Institute for Human and Industrial Engineering, University of Karlsruhe, Karlsruhe, Germany

Preface

The present volume is intended to provide a comprehensive state-of-the-art overview of current research on cognitive and applied aspects of eye movement research. It includes contributions that originated from papers presented at the 11th European Conference on Eye Movements (August 22–25, 2001, Turku, Finland), supplemented by invited chapters and commentaries. The ECEM series of conferences was commenced in 1981 by Professor Rudolf Groner in Bern. It brings together researchers from various disciplines with an interest to study behavioral, cognitive, neurobiological and clinical aspects of eye movements. This book presents a selection of contributions addressing cognitive and applied aspects of oculomotor research. All chapters were reviewed by two referees, in most cases fellow contributors, but in addition a number of outside referees were also consulted.

The book is divided into five sections: I Visual information processing and saccadic eye movements; II Eye movements in reading and language processing; III Computational models of eye movement control in reading; IV Eye movements in human–computer interaction; and V Eye movements in media applications and communication. Each section ends with a commentary chapter written by a distinguished scholar, aimed at discussing and integrating the empirical contributions and providing an expert view on the most significant present and future developments in the respective area. A sister book is also published (Hyönä, Munoz, Heide & Radach (eds), *The Brain's Eye: Neurobiological and Clinical Aspects of Oculomotor Research*, Amsterdam: Elsevier Science), where the emphasis is on lower-level and clinical aspects of eye movements.

Section I explores various aspects of the relationship between visual information processing and saccadic eye movements. Not surprisingly, a major theme of this section concerns selective visual attention, which is generally seen as a mechanism that bridges vision and eye movements. The section starts with a review by Godijn and Theeuwes on what is currently known of the relationship between endogenous and exogenous attention and saccades. The authors then present some of their own experiments that study this relationship by means of a new oculomotor capture paradigm. To account for their empirical findings, Godijn and Theeuwes propose a competitive integration model for the interaction between attention and saccades. The chapter by Kean and Lambert also explores the relationship between covert and overt attention shifts, however, with specific emphasis on the role of peripheral information in attentional orientation. The authors review several recent experiments that use eye movements as an indicator of covert attention, and provide demonstrations that the information conveyed by peripheral stimuli is capable of modulating our attentional orienting behavior. Hodgson and Golding's work, in normal subjects and in neurological patients, studies more complex situations in which saccades are made on the basis of changing task rules. They demonstrate, in a number of different tasks, how humans are able to implement the very flexible control of behavior required to account for changing demands from the environment. Shen, Reingold, Pomplun and Williams explore the interaction between a central discrimination task and the selection of the next target based on peripheral information. An important result is that they find that saccade selectivity is rather unaffected by the difficulty of the discrimination task, suggesting that the central discrimination and the peripheral analysis do not share the same pool of attentional resources. The chapter by Walker and Doyle introduces an important, though frequently neglected, aspect of eye movement control, namely that saccade target selection must normally occur in a multisensory environment. The authors start with a review of a number of studies on saccade reorienting to a combination of visual and auditory targets. They also present a recent experimental study of the effect of auditory and tactile distractors on saccade trajectories, providing further evidence that interactions between different sensory stimuli can indeed modify saccadic responses. The chapter by Engbert and Kliegl looks at the pattern of microsaccades that occur under the condition of different changes of the display. To analyze oculomotor behavior, the authors introduce a new algorithm for the detection of very small saccades, based on an analysis in 2D velocity space. In contrast to the current view, they find that some microsaccades can be monocular, with different statistical properties of the movements of each eye. In the last empirical chapter of this section, de'Sperati demonstrates, with a number of striking experiments, how eye movements can be successfully utilized to recover the kinematics of purely internal mental processes related to mental imagery. The section ends with a commentary by Findlay presenting a critical view on how the theme of attention is used in the various section contributions. As a viewpoint alternate to rashly accepting an outstanding role of attention in oculomotor control, he proposes that oculomotor workers should adopt a perspective in which the movements of the eyes are regarded as a primary feature of vision.

Section II includes empirical contributions dealing with reading and language processing. Over the years, eye-tracking has become increasingly popular to study written language processing, particularly word identification and syntactic parsing. The chapters by Tsai and McConkie, Rayner *et al.* and Liversedge build directly on this well-established research tradition. Tsai and McConkie study reading behavior in Chinese to examine whether single characters are more influential than words in guiding the eyes of Chinese readers. Rayner, White, Kambe, Miller and Liversedge provide a review of recent work examining parafoveal information acquisition and the extent to which parafoveally visible words can influence foveal processing. Liversedge's chapter deals with an important aspect of sentence processing – assigning thematic roles to clause constituents. Kennedy, Brooks, Flynn and Prophet apply the eye-tracking technique to the investigation of spatial coding in reading. The critical question is to what extent readers represent the exact spatial location of words to guide the eyes when rechecking or when additional processing becomes necessary. Hyönä,

Lorch and Rinck propose the use of eye-tracking to study comprehension processes when reading long texts – a research area where oculomotor analyses may significantly contribute to future progress. The chapter by Vonk and Cozijn offers a methodological contribution, suggesting that the time spent in saccadic movements should be added to the processing time measures; moreover, they argue that regressive and progressive fixation cycles need to be analysed separately. Section II also comprises studies in areas of research to which the eye-tracking methodology has not yet been extensively applied. Gilman and Underwood examine in their contribution how musical notation is processed by expert and less expert musicians. They are particularly interested in defining the extent of the perceptual span in music reading. Morris and Williams demonstrate how eye-tracking can be successfully applied to study the process by which readers infer meaning for new vocabulary items. Meyer and Dobel in turn discuss a new eye movement paradigm to study the time course of spoken word production. In their commentary chapter, Inhoff and Weger consider the section a good example of the "methodological middle-ground" that they would like to advocate, emphasizing that performance is being studied in relatively natural task environments with rigorous experimental controls. They conclude by making suggestions on how to go about employing the plethora of eye movement measures already introduced in the field.

Over the last two decades oculomotor research on reading has reached a state of development that now makes it possible to develop realistic computational models of eye movement control in this complex task. These models attempt to account for the dual nature of written language processing, "moving eyes" and "reading words", as Grainger puts it in his commentary chapter for Section III. They also seek to incorporate principles of visuomotor control as found in basic research, one being the relation between visual selection and the generation of saccades. This is the focus of the chapter by Reingold and Stampe, who found a marked decrease in the frequency of saccades due a task irrelevant flicker that was more pronounced when the flicker occurred in the direction of the next saccade. On the basis of this saccade inhibition effect, they argue in favor of a close relation between the allocation of visual attention and saccade programming. Sequential shifts of visual attention play a central role in the E-Z Reader model presented in the chapter by Pollatsek, Reichle & Rayner. They provide a synopsis of relevant empirical findings and then discuss the major design principles and core ideas of their modeling approach. Among the new features implemented in the current version are a pre-attentive stage of word processing and a new mechanism for the generation of refixations. The chapter concludes with a study on the reading of compounds words, the first attempt to explicitly address the processing of morphologically complex words in a computational model. The general architecture of the SWIFT model described by Kliegl and Engbert is similar in some ways to attention-based sequential processing models but its principles of operation differ in several respects. Critically, lexical processing is seen as spatially distributed and the timing of saccades is co-determined by autonomous saccade generation and processing-based inhibition of foveal targets. In their chapter, Kliegl and Engbert demonstrate how the model can predict the outcome of typical manipulations of parafoveal information using gaze contingent display changes in sentence reading.

Yang and McConkie propose a competition-inhibition theory of eye movement control in reading that is grounded in neurobiological research on saccade generation and also includes a mechanism for the autonomous triggering of saccades. In experiments using different classes of saccade contingent display changes they found that only saccades that are initiated after relatively long latencies are delayed by such changes. Based on these results they argue against direct cognitive control of most saccades in reading and propose several mechanisms how linguistic processing may affect saccade generation in a more indirect way. The Glenmore model proposed in the chapter by Reilly and Radach introduces the notion of a letter-based spatial saliency representation as a means to integrate visual and linguistic processing in the generation of oculomotor behavior in reading. An interactive activation network of letter and word processing provides top down feedback that, together with visual information, co-determines the selection of parafoveal words as saccade targets. Grainger, in his commentary on Section III emphasizes the importance of relating models that account for the dynamics of reading to the tradition of computational models of single word processing. Taken together, the developments presented in this section cannot only be expected to contribute to progress in reading research, but they may also serve to further our understanding of the nature of vision-based cognitive information processing in general.

The ascent of new information technologies has a double effect on eye movement research: First, it paves the way for precise measurement and straightforward analysis of oculomotor behavior via affordable and robust equipment. Second, the use of new information technologies itself becomes a new subject of applied oculomotor research. The chapters in Section IV consider both the application of eye movements as a medium of control in human-computer interaction and as a tool in studying the usability of interfaces. Surakka, Illi and Isokoski provide an introduction into the use of voluntary eye movements for interaction with computers. This requires the real-time tracking of eye movements and the implementation of eye-based feedback signals into multimodal interfaces. Discussing results of two recent studies, they demonstrate the feasibility of combining eye movements with voluntary facial muscle activity as a way to execute mouse pointing operations. Goldberg and Wichansky provide in their chapter a guide to eye-tracking in usability evaluation that will be especially useful for new researchers entering the field. This review provides information for both software usability specialists, who consider the potential merits of eye-tracking, and for eye movement researchers, who are interested in usability evaluation as a field of application. One important problem in the development of computer interfaces is how to present spatial configurations in computer interfaces. This issue is addressed in experiments reported in the chapter by Crosby and Sophian, who propose several ways in which presentation of data can facilitate the integration of quantitative information. The application of eye-tracking in the evaluation of industrial manufacturing software settings is discussed in the contribution by Zülch and Stowasser. Using the example of scheduling problems in a complex shop floor system, they show the usefulness of oculomotor analyses in industrial human–computer interaction and draw conclusions for the design of control systems and the visualization of object-oriented data. O'Sullivan, Dingliana and Howlett report a series of studies that used eye movements in the area of interactive computer graphics and virtual reality. They show how techniques like gaze directed rendering and collision handling can be used to enhance the perceived realism of images and graphical simulations in dynamic virtual environments. The commentary to this section by Jacob and Karn provides a detailed review of the field, from the history of eye-tracking in human–computer interaction to a discussion of the most prominent issues in theory and methodology. Their look into the future suggests that the application of eye movements will be an important ingredient of innovative developments in several branches of modern information technologies.

A second important area of applied oculomotor research is in the field of visual communication and media. Here it is of particular value to develop eye-tracking applications that can be used in real-life environments without considerable difficulty. Section V comprises studies in which eye-tracking is applied to study social interaction and the intake of information from modern media. Radach, Lemmer, Vorstius, Heller and Radach demonstrate that advertisements in which the relationship between the text and the picture is not transparent do a better job in attracting visual processing and inducing retention in memory than those in which the content of the two types of information explicitly converges. Stenfors, Morèn and Balkenius recorded viewers' eye movements while they scanned internet web pages; they demonstrate (among other things) that web advertisements can easily be ignored. Kitamura, Horii, Takeuchi, Kotani and d'Ydewalle set out to determine the optimal speed for reading scrolling text (i.e. one character is added to the right-hand corner of the text window while the leftmost character is deleted). Holmqvist, Holsanova, Barthelson and Lundqvist compare newspaper reading to netpaper reading and conclude that netpaper reading involves more scanning in comparison to ordinary newspaper reading. De Bruycker and d'Ydewalle examine the factors that determine the extent to which adults and children visually process subtitles while watching subtitled movies. Gullberg reports a study on gaze patterns during social interaction with a particular focus on the factors that determine when the speaker's gestures are gazed upon by the listener. Finally, Tchalenko, Marco, Dempere, Hu and Yang provide a summary of studies examining the eye movement patterns of artists while they draw portraits. In his commentary, Gale welcomes the fact that applied eye movement research in the context of visual media has gained increased popularity; at the same time, however, he sees a clear need to extend good scientific rigor to research in this area.

All in all, the present volume demonstrates the fruitfulness of the eye tracking methodology in tackling both theoretically motivated and applied issues. We expect that oculomotor research will become increasingly influential, as it provides a privileged means for addressing fundamental questions about the workings of the mind as well as a tool to determine human performance in many applied settings. The novel and diverse approaches taken in the empirical contributions presented in the present volume are likely to trigger new lines of research and to provoke lively theoretical debates within and beyond the eye movement research community. We also hope that the commentary and review chapters will contribute to the solid theoretical grounding and scientific coherence that, as we believe, is particularly valuable for a dynamically developing field such as eye movement research.

Finally we would like to express our gratitude to all the contributors and reviewers, and to our partners at Elsevier Science, Oxford, whose unrelenting efforts made this book possible.

Jukka Hyönä, Ralph Radach and Heiner Deubel

Section 1

Visual Information Processing and Saccadic Eye Movements

Section Editor: Heiner Deubel

Chapter 1

The Relationship Between Exogenous and Endogenous Saccades and Attention

Richard Godijn and Jan Theeuwes

Visual scenes typically contain many objects that compete for the control of attention and saccades, on the basis of their intrinsic salience (exogenous control) or their relevance for the goals of the observer (endogenous control). The present chapter reviews the evidence regarding the relationship between endogenous and exogenous attention and saccades. Furthermore, a competitive integration model is presented, which provides a framework for understanding exogenous and endogenous control of saccades as well as the relationship between attention and saccades.

Introduction

During our everyday lives we are continuously confronted with a visual environment containing a vast amount of information. In order to interact adaptively with the environment we need to select the information that is relevant for our goals and to ignore what is irrelevant. Selection may be achieved by overt orienting (saccades), which allows the high acuity fovea to be directed to the focus of interest, or by covert orienting (attention), which facilitates the processing of selected objects without shifting the gaze direction. A fundamental research question concerns the mechanisms that control what parts of the visual scene are selected. On the one hand, selection may be controlled by stimulus properties, irrespective of the goals of the observer. For example, a salient new object suddenly appearing in the visual field will capture our attention (e.g. Yantis & Jonides, 1984; Jonides & Yantis, 1988; Theeuwes, 1994, 1995) and our eyes (e.g. Theeuwes *et al.*, 1998, 1999) even if it is irrelevant for the task at hand. This control mode is known as exogenous control (or stimulus-driven, involuntary, bottom-up). On the other hand, selection may be controlled by our goals and expectations. For example, when we are searching for a specific target object we will

The Mind's Eye: Cognitive and Applied Aspects of Eye Movement Research

ISBN: 0–444–51020–6

tend to select objects that share one or more feature with the target (e.g. Treisman & Gelade, 1980; Wolfe *et al.*, 1989; Findlay, 1997). This control mode is known as endogenous control (or goal-directed, voluntary, top-down).

Over the past 20 years a great deal of research has been conducted to determine the relationship between (endogenous and exogenous) shifts of attention and saccades. Since attention and saccades both have the goal of selecting the relevant portions of a visual scene, the idea that attention and saccades are to a certain extent related is intuitively appealing. In this view attention and saccades are related on the basis of their common function. That is, in order to further process and respond to an object, both orienting systems are typically directed to the same object, although in principle their focus may be dissociated. An alternative view which assumes a tighter relationship between attention and saccades is the efference view (Posner, 1980). According to this view attention is required at the saccade destination in order to program a saccade. Furthermore, attention shifts are accomplished by preparing an eye movement to that location (e.g. Klein, 1980; Rizzolatti *et al.*, 1987). This does not imply that whenever attention moves the eyes must follow. It is assumed that attention and saccade programming are causally related, but a separate go-signal is required to trigger the saccade that has been programmed (e.g. Deubel & Schneider, 1996; Rizzolatti *et al.*, 1987). Therefore, attention may move while the eyes remain fixated (e.g. Posner, 1980).

The central goal of this chapter is to provide an overview of the evidence regarding the nature of the relationship between attention and saccades for exogenous and endogenous orienting. Furthermore, a competitive integration model will be presented (Godijn & Theeuwes, in press-a; in press-b) in which exogenous and endogenous saccades are programmed within a common saccade map. This model provides a framework within which the relation between attention and saccades may be understood.

Although there have been quite a few studies examining the relationship between attention and saccades during reading (e.g. Morrison, 1984; Henderson & Ferreira, 1990; Henderson, 1992; Reichle *et al.*, 1998) we consider this a separate issue with quite specific demands on the attentional and oculomotor systems. Therefore these studies will not be discussed.

Attention and Endogenous Saccades

The most common method to examine the relationship between attention and saccades has been the dual-task paradigm (e.g. Deubel & Schneider, 1996; Hoffman & Subramaniam, 1995; Kowler, *et al.*, 1995; Shepherd *et al.*, 1986). Typically, the primary task is an eye movement task in which the participant is required to execute a saccade to a peripheral saccade goal, which is indicated by a symbolic cue (e.g. an arrow presented in the centre of the visual field). The secondary task is usually a manual response task toward a probe stimulus which is either presented at the saccade target location or somewhere else in the visual field. The rationale behind this procedure is that, if attention precedes the eyes, identification and detection of a probe stimulus should be facilitated when it is presented at the saccade destination compared to when it is presented at a different location. There is quite some evidence that when an endogenous saccade is executed towards a particular location, performance on the secondary manual response task is better when the probe stimulus is presented at the saccade target location than when it is presented somewhere else in the visual field. For example, Kowler *et al.* (1995) conducted a series of experiments in which participants were required to make a saccade while performing a letter identification task. Participants viewed displays containing eight pre-masks on a circular array surrounding a central fixation cross. A saccade was executed to one of the peripheral objects as indicated by a central arrow. Simultaneously with the onset of the arrow, the pre-masks were replaced by letters, which were masked 200 ms later. In the random report condition the letter "Q" appeared randomly at one of the display locations at the end of the trial and participants were instructed to report the letter that had appeared there. In the fixed report condition participants were always required to report the letter at a fixed location. The results showed that in the random report condition identification accuracy was best when the saccade goal and letter target were at the same location, which indicated that attention had preceded the eyes to the saccade goal. However, in the fixed report condition identification accuracy was always very good and did not depend on the location of the saccade goal. Kowler *et al.* suggested that in the fixed report condition participants might have focused on the letter identification task before programming the saccade. Indeed, saccade latencies were 50–75 ms higher in the fixed report condition than in the random report condition. In a subsequent experiment Kowler *et al.* examined the possible interference between the letter identification task and the eye movement task by requiring participants to perform the tasks either alone or together under varying priority instructions. The results revealed a trade-off between the two tasks, but in this experiment an advantage for letters presented at the saccade destination was found in the fixed report condition as well as in the random report condition.

Other dual-task studies that have used discrimination tasks to asses attention allocation have also found perceptual benefits at the saccade destination (e.g. Deubel & Schneider, 1996; Hoffman & Subramaniam, 1995). Furthermore, the results of these studies indicate that the coupling between selective attention and saccades is obligatory, since the perceptual benefits were still found when participants had preknowledge of the discrimination target location.

Even though the studies discussed above indicate a strong relationship between attention and endogenous saccades, it appears that this relationship only exists when an identification task or a discrimination task is used as a secondary task. If the secondary task is a detection task the results are less clear-cut. For example, Remington (1980) used a threshold detection task in which on half of the trials a brief luminance increment occurred at one of three positions (left, center, right). The luminance increment occurred at varying stimulus-onset-asynchronies (SOAs) relative to the presentation of a central arrow cue, which indicated the saccade goal. The results revealed no effect of saccade goal on detection accuracy when the luminance increment occurred during the interval between presentation of the saccade cue and saccade execution. Remington interpreted these results as evidence that attention did not precede endogenous saccades (see also Klein, 1980). There appears to be only one study that did find perpetual benefits at the saccade destination using a detection task (Shepherd *et al.*, 1986). However, as pointed out by Hoffman (1998) the results of Shepherd *et al.* are hard to interpret, since in their study the detection probe remained visible until after the manual response. Since the manual response often occurred after the eyes had moved, the perceptual benefits at the saccade destination could have been due to the facilitated processing of foveated probes.

Another task that requires the detection of stimuli is the temporal order judgment task (TOJ task; e.g. Stelmach & Herdman, 1991; Maylor, 1985). In a TOJ task two stimuli are presented in the periphery and participants are required to judge which stimulus was presented first. This task has been found to be sensitive to attentional allocation (e.g. Stelmach & Herdman, 1991; Maylor, 1985). That is, if one of the TOJ stimuli is presented at the attended location, this stimulus is judged to have appeared first on the majority of trials on which both TOJ stimuli are presented simultaneously. Stelmach *et al.* (1997) used this task to examine the relationship between attention and endogenous saccades. Participants executed a saccade in the direction indicated by an auditory cue ("left" or "right") or they maintained fixation when the auditory cue "center" was presented. After a varying SOA relative to the saccade cue the TOJ stimuli were presented. The rationale was that if attention preceded the eyes the stimulus at the saccade destination should have been judged to have appeared first on the majority of trials. The results revealed no effect of saccade destination, suggesting that attention did not precede the eyes.

When examining these results it is not immediately clear why the relationship between attention and endogenous saccades cannot be established when a probe detection task or a TOJ task is used. One possible explanation is provided by Stelmach *et al.* (1997). They argued that there may be two types of attention referred to as selective and preparatory attention. According to their view, identification tasks assess selective attention while detection tasks assess preparatory attention (Stelmach *et al.*, 1997; also see LaBerge, 1995). When a number of objects are present in a visual scene selective attention allows processing operations to be performed on the selected object while the other objects are ignored (or "filtered out"). Preparatory attention refers to an attentional shift to a specific location in anticipation of a specific object. That is, preparatory attention operates on the basis of expectations concerning when and where a task-relevant object will appear. Stelmach *et al.* argued that selective attention may operate in the order of tens of milliseconds while preparatory attention may operate in the order of hundreds of milliseconds. This would explain why tasks that have assessed selective attention (e.g. identification and discrimination tasks) have provided evidence for a strong link between attention and saccade programming while tasks that have assessed preparatory attention (e.g. detection and TOJ tasks) have not. Since endogenous saccade latencies range from 200–400 ms, preparatory attention may be too slow to precede a saccade. Furthermore, since attention and saccades have the common function of selecting the relevant information in a visual scene, it makes intuitive sense that selective attention precedes saccades, but preparatory attention does not. In fact, according to Schneider and Deubel (Schneider & Deubel, 2002; Schneider, 1995; Deubel & Scheider, 1996) attention is a common selection mechanism for saccades (or actions in general) as well as for perception.

Another speculation might be that a detection task does not need the same focused attention as identification tasks do. There is evidence that the detection of simple features (i.e., a probe onset) can proceed pre-attentively without the need for focused attention. Indeed, participants may "know" that something was presented but may not know where in the visual field this occurred (Sagi & Julesz, 1985). The detection of a basic feature may occur without the necessity of allocating spatial attention and the generation of a spatial code for the execution of the endogenous saccade and the (pre-attentive) detection of the probe may therefore proceed without any interference. In other words, detection of probes may not benefit from the allocation of spatial attention and therefore it may appear that there is no relation between attention and endogenous saccades when a detection task is used.

Attention and Exogenous Saccades

The evidence reviewed in the previous section suggests that attention does precede endogenous saccades. An equally fundamental question is whether attention also precedes exogenous saccades. Exogenous saccades are saccades that are triggered by some event in the environment and are executed independent of the goal of the observer. It is well known that abrupt onsets have the ability to capture attention independent of the goals of the observer (Yantis & Jonides, 1984; Theeuwes, 1991, 1994, 1995). Therefore, it is likely that abrupt onsets may also elicit exogenous saccades. Thus, similar to the paradigms employed to investigate endogenous saccades, in a number of dual-task studies abrupt onsets were presented in the periphery to elicit "exogenous" saccades and performance on a secondary manual response task, sensitive to attentional allocation, was examined. With this type of set-up, several studies have indeed found a consistent relationship between attention and exogenous saccades (e.g. Remington, 1980; Posner, 1980; Stelmach *et al.*, 1997; Schneider & Deubel, 2002). However, there is a fundamental methodological issue to consider. The question is whether in these studies the saccades made toward the onset were truly "exogenous". Participants received the instruction to make an eye movement toward the abrupt onset and this added an endogenous component to the exogenous properties of the abrupt onset. The only way to investigate genuine exogenous saccades is to provide an explicit endogenous saccade goal that is different from the location of the abrupt onset. If a saccade is made to the onset even though the observer had the intention to execute a saccade toward another goal only then one can speak of an exogenous saccade.

In the literature there appear to be only two paradigms that fulfil these requirements and are able to generate genuine exogenous saccades. First, in the anti-saccade task (e.g. Hallet, 1978; Mokler & Fischer, 1999) an abrupt onset is presented in the periphery and participants have the task of executing a saccade in the opposite direction. Saccades that are correctly executed toward the opposite location (so-called anti-saccades) are considered to be endogenous while saccades that are erroneously executed toward the onset (pro-saccades) are considered exogenous (e.g. Klein & Shore, 2000). Typically participants make these erroneous pro-saccades on about 10 to 30% of the trials, depending on the specific characteristics of the task (e.g. Fischer & Weber, 1992, 1996; Mokler & Fischer, 1999). For example, when the fixation point is removed shortly before the onset appears the proportion of erroneous pro-saccades is increased (e.g. Fischer & Weber, 1992; Reuter-Lorenz *et al.*, 1995). Latencies of exogenous pro-saccades are generally significantly shorter than latencies of endogenous anti-saccades (e.g. Hallet, 1978).

A second paradigm that is assumed to generate genuine exogenous saccades is the oculomotor capture paradigm (e.g. Theeuwes *et al.*, 1998, 1999). In this task, participants viewed displays containing six equi-spaced grey circles presented on an imaginary circle around a central fixation point. After one second all of the circles but one changed to red. Participants had the explicit instruction to make a saccade towards the only grey element in the display. On some trials, an irrelevant red circle, presented with an abrupt onset, was added to the display. In Theeuwes *et al.* (1999) a control condition was used in which an additional non-onset distractor was added to the display at the beginning of the trial. In Theeuwes *et al.* (1998) there was no additional non-onset distractor on trials without an onset. The results of both studies showed that when no onset was added to the display, observers generated endogenous saccades that went directly towards the uniquely coloured circle. However, on those trials on which an onset was added to the display, the eyes went in the direction of the onset on about 30 to 40% of these trials, stopped briefly, and then went on to the target. Figure 1.1 shows the results. The graphs on the left side depict the control condition without the onset; the graphs on the right side depict the condition in which an onset was presented. Note that in the condition with the onset, the eyes often went to the onset. This occurred even when the onset appeared on the opposite side of the target circle, although the proportion of saccades to the onset was greater at a 90° separation than at a 150° separation.

Since participants were required to execute a saccade to the uniquely coloured elements they had a clear top-down goal. However, despite this endogenous goal, on about 30 to 40% of the trials on which an onset was presented a saccade was executed toward the abrupt onset (see Figure 1.1). These saccades can be considered as genuinely exogenous, since they are completely irrelevant for the task at hand and were executed even though there was an explicit instruction to move the eyes to another location. Saccades made to the colour singleton are considered endogenous (e.g. Theeuwes *et al.*, 1998, 1999; Klein & Shore, 2000).

The oculomotor capture paradigm has a few advantages over the anti-saccade task (see also Klein & Shore, 2000). First, the location of the target is completely independent from that of the onset. In contrast, in the anti-saccade task the anti-saccade location is defined in terms of the location of the onset (i.e. the anti-saccade location is always opposite the onset). Second, in the oculomotor capture paradigm the competition between exogenous and endogenous saccades can be examined by comparing saccade behaviour on trials on which an onset is presented with trials on which no onset is presented. This is obviously impossible in the anti-saccade task. Third, in the anti-saccade task there are often not enough erroneous pro-saccades to allow a reliable statistical analysis. Several additional manipulations have been used to increase the number of pro-saccades. For example, by removing the fixation point and by

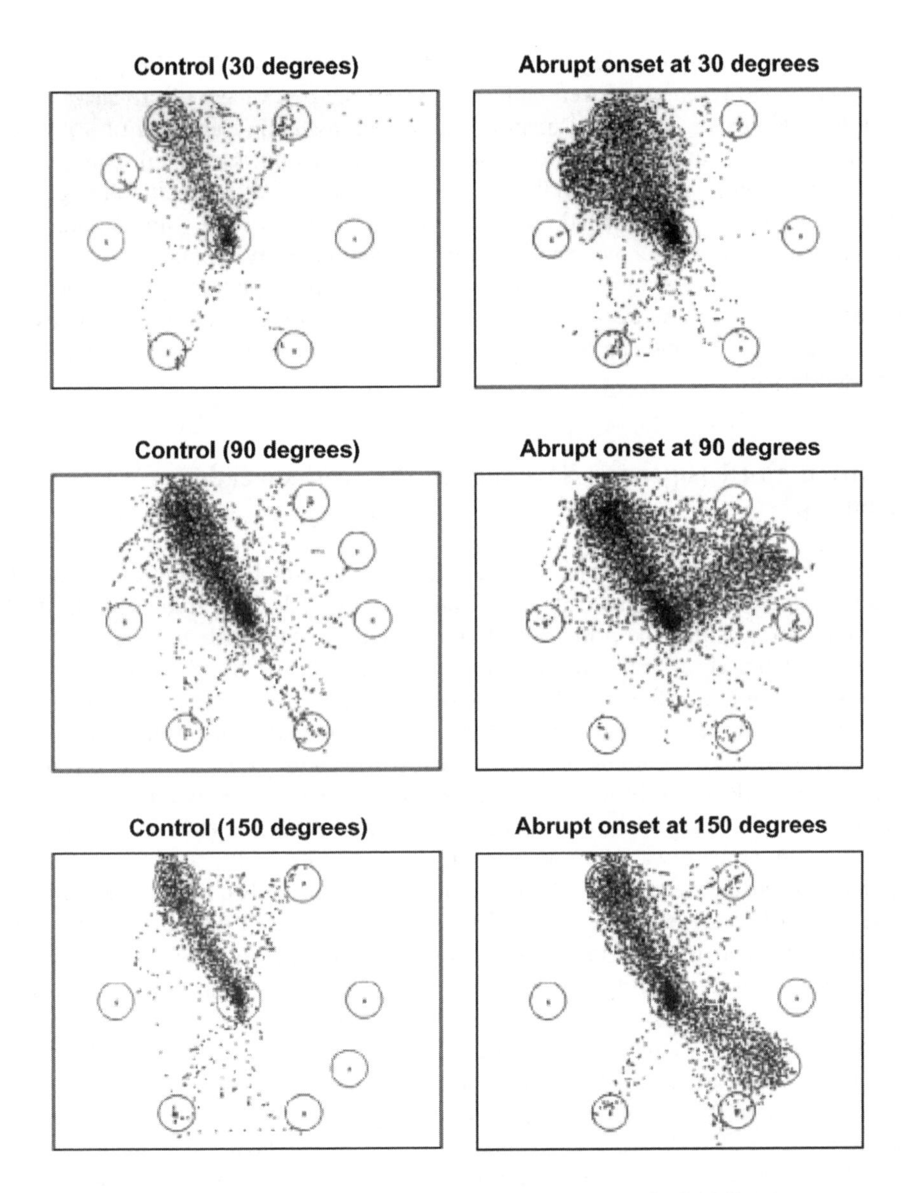

Figure 1.1: Oculomotor capture. Data from Theeuwes *et al.* (1999). Eye movement behaviour in the condition in which an abrupt onset was presented simultaneously with the target. The results are collapsed over all eight participants and normalised with respect to the position of target and onset. Sample points (every 4 ms) were only taken from the first saccade. Left panels: Eye movement behaviour in the control condition in which no abrupt onset was presented. Right panels: Eye movement behaviour in the condition in which an abrupt onset was presented; Either close to the target (TOP), somewhat away from the target (MIDDLE) or at the opposite side from the target (BOTTOM). From Theeuwes *et al.* (1999).

presenting a pre-cue at the saccade location before the imperative stimulus (e.g. Mokler & Fischer, 1999) researchers were able to increase the percentage of pro-saccades to about 20%. However, additional manipulations such as the presentation of a pre-cue may affect the allocation of attention when an exogenous pro-saccade or endogenous anti-saccade is programmed. Note that in the oculomotor capture paradigm exogenous saccades to the onset are elicited on about 30% of the trials even without additional manipulations such as pre-cues and offset of the central fixation point.

Now that we have introduced two paradigms in which genuine exogenous saccades are generated we can address the question whether there is a relationship between attention and exogenous saccades. To date there are not many studies that have addressed this issue.

Attention and Exogenous Saccades in the Anti-saccade Paradigm

Recently, two (unpublished) studies have examined the attentional allocation prior to exogenous and endogenous saccades in the anti-saccade task. Both studies used a dual-task paradigm in which the primary task was an anti-saccade task and the secondary task was a letter discrimination task. In Mokler *et al.* (2000) participants viewed displays containing two figure-eight pre-masks left and right of a central fixation point. After one second there was an onset of the anti-saccade stimulus around one of the pre-masks. In order to raise the probability that participants would generate erroneous pro-saccades to the onset of the anti-saccade stimulus the fixation point was removed 200 ms before the onset of the anti-saccade stimulus (e.g. Fischer & Weber, 1992) and a 100% valid pre-cue was presented at the location to which a saccade would be required (e.g. Fischer & Weber, 1996; Mokler & Fischer, 1999). Some line segments of the pre-masks were removed 20 or 120 ms after the onset of the anti-saccade stimulus, revealing two letters. One of the letters was the discrimination target ("E" or "reversed E"). Another 100 ms later the letters were masked. Participants were required to execute a saccade in the opposite direction of the anti-saccade stimulus and to determine the identity of the discrimination target. Furthermore, in addition to the typical secondary manual response task, participants were required to indicate whether they thought they had made an erroneous pro-saccade to the onset of the anti-saccade stimulus. The results showed that erroneous pro-saccades were made on 20% of the trials. On 57% of these trials participants were unaware that they had made a pro-saccade. When the eyes went directly to the saccade target location discrimination accuracy was highest when the discrimination target was presented at the saccade target location, suggesting that attention preceded endogenous anti-saccades. When erroneous pro-saccades were perceived discrimination accuracy was highest when the discrimination target was presented at the location of the anti-saccade stimulus. However, when erroneous pro-saccades were unperceived discrimination accuracy was highest when the discrimination target was presented at the saccade target location. According to Mokler *et al.* (2000) this suggested that unintended saccades could be executed without a presaccadic shift of attention. However, this lacked an adequate control condition. Since there was no non-attended control location it could not be determined whether attention moved to both locations or to just one (see Godijn & Theeuwes, in prep. showing evidence for the parallel allocation of attention to two non-contiguous locations when preparing two saccades).

Another recent study by Irwin *et al.* (submitted) was very similar to Mokler *et al.* In Irwin *et al.* a pre-cue was also used to induce erroneous pro-saccades. However, the pre-cue was completely non-predictive of the saccade target location, whereas in Mokler *et al.* (2000) the precue was 100% valid. Another difference compared to Mokler *et al.* (2000) was that the discrimination letters were briefly flashed in Irwin *et al.* but not masked as in Mokler *et al.* In contrast to the results of Mokler *et al.* (2000) discrimination accuracy was always higher at the location to which the eyes moved compared to the opposite location. Furthermore, discrimination accuracy did not differ between trials on which the erroneous pro-saccades were perceived and those on which they were not. These results suggest that attention preceded endogenous saccades to the saccade target location as well as exogenous saccades to the anti-saccade stimulus.

One factor that may have affected the results in both Mokler *et al.* and Irwin *et al.* is the presentation of the pre-cue. In Mokler *et al.* it always indicated the saccade target location, but in Irwin *et al.* it was non-predictive of the saccade target location. Previous research has shown that when peripheral cues are non-predictive of the target location, responses to a target presented at least a few hundred milliseconds later at the cued location are delayed (e.g. Posner & Cohen, 1984; Rafal *et al.*, 1989) and discrimination performance at that location is impaired (Lupiáñez *et al.*, 1997; Pratt *et al.*, 1997). This phenomenon, known as inhibition-of-return (Posner & Cohen, 1984) will be discussed further in the following section. Since the vast majority of erroneous pro-saccades in Irwin *et al.* were made when the anti-saccade stimulus appeared at the location opposite the pre-cue it is possible that discrimination accuracy was higher at the location of the anti-saccade stimulus than at the saccade target location, because of inhibition-of-return (IOR) to the location of the precue.

Taken together, on the basis of the studies by Mokler *et al.* (2000) and Irwin *et al.* (submitted) it is unclear whether attention precedes exogenous saccades in the anti-saccade paradigm.

Attention and Exogenous Saccades in the Oculomotor Capture Paradigm

Theeuwes *et al.* (1999, Experiment 2) addressed the question whether exogenous saccades to the onset in the oculomotor capture paradigm are preceded by a shift of attention. Participants had to make a saccade to the uniquely coloured grey circle in order to identify a small letter (C or reversed C) contained within it. In addition, a large irrelevant letter was presented at the location of the onset. This letter could be either congruent or incongruent with the letter inside the uniquely coloured grey circle. Due to the relatively large size of the letter inside the onset its identity could be determined by a shift of covert attention. In other words, it was not necessary to execute a saccade to the location of the onset to identify the letter inside it. On congruent trials, the large letter was identical to the letter appearing in the singleton target, so that the letter inside the onset activated the same response as the letter in the singleton target. On incongruent trials, the large letter was different from the letter appearing in the singleton target, so that the large letter activated the inappropriate response. Theeuwes *et al.* reasoned that if congruency of the letter inside the onset would have an effect on responding to the letter contained in the grey target circle, attention must have shifted to the location of the onset. The results indicated a reliable congruency effect on trials on which the eyes went directly to the target. Theeuwes *et al.* concluded that regardless of whether an exogenous saccade was executed toward the onset, attention always went to the location of the onset.

However, the method employed by Theeuwes *et al.* (1999) was problematic for a number of reasons. First, because a large response relevant letter was present at the location of the onset, this location may have received attention, not because an irrelevant element was presented at that location, but because an element was presented that contained a large and response-relevant letter. If the response-relevant identity of the letter in the onset attracted attention to that location, it is hard to claim that it was the onset that captured attention. A second reason for questioning thet congruency manipulation of Theeuwes *et al.* (1999) was offered by Folk and Remington (1998). In line with the notion of perceptual load (Lavie, 1995), they suggested that when the number of objects is small, identity information can influence response mechanisms in parallel. In other words, they claimed that attention may have gone in parallel to both the singleton target and the onset containing the large letter causing a congruency effect on responding.

A recent study by Godijn and Theeuwes (in press-a) examined whether inhibition-of-return (IOR) occurs at the location of the onset in the oculomotor capture paradigm. This study is relevant for this issue, since the standard claim underlying inhibition-of-return (IOR) is that attention is inhibited from returning to a previously attended location (e.g. Posner & Cohen, 1984; Pratt *et al.*, 1997). Furthermore, saccades are also inhibited from moving to a previously attended location (e.g. Abrams & Dobkin, 1994). Similar to Theeuwes *et al.* (1999) participants were instructed to execute an endogenous saccade to a uniquely coloured target while a task-irrelevant onset was presented at a different location. After fixating the initial target another object became the next saccade target. This was done by changing the colour of one of the distractor circles into the target colour (from red to grey). The new target was presented at the location at which the onset had previously appeared or at the location of one of the non-onset distractors. The results confirmed earlier findings of Theeuwes *et al.* (1999) that attention was captured by the onset even when the eyes went directly to the target. Saccade latencies to the second target were longer when the new target appeared at the location at which the onset had appeared than when it appeared at another location. The size of this IOR effect was the same regardless of whether the eyes first went to the onset or to the initial target location.

The evidence from Godijn and Theeuwes (in press-a) that onsets capture attention even when the eyes go directly to the target is indirect and based on the assumption that IOR reflects the previous allocation of attention. No study has directly examined the allocation of attention in the oculomotor capture paradigm. Therefore, further research is needed to address this issue. Using a dual-task paradigm with the oculomotor capture paradigm instead of the anti-saccade task would have a number of distinct advantages. First, discrimination targets may also be presented at non-onset distractors at which attention is typically not directed (e.g. Theeuwes *et al.*, 1998, 1999). This would provide a suitable control condition with which discrimination accuracy of targets at the saccade target and onset may be compared. Second, in the oculomotor capture paradigm a condition in which an onset is presented may be compared with a condition in which no onset is presented. If attention is captured by the onset on trials on which the eyes directly move to the saccade target it may be expected that discrimination accuracy of letters presented at the saccade target location is higher when the eyes go to the saccade target on no-onset trials than on onset trials.

Exogenous Saccades and Awareness

The question whether attention precedes exogenous saccades has also been addressed by examining observers' conscious awareness (e.g. Mokler & Fischer, 1999). In fact, Mack and Rock (1998) have argued that conscious awareness is a prerequisite of attention. Thus, if a specific object did not reach conscious awareness it could not have captured attention. Likewise, it has been suggested that the awareness of gaze position depends on attentional allocation (e.g. Deubel *et al.*, 1999; Mokler & Fischer, 1999). This idea is supported by a study by Deubel *et al.,* (1999). In this study participants had the task of executing a saccade to a target location. Furthermore, a probe stimulus was presented at varying SOAs relative to the signal to move the eyes. At the end of the trials participants were required to indicate where they were looking when the probe stimulus appeared. The results showed that they tended to judge that they were looking at the saccade location well before they actually moved their eyes. Since it is assumed that attention precedes the eyes to the saccade destination, Deubel *et al.* concluded that the attended location was typically misinterpreted as the gaze position.

If it is assumed that the perception of gaze position requires attention then it is possible that saccades that go unnoticed by observers are not preceded by attention. A number of studies have in fact shown that observers are often unaware that they execute exogenous saccades toward onsets in the anti-saccade task (e.g. Mokler & Fischer, 1999; Mokler *et al.*, 2000) and in the oculomotor capture paradigm (Theeuwes *et al.*, 1998, 1999). Mokler and Fischer (1999) interpreted this finding as evidence that involuntary (exogenous) saccades are often not preceded by attention.

An alternative possibility is that exogenous saccades are always preceded by attention, but whether the change in gaze position reaches awareness depends on the speed of disengagement. It has been suggested that conscious awareness requires a certain degree of sustained allocation of attention (e.g. Most & Simons, 2001; Neisser, 1967). Therefore, it is possible that when attentional allocation on the location of the onset is too brief the erroneous saccade to the onset will not be perceived. Moreover, a number of studies have found evidence that attention and awareness can be dissociated (e.g. McCormick, 1997; Kentridge *et al.*, 1999; Lambert *et al.*, 1999).

Speed of Attentional Disengagement, Awareness and Oculomotor Capture

As an alternative to the view that in the anti-saccade task and oculomotor capture paradigm unperceived erroneous saccades to the onset are not preceded by attention, we propose that attention is always captured by the onset and whether the eyes move to the onset and whether these erroneous saccades are perceived depends on the speed of attentional disengagement from the location of the onset. This proposal is based on the following four assumptions. First, we assume that onsets capture attention, at least when attention is not already engaged on an object when the onset appears (e.g. Yantis & Jonides, 1990; Theeuwes, 1991). Second, erroneous saccades to the onset are only perceived when attention is directed to the onset for a sufficient amount of time. Third, attention is required at the saccade destination in order to program a saccade (e.g. Deubel & Schneider, 1996; Kowler *et al.*, 1995; Rizzolatti *et al.*, 1987). According to this assumption short fixation durations on the onset prior to a saccade to the target location (e.g. Theeuwes *et al.*, 1998, 1999; Mokler & Fischer, 1999) can only occur if attention shifts to the target location prior to the execution of the saccade to the onset. Furthermore, saccade programming will not be completed if attention disengages too soon. Fourth, shifts of exogenous attention are faster than shifts of endogenous attention (e.g. Theeuwes *et al.*, 2000). Therefore, in the oculomotor capture paradigm and the anti-saccade task the attention shift to the onset precedes the attention shift to the target location.

Given these assumptions the saccade behaviour and the awareness of the gaze position critically depend on the attentional capture by the onset and the speed of attentional disengagement from the location of the onset. If attentional disengagement from the onset is extremely fast, no saccade will be executed to the onset. In this case, attention will move from the onset location to the target location. Attention is then engaged on the target location for a sufficient amount of time to allow saccade programming to the target location. If attentional disengagement from the onset location is slower a saccade will be executed to the onset. These erroneous saccades to the onset can be distinguished on the basis of whether or not participants are aware of the erroneous saccade. This also depends on the speed of disengagement. Thus, when attentional disengagement from the onset location is relatively fast (but not fast enough to prevent an erroneous saccade) participants will not notice they moved their eyes toward the onset. If attentional disengagement is relatively slow participants will become aware of the erroneous saccade. Obviously, these are speculations which need to be tested in future research.

The Competitive Integration Model

Godijn and Theeuwes (in press-a, in press-b) developed a competitive integration model to account for the competition between exogenous and endogenous saccade programming. This model is primarily based on saccade behaviour from a variety of different tasks, but it also provides a framework within which the relationship between attention and saccades and phenomenon such as IOR may be understood. We first discuss the assumptions concerning saccade programming and from there we turn to the control signals that precede the final saccade programming stage.

Architecture

The competitive integration model (e.g. Godijn & Theeuwes, in press-a, in press-b) assumes that exogenous and endogenous saccades are programmed in a single oculomotor system. Similar to a number of previous models (e.g. Kopecz, 1995; Findlay & Walker, 1999; Trappenberg *et al.*, 2001) the competitive integration model assumes that saccade programming occurs on a common saccade map, in which information from different sources (e.g. endogenous and exogenous) is integrated. Figure 1.2 illustrates the basic idea. Saccade-related activation at one location spreads to neighbouring locations, but inhibits distant locations (Figure 1.2a). Thus, saccade programming is a competition between activations at locations represented in the saccade map. When two relatively distant locations are activated this activation is mutually inhibitory (Figure 1.2b), but when two nearby locations are activated, this activation is mutually excitatory (Figure 1.2c).

Temporal Trigger

It is assumed that a saccade is executed when a certain activation threshold is reached in the saccade map. In contrast to Findlay and Walker's (1999) model there are no separate "fixate" and "move" centres. Instead, the fixation location is part of the saccade map. Fixation-related activity is a critical aspect for the temporal trigger. When observers are actively fixating a specific location, the central portion of the saccade map is strongly activated (Krauzlis *et al.,* 1997; also see Kopecz, 1995).There is lateral inhibition between the fixation location and peripheral locations precisely like the lateral inhibition between distant peripheral locations. Therefore, when the fixation location is strongly activated this prevents the threshold from being reached at peripheral locations. When a saccade is required the fixation-related activation may be inhibited (typically referred to as oculomotor disengagement) releasing peripheral locations from the lateral inhibition from the fixation location.

Saccade Destination

The competitive integration model assumes that the saccade is directed to the mean vector of activity in the saccade map once the threshold is reached (e.g. Tipper *et al.*, 2000, 2001). Therefore, when other locations are activated when a threshold is reached, deviations of the saccade trajectory from the threshold location will occur. When two nearby locations are strongly activated the eyes will typically land somewhere between the two locations (e.g. Coren & Hoenig, 1972; Findlay, 1982; Godijn & Theeuwes, in

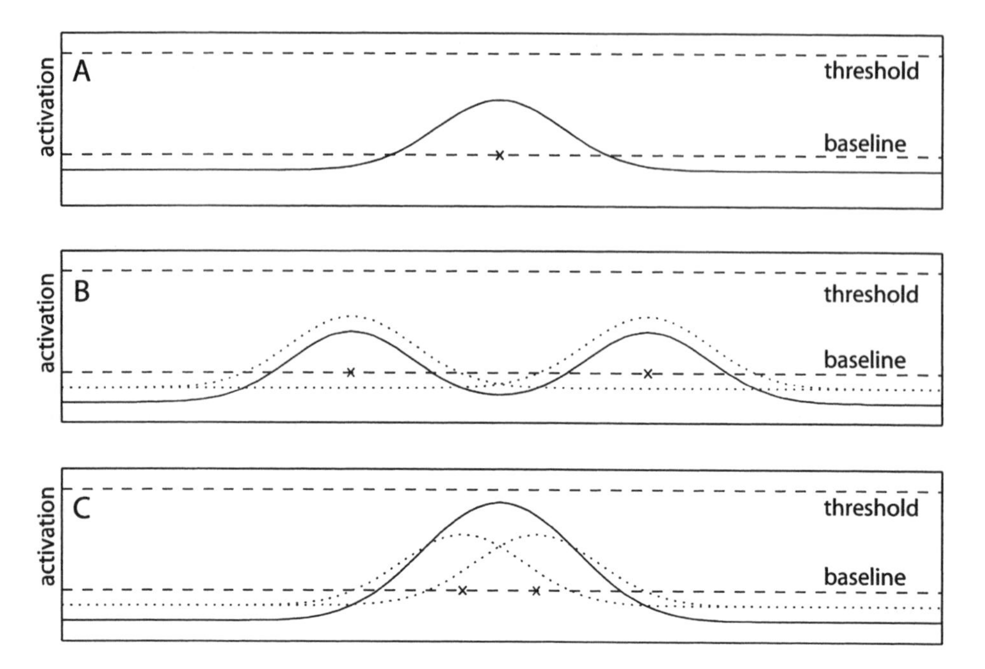

Figure 1.2: Activation patterns in the saccade map. (A) When a saccade is programmed to a certain location "x" in the saccade map, representing a location in the visual field, the activation spreads out to neighbouring locations, but inhibits distant locations. (B) When two saccades are programmed in parallel, activation related to both goals (the broken lines) is combined (continuous line) and when the two locations are relatively far apart activation is mutually inhibitory. (C) On the other hand, when two locations are relatively close together, the combined activation may result in a high activation peak somewhere between the two locations. From Godijn and Theeuwes (in press-b).

press-b). However, when two distant locations are strongly activated, the mutual inhibition will prevent the threshold from being reached.

Location-specific Inhibition

In order to execute a saccade to a target location, activation at other locations that are activated (e.g. the location of a task-irrelevant onset) needs to be inhibited. In addition to the lateral inhibition in the saccade map another inhibitory mechanism is assumed, which acts directly on the activation at a specific location (e.g. Tipper *et al.*, 2001). This allows the conflict between two locations to be resolved and biases saccade programming toward desired locations. This location-specific inhibition may result in a sub-baseline activation level causing the eyes to deviate away from the inhibited location (e.g. Tipper *et al.*, 2000, 2001; also see Doyle & Walker, 2001; Sheliga *et al.*, 1994, 1995), because the mean vector of activity will be shifted away from the inhibited location.

Attentional Control Signals

Before a saccade can be programmed to a specific location, control signals must be delivered to the saccade programming map. In a visual scene there are typically many potential targets for a saccade and therefore selection is required. In accordance with Schneider (1995) we propose that there is a common attentional selection mechanism for saccades and sensory processing. Thus, when an attentional control signal is applied to the saccade map in order to program a saccade the control signal is also applied to the visual system in which object features are processed (see also Chelazzi & Corbetta, 2000). Therefore, processing of object features is facilitated at the saccade destination. In addition, when a location in the saccade map is strongly activated, control signals may be applied to that location preventing the saccade. This inhibitory control signal is also passed on to the visual system responsible for the processing of object features resulting in impaired processing of object features at the saccade destination.

Neural Correlates

The superior colliculus (SC) is typically considered the locus of the final stage of saccade programming (for reviews see Schall, 1991; Wurtz *et al.*, 2000; Sparks & Mays, 1981). The architecture of the SC is consistent with the competitive integration model (e.g. Olivier *et al.*, 1998; Munoz & Istvan, 1998) and it receives cortical input from a number of areas such as the frontal eye fields (FEF) and the lateral intraparietal region (LIP) in the posterior parietal cortex (PPC). It has been suggested that a fronto-parietal circuit involving these areas is responsible for delivering the control signal required for saccade programming (e.g. Chelazzi & Corbetta, 2000). Furthermore lesion studies have suggested that the FEF is involved in inhibiting saccades (e.g. Guitton *et al.,* 1985; Henik *et al.*, 1994; Rafal *et al.*, 2000). The LIP not only projects to the SC, but also to areas in the temporal and visual cortex, in particular IT and V4. Area V4 is specialized in processing object features, while IT is specialised in processing complex objects (Tanaka, 1993). According to LaBerge (1995) when a location is selected activation flows from the PPC to V4 and on to IT and enhances activation at the selected area while suppressing activation at surrounding areas. In the context of the competitive integration model it may be proposed that the attentional control signals arise from a fronto-parietal circuit including the FEF and LIP and are expressed in the SC for saccade programming and in visual areas such as V4 and IT in order to facilitate the perceptual processing at the selected location.

An Oculomotor Suppression Hypothesis of IOR

Within the framework of this competitive integration model Godijn and Theeuwes (in press-a) proposed an account of IOR based on oculomotor suppression. According to this account IOR is a result of a location-specific inhibition in the saccade map. This inhibition is applied to a specific peripheral location in order to prevent the execution of a saccade. Furthermore, when a saccade is required activation at the fixation location is inhibited, releasing peripheral locations from the lateral inhibition from the fixation location. The location-specific inhibition causes a sub-baseline level of activation of the inhibited location. Within the framework of the competitive integration model the consequence of this inhibition is twofold: Subsequent saccades to the inhibited location are delayed (e.g. Abrams & Dobkin, 1994; Klein & MacInnis, 1999; Godijn & Theeuwes, in press-a). That is, it takes longer for the threshold to be reached at that location. Furthermore, saccade trajectories deviate away from the previously inhibited location (e.g. Tipper *et al.*, 2000, 2001; Sheliga *et al.*, 1994, 1995; Doyle & Walker, 2001), since the mean vector of activity shifts away from the inhibited location. Figure 1.3 illustrates this idea. Shown is the saccade map representing activations that occur during a typical trial on which the eyes move to the target. At the start of the trial there is strong activation around the central fixation location (Figure 1.3a). After the presentation of the target display the fixation location receives top-down inhibition as the observer prepares to make a saccade (1.3b). Before the activation at the onset location can reach threshold target-related input reaches the saccade map and top-down inhibition acts on the location of the onset distractor (1.3d). Due to the inhibition at the location of the onset distractor the eyes move to the target, but with a slight deviation in the trajectory away from the onset distractor. If the activation at the onset distractor sets in too late to prevent the threshold from being reached at the location of the onset distractor the eyes first move to the onset distractor. If the activation of the onset distractor is inhibited shortly after the threshold is reached the saccade may fall short of the onset distractor and the reduced activation at that location will allow the threshold to be reached at the target location relatively quickly. This is consistent with the findings of Godijn and Theeuwes (in press-a, in press-b). In addition, subsequent saccades to the location of the onset distractor will be delayed (Godijn & Theeuwes, in press-a).

Relation to Other Models

Relations to Other Models of Saccade Programming

Previous models of saccade programming have typically assumed separate systems for the spatial and temporal aspects ("WHEN" and "WHERE") of saccade control (e.g. Findlay & Walker, 1999; Becker & Jürgens, 1979). These models are consistent with findings from double-onset studies (e.g. Walker *et al.*, 1995, 1997; Ottes *et al.*, 1985; Lévy-Schoen, 1969). In these studies participants were required to execute a saccade to a target presented with an abrupt onset. On some trials an onset distractor was also presented simultaneously with the target. These studies showed that distractors that were presented near a saccade target affected the endpoint of the saccade, but not the latency, while distractors presented far from the saccade target affected the latency, but not the endpoint. Specifically, when a distractor was presented near the saccade target

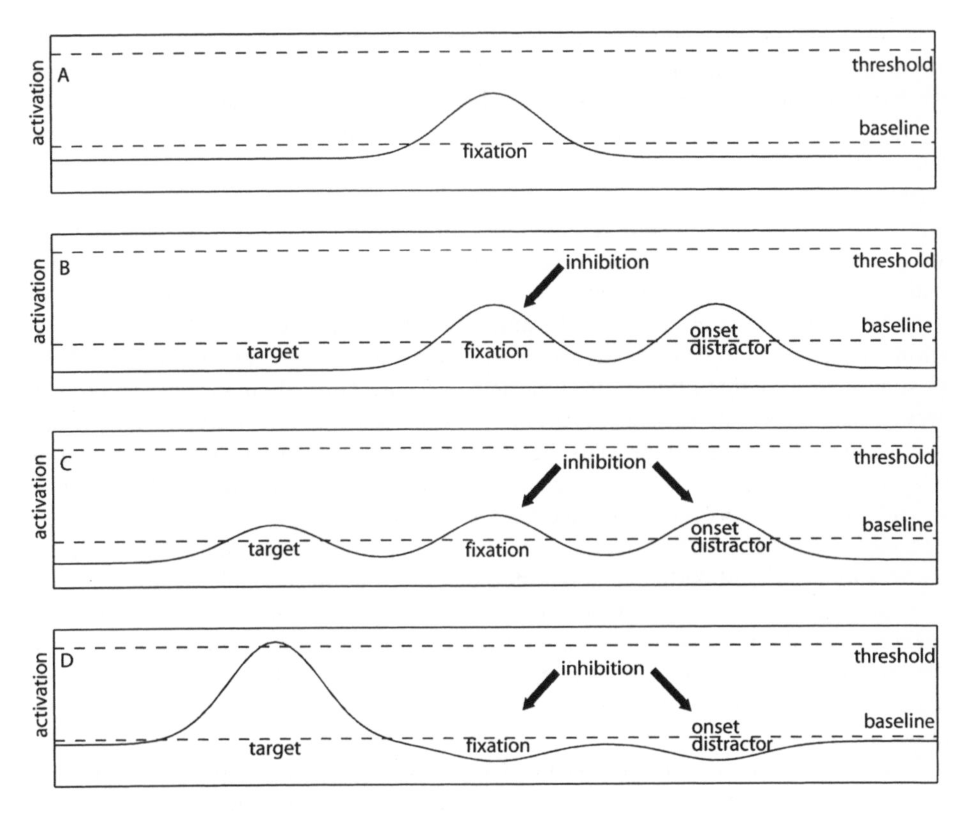

Figure 1.3: An illustration of the time course of activation in the saccade map according to the competitive integration model on a typical trial in the oculomotor capture paradigm. See text for details.

the eyes often landed somewhere between the saccade target and the distractor (global effect; e.g. Coren & Hoenig, 1972; Findlay, 1982). However, when the distractor was presented far from the saccade target the eyes moved to one or the other (no global effect), but with an increased latency relative to a no-distractor control condition (e.g. Walker *et al*., 1995, 1997; Lévy-Schoen, 1969). According to models assuming separate WHEN and WHERE processes remote distractors affect the WHEN process, while neighbouring distractors affect the WHERE process (e.g. Walker *et al*., 1997). Although the results of double-onset studies are consistent with models assuming a separation between the temporal and spatial aspects of saccade control, models in which the spatial and temporal aspects are integrated can also explain these findings. For example, the latency increase caused by distractors presented far from the onset target may be explained by the competitive integration model through its lateral inhibition architecture (see Figure 1.2b). Distractors presented close to the onset target do not increase latency, because there is mutual excitation between two nearby locations (see Figure 1.2c). Furthermore, a global effect is not found with far distractors, since distractor-related activity is inhibited in order to resolve the conflict between the target and distractor locations (see Figure 1.3). That is, the lateral inhibition between the locations prevents the threshold from being reached. Top-down inhibition of the distractor location is needed to release the activation at the target location from the lateral inhibition from the distractor location. When the threshold is reached the distractor location is no longer strongly activated so that no global effect occurs. The question remains how the competitive integration model can explain that saccade latencies are not reduced in double-onset studies on trials with a near distractor relative to trials without a distractor. Given the architecture of the competitive integration model one would expect such an effect, since the mutual excitation between two locations should allow the threshold to be reached faster than when no distractor is presented. However, as suggested by Ottes *et al.* (1985) participants may delay their saccade in order to prevent fast inaccurate saccades that land between target and distractor. Thus, it may well be that on near distractor trials the latency distributions are mixed. On some trials relatively fast global saccades are executed, while on other trials participants delay their saccade in order to prevent fast inaccurate saccades, resulting in relatively slow accurate saccades. The net effect would depend on the proportion of delayed saccades. In Ottes *et al.* (1985) all participants reported that delaying the saccade required a deliberate strategy. In the context of the competitive integration model a delay in saccade execution would be achieved by top-down activation of the fixation location. Due to the lateral inhibition from the fixation location activation at both the target and distractor location would be reduced, allowing more time for distinguishing between the target and distractor.

The competitive integration model of Godijn and Theeuwes (in press-a; in press-b) is similar to a number of other models that also assume a competitive integration structure (e.g. Trappenberg *et al.*, 2001; Kopecz, 1995; Findlay & Walker, 1999). Trappenberg *et al.* (2001) developed a neural model of saccade programming based on competitive integration of exogenous and endogeneous signals in the superior colliculus. Their model produced activity patterns very similar to activity patterns of cells in the superior colliculus. Furthermore, the initial saccade latencies of the model fitted well with a range of oculomotor effects. However, the model is only concerned with the initial saccade latencies and therefore does not account for effects on saccade amplitude, trajectories and fixation durations between saccades.

Modelling the Relationship Between Attention and Saccades

The competitive integration model assumes a strong relationship between attention and saccades. Whenever a location is selected as the goal for a saccade (exogenously or endogenously) perceptual processing of objects at the saccade destination is facilitated. Thus, there is a common attentional selection for the saccades and for object recognition. This is similar to Schneider's (1995) visual attention model (VAM). In this model selection is achieved in two parallel processing streams, a dorsal and a ventral stream. It is assumed that the dorsal stream (from V1 to PPC) selects information from an object for actions, while the ventral stream (from V1 to IT) selects information from an object for recognition. Attentional selection occurs downstream in V1 and is passed through the dorsal and ventral processing streams. This selection is coupled to a single object. That is, when an object is selected, actions such as a saccades are programmed to that object and recognition of that object is facilitated. This does not mean that when attention is directed to an object a saccade is always executed towards it. It is assumed that a separate GO-signal is required for saccade execution.

Another model that assumes a strong relationship between attention and saccades is the premotor theory of attention of Rizzolatti and colleagues (1987, 1994). According to the premotor theory the mechanisms responsible for spatial attention are localised in the spatial pragmatic maps. Thus, attention shifts are accomplished by programming an action to a specific location. This is similar to VAM, but one major difference between the two theories is the direction of the causality: VAM assumes that a saccade program is a consequence of an attention shift, while the premotor theory assumes that programming a saccade causes an attention shift. Irrespective of the direction of the causality between attention and saccades, a one-to-one relationship between attention and saccades is predicted by both theories. That is, attention cannot be directed to an object without programming a saccade. Therefore, VAM and the premotor theory both represent efference theories of the relationship between attention and saccades (Posner, 1980).

Compared with VAM and the premotor theory the competitive integration model of Godijn and Theeuwes (in press-a; in press-b) is more concerned with the mechanisms responsible for saccade programming. First and foremost it is a model of saccade control. Although it is assumed that the control signals required for saccade programming are also applied to the (ventral) visual system, responsible for processing object features, the mechanisms responsible for object recognition are in its present form not specified. Future efforts in the development of the competitive integration model should shift the focus from the mechanisms responsible for saccade programming to the mechanisms responsible for object recognition.

References

Abrams, R. A., & Dobkin, R. S. (1994). Inhibition of return: Effect of attentional cuing on eye movement latencies. *Journal of Experimental Psychology: Human Perception and Performance, 20*, 467–477.

Becker, W., & Jürgens, R. (1979). An analysis of the saccadic system by means of double step stimuli. *Vision Research, 19*, 967–983.

Chelazzi, L., & Corbetta, M. (2000). Cortical mechanisms of visuospatial attention in the primate brain. In: M. S. Gazzaniga (ed.), *The New Cognitive Neurosciences* (pp. 667–686). Cambridge, MA: MIT Press.

Coren, S., & Hoenig, P. (1972). Effect of non-target stimuli on the length of voluntary saccades. *Perceptual and Motor Skills, 34*, 499–508.

Deubel, H., Irwin, D. E., & Schneider, W. X. (1999). The subjective direction of gaze shifts long before the saccade. In: W. Becker, H. Deubel and Th. Mergner (eds), *Current Oculomotor Research: Physiological and Psychological Aspects* (pp. 65–70). New York, London: Plenum.

Deubel, H., & Schneider, W. X. (1996). Saccade target selection and object recognition: Evidence for a common attentional mechanism. *Vision Research, 36*, 1827–1837.

Doyle, M., & Walker, R. (2001). Curved saccade trajectories: Voluntary and reflexive saccades curve away from irrelevant distractors. *Experimental Brain Research, 139*, 333–344.

Findlay, J. M. (1982). Global processing for saccadic eye movements. *Vision Research, 22*, 1033–1045.

Findlay, J. M. (1997). Saccade target selection during visual search. *Vision Research, 37*, 617–631.

Findlay, J. M., & Walker, R. (1999). A model of saccade generation based on parallel processing and competitive integration. *Behavioral and Brain Sciences, 22*, 661–721.

Fischer, B., & Weber, H. (1992). Characteristics of "anti" saccades in man. *Experimental Brain Research, 89*, 415–424.

Fischer, B., & Weber, H. (1996). Effects of precues on error rate and reaction times of anti-saccades in human subjects. *Experimental Brain Research, 109*, 507–512.

Folk, C. M., & Remington, R. W. (1998). Selectivity in distraction by irrelevant feaural singletons: Evidence for two forms of attentional capture. *Journal of Experimental Psychology: Human Perception and Performance, 24*, 847–858.

Godijn, R., & Theeuwes, J. (2002a). Oculomotor capture and inhibition of return: Evidence for an oculomotor suppression account of IOR. *Psychological Research, 66*, 234–246.

Godijn, R., & Theeuwes, J. (2002b). Programming of exogenous and endogenous saccades: Evidence for a competitive integration model. *Journal of Experimental Psychology: Human Perception and Performance, 28*, 1039–1054.

Godijn, R., & Theeuwes, J. (submitted). Programming saccade sequences requires parallel allocation of attention.

Guitton, D., Buchtel, H. A., & Douglas, R. M. (1985). Frontal lobe lesions in man cause difficulties in suppressing reflexive glances and in generating goal directed saccades. *Experimental Brain Research, 58*, 455–472.

Hallet, P. E. (1978). Primary and secondary saccades to goals defined by instructions. *Vision Research, 18*, 1279–1296.

Henderson, J. M. (1992). Visual attention and eye movement control during reading and picture viewing. In: Rayner, K. (ed.), *Eye Movements and Visual Cognition* (pp. 261–283). Berlin: Springer.

Henderson, J. M., & Ferreira, F. (1990). Effects of foveal processing difficulty on the perceptual span in reading: Implications for attention and eye movement control. *Journal of Experimental Psychology: Learning, Memory, and Cognition, 16*, 417–429.

Henik, A., Rafal, R., & Rhodes, D. (1994). Endogenously generated and visually guided saccades after lesions of the human frontal eye fields. *Journal of Cognitive Neuroscience, 6*, 400–411.

Hoffman, J. E. (1998). Visual attention and eye movements. In: Pashler, H. (ed.), *Attention* (pp. 119–153). Erlbaum.

Hoffman, J. E., & Subramaniam, B. (1995). The role of visual attention in saccadic eye movements. *Perception & Psychophysics, 57*, 787–795.

Irwin, D. E., Brockmole, J. R., & Kramer, A. F. (submitted). Attention precedes involuntary saccades.

Irwin, D. E., Colcombe, A. M., Kramer, A. F., & Hahn, S. (2000). Attention and oculomotor capture by onset luminance and colour singletons. *Vision Research, 40*, 1443–1458.

Jonides, J., & Yantis, S. (1988). Uniqueness of abrupt visual onset in capturing attention. *Perception & Psychophysics, 43*, 346–354.

Kentridge, R. W., Heywood, C. A., & Weiskrantz, L. (1999). Attention without awareness in blindsight. *Proceedings of the Royal Society of London — B, 266*, 1805–1811.

Klein, R. M. (1980). Does oculomotor readiness mediate cognititve control of visual attention? In: R. Nickerson (ed.), *Attention and Performance VIII* (pp. 259–275). New York: Academic Press.

Klein, R. M., & MacInnes, W. J. (1999). Inhibition of return is a foraging facilitator in visual search. *Psychological Science, 10*, 346–352.

Klein, R. M., & Shore, D. I. (2000). Relations among modes of visual orienting. In: S. Monsell and J. Driver (eds), *Attention and Performance XVIII* (pp. 196–208). Cambridge, MA: MIT Press.

Kopecz, K. (1995). Saccadic reaction times in gap/overlap paradigm: A model based on integration of intentional and visual information on neural dynamic fields. *Vision Research, 35*, 2911–2925.

Kowler, E., Anderson, E., Dosher, B., & Blaser, E. (1995). The role of attention in the programming of saccades. *Vision Research, 35*, 1897–1916.

Krauzlis, R. J., Basso, M. A., & Wurtz, R. H. (1997). Shared motor error for multiple eye movements. *Science, 279*, 1693–1695.

LaBerge, D. (1995). *Attentional processing: The brain's art of mindfulness.* Cambridge, MA: Harvard University Press.

Lambert, A., Naikar, N., McLachlan, K., & Aitken, V. (1999). A new component of visual orienting: Implicit effects of peripheral information and subthreshold cues on covert attention. *Journal of Experimental Psychology: Human Perception and Performance, 25*, 321–340.

Lavie, N. (1995). Perceptual load as a necessary condition for selective attention. *Journal of Experimental Psychology: Human Perception and Performance, 21*, 451–468.

Lévy-Schoen, A. (1969). Détermination et latence de la résponse oculomotrice à deux stimulus simultanés ou successifs selon leur excentricité relative. *L'Année Psychologique, 69*, 373–392.

Lupiáñez, J. Milán, E. G., Tornay, F. J., Madrid, E., & Tudela, P. (1997). Does IOR occur in discrimination tasks? Yes, it does, but later. *Perception & Psychophysics, 59*, 1241–1254.

Mack, A., & Rock, I. (1998). *Inattentional Blindness.* Cambridge, MA: MIT Press.

McCormick, P. A. (1997). Orienting attention without awareness. *Journal of Experimental Psychology: Human Perception and Performance, 23*, 168–180.

Maylor, E. A. (1985). Facilitory and inhibitory components of orienting in visual space. In: M. I. Posner and O. S. M. Marin (eds), *Attention and Performance XI* (pp. 189–204). Hillsdale, NJ: Erlbaum.

Mokler, A., Deubel, H., & Fischer, B. (2000). Unintended saccades can be executed without a presaccadic attention shift. Poster presented at ECVP 2000, Groningen, The Netherlands.

Mokler, A., & Fischer, B. (1999). The recognition and correction of involuntary prosaccades in an anti-saccade task. *Experimental Brain Research, 125*, 511–516.

Morrison, R. E. (1984). Manipulation of stimulus onset delay in reading: Evidence for parallel programming of saccades. *Journal of Experimental Psychology: Human Perception and Performance, 10*, 667–682.

Most, S. B., & Simons, D. J. (2001). Attention capture, orienting, and awareness. In: C. Folk and B. Gibson (eds), *Attraction, Distraction, and Action: Multiple perspectives on attentional capture* (pp. 151–173). North-Holland: Elsevier.

Munoz, D. P., & Istvan, P. J. (1998). Lateral inhibitory interactions in the intermediate layers of the monkey superior colliculus during interrupted saccades. *Journal of Comparitive Neurology, 276*, 169–187.

Neisser, U. (1967). *Cognition and Reality: Principles and Implications of Cognitive Psycholgy.* San Francisco, CA: W. H. Freeman.

Olivier, E., Porter, J. D., & May, P. J. (1998). Comparison of the distribution and somatodendritic morphology of tectotectal neurons in the cat and monkey. *Visual Neuroscience, 15*, 903–922.

Ottes, F. P., Van Gisbergen, J. A. M., & Eggermont, J. J. (1985). Latency dependence of colour-based target vs. nontarget discrimination by the saccadic system. *Vision Research, 25*, 849–862.

Posner, M. I. (1980). Orienting of attention. *Quarterly Journal of Experimental Psychology, 32*, 3–25.

Posner, M. I., & Cohen, Y. A. (1984). Components of visual orienting. In: H. Bouma and D. G. Bouwhuis (eds), *Attention and Performance X* (pp. 531–556). Hillsdale, NJ: Lawrence Erlbaum Associates.

Pratt, J., Kingstone, A., & Khoe, W. (1997). Inhibition of return in location- and identity-based choice decision tasks. *Perception & Psychophysics, 59*, 964–971.

Rafal, R. D., Calabresi, P. A., Brennan, C. W., & Sciolto, T. K. (1989). Saccade preparation inhibits reorienting to recently attended locations. *Journal of Experimental Psychology: Human Perception and Performance, 15*, 673–685.

Rafal, R. D., Machado, L. J., Ro, T., & Ingle, H. W. (2000). Looking forward to looking: Saccade preparation and control of the visual grasp reflex. In: S. Monsell and J. Driver (eds), *Attention and Performance XVIII* (pp. 155–174). Cambridge, MA: MIT Press.

Reichle, E. D., Pollatsek, A, Fischer, D. L., & Rayner, K. (1998). Toward a model of eye movement control in reading. *Psychological Review, 105*, 125–157.

Remington, R. W. (1980). Attention and saccadic eye movements. *Journal of Experimental Psychology: Human Perception and Performance, 6*, 726–744.

Reuter-Lorenz, P. A., Oonk, H. M., Barnes, L. L., & Hughes, H. C. (1995). Effects of warning signals and fixation point offsets on the latencies of pro- versus anti-saccades: Implications for an interpretation of the gap effect. *Experimental Brain Research, 103*, 287–293.

Rizzolatti, G., Riggio, L., Dascola, I., & Umiltà, C. (1987). Reorienting attention across the horizontal and vertical meridians: Evidence in favor of a premotor theory of attention. *Neuropsychologica, 25*, 31–40.

Rizzolatti, G., Riggio, L., & Sheliga, B. M. (1994). Space and selective attention. In: C. Umiltà and M. Moscovitch (eds), *Attention and Performance XV* (pp. 231–265). Cambridge, MA: MIT Press.

Sagi, D., & Julesz, B. (1985). "Where" and "what" in vision. *Science, 228*, 1217–129.

Schall, J. D. (1991). Neuronal basis of saccadic eye movements in primates. In: A. G. Leventhal (ed.), *Vision and Visual Dysfunction: Vol. 4. The Neural Basis of Visual Function* (pp. 388–442). London: Macmillan.

Schneider, W. X. (1995). VAM: A neuro-cognitive model for for visual attention control of segmentation, object-recognition and space-based motor actions. *Visual Cognition, 2*, 331–376.

Schneider, W. X., & Deubel, H. (2002). Selection-for-perception and selection-for-spatial-motor-action are coupled by visual attention: A review of recent findings and new evidence from stimulus-driven saccade control. In: W. Prinz and B. Hommel (eds), *Attention and Performance XIX*, Oxford: Oxford University Press.

Sheliga, B. M., Riggio, L., & Rizzolatti, G. (1994). Orienting of attention and eye movements. *Experimental Brain Research, 98*, 507–522.

Sheliga, B. M., Riggio, L., & Rizzolatti, G. (1995). Spatial attention and eye movements. *Experimental Brain Research, 105*, 261–275.
Shepherd, M., Findlay, J. M., & Hockey, R. J. (1986). The relationship between eye movements and spatial attention. *Quarterly Journal of Experimental Psychology, 38A*, 475–491.
Sparks, D. L., & Mays, L. E. (1981). The role of the superior colliculus in the control of saccadic eye movements. A current perspective. In: A. F. Fuchs and W. Becker (eds), *Progress in Oculomotor Research* (pp.137–144). North-Holland: Elsevier.
Stelmach, L. B., Campsall, J. M., & Herdman, C. M. (1997). Attentional and ocular movements. *Journal of Experimental Psychology: Human Perception and Performance, 23*, 823–844.
Stelmach, L. B., & Herdman, C. M. (1991). Directed attention and perception of temporal order. *Journal of Experimental Psychology: Human Perception and Performance, 17*, 539–550.
Tanaka, K. (1993). Neuronal mechanisms of object recognition. *Science, 262*, 685–688.
Theeuwes, J. (1991). Exogenous and endogenous control of attention: The effects of visual onsets and offsets. *Perception & Psychophysics, 49*, 83–90.
Theeuwes, J. (1994). Stimulus-driven capture and attentional set: Selective search for color and visual abrupt onsets. *Journal of Experimental Psychology: Human Perception and Performance, 20*, 799–806.
Theeuwes, J. (1995). Abrupt luminance change pops out: Abrupt color change does not. *Perception & Pyschophysics, 57*, 637–644.
Theeuwes, J., Atchley, P., & Kramer, A. F. (2000). On the time course of top-down and bottom-up control of visual attention. In: S. Monsell and J. Driver (eds), *Attention & Performance XVIII.* Cambridge: MIT Press.
Theeuwes, J., Kramer, A. F., Hahn, S., & Irwin, D. E. (1998). Our eyes do not always go where we want them to go: Capture of the eyes by new objects. *Psychological Science, 9*, 379–385.
Theeuwes, J., Kramer, A. F., Hahn, S., Irwin, D. E., & Zelinsky, G. J. (1999). Influence of attentional capture on oculomotor control. *Journal of Experimental Psychology: Human Perception and Performance, 25*, 1595–1608.
Tipper, S. P., Howard, L. A., & Houghton, G. (2000). Behavioral consequences of selection from population codes. In: S. Monsell and J. Driver (eds), *Attention and Performance XVIII* (pp. 223–245). Cambridge, MA: MIT Press.
Tipper, S. P., Howard, L. A., & Paul, M. A. (2001). Reaching affects saccade trajectories. *Experimental Brain Research, 136*, 241–249.
Trappenberg, T. P., Dorris, M. D., Munoz, D. P., & Klein, R. M. (2001). A model of saccade initiation based on the competitive integration of exogenous and endogenous signals in the superior colliculus. *Journal of Cognitive Neuroscience, 13*, 256–271.
Treisman, A., & Gelade, G. (1980). A feature integration theory of attention. *Cognitive Psychology, 12*, 97–136.
Walker, R., Kentridge, R. W., & Findlay, J. M. (1995). Independent contributions of the orienting of attention, fixation offset and bilateral stimulation on human saccadic latencies. *Experimental Brain Research, 103*, 294–310.
Walker, R., Deubel, H., Schneider, W. X., & Findlay, J. M. (1997). Effect of remote distractors on saccade programming: Evidence for an extended fixation zone. *Journal of Neurophysiology, 78*, 1108–1119.
Wolfe, J. M., Cave, K. R., & Franzel, S. L. (1989). Guided search: An alternative to the feature-integration model for visual search. *Journal of Experimental Psychology: Human Perception and Performance, 15*, 419–433.

Wurtz, R. H., Basso, M. A., Paré, M., & Sommer, M. A. (2000). The superior colliculus and the cognitive control of movement. In: M. S. Gazzaniga (ed.), *The New Cognitive Neurosciences* (pp. 573–587). Cambridge, MA: MIT Press.

Yantis, S., & Jonides, J. (1984). Abrupt visual onsets and selective attention: Evidence from visual search. *Journal of Experimental Psychology: Human Perception and Performance, 10*, 601–621.

Yantis, S., & Jonides, J. (1990). Abrupt visual onsets and selective attention: Voluntary versus automatic allocation. *Journal of Experimental Psychology: Human Perception and Performance, 16*, 121–134.

Chapter 2

Orienting of Visual Attention Based on Peripheral Information

Matthew Kean and Anthony Lambert

This chapter is concerned with peripheral stimuli from which useful information may be extracted and used to direct attention. We describe two recent spatial cueing studies which, using saccadic eye movements as the response, investigated the orienting of visual attention towards and away from a salient peripheral stimulus. In addition, we review other recent work from our laboratory which has been concerned with information conveyed by peripheral cues. Collectively, the findings indicate that peripheral information can elicit extremely rapid shifts of visual attention. We argue that attention capture is not the only process that is capable of activating the attentional response system via a bottom-up pathway.

Introduction

At any given moment, peripheral vision accounts for the vast majority of our internal visual representation of the environment before us. The fovea is only relatively tiny, occupying approximately the central two degrees of the retina (Liversedge & Findlay, 2000) — with its resulting image being roughly equivalent to the size of a full moon (Goldberg, 2000). Typically, when we orient our attention to different locations within the visual field we execute an *overt* attention shift, by moving the eyes, head, or entire body to align the fovea with a new object of interest. However, while the focus of attention may, in this way, coincide with the area of the visual field to which the fovea is directed, the two are also potentially dissociable. A *covert* visual attention shift (Posner, 1980) occurs when the focus of attention moves to an area of the peripheral or parafoveal visual field independent of any overt movements — this process is commonly referred to as "looking out of the corner of your eye".

The Mind's Eye: Cognitive and Applied Aspects of Eye Movement Research

ISBN: 0–444–51020–6

Research into covert visual attention has flourished since Posner and colleagues developed the spatial cueing task (Posner *et al.*, 1978; Posner, 1980). In this task, subjects are required to respond to a peripherally presented target, which is preceded by a cue that serves to direct covert visual attention to a particular location. A typical finding is one of more efficient processing of targets appearing in the cued location relative to uncued locations. A processing advantage at the cued location, such as faster detection of the target, is thought to reflect a shift of covert visual attention (e.g. Posner, 1980).

A variety of responses to the target have been required of observers in different versions of the spatial cueing task, including simple detection (e.g. Posner, 1980), discrimination (e.g. Müller & Rabbitt, 1989), and temporal order judgement (e.g. Zackon *et al.*, 1999). The majority of spatial cueing studies have employed manual (i.e., key-press) responses. An alternative to manual responses — and one which has been utilised in comparatively few spatial cueing studies — is saccadic eye movements (e.g. Crawford & Müller, 1992; Briand *et al.*, 2000; Lambert *et al.*, 2000). In such studies, participants are required to respond to the appearance of the target by executing a voluntary saccade in order to fixate it.

It would be a mistake to view this methodology as measuring overt, *as opposed to* covert, orienting. While responding is performed by executing a saccadic eye movement (an overt indicator of an attentional movement), the latency to initiate a saccadic response comprises both overt and covert components (cf. Crawford & Müller, 1992). In both manual and saccadic response situations, attention is covertly allocated to a peripheral cue, since subjects are required to maintain central fixation until target appearance; a response can only be initiated after the target has been detected in the subject's peripheral vision. In other words, even with saccades as the mode of response, subjects must covertly attend to the cued location prior to executing a response.

Eye movements, as a mode of response, have a number of advantages over manual responses. Firstly, a saccadic eye movement is the primary overt behavioural indicator of the orienting of visual attention in everyday life. It could be argued, then, that eye movements provide an index of orienting with a higher degree of face validity than manual responses, such as space-bar presses. A related benefit of employing saccades as a response measure is that the findings may be informative with respect to the nature of the relationship between shifts of covert attention and overt attention (in the form of eye movements). This issue has been of theoretical interest for some time, and remains so today (Posner, 1980; Rizzolatti *et al.*, 1987; Hoffman & Subramaniam, 1995; Kowler *et al.*, 1995; Sheliga *et al.*, 1995; Deubel & Schneider, 1996; Corbetta, 1998). Given the intimate link between eye movements and covert attention in everyday orienting behaviour, exploring the influence of peripheral stimuli on saccades is a valuable tool for advancing our understanding of the capabilities — and limitations — of our ability to orient visual attention. Finally, in the case of eye movement responses, directional and accuracy data may be collected in addition to latency data.

A useful distinction has been made between two fundamentally different processes that are capable of effecting attention shifts; one requiring conscious direction (i.e., a top-down, voluntary process), and another which acts independent of conscious control (i.e., a bottom-up, or stimulus-driven process). These two theoretical constructs are known as the endogenous and exogenous orienting mechanisms, respectively (Posner, 1980). For Posner (1980), the endogenous orienting mechanism is employed when attention is "directed by a central decision" (p. 19). As the distinction between the two processes is a functional dichotomy, the exogenous orienting mechanism is thus responsible for attention shifts that "occur involuntarily" (i.e., without conscious intent) (Folk *et al.*, 1992, p. 1030).

The notion of two separate orienting systems — delineated with respect to independence from, versus reliance on, conscious direction — is supported by evidence of different neural underpinnings associated with the respective processes. Converging evidence suggests that certain subcortical areas of the brain — most notably the superior colliculus — are more involved in exogenous shifts of attention, while activation of particular cortical areas — such as the dorsolateral pre-frontal cortex — is associated with endogenous orienting (e.g. Corbetta *et al.*, 1993; Robinson & Kertzman, 1995; Rosen *et al.*, 1999; Zackon *et al.*, 1999; Robertson & Rafal, 2000).

The endogenous orienting mechanism is — by definition — consciously directed, and its execution is consequently relatively slow. For example, Müller and Rabbitt (1989) showed that optimal effects of attention shifts driven by the endogenous orienting mechanism are manifested at cue–target onset delays (i.e., stimulus onset asynchronies; SOAs) of between 275 and 400 ms. The exogenous orienting mechanism, on the other hand, is capable of acting rapidly. For example, *attention capture* — one process which operates independently of conscious direction (i.e., via the exogenous orienting mechanism) — exerts its maximal effects at brief SOAs (i.e., around 100 ms) (Wright & Ward, 1994). Attention capture is the reflexive allocation of attention to a luminance change in peripheral vision, whether the change is a sudden onset or offset (Theeuwes, 1991), or a change in luminance of an already-present stimulus (Posner & Cohen, 1984). This effect has been observed when the eliciting stimulus (i.e., the *cue* in spatial cueing studies) is irrelevant to the task at hand (Jonides, 1981), and even, in many instances, when it is known that orienting attention to the cued location will be detrimental to performance (Posner *et al.*, 1982).

An Informative Peripheral Cue

One methodological technique, which has been used to investigate the influence of top-down factors on exogenous orienting, is to use a peripheral cue that is informative with respect to the likely location of the impending target. The cue may indicate that the target is most likely to appear close by (e.g. Müller & Rabbitt, 1989; Crawford & Müller, 1992) or in an uncued location (e.g. Posner *et al.*, 1982; Tepin & Dark, 1992; Danckert & Maruff, 1997). It is thought that shifting visual attention towards or away from an informative peripheral cue involves both the exogenous mechanism, which is activated in response to the peripheral stimulation provided by the cue, and the endogenous mechanism, which acts to align attention with the most likely target-location as indicated by the cue's location (e.g. Müller & Rabbitt, 1989; Danziger & Kingstone, 1999). There is ecological validity to the use of an informative peripheral cue, as a sudden peripheral event may well signal that it is worth attending to a particular location in the visual field. Situations in which a peripheral event serves as a stimulus to attend to an alternative location are perhaps less common than those in which we deliberately attend to the location of the peripheral event itself. However, some such scenarios do occur in everyday life; for example, when a traffic light changes to green, before proceeding one might subsequently attend to adjoining streets to check whether another vehicle was running a red light.

We carried out a study addressing this issue using a salient sudden-onset peripheral cue, which was informative with respect to the likely location of the potential target. The most important findings are presented here; full details are described in Kean and Lambert (2001). In Experiment 1, each observer participated in two conditions: in the "Ipsilateral" condition, the target was most likely (p=0.8) to appear near the cued location, while in the "Contralateral" condition the target usually (p=0.8) appeared in the visual hemi-field opposite the cue. Prior to undertaking each condition, participants were informed of the contingent relationship between cue and target-locations. The sudden-onset cue and target stimuli were readily distinguishable from each other, and were presented at different eccentricities (7.3° and 5.5° from fixation, respectively). Participants responded as quickly as possible to the appearance of the target by executing a saccadic eye movement in order to fixate it. In each condition, at each of four SOAs (zero, 100, 300 and 600 ms), saccadic latencies on trials in which the target appeared in the expected location were compared with latencies on trials in which the target appeared in the unexpected location. This study was the first to have examined this issue (i.e., expectancy-based orienting of attention towards and away from an informative peripheral cue) using saccadic latencies as the response measure.

Results for the "Ipsilateral" condition are shown in Figure 2.1a.[1] An ANOVA revealed that, overall, mean saccadic latency was faster for trials on which the target appeared at the cued and expected location (325 ms) than at the uncued and unexpected location (343 ms). However, the magnitude of this cueing effect varied according to SOA (i.e., the interaction between target-location and SOA was significant), with the difference reaching significance only at the two brief SOAs (zero and 100 ms). This suggests that observers were able to rapidly direct visual attention towards the side on which the cue was presented and the target was most likely to appear. Subjects were, however, unable to maintain their attentional focus on the side where the cue had appeared: at the SOAs of 300 and 600 ms the cueing effect which was observed at the short SOAs disappeared, with saccadic latencies now being no faster for trials on which the target appeared in the expected location than in the unexpected location.

Given that subjects were informed of the high probability of ipsilateral target appearance, one might expect to see faster latencies at the cued location across the two longest SOAs, due to the operation of the endogenous orienting mechanism. Indeed, such a finding is often observed in studies using manual responses (e.g. Müller & Rabbitt, 1989; Danziger & Kingstone, 1999). While our finding, of no latency advantage at a stimulated and expected location with long cue–target delays, is uncommon, it is not without precedent. In a study which compared key-press responses with saccadic eye movement responses, and which also employed a peripheral cue that signalled a high probability of ipsilateral target presentation, Crawford and Müller

(1992) found a cueing effect at an SOA of 500 ms with manual responses, but not with eye movement responses. To explain this disparity between the two modes of response, these authors invoked the notion of inhibition of return (IOR) — a process which influences attentional orientation at long SOAs, by temporarily biasing either attention (Posner & Cohen, 1984; Posner *et al.*, 1985) or oculomotor responses (Abrams & Dobkin, 1994) against returning to a previously stimulated location. Crawford and Müller (1992) suggested that

> inhibition of return may be more difficult to suppress on saccadic eye movements than on simple manual responses with stationary eyes. (p. 303)

In the "Ipsilateral" condition of Experiment 1 by Kean and Lambert (2001), the order of participating in the two conditions did not impact on the results; there was no evidence that the previously learned "Contralateral" cue–target contingency impacted on the acquisition of the new "Ipsilateral" rule.[2] An ANOVA revealed that in the "Contralateral" condition, however, the effect of cueing/expectancy across SOA varied according to whether subjects undertook the "Contralateral" condition first or second (see Figure 2.1b and 2.1c). Further analysis revealed a significant latency advantage (of 46 ms) for the expected, but *uncued*, location at the SOA of 100 ms for those who undertook this condition first,[3] but not for those who had recently participated in the "Ipsilateral" condition.

Provided attention is not focused in advance, attention is *initially* reflexively drawn to a sudden-onset peripheral stimulus (Yantis & Jonides, 1990). Successfully shifting attention away from the stimulus (e.g. a cue) and to an alternative location involves, not just orienting, but *re*direction or *re*orienting of attention. The process of reorienting visual attention away from a peripheral stimulus is thought to involve three components: disengagement of attention from the initial location, moving the focus of attention to the new position, and reengagement of attention at the new location (Posner, 1988). The significant expectancy effect at the SOA of 100 ms for the subjects who participated in the "Contralateral" condition first suggests that visual attention can be rapidly reoriented away from a salient sudden-onset peripheral cue — at least under simple target detection conditions when the method of responding is by saccadic eye movements. The group of subjects who participated in the "Contralateral" condition first provides a more valid indication of normal subjects' capacity to rapidly redirect attention away from a salient peripheral stimulus based on expectancy. The task and stimuli were novel to this group of subjects; any potential effects of fatigue, familiarity, or bias towards the cue were absent.

In order to investigate the time course of this rapid redirection of attention in more detail, we (Kean & Lambert, 2001) conducted a further experiment (Experiment 2) which was similar to the "Contralateral" condition of Experiment 1, except only short SOAs (33, 67 and 100 ms) were employed. Figure 2.2 shows the saccadic latency data from this experiment. A significant latency advantage (of 31 ms) for the expected, but uncued, location was again observed at an SOA of 100 ms — the new subjects were also able to rapidly shift their attention away from the cue and towards the more likely

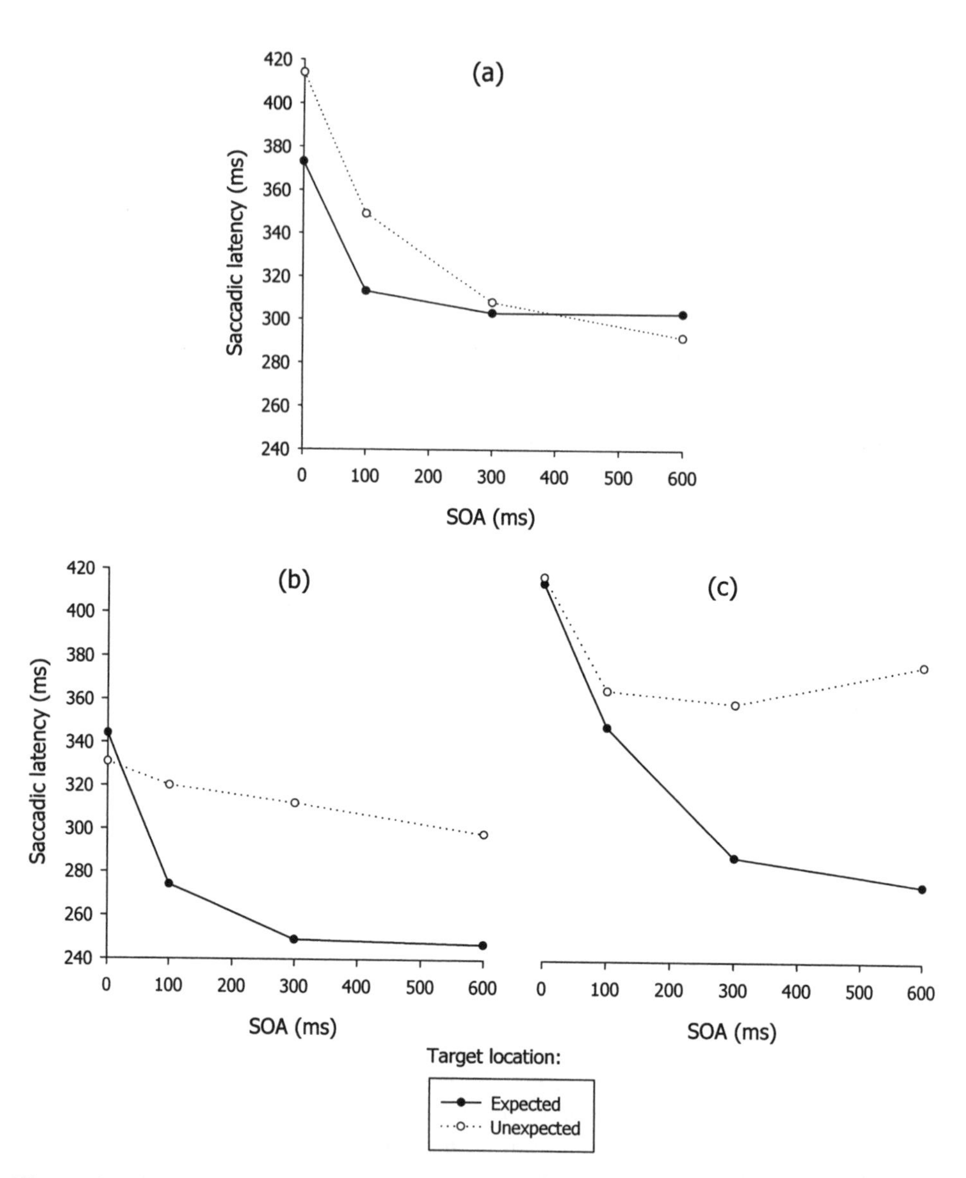

Figure 2.1: Mean saccadic latencies to expected versus unexpected target locations, across four SOAs, in Experiment 1 of Kean and Lambert (2001). (a): "Ipsilateral" condition. (b): "Contralateral" condition — subjects who participated in this condition first. (c): "Contralateral" condition — subjects who participated in this condition second. (For the "Ipsilateral" condition, Expected = target presented ipsilateral to the cue, Unexpected = target presented contralateral to the cue. For the "Contralateral" condition, Expected = target presented contralateral to the cue, Unexpected = target presented ipsilateral to the cue. SOA = stimulus onset asynchrony).

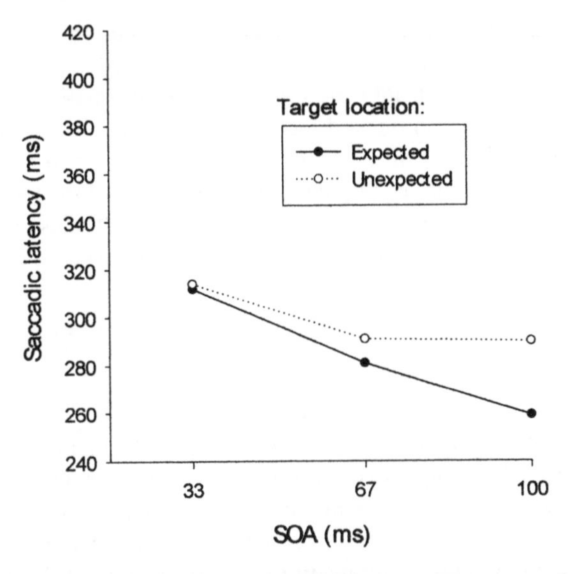

Figure 2.2: Mean saccadic latencies to expected versus unexpected target locations, across three SOAs, in Experiment 2 of Kean and Lambert (2001). (Expected = target presented contralateral to the cue, Unexpected = target presented ipsilateral to the cue. SOA = stimulus onset asynchrony).

target-location. However, at the shorter SOAs of 33 and 67 ms, there was no significant difference between latencies to the expected versus unexpected locations. This suggests that the process of reorienting attention away from the cue, and towards the alternative location, is incomplete with these brief cue–target delays.

Evidence for reorienting away from a stimulated location at an SOA of 100 ms or less is a rare outcome. The majority of studies in which an informative peripheral cue has signalled likely remote target appearance have failed to show a processing advantage for a predicted (and nonstimulated) location, relative to the cued (and unlikely) location, at short SOAs (e.g. Posner *et al.*, 1982; Lambert *et al.*, 1987; Warner *et al.*, 1990, Experiment 1; Tepin & Dark, 1992. Prior to the current work, only two spatial cueing studies (Warner *et al.*, 1990, Experiment 2; Danziger & Kingstone, 1999) have found a processing advantage at an uncued/expected location, relative to a cued/unexpected location, at an SOA of 100 ms or less. It is noteworthy that, compared to both of these studies, the current experiments consisted of a relatively small number of trials. In Warner *et al.,* (1990), for example, evidence that attention could be oriented, at an SOA of 100 ms, from a cue to an alternative location was only observed after extensive practice (approximately 4600 trials). In the study by Danziger and Kingstone (1999), in which subjects were able to direct attention away from a stimulated location to an alternative location at the brief SOA of 50 ms, subjects were exposed to 660 trials in each of three conditions. In our study (Kean & Lambert, 2001), subjects were presented with only 176 trials in each condition of Experiment 1, and 198 trials in Experiment 2. Aspects of the methodology employed in our study might contribute to an explanation of why such rapid redirection of attention was found with comparatively few trials. The most obvious distinction — between the methodologies employed in previous research into the effects of an informative peripheral cue predicting target appearance in an uncued location, and our study — relates to the mode of response (i.e., saccadic eye movements versus manual responses). If it is accepted that eye movements are a more ecologically valid indicator of the locus of visual attention than key-press responses, then it is possible that studies employing manual responses may have disguised the extent of normal observers' capacity to rapidly reorient visual attention.

Summary

With only brief cue–target onset asynchronies, subjects in the Kean and Lambert (2001) study exhibited an impressive degree of flexibility in the orienting of spatial attention. Under the current experimental conditions, subjects were able to rapidly (at the SOA of 100 ms) shift visual attention either towards or away from a salient peripheral stimulus, depending on the contingent relationship between the cue's location and the likely location of the impending target.

Paradoxically, observers showed less flexibility in orienting with long cue–target delays. At the relatively long SOAs of 300 and 600 ms, participants could readily orient their attention away from the peripheral cue when they were aware that the target was unlikely to appear at the cued location. In contrast, however, in the "Ipsilateral" condition of Experiment 1, the expected location (ipsilateral to the cue) was not associated with attentional orientation at these long SOAs; saccadic eye movements to the target were initiated just as quickly when the target appeared in the uncued (and unexpected) location as when it appeared in the cued (and expected) location. Observers responding with eye movements appear to exhibit a limited capacity to endogenously maintain their attentional focus at the location of a peripheral cue indicating likely nearby target appearance. One interpretation of this finding is that IOR nullifies the effect of endogenous orienting towards a cued location when observers respond with a saccadic eye movement to the target (cf. Crawford & Müller, 1992).

Distinct Bilateral Cues

The most interesting finding from Kean and Lambert (2001) was at the SOA of 100 ms. At this brief cue–target onset delay, the visual orienting system exhibits a remarkable degree of flexibility, at least when subjects respond to the target with saccadic eye movements, and when the task environment is kept simple. We were interested in whether observers would also show evidence for rapid orienting, on the basis of expectancy information conveyed by informative peripheral cues, when the task environment was made slightly more complex.

To address this issue, we conducted a spatial cueing study that involved the presentation of sudden-onset peripheral cue stimuli that, across three conditions, either conveyed useful information regarding the likely location of the impending target (i.e., informative cues) or did not (i.e., uninformative cues). (Again, the main findings are reported here; full details will appear in Kean and Lambert (in press)). However, while the majority of studies involving peripheral cueing have employed a single cue, the present experiment investigated the effects on orienting of two bilaterally- and simultaneously-presented cues. The cues were perceptually identical to each other except for their respective levels of luminance: i.e., on a given trial, one cue was brighter than the other. The luminance levels of the target stimulus, the bright cue, and the dim cue were 15.5 candelas per square metre (cd/m^2), 11.7 cd/m^2, and 2.0 cd/m^2, respectively. Participants again responded to the appearance of the target by executing a saccadic eye movement in order to fixate it. In each condition of the experiment, at each of four brief SOAs (zero, 50, 100 and 150 ms), saccadic latencies for trials on which the target appeared on the side of the bright cue were compared with latencies for trials on which the target appeared near the dim cue.

There were three conditions in this experiment: a "Neutral" condition, in which the target was equally likely to appear on the side of the bright cue or the dim cue, as well as "Dim side likely" and "Bright side likely" conditions, in which the target usually (p=0.8) appeared ipsilateral to the dim or bright cue, respectively. Subjects were informed of the cue–target relationship in advance. An ANOVA revealed a significant three-way interaction between condition, target-location (i.e., ipsilateral to the bright versus dim cue), and SOA. Given this interaction, the conditions were analysed separately.

Figure 2.3a shows the saccadic latency data from the "Neutral" condition. An ANOVA revealed a significant main effect of target-location; mean saccadic latency was, overall, faster to the target when it was presented on the side of the bright (306 ms) versus the dim cue (320 ms). This advantage for the bright cue's location was evident even at the brief cue–target onset delay of 50 ms: there was a significant saccadic latency advantage for the location of the bright cue across the SOAs of zero and 50 ms. These results point to the operation of the exogenous process of attention capture; i.e., it appears that attention was reflexively — and rapidly — drawn to the more salient cue.

Results for the "Dim side likely" condition are shown in Figure 2.3b. An ANOVA revealed no main effect of target-location, and no interaction between target-location and SOA. The data from this condition were compared directly with those from the "Neutral" condition. Although the interaction between condition and target-location failed to reach significance, the pattern of results suggests that attention capture was suppressed in the "Dim side likely" condition at an early stage. Specifically, while the bright cue exerted an attention-capturing effect across the SOAs of zero and 50 ms in the "Neutral" condition, a nonsignificant latency advantage for the location of the dim cue was observed within this brief time course in the "Dim side likely" condition. One interpretation of this finding is that disengagement of attention (Posner, 1988) from the location of the bright cue was rapidly accomplished. However, given that there was no latency advantage for targets appearing on the side of the dim cue, it would appear that subjects were unable to reverse the effect of attention capture and align their attention with the least perceptually salient, but more strategically useful, of the two stimuli.

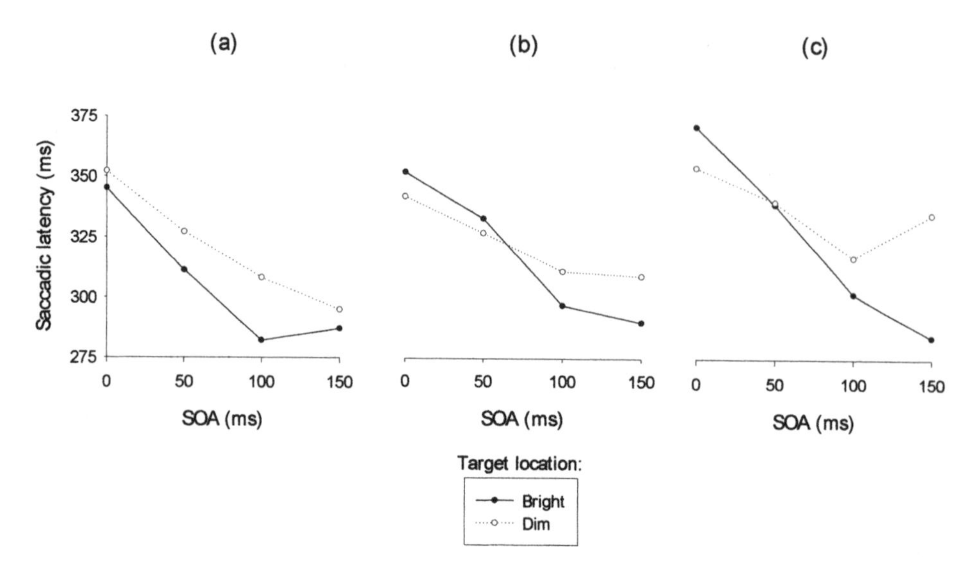

Figure 2.3: Mean saccadic latencies to the target presented ipsilateral to the bright versus dim cue, across four SOAs, in Kean and Lambert (in press). (a): "Neutral" condition. (b): "Dim side likely" condition. (c): "Bright side likely" condition. (Bright = target presented ipsilateral to the bright cue, Dim = target presented ipsilateral to the dim cue. SOA = stimulus onset asynchrony).

Results for the "Bright side likely" condition are shown in Figure 2.3c. An ANOVA revealed no main effect of target-location, but a significant interaction between target-location and SOA. Again, in contrast to the "Neutral" condition, a nonsignificant latency advantage for the location of the *less* salient cue was observed across the SOAs of zero and 50 ms. This finding was unexpected; the bright cue's attention-capturing property, which was evident in the "Neutral" condition, appeared to have been attenuated now that the target was four times more likely to appear near the bright cue. One possible explanation for this finding is that attention was rapidly disengaged from the bright cue's location in the "Bright side likely" condition, just as it appeared to have been in the "Dim side likely" condition. It may be that, in order to initiate strategic orienting to one of two bilateral stimuli, attention is initially disengaged from the more salient, attention-capturing stimulus, even if that same location is subsequently favoured for attentional allocation. However, further research is needed to evaluate this post-hoc speculation.

Further analysis of the "Bright side likely" condition revealed a significant saccadic latency advantage (of 51 ms) for the location of the bright cue at the SOA of 150 ms. This finding differs from the "Neutral" condition, where the latency advantage for targets appearing near the bright cue had dissipated to 8 ms by this SOA. The disparity between the two conditions suggests that a process distinct from reflexive attention capture underlies orienting to the bright cue at the SOA of 150 ms in the "Bright side likely" condition. One interpretation of this finding is that it may represent the leading edge of endogenous orienting to the most likely target-location. However, this explanation, in its simplest form, is problematic for the following reason: If the *endogenous* orienting mechanism *alone* could allocate attentional resources to the expected target-location at this SOA, we would also expect to see evidence for its operation in the "Dim side likely" condition at the same SOA. One possible resolution to this problem is that the effect of attention capture, having been momentarily suppressed in order to initiate strategic orienting, might resurface after a delay of around 100 to 150 ms. At the SOA of 150 ms in the "Bright side likely" condition, the exogenous and endogenous orienting mechanisms may act together to facilitate orienting to the location which had been previously stimulated by the bright cue, and at which the target is also expected. In contrast, at this SOA in the "Dim side likely" condition, any residual influence of attention capture (by the bright cue) would exert a detrimental effect on the attempt to strategically direct attention to the dim cue's location.

Summary

At the brief SOAs that we (Kean & Lambert, in press) employed in this experiment, it is evident that orienting of visual attention in response to bilateral cues can be affected by cue informativeness. The rapid reflexive orienting to the more salient of the two cue stimuli, which was observed in the "Neutral" condition, was suppressed in both the "Bright side likely" and "Dim side likely" conditions. We propose that the *attempt* to strategically orient to one of two stimuli momentarily disengages the reflexive process of attention capture, even when this is detrimental to performance.

While the results point to rapid expectancy-based interference in the reflexive process of attention capture, there are clearly limitations to the ability to strategically shift attention towards one of two bilaterally- and simultaneously-appearing stimuli. Not only did informativeness exert a somewhat detrimental influence on orienting in the "Bright side likely" condition, but in the "Dim side likely" condition there was no latency advantage at the dim cue's location, suggesting that these subjects were unable to completely avoid attention capture by the more perceptually salient stimulus.

Derived cueing

The experiments just described were investigations into the orienting of visual attention based on *elementary* information conveyed by peripheral stimuli. In Kean and Lambert (2001), for example, observers attempted to shift their attention based on only one feature of the peripheral cue — its location. In Kean and Lambert (in press), some discrimination between the two bilateral cue stimuli was necessary before attention could be strategically shifted to the appropriate location.

In everyday life, more information is usually encapsulated in peripherally perceived objects — to which we may or may not also overtly attend — than simple luminance variations. Stimulus features that relate to the *identity* of an object in peripheral vision, aside from its gross physical characteristics like location or luminance, may be critical factors in determining the next area of the visual field to which attention will be directed.

Recent studies by Lambert and colleagues (e.g. Lambert & Sumich, 1996; Lambert *et al.*, 1999; Lambert *et al.*, 2000; Lambert & Roser, 2001; Lambert & Duddy, 2002) have explored the influence of a variety of stimulus attributes on orienting behaviour; in some instances using target-detection manual responses, and in others target-detection saccades. In some of the experiments reported in these papers, extremely rapid effects of orienting on the basis of peripheral information — conveyed, in these instances, by features relating to the *identities* of the cues — have been observed.

Lambert and Duddy (2002), for example, investigated whether a probabilistic relationship between the arrangement of bilateral cues and the target-location could exert an influence on orienting behaviour. As in the experiment described above by Kean and Lambert (2001), the two bilaterally- and simultaneously-presented cues differed from each other, but the distinction between Lambert and Duddy's (2002) cues rested on their respective identities — specifically, one of these cues was the letter X and the other T. For each subject a sudden-onset target either usually (p=0.8) appeared near the X and opposite the T, or vice versa, and subjects were informed of this relationship. In the four experiments that utilised this same basic design, a target detection advantage (as indexed by manual response latencies) for the expected target-location was observed across a range of SOAs — even when these cue–target onset delays were very brief (i.e., zero, 33, 66, and 100 ms).

Distinct bilateral letter cues were also employed by Lambert *et al.* (2000, Experiment 1), in an experiment that used saccadic eye movements as the response measure. As for the letter cues in the Lambert and Duddy (2002) study, there was a predictive relationship between the arrangement of the cues and the likely target-location. In this experiment, however, subjects were not informed of the cue–target relationship. Despite the fact that most subjects were not aware of the existence of any cue–target relationship, a saccadic latency advantage was observed at the likely target-location after a relatively brief practice period. This effect had a rapid time course (it was apparent at the SOA of 100 ms), suggesting that attention was rapidly oriented in response to the predictive information carried by the letters. Lambert *et al.* (2000) concluded that the cue–target relationship had been implicitly acquired.

Orienting on the basis of a cue–target contingency of which subjects remained unaware was also found by Lambert *et al.* (1999, Experiment 1). In this study, in which manual responses to the target were employed, a significant latency advantage for the most likely target-location — as indicated by the arrangement of bilateral letter cues — was again observed at the brief SOA of 100 ms. Lambert and Roser (2001), in a study which also employed manual responses to the target, found an analogous effect with distinct bilateral cues that were subjectively similar colours. Once again, the effect was rapid — it was observed at the SOA of 100 ms — and was independent of cue–target contingency awareness.

In addition to these rapid attentional effects — based on the processing of relatively low-level stimulus features — *implicit* orienting on the basis of an undeniably complex attribute of a peripheral stimulus has also been observed, although after a longer SOA (Lambert & Sumich, 1996). Lambert and Sumich (1996) investigated the influence of peripherally presented semantic information on orienting. In three experiments, which all shared the same basic design, an SOA of 600 ms elapsed between the appearance of a cue and the presentation of a target, to which participants made a simple key-press detection response. The cue in these experiments was a single, peripherally presented word, which participants were instructed to ignore. Half of the words referred to living things (e.g. frog), while the remaining half referred to inanimate objects (e.g. bike). For half of the participants, the target usually (p=0.8) appeared on the same side as the cue word if it referred to something living, and on the opposite side if it referred to something non-living; the reverse contingency applied for the remaining participants. In all three experiments, participants responded more rapidly to targets that appeared at the likely location, as indicated by the semantic category of the purportedly unattended cue words. This was in spite of the fact that the majority of subjects (26/32) were unaware of the existence of any cue–target relationship, and that none of those who claimed contingency awareness mentioned word category in their descriptions of the perceived relationship. Unfortunately, no shorter SOAs were tested in this study, so it is uncertain whether any effects might be observed at brief SOAs. Presumably, though, the processing required to interpret the cue is a time-consuming one in the case of such relatively complex stimuli. Nevertheless, the capacity to orient on the basis of such high-level attributes of peripheral stimuli — when subjects were unaware of the cue-identity/target-location relationship — reveals

> an impressive sensitivity of the human perceptual system to abstract characteristics of ostensibly irrelevant visual objects. (Lambert & Sumich, 1996, p. 516)

Summary

William James viewed learning and experience as important factors in the control of attentional behaviour. James suggested, over a century ago, that shifts of attention could be "derived", whereby the stimulus "owes its interest to association with some other immediately interesting thing" (James, 1890/1983, p. 393). This notion has only recently been tested in spatial cueing studies. The work considered in the section above points to the following conclusion: *The identity of a cue is able to influence attentional orientation by virtue of its association with a target event.* Essentially, contingent relationships between features of a cue (or cues) and the most likely target-location may be acquired, and such learned associations are capable of influencing the attentional response system. Lambert and colleagues (e.g. Lambert *et al.*, 2000) have referred to the process responsible for this type of attentional effect as "derived cueing", after James's notion of "derived attention".

Associations between cue-identity and target-location can be explicitly held — an observer may, for example, be informed of the relationship prior to participation, or they may become aware of it during the course of the task. In addition, a cue-identity/target-location contingency may be acquired — or "derived" — implicitly (Lambert & Sumich, 1996; Lambert *et al.*, 1999, Experiment 1; Lambert *et al.*, 2000, Experiment 1; Lambert & Roser, 2001); in other words, it appears that the learning which underlies derived cueing can proceed independently of conscious processes.

General Discussion

In normal circumstances, humans make between three to five saccadic eye movements per second, interspersed with periods of 200–300 ms during which the eyes do not move to any marked extent (Fischer & Weber, 1993). Saccades — the most common means of overtly orienting visual attention — are intricately tied to shifts of covert attention (Rizzolatti *et al.*, 1987; Schneider, 1995; Corbetta, 1998). Our eyes do not (in general, at least) jump randomly from one point in the visual scene to another, by chance happening to focus on an interesting object. We must inspect the large area of the visual field that lies outside the immediate range of the fovea so that objects of interest at parafoveal or peripheral locations, which are currently receiving sub-optimal processing, may be overtly attended.

But to what does visual attention shift? What features of an "object of interest" might be processed by the observer and used to guide orienting? Many studies have shown that attention is rapidly drawn to — or "captured by" — a luminance change in the periphery. However, in everyday life, a peripheral event is seldom merely a luminance change; usually it comprises *information*, which may — or may not — be useful to the organism.

The evidence reviewed in this chapter demonstrates that information conveyed by peripheral stimuli contributes to attentional orientation. This has been found across a broad range of cue–target onset asynchronies: it appears that informative peripheral cues can elicit rapid, as well as delayed effects. We will focus here primarily on observations at brief SOAs, as some of the most interesting findings described above have involved rapidly acting orienting effects. Of course, orienting of visual attention based on peripheral information at longer SOAs — e.g. greater than approximately 275 ms (Müller & Rabbitt, 1989) — is also of interest. Processes that operate within this range of the attentional time course include the endogenous orienting mechanism and IOR. At long SOAs, we can observe interactions between these concurrently active processes, as they each compete for directional control over the "spotlight" of attention.

In the study by Kean and Lambert (2001) described above, findings at the long SOAs of 300 and 600 ms in the "Ipsilateral" condition of Experiment 1 are especially relevant to this notion of a dual-process interaction. In this situation, even though subjects might be said to be endogenously attempting to orient to the previously cued location, it appeared that IOR nullified this effort. A similar finding was reported by Crawford and Müller (1992) for subjects who responded with a saccade to the target, but not for subjects who responded manually. One lesson to be derived from Crawford and Müller's (1992) data — one worth bearing in mind when considering findings from the field of visual attention — is that saccadic eye movement responses and key-press responses may yield different results (see also Maylor, 1985; Briand *et al.*, 2000).

Rapid attentional orienting effects (i.e., effects observed at an SOA of around 100 ms or less) occur in response to a diverse range of peripheral information. In its simplest form, the mere stimulation provided by a single peripheral event can be all the "information" that is necessary for a rapid shift of visual attention to occur (i.e., in the case of attention capture). In a similar vein, we (Kean & Lambert, in press) have shown that the more salient of two, otherwise identical, stimuli exerts a rapidly acting influence on attention, in a manner closely resembling — or possibly identical to — that of attention capture. Some spatial cueing studies have found, though, that if a salient cue conveys information predicting target appearance in an alternative location, rapid suppression of attention capture can occur (e.g. Lambert *et al.*, 1987; Tepin & Dark, 1992; Kean & Lambert, in press). For example, in the Kean and Lambert (in press) study described above, the more salient of two bilateral cue stimuli appeared to capture attention when the arrangement of these cues was uninformative with respect to the impending target location, but when the target was four times more likely to appear ipsilateral to the dimmer cue, this effect was attenuated.[4] It has also been shown, in a small number of studies, that on detection of a unilateral peripheral event which signals "attend elsewhere", attention can be rapidly and successfully redirected to an alternative location (Warner *et al.*, 1990; Danziger & Kingstone, 1999; Kean & Lambert, 2001). Furthermore, surprisingly rapid[5] derived cueing effects on the basis of peripheral information have also been observed (Lambert *et al.*, 1999; Lambert *et al.*, 2000; Lambert & Roser, 2001; Lambert & Duddy, 2002).

It seems reasonable to surmise that distinct processes may subserve rapid orienting in response to different attributes of peripheral cues. And it appears, indeed, that the mechanism underlying rapid derived cueing is distinct from — at least — the process responsible for attention capture. Lambert *et al.* (2000, Experiment 2), for example, using saccadic eye movements as the response measure under monocular viewing conditions, showed that derived cueing and attention capture yielded distinctive patterns of naso-temporal asymmetry; specifically, the influence of derived cueing was confined exclusively to the nasal visual field, while a greater temporal field effect was seen in the case of attention capture. This finding indicates that distinct neural processes most likely subserve the two types of effects.

Certain processes can act rapidly to shift attention *away* from a peripheral stimulus. Attention capture can be suppressed, for example, or the peripheral event may be used as a stimulus to attend elsewhere. This outcome lies in stark contrast to attention capture, whereby attention is swiftly directed to a stimulated location — suggesting that separate processes account for the different effects. An attention shift away from — and, for that matter, towards — a source of peripheral stimulation, based on information conveyed by its location (e.g. Kean & Lambert, 2001), is widely thought to involve the operation of the endogenous orienting mechanism (Posner *et al.*, 1982; Müller & Rabbitt, 1989; Tepin & Dark, 1992; Danziger & Kingstone, 1999). Expanding on a model proposed by Yantis and Jonides (1990), for example, Tepin and Dark (1992) hypothesised that the attentional response mechanism which responds to a peripheral cue comprises a "detection" and a "prioritisation" component. The detection function is concerned with registration of a peripheral event, while the prioritisation function "specifies what should be done given the occurrence of the event" (Tepin & Dark, 1992, p. 128). These authors proposed that detection of a cue is an automatic process, while the prioritisation function is endogenously programmable depending on the demands of the task.

The idea that the prioritisation function is endogenously programmable would be hard to deny. Evidence certainly suggests that, given enough time, we are able to voluntarily direct our attention based on information conveyed by a peripheral stimulus concerning the likely location of an impending target (e.g. Posner *et al.*, 1982). However, the orienting of visual attention based on information conveyed by a peripheral cue's location can be extremely rapid, even when this involves redirecting attention away from the cue (Warner *et al.*, 1990; Danziger & Kingstone, 1999; Kean & Lambert, 2001). Indeed, Tepin and Dark (1992) themselves showed that attention capture could be suppressed at the brief SOA of 33 ms. Such brief SOAs (i.e., around 100 ms and below) represent a time course in which — according to conventional wisdom — the influence of the endogenous orienting mechanism is *not normally expected* (e.g. Müller & Rabbitt, 1989).

One possible explanation for rapid reorienting of attention (as reported in Kean & Lambert, 2001) is that consciously directed (i.e., endogenous) processes are capable of operating within a substantially shorter time frame than is typically thought possible. Perhaps, for example, a sudden-onset peripheral cue that is predictive of another location is more conducive to eliciting rapid voluntary attention shifts than, say, an arrow presented at fixation. However, we should not be too hasty in reaching this conclusion. An alternative possibility is that Tepin and Dark's (1992) "prioritisation function" may be activated — or "programmed" — via *implicit*, as well as via explicit sources. In other words, it could be the case that the process of shifting attention away from a cue and to an alternative location need not rely on the endogenous orienting mechanism. An underlying mechanism's independence from conscious input would account for the rapidity of reorienting observed in our study (Kean & Lambert, 2001). This speculation is most plausible with respect to a simple task environment — such as a scenario in which only two possible target-locations exist — because the programming of the appropriate cue–target contingency is presumably readily achievable, potentially via simple associative learning.

In broad agreement with the hypothesis that orienting away from a peripheral stimulus might occur independently of conscious influence, Kentridge *et al.* (1999) proposed that

> a simple cue–target contingency may be encoded in the collicular-parietal, eye-movement control system and may not necessarily require the involvement of frontal language and memory systems. (p. 1810)

The superior colliculus is known to be involved in the orienting of covert visual attention, as well as the preparation and generation of saccadic eye movements. It is possible, then, that the ability to rapidly redirect visual attention is particularly evident when individuals respond with a saccadic eye movement because this type of simple reorienting of covert attention, as well as the actual response to the target, are both mediated via closely-related neural pathways. It is worth noting here, parenthetically, that spatial cueing studies in which a peripheral cue signals likely contralateral target appearance are similar to studies employing the antisaccade task (see, e.g., Everling & Fischer, 1998; Zhang & Barash, 2000). An exploration of the relationship between these closely related paradigms would be an interesting avenue of investigation.

In the Kean and Lambert (2001) study, the possibility that endogenous processes played a role in *rapid* reorienting away from the cue cannot be excluded, because subjects were aware of the likelihood of contralateral target appearance. Furthermore, the co-occurrence of explicit and implicit processes may, in at least some instances, contribute to rapid reorienting effects. Both top-down and bottom-up pathways might independently encode the association between the cue's location and the likely target-location, and separately call on the attentional response system for resource allocation. However, although our subjects possessed explicit knowledge of the cue–target contingency, it does not necessarily follow that top-down processes were responsible for reorienting effects observed *across all SOAs*. Even when an observer is aware that they should attend to a certain location, the processes governing the allocation of attention to that location might *only* involve an implicit (unconscious) pathway. Further research is needed to clarify these issues.

Notwithstanding these considerations, though, a rapid time course of attentional orientation is consistent with the involvement of non-voluntary processes, but problematic for an interpretation resting solely on the endogenous orienting mechanism. We suggest there is reason to doubt the widely held assumption that attention capture is the only process that acts via the exogenous orienting mechanism to exert an influence on attentional allocation. There might, instead, be a number of processes — such as derived cueing, suppression of attention capture, and reorienting of attention — that can each operate independently of conscious direction (i.e., in a bottom-up, stimulus-driven fashion), ultimately impacting on attentional orientation. (Alternatively, one could view suppression of attention capture and reorienting of attention as components of the derived cueing process). Essentially, there may be multiple routes by which the exogenous orienting mechanism may be activated.

Much of the evidence considered in this chapter points to the conclusion that information conveyed by peripheral stimuli is capable of influencing our attentional orienting behaviour. However, this research has undoubtedly only scratched the surface of such effects. There may well exist a cornucopia of stimulus attributes that exert an influence on visual attention in our daily lives. For instance, might attention be rapidly allocated to one of two onset cues which is less perceptually salient but more critical to immediate survival? Further, recent evidence has shown that attention can be drawn to a peripheral event even when the observer is not subjectively aware of the source of stimulation (McCormick, 1997; Danziger *et al.*, 1998; Kentridge *et al.*, 1999; Lambert *et al.*, 1999). Is it possible that the *nature* of a peripheral stimulus, which is not represented in conscious awareness, might exert an influence on the attentional response system? These questions await attention.

Acknowledgements

Some of the research reported here (Kean & Lambert, 2001; Kean & Lambert, in press) was undertaken as partial fulfilment of a Ph. D. degree at the University of Auckland, and was supported by the University of Auckland Research Committee grant to the first author. We are grateful to two anonymous reviewers for helpful comments on an earlier draft of this chapter.

Notes

1 Mean saccadic latencies were relatively long at the SOA of zero ms. An SOA affords the opportunity to prepare for the target's appearance, and with no cue–target delay this warning-signal effect is absent. In addition, participants knew that, given a cue–target delay, the second onset would be the target. This temporal information, which could be used to discriminate between the two stimuli, is absent at zero ms SOA.
2 See Appendix A (Table 2.1) for the data from the "Ipsilateral" condition broken down according to the order in which subjects participated in this condition.
3 An anonymous reviewer suggested that stroboscopic (beta) apparent motion might account for this finding. However, the parameters employed here (i.e., an SOA of 100 ms, and a separation of 12.8° between the two stimuli when they appear opposite each other) are incompatible with the spatiotemporal limits of apparent motion (see, e.g., Caelli & Finlay, 1979).
4 Curiously, this "suppression of attention capture" effect was also observed in the "Bright side likely" condition: it seems that the attempt to orient *towards* the brighter stimulus also diminished its attention-capturing property.
5 This is not meant to imply that derived cueing effects are manifested exclusively at brief SOAs. See Lambert and Sumich (1996), for example, for evidence that an implicitly derived — and complex — cue–target contingency is capable of exerting an influence on orienting at an SOA of 600 ms.

References

Abrams, R. A., & Dobkin, R. S. (1994). Inhibition of return: Effects of attentional cuing on eye movement latencies. *Journal of Experimental Psychology: Human Perception and Performance, 20,* 467–477.

Briand, K. A., Larrison, A. L., & Sereno, A. B. (2000). Inhibition of return in manual and saccadic response systems. *Perception and Psychophysics, 62,* 1512–1524.

Caelli, T., & Finlay, D. (1979). Frequency, phase, and colour coding in apparent motion. *Perception, 8,* 59–68.

Corbetta, M. (1998). Frontoparietal cortical networks for directing attention and the eye to visual locations: Identical, independent, or overlapping neural systems? *Proceedings of the National Academy of Sciences: USA, 95,* 831–838.

Corbetta, M., Miezin, F. M., Shulman, G. L., & Petersen, S. E. (1993). A PET study of visuospatial attention. *The Journal of Neuroscience, 13,* 1202–1226.

Crawford, T. J., & Müller, H. J. (1992). Spatial and temporal effects of spatial attention on human saccadic eye movements. *Vision Research, 32,* 293–304.

Danckert, J., & Maruff, P. (1997). Manipulating the disengage operation of covert visual spatial attention. *Perception and Psychophysics, 59,* 500–508.

Danziger, S., & Kingstone, A. (1999). Unmasking the inhibition of return phenomenon. *Perception and Psychophysics, 61,* 1024–1037.

Danziger, S., Kingstone, A., & Rafal, R. D. (1998). Orienting to extinguished signals in hemispatial neglect. *Psychological Science, 9,* 119–123.

Deubel, H., & Schneider, W. X. (1996). Saccade target selection and object recognition: Evidence for a common attentional mechanism. *Vision Research, 36,* 1827–1837.

Everling, S., & Fischer, B. (1998). The antisaccade: A review of basic research and clinical studies. *Neuropsychologia, 36,* 885–899.

Fischer, B., & Weber, H. (1993). Express saccades and visual attention. *Behavioral and Brain Sciences, 16,* 553–610.

Folk, C. L., Remington, R. W., & Johnston, J. C. (1992). Involuntary covert orienting is contingent on attentional control settings. *Journal of Experimental Psychology: Human Perception and Performance, 18,* 1030–1044.

Goldberg, M. E. (2000). The control of gaze. In: E. R. Kandel, J. H. Schwartz, and T. M. Jessell (eds), *Principles of Neural Science* (4th edn) (pp. 782–800). New York: McGraw-Hill.

Hoffman, J. E., & Subramaniam, B. (1995). The role of visual attention in saccadic eye movements. *Perception and Psychophysics, 57,* 787–795.

James, W. (1983). *The Principles of Psychology*. Cambridge, MA: Harvard University Press. (Original work published 1890).

Jonides, J. (1981). Voluntary versus automatic control over the mind's eye's movement. In: J. B. Long and A. D. Baddeley (eds), *Attention and Performance IX* (pp. 187–203). Hillsdale, NJ: Erlbaum.

Kean, M., & Lambert, A. (2001). The influence of an informative sudden-onset peripheral cue on the latency of target-detection saccades. Unpublished manuscript, University of Auckland.

Kean, M., & Lambert, A. (In press). The influence of a salience distinction between bilateral cues on the latency of target-detection saccades. Manuscript submitted for publication to British Journal of Psychology.

Kentridge, R. W., Heywood, C. A., & Weiskrantz, L. (1999). Attention without awareness in blindsight. *Proceedings of the Royal Society of London B, 266,* 1805–1811.

Kowler, E., Anderson, E., Dosher, B., & Blaser, E. (1995). The role of attention in the programming of saccades. *Vision Research, 35,* 1897–1916.

Lambert, A., & Duddy, M. (2002). Visual orienting with central and peripheral precues: Deconfounding the contributions of cue eccentricity, cue discrimination and spatial correspondence. *Visual Cognition, 9,* 303–336.

Lambert, A., Naikar, N., McLachlan, K., & Aitken, V. (1999). A new component of visual orienting: Implicit effects of peripheral information and subthreshold cues on covert attention. *Journal of Experimental Psychology: Human Perception and Performance, 25,* 321–340.

Lambert, A., Norris, A., Naikar, N., & Aitken, V. (2000). Effects of informative peripheral cues on eye movements: Revisiting William James' "derived attention". *Visual Cognition, 7,* 545–569.

Lambert, A., & Roser, M. (2001). Effects of bilateral colour cues on visual orienting: Revisiting William James' "derived attention". *New Zealand Journal of Psychology, 30,* 16–22.

Lambert, A., Spencer, E., & Mohindra, N. (1987). Automaticity and the capture of attention by a peripheral display change. *Current Psychological Research and Reviews, 6,* 136–147.

Lambert, A., & Sumich, A. (1996). Spatial orienting controlled without awareness: A semantically based implicit learning effect. *Quarterly Journal of Experimental Psychology, 49A,* 490–518.

Liversedge, S. P., & Findlay, J. M. (2000). Saccadic eye movements and cognition. *Trends in Cognitive Sciences, 4,* 6–14.

Maylor, E. A. (1985). Facilitatory and inhibitory components of orienting in visual space. In: M. I. Posner and O. S. M. Marin (eds), *Attention and Performance XI* (pp. 189–204). Hillsdale, NJ: Erlbaum.

McCormick, P. A. (1997). Orienting attention without awareness. *Journal of Experimental Psychology: Human Perception and Performance, 23,* 168–180.

Müller, H. J., & Rabbitt, P. M. A. (1989). Reflexive and voluntary orienting of visual attention: Time course of activation and resistance to interruption. *Journal of Experimental Psychology: Human Perception and Performance, 15,* 315–330.

Posner, M. I. (1980). Orienting of attention. *Quarterly Journal of Experimental Psychology, 32,* 3–25.

Posner, M. I. (1988). Structures and functions of selective attention. In: T. Boll and B. Bryant (eds), *Clinical Neuropsychology and Brain Function: Research, Measurement, and Practice* (pp. 169–202). Washington, DC: American Psychological Association.

Posner, M. I., & Cohen, Y. (1984). Components of visual orienting. In: H. Bouma and D. G. Bouwhuis (eds), *Attention and Performance X* (pp. 531–556). Hillsdale, NJ: Erlbaum.

Posner, M. I., Cohen, Y., & Rafal R. (1982). Neural systems control of spatial orienting. *Philosophical Transactions of the Royal Society of London B, 298,* 187–198.

Posner, M. I., Nissen, M. J., & Ogden, W. C. (1978). Attended and unattended processing modes: The role of set for spatial location. In: H. Pick and E. Saltzman (eds), *Modes of Perceiving and Processing Information* (pp. 137–157). Hillsdale, NJ: Erlbaum.

Posner, M. I., Rafal, R. D., Choate, L. S., & Vaughan, J. (1985). Inhibition of return: Neural basis and function. *Cognitive Neuropsychology, 2,* 211–228.

Rizzolatti, G., Riggio, L., Dascola, I., & Umilta, C. (1987). Reorienting attention across the horizontal and vertical meridians: Evidence in favor of a premotor theory of attention. *Neuropsychologia, 25,* 31–40.

Robertson, L. C., & Rafal, R. (2000). Disorders of visual attention. In: M. S. Gazzaniga (ed.), *The New Cognitive Neurosciences* (pp. 633–649). Cambridge, MA: MIT Press.

Robinson, D. L., & Kertzman, C. (1995). Covert orienting of attention in macaques. III. Contributions of the superior colliculus. *Journal of Neurophysiology, 74,* 713–721.

Rosen, A. C., Rao, S. M., Caffarra, P., Scaglioni, A., Bobholz, J. A., Woodley, S. J., Hammeke, T. A., Cunningham, J. M., Prieto, T. E., & Binder, J. R. (1999). Neural basis of endogenous and exogenous spatial orienting: A functional MRI study. *Journal of Cognitive Neuroscience, 11,* 135–152.

Schneider, W. X. (1995). VAM: A neuro-cognitive model for visual attention control of segmentation, object recognition, and space-based motor action. *Visual Cognition, 2,* 331–376.

Sheliga, B. M., Riggio, L., & Rizzolatti, G. (1995). Spatial attention and eye movements. *Experimental Brain Research, 105,* 261–275.

Tepin, M. B., & Dark, V. J. (1992). Do abrupt-onset peripheral cues attract attention automatically? *Quarterly Journal of Experimental Psychology, 45A,* 111–132.

Theeuwes, J. (1991). Exogenous and endogenous control of attention: The effect of visual onsets and offsets. *Perception and Psychophysics, 49,* 83–90.

Warner, C. B., Juola, J. F., & Koshino, H. (1990). Voluntary allocation versus automatic capture of visual attention. *Perception and Psychophysics, 48,* 243–251.

Wright, R. D., & Ward, L. M. (1994). Shifts of visual attention: An historical and methodological overview. *Canadian Journal of Experimental Psychology, 48,* 151–166.

Yantis, S., & Jonides, J. (1990). Abrupt visual onsets and selective attention: Voluntary versus automatic allocation. *Journal of Experimental Psychology: Human Perception and Performance, 16,* 121–134.

Zackon, D. H., Casson, E. J., Zafar, A., Stelmach, L., & Racette, L. (1999). The temporal order judgment paradigm: Subcortical attentional contribution under exogenous and endogenous cueing conditions. *Neuropsychologia, 37,* 511–520.

Zhang, M., & Barash, S. (2000). Neuronal switching of sensorimotor transformations for antisaccades. *Nature, 408,* 971–975.

Appendix A

Table 2.1: Mean saccadic latencies (ms) for the "Ipsilateral" condition of Kean and Lambert (2001; Experiment 1) – according to the order in which subjects participated in this condition.

Target-Location	**Subjects who undertook this condition first**			
	SOA (0 ms)	**SOA (100 ms)**	**SOA (300 ms)**	**SOA (600 ms)**
Expected	415	361	333	341
Unexpected	451	394	342	322
	Subjects who undertook this condition second			
	SOA (0 ms)	**SOA (100 ms)**	**SOA (300 ms)**	**SOA (600 ms)**
Expected	336	270	276	269
Unexpected	381	310	279	266

(Expected = target presented ipsilateral to the cue. Unexpected = target presented contralateral to the cue. SOA = stimulus onset asynchrony).

Chapter 3

Executive Contributions to Eye Movement Control

Timothy L. Hodgson and Charlotte Golding

This paper addresses the question of how humans perform in situations where task rules change dynamically from moment to moment. Using several tasks requiring the control of saccades based on changing rules, we show that the spatial and temporal properties of saccades are influenced by previously active associations. Control subjects are usually able to overcome these influences and execute the correct response. In contrast, patients with frontal damage are unable to control saccades efficiently when no external cues are provided to instruct the current stimulus-response association. We propose that frontal cerebral cortex monitors conflict between actual and optimum neural states and biases competitive neural interactions to meet high-level goals.

Introduction

Active vision is a dynamic process involving the flexible co-ordination of different search strategies to meet task goals (Yarbus, 1967; Land *et al.*, 1999; Hayhoe *et al.*, 1997; Hodgson *et al.*, 2000a). The function of switching between strategies is often attributed to "executive" or "supervisory" control processes (Baddeley & Della Salla, 1996; Shallice & Burgess, 1998), yet the details by which such control might be implemented remains poorly understood. Here we describe the characteristics of eye movements made by normal participants and neurological patients during several tasks which require cognitive flexibility. The results provide important insights into the demands placed on eye movement control during complex cognitive tasks, as well as characteristics of executive control and its neural substrates.

The Mind's Eye: Cognitive and Applied Aspects of Eye Movement Research

ISBN: 0–444–51020–6

Switching Between Simple Saccade Tasks

The majority of investigations into saccadic eye movements have used very simple paradigms in which participants execute a single saccadic response on each trial. Typically, participants might be asked to make a movement towards (pro-saccade) or away from (anti-saccade) a salient stimulus onset in the periphery. In the majority of studies the task to be performed remains constant throughout a block of trials; however a few studies have examined the flexible control of saccades by requiring participants to switch between these two tasks within a block of trials.

In Hallett and Adams' (1980) original paper on anti-saccades it was found that there was no difference in response latencies when pro- and anti-saccade trials were mixed within a block. The authors concluded that the "psychological details" under which eye movement tasks were presented had little influence on saccadic latency. In contrast, another study by Weber (1995) reported an increase in both saccadic latency and error rates on the first trial following a switch between the two tasks.

An important difference between these two studies was that in Hallett's design the task was precued before the start of a trial, whilst in Weber's study the shape of the saccade target instructed either a pro- or an anti-saccade. The fact that participants had more time to prepare themselves for the new task in Hallett and Adams' study might explain the absence of an interference effect when switching between the two tasks.

Recently, Hunt and Klein (in press) varied the interval between the task cue and the onset of a target stimulus. Using an analogous pro/anti manual response task, they found a significant increase in manual response times on the first trial of a new run of pro or anti responses. This "task switch cost" persisted even at long cue–target intervals (CTIs), where participants had time to prepare themselves for the switch in task. In contrast, when saccades were used as the dependent motor response, significant switch costs only arose at the shortest cue–target intervals.

The absence of switch costs on saccade latencies at long CTIs is surprising in the light of the robustness of task switch costs in other paradigms (Pashler, 2000; Allport *et al.*, 1994; Allport & Wylie, 2000; Rogers & Monsell, 1995). However, one plausible explanation for the absence of costs under these conditions is that reflexive pro-saccades are so automatic that they are not susceptible to task switch costs. It may be for this reason that participants are able to fully enable and disable reflexive responding prior to the start of a new run of anti-saccades without incurring any behavioural cost (Hunt & Klein, 2002; Hodgson *et al.,* submitted manuscript).

Symbolic Control of Eye Saccades

As well as coordinating behaviour based on peripheral target onsets, humans and other animals are also able to direct eye movements and attention using more abstract cues in their environment (Wise & Murray, 2000; Hodgson & Müller, 1999). Recently we have examined a task which probes how humans learn and flexibly control these symbolic associations. In the "rule reversal" task a computer generates a rule linking the colour of a central cue with the direction of a saccade. At the start of the experiment the participant has to guess which direction is associated with the coloured cue by looking towards either the left or the right response box. Following a fixation longer than 800 ms on one of the response boxes, feedback is given to indicate whether the participant's decision was correct or incorrect (Figure 3.1a). Following runs of between 9 and 13 consecutive correct responses the computer reverses the rule linking the colour of the cue with the direction of response (rule changes are cued by the occcurence of an unexpected error feedback). Up to eight of these rule reversals can occur in a block of 100 trials.

On the first trial following the unexpected error which cues the rule change, saccadic latencies are found to be increased in control participants. Furthermore, an increased number of erroneous saccades are observed after the rule change, suggesting that controls have a tendency to execute movements based on the preceding, but now incorrect, rule mapping. The majority of these errors are quickly corrected, so that gaze is redirected towards the appropriate response location and a correct feedback is given by the computer (Figure 3.1b, Figure 3.2a).

We have also tested six patients who had brain damage affecting the frontal cerebral cortex (Hodgson *et al.*, 2000b). All the patients had damage in the right lateral frontal region; none of their lesions involved areas in the medial wall of the frontal cortex (e.g. anterior cingulate) or primary oculomotor areas such as the frontal and supplementary eye fields (mean age: 58±12, range 36–66).

The patient group was found to make an increased rate of uncorrected saccade errors and elevated response latencies compared to a group of nine mixed age control subjects (mean age: 54.5±19, range 29–75) (Figure 3.2). Notably, they also made significantly *fewer* corrective saccadic responses than age-matched controls on the first trial after a rule reversal (significant interaction between subject group and trial after rule change: $F(4,56) = 10.42$, $p < 0.001$). On debriefing the patients reported being all too aware that they were performing poorly in the task; yet despite this insight they found it difficult to acquire new stimulus-response mappings.

It is interesting to consider the relationship between the performance of patients in the rule reversal task and the more commonly reported deficit in anti-saccade performance following frontal lobe damage (Everling & Fischer, 1998; Fukushima *et al.*, 1988; Guitton *et al.*, 1985; Lasker *et al.*, 1987; Pierrot-Deseilligny *et al.*, 1991). Previous studies in humans and primates suggest that frontal cortical regions, including those lying outside primary oculomotor centres are important in suppressing reflexive saccades towards sudden onset targets (Schlag-Rey *et al.*, 1992; Schlag-Rey *et al.*, 1997; Everling *et al.*, 1999; Everling & Munoz, 2000; Everling *et al.*, 1998; O'Driscoll *et al.*, 1995; Sweeney *et al.*, 1996; Reingold & Stampe, 2002). A possible interpretation of the present results is that frontal patients are unable to suppress errors in oculomotor output based on strongly reinforced symbolic mappings as well as peripheral onsets.

One of the patients tested on the rule reversal test was the subject of a more detailed study of anti-saccade performance (Walker *et al.*, 1998). This patient made close to 100% errors in the anti-saccade task, but invariably corrected his mistakes such that he ended each trial fixating the correct location. This contrasts with the performance of the same patient in the rule reversal task where saccade errors were not corrected. In fact, total error rates (corrected plus uncorrected errors) were not significantly

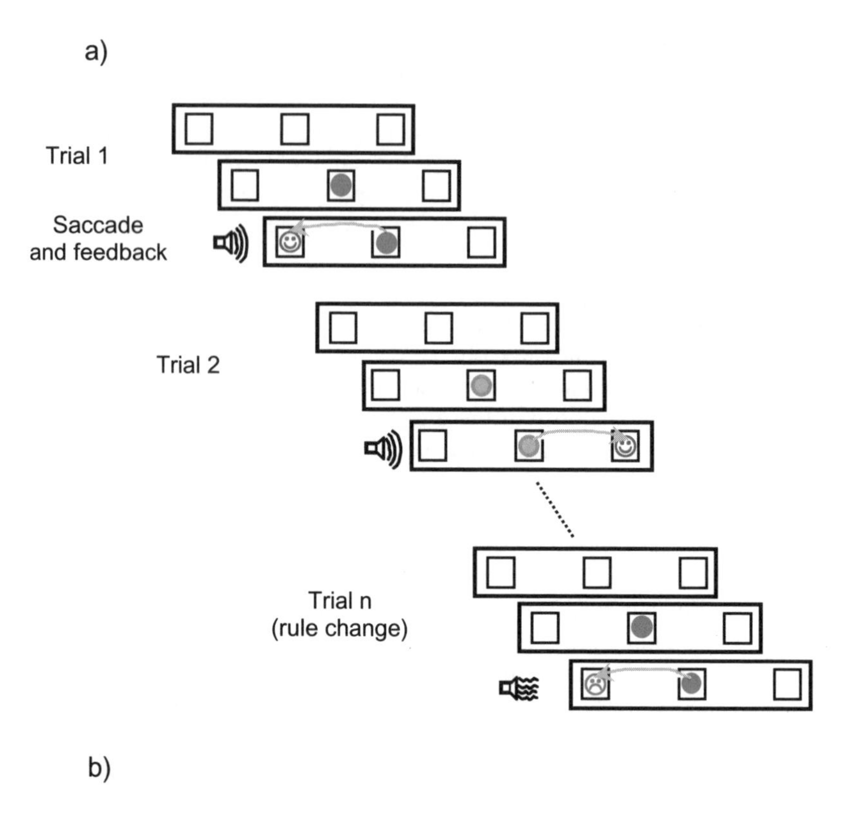

b)

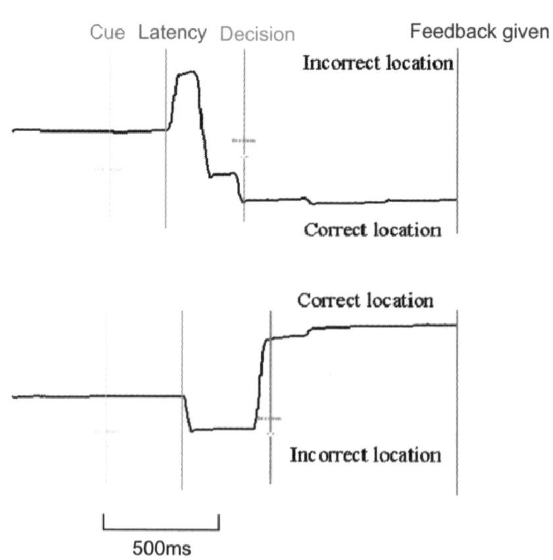

Figure 3.1: (a) Rule reversal task. Subjects learn a rule linking a coloured shape with a movement to either the left or the right. Feedback is given following a fixation >800ms

increased for frontal patients relative to controls on the first trial after a rule reversal. It was only a comparison of corrected errors that revealed a significant difference between the two groups, suggesting that an isolated deficit in response suppression is unlikely to explain the result.

In the light of the comparative data on anti-saccade performance the problems experienced by frontal patients in the rule reversal task might be best described as reflecting an impairment in the behavioural monitoring functions necessary to learn arbitrary stimulus–saccade associations rather than a failure to suppress saccade execution. Information concerning recent actions (e.g. saccade direction), their antecedents (e.g. colour) and associated feedback must be integrated over time to establish new rule mappings. It may be a deficit in this integrative process which underlies lateral frontal patients' performance deficit in this task.

Card Sorting

One question arising from the performance of controls and patients in the rule reversal task is how the findings might generalise to more complex tasks used in standard clinical neuropsychological assessment. Many other tasks considered to be sensitive to "executive dysfunction" are not explicitly oculomotor in nature, but neverthelss demand the coordination of sophisticated eye movement strategies (Hodgson *et al.*, 2000a).

Recently we have examined the spontaneous eye movements made whilst performing a modification of the Wisconsin card sort task, a test commonly used to assess behavioural flexibility in neuropsychological patients. In our version of the task, each trial involves the presentation of three "Response" cards depicting different coloured symbols at the top of a computer touch screen. A fourth "Stimulus" card is also displayed at the bottom of the screen. The participants' task is to point to the Response card that matches the Stimulus card according to the dimension of either shape or colour (Figure 3.3a) (Golding *et al.*, 2001). In some blocks the dimension to sort by is indicated at the start of the trial by a *direct* cue: "SHAPE" or "COLOUR'. Performance on these blocks is contrasted with blocks in which the dimension to sort by was instructed by an *indirect* cue: "SAME" or "DIFFERENT" (relative to the last trial).[1]

After a change in sorting dimension, a behavioural cost is seen on a number of measures in control subjects. Total response times as well as the latency of the first

---

on one of the response boxes allowing several eye movements to be made before a decision is recorded. After a random number of trials the rule can reverse (indicated by the occurrence of an unexpected error on the rule change trial). The task is self-paced with at least 1500ms elapsing between each trial. (b) Eye position traces showing example corrective saccades. Labelled vertical lines superimposed on eye movement trace indicate timing of discrete stimulus and behavioural events within a trial. On a proportion of trials participants make saccade errors followed by a corrective movement towards the correct response box.

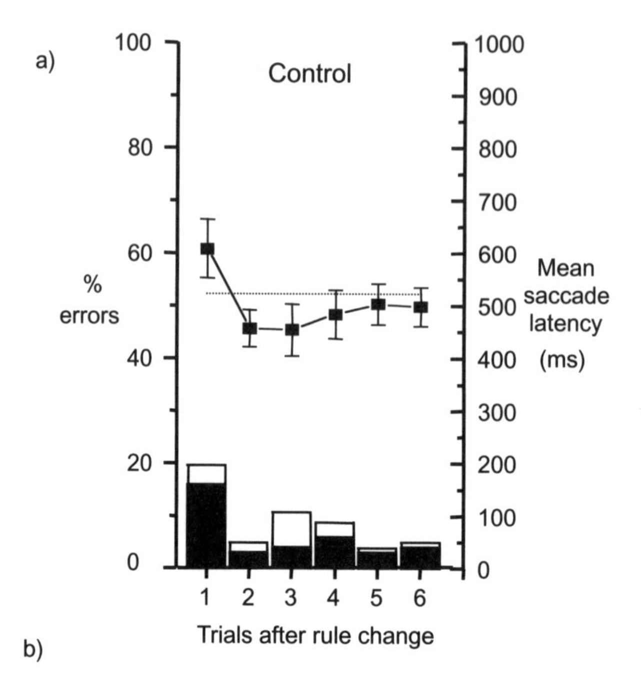
a)
Control
100
80
60
40
20
0
% errors
1000
900
800
700
600
500
400
300
200
100
0
Mean saccade latency (ms)
1 2 3 4 5 6
Trials after rule change

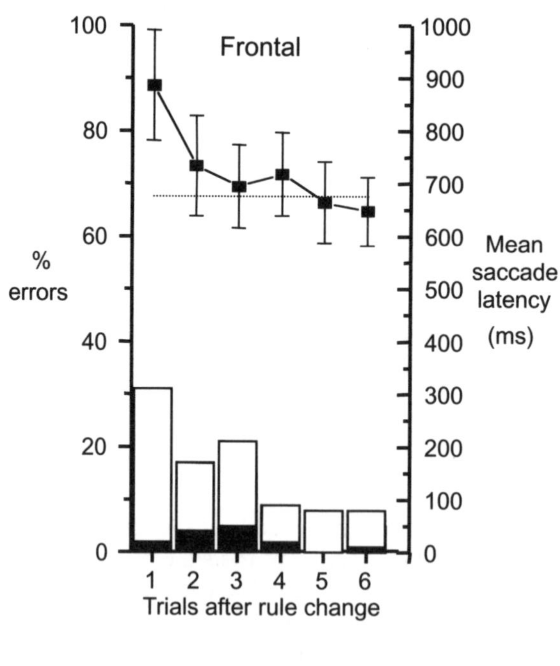
b)
Frontal
100
80
60
40
20
0
% errors
1000
900
800
700
600
500
400
300
200
100
0
Mean saccade latency (ms)
1 2 3 4 5 6
Trials after rule change
Saccade latency
Errors
Corrected errors

saccade are increased (Figure 3.3b).[2] Most participants make large saccades directly towards the correct Response card before pointing to it. However, following a rule change more extraneous saccades are seen.

We also observe an effect of dimension changes on the dynamics of saccades made between the Stimulus and Response cards. The trajectories of saccades are subtly curved towards the location of non-target cards which share a common feature with the Stimulus card (Figure 3.3c). The effect is particularly marked on the first two trials after a rule switch when colour is the distracting stimulus dimension (Figure 3.3a and b) (interaction effect on initial direction of saccade, trial type (switch/non-switch) × distractor location F (1, 9) = 23.19, P = 0.001).

Despite this competing influence of distracting stimulus features on eye movements, control participants correctly selected the target card on 99% of trials. It made little difference whether direct or indirect cues were used to instruct the sorting rules (Figure 3.4).

We have also tested a single patient with a well circumscribed lesion in the right lateral prefrontal cortex using this task (Golding *et al.,* 2001). We found that this patient was able to perform the task well when the dimension was directly cued (albeit with more eye movements and longer response times than controls). However, when indirect cues were used, the patient found it very difficult to keep track of the current sorting dimension and continued to sort stimuli on the basis of the shape dimension throughout most of the test (Figure 3.4). As with the data from the rule reversal task described above, this pattern of performance is consistent with a role for frontal structures in self-monitoring, rather than a specific role in initiating attentional switches across perceptual dimensions (Rogers *et al.*, 2000; Pollmann *et al.*, 2000).

Cognitive Planning

The card sorting task and rule reversal task described above explicitly test the ability to implement flexible behavioural control. But in other "executive" tasks this demand is not so transparent. An example of such a test is the "Tower of London" (TOL).[3] In this task participants are presented with two arrangements of coloured billiard balls on top of each other in pockets (Figure 3.5a). The aim is to rearrange the lower set of balls to match the upper set using the shortest possible number of moves (N.B. balls always fall to the bottom of pockets and cannot be taken out of pockets without removing balls positioned above them).

Superficially, the task seems to require an ability to hold a sequence of actions in mind and then replay them during problem execution. However, an alternative

---

Figure 3.2: Performance characteristics of (a) controls and (b) patients with right lateral frontal cortex damage in the rule reversal task, showing saccadic latency, corrected and uncorrected saccade errors following a reversal in stimulus-response mappings (dotted line shows latency in control condition for which mappings remained constant throughout the block).

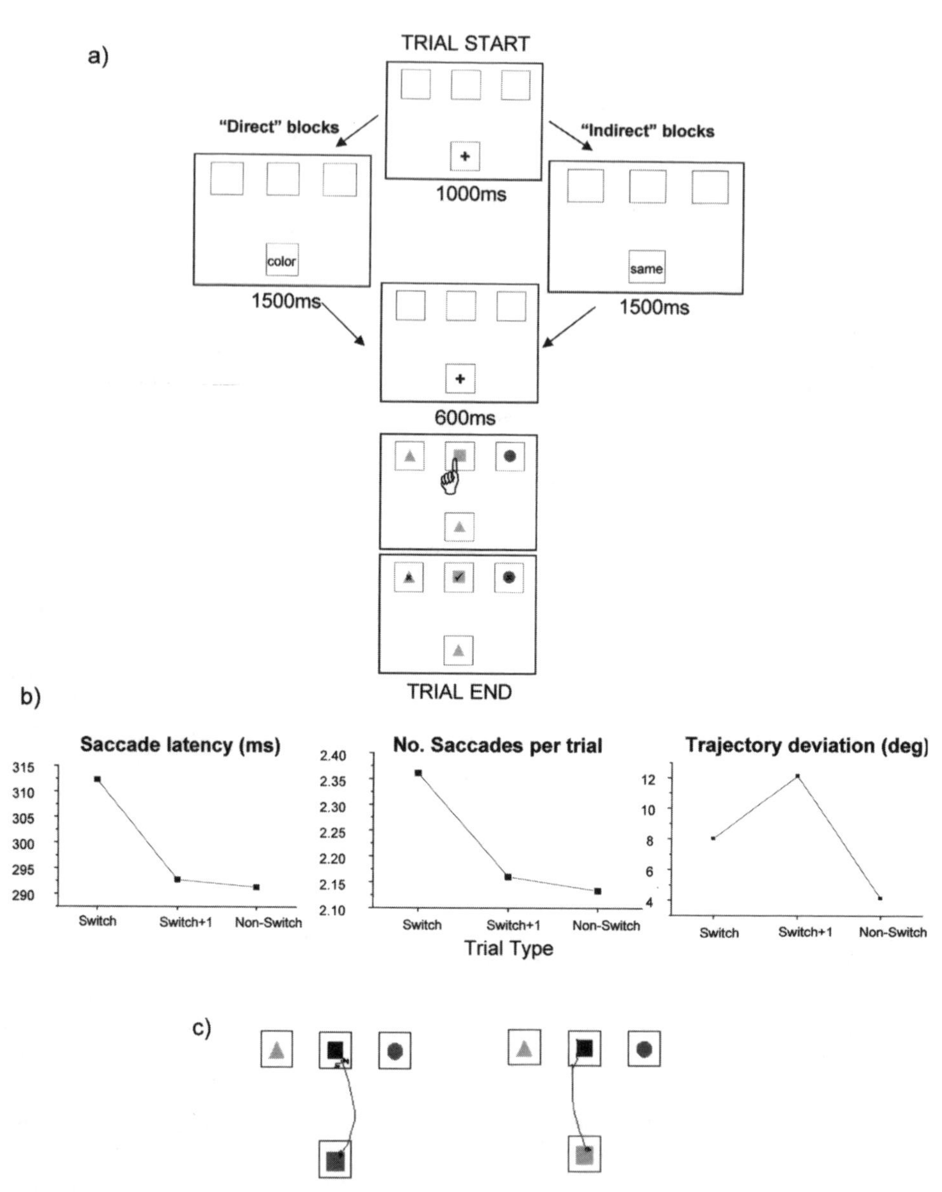

Figure 3.3: (a) Modified card sorting task. Subjects are required to point to one of three "response" cards, which matches a "stimulus" card according to one of two possible stimulus dimensions. The dimension to sort by is indicated by either a *direct* cue ("shape"/"colour") or an *indirect* cue ("same"/"different" relative to last trial) in different trial blocks. (b) Graphs showing switch costs following a change in sorting dimensions on latency of first saccade, total number of saccades and direction of first saccade. (c) Examples of eye movement traces showing curvature of trajectories towards salient distractors on shape matching trials.

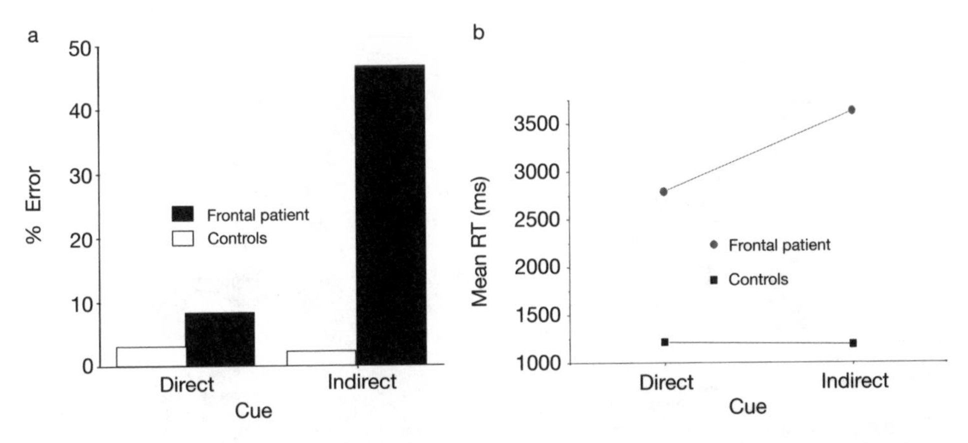

Figure 3.4: Comparison of (a) error rates and (b) reaction times in the card sorting task between control subjects and a single patient with right prefrontal cortex damage.

view of cognitive planning is that only a few key features need to be encoded during advanced planning resulting in a much lower load on short-term memory during problem solving (Ward & Allport, 1997; Phillips *et al.*, 2001).

In one recent study we measured the eye movements made by control participants while they planned solutions to TOL problems (Hodgson *et al.*, 2000a). Participants were shown pictures of TOL problems and were asked to solve them "in their heads" without moving the balls. When they thought they had worked out the solution they pressed a response key and told the experimenter how many moves it would take to solve the problem (i.e. the "one touch" tower of London task, Owen *et al.*, 1995).

We wanted to establish whether eye movements were used in a strategic way during the task or were randomly sampling the display regardless of what moves were required to solve a problem. We therefore designed problems to be visually very similar to one another whilst requiring quite different strategies for solution. On some trials successful solution required a shunting manoeuvre using the central blue ball, positioning it into a temporary location whilst making several other moves. Failure to appreciate this key feature of these "blue ball" problems leads to an impasse in which the participant must back track and undo moves in order to reach the correct solution. In contrast, "non-blue ball" problems do not require this sub-goal move; on these trials balls from the left or the right pockets must be moved first (Figure 3.5a). Both these problem types were presented in two "isomer" forms, requiring either predominantly left to right or right to left ball moves.

The task was made sufficiently difficult to induce high error rates in some control subjects. When we compared the pattern of eye movements made by these "error makers" with those who performed the task efficiently, we found that error makers spent longer fixating problem-irrelevant balls and locations. In contrast, efficient problem solvers were more focused on the central blue ball location on blue ball trials and lateral balls on non-blue ball trials (Figure 3.5b) (Location × problem type ×

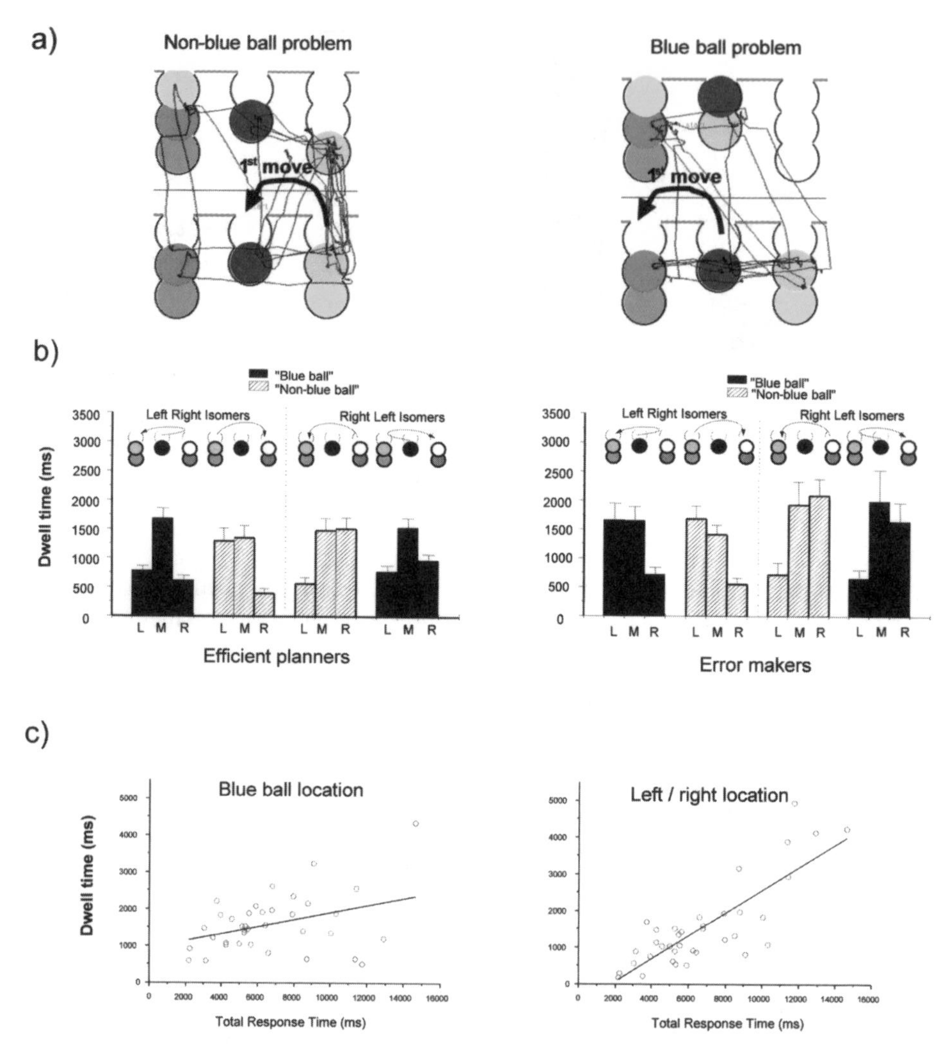

Figure 3.5: Eye movement recordings during cognitive planning. Participants were shown pictures of "Tower of London" problems. Their task was to mentally plan the shortest sequence of moves which would rearrange the bottom set of balls to match the top. The problems used were designed to be visually similar, but required different strategies for solution. (a) example eye movement traces on problems which required either "blue ball" or "non-blue ball" solution strategies (right to left problem "isomers" shown). (b) Mean gaze times per trial on lower array of balls for participants who performed the task efficiently and those who made >5% errors in the task, showing mean fixation times on left, middle and right ball locations for both left to right and right to left problem isomers. (c) Correlation of fixation time on different ball locations with total response time for blue ball problems (data from all relevant trials collapsed across subjects).

isomer type interactions for efficient problem solvers: $F(2,8) = 19.14$, $p < 0.001$ and error makers: $F(2,8) = 0.34$).

It was also found that the total time taken to solve blue ball problems correlated positively with the time spent fixating the lateral ball locations ($R^2 = 0.67$, $F(1,35)=70.71$, $p < 0.0001$). In contrast, the time spent looking at the blue ball itself correlated only weakly with solution time ($R^2 = 0.14$, $F(1,35) = 5.78$, $p < 0.05$) (Figure 3.5c). In other words, the total time taken to solve a problem was more dependent upon the time inspecting *irrelevant* rather than problem relevant locations. This result suggests that rather than rehearsing and recalling a fully formed action sequence, mental planning primarily involves locating the critical feature of a particular problem (e.g. the blue ball manoeuvre). Once this search process has been completed, the rest of the solution easily falls into place.

Other studies have reached similar conclusions about the nature of cognitive planning using different analyses (Ward & Allport, 1997; Phillips *et al.*, 2001). However, our study additionally demonstrated the importance of behavioural flexibility in solving the task. When we analysed total planning times on consecutive trials we found that participants who had high error rates were quicker to press the response key when the solution strategy of the current problem matched that of the previous trial (i.e. blue ball followed by a blue ball problem). But for those individuals who made relatively few errors in the task, it made no difference whether the previous problem was of the same or of a different type to the preceding trial (Figure 3.6) (trial type × preceding trial type × subject group: $F(1,6) = 11.74$, $p < 0.01$).

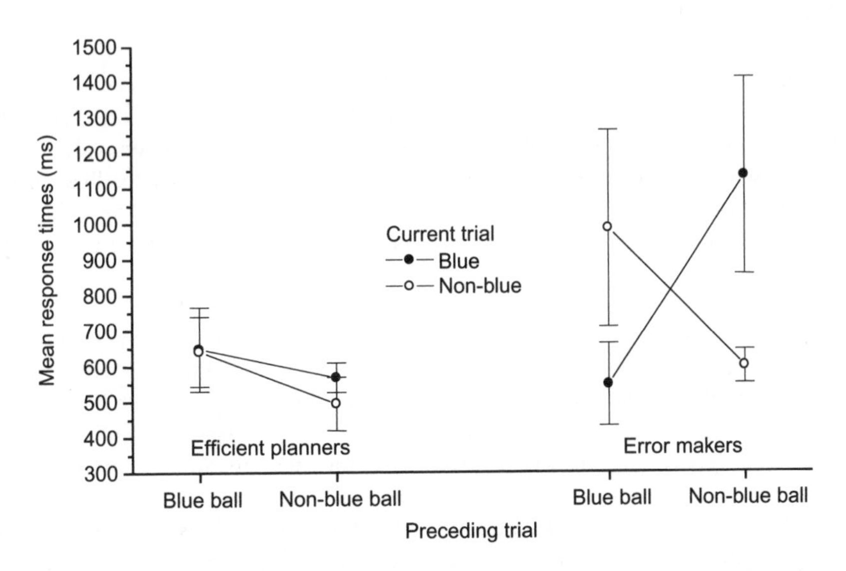

Figure 3.6: Total response times on blue and non-blue ball problems plotted against the problem type presented on the preceding trial. Efficient planners showed no effect of previous problem type. In contrast, for error prone subjects, planning times were strongly modulated by the strategy required on the preceding problem.

We conclude from this work that the TOL puts similar demands on the control of attention and eye movements as the other "executive" tasks described above. Although the stimulus presented to the subject remains relatively constant from trial to trial, attention must be shifted in a fluid manner between different objects, actions and stimulus–response associations. As well as requiring an ability to suppress previously reinforced behaviour, the TOL task also relies on the capacity to monitor recent actions, the sensory context in which they were executed and their success in achieving the task goals.

Discussion and Conclusions

We have examined the performance of neurologically normal participants during three tasks which require flexible control of eye movement strategies. In all three tasks previously reinforced gaze strategies influenced current response execution. In a symbolically cued eye movement task incorrect gaze shifts were executed following a rule reversal. During card sorting, more subtle effects were observed during on-line control, including deviation of movement trajectories following a change in the sorting dimension. In both these tasks, interference effects lasted at most two trials after a change in rules, with performance returning to baseline levels thereafter. Finally, in the more complex Tower of London test, analysis of eye movements revealed the importance of directing attention towards different problem features on visually similar problems. Some controls failed to direct attention in this fluid manner and showed interference from the preceding trial strategy.

Rather than an all controlling "homunculus", our studies suggest that executive control processes are a sleeping partner to more automatic mechanisms which take over once environment–behaviour associations have been learned. The function of executive control is to monitor this automatic system to ensure that behaviour remains compatible with high-level goals. It is only when motor and sensory feedback conflict with expectations based on these goals that direct top-down control must intervene. But can we ever hope to understand how such control might be implemented at the neural level?

Recent neurophysiological evidence suggests that attention can be viewed as a "race" between the firing rates of different neuronal populations selectively tuned for different attributes of an organism's environment (Desimone, 1999; Duncan, 1999; Chelazzi *et al.*, 1993). In this account frontal areas are assumed to provide a tonic biasing signal onto neurons coding for a sensory feature of interest (e.g. a particular colour, shape or location). This gives neurons with a preference for these features a small head start in the race (i.e. an increase in their base-line firing rates). Following stimulus onset, this small lead is exaggerated by mutually inhibitory connections between competing neural populations, such that neurons which represent objects with the attended visual features end up winning the race (i.e. reaching a criterion level of firing). In this way a distributed mosaic of activation emerges across posterior brain regions corresponding to the sensory features of a single visual object. Objects selected in this manner are then more likely to become the target for actions such as eye movements.

This "biased competition" model provides a compelling picture of the visual brain in action. But it also provides a framework for understanding how attentional mechanisms select not only between objects and locations but also alternative stimulus–response mappings and behavioural schema (Miller, 1999; Miller & Cohen, 2001).

For example, in the rule reversal task described above (Figure 3.1a) the presentation of a coloured shape can instruct two alternative eye movement responses on every trial. According to the biased competition view of task selection, pools of neurons coding for mutually exclusive colours, locations and colour-location mappings are simultaneously activated at cue onset. When an association is well established, neurons representing the correct association have the advantage and quickly suppress firing in neurons representing the incorrect mapping. However, following a reversal in reinforced stimulus-response mappings, enhanced competition arises between these different groups of neurons (see Asaad *et al.*, 1998). As a result the race becomes so close that even intact brains execute the incorrect behaviour on a proportion of trials.

Our investigations in patients with frontal lobe damage suggest that the frontal cerebral cortex is important for monitoring these competitive interactions as well as exerting top-down control. We suggest that frontal areas monitor recent behaviour and use this information to make predictions about future actions and expected sensory and motor feedback. It is only when these predictions are breached (e.g. by an unexpected penalising feedback or an error in motor output) that frontal regions adaptively bias neural competition towards new associative mappings (Miller & Cohen, 2000; Miller, 1999; Damasio, 1996; Hasegawa *et al.*, 2000; Fletcher *et al.*, 2001). Importantly, this account suggests that it makes little sense to characterize attentional selection processes as being exclusively oculomotor (Rizzolatti *et al.,* 1987; Rizzolatti *et al.,* 1994), perceptual or even based on retrieval of stimulus-response mappings from long term memory (Meiran, 1996). Executive contributions to attentional selection must operate across multiple domains and processing levels to ensure that behaviour continues to match expectations based on high-level goals.

Acknowledgements

This study was supported by Wellcome Trust Programme Grant Number: 037222/2/98/A. Charlotte Golding is supported by a Wellcome prize studentship. We would like to express our thanks to Marcia Chamberlain, Dominic Mort, Masud Husain, Clive Rosenthal and Prof. Christopher Kennard for their contributions to the work described in this chapter.

Notes

1 The first sorting dimension was instructed by the Experimenter at the start of the block.
2 Error rates are also increased on switch trials, making it unlikely that the cost reflects a straightforward change in decision criterion or response bias.
3 So called because it is a variant of the Tower of Hanoi puzzle and was devised at University College London (Shallice, 1982).

References

Allport, A., & Wylie, G. (2000). Task switching, stimulus-response bindings, and negative priming. In S. Monsell and J. Driver (eds), *Attention and Performance XVIII: Control of Cognitive Processes*. Cambridge: MIT Press.

Allport, D. A., Styles, E. A., & Hseih, S. (1994). Shifting intentional set: Exploring the dynamic control of tasks. In C. Umilta and M. Moscovitch (eds), *Attention and Performance XV: Conscious and Non-conscious Information Processing*. Cambridge, MA: MIT Press.

Asaad, W. F., Rainer, G., & Miller, E. K. (1998). Neural activity in the primate prefrontal cortex during associative learning. *Neuron*, *21*, 1399–1407.

Baddeley, A., & Della Sala, S. (1996). Working memory and executive control. *Philosophy Transactions of the Royal Society of London B, Biological Sciences 351*, 1397–1404.

Chelazzi, L., Miller, E. K., Duncan, J., & Desimone, R. (1993). A neural basis for visual search in inferior temporal cortex. *Nature*, *363*, 345–347.

Damasio, A. R. (1996). The somatic marker hypothesis and the possible functions of the prefrontal cortex. *Philosophy Transactions of the Royal Society of London B, Biological Sciences 351*, 1413–1420.

Desimone, R. (1999). Visual attention mediated by biased competition in extrastriate cortex. In: G. W. Humphreys, J. Duncan and A. Triesman (eds), *Attention, Space and Action* (pp. 13–31). Oxford: Oxford University Press.

Duncan, J. (1999). Converging levels of analysis in the cognitive neuroscience of visual attention. In G. W. Humphreys, J. Duncan and A. Triesman (eds), *Attention, Space and Action* (pp. 112–129). Oxford: Oxford University Press.

Everling, S., Dorris, M. C., Klein, R. M., & Munoz, D. P. (1999). Role of primate superior colliculus in preparation and execution of anti-saccades and pro-saccades. *Journal of Neuroscience*, *19*, 2740–2754.

Everling, S., & Fischer, B. (1998). The antisaccade: A review of basic research and clinical studies. *Neuropsychologia*, *36*, 885–899.

Everling, S., & Munoz, D. P. (2000). Neuronal correlates for preparatory set associated with pro-saccades and anti-saccades in the primate frontal eye field. *Journal of Neuroscience*, *20*, 387–400.

Fletcher, P. C., Anderson J. M., Shanks D. R., Honey R., Carpenter, T. A., Donovan T, Papadakis, N., & Bullmore, E. T., (2001). Responses of human frontal cortex to surprising events are predicted by formal associative learning theory. *Nature Neuroscience*, *4*(10), 1043–1048.

Fukushima, J., Fukushima, K., Chiba, T., Tanaka, S., Yamashita, I., & Kato, M. (1988). Disturbances of voluntary control of saccadic eye movements. *Biological Psychiatry*, *23*, 670–677.

Golding, C., Hodgson, T. L., & Kennard, C. (2001). Eye movements and set inhibition during a card sorting task. *Journal of Cognitive Neuroscience*, (suppl.), 48.

Guitton, D., Buchtel, H. A., & Douglas, R. M. (1985). Frontal lobe lesions in man cause difficulties in suppressing reflexive glances and in generating goal-directed saccades. *Experimental Brain Research*, *58*, 455–472.

Hallett, P. E., & Adams, B. D. (1980). The predictability of saccadic latency in a novel voluntary oculomotor task. *Vision Research*, *20*, 329–339.

Hasegawa, R. P., Blitz, A. M., Geller, N. L., & Goldberg, M. E. (2000). Neurons in monkey prefrontal cortex that track past or predict future performance. *Science*, *290*, 1786–1789.

Hayhoe, M. H., Bensinger, D. G., & Ballard, D. H. (1997). Task constraints in visual working memory. *Vision Research*, *38*(1), 125–137.

Hodgson, T. L., Bajwa, A., Owen, A. M., & Kennard, C. (2000a). Strategic control of gaze direction during the Tower of London task. *Journal of Cognitive Neuroscience, 12*(5), 894–907.

Hodgson, T. L., Chamberlain, M. M., & Kennard, C. (2000b). Gaze control, set switching and the prefrontal cerebral cortex. *Society for Neuroscience Abstracts*. New Orleans.

Hodgson, T. L., Golding, C., Mort, D., Molyva, D., Rosenthal, C. & Kennard, C. (submitted manuscript). Eye movements during task switching: Reflexive, symbolic and affective contributions to response selection.

Hodgson, T. L., & Müller, H. J. (1999). Attentional orienting in two-dimensional space. *Quarterly Journal of Experimental Psychology, 52A*, 615–648.

Hunt, A. R., & Klein, R. M. (2002). Eliminating the cost of task set reconfiguration. *Memory & Cognition, 30*(4) 529–539.

Land, M. F., Mennie, N., & Rusted, J. (1999). The roles of vision and eye movements in the control of activities of daily living. *Perception, 28*, 1311–1328.

Lasker, A. G., Zee, D. S., Hain, T. C., & Folstein, S. E. (1987). Saccades in Huntington's disease: Initiation defects and distractibility. *Neurology, 44*, 2285–2289.

Meiran, N. (1996). Reconfiguration of processing mode prior to task performance. *Journal of Experimental Psychology: Learning, Memory and Cognition, 22*, 1423–1442.

Miller, E. K. (1999). The prefrontal cortex: Complex neural properties for complex behavior. *Neuron, 22*, 15–17.

Miller, E. K., & Cohen, J. D. (2001). An integrative theory of prefrontal function. *Annual Review of Neuroscience, 24*, 167–202.

O'Driscoll, G. A., Alpert, N. M., Matthysse, S. W., Levy, D. L., Rauch, S. L., & Holzman, P. S. (1995). Functional neuroanatomy of antisaccade eye movements investigated with positron emission tomography. *Proceedings of the National Academy of Sciences, USA, 92*, 925–929.

Owen, A. M., Sahakian, B. J., Hodges, J. R., Summers, B. A., Polkey, C. E., & Robbins, T. W. (1995). Dopamine-dependent fronto-striatal planning deficits in early Parkinson's disease. *Neuropsychology, 9*, 126–140.

Pashler, H. (2000). Task switching and multi-task performance. In: S. Monsell and J. Driver (eds), *Attention and Performance XVIII: Control of Cognitive Processes* (pp. 277–309). Cambridge: MIT Press.

Pierrot-Deseilligny, C. P., Rivaud, S., Gaymard, B., & Agid, Y. (1991). Cortical control of reflexive visually-guided saccades. *Brain, 114*, 1472–1485.

Philips, L. H., Wynn, V. E., McPherson, S. and Gilhooly, K. J. (2001). Mental planning and the Tower of London task. *Quarterly Journal of Experimental Psychology, 54A*(2), 579–598.

Pollmann, S., Weidner, R., & Müller, H. J., von Cramon, D. Y. (2000). A fronto-posterior network involved in visual dimension changes. *Journal of Cognitive Neuroscience*, 12(3), 480–494.

Reingold, E. M., & Stampe, D. M. (2002). Saccadic inhibition in voluntary and reflexive saccades. *Journal of Cognitive Neuroscience, 14*(3), 371–388.

Rizzolatti, G., Riggio, L., Dascola, I., & Umiltá, C. (1987). Reorienting attention across the horizontal and vertical meridians: Evidence in favour of a premotor theory of attention. *Neuropsychologia, 25*, 31–40.

Rizzolatti, G., Riggio, L., & Sheliga, B. M. (1994). Space and selective attention. In: C. Umiltá and M. Moscovitch (eds), *Attention and Performance XV* (pp. 231–265). Cambridge: MIT Press.

Rogers, R. D., & Monsell, S. (1995). Costs of a predictable switch between simple cognitive tasks. *Journal of Experimental Psychology: General, 124*, 207–231.

Rogers, R. D., Andrews, T. C., Grasby, P. M., Brooks, D. J., & Robbins, T. W. (2000). Contrasting cortical and subcortical activations produced by attentional-set shifting and reversal learning in humans. *Journal of Cognitive Neuroscience*, *12*(1), 142–162.

Schlag-Rey, M., Amador, N., Sanchez, H., & Schlag, J. (1997). Anti-saccade performance predicted by neuronal activity in the supplementary eye field. *Nature*, *390*, 398–401.

Schlag-Rey, M., Schlag, J., & Dassonville, P. (1992). How the frontal eye field can impose a saccade goal on superior colliculus neurons. *Journal of Neurophysiology*, *67*, 1003–1005.

Shallice, T. (1982). Specific impairments of planning. *Philosophical Transactions of the Royal Society of London B, Biological Sciences*, *298*, 199–209.

Shallice, T., & Burgess, P. (1998). The domain of supervisory processes and the temporal organization of behaviour. In: A. C. Roberts, T. W. Robbins and L. Weiskrantz (eds), *The Prefrontal Cortex* (pp. 22–35). Oxford: Oxford University Press.

Sweeney, J. A,. Mintun, M. A., Kwee, S., Wiseman, M. B., Brown, D. L., Rosenberg, D. R., & Carl, J. R. (1996). Positron emission tomography study of voluntary saccadic eye movements and spatial working memory. *Journal of Neurophysiology*, *75*, 454–468.

Walker, R., Husain, M., Hodgson, T. L., Harrison, J., & Kennard, C. (1997). Saccadic eye movement and working memory deficits following damage to human prefrontal cortex. *Neuropsychologia*, *36*(1), 1141–1159.

Ward, G., & Allport, A. (1997). Planning and problem-solving using the five-disc Tower of London task. *The Quarterly Journal of Experimental Psychology*, *50A*(1), 49–78.

Weber, H. (1995). Presaccadic processes in the generation of pro and anti saccades in human subjects — a reaction-time study. *Perception*, *25*, 1265–1280.

Wise, S. P., & Murray, E. A. (2000). Arbitrary associations between antecedents and actions Trends. *Neuroscience*, *23*, 271–276.

Yarbus, A. (1967). *Movements of the Eyes*. New York: Plenum.

Chapter 4

Saccadic Selectivity During Visual Search: The Influence of Central Processing Difficulty

Jiye Shen, Eyal M. Reingold, Marc Pomplun and Diane E. Williams

The current study examined the relation between the difficulty of central discrimination and the efficiency of peripheral selection in visual search tasks. Participants were asked to search for a target among high-, medium-, and low-similarity distractors. In Experiment 1, while the duration of current fixations increased with increasing target-distractor similarity, there was no evidence that saccadic selectivity was influenced by the target-distractor similarity of the previously fixated item or by the duration of the previous fixation. In addition, we manipulated the difficulty of the central discrimination by introducing a concurrent visual task (Experiment 2) and by presenting a gaze-contingent moving mask (Experiment 3). Although both manipulations substantially degraded the overall visual search performance, the magnitude of peripheral selection was not affected. Results from the current study suggest that peripheral selection is a robust process, largely independent of the central processing difficulty.

Introduction

Visual search is one of the dominant paradigms used for investigating visual attention. In a typical visual search task, participants have to decide whether a search display contains a designated target among distractors (nontarget elements). In most studies, response times (RTs) and error rates are analyzed as a function of the number of items in the display (display size). Based on such data, several theories of visual search have been suggested (e.g., Duncan & Humphreys, 1989, 1992; Treisman & Gelade, 1980;

The Mind's Eye: Cognitive and Applied Aspects of Eye Movement Research

ISBN: 0-444-51020-6

Treisman & Sato, 1990; Treisman, Sykes & Gelade, 1977; Wolfe 1994; Wolfe, Cave & Franzel, 1989). By monitoring participants' eye movements during the search process, fine-grained temporal and spatial measures (such as fixation duration, initial latency to move, saccadic amplitude, saccadic error, etc.) could be provided to supplement global performance indicators such as RT and error rate (Bertera & Rayner, 2000; Binello, Mannan & Ruddock 1995; Gould, 1967; Jacobs, 1986; Motter & Belky, 1998a; Rayner & Fisher, 1987; Viviani & Swensson, 1982; D. E. Williams, Reingold, Moscovitch & Behrmann, 1997; Zelinsky & Sheinberg, 1997; see Rayner, 1998 for a review).

The current paper illustrates one important way in which the eye movement data could provide unique insights into the search process by quantifying the extent of saccadic selectivity — the bias in the spatial distribution of saccadic endpoints towards or away from certain distractor types. Accordingly, a brief review of several visual search theories and their predictions concerning saccadic selectivity are provided, followed by a review of empirical studies on saccadic selectivity. Then, we report three experiments investigating the relation between the difficulty of the central discrimination and the magnitude of saccadic selectivity, showing that the guidance of eye movements is robust and largely independent of central processing difficulty.

Theories of Visual Search and Predictions Concerning Saccadic Selectivity

In a complex search display containing several types of distractors with different levels of target–distractor similarity, participants will typically have to make a few saccades before a decision on target presence can be made. During this process, will they exhibit saccadic selectivity by preferentially directing their eye movements towards one type of distractors over others? Different predictions could be derived from current theoretical frameworks on visual search.

An early theory of visual search is the original feature integration theory by Treisman and her colleagues (Treisman & Gelade, 1980; Treisman, Sykes & Gelade, 1977). This theory proposes the existence of preattentive feature maps, one for each stimulus dimension (such as color, shape, orientation, etc.). Information from parallel preattentive processes could only mediate performance in a visual search task if the target was defined by the presence of a unique feature (i.e., feature search), such as searching for a green *X* among red and blue *X*s. However, if the target is defined by a specific combination of features (i.e., conjunction search), such as searching for a green *X* among red *X*s and green *O*s, attention is necessary to locally combine the information from the corresponding feature maps. As a result, participants have to inspect the search display in a serial item-by-item fashion until target detection or exhaustive search. Given the nature of serial item-by-item processing, this theory predicts that in a conjunction search task that allows free eye movements, each type of distractors has an equal probability of being fixated for inspection and therefore there will be no selectivity in the distribution of saccades.

The original feature integration theory was inconsistent with the findings from many subsequent studies. For example, parallel or highly efficient performance has been found in a variety of conjunction search tasks (e.g., McLeod, Driver & Crisp, 1988; Nakayama & Silverman, 1986; Theeuwes & Kooi, 1994; Wolfe *et al.*, 1989; Zohary & Hochstein, 1989). This is inconsistent with the notion of serial item-by-item search proposed by that theory. Furthermore, some feature search tasks were found to induce serial or inefficient performance (e.g., Nagy & Sanchez, 1990; Wolfe, Friedman-Hill, Stewart & O'Connell, 1992). This indicates that search efficiency in both feature and conjunction search tasks may vary along a continuum (see Wolfe, 1998, for review).

Several other theories have been proposed to explain variations in search efficiency. For example, the guided search model by Wolfe and his colleagues (e.g., Cave & Wolfe, 1990; Wolfe 1994; Wolfe *et al.*, 1989) argues that in a visual search task participants selectively use peripheral information to guide the search process. In an initial processing stage, a parallel analysis is carried out across all locations of a search display; preattentive information is extracted to segment the search display and to create an "activation map'. The overall activation at each stimulus location consists of a top-down component, reflecting the similarity to the target, and a bottom-up component, quantifying the similarity to the other distractors. The activation map is then used to guide shifts of attention in a subsequent stage of serial search (the focus of attention is directed serially to the locations with the highest activation until the target is found or the criterion to make a negative response is reached. One prediction based on this model is that those distractors that are more similar to the target item would be more likely to be fixated than would the less similar ones and therefore there would be a bias in the distribution of saccadic endpoints (i.e., saccadic selectivity).

A similar prediction of search efficiency and saccadic selectivity can be derived from the attentional engagement theory proposed by Duncan and Humphreys (1989, 1992). These researchers argue that in a search task, display inputs must be entered into visual short-term memory before accessing awareness and becoming the focus of current behavior. Due to the limited capacity of the visual short-term memory system, information is admitted competitively following a process of selection on the basis of both target–distractor similarity and distractor–distractor similarity. As a result, selective processing of visual information will likely occur in a complex search task. Similarly, the revised feature integration theory (feature-inhibition hypothesis: Treisman & Sato, 1990) also predicts selective processing of distractors by proposing that individual feature maps can inhibit nontarget features. If the features are sufficiently distinct and separable, such a mechanism might eliminate the activity generated in the master map by distractor items, allowing the target to pop out. However, if the inhibition is incomplete, a serial scan is necessary through the master map, in which the locations differ only in their levels of activation. Given the mechanism of distractor inhibition, those locations containing items that are more similar to the target or share feature(s) with the search target will be more likely to "survive" the inhibition for further inspection.

Empirical Studies on Saccadic Selectivity

During the past few decades, the guidance of eye movements have been examined in several visual search studies. In one of the earliest studies, L. G. Williams (1967) employed search arrays of 100 simple geometric forms, which were defined by unique combinations of color, shape, and size. In the center of each display item, there was a two-digit number and participants were instructed to search for a particular number. In most of the trials, additional information concerning one or more attributes (color, shape, or size) of the search target was prespecified. Williams found that providing participants with prior color information greatly shortened the search time. Participants could effectively restrict their search within the color dimension, directing most of the saccades towards those items sharing the target color. This was true regardless whether the target color was specified alone or together with additional shape and/or size information. Providing participants with size information, however, was much less effective whereas specifying the shape of the target yielded little evidence of saccadic selectivity. In a subsequent study, Luria and Strauss (1975) asked participants to look for the only dial in an array of 16 that had not been rotated from the starting position. They examined the efficiency of coding dials by color, shape, and a combination of the two. They similarly found that, when provided with color information, participants showed a marked tendency to direct saccades towards target-color items. Unlike Williams (1967), they found that participants could use the shape information to guide search, although to a lesser extent compared to the color information. When provided with both the color and shape information, participants' saccadic endpoints were guided by both dimensions.

More recent studies on saccadic selectivity have yielded conflicting results. Zelinsky (1996) had participants search through displays containing two subsets of distractors. A subset of "similar" distractors was chosen to share either color or orientation with the search target whereas another subset of "dissimilar" distractors did not share color or orientation with the target item. Zelinsky reasoned that if the search process were guided (e.g., Wolfe, 1994), participants should make more saccades towards the similar distractors than towards the dissimilar ones. He examined the distribution of saccadic endpoints, only to find very weak evidence for guidance. Of all valid saccades, 55% were directed to the similar distractors and 45% were directed to the dissimilar ones. Instead of biasing saccadic endpoints towards one specific type of distractors, Zelinsky argued that participants adopted an oculomotor strategy to aid the search process in a multi-element display. The oculomotor strategy involves programming a series of fixations in an orderly fashion such that every display element has an equal chance of being recognized correctly. In accordance with this argument, he found that the endpoints of the first saccades were systematically biased towards the top-left quadrant of the display.

However, subsequent studies have shown strong evidence of selectivity in the distribution of saccadic endpoints in visual search tasks (e.g., Bichot & Schall, 1998; Findlay, 1997; Findlay, Brown & Gilchrist, 2001; Findlay & Gilchrist, 1998; Hooge & Erkelens, 1999; Motter & Belky, 1998b; Pomplun, Reingold & Shen, 2001a, submitted; Pomplun, Reingold, Shen & Williams, 2000; Scialfa & Joffe, 1998; Shen & Reingold, 1999; Shen, Reingold & Pomplun, 2000; Williams & Reingold, 2001). Stimulus dimensions such as color, shape, contrast polarity, and size have been shown to guide the search process. In light of these findings, the failure to demonstrate saccadic selectivity by Zelinsky (1996) seems to be anomalous. This discrepancy across studies might be attributable to several factors. First, although the differences observed in the experiment by Zelinsky were relatively small and sometimes nonsignificant, saccades were still more likely to be directed towards "similar" distractors than towards "dissimilar" distractors (55% vs. 45%). Thus, Zelinsky's findings do provide weak evidence for selectivity, as he acknowledges. Second, to examine guidance, Zelinsky compared "similar" and "dissimilar" distractors, with "similar" distractors consisting of two types (target-color vs. target-orientation). When calculating the proportion of saccades directed towards "similar" distractors, he did not distinguish between saccades directed towards target-color distractors and those directed towards target-shape distractors. Given that participants heavily rely on color information to guide visual search but make little use of orientation information (e.g., Motter & Belky, 1998b; Williams & Reingold, 2001), lumping these two types of distractors together may have led to an underestimation of the strength of guidance in his study.

Effect of Central Processing Difficulty on Peripheral Selection

Most of the above-mentioned studies employed multi-fixation search tasks. In such tasks, during each fixation information may be extracted from foveal, parafoveal, and peripheral regions of the visual field (Findlay, 1997; Hooge & Erkelens, 1999; Lévy-Schoen, 1981; Rayner & Fisher, 1987). Such information subserves both the discrimination of the foveated object from the search target (henceforth the central discrimination task) as well as the selection of the next object to be foveated (henceforth the peripheral selection task). The main goal of the current study was to investigate whether the difficulty of the central discrimination task affects the efficiency of the peripheral selection task (i.e., saccadic selectivity). This issue has been recently studied by Hooge and Erkelens (1999). These investigators examined saccadic selectivity in a search task by manipulating the difficulty of the central discrimination and peripheral selection tasks separately. They found that the difficulty of the central discrimination task influenced the peripheral analysis and saccadic selectivity indirectly via fixation duration. Specifically, they demonstrated that saccadic selectivity was more pronounced when the difficulty of the central discrimination task was increased. They argued that increasing the difficulty of the central discrimination task led to longer fixation duration, which in turn permitted more time for peripheral analysis, consequently leading to a better selection of the next saccadic target.

Hooge and Erkelens's (1999) interpretation can be illustrated by a variant of the waiting-room metaphor proposed by Navon and Pearl (1985; henceforth the "waiting-room" model). Let us imagine the central discrimination of the foveated stimulus as the doctor examining one patient while at the same time, peripheral selection is likened to a nurse screening other patients in the waiting room, attempting to select the next patient to be examined by the doctor based on the severity of the symptoms. The amount of information the nurse can gather and consequently the quality of the selection depends on the availability of the doctor — the longer the doctor is preoccupied by the previous patient, the more information the nurse can gather from other patients.

Note that in this metaphor the waiting time is solely determined by the availability of the doctor rather than the complexity of the procedures carried out by the nurse (such as retrieving the patients' files, taking temperature, interviewing patients, etc.). This parallels the findings by Hooge and Erkelens that fixation duration in a multi-fixation search task was influenced by the difficulty of the foveal discrimination task but not by the difficulty of the peripheral selection task (see also Findlay, 1997; Gould, 1967; Jacobs, 1986; Nazir & Jacobs, 1991).

Experiment 1

The goal of Experiment 1 was to evaluate the waiting-room model concerning the relation between the difficulty of the central discrimination and peripheral selection. The manipulation of central discrimination difficulty was accomplished by adopting a set of stimuli with varying degrees of target–distractor similarity (high, medium, or low similarity). We examined the proportions of fixations on the high-, medium-, and low-similarity distractors as a function of the target–distractor similarity of the previously fixated display item and of the duration of the previous fixation.

Method

Participants Eight participants were tested individually in a single one-hour session. All participants had normal or corrected-to-normal visual acuity. They were naïve with respect to the purpose of the experiment and received course credits for their participation.

Apparatus The eyetracker employed in the current study was the SR Research Ltd. EyeLink system. This system has a sampling rate of 250 Hz (4 ms temporal resolution) and an average error of less than 0.5° of visual angle in the computation of gaze position. The EyeLink headband has three cameras, allowing simultaneous tracking of both eyes and of head position for head-motion compensation. By default, only the participant's dominant eye was tracked. In the present investigation, the configurable acceleration and velocity thresholds were set to detect saccades of 0.5° or greater. Stimulus displays were presented on two monitors, one for the participant (a 17-inch ViewSonic 17PS) and the other for the experimenter. The experimenter monitor was used to give feedback in real-time about the participant's computed gaze position.

Stimuli As illustrated in Figure 4.1, four types of display items — search target, high-, medium-, and low-similarity distractors — were created based on Hooge and Erkelens (1999). Each individual item had a diameter of 1.37°. The target was a circle with a line width of 0.17° and the distractors were *C*s with a gap of 0.1°. The high-, medium-, and low-similarity distractors had a line width of 0.17°, 0.26°, and 0.34°, respectively. The orientation of the gap of individual *C*s was chosen randomly from facing up, left, down, and right.

Design Search displays were created by using an imaginary matrix of 6 × 6 cells, which subtended 13.2° × 13.2° at a viewing distance of 60 cm. For all trials, the total number of items presented in a display (display size) was held constant at 18. Participants were asked to search for the target item in displays containing all three types of distractors (high-, medium-, and low-similarity distractors). In a target-absent display, there were six distractors of each type. In target-present trials, a target-absent display was first created and then one of the distractors at 16 possible target locations (four cells at each corner of the grid) was randomly chosen to be replaced by the target item.

Each participant performed 480 trials in five blocks of 96 trials. An equal number of target-present and target-absent trials were used. The order of stimulus displays was randomized with a restriction that no more than four consecutive displays of a given type would occur. At the beginning of the experiment, participants received 24 practice trials.

Procedure A 9-point calibration procedure was performed at the beginning of the experiment, followed by a 9-point calibration accuracy test. Calibration was repeated if any point was in error by more than 1° or if the average error for all points was greater than 0.5°. Each trial started with a drift correction in the gaze position. Participants were instructed to fixate on a black dot in the center of the computer screen and then press a start button to initiate a trial. They were asked to search for the target item and indicate whether it was in the display or not by pressing an appropriate button

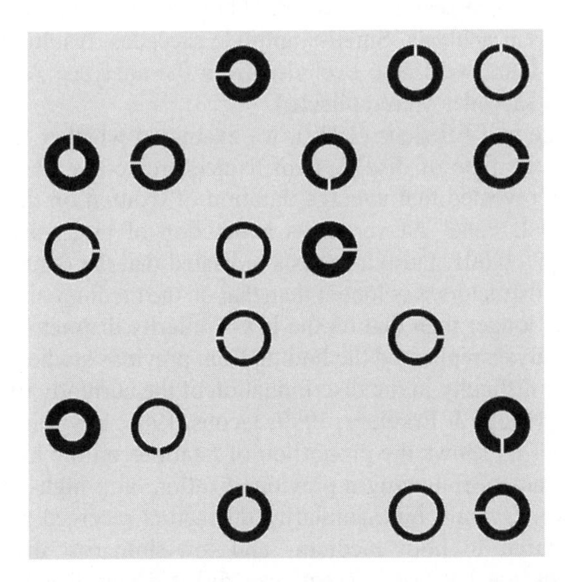

Figure 4.1: Sample search displays used in Experiment 1. The target was a filled circle **O** and the distractors were open circles with different target–distractor similarity (High: O; Medium: O; Low: O).

as quickly and as accurately as possible. The trial terminated if participants pressed one of the response buttons or if no response was made within 20 seconds. The time between display onset and the participant's response was recorded as the response time. The particular buttons used to indicate target presence were counterbalanced across participants.

Results and Discussion

Trials with a saccade or a blink overlapping the onset of the search display, or with an incorrect response, were excluded from analysis. These exclusions accounted for 2.3% and 3.6% of total trials respectively. Following the convention of visual search literature, for each participant, an outlier analysis was performed on the target-absent and target-present trials separately to eliminate those trials with response times more than 3.0 standard deviations above or below the mean. This resulted in the removal of 2.6% of trials from further analysis. The average response times were 2456.6 ms for the remaining target-absent trials and 1729.6 ms for the target-present trials.

For each trial, the distance was calculated between the fixation following each saccade and every item in the display. The item closest to the fixation was taken to be the target of that saccade. The number of saccades towards each type of distractor (high-, medium-, and low-similarity distractors) was then summed to assess saccadic selectivity. As pointed out by Zelinsky (1996), results from target-absent trials can be interpreted more clearly than those from target-present trials where the presence of the target item may influence search behavior. Therefore, only target-absent trials were included in the current analysis. Small-amplitude saccades, resulting in no change in the fixated display item, were also excluded from the analysis. Across eight participants, 12,722 valid saccades were collected.

Following Hooge and Erkelens (1999), we examined whether saccadic selectivity was influenced by the type of display item fixated previously. A one-way repeated-measures ANOVA revealed that average duration of fixation on the currently fixated item (see Figure 4.2, Panel A) varied as a function of target-distractor similarity, $F(2, 14) = 57.45, p < 0.001$. Pairwise *t*-tests indicated that the duration of fixations on the high-similarity distractors was longer than that on the medium-similarity distractors, which, in turn, was longer than that on the low-similarity distractors, all $ts\ (7) > 6.36$, $ps < 0.001$. This analysis replicated the finding from previous studies that fixation duration varied with the difficulty in the discrimination of the currently fixated display item (e.g., Gould, 1967; Hooge & Erkelens, 1999; Jacobs, 1986; Lévy-Schoen, 1981).

Figure 4.2 (Panel B) shows the proportion of fixations on the high-, medium-, and low-similarity distractors following a previous fixation on a high-, medium-, or low-similarity distractor. Overall, high-similarity distractors received a larger percentage of fixations compared to both medium- and low-similarity distractors; saccadic frequencies towards the latter two types also differed significantly, all $Fs\ (1, 7) > 48.01$, $ps < 0.001$. Thus, this analysis clearly indicates that search process is guided by the overall target–distractor similarity: the more closely the distractors resemble the search target, the more likely they will be fixated during the search process. However,

a)

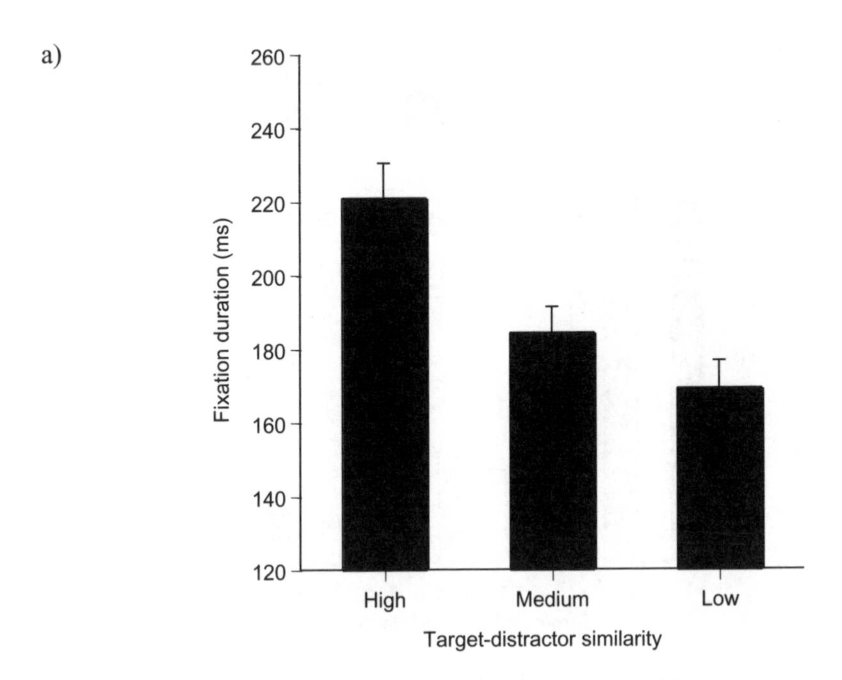

b)

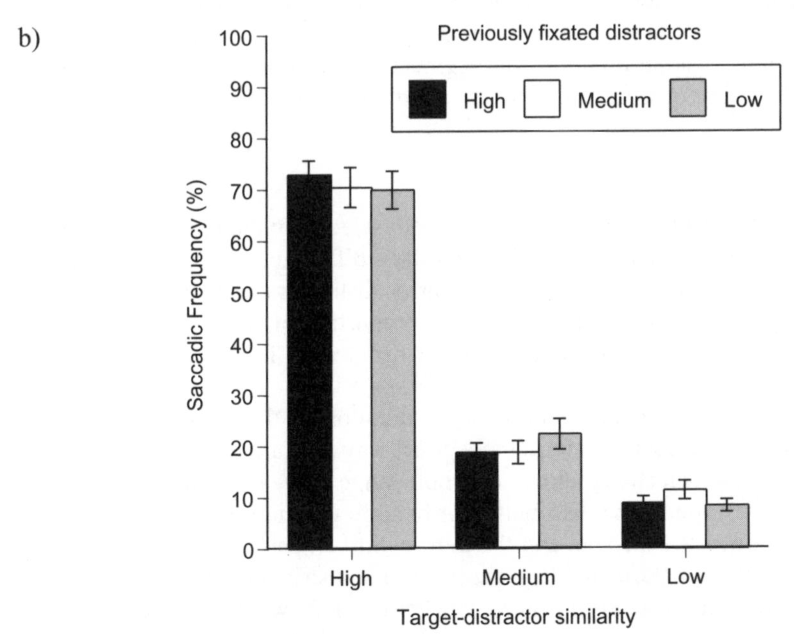

Figure 4.2: Panel A: Average duration of fixations as a function of target–distractor similarity of the currently fixated stimulus. Panel B: Frequency of saccades towards the high-, medium-, and low-similarity distractors as a function of the type of distractor (high-, medium-, or low-similarity) fixated previously.

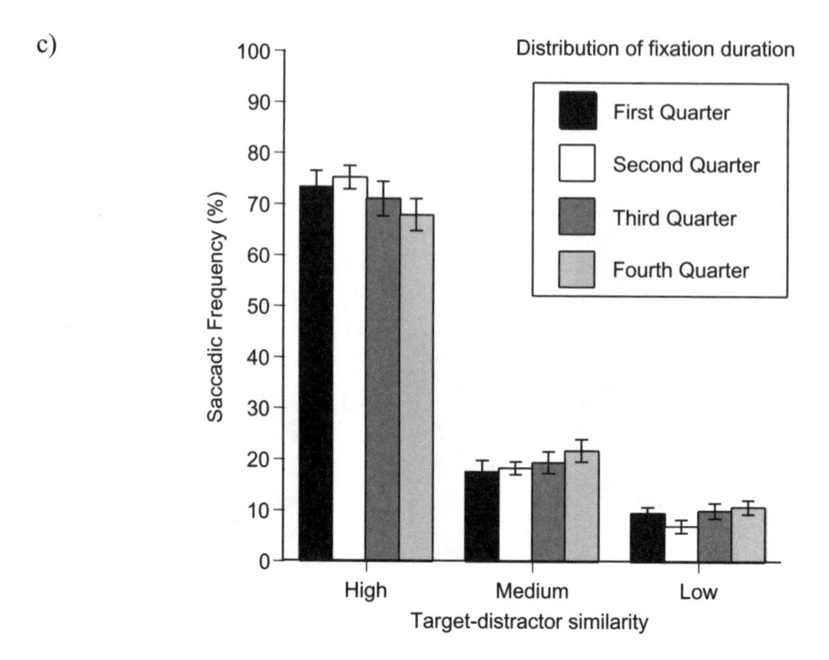

Figure 4.2: Panel C: Frequency of saccades directed towards the high-, medium-, and low-similarity distractors as a function of the previous fixation duration. The distribution of the previous fixation duration was segmented into four quarters by using the first quartile, the median, and the third quartile.

inconsistent with the waiting-room model, saccadic selectivity did not vary as a function of the central discrimination difficulty. Despite a difference of 36.6 ms in fixation duration between the high- and medium-similarity distractors and 15.0 ms between the medium- and low-similarity distractors, the proportion of fixations on the high-, medium-, and low-similarity distractors did not differ as a function of the type of distractor fixated previously, all Fs $(2, 14) < 3.76$, $ps > 0.05$.

If peripheral analysis benefits from a longer duration of the previous fixation as suggested by the waiting-room model, a more direct way of examining this issue would be to measure saccadic selectivity across the whole range of the fixation duration distribution. Thus, we segmented the distribution of fixation duration into four quarters by using the first quartile, the median, and the third quartile. (The average fixation durations for the first, second, third, and last quarters of the distribution were 119.3, 172.8, 214.7, and 306.4 ms respectively.) Figure 4.2 (Panel C) shows that the selectivity of a subsequent saccade became slightly weaker following a longer fixation duration. The percentage of saccades directed towards the high-similarity distractors decreased from 73.2% in the first quarter of the fixation duration distribution to 67.8% in the fourth quarter of the distribution, $t\ (7) = 2.81$, $p < 0.05$. This finding is clearly inconsistent with the prediction of the waiting-room model.

Although the current experiment attempted to replicate the findings by Hooge and Erkelens (1999), there are some methodological differences between the two studies that may explain the inconsistency in findings. Specifically, Hooge and Erkelens adopted a blocked design (using different combinations of line width and gap size of *C*s in different sessions). In contrast, in the current study a combination of high-, medium-, and low-similarity distractors were presented in each display. In addition, unlike in the present study, Hooge and Erkelens manipulated the difficulty of peripheral selection (line-width of *C*s). Accordingly, one could argue that these methodological differences are responsible for the sizeable difference in fixation duration between the high- and low-similarity distractors as well as the effect of central processing difficulty on peripheral selection found in the study by Hooge and Erkelens. One could further argue that a difference of 50 ms in fixation duration observed in the current study might be too small to influence saccadic selection. However, our examination of saccadic selectivity across the whole range of the fixation duration distribution (see Figure 4.2C) did not provide evidence for this argument.

A recent study by Findlay *et al.* (2001) may be relevant to the current findings. These investigators examined the relationship between the duration of a previous fixation and the precision of target acquisition in a subsequent saccade. They similarly found that saccades following brief fixations had the same probability of reaching the target as those following longer fixations. Thus, results from the current experiment and Findlay *et al.* indicate that saccadic selectivity is not strongly influenced by the duration of previous fixation as proposed by the waiting-room model.

Experiment 2

The underlying assumption of the waiting-room model is that the peripheral analysis is an independent process, which is carried out in parallel to the central discrimination task but does not share resources with the latter. If the central discrimination and peripheral selection share resources, a different prediction can be derived when increasing the difficulty of the central discrimination task. This can be illustrated considering prior research on foveal load (henceforth the "foveal-load" model). Those studies (e.g., Ikeda & Takeuchi, 1975; Mackworth, 1965) typically employed two concurrent visual tasks, one in the center of the display and the other in the periphery. L. J. Williams (1989), for example, had participants perform a central letter discrimination task that either induced a high or low foveal load. In a simultaneous peripheral task, participants named a one-digit number shown in the periphery. It was found that a more difficult central task (i.e., higher foveal load) decreased participants' performance in the peripheral detection task. Similar dependence of peripheral analysis on the difficulty of the central processing has also been found in reading research. Henderson and Ferreira (1990) investigated the influence of foveal task difficulty on the benefit of parafoveal previewing by manipulating the lexical frequency and syntactic difficulty of the foveated word. They found that less parafoveal information was acquired when the foveal processing was difficult, despite the fact that in the difficult condition the parafoveal word was available for a longer amount of time than in the easy condition.

In the context of visual search, this model would predict that peripheral analysis is inversely related to the demands of central processing and saccade selection will be less efficient if the central discrimination task becomes more difficult.

In Experiment 2, we further examined whether participants' performance of peripheral selection is influenced by the difficulty of the central discrimination with a dual-task manipulation. In a dual-task condition, the central processing was rendered more difficult with the introduction of a concurrent visual task — besides detecting the presence of the search target, participants also had to find and memorize the largest number presented within a gaze-contingent moving window (see Figure 4.3 for an example). Gaze-contingent techniques have been widely used in reading, scene perception, and recently in visual search studies (e.g., Bertera & Rayner, 2000; Murphy & Foley-Fisher, 1988; Pomplun, Reingold & Shen, 2001a, b; Rayner & Fisher, 1987; Rayner, Inhoff, Morrison, Slowiaczek & Bertera, 1981; Reingold, Charness, Pomplun & Stampe, 2001; van Diepen, De Graef & d'Ydewalle, 1995; see Rayner, 1998 for a review). In a single-task condition, participants were asked to concentrate on the visual search task only while ignoring the numbers presented. We examined whether saccadic selectivity differ across the two conditions — the foveal-load model predicts a decrease in saccadic selectivity in the dual-task condition whereas the waiting-room model makes an opposite prediction.

Method

Participants Eight participants were tested in a single one-hour session. None of them had participated in the previous experiments. All participants had normal or corrected-to-normal vision and were paid $10 for their participation. They were not aware of the purpose of the experiment.

Stimuli and design Four types of display items — search target, high-, medium-, and low-similarity distractors — were constructed by using a matrix of 4 × 4 squares, with eight of them filled with black and the rest remaining white (see von Grünau, Dubé & Galera, 1994, Experiment 5). The high-, medium-, and low-similarity distractors had a physical difference of 1, 3, and 6 with respect to the target item. The physical difference was defined as the number of black squares that had to be moved from one location in the 4 × 4 matrix to another in order to change the target into the respective distractor. This construct of similarity by physical difference was verified by a subjective rating of stimulus similarity as well as search efficiency data (von Grünau *et al.*, 1994). Each individual element subtended 1.0° × 1.0° on a white background of 15.2° × 15.2°.

The current experiment examined whether participants' performance of peripheral selection is influenced by the difficulty of the central discrimination with a dual-task manipulation. This was achieved by introducing a gaze-contingent moving window and by instructing participants to attend, or not to attend, to information presented within the window across conditions. In each trial, a circular gaze-contingent moving window of 4.8° in diameter was presented (see Figure 4.3 for an example). The moving

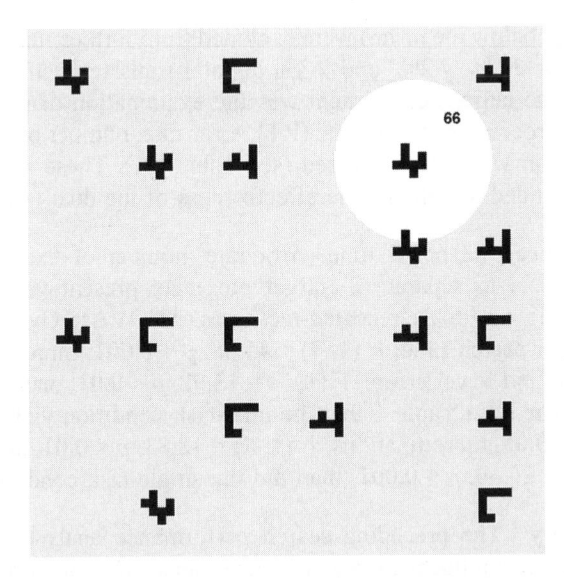

Figure 4.3: Sample search displays overlaid with a gaze-contingent moving window (Experiment 2). Target was and distractors had different levels of target-distractor similarity (High: ; Medium: ; Low:). The window center was aligned with the current gaze position.

window, constantly centered on the participants' fixation point, unveiled one 2-digit number from 10 to 99 during each fixation. The numbers were placed at a distance of 1.8° from any neighboring display item. Each individual number extended 0.4° horizontally and 0.3° vertically. Display items falling outside the window were clearly visible to the participants. In the single-task condition, participants were asked to make a decision regarding the presence of the search target as quickly and as accurately as possible. They were instructed to ignore any number presented within the window. In the dual-task condition, once participants made response regarding the presence of the search target, they also had to report the largest number they had seen in that trial.

The current experiment included eight blocks of 36 test trials, with half of the trials in the single-task load condition and the other half in the dual-task condition. This amounted to 72 trials in each cell of the design (target presence by task manipulation). Single-task and dual-task conditions were tested in alternating blocks with the order of conditions counterbalanced across participants. At the beginning of the experiment, participants received 48 practice trials.

Results and Discussion

Trials with a saccade or a blink overlapping the onset of a search display, with an incorrect response, or with an excessively long or short response time (3.0 standard deviations above or below the mean) were excluded from further analysis. These exclusions accounted for 2.6%, 7.2%, and 2.3% of total trials respectively. Although the primary focus of the current experiment was the examination of saccadic selectivity during the search process, response time (RT), error rate, number of fixations per trial, and fixation duration were also analyzed (see Table 4.1). These search performance measures were included to validate the effectiveness of the dual-task manipulation.

Search performance Response time, error rate, number of fixations, and fixation duration were subject to separate 2 (target presence: present vs. absent) × 2 (task manipulation: single vs. dual) repeated-measures ANOVAs. Overall, target-absent trials yielded longer search time, $F\ (1,\ 7) = 45.88$, $p < 0.001$, more fixations, $F\ (1,\ 7) = 68.08$, $p < 0.001$, and fewer errors, $F\ (1,\ 7) = 13.92$, $p < 0.01$, than did target-present trials. It is also clear from Table 1 that the dual-task condition yielded longer RT, $F\ (1,\ 7) = 10.33$, $p < 0.05$, more fixations, $F\ (1,\ 7) = 12.83$, $p < 0.01$, and longer fixation duration, $F\ (1,\ 7) = 26.64$, $p < 0.001$, than did the single-task condition.

Saccadic selectivity The preceding search performance analyses suggest that the dual task manipulation influenced search performance. It is important to determine whether saccadic selectivity was similarly affected. For each individual participant, the proportions of saccades directed to each type of distractor were determined in both the single-task condition and the dual-task condition (see Figure 4.4).

As can be seen from Figure 4.4, saccadic frequency towards high-similarity distractors was higher than that towards low-similarity distractors, which, in turn, was higher than that towards the low-similarity distractors, all Fs $(1,\ 7) > 143.78$, $ps < 0.001$. More importantly, although the dual-task manipulation decreased search efficiency, it had very little influence on the pattern of saccadic selectivity. Proportions of saccades directed towards the high-, medium-, and low-similarity distractors did not differ across

Table 4.1: Search performance as a function of target presence and task manipulation in Experiment 2.

	Single task		**Dual task**	
	Absent	**Present**	**Absent**	**Present**
Response time (ms)	2157.7	1289.2	2788.2	1665.6
	(182.4)	(84.7)	(331.8)	(204.3)
Error rate (%)	1.7	13.4	2.7	10.8
	(0.6)	(3.3)	(0.5)	(1.6)
Number of fixations per trial	8.7	4.3	10.4	5.5
	(0.7)	(0.2)	(1.1)	(0.5)
Fixation duration (ms)	203.8	193.8	221.6	232.6
	(7.6)	(8.7)	(6.5)	(9.0)

Note: Values in parentheses represent standard errors.

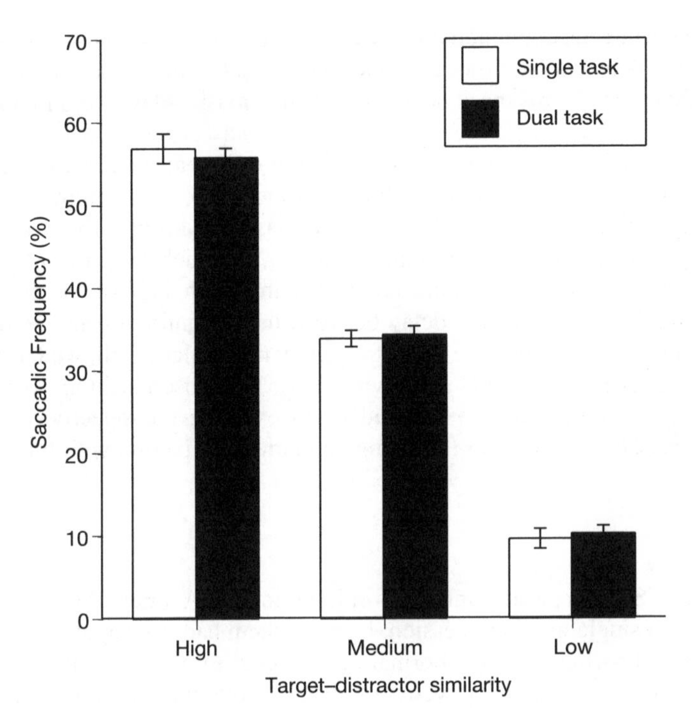

Figure 4.4: Frequency of saccades directed to the high-, medium-, and low-similarity distractors in the single-task and dual-task conditions.

the single-task and dual-task conditions, all *F*s <1. This pattern of saccadic selectivity did not provide support for either the waiting-room model or the foveal-load model.

Experiments 3A and 3B

Experiment 3 was designed to further investigate the relation between the difficulty of the central discrimination and peripheral selection by employing a more powerful manipulation of the central discrimination difficulty. We adopted a gaze-contingent moving-mask paradigm, in which a mask was presented centered on the point of gaze at a certain delay following the beginning of a fixation. With this manipulation, we selectively masked information required for the central discrimination while leaving information vital for the peripheral selection relatively intact. If the central discrimination and peripheral selection are independent processes, interference with the central discrimination should have no effect on peripheral selection. We should predict that saccadic selectivity remains constant even if the information supporting the central discrimination is severely degraded. However, if the two processes are interdependent as suggested by the foveal-load model, a decrease in saccadic selectivity is predicted when the central discrimination becomes more difficult.

Two versions of mask manipulation were adopted. In Experiment 3A, we manipulated the delay between fixation onset and mask onset: a no-mask condition, and two masking conditions (50-ms mask delay and 117-ms mask delay) were included. Based on previous studies (e.g., Rayner *et al.*, 1981), it was expected that a shorter mask delay (50 ms) would lead to a greater degradation of foveal processing and add to the difficulty of the discrimination task. In Experiment 3B, we manipulated the mask frequency by including three conditions: a no-mask condition, in which none of the fixations was masked, a sparse-masking condition, in which a quarter of the fixations were masked, and a dense-masking condition, in which half of the fixations were masked. In masked fixations, the delay between the beginning of the fixation and the onset of the mask was 50 ms. The mask frequency manipulation allowed us to compare search performance and saccadic selectivity across the dense-masking, sparse-masking and no-mask conditions. It also allowed us to examine the selectivity of saccades following a masked fixation and following an unmasked fixation within the same trial.

Method

Participants Sixteen participants (half in Experiment 3A and half in Experiment 3B) were tested in a single one-hour session. None of them had participated in the previous experiments. All participants had normal or corrected-to-normal vision and were paid $10 for their participation. They were not aware of the purpose of the experiment.

Stimuli and design The same set of display items as in the previous experiment was used. Each individual item subtended 1.37° both horizontally and vertically. One major change implemented in the current experiment was the introduction of a circular gaze-contingent moving mask in some trials. The mask, 4° of visual angle in diameter, was displayed 50 ms (50-ms mask delay condition in Experiment 3A and all mask conditions in Experiment 3B) or 117 ms (117-ms mask delay condition in Experiment 3A) following the onset of a fixation, and remained centered on the gaze position. The mask, composed of random black-and-white patches, replaced display items or fragments of display items that were within 2.0° radius from the gaze position. The pattern of the moving mask varied from trial to trial (see Figure 4.5 for an example). In other trials, no mask was displayed (no-mask condition).

In Experiment 3A, participants were tested in six blocks of 48 trials, with half of the trials in the no-mask condition and the remaining trials divided evenly between the 50-ms and 117-ms mask delay conditions. In Experiment 3B, participants received 288 test trials, with 48 trials in each cell of the design (mask condition by target presence). In both experiments, participants also received 48 practice trials at the beginning of the experiment.

Results and Discussion

Trials with a saccade or a blink overlapping the onset of a search display, with an incorrect response, with an excessively long or short response time, or with no

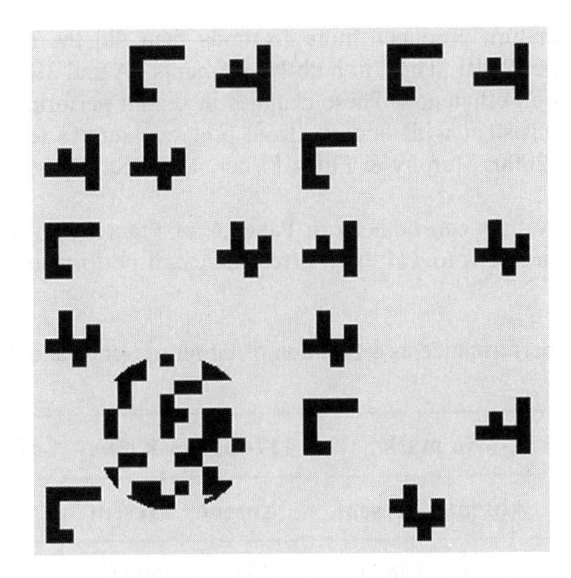

Figure 4.5: Sample search displays used in Experiments 3A and 3B. The gaze-contingent moving mask was centered on the participant's current gaze position.

response (timed-out) were excluded from further analysis. These exclusions accounted for 1.5%, 4.8%, 0.6%, and 2.7% of total trials respectively for Experiment 3A and 2.0%, 4.8%, 1.4%, 0% respectively for Experiment 3B.

Search performance In order to validate the effectiveness of the mask manipulation, response time, error rate, and number of fixations per trial were analyzed by separate 2 (target presence: present vs. absent) × 3 (mask condition: no mask, 50-ms mask delay, or 117-ms mask delay in Experiment 3A; no mask, dense masking, or sparse masking in Experiment 3B) repeated-measures ANOVAs. Tables 4.2 and 4.3 summarize the results of these analyses. As can be seen in Table 4.2, in Experiment 3A response times were longest in the 50-ms mask delay condition, and shortest in the no-mask condition. This difference was more pronounced in target-absent trials than in target-present trials, as indicated by a significant interaction between target presence and mask condition, $F\ (2, 14) = 8.19$, $p < 0.01$. This pattern was similarly exhibited in the fixation-number data, $F\ (2, 14) = 5.36$, $p < 0.01$. In addition, the error-rate data suggest that the presence of a foveal mask was detrimental to the search performance, as the 50-ms mask delay condition was more error-prone than were the other two conditions, $F\ (2, 14) = 5.24$, $p < 0.01$.

Similarly, as can be seen in Table 4.3, search performance in Experiment 3B varied as a function of mask condition. Response times were longest in the dense-masking condition and shortest in the no-mask condition. This difference was more pronounced in target-absent trials than in target-present trials, as indicated by a significant interaction between target presence and mask condition, $F\ (2, 14) = 4.29$, $p < 0.05$. The dense-masking condition also produced more fixations than did the sparse-masking condition, which in turn produced more fixations than did the no-mask condition, $F(2, 14) = 26.63, p < 0.001$. Thus, in both Experiments 3A and 3B, masking substantially degraded search efficiency. These changes in search performance induced by a foveal mask are consistent with findings from previous studies (e.g., Bertera, 1988; Bertera & Rayner, 2000; Murphy & Foley-Fisher, 1988; Rayner *et al.*, 1981).

Saccadic selectivity As can be seen in Panel A of Figure 4.6, in Experiment 3A, although the presence of a foveal mask affected search performance substantially, it

Table 4.2: Search performance as a function of target presence and mask condition in Experiment 3A.

	No mask		117-ms mask delay		50-ms mask delay	
	Absent	**Present**	**Absent**	**Present**	**Absent**	**Present**
Response time (ms)	3036.2 (363.2)	1467.0 (167.5)	5191.3 (811.9)	2609.9 (467.3)	7711.9 (1012.1)	3894.7 (619.6)
Error rate (%)	1.0 (0.3)	3.8 (1.2)	1.7 (1.0)	7.6 (1.6)	6.6 (2.8)	12.5 (2.9)
Number of fixations/trial	12.4 (1.4)	5.8 (0.8)	19.5 (3.2)	9.3 (1.9)	27.2 (3.6)	13.3 (2.7)
Fixation duration (ms)	194.7 (7.5)	194.2 (8.2)	225.1 (9.1)	234.7 (12.6)	241.2 (13.2)	251.3 (20.4)

Note: Values in parentheses represent standard errors.

Table 4.3: Search performance as a function of target presence and mask condition in Experiment 3B.

	No mask		Sparse masking		Dense masking	
	Absent	**Present**	**Absent**	**Present**	**Absent**	**Present**
Response time (ms)	2529.4 (301.5)	1359.1 (163.1)	3081.3 (392.5)	1675.0 (237.5)	3655.7 (374.7)	2161.5 (298.1)
Error rate (%)	0.5 (0.3)	6.5 (2.4)	2.1 (0.7)	6.8 (2.2)	2.9 (2.0)	10.2 (2.7)
Number of fixations/trial	10.8 (1.2)	5.0 (0.5)	12.7 (1.5)	6.2 (0.7)	14.3 (1.4)	7.6 (1.0)
Fixation duration (ms)	203.54 (9.0)	195.3 (11.9)	218.1 (9.7)	222.8 (10.5)	227.2 (9.6)	236.4 (12.1)

Note: Values in parentheses represent standard errors.

had very little effect on the performance of peripheral selection. Proportions of saccades directed towards the high-, medium- and low-similarity distractors did not differ across the no-mask, 50-ms mask delay, and 117-ms mask delay conditions, all $Fs\ (2, 14) < 1.99$, $ps > 0.05$.

In Experiment 3B, we first categorized five types of saccade (saccade type): those in the no-mask condition, those following a masked fixation in the dense-masking condition, those following an unmasked fixation in the dense-masking condition, those following a masked fixation in the sparse-masking condition, and those following an unmasked fixation in the sparse-masking condition. For each individual participant, the proportions of fixations on the high-, medium-, and low-similarity distractors were calculated (see Figure 4.6, Panel B). Despite the fact that the central discrimination was made more difficult by the presence of a gaze-contingent moving mask, saccadic selectivity was not influenced by the mask manipulation. The proportion of saccades directed towards high-similarity distractors remained the same across the five types of saccades, $F\ (4, 28) = 1.36$, $p = 0.274$.

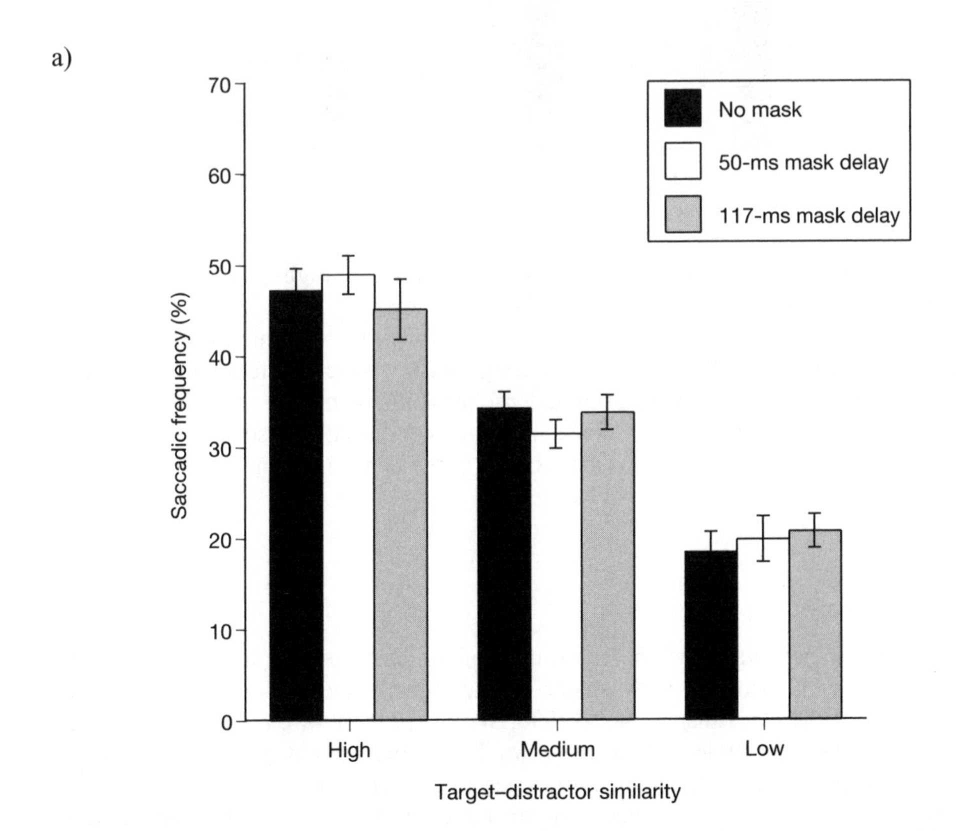

Figure 4.6: Panel A: Frequency of saccades directed to the high-, medium-, and low-similarity distractors in the no-mask, 50-ms mask delay, and 117-ms mask delay conditions in Experiment 3A.

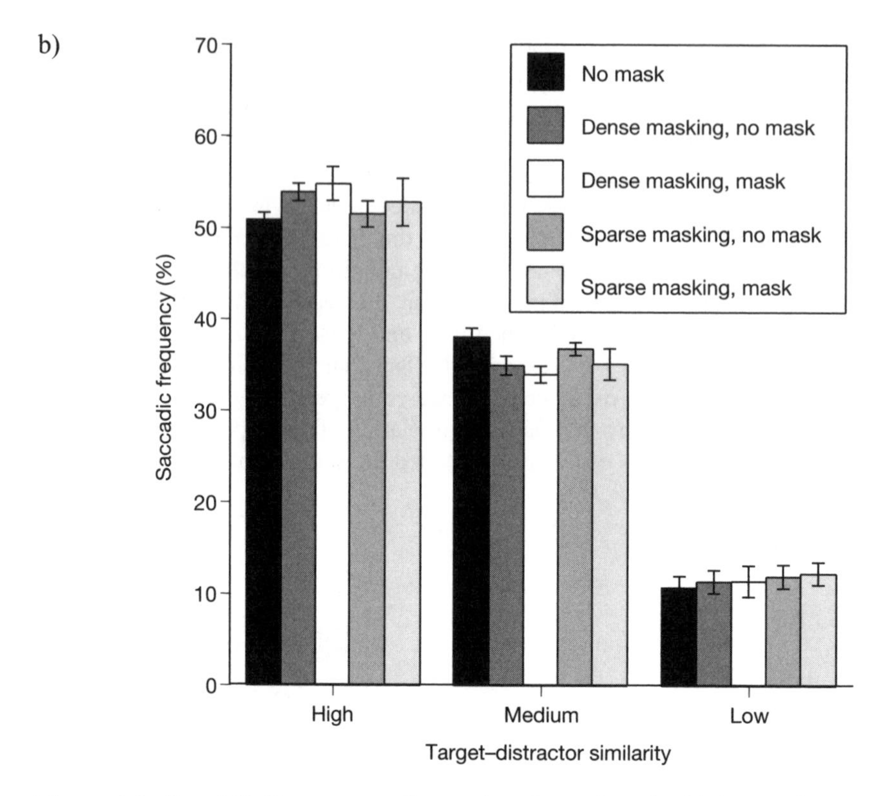

Figure 4.6: Panel B: Frequency of saccades directed to the high-, medium-, and low-similarity distractors in Experiment 3B. Note: No mask = saccades in the no-mask condition; sparse masking, no mask = saccades following an unmasked fixation in the sparse-masking condition; sparse masking, mask = saccades following a masked fixation in the sparse-masking condition; dense masking, no mask = saccades following an unmasked fixation in the dense-masking condition; dense masking, mask = saccades following a masked fixation in the dense-masking condition.

Summary

The current paper examined the robustness of guidance of eye movements during visual search. Consistent with major visual search theories (e.g., Duncan & Humphreys, 1989; Treisman & Sato, 1990; Wolfe, 1994; Wolfe *et al.*, 1989; but see Treisman & Gelade, 1980; Treisman *et al.*, 1977), the present experiments demonstrated that participants direct their saccades selectively during the search process, revealing a strong correspondence between target–distractor similarity and saccadic frequency towards the respective distractors. This adds to a growing literature on the guidance of eye movements in visual search tasks (e.g., Bichot & Schall, 1998; Findlay, 1997; Findlay *et al.*, 2001; Findlay & Gilchrist, 1998; Hooge & Erkelens,

1999; Luria & Strauss, 1975; Motter & Belky, 1998b; Pomplun *et al.*, 2001a, submitted; Pomplun *et al.*, 2000; Scialfa & Joffe, 1998; Shen & Reingold, 1999; Shen, Reingold & Pomplun, 2000; D. E. Williams & Reingold, 2001; Williams, 1967)

We studied the robustness of visual guidance by examining whether the magnitude of saccadic selectivity is influenced by the difficulty of the central discrimination task. In Experiment 1, while the duration of current fixations increased with increasing target–distractor similarity, there was no evidence that saccadic selectivity was influenced by the target–distractor similarity of the previously fixated display item or by the duration of the previous fixation (see also Findlay *et al.*, 2001). These findings are inconsistent with the predictions by the waiting-room model (Hooge & Erkelens, 1999). In addition, we manipulated the difficulty of the central discrimination by introducing a concurrent visual task (Experiment 2) and by presenting a gaze-contingent moving foveal mask (Experiments 3A and 3B). Although both manipulations substantially degraded the overall visual search performance, the magnitude of peripheral selection was not affected. This is not consistent with the notion that the central discrimination and the peripheral analysis share the same pool of attentional resources as suggested by the foveal load model. Thus, the current series of experiments provide convergent evidence that peripheral selection is a robust process, largely independent of the central processing difficulty.

The dissociations between the strong effects of the present manipulations on the central processing difficulty and the lack of impact on saccadic selectivity suggest that different processes may underlie the performance of the central discrimination and peripheral selection tasks. Specifically, peripheral selection may be supported by a preattentive process, which is carried out in a spatially parallel manner. Although the preattentive processing successfully supports and biases the selection of the next display item to be fixated, it does not enable the observer to determine the exact form of the item or to bind the individual features into a complete object (see Rayner & Fisher, 1987; Treisman, 1996; Wolfe, 1998; Wolfe & Bennett, 1997). In contrast, the post-selection central discrimination requires focal attention and involves extracting critical features of the fixated item or integrating individual features into a complete object (e.g., Navon & Pearl, 1985; Rayner & Fisher, 1987; Treisman, 1988; Wolfe, 1998; Wolfe & Bennett, 1997). Thus, the present study provides important convergent evidence that saccadic selectivity in visual search is a form of preattentive guidance.

During the past two decades, several studies have investigated the relation between foveal processing and peripheral analysis in reading research, and different theoretical models (e.g., Henderson, 1992; Henderson & Ferreira, 1990, 1993; Morrison, 1984) have been proposed to explain the interplays between the two processes in reading tasks. The current study, along with Hooge and Erkelens (1999) and Findlay *et al.* (2001), began to provide similar data in the context of visual search. The picture emerging from these recent studies is complex and suggests that the components of visual processing are multi-determined by variables such as task context, stimulus materials, and attentional factors. The present investigation highlights the need for further explorations of the relation between foveal processing and peripheral analysis in complex visual tasks and provides a theoretical and methodological framework for this line of research.

Acknowledgements

Preparation of this manuscript was supported by a grant to Eyal M. Reingold from the Natural Science and Engineering Research Council of Canada (NSERC) and a grant to Marc Pomplun from the Deutsche Forschungsgemeinschaft (DFG). We would like to thank Heiner Deubel, Raymond Klein, Jan Theeuwes, and Ignace Hooge for their helpful comments on an earlier draft of this paper.

References

Bertera, J. H. (1988). The effect of simulated scotomas on visual search in normal subjects. *Investigative Ophthalmology & Visual Science, 29*, 470–475.

Bertera, J. H., & Rayner, K. (2000). Eye movements and the span of effective stimulus in visual search. *Perception & Psychophysics, 62*, 576–585.

Bichot, N. P., & Schall, J. D. (1998). Saccade target selection in macaque during feature and conjunction visual search. *Visual Neuroscience, 16*, 81–89.

Binello, A., Mannan, S., & Ruddock, K. H. (1995). The characteristics of eye movements made during visual search with multi-element stimuli. *Spatial Vision, 9*, 343–362.

Cave, K. R., & Wolfe, J. M. (1990). Modeling the role of parallel processing in visual search. *Cognitive Psychology, 22*, 225–271.

Duncan, J., & Humphreys, G. W. (1989). Visual search and stimulus similarity. *Psychological Review, 96*, 433–458.

Duncan, J., & Humphreys, G. W. (1992). Beyond the search surface: Visual search and attentional engagement. *Journal of Experimental Psychology: Human Perception & Performance, 18*, 578–588.

Findlay, J. M. (1997). Saccade target selection during visual search. *Vision Research, 37*, 617–631.

Findlay, J. M., Brown, V., & Gilchrist, I. D. (2001). Saccade target selection in visual search: The effect of information from the previous fixation. *Vision Research, 41*, 87–95.

Findlay, J. M., & Gilchrist, I. D. (1998). Eye guidance and visual search. In: G. Underwood (ed.), *Eye Guidance in Reading, Driving and Scene Perception* (pp. 295–312). Oxford: Elsevier.

Gould, J. D. (1967). Pattern recognition and eye-movement parameters. *Perception & Psychophysics, 2*, 399–407.

Henderson, J. M. (1992). Visual attention and eye movement control during reading and picture viewing. In: K. Rayner (ed.), *Eye Movements and Visual Cognition: Scene Perception and Reading* (pp. 260–283). New York: Springer-Verlag.

Henderson, J. M., & Ferreira, F. (1990). Effects of foveal processing difficulty on the perceptual span in reading: Implications for attention and eye movement control. *Journal of Experimental Psychology: Learning, Memory, & Cognition, 16*, 417–429.

Henderson, J. M., & Ferreira, F. (1993). Eye movement control during reading: Fixation measures reflect foveal but not parafoveal processing difficulty. *Canadian Journal of Experimental Psychology, 47*, 201–221.

Hooge, I. T., & Erkelens, C. J. (1999). Peripheral vision and oculomotor control during visual search. *Vision Research, 39*, 1567–1575.

Ikeda, M., & Takeuchi, T. (1975). Influence of foveal load on the functional visual field. *Perception & Psychophysics, 18*, 255–260.

Jacobs, A. M. (1986). Eye-movement control in visual search: How direct is visual span control? *Perception & Psychophysics, 39*, 47–58.

Lévy-Schoen, A. (1981). Flexible and/or rigid control of visual scanning behaviour. In: D. F. Fisher, R. A. Monty and J. W. Senders (eds), *Eye Movements: Cognition and Visual Perception* (pp. 299–314). Hillsdale, NJ: Lawrence Erlbaum.

Luria, S. M., & Strauss, M. S. (1975). Eye movements during search for coded and uncoded targets. *Perception & Psychophysics*, *17*, 303–308.

Mackworth, N. H. (1965). Visual noise causes tunnel vision. *Psychonomic Science*, *3*, 67–68.

McLeod, P., Driver, J., & Crisp, J. (1988). Visual search for a conjunction of movement and form is parallel. *Nature*, *332*, 154–155.

Morrison, R. E. (1984). Manipulation of stimulus onset delay in reading: Evidence for parallel programming of saccades. *Journal of Experimental Psychology: Human Perception & Performance*, *10*, 667–682.

Motter, B. C., & Belky, E. J. (1998a). The zone of focal attention during active visual search. *Vision Research*, *38*, 1007–1022.

Motter, B. C., & Belky, E. J. (1998b). The guidance of eye movements during active visual search. *Vision Research*, *38*, 1805–1815.

Murphy, K. St. J., & Foley-Fisher, J. A. (1988). Visual search with non-foveal vision. *Ophthalmic & Physiological Optics*, *8*, 345–348.

Nagy, A. L., & Sanchez, R. R. (1990). Critical color differences determined with a visual search task. *Journal of the Optical Society of America — A*, *7*, 1209–1217.

Nakayama, K., & Silverman, G. H. (1986). Serial and parallel processing visual feature conjunctions. *Nature*, *320*, 264–265.

Navon, D., & Pearl, D. (1985). Preattentive processing or prefocal processing? *Acta Psychologia*, *60*, 245–262.

Nazir, T., & Jacobs, A. M. (1991). The effects of target discriminability and retinal eccentricity on saccade latencies: Analysis in terms of variable criterion theory. *Psychological Research*, *53*, 287–299.

Pomplun, M., Reingold, E. M., & Shen, J. (2001a). Peripheral and parafoveal cueing and masking effects on saccadic selectivity. *Vision Research*, *41*, 2757–2769.

Pomplun, M., Reingold, E. M., & Shen, J. (2001b). Investigating the visual span in comparative search: The effects of task difficulty and divided attention. *Cognition*, *81*, B57-B67.

Pomplun, M., Reingold, E. M., & Shen, J. (submitted). Area activation: A computational model of saccadic selectivity in visual search. *Cognitive Science*.

Pomplun, M., Reingold, E. M., Shen, J., & Williams, D. E. (2000). The area activation model of saccadic selectivity in visual search. In: L. R. Gleitman and A. K. Joshi (eds), *Proceedings of the 22nd Annual Conference of the Cognitive Science Society* (pp. 375–380). Mahwah, NJ: Elrbaum.

Rayner, K. (1998). Eye movements in reading and information processing: 20 years of research. *Psychological Bulletin*, *124*, 372–422.

Rayner, K., & Fisher, D. L. (1987). Letter processing during eye fixations in visual search. *Perception & Psychophysics*, *42*, 87–100.

Rayner, K., Inhoff, A. W., Morrison, R. E., Slowiaczek, M. L., & Bertera, J. H. (1981). Masking of foveal and parafoveal vision during eye fixation in reading. *Journal of Experimental Psychology: Human Perception & Performance*, *7*, 167–179.

Reingold, E. M., Charness, N., Pomplun, M., & Stampe, D. M. (2001). Visual span in expert chess players: Evidence from eye movements. *Psychological Science*, *12*, 48–55.

Scialfa, C. T., & Joffe, K. (1998). Response times and eye movements in feature and conjunction search as a function of target eccentricity. *Perception & Psychophysics*, *60*, 1067–1082.

Shen, J., & Reingold, E. M. (1999). Saccadic selectivity during visual search: The effects of shape and stimulus familiarity. In: M. Hahn and S. C. Stoness (eds), *Proceedings of the 21st Annual Conference of the Cognitive Science Society* (pp. 649–652). Mahwah, NJ: Erlbaum.

Shen, J., Reingold, E. M., & Pomplun, M. (2000). Distractor ratio influences patterns of eye movements during visual search. *Perception, 29*, 241–250.
Theeuwes, J., & Kooi, F. L. (1994). Parallel search for a conjunction of contrast polarity and shape. *Vision Research, 34*, 3013–3016.
Treisman, A. (1988). Features and objects: The fourteenth Bartlett memorial lecture. *The Quarterly Journal of Experimental Psychology*, 40A, 201–237
Treisman, A. (1996). The binding problem. *Current Opinion in Neurobiology, 6*, 171–178.
Treisman, A., & Gelade, G. (1980). A feature integration theory of attention. *Cognitive Psychology, 12*, 97–136.
Treisman, A., & Sato, S. (1990). Conjunction search revisited. *Journal of Experimental Psychology: Human Perception & Performance, 16*, 459–478.
Treisman, A., Sykes, M., & Gelade, G. (1977). Selective attention and stimulus integration. In: S. Dornic (ed.), *Attention and Performance III* (pp. 280–292). Amsterdam: North-Holland.
van Diepen, P. M. J., De Graef, P., & d'Ydewalle, G. (1995). Chronometry of foveal information extraction during scene perception. In: J. M. Findlay, R. Walker and R. W. Kentridge (eds), *Eye Movement Research: Mechanism, Processes and Applications* (pp. 349–362). Elsevier: North-Holland.
Viviani, P., & Swensson, R. G. (1982). Saccadic eye movements to peripherally discriminated visual targets. *Journal of Experimental Psychology: Human Perception & Performance, 8*, 113–126.
von Grünau, M., Dubé, S., & Galera, C. (1994). Local and global factors of similarity in visual search. *Perception & Psychophysics, 55*, 575–592.
Williams, D. E., & Reingold, E. M. (2001). Preattentive guidance of eye movements during triple conjunction search tasks: The effects of feature discriminability and saccadic amplitude. *Psychonomic Bulletin & Review, 8*, 476–488.
Williams, D. E., Reingold, E. M., Moscovitch, M., & Behrmann, M. (1997). Patterns of eye movements during parallel and serial visual search tasks. *Canadian Journal of Experimental Psychology, 51*, 151–164.
Williams, L. G. (1967). The effect of target specification on objects fixated during visual search. *Perception & Psychophysics, 1*, 315–318.
Williams, L. J. (1989). Foveal load affects the functional field of view. *Human Performance, 2*, 1–28.
Wolfe, J. M. (1994). Guided search 2.0: A revised model of visual search. *Psychonomic Bulletin & Review, 1*, 202–238.
Wolfe, J. M. (1998). Visual search. In H. Pashler (ed.), *Attention* (pp. 13–71). London: Psychology Press.
Wolfe, J. M., & Bennett, S. C. (1997). Preattentive object files: Shapeless bundles of basic features. *Vision Research, 37*, 25–43.
Wolfe, J. M., Cave, K. R., & Franzel, S. L. (1989). Guided search: An alternative to the feature integration model for visual search. *Journal of Experimental Psychology: Human Perception & Performance, 15*, 419–433.
Wolfe, J. M., Friedman-Hill. S. R., Stewart, M. I., & O'Connell, K. M. (1992). The role of categorization in visual search for orientation. *Journal of Experimental Psychology: Human Perception & Performance, 18*, 34–49.
Zelinsky, G. J. (1996). Using eye saccades to assess the selectivity of search movements. *Vision Research, 36*, 2177–2187.
Zelinsky, G. J., & Sheinberg, D. L. (1997). Eye movements during parallel-serial visual search. *Journal of Experimental Psychology: Human Perception & Performance, 23*, 244–262.
Zohary, E., & Hochstein, S. (1989). How serial is serial processing in vision? *Perception, 18*, 191–200.

Chapter 5

Multisensory Interactions in Saccade Generation

Robin Walker and Melanie Doyle

Saccadic eye movements may be guided by salient objects and events within the environment or directed at will in the absence of an external stimulus. Selection is necessary to select just one target from the vast array of competing stimuli in the environment (Schall, 1995) and, as saccades may be directed to visual, auditory and tactile stimuli, the processes of target selection must operate in a multisensory context. Do multisensory interactions have the capacity to influence saccadic eye movements? In the following chapter we will discuss neurophysiological and behavioural evidence of multisensory interactions in saccade generation. Particular focus will be given to a recent study examining the effect of multisensory interactions on saccade trajectory.

Neurophysiological Basis of Multisensory Interactions

For multisensory interactions to occur at neural level different sensory stimuli must have access to the same neurons. There is a great deal of neurophysiological evidence that visual, auditory and tactile stimuli have access to the same "multisensory" neurons in a number of areas (Stein & Meredith, 1993). One such area, the superior colliculus, has been studied in great depth and is of particular relevance here as it is involved in eye movement generation. Multisensory neurons are found only in the deep layers of the superior colliculus (DLSC) and it is the DLSC that are involved in eye movement generation (Stein & Meredith, 1993). It has been estimated that about half of DLSC neurons are multisensory and respond to some combination of visual, auditory and tactile/somatosensory stimuli (Meredith & Stein, 1986a; Wallace *et al.*, 1998).

Within the DLSC, individual multisensory neurons tend to respond to stimuli from the same region of external space so their receptive fields are broadly aligned for different modalities. Overall, neurons are arranged so that the receptive field locations

The Mind's Eye: Cognitive and Applied Aspects of Eye Movement Research

ISBN: 0–444–51020–6

vary systematically across the structure to form a topographic map of external space. The vertical representation of space runs medio-laterally with superior space represented medially and inferior space laterally, whilst the horizontal representation of space runs rostro-caudally, with medial space represented rostrally and lateral space caudally (see Stein & Meredith, 1993). Though there are some variations between modalities, these maps are roughly aligned for different modalities so that a particular region of the map will be responsive to visual, auditory and somatosensory stimuli within the same region of external space.

The superior colliculus is known to be involved in the generation of saccadic eye movements. Individual neurons are associated with a particular movement vector (or movement of a particular amplitude and direction). Saccadic eye movements are brought about by the combined activity of a large group of neurons (McIlwain, 1991) — up to 25% of SC neurons may be involved in a single eye movement (Munoz & Wurtz, 1995) — and each neuron may be involved in a range of eye movements. The *movement field* of a neuron delimits a region of space to which the eye must move for that neuron to take part in the movement (Stein & Meredith, 1993). Movement fields vary systematically across the population of neurons and, as with sensory receptive fields, the movement fields form a topographical map of external space. Furthermore, these motor maps are aligned with sensory maps such that stimulation of motor neurons at the site of a given sensory stimulus will result in an eye movement to that stimulus (Stein & Meredith, 1993).

Within the DLSC, individual multisensory neurons may exhibit an enhanced response to particular combinations of stimuli provided that certain criteria are satisfied. Firstly, the neuron must be capable of responding to the particular combination of modalities (Stein & Meredith, 1993). Secondly, stimulus location is important as the stimuli must lie within their respective receptive fields, and must therefore be aligned in external space for facilitation to occur (Meredith & Stein, 1986b). Thirdly, the relative timing of the stimuli must be such that the neural response to the two stimuli overlaps in time — this does not necessarily require that the stimuli occur simultaneously (Meredith *et al.*, 1987). In essence, the neurons appear to favour combinations of stimuli that are broadly aligned in space and time and likely to originate from the same object or event. A final factor relates to the efficacy of the individual stimuli: within limits, the magnitude of facilitation increases as the efficacy of the individual stimuli decreases (Meredith & Stein, 1986a). Similar rules have been found to apply to multisensory integration in other areas, such as the cortex (Stein & Wallace, 1996).

The rules of multisensory integration have been studied in depth in the superior colliculus and it is in this area that one can most readily associate behavioural outcomes of multisensory integration with underlying neural structures. Stein and Meredith (1993) suggest that multisensory integration in different neural areas subserves different roles: whilst integration in the superior colliculus may influence orienting behaviours, integration in the cortex may play a greater role in perception. The present chapter will focus upon the superior colliculus in the later discussion of the neural correlate of the observed behavioural effects as it is in this area that the neural basis and outcome of multisensory integration is best understood.

Behavioural Evidence of Multisensory Interactions

In a multisensory environment two scenarios are possible: different sensory stimuli may arise from the *same* object or event, or the stimuli may originate from *different* objects or events. Where stimuli arise from the *same* object it is possible that both will facilitate selection of that object and may therefore interact in a co-operative fashion. In contrast, different sensory stimuli arising from *different* objects are likely to favour the selection of different objects and therefore to compete as targets for overt behaviours, such as saccades.

In a series of studies Stein and colleagues (Stein *et al.*, 1988, 1989) examined the accuracy of orienting behaviour in cats following visual targets with spatially aligned and disparate auditory stimuli. Accuracy was increased following spatially congruent sound and reduced with sound at distant locations. It was noted that "the rules that define the integration of multisensory inputs at the level of the single superior colliculus neuron also predict the overt behaviour of the animal when it is required to attend and orient to multisensory stimuli" (Stein *et al.*, 1988, p. 358). It appears that this is also true for facilitation of overt orienting in human observers (Costin *et al.*, 1991). More recent studies have examined the impact of multisensory interactions on human saccadic eye movements. A number of these have focused upon reductions in saccade latency associated with multisensory stimuli, whilst others have examined the effects of multisensory competition on saccade amplitude and trajectory.

Saccade Latency

Saccade latencies may be reduced when a visual target is accompanied by a sound (Hughes *et al.,* 1994; Konrad *et al.*, 1989; Lee *et al.*, 1991; Frens *et al.*, 1995). Hughes *et al.* (1994) examined the impact of sound upon manual and saccadic responses to visual stimuli to determine whether the facilitation was due to statistical facilitation (Miller, 1982), or was the result of multisensory interaction at neural level based on additive (Schwartz, 1989) or multiplicative (Stein & Meredith 1993) processes. Visual and auditory stimuli were presented alone or in combination on the left or right of a display, 15 degrees from the vertical midline. Latencies of manual and saccadic responses were reduced when spatially congruent visual and auditory stimuli were presented. The pattern of responses with saccades is particularly interesting. When visual and auditory stimuli were spatially congruent saccade latencies were reduced by up to 50 ms. However, when the two stimuli were on opposite sides of the midline, latencies were increased relative to the visual baseline. The latter effect mirrored the "remote distractor" effect observed with paired visual stimuli presented some distance apart (Walker *et al.*, 1997; Findlay & Walker, 1999). The degree of facilitation observed with spatially congruent stimuli was too large to be explained by statistical facilitation or by additive neural processes. Rather, it was suggested that facilitation reflected multiplicative neural processes, possibly within the superior colliculus. This is consistent with the single cell response reported by Stein and colleagues (see e.g., Stein & Meredith, 1993).

Saccade Amplitude

Saccade amplitude may also be modified by multisensory interactions. Frens *et al.* (1995) presented visual targets and auditory distractor stimuli in close proximity and observed that saccades were directed to the average location of the two stimuli. (Incidentally, this averaging effect was accompanied by a reduction in saccade latency of approximately 40 ms.) Similarly, Lueck *et al.* (1990) found that the endpoint of saccades to auditory targets shifted towards a visual distractor. As with the remote latency effect noted previously, these multisensory amplitude effects resemble those observed for paired stimuli from the same modality. The "global" effect occurs when target and distractor stimuli are in close proximity and the saccade is directed to the weighted average location of the two stimuli rather than to the target (Findlay, 1982; Deubel *et al.,* 1984).

The global effect has been attributed to neural activity, specifically the failure to form distinct peaks of activity associated with target and distractor stimuli presented in close proximity. Briefly, target selection should result in the selection of a single peak of activity that serves as the target for the upcoming saccade. If target and distractor stimuli are presented in close proximity it may be difficult, or impossible, to resolve target and distractor activity and instead a single peak is formed associated with the activity related to both stimuli. As a result, the saccade is directed to the average location of the two stimuli. Multisensory global effects are likely to require a similar explanation. Indeed, Frens *et al.* (1995) asserted that "the averaging responses between visual and auditory targets . . . may . . . be caused by the same mechanism that is thought to be responsible for the averaging responses towards visual targets" (p. 814). This does, however, require that target selection occurs within a multisensory context and that different sensory stimuli produce activity in a common topographical map.

Another effect that may be explained by such selection processes operating within a multisensory context is the remote latency effect, that is, response times to a target are increased by the presence of a remote distractor (Lévy-Schoen, 1969; Findlay, 1983; Walker *et al.*, 1997). In this effect, target and distractor stimuli presented some distance apart produce separate peaks of activity and, relative to when only one peak of activity is present, extra time is required to select one of the two peaks as the target for the next saccade. A "remote distractor" latency effect was reported by Hughes *et al.* (1994) when the visual target was accompanied by an auditory distractor on the opposite side of the midline, 30 degrees from the target.

Saccade Trajectory

Saccade trajectories may be modified by competition between target and non-target stimuli. For example, Sheliga *et al.* (1994, 1995) observed that saccades to visual targets curved away from visual distractors to which the observer had previously oriented attention. In such studies the distractor was used to provide information about the task, e.g., the distractor was a directional cue indicating the location of the target, and observers had to attend to it before they could make a saccade. When the distractor was on the right, saccades curved to the left and when the distractor was on the left, saccades curved to the right.

Given that distractor effects, such as the global and remote effect, may be observed for combinations of stimuli from the same or different modalities, it was of interest to determine whether trajectory effects could also be observed with stimuli from different modalities. As noted above, different sensory stimuli would need access to the same topographical maps — or there would need to be a degree of interplay between topographical maps for different sensory stimuli — for multisensory amplitude effects to occur. Multisensory trajectory effects would also require sensory convergence at the level of topographic maps, but would also require that such interactions had the capacity to modify a saccade in progress rather than just the endpoint of the saccade. Although neurophysiogical studies have demonstrated neural interactions in the superior colliculus with visual, auditory and somatosensory stimuli behavioural studies have examined multisensory interaction effects with visual and auditory stimuli only.

In a series of studies we examined the impact of visual, auditory and tactile distractors upon the trajectories of saccades to visual targets (see Doyle & Walker, 2002 for further details). In the present discussion we will focus upon multisensory effects with relevant targets (but see Doyle & Walker, 2001 for curvature effects with irrelevant visual targets).

Experiment: Multisensory Effects in Saccade Trajectory

The primary aim of this experiment was to examine the trajectories of saccades initiated to visual targets when accompanied by visual, auditory and tactile distractors. It was expected that saccades would curve away from the distractor stimuli due to multisensory interactions between target and distractor stimuli. This hypothesis was supported by results of previous studies that reported such curvature with visual distractors (Sheliga *et al.*, 1994; Sheliga *et al.,* 1995) and by evidence that other multisensory distractor effects have resembled those observed in unimodal contexts (Frens *et al.*, 1995; Hughes *et al.*, 1994).

The display used is shown in Figure 5.1a. The stimulus framework consisted of two horizontal and two vertical metal bars, each of which had a stimulus generator attached. Individual stimulus generators incorporated visual (yellow LED, 5 mm diameter × 9 mm depth), auditory (white noise delivered through Piezo sounders, Kingstate (supplied by RS components), 25 mm diameter, 5 V input, 250 Hz — 20 kHz frequency response) and tactile (vibratory stimuli delivered to index finger from a linear actuator, TransDimensional International Co., TDITAC) stimuli. The three stimuli were spatially aligned within each stimulus generator but the relative location of the four generators could be readily altered. Targets were visual stimuli above and below fixation whilst distractors were visual, auditory or tactile stimuli to the left or right of the midline: in order that the distractor location did not need to be reset during each session distractors were in the upper half of the display for half of the observers

a

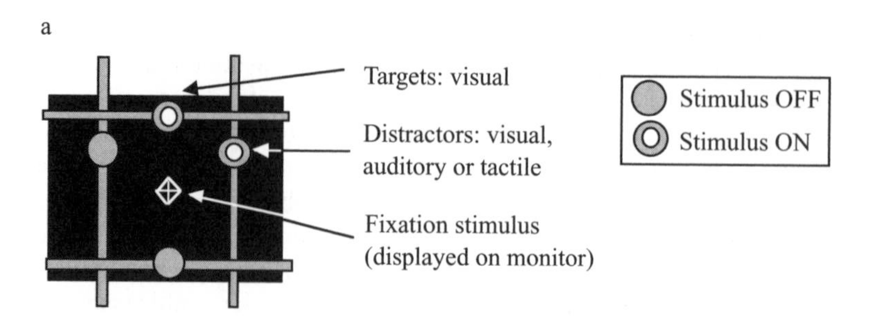

b

Fixation display (800-1300 ms) → Test display (1000 ms) → Intertrial interval (750 ms)

Distractor trials: saccade direction based on distractor location

Control trials: saccade when fixation stimulus changes, direction alternates from trial to trial

Figure 5.1: Schematic representation of display (a) and display used in test and control trials in the experiment (b). The display consisted of a central fixation stimulus (shown on the monitor), two visual targets (above and below fixation) and a single visual, auditory or tactile distractors (to the left or right of the vertical midline). In test trials distractors provided information about the task — the location of the distractor indicated the direction of the saccade. In control trials no distractor was shown and saccade direction was based on prior instruction.

and in the lower half for the remaining observers. The fixation stimulus was shown on a monitor located directly behind the framework. Observers sat with their head in a chin-rest 40 cm from the framework, and their arms outstretched so that their index fingers were in contact with the tactile stimuli on each side of the framework.

Each observer completed three test blocks containing 120 trials (20 trials for each combination of distractor modality and target location) and a control block of 40 trials, each preceded by a short practice block. Trials were defined on the basis of distractor modality (visual, auditory or tactile), distractor location (left or right) and distractor hemifield (same, opposite). Equal numbers of trials from each combination of distractor modality (visual, auditory, tactile), distractor location (left, right) and distractor hemifield (same, opposite) were combined with a single block and were interleaved in random order. Distractor hemifield refers to the relative location of target and distractor stimuli, specifically whether they were in the same half (upper or lower) or opposite halves of the display: this variable had no effect and will not be discussed further (but see Doyle & Walker, 2002 for details).

In test, or "distractor", trials the fixation stimulus and target stimuli were both visible at the start of each trial. Following a random fixation period, between 800 and 1300 ms, a distractor appeared to the left or right of fixation. Target direction was based on the location (left or right) of the distractor, e.g., observers were instructed to make a saccade to the upper target if the distractor was on the left and to the lower target if the distractor was on the right. Visual, auditory and tactile distractors were interleaved within a single block and the same instructions applied regardless of distractor modality. The target and distractor stimuli were presented for a further 1000 ms and before the display was cleared and an inter-trial interval of 750 ms occurred. Observers were asked to fixate the central stimulus and to maintain fixation until the distractor had appeared. Following distractor onset, they had to localise the distractor *without* moving their eyes and then to make a saccade, as quickly and as accurately as possible to the relevant target.

In a control block observers were shown two targets and asked to make saccades from fixation to the centre of the target when the fixation stimulus changed, alternating between upper and lower targets on successive trials. Although this did not provide a suitable control for saccade latency it provided a baseline for saccade curvature and allowed us to assess curvature when distractors were present relative to that obtained when no distractor was present.

Measuring Curvature

Saccade curvature is measured by comparing the path of the saccade to the direct route from the start to the end of the saccade, or from the initial fixation point to the target (see Doyle & Walker, 2001 for details). The saccade path itself may be assessed by using a single point on the saccade, by using more than one point or by assessing the area under the entire curve. In our study we have measured the distance between the actual path and the direct route at a single point on the saccade, the point at which the saccade deviated most from the direct route. This distance is measured perpendicular to saccade direction, e.g., deviation is measured horizontally for vertical saccades and vertically for horizontal saccades. To prevent changes in saccade curvature related to saccade amplitude from biasing the results the measure of deviation was divided by saccade amplitude to obtain a ratio of curvature per unit amplitude. Finally, a sign was used to indicate the absolute direction of curvature, e.g., for vertical saccades; negative values indicated leftwards curvature and positive values indicated rightwards curvature.

Results

The saccade data from one observer is shown in Figure 5.2.

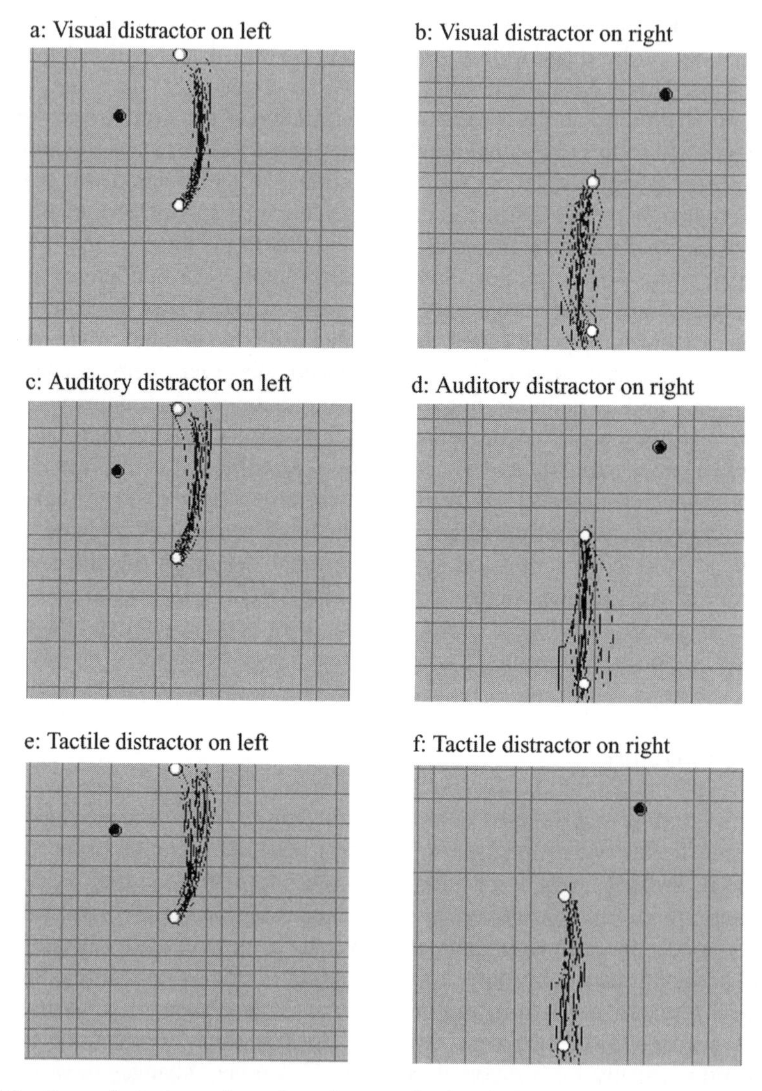

Figure 5.2: Saccade trajectories taken from a single observer. In this block of trials the observer was instructed to make a saccade to the upper target if the distractor was on the left and to the lower target if the distractor was on the right.

The mean signed curvature data for all six observers is shown in Figure 5.3.

It appears that when distractors are presented on the right strong leftwards curvature is observed for all distractor modalities. In contrast, saccade trajectories curve to the right when visual distractors are on the left, but show a small degree of leftwards curvature when auditory and tactile distractors occur in the same location. (N.B. individual differences are found in trajectory effects and the group data reveals a different trend to that apparent in the data for the single subject in Figure 5.2.) The magnitude of curvature is relatively small and, at least in the case of auditory and tactile distractors, may not be statistically significant from zero: significant changes in direction of curvature are assessed by comparing curvature with right side and left side distractors within each modality.

A three-way ANOVA with the factors distractor modality (visual, auditory, tactile), distractor hemifield (same, opposite) and distractor location (left, right) was used to analyse the curvature data. Although ANOVA revealed no significant main effects of distractor modality, hemifield or location there was a significant interaction of distractor modality and distractor location ($F(2,10) = 11.06, p < 0.01$). Post-hoc tests showed a significant effect of distractor location upon signed curvature in all three modalities (Tukey HSD: visual — $p < 0.01$, auditory — $p < 0.05$, tactile — $p < 0.05$). For distractors in all three modalities curvature was more positive (indicating increased rightward curvature) when the distractor was on the left relative to when it was on the right. In other words, for distractors in all three modalities, the modulation of saccade curvature was greater when the distractor was on the right relative to when it was on the left.

Discussion

These results indicate that multisensory interactions can alter the path of a saccade such that saccades to visual target curve away from visual, auditory and tactile distractors.

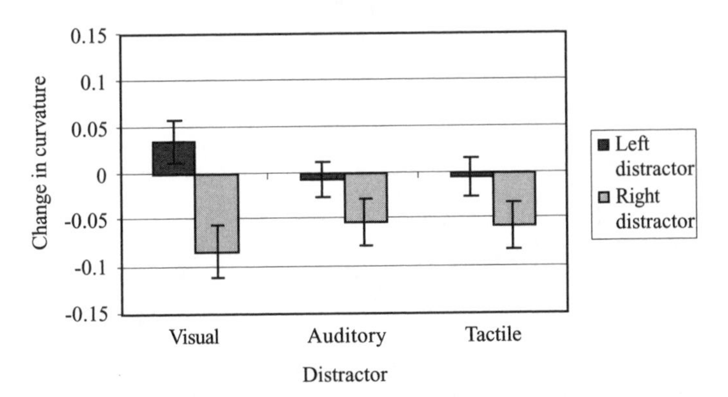

Figure 5.3: The mean curvature data for six observers in shown. Change in curvature (in the test relative to the control condition) is shown for conditions in which visual, auditory and tactile distractors were presented to the left or right of fixation. Negative values indicate leftwards curvature, whilst positive values indicate rightwards curvature.

Furthermore, the trajectory effects observed in multisensory contexts resembled those obtained with combinations of stimuli from the same modality. Like the global and remote effects described earlier, it is likely that multisensory trajectory effects will demand similar explanations to trajectory effects observed for visual stimuli. The most plausible explanations of curvature come from inhibition-based models developed to explain curvature of saccades (Sheliga *et al.*, 1994, 1995) and reach trajectories (Tipper *et al.*, 1997, 2000).

According to inhibition-based models (Sheliga *et al.*, 1994; Sheliga *et al.*, 1995; Tipper *et al.*, 1997; Tipper *et al.*, 2000; Tipper *et al.*, 2001) curvature arises during the process of target selection due to inhibition of distractor-related activity. In these hypotheses it is proposed that the overt response to the distractor stimulus is programmed, either to achieve covert orienting (Sheliga *et al.*, 1994) or automatically (Tipper *et al.*, 1997), and must then be inhibited for a saccade to be made to the target. Tipper's model is shown in Figure 5.4, which shows a schematic representation of the neural activity and overt response associated with the target and distractor stimuli alone and with the target and distractor presented together. Because movements such as eye movements and reach movements rely upon a large number of neurons, there is likely

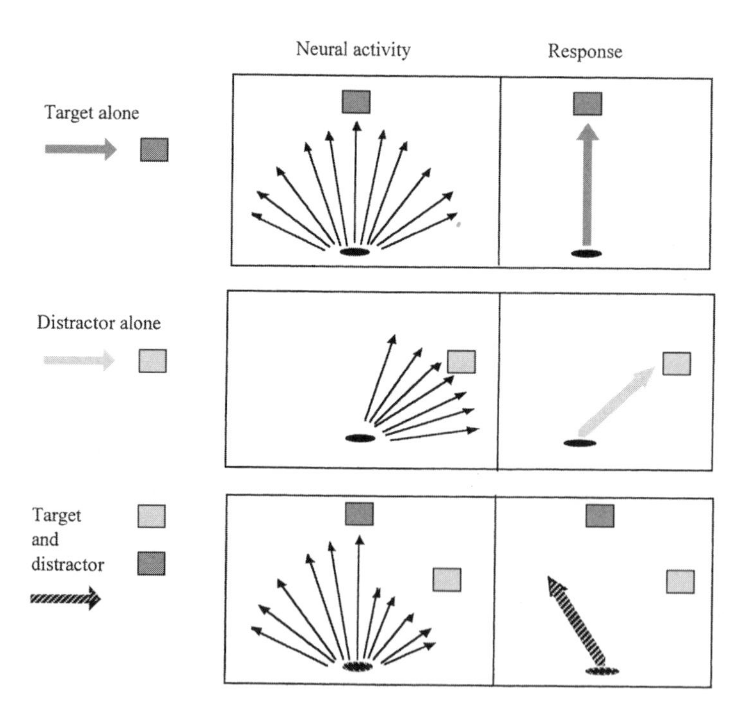

Figure 5.4: Tipper's model of curved trajectory effects. This is a schematic representation of the neural activity and overt response associated with a target, a distractor and a target with a competing distractor present. It illustrates the manner in which curvature might occur as outlined in the inhibition-based hypotheses (see text for description).

to be overlap between the populations of neurons involved in movements to target and distractor stimuli (Figure 5.4) and a single neuron may potentially be involved in both responses. Inhibition of the distractor response reduces the activity in neurons associated with the response to the distractor but may also reduce activity in a subset of neurons involved in programming the response to the target. Due to the imbalance in activity in neurons involved in the target response the overt response to the target, which depends upon the combined activity of a group of neurons (McIllwain, 1991), then deviates away from the inhibited region.

One limitation of the inhibition-based theories is that they do not explain why saccades, after deviating away from distractor stimuli, then curve back towards the target. Recent studies have suggested that curvature is an inherent property of saccades (Aizawa & Wurtz, 1998) or that deviations in saccade trajectory may be corrected during the course of the saccade (Quaia *et al.*, 2000). A further shortcoming is that the theories, at present, do not relate trajectory effects to the global and remote effects observed earlier. In this respect two factors appear important. Firstly, activity associated with target and distractor stimuli must be resolved for curvature to occur: in other words, it is necessary to differentiate peaks of activity for the two stimuli to select one and inhibit the other. Secondly, activity associated with the distractor stimulus must then be inhibited (Sheliga *et al.*, 1994, 1995). The competition between target and distractor stimuli leads to curvature that closely resembles that obtained with remote distractors. Indeed, latency increases have been found in association with saccade curvature (Doyle & Walker, 2001) when distractors were presented in the opposite half of the display to the target stimuli, but not when distractors and targets were in the same half of the display. Though a clear relationship between curvature and global or remote effects has not been identified, the effects appear to have some common elements, e.g., latency effects.

General Discussion

Evidence of multisensory trajectory, global and remote effects supports the view that target selection must take place in a multisensory context and also suggests that such selection occurs within an environment where saccade programming is achieved through population coding. The latency facilitation effects observed with multisensory stimuli add further support to the view that multisensory interactions can influence saccade generation. Though the superior colliculus is by no means the only area likely to be of influence, it does support multisensory responses and controls saccade generation through population encoding. It therefore seems reasonable to consider the response properties of this area when developing theories of saccade target selection in multisensory contexts.

Although almost half of DLSC neurons respond to only one sensory modality it appears that the *multisensory* neurons are heavily involved in eye movement generation. A study of the sensory properties of eye movement related neurons reported that 74% responded to stimuli of more than one modality (Meredith & Stein, 1985). Another factor that highlights the importance of multisensory neurons in this context is their connections with the broader network of areas involved in eye movement control. The frontal eye fields (FEF) are involved in eye movement generation and have topographical excitatory and inhibitory links with the SC that allow the FEF to impose a saccade goal on the SC (Schlag-Rey *et al.*, 1992). The connections from the FEF may also have the capacity to influence multisensory interactions as they target multisensory neurons in the SC (Meredith, 1999). Further research will be needed to ascertain the role that the FEF and other cortical areas may play in resolving multisensory competition.

In sum, interactions between different sensory stimuli can modify saccade responses. Co-operative multisensory interactions may facilitate responses resulting in faster eye movements to multisensory targets (e.g., Hughes *et al.*, 1994). In contrast, competition between stimuli of different modalities may increase eye movement latency (Hughes *et al.*, 1994) or alter the amplitude (Frens *et al.*, 1995) or path that the eye follows (Doyle & Walker, 2002). Neurophysiological studies have revealed a potential neural basis for such multisensory effects and theories used to explain unimodal distractor effects appear to apply at multisensory level. As the external environment provides a vast array of potential targets for saccadic eye movements, it is important to gain a better understanding of how multisensory interactions influence saccades and to develop theories that relate these effects to unimodal effects and the underlying neural circuitry.

Acknowledgements

This work was supported by a grant from the Wellcome Trust. The authors would like to thank Andy Haswell for developing and building the stimulus presentation framework and Derek Neil for designing and programming the saccade analysis program.

References

Aizawa, H., & Wurtz, R. H. (1998). Reversible inactivation of monkey superior colliculus. I. Curvature of saccadic trajectory. *Journal of Neurophysiology*, *79*, 2082–2096.

Costin, D., Neville, H. J., Meredith, M. A., & Stein, B. E. (1991). Rules of multisensory integration and attention: ERP and behavioural evidence in humans. *Society for Neuroscience Abstracts*, *17*, 656.

Deubel, H., Wolf, W., & Hauske, G. (1984). The evaluation of oculomotor error signals. In: A. G. Gale and F. Johnson (eds), *Theoretical and Applied Aspects of Oculomotor Research* (pp. 55–62). Amsterdam: Elsevier.

Doyle, M. C., & Walker, R. (2001). Curved saccade trajectories: Voluntary and reflexive saccades curve away from irrelevant distractors. *Experimental Brain Research*, *139*, 333–344.

Doyle, M. C., & Walker, R. (2002). Multisensory interactions in saccade target selection: Curved saccade trajectories. *Experimental Brain Research*, *142*, 116–130.

Findlay, J. M. (1982). Global processing for saccadic eye movements. *Vision Research*, *22*, 1033–1045.

Findlay, J. M. (1983). Visual information processing for saccadic eye movements. In: A. Hein and M. Jeannerod (eds), *Spatially Oriented Behaviour* (pp. 281–303). Springer-Verlag, New York.

Findlay, J. M., & Walker, R. (1999). A model of saccade generation based on parallel processing and competitive inhibition. *Behavioural and Brain Sciences*, *22*, 661–721.

Frens, M. A., Van Opstal, A. J., & Van der Willigen, R. F. (1995). Spatial and temporal factors determine auditory-visual interactions in human saccadic eye movements. *Perception and Psychophysics*, *57*, 802–816.

Hughes, H. C., Reuter-Lorenz, P. A., Nozawa, G., & Fendrich, R. (1994). Visual-auditory interactions in sensorimotor processing: saccades versus manual responses. *Journal of Experimental Psychology: HPP*, *20*, 131–153.

Konrad, H. R., Rea, C., Olin, B., & Colliver, J. (1989). Simultaneous auditory stimuli shorten saccade latencies. *Laryngoscope*, *99*, 1230–1232.

Lee, C., Chung, S., Kim, J., & Park, J. (1991). Auditory facilitation of visually guided saccades. *Society for Neuroscience Abstracts*, *17*, 862.

Lévy-Schoen, A. (1969). Determination et latence de la reponse oculomotrice a deux stimulus. *L'Annee Psychologique*, *69*, 373–392.

Lueck, C. J., Crawford, T. J., Savage, C. J., & Kennard, C. (1990). Auditory-visual interaction in the generation of saccades in man. *Experimental Brain Research*, *82*, 149–157.

McIlwain, J. T. (1991). Distributed spatial coding in the superior colliculus: a review. *Visual Neuroscience*, *6*, 3–13.

Meredith, A. M. (1999). The frontal eye fields target multisensory neurons in the cat superior colliculus. *Experimental Brain Research*, *128*, 460–470.

Meredith, M. A., Nemitz, J. W., & Stein, B. E. (1987). Determinants of multisensory integration in superior colliculus neurons. I. Temporal factors. *Journal of Neuroscience*, *7*, 3215–3229.

Meredith, M. A., & Stein, B. E. (1986a). Visual, auditory, and somatosensory convergence on cells in superior colliculus results in multisensory integration. *Journal of Neurophysiology*, *56*, 640–662.

Meredith, M. A., & Stein, B. E. (1986b). Spatial factors determine the activity of multisensory neurons in the cat superior colliculus. *Brain Research*, *365*, 350–354.

Munoz, D. P., & Wurtz, R. H. (1992). Role of the rostral superior colliculus in active visual fixation and execution of express saccades. *Journal of Neurophysiology*, *67*, 1000–1002.

Quaia, C., Pare, M., Wurtz, R. H., & Optican, L. M. (2000). Extent of compensation for variations in monkey saccadic eye movements. *Experimental Brain Research*, *132*, 39–51.

Schall, J. D. (1995). Neural basis of saccade target selection. *Reviews in the Neurosciences*, *6*, 63–85.

Schlag-Rey, M., Schlag, J., & Dassonville, P. (1992). How the frontal eye field can impose a saccade goal on superior colliculus neurons. *Journal of Neurophysiology*, *67*, 1003.

Schwartz, W. (1989). A new model to explain the redundant-signals effect. *Perception and Psychophysics*, *46*, 498–500.

Sheliga, B. M., Riggio, L., & Rizzolatti, G. (1994). Orienting of attention and eye movements. *Experimental Brain Research*, *98*, 507–522.

Sheliga, B. M., Riggio, L., & Rizzolatti, G. (1995). Spatial attention and eye movements. *Experimental Brain Research*, *105*, 261–275.

Stein, B. E., Huneycutt, W. S., & Meredith, M. A. (1988). Neurons and behaviour: The same rules of multisensory integration apply. *Brain Research*, *448*, 355–358.

Stein, B. E., & Meredith, M. A. (1993). The merging of the senses. Cambridge MA: MIT Press.

Stein, B. E., Meredith, M. A., Huneycutt, W. S., & McDade, L. (1989). Behavioural indices of multisensory integration: Orientation to visual cues is affected by auditory stimuli. *Journal of Cognitive Neuroscience*, *1*, 12–24.

Stein, B. E., & Wallace, M. T. (1996). Comparisons of cross-modality integration in midbrain and cortex. *Progress in Brain Research*, *112*, 289–299.

Tipper, S. P., Howard, L. A., & Houghton, G. (2000). Behavioural consequences of selection from neural population codes. In S. Monsell and J. Driver (eds), *Control of Cognitive Processes: Attention and Performance XVIII*, Cambridge, MA: MIT Press.

Tipper, S. P., Howard, L. A., & Jackson, S. J. (1997). Selective Reaching to Grasp: Evidence for Distractor Interference Effects. *Visual Cognition*, *4*, 1–38.

Tipper, S. P., Howard, L. A., & Paul, M. (2001). Reaching affects saccade trajectories. *Experimental Brain Research*, *136*, 241–249.

Walker, R., Deubel, H., Schneider, W., & Findlay, J. M. (1997). Effect of remote distractors on saccade programming: Evidence for an extended fixation zone. *Journal of Neurophysiology*, *78*, 1108–1119.

Wallace, M. T., Meredith, M. A., & Stein, B. E. (1998). Multisensory interaction in the superior colliculus of the alert cat. *Journal of Neurophysiology*, *80*, 1006–1010.

Chapter 6

Binocular Coordination in Microsaccades

Ralf Engbert and Reinhold Kliegl

When viewing a stationary object, our eyes perform miniature eye movements, which are classified as drift, tremor and microsaccades. While drift and tremor are irregular movements with a low correlation between the eyes, microsaccades are ballistic movements, traditionally defined as binocular phenomena. Here we propose that microsaccades can be both binocular and monocular with different statistical properties, in particular with different mean orientations. For this analysis, we have developed a new algorithm for the detection of microsaccades based on a transformation to 2D velocity space. Based on our algorithm, we investigate oculomotor responses to display changes. We show that the rate of binocular microsaccades is strongly modulated by changes in visual input. These results suggest that the occurrence of microsaccades is a dynamically rich phenomenon, which may give new insights into the specific functional relevance of microsaccades and may be used to study binocular aspects of oculomotor control.

Introduction

While we attempt to fixate on a stationary object, the eyes do not remain perfectly motionless. Instead, small-amplitude eye movements can be observed (Ratliff & Riggs 1950; Ditchburn, 1955). Therefore, the term "fixation" seems to be misleading, but it turns out that these miniature eye movements or micromovements can be neglected for many behaviourally relevant tasks, e.g. if we study oculomotor control in image processing or reading (Rayner, 1998). Consequently, the term *fixational eye movements* has been created to express both the fact that we are intending to keep our eyes on a stationary object, while the eyes perform small-amplitude movements (Yarbus, 1967; Ditchburn, 1973; for a review see: Ciuffreda & Tannen, 1995).

If we suppress fixational eye movements experimentally in the paradigm of retinal stabilization (Riggs *et al.*, 1953), participants see parts of an object (e.g., line segments

The Mind's Eye: Cognitive and Applied Aspects of Eye Movement Research

ISBN: 0–444–51020–6

in a letter) disappear within a few seconds. The underlying psychological mechanism is *retinal adaptation*, which may have evolved as an elegant property of our visual system to force rapid detection of moving objects (e.g., an approaching predator). When we fixate on a stationary object under retinal stabilization, however, retinal adaptation is disastrous and causes the image to fade from perception. In this respect, fixational eye movements serve an important purpose to maintain the perception of stationary visual scenes by providing small random displacements.

Miniature eye movements or micromovements can be subdivided into three distinct components (Ciuffreda & Tannen, 1995). The first component is *tremor*, which is a high-frequency oscillatory component ranging from 30 to 100 Hz. Tremor is superimposed on *drift*, the second component of fixational eye movements. Drift is a low-velocity movement with a peak velocity below 30-min arc per second. Neither drift nor tremor are correlated between the two eyes. Thus, these types of micromovements have been interpreted as noise in the oculomotor system, which originates from random firing of brainstem motor neurons. Due to its noisy origin, drift and tremor are believed to be error-producing; however, these micromovements may also be error-correcting at a time (Nachmias, 1959, Kowler, 1991).

The third component is the *microsaccade*. Microsaccades are rapid small-amplitude movements, which occur at a typical mean rate of 1 to 2 per second. Amplitudes of microsaccades are rarely larger than 30-min arc. With respect to binocular coordination in microsaccades, Ciuffreda and Tannen (1995: 12) conclude that

> microsaccades are always binocular and have a high amplitude correlation (0.6 to 0.9) between the eyes, which suggests that they are under central neurological control.

Functionally, all types of micromovements serve to counteract retinal adaptation. To find the specific functions of the three types of micromovements, however, is a long-standing problem in eye movement research (Kowler & Steinman, 1980).

Microsaccades are traditionally believed to be error-correcting (Ditchburn, 1980). If we assume that tremor and drift are uncorrelated between the eyes, microsaccades would have to be error-correcting in order to maintain binocular coordination. However, there are two observations rejecting this function for microsaccades. First, microsaccades can be suppressed voluntarily in high-acuity observation tasks (Steinman *et al.*, 1967; Findlay, 1976; Winterson & Collewijn, 1976; Bridgeman & Palca, 1980). Consequently, if error-correction were the primary function, we would — paradoxically — observe increasing binocular disparities during high-acuity observations tasks with substantial suppression of microsaccades because of rapidly accumulating deviations caused by drift and tremor. Second, when microsaccades are suppressed, slow eye movements, i.e. drift and tremor, maintain both fixation position and binocular coordination (Steinman *et al.*, 1973). Putting together these findings, Kowler and Steinman (1980) concluded that "microsaccades serve no useful purpose."

It has been reported recently by Martinez-Conde *et al.* (2000) that microsaccades are correlated with bursts of spikes of single cells in primary visual cortex (V1) of macaque monkeys. This new result may help explain the functional significance of microsaccades. Nevertheless, generation of microsaccades may still primarily represent a noise-producing low-level oculomotor phenomenon, which counteracts retinal adaptation.

Our concern with binocular microsaccades grew out of a more general interest in binocular coordination in fixational eye movements and its interplay with visual attention (Engbert & Kliegl, 2002). The slow components of miniature eye movements (drift and tremor) are rather irregular and show statistical properties of a random walk (van Kampen, 1981). Microsaccades, however, create more linear movement segments embedded in the eyes' trajectories during fixational movements. This property can be exploited to characterize binocular coordination. Furthermore, microsaccades have well-defined duration, amplitude, peak velocity and angular orientation — several parameters for a detailed quantitative analysis. Motivated by these advantages, we restrict our study of binocular coordination to the analysis of microsaccades here.

Detection of Microsaccades

Microsaccades can be observed by visual inspection in eye movement recordings when a participant is fixating a stationary object (see: Methods). They are ballistic and appear as linear parts of the eye movement trajectory (Figure 6.1a). While the peak velocity during saccades is significantly higher than the maximum velocity during drift and tremor, starting and end points of microsaccades are very difficult to determine, in particular because of noise arising from the slow components of micromovements and from noise generated by the recording technique. Therefore, a critical problem of a detection algorithm for microsaccades is the estimation of robust velocity thresholds.

We developed a new algorithm for the detection of microsaccades in 2D velocity space to directly address the problem of stochastic fluctuations of the velocity. In a first step, we transform the recorded time series (N data samples) of eye positions, $\{\vec{x}_n\} : \vec{x}_1, \vec{x}_2, \vec{x}_3, \ldots, \vec{x}_{N-1}, \vec{x}_N$, to velocities. The velocity $\vec{v}$ is generally defined as the time derivative of the position $\vec{x}$ and can be approximated by computing finite differences to estimate the derivative,

$$\vec{v} = \frac{\mathrm{d}\vec{x}}{\mathrm{d}t} \approx \frac{\Delta\vec{x}}{\Delta t} \tag{6.1}$$

where $1/\Delta t$ is the sampling rate of the eye tracker ($\Delta t = 4$ ms in our experiments). To obtain a symmetric estimate of the velocity of the nth data sample, we use the arithmetic mean of the two differences between $\vec{x}_n$ and the neighboring samples $\vec{x}_{n-1}$ and $\vec{x}_{n+1}$,

$$\hat{\vec{v}}_n = \frac{1}{2}\left(\frac{\vec{x}_n - \vec{x}_{n-1}}{\Delta t} + \frac{\vec{x}_{n+1} - \vec{x}_n}{\Delta t}\right) = \frac{\vec{x}_{n+1} - \vec{x}_{n-1}}{2\Delta t} \tag{6.2}$$

Next, to reduce noise in this estimate of the velocity, we compute a moving average of three successive data samples to obtain the time series of the velocity vector for our analysis,

$$\vec{v}_n = \frac{1}{3}(\hat{\vec{v}}_{n+1} + \hat{\vec{v}}_n + \hat{\vec{v}}_{n-1}) = \frac{\vec{x}_{n+2} + \vec{x}_{n+1} - \vec{x}_{n-1} - \vec{x}_{n-2}}{6\Delta t} \qquad (6.3)$$

which is an estimate based on four data samples.[1] As a consequence of the randomly distributed orientations of the velocity vectors during fixational eye movements, the resulting mean value is effectively zero. A plot of the corresponding trajectory in 2D velocity space shows an erratic trajectory around the origin of the coordinate system (Figure 6.1b), while microsaccades can be identified by excursions of the trajectory to higher values of the velocities.

The second step is the computation of robus velocity thresholds, which is crucial for our detection algorithm. As mentioned before, peak values of velocities during microsaccades are typically higher than during tremor and drift. A more difficult problem is to determine onsets and end points of microsaccades, because the velocities in these points are indistinguishable from velocity values for tremor and drift.

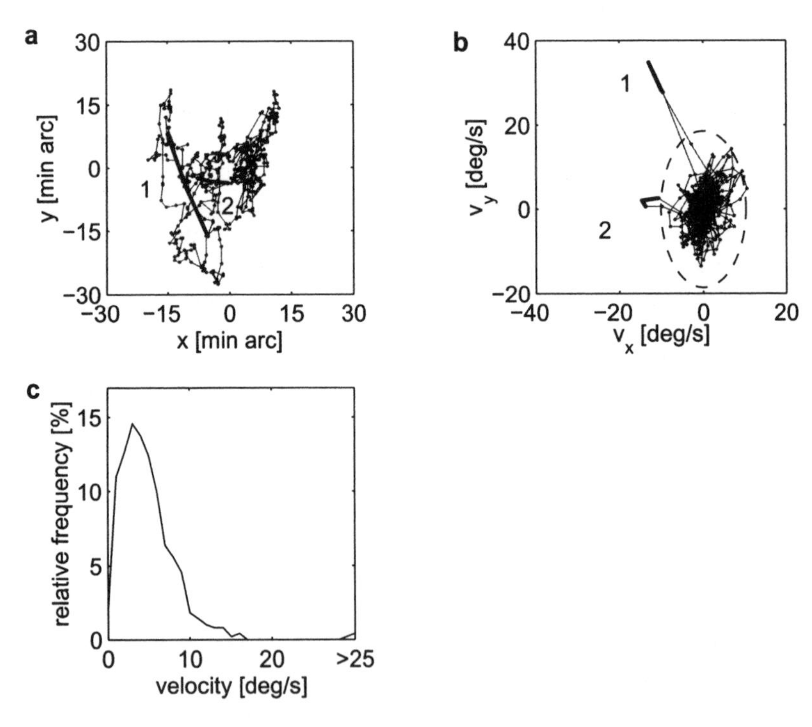

Figure 6.1: Illustration of the detection algorithm. (a) The eye movement trajectory of the first 500 samples (or 2 s) of a fixation. The numbers denote the microsaccades detected by our algorithm. (b) The same trajectory plotted in 2D velocity space. Velocities are computed using Eq. (6.3). The ellipse is defined by the detection thresholds, Eq. (6.4–6.5). If there are more than three velocity samples falling outside this ellipse, we define these sequences as microsaccades. (c) Velocity distribution computed from all data of the fixation (952 samples or 3808 ms).

In a histogram of the velocities (Figure 6.1c) of fixational eye movements only few samples of the velocity lie above 25 min arc/s. To compute an adaptive threshold varying with the amount of stochasticity observed over different trials and participants, we use a measure in units of the standard deviation of the velocity in a given trial, i.e. a relative threshold, for both horizontal η_x and vertical η_y components,

$$\eta_{x,y} = \lambda\sigma_{x,y} \tag{6.4}$$

where λ is a dimensionless parameter for the velocity threshold. To obtain a robust threshold value for $\eta_{x,y}$, we apply a median estimator to the time series $\{\vec{v}_n\}$ of velocities for the computation of the mean values,

$$\sigma_{x,y} = \sqrt{\langle v^2_{x,y} \rangle - \langle v_{x,y} \rangle^2} \tag{6.5}$$

where $\langle . \rangle$ denotes the median here. The two components of the resulting thresholds determine the axes of an ellipse (Figure 6.16). A microsaccade is obtained, if velocity values fall outside this ellipse. The criterion used in our algorithm can be formulated as

$$\left(\frac{v_{x,n}}{\eta_x}\right)^2 + \left(\frac{v_{y,n}}{\eta_y}\right)^2 > 1 \tag{6.6}$$

A final problem is related to the duration of the microsaccade. Obviously, in the case that only one sample fulfils Eq. (6.6) we do not intend to assign a microsaccade to this event. Rather, we use a threhold of three data samples (12 ms) as a minimum duration of a microsaccade.

In our algorithm, the detection threshold is varied by the free parameter λ, Eq. (6.4), which determines the threshold as a multiple of the median-based standard deviation of the velocity components. By this definition the threshold can be easily adapted to different noise levels produced by interindividual differences in participants or by different eye tracking technologies. Therefore, our algorithm seems promising for many applications in eye movement research.[2]

Binocular Coordination of Microsaccades

A starting point for our analysis of binocular coordination in terms of microsaccades is to define binocular microsaccades. According to our definition binocular microsaccades are saccades occurring in left and right eyes with a temporal overlap. This definition can be used to derive a criterion, which can easily be implemented on a digital computer. If we observe a microsaccade in the right eye starting at time r_1 and ending at time r_2 and a microsaccade in the left eye beginning at time l_1 and stopping at time l_2, the criterion for temporal overlap can be implemented by the conditions:

$$r_2 > l_1 \quad \text{and} \quad r_1 < l_2 \tag{6.7}$$

Since the labels "left" and "right" are in a sense arbitrary, such a criterion must be invariant (or symmetric) under exchange of the eyes, which can be easily verified by exchanging the letters "*r*" and "*l*" in Eq (6.7). Examples for binocular and monocular microsaccades are given in Figure 6.2a,b. We detect three microsaccades in the left eye, but we compute only a single microsaccade in the right eye using our algorithm based on Eqs. (6.3–6.6). In applying the temporal overlap criterion for binocular microsaccades, Eq. (6.7), we find that the microsaccade produced by the right eye is a binocular one, since there is overlap with a microsaccade occurring in the left eye (Figure 6.2a–d).

Microsaccades are ballistic movements like long-distance saccades, which is strongly suggested by the observed fixed relation between microsaccade amplitudes and peak velocities (Zuber *et al.*, 1965). Since the fixed relation is characteristic of microsaccades, we can use this fact to test the plausibility of our algorithm. If we assume that monocular microsaccades occur only as an artifact of our algorithm, the fixed relation between peak velocity and amplitude should no longer hold for

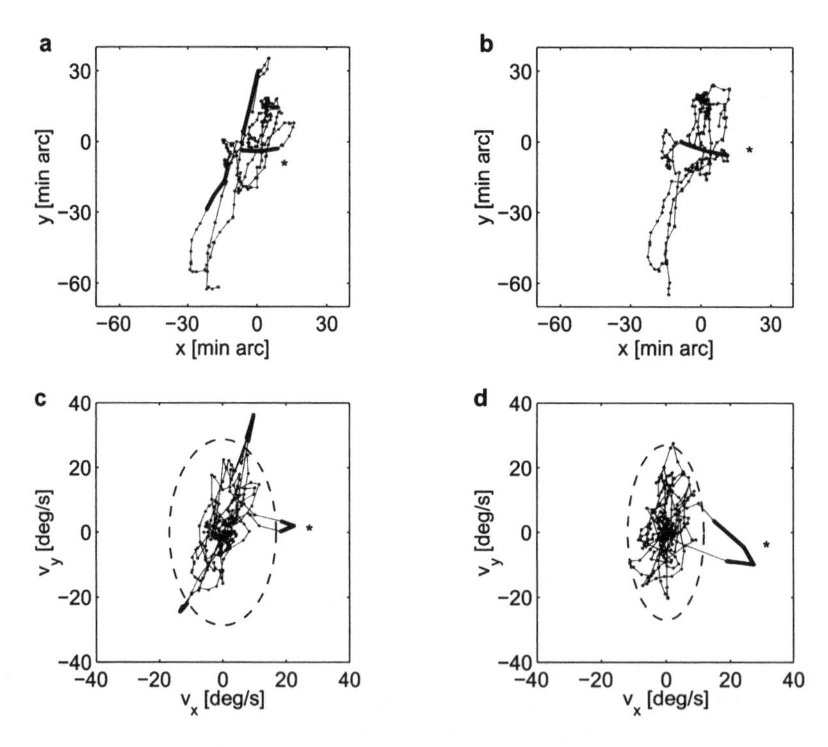

Figure 6.2: Examples of binocular microsaccades. (a) Irregular miniature eye movements are interrupted by small but rapid microsaccades, which can be defined as linear parts of the trajectory (bold lines with numbers). Data were recorded from the left eye during a fixation of 960 ms. (b) Position data plotted for the right eye. (c,d) Plot of the same trajectory as in (a,b) in 2D velocity space, described in Figure 6.1c. The star (★) marks the binocular microsaccade detected using Eq. (6.7).

monocular microsaccades — becase there is no reason why peak velocity and amplitude should be in a fixed relation for drift and tremor components. However, a scatter plot demonstrates that we obtain the same relation between peak velocity and amplitude for binocular and monocular microsaccades (Figure 6.3a–c). Peak velocity and amplitude of a binocular microsaccade are defined as the maxima of the corresponding parameters of saccades in the left and right eyes. This definition causes the small shift in Figure 6.3c towards greater amplitudes and higher peak velocities.

Because microsaccades are defined as intrafixational movements, an overlap of presaccadic and postsaccadic foveal areas is required. This constraint limits the maximum amplitude of microsaccades to about 1°. Since we observe amplitudes up to 2°, we cannot rule out the presence of some small voluntary saccades in our data (Figure 6.3a–c). There are less than 10% of the binocular microsaccades in Figure 6.3c

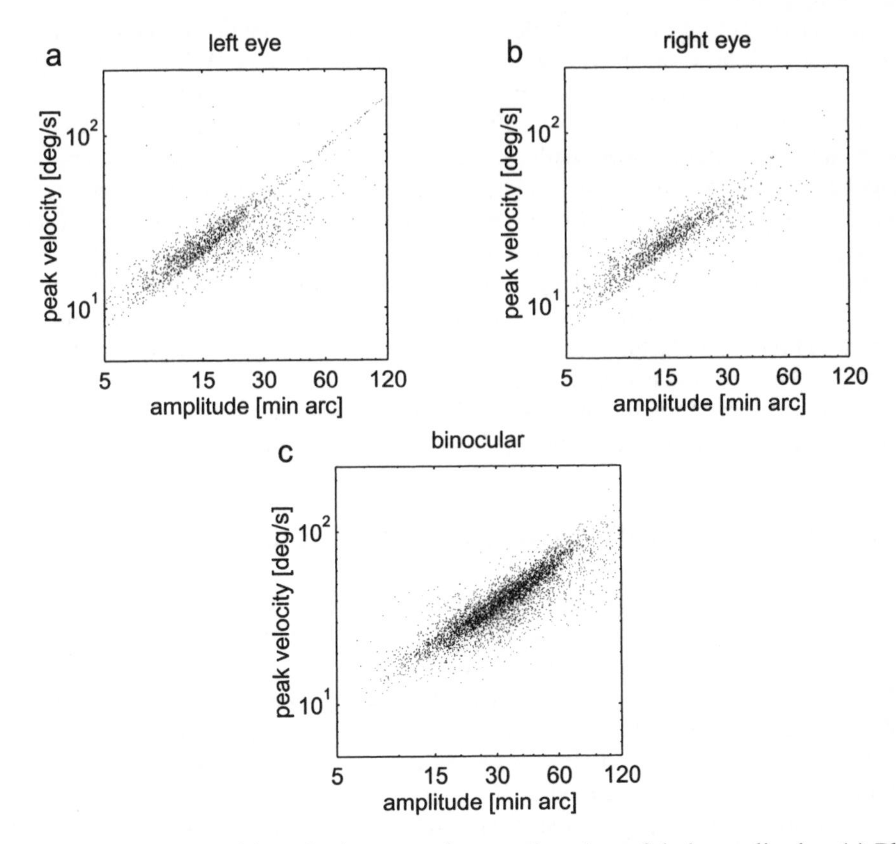

Figure 6.3: Peak velocities of microsaccades as a function of their amplitudes. (a) Plot for monocular microsaccades obtained for the left eye (1890 microsaccades, detected during fixation of 0.8 s on a stationary +-symbol, before the display change). (b) Same as in (a), but for the data from the right eye (1449 microsaccades). (c) Scatter plot for all binocular microsaccades (4894), which were determined using the temporal overlap criterion, Eq. (6.7).

with amplitudes greater than 1°. We have checked, however, that the results reported in this study did not change qualitatively after introducing an amplitude cutoff of 1°.

An interesting parameter of saccadic eye movements is their orientation angle. If we observe a horizontal component Δx and a vertical component Δy in a microsaccade, the orientation angle ϕ can be calculated from the relation $\tan \phi = \Delta y/\Delta x$.[3] For our data we find large differences in the angular orientation between monocular and binocular microsaccades (Figure 6.4a). While binocular microsaccades show an angular distribution, which is dominated by horizontal orientations (thin line in Figure 6.4), monocular microsaccades have a preferred vertical orientation (bold line). This result provides further evidence against the hypothesis that monocular microsaccades represent noisy artifacts of our detection algorithm, since the clear separation of the angular distributions between binocular and monocular microsaccades cannot be created by noise in the detection process.

An analysis of all microsaccades (Figure 6.4b), i.e. binocular and monocular microsaccades, shows a superposition of the two distributions. While there is still a majority of microsaccades with horizontal orientation, the combined distribution is clearly more symmetric than the distributions of binocular and monocular microsaccades alone. We interpret this result as a functional division. Binocular coordination is trained for vergence movements, which are in most cases horizontal. As a consequence, binocular microsaccades might be orientated with a horizontal preference. It follows directly from this hypothesis that monocular microsaccades may be essential in order to produce a symmetrical angular distribution over all microsaccades. Our interpretation is, however, preliminary, and awaits to be tested with different experimental paradigms and recording techniques. It is important to note that the combined distribution (Figure 6.4b) of all microsaccades would have been the result of a monocular analysis. Therefore, our results suggest that a binocular analysis may provide a new approach to investigate the specific functions for binocular and monocular microsaccades.

Modulation of Binocular Microsaccades by Visual Stimuli

Microsaccades might be experimentally indistinguishable from small error-correcting saccades following long-distance inspection saccades, although these error-correcting saccades are functionally different from microsaccades during fixation. Therefore, for a first investigation on the statistics of monocular and binocular microsaccades, we excluded the interaction of microsaccades with inspection saccades and used an experimental set-up, in which participants fixated a stationary target, which was replaced by a different symbol during fixation.

Participants were required to fixate on a cross (+). After 1.5 to 2 s, the fixation cross was replaced by a different symbol (except for a control condition, in which the fixation cross was unchanged). After another randomised time interval the symbol disappeared and participants were instructed to respond with a key-press as fast as possible (see: Methods). The symbols used in the experiment were adopted from typical attention experiments, i.e. three different arrow symbols (<, >, <>) or a fixation cross (+) with a different color (green or red).

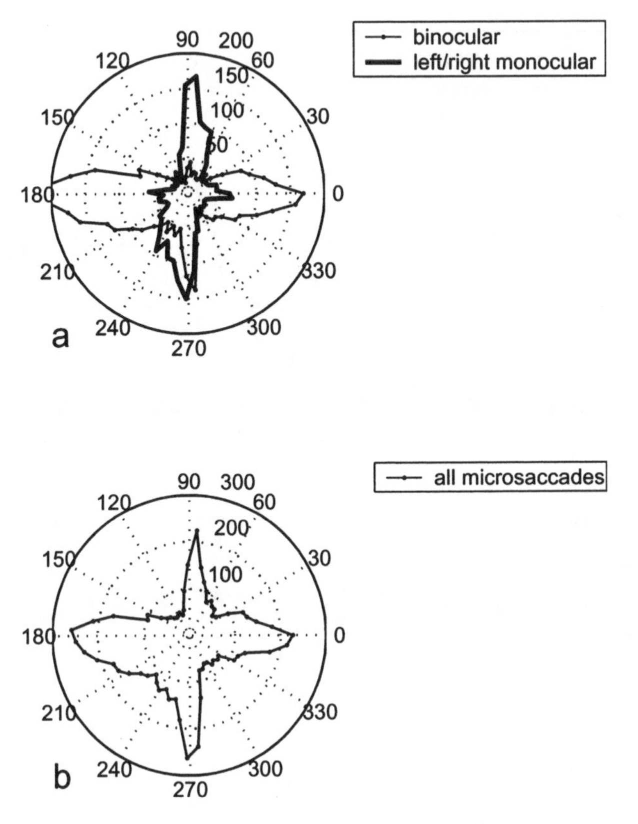

Figure 6.4: Angular distribution of microsaccades. (a) The distributions of the orientation of binocular (thin line) and monocular microsaccades (bold line) show a pronounced difference with a vertical preference for monocular microsaccades and a horizontal preference for binocular orientation. The distributions were calculated from the same data as in Figure 6.3. (a time window of 0.8 s, starting 800 ms before the display change). The radial axis is frequency of cases, computed over 63 bins (width approx. 5.7°). (b) A combined plot of monocular and binocular microsaccades shows a more symmetric distribution (compared to the separate analysis in (a)) between vertical and horizontal components. It is important to note that this plot would be the result from an analysis of microsaccades in one eye only.

As a statistical measure of binocular microsaccades we computed their rate of occurrence in a time window of 80 ms for all trials of 20 participants. During the initial fixation, the rate of microsaccades settled to a baseline level of approximately 1.5 per second (Figure 6.5a). There are considerable interindividual differences with a minimum of 0.71 s^{-1} and a maximum of 2.45 s^{-1} (Table 6.1). The display change occurred at time $t = 0$ ms on the time axis. Induced by the display change, we observed a drop in the microsaccade rate to less than 0.5 s^{-1} for the arrow symbols as well as for the color changes. The minimum of the microsaccade rate is reached around

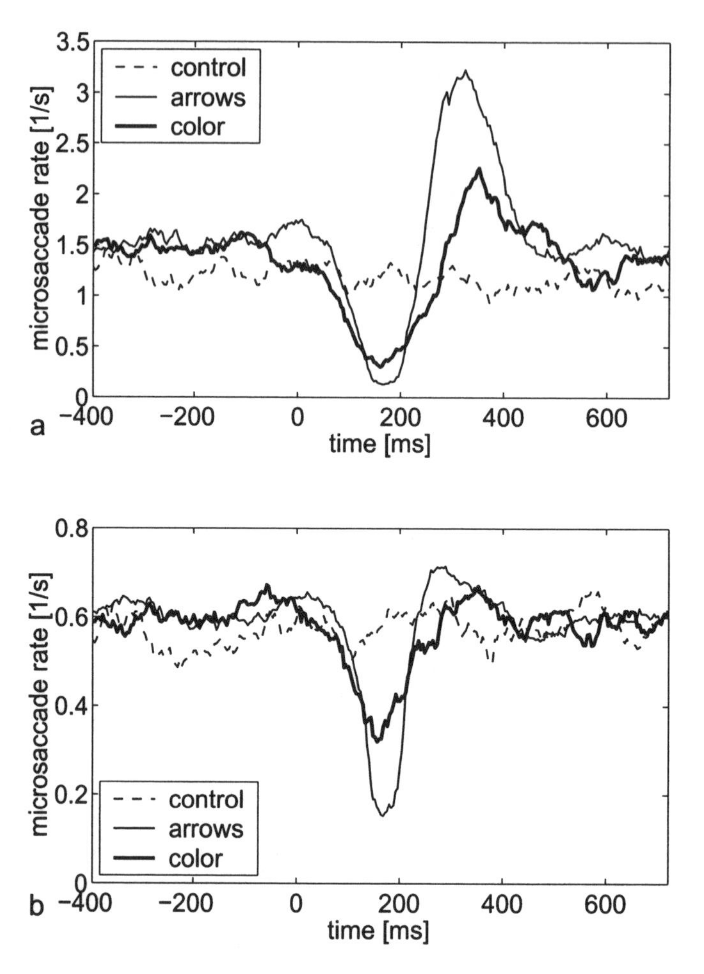

Figure 6.5: Modulation of the statistics of binocular microsaccades by display changes. The microsaccade rate was computed in a moving window of width 80 ms. (a) The time-course of the rate of microsaccades in the experiment showed a strong response to the exchange of the central stimulus compared to the baseline level of 1.5 microsaccades per second. The arrow symbols induced a stronger response (maximum rate approximately 3 s^{-1}). (b) The binocular fraction, computed as the ratio of number of binocular microsaccades to all microsaccades). As in (a) the drop in the binocular fraction exhibited a more pronounced response for the arrow symbols (minimum fraction around 0.2) than for the color symbols (minimum approximately 0.4).

Table 6.1: Mean rate of microsaccades per second.

Participant	**Binocular**	**Monocular, right eye**	**Monocular, left eye**
1	0.71	0.50	0.34
2	0.73	0.44	0.56
3	0.74	0.26	0.28
4	0.77	0.23	0.38
5	1.03	0.51	0.46
6	1.13	0.24	0.54
7	1.17	0.49	0.80
8	1.23	0.91	0.50
9	1.24	0.25	0.66
10	1.33	0.42	0.57
11	1.39	0.82	0.07
12	1.46	0.32	0.27
13	1.59	0.52	0.81
14	1.64	0.52	0.92
15	1.72	0.15	0.55
16	1.89	0.63	0.91
17	1.99	1.20	1.17
18	2.13	0.52	0.30
19	2.21	0.43	0.31
20	2.45	0.50	0.26
Mean	1.43	0.49	0.53

150 ms after the display change. After this drop, the rate of microsaccades increases to about 3 s^{-1} in the case of the arrow symbols and to approximately 2 s^{-1} in the case of the color change, where the maximum is reached around 370 ms after the display change. The microsaccade rate drops to the baseline level around 500 ms after the display change. Our finding of the decrease in the rate of microsaccades replicates the suppression effect for microsaccades, which was first observed by Winterson and Collewijn (1976), Findlay (1976), and Ridgeman and Palca (1980) in high-acuity observation tasks. Interestingly, a similar type of inhibition has recently been reported for long-distance saccades (Reingold & Stampe 2000, in press; Pannasch *et al.,* 2001). Thus, our results provide a further example for similar properties between micro- and long-distance saccades.

In addition to the inhibition of microsaccades — a well-known phenomenon — we find microsaccadic enhancement, which starts directly after the microsaccadic inhibition. While the drop in microsaccade rate is quantitatively similar between arrow and color conditions, the increase is stronger in the case of arrow symbols. However, a more detailed analysis, which highlights the binocular effects in microsaccades, shows that there are already differences between the experimental conditions in the inhibition stage of the microsaccadic modulation.

As a simple measure for binocular coordination in microsaccades we compute the binocular fraction, i.e. the proportion of binocular microsaccades divided by the number of all microsaccades,

$$\gamma\tau(t) = \frac{N_B\,(\tau,t)}{N_B(\tau,t) + N_L(\tau,t) + N_R(\tau,t)} \tag{6.8}$$

Where $N_B(\tau,t)$ is the number of binocular microsaccades in a given time window of length τ centered at time t and $N_L(\tau,t)$, $N_R(\tau,t)$ are the corresponding numbers of monocular left and right microsaccades. Our analysis indicates that the binocular fraction changes significantly during the display change (Figure 6.5b). The most dramatic effect is a drop in the binocular fraction during the inhibition epoch, where the minimum occurs around 170 ms after the display change.

These results show that binocular coordination is strongly modulated by display changes. The selective response of binocular and monocular microsaccades is an interesting phenomenon, which may help to understand the role of microsaccades in the maintenance of perception of stationary scenes.

Discussion

In an exploratory analysis, we investigated the dynamics of binocular coordination in terms of the statistics of microsaccades. We showed that a distinction between monocular and binocular microsaccades yields interesting differences in angular orientations, which are inaccessible in a monocular analysis. Based on this result we strongly recommend binocular recordings in experiments on fixational eye movements, since monocular microsaccades may only provide partial knowledge of the oculomotor system.

It has been shown in 1960 that almost all microsaccades are binocular (Krauskopf *et al.*, 1960; Riggs & Niehl, 1960). Here we used a video-based technology for the analysis of micromovements in combination with a refined algorithm for the detection of microsaccades. Therefore, our results are preliminary and need to be replicated with different eye tracking techniques (e.g., search coil). Recent experimental and theoretical studies suggest, however, that binocular synchronization in (long-distance) saccades may be less robust than previously assumed (Zhou & King 1997, 1998; King & Zhou, 2000).

Our investigation was based on a new algorithm for the detection of microsaccades. In the algorithm, a transformation to 2D velocity space is used to disentangle microsaccades from slower components of fixational eye movements, i.e. drift and tremor. Due to their ballistic nature, microsaccades can — like long distance saccades — be characterized by a more linear movement sequence embedded in the rather irregular time series of micromovements during fixation. In velocity space, microsaccades can be identified by excursions of the eye's trajectory to values, which lie clearly above a threshold. This threshold is computed as a multiple of a median-based estimator of the standard deviation of the velocity. Because of this procedure, our detection algorithm is capable to tackle the problem of noise in the analysis of microsaccades.

Based on these methodological foundations, we investigated the modulation of microsaccades in response to display changes. We observed a fast response with a drop of the rate of microsaccades, where a minimum value is reached about 170 ms after the display change, and a slower response, which leads to an increase of the microsaccade rate with a maximum value about 320 ms (arrow symbols) to 350 ms (color symbols) after the display change. While there are similar results for both types of exchanged symbols (arrows vs. color) in the binocular microsaccade rate, a more detailed analysis of the binocular coordination emphasizes differences for the two conditions even in the fast response. To this end we computed the binocular fraction of microsaccades. Our findings showed that a binocular analysis of micromovements can provide new and interesting variables for the characterization of miniature eye movements.

In summary, the modulation of binocular microsaccade statistics turned out to be a dynamically rich phenomenon, which may give new insights into both the specific functions of microsaccades and the principles of binocular coordination in the oculomotor system.

Acknowledgements

We thank Christian Lakeberg for implementing the experiment and Jochen Laubrock for valuable discussions, and Bruce Bridgeman (Santa Cruz), John Findlay (Durham) and Boris Velichkovsky (Dresden) for valuable comments on the manuscript. This work was supported by Deutsche Forschungsgemeinschaft (DFG, grant KL 955/3).

Notes

1 See also: Eyelink System Docuemntation, Version 1.2, SensoMotoric Instruments, 1999, p. 352.
2 Our algorithm will be made available via the internet at: http://www.agnld.uni-potsdam.de/~ralf/micro/.
3 To obtain the angle ϕ from this relation, most programming languages (e.g., MATLAB or C) provide the inverse function atan2(Δy, Δx).

References

Bridgeman, B., & Palca, J. (1980). The role of microsaccades in high acuity observational tasks. *Vision Research*, *20*, 813–817.

Ciuffreda, K. J., & Tannen, B. (1995). *Eye Movement Basics for the Clinician*. St. Louis: Mosby.

Ditchburn, R. W. (1955). Eye movements in relation to retinal action. *Optica Acta*, *1*, 171–176.

Ditchburn, R. W. (1973). *Eye Movements and Visual Perception*. Oxford: Clarendon Press.

Ditchburn, R. W. (1980). The function of small saccades. *Vision Research, 20*, 271–272.
Engbert, R., & Kliegl, R. (2002). Microsaccades uncover the orientation of covert attention (submitted).
Findlay, J. M. (1976). Direction perception and fixation eye movements. *Vision Research, 14*, 703–711.
van Kampen, N. G. (1981). *Stochastic Processes in Psychics and Chemistry*. Amsterdam: North-Holland.
King, W. M., & Zhou, W. (2000). New ideas about binocular coordination of eye movements: Is there a chameleon in the primate family tree? *The Anatomical Record, 261*, 153–161.
Kowler, E. (1991). The stability of gaze and its implications for vision. In: J. Cronly-Dillon (ed.), *Vision and Visual Dysfunction (Volume 8, Eye movements)*, Boca Raton: CRC.
Kowler, E., & Steinman, R. M. (1980). Small saccades serve no useful purpose: Reply to a letter by R. W. Ditchburn. *Vision Research, 20*, 273–276.
Krauskopf, J., Cornsweet, T. N., & Riggs, L. A. (1960). Analysis of eye movements during monocular and binocular fixation. *Journal of the Optical Society of America, 50*, 572–578.
Martinez-Conde, S., Machnik, S. L., & Hubel, D. H. (2000). Microsaccadic eye movements and firing of single cells in the striate cortex of macaque monkeys. *Nature Neuroscience, 3*, 261–258.
Nachmias, J. (1959). Two-dimensional motion of the retinal image during monocular fixation. *Journal of the Optical Society of America, 49*, 901.
Pannasch, S., Dornhoefer, S. M., Unema, P. J., & Velichkovsky, B. M. (2001). The omnipresent prolongation of visual fixations: Saccades are inhibited by changes in situation and in subject's activity. *Vision Research, 41*, 3345–3351.
Ratliff, F., & Riggs, L. A. (1950). Involuntary motions of the eye during monocular fixation. *Journal of Experimental Psychology, 40*, 687–701.
Rayner, K. (1998). Eye movements in reading and information processing: 20 years of research. *Psychological Bulletin, 124*, 372–422.
Reingold, E. M., & Stampe, D. M. (2000). Saccadic inhibition and gaze contingent research paradigms. In: A. Kennedy, R. Radach, D. Heller and J. Pynte (eds), *Reading as a Perceptual Process* (pp. 119–145). Oxford: North-Holland.
Reingold, E. M., & Stampe, D. M. (in press). Saccadic inhibition in reading. *Journal of Experimental Psychology Human Perception and Performance.*
Riggs, L. A., & Niehl, E. W. (1960). Eye movements recorded during convergence and divergence. *Journal of the Optical Society of America, 50*, 913–920.
Riggs, L. A., Ratliff, F., Cornsweet, J. C., & Cornsweet T. N. (1953). The disappearance of steadily fixated test objects. *Journal of the Optical Society of America, 43*, 495–501.
SMI Eyelink System Documentation, (1999), Senso-Motoric Instruments.
Steinman, R. M., Cunitz, R. J., Timberlake, G. T., & Herman, M. (1967). Voluntary control of microsaccades during maintained monocular fixation. *Science, 155*, 1577.
Steinman, R. M., Haddad, G. M., Skavenski, A. A., & Wyman, D. (1973). Miniature eye movements. *Science, 181*, 810–819.
Winterson, B. J., & Collewijn, H. (1976). Microsaccades during finely guided visuomotor tasks. *Vision Research, 16*, 1387–1390.
Yarbus, A. L. (1967) *Eye Movements and Vision*. New York: Plenum.
Zhou, W., & King, W. M. (1997). Binocular eye movements are not coordinated during REM sleep. *Experimental Brain Research, 117*, 153–160.
Zhou, W., & King, W. M. (1998). Premotor commands encode monocular eye movements. *Nature, 393*, 692–695.
Zuber, B. L., Stark, L., & Cook, G. (1965). Microsaccades and the velocity-amplitude relationship for saccadic eye movements. *Science, 150*, 1459–1460.

Appendix: Methods

Twenty participants (undergraduates of the University of Potsdam) performed 240 trials each. Experiments were presented on a 21-inch EYE-Q 650 Monitor (832 × 624 resolution; frame rate 75 Hz) controlled by an Apple Power Macintosh G3 Computer. Eye movements were recorded using a video-based SMI Eyelink System (Senso-Motoric Instruments) with a sampling rate of 250 Hz and an eye position resolution of 20-sec arc. A trial started with a fixation cross (size 2.3°; white on dark background), which was presented centrally for 1500 to 2000 ms. Time intervals were randomised within the given range in steps of approximately 100 ms. In the control condition, the fixation cross (+) remained unchanged. In all other conditions, the fixation cross was exchanged by one of three arrow symbols (<; >; <>) or the fixation cross changed its color to green or red. After another interval of 1500 to 2000 ms, the symbol disappeared. Participants were instructed to respond as fast as possible, when the fixation symbol disappeared, but to ignore the display changes, which occurred before (except for the control condition). After preprocessing of the data we retained 4040 trials, containing 35,619 microsaccades (binocular: 20,480; monocular: 15,139).

Chapter 7

The Inner Working of Dynamic Visuo-spatial Imagery as Revealed by Spontaneous Eye Movements

Claudio de'Sperati

Mental imagery is a basic form of cognition central to activities such as problem-solving or creative thinking. However, phenomena like mental rotation, in which mental images undergo spatial transformations; and motion imagery, in which we imagine objects in motion, are very elusive. Although widely investigated, mental rotation in humans has never been observed directly in its instantaneous evolution. This chapter is aimed at showing that, as these processes give rise to organized patterns of eye movements, the underlying mental kinematics can be successfully recovered. In a visuo-spatial variant of mental rotation, sequences of saccades repeatedly and spontaneously reproduced the circular trajectory supposed to be mentally covered. The average gaze rotation had an ordered angular progression, with an asymmetrical angular velocity profile whose peak and mean velocity increased with the amount of rotation, as in goal-directed movements. In addition, spontaneous saccades during instructed rotary motion imagery revealed a considerable capability of the eyes to reproduce the imagined motion without introducing important distortions. Thus the recording of spontaneous eye movements can be a valuable tool to objectively measure the time-course of visuo-spatial dynamic thinking.

Eye Movements and Visual Imagery

The history of science has plenty of reports of vivid mental images that have appeared more or less suddenly to eminent scientists and that have led to fortunate intuitions of important discoveries. In spite of being faint compared to visual perception, visual

The Mind's Eye: Cognitive and Applied Aspects of Eye Movement Research

ISBN: 0–444–51020–6

imagery is a basic form of cognition that enters deeply into our everyday life in a number of different circumstances, from pure aesthetic contemplation to problem-solving and creative thinking. Especially with imagery, scientists are confronted with the fundamental problem of how to measure mental activities. In the past thirty years, a large body of experimental observations has determined a substantial improvement in our understanding of visual imagery (see Kosslyn, 1980, 1994; Finke & Shepard, 1986; Finke, 1989; Kosslyn *et al.*, 2001a). Certainly, a key step has been to consider this intrinsically subjective activity as an internal process that can be measured through its overt, objective consequences.

Mental chronometry has been extensively used to characterize visual imagery. In a famous experiment, Stephen Kosslyn (1973, 1980) showed that the time to mentally inspect an island reflected the covered distances. The proportionality between the response time (RT) and the distance required to be covered in the mind was taken as a sign that "viewing" a mental image is somewhat equivalent to actually viewing that image. Therefore, a mental visual image would possess many characteristics of an actual visual image, including spatial extent (Finke 1989). Several subsequent studies confirmed this basic tenet (see Kosslyn, 1994).

Mental images are not just static representations, as they can undergo dynamic[1] spatial transformations. The introduction of the mental rotation paradigm by Roger Shepard and colleagues is considered a milestone in this regard (Shepard and Metzler, 1971; see Shepard & Cooper, 1986). In a prototypical mental rotation task, subjects have to decide about the sameness of two objects presented at different orientations. The decision takes more time as the orientation disparity increases, as if images were being rotated in the mind in order to match their orientation before giving the response. Thus, mental rotation can be regarded as a dynamic, time-consuming internal process that performs gradually angular transformations at a certain speed.

Although RTs have been instrumental in characterizing a number of mental processes and sub-processes, they cannot expose directly the continuous spatio-temporal evolution of a mental process. In the case of mental rotation, this restriction is important: The existence of a gradual rotation is usually inferred from the proportionality between RT and the presumed angle of rotation (Finke & Shepard, 1986; see also Cooper, 1976 for additional evidence still based on RTs), but in humans nobody has ever directly observed the instantaneous progression of mental rotation.

It is a common belief that the eyes provide us a window into mental life. Eye movements might indeed contain just that piece of information that RT lack, as they seem to have the characteristics to continuously or almost continuously track both in space and time the evolution of mental events, especially those with a significant visuo-spatial content. Many insights have been gained from the analysis of spontaneous saccades in a variety of conditions (for reviews, see Yarbus, 1967; Viviani, 1990; Kowler, 1995; Rayner, 1998; Liversedge & Findlay, 2000), including visual inspection and search (Noton & Stark, 1971a,b; Zelinsky *et al.*, 1997), memory and recall tasks (Ballard *et al.*, 1995; Brandt & Stark, 1997), attention tasks (Klein, 2000), visual illusions (Coren & Girgus, 1978; Ellis & Stark, 1978; Coren, 1986; Scotto *et al.*, 1990), language comprehension (Tanenhaus *et al.*, 1995), and problem-solving (Ballard *et al.*, 1997; Epelboim & Suppes, 2001). The oculomotor activity is a complex mixture of bottom-up (of retinal origin) and top-down (of mental origin) components, whose relative contribution depends upon the particular activity one is involved in. This point can be appreciated in Figure 7.1, which illustrates both a "default" pattern of fixation, in which the gaze is attracted by salient points of a face, and, conversely, the very different sequences of saccades occurring while looking at the same scene but under different cognitive demands. This finding simply reflects ordinary visual inspection, which is almost never purely guided by the characteristics of the visual stimulus but is often directed to disambiguate our own a priori hypotheses about the visual world, in a continuous action-perception loop (Neisser, 1976). Ideally, mental imagery in darkness offers the opportunity to open the loop by functionally disconnecting the eyes from the retina while maintaining their functional connection with the mental input. In this way, the oculomotor circuitry would work in a pure top-down mode.

Although apparently promising to unravel pure mental processes, recording free eye movements in imagery may face us with huge interpretative problems. The extraordinary complexity of the patterns of saccades and fixations that may open out in a tenth of a second, together with a poor experimental control of the mental content itself (think of eye movements in dreams), oblige us to focus on very circumscribed hypotheses or to limit the analyses to some statistical properties of the whole pattern. An example of a successful confirmation of a hypothesis concerning the nature of mental images, at the cost of renouncing to comprehend the mental operations involved, is a study on mental recall (Brandt & Stark, 1997). Anticipating the above-mentioned idea of a functional equivalence between perception and imagery, Hebb (1968) suggested that, for a mental image to be an equivalent of a visual image, eye movements while visually imagining a scene should be the same as those during visual inspection of the same scene. Brand and Stark addressed this issue by asking subjects to view checkered diagrams and subsequently to recall them, while concurrently recording their eye movements. By using the "string editing analysis" (Hacisalihzade, *et al.*, 1992) to quantify the degree of similarity of the sequences of saccades and fixations (the so-called scan-path, Noton & Stark, 1971a,b), they were able to show that, apart from some idiosyncratic differences, the patterns of eye movements in the two conditions had a similar structure. No attempt to understand the meaning of the particular scan-paths was made.

Can free eye movements reveal the inner working of dynamic visual imagery such as mental rotation? In an earlier study, Carpenter and Just (1978) asked subjects to decide whether two objects displayed in a different orientation were the same or mirror-copies. Concurrent recording of eye movements showed that fixations tended to concentrate on those parts of the stimuli that were presumably more informative to solve the task. Accordingly, subjects moved the eyes back and forth between corresponding parts of the two objects. In that experiment eye movements were of not much help in clarifying the process supposed to be at the core of mental rotation, that is, the progressive re-orientation of the mental image. This was probably because the task involved objects whose visual features, although simple, had the potential to attract the gaze, resulting in a sequence of saccades that may have predominantly reflected this aspect. For this reason, it would be interesting to test the mental rotation hypothesis with tasks that do not involve objects but only "pure" motion in visual space.

a)

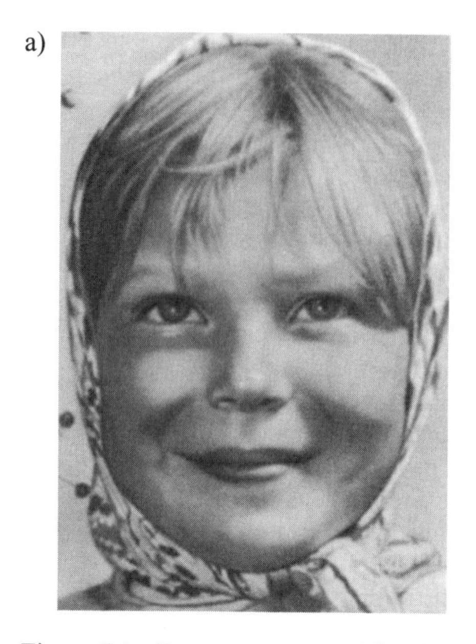

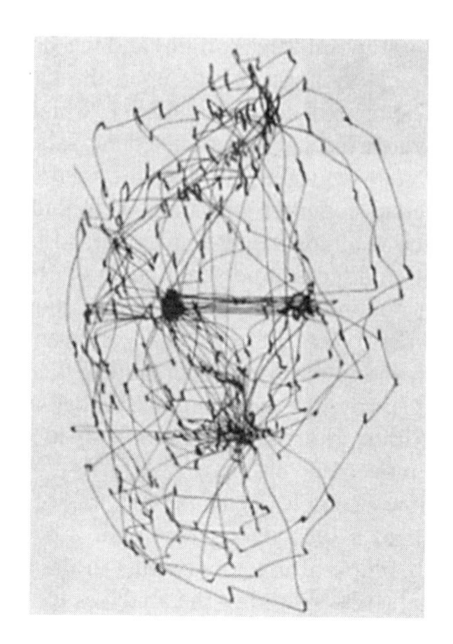

Figure 7.1: Eye movements while observing a scene. In **a** the gaze clusters at salient points of the face (e.g., eyes, mouth). In **b**, a picture is viewed after the experimenter asked the following questions (from 1 to 7): "View the picture", "Pay attention to the environment", "Attribute an age to the characters", "Guess what the family was doing before the visitor arrival", "Recall the characters' clothing", "Recall the position of the characters and of the objects in the room", "Guess how long the visitors had been away". The pattern of exploratory saccades depends heavily on the cognitive demand (from Yarbus, 1967).

Visuo-motor and Visuo-spatial Mental Rotation

Mental rotation is usually regarded as a perceptual-like process. In the last twenty years, converging evidence has highlighted the involvement of motor processes in mental rotation (Sekiyama, 1982; Parsons, 1987, 1994; Parsons *et al.*, 1995; Cohen *et al.*, 1996; de'Sperati & Stucchi, 1997, 2000; Kosslyn *et al.*, 1998, 2001b; Wohlschläger & Wohlschläger, 1998; Wexler *et al.*, 1998; Ganis *et al.*, 2000; Sirigu & Duhamel, 2001). Among the various tasks in which it has been shown that the motor system participates in mental rotation (including classic, perceptual-like mental rotation tasks), the so-called visuo-motor mental rotation task is of particular interest here.

This task was introduced in the mid-1980s by Apostolos Georgopoulos and colleagues as a motor variant of mental rotation (Georgopoulos & Massey, 1987; Pellizzer & Georgopoulos, 1993). They required subjects to make arm movements at various angles from a stimulus direction. As the angle increased, the latency of the movement increased too. The increase in latency with increasing stimulus–response angular differences has also been found recently in the case of saccades made to

b)

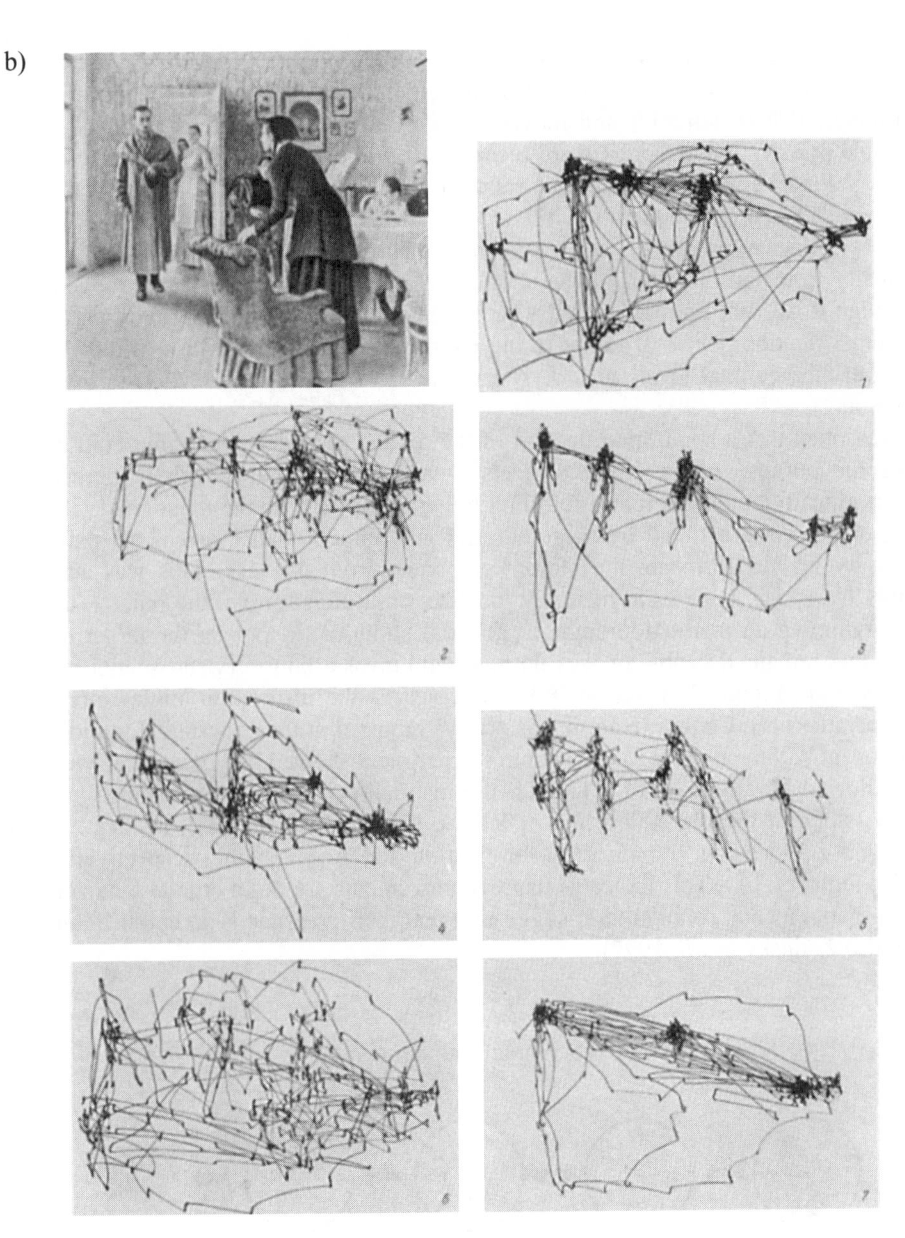

Figure 7.1: continued.

mentally rotated targets, when the visual stimulus was a cue and not the target of the saccadic eye movement (rotated saccades, de'Sperati, 1999; Fischer *et al.*, 1999). These observations were interpreted as the consequence of a mental rotation process, as if subjects had mentally performed a gradual rotation from the stimulus direction to the movement direction. The mental rotation hypothesis received substantial support from a well-known neurophysiological finding (Georgopoulos *et al.*, 1989). Two monkeys were trained to make arm movements perpendicular to the direction of a stimulus, that is, stimulus and movement were always 90 degrees apart. Neurons from the primary motor cortex were recorded. It was found that, in the course of the lapse of time between stimulus presentation and movement onset, the direction of the so-called neuronal population vector, i.e., the direction of the impending movement as collectively coded by an ensemble of neurons (Georgopoulos, 1990), passed gradually from the direction of the stimulus to the direction of the arm movement. Although it was not possible to tell whether the observed neuronal activity pertained to mental rotation *per se* or rather to the motor component of the task, it did clearly show, at the neuronal level, an orderly passage from the initial direction to the final direction.

I reasoned that a slight modification of the rotated saccades paradigm could easily serve the purpose of assigning subjects a task that is neither strictly perceptual nor specifically motor, therefore fulfilling rather closely the above-mentioned requirement of a pure directional transformation in an "empty" visual space. All that was necessary was to eliminate the motor component from the task. This was attained simply by asking subjects to mentally localize, on a target circle, the point resulting from summing an instruction angle to a visual stimulus (a spot on the target circle, that either remained visible or was flashed), and press a button (visuo-spatial mental rotation task, Figure 7.2). Again, RTs increased as the instruction angles increased (de'Sperati 1999). Central fixation was always required in the experiment, although a subsequent within-subjects ($N = 5$) control experiment showed that, when free viewing was allowed, the increase of RTs with the instruction angle did not change significantly (slope of the regression line of RT vs. instruction angle: 2.01 ms/deg ± 0.66 S.D. and 2.59 ms/deg ± 0.45 S.D., for fixation and free viewing, respectively, $p = 0.277$; Figure 7.3A). This indicates that eye movements are not a crucial factor in the visuo-spatial mental rotation task, suggesting that their presence is an epiphenomenon (see also Zelinsky *et al.*, 1997).

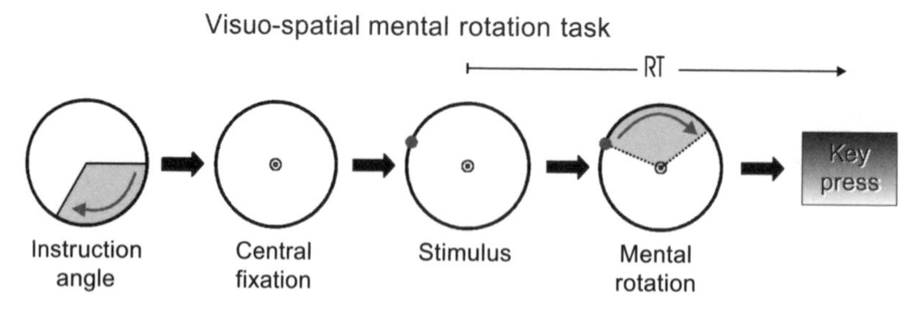

Figure 7.2: The visuo-spatial mental rotation task. Subjects view an instruction angle for a few seconds, then fixate a central spot. After a random pause a spot (stimulus) appears somewhere on the target circle. Subjects have to mentally add the instruction angle to the stimulus in the clockwise direction and press a button upon localization of its (imagined) end point. RT = response time (adapted from de'Sperati, 1999).

RTs did not depend on the radius of the target circle in the range 2.5–10.0 degrees (Figure 7.3B), confirming that the crucial variable is the amount of the angular transformation interposed between the stimulus and the response. However, for the canonical instruction angle of 180°, RTs were significantly lower than could be expected by extrapolating RT data from the other non-canonical instruction angles (951 ms ± 157 S.D. vs. 2131 ms, $p<0.001$; Figure 7.3C). This result suggests that different strategies are at play: The 180° instruction angle might require simply a mirror operation in a one-shot fashion, and not a gradual reorientation up to the desired direction (see Fischer *et al.*, 1999). The former operation is supposed to underlie anti-saccades, i.e., saccades made in the direction opposite to the visual stimulus (Hallett, 1978). In fact, the latency of anti-saccades is also lower than that of saccades made at non-canonical angles from a visual stimulus (Fischer *et al.*, 1999). Thus, the visuo-spatial task with non-canonical instruction angles is an example of a geometrical problem-solving task whose solution strategy, accordingly to RT measures, seems to be based upon a mental rotation process starting at the stimulus direction and ending at the to-be-computed direction.

In the sections that follow I will overview some recent findings in which free eye movements have been recorded during visuo-spatial rotation and during rotary motion imagery, with the aim of showing that, in spite of the previously evoked difficulties, the eyes can reveal many aspects of our visuo-spatial thinking.

Spontaneous Eye Movements in the Visuo-Spatial Mental Rotation Task

I replicated the visuo-spatial task experiments with three instruction angles (70, 115 and 160°) and without requiring central fixation. The spontaneous eye movements of fourteen adult subjects, *naive* to the purpose of the experiment, were recorded in head-fixed conditions throughout the stimulus-response time lapse (sampling frequency: 500 Hz). The scleral coil method was used for two participants, while for the remaining twelve the more comfortable infrared oculometry technique was adopted (Dr Bouis Oculometer, nominal resolution: <0.3°). No mention of eye movements was made to the subjects, except to say that the gaze had to be maintained on the central fixation spot from the disappearance of the instruction angle to the stimulus presentation. They simply had to respond by pressing a button. Mean RTs ranged from 1684 ms to 2056 ms and increased linearly with the instruction angle (slope = 4.1 ms/deg, intercept = 1400 ms). Four subjects were also tested for spatial accuracy and precision: After the button press, they moved the mouse cursor to the estimated point on the target circle. The mean error was less than 6° (on average, 2.4% of the instruction angle) in the counterclockwise direction, with an average circular standard deviation of 20.3°, confirming the rather good performance attained in this task.

The eye movement data were beyond any expectations: In the short time-span between stimulus onset and the button press, it was not uncommon that a sequence of saccades bring the gaze from the stimulus direction to the desired direction. Excerpts of spontaneous eye movements during the stimulus-response time lapse are shown in

A

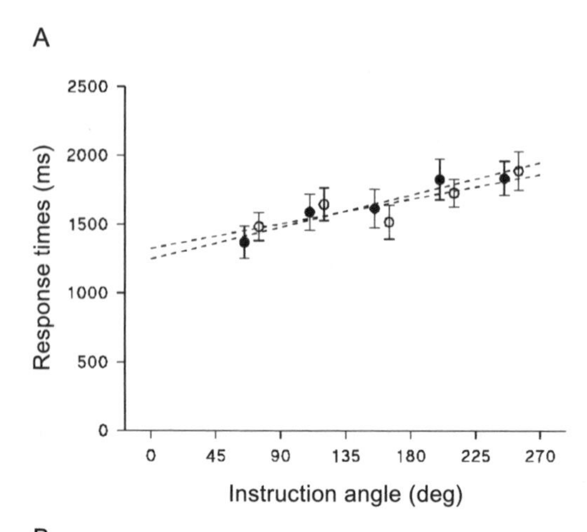
Response times (ms)
2500
2000
1500
1000
500
0
0
45
90
135
180
225
270
Instruction angle (deg)

B

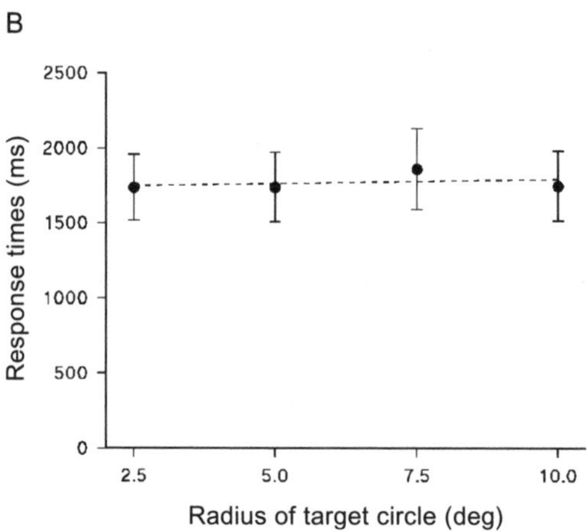
Response times (ms)
2500
2000
1500
1000
500
0
2.5
5.0
7.5
10.0
Radius of target circle (deg)

C

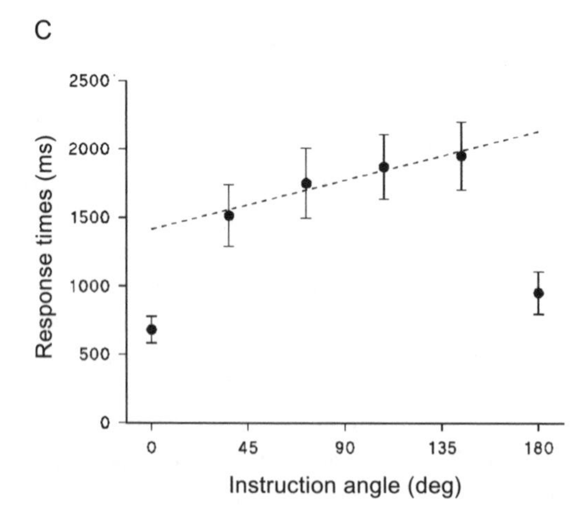
Response times (ms)
2500
2000
1500
1000
500
0
0
45
90
135
180
Instruction angle (deg)

Figure 7.4. Types of eye movements ranged from a single-saccade pattern (a), to cases in which a complete, almost circular trajectory composed of up to 78 saccades was clearly detectable (f). Patterns of intermediate complexity are illustrated in Figure 7.4b–e. In one fourth of the trials there were more than three saccades.

The experimenter could even detect a few erroneous counterclockwise rotations during the recording session just by looking at the raw signal displayed on the oscilloscope (Figure 7.4g)! Subjects confirmed these errors in the subsequent debriefing. Saccades shifting the gaze in the counterclockwise direction were very rare, of small amplitude and mostly confined at the end of the trajectory. Around 40% of the saccades were "saccades to nowhere", ending at least 1.5 degrees away from both the central position and the target circle. At the end of the experiment, most subjects could not recall where they had been looking during the task.

Figure 7.5a–c shows the superimposed raw eye position records for the three instruction angles for one subject. The gaze is clearly clustered within the space delimited by the instruction angle and is almost never directed to other parts of the visual field. The color code (see caption) shows that the closer the gaze direction to the desired direction, the later this occurred from the time of stimulus presentation. This means that the gaze, initially directed towards the stimulus, was approaching the desired direction. The averaging of eye position data points over all repetitions and subjects forms a single average trajectory for each instruction angle (Figure 7.5d–f), which shows an orderly progression from the stimulus direction to the desired direction, passing through the intermediate directions, as expected on the basis of the mental rotation hypothesis.

It is important to emphasize that the crucial aspect of this phenomenon is the ordered angular progression of the gaze, rather than its particular spatial pattern (recall that the problem is formulated only in angular terms, i.e., without explicit trajectory constraints, and that RTs are independent of the radius of the target circle). For example, many raw sequences of saccades did not even partially resemble the average trajectories illustrated in Figure 7.5d–f. Yet, even saccadic patterns apparently uninformative if taken in isolation (e.g. Figure 7.5h) may be just the idiosyncratic overt expression of a covert vector rotating gradually. That this is indeed the case is suggested by the significant correlation ($r = 0.686$, $p < 0.001$, $N = 300$) between

---

Figure 7.3: Response times in the visuo-spatial mental rotation task. **A** illustrates the comparison between a condition in which subjects maintain central fixation throughout the task (filled symbols) and a condition in which the same subjects are allowed free viewing (open symbols). No differences emerged between the two conditions. In **B** it is shown that the response times do not change for different amplitudes of the target circle. In **C** the response times with non-canonical instruction angles are compared to those with the 0° (no rotation) and the 180° (mirror rotation) instruction angles. The 180° instruction angle determines response times that are markedly lower than those with the other non-canonical instruction angles. The dashed lines are the regression lines (in **C** the responses for the 0° and the 180° instruction angles have not been considered). Bars = standard deviation.

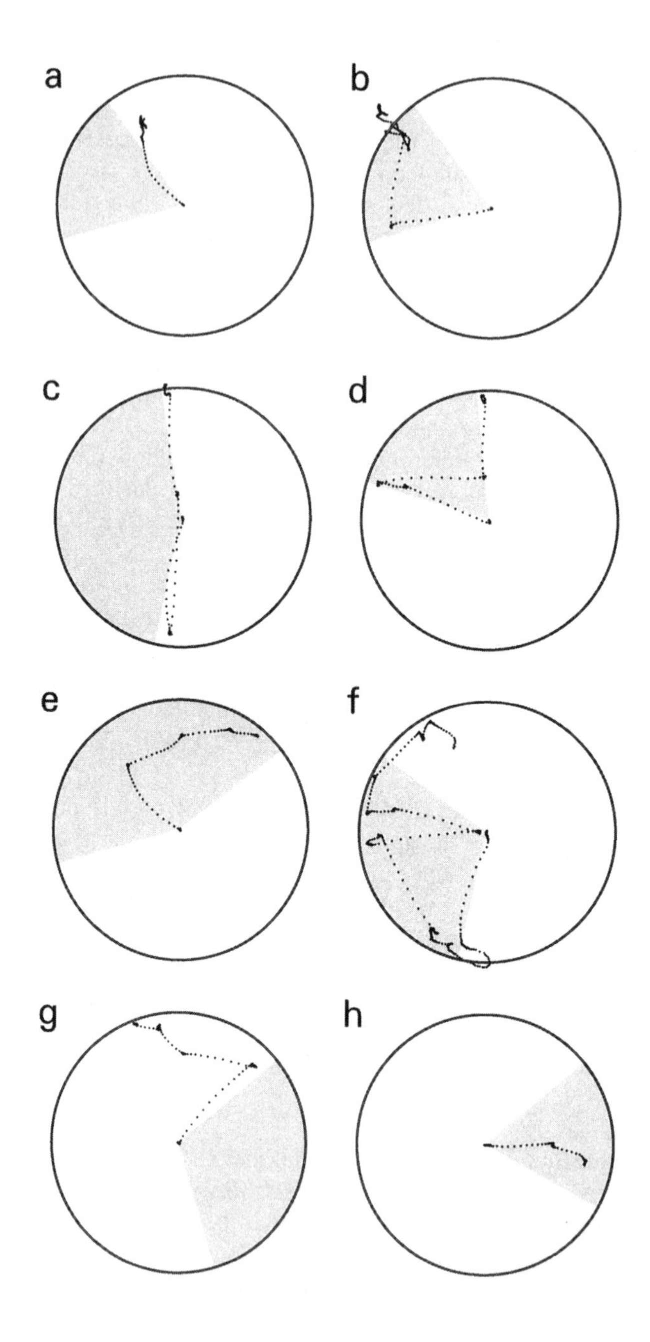

Figure 7.4: Spontaneous eye movements in the visuo-spatial mental rotation task, from stimulus onset to the button press. The gaze is initially in the central position. The gray areas indicate the angles supposed to be covered by mental rotation. In several cases the eyes reproduce the entire trajectory expected on the basis of the mental rotation hypothesis.

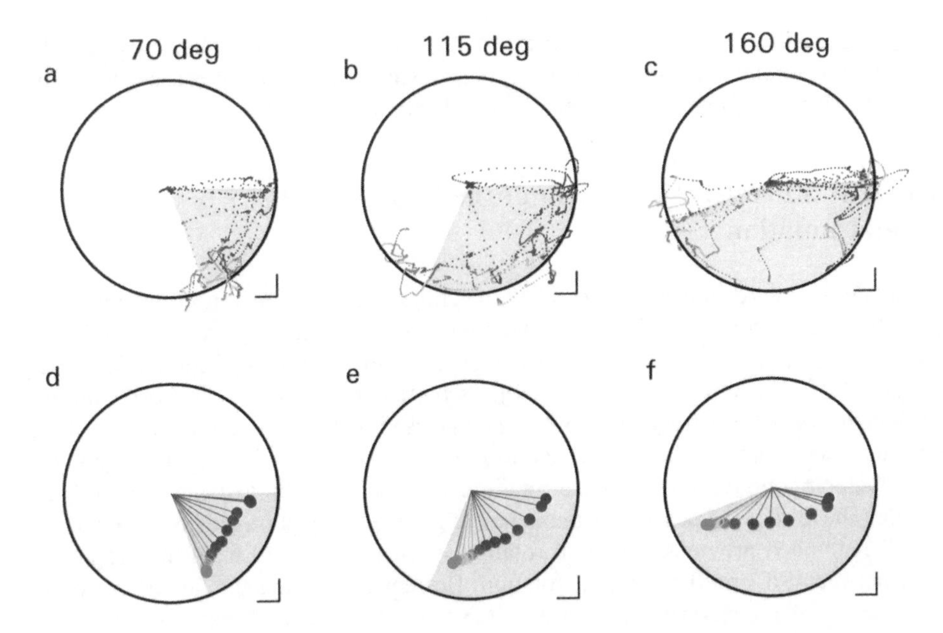

Figure 7.5: Rotation of the gaze: Superimposed raw eye position traces from a single subject (**a–c**) and average gaze trajectories computed over all subjects (**d–f**). In order to time-lock the eye position traces to both the stimulus and the response, each individual recording has been subdivided into 16 bins, starting from stimulus presentation up to the time of the manual response for that trial, so that each bin represents always 1⁄16 of RT. A color has been assigned to each bin, deep blue standing for the time of stimulus presentation (1st bin) and red standing for the time of the manual response (16th bin), with intermediate colors standing for intermediate times. In **d–f**, eye position data in each *n*th bin have been averaged both between- and within-subjects for each instruction angle and the resulting mean gaze position is plotted (colored circles, connected to the central position for graphical purposes). Gray areas are as in Figure 7.6. All stimulus directions have been aligned at 3 o'clock. Calibration bars: 1°.

direction and latency of the first saccade (that is, the saccade shifting the gaze away from central fixation position): the higher the saccadic latency, the closer the saccadic direction to the desired direction. This holds also by selecting those trials (19%) in which only one saccade is present ($r = 0.601$, $p < 0.01$), further confirming that the internal trajectory evolves regardless the specific pattern of overt eye movements, even in the extreme case of single-saccade trials. This is also in line with the above-mentioned idea that eye movements are just an epiphenomenon of visuo-spatial mental rotation, and not a necessary condition for its evolution.

The smooth appearance of the mean trajectory derives from having averaged many saccades. This is not to say that mental rotation is necessarily a saccadic-like process but just that the working of the saccadic circuitry constrains the final output. We will come back to this point. For now suffice to say that mental rotation has been found to have a behavioral correlation in the form of sequences of spontaneous saccades whose angular evolution complies with the supposed internal process.

Towards a Precise Assessment of the Evolution of Mental Rotation

A number of interesting properties of the average gaze rotation can be quantitatively measured. In order to explore this phenomenon more thoroughly, particularly its kinematical aspects, the experiment was repeated with eight instruction angles (0, 20, 40, 60, 80, 100, 120 and 140°). Ten subjects participated in this experiment. Infrared oculometry (sampling frequency: 500 Hz) was used to record eye movements. Three questions stand out for their relevance to mental rotation. The first is the estimate of mean velocity and duration of gaze rotation. The second is its instantaneous velocity profile. The third is the possible dependency of gaze velocity upon the instruction angle.

As mentioned previously, mental rotation is usually inferred from the proportionality between RT and the angle of rotation. Besides supporting the qualitative notion of existence of a mental rotation process, RTs have allowed to estimate its speed, taken as the inverse of the slope of the regression line of RTs over the rotation angles. From the regression line, in this experiment the estimate of mental rotation speed in the visuo-spatial mental rotation task was 238 deg/s ± 40 S.D. However, the mean rotation speed of the gaze (i.e., its mean angular velocity) was significantly lower (56.7 deg/s ± 17.4 S.D., $p < 0.001$). Analogous reasoning applies to the estimate of the average mental rotation duration, which, if estimated through RTs, was 360 ms ± 260 S.D. The mean duration of gaze rotation (computed as the period in which angular velocity exceeded 5 deg/s) was significantly ($p < 0.001$) higher, being 1250 ms ± 500 S.D. Thus, as compared to the supposed mental rotation process, gaze rotation appears to be slower and lasting longer.

As for the instantaneous velocity profile, it is usually assumed that mental rotation speed is constant. According to Kosslyn (1980: p. 56), in mental rotation "the mind seems to have some kind of process starting and moving at constant velocity". Yet, it is easily recognizable from the profiles of instantaneous angular position, velocity and acceleration (Figure 7.6) that gaze rotation was far from being a constant velocity process. Rather, its mean angular progression is well captured by an asymmetrical sigmoidal function, which in turn results in an asymmetrical velocity and acceleration profile. The asymmetry was not dependent upon the particular averaging procedure adopted, because it was already present in the individual trials.

It is tempting to speculate that the asymmetry of gaze rotation depends on the presence of a double mechanism in mental rotation, consisting of an initial open-loop, propelling phase and a final closed-loop, adjustment phase. This is reminiscent of the asymmetry of goal-directed movements, a characteristic already recognized in the nineteenth century (Woodworth, 1899; see Jeannerod, 1988). Basically, in reaching a target the last part of the movement decelerates because in passing to a closed-loop control, there is a trade-off between better spatial precision and slowness of the

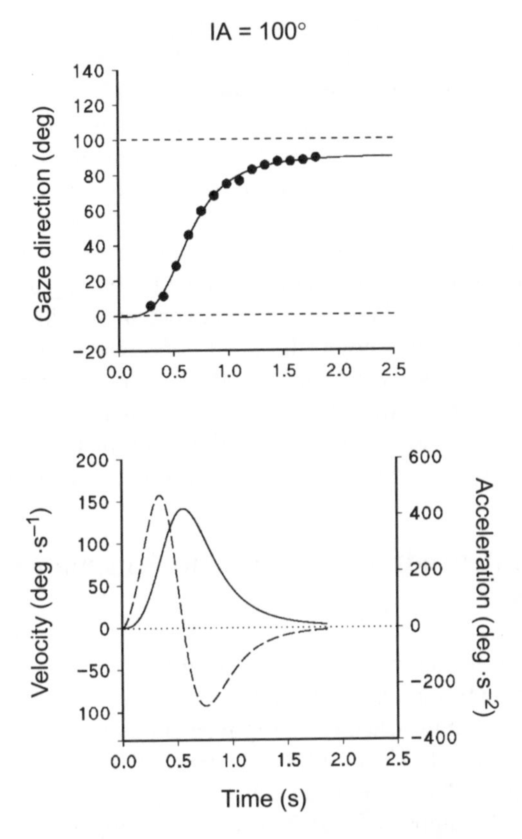

Figure 7.6: Rotation of the gaze: Mean gaze direction, angular velocity and angular acceleration as a function of time for one representative instruction angle (IA). In the top panel, the binned mean gaze direction (closed circles) has been fitted with a Morgan-Mercier-Florin growth model (continuous line). Bin duration has been re-normalized to the average response time. The expected initial and final directions are represented as horizontal dashed lines. In the bottom panel the angular velocity (continuous line) and acceleration (dashed line) have been obtained as the first and second derivative, respectively, of the direction-fitting function. It can be seen that gaze rotation does not evolve uniformly in time.

system. Applied to the present case, this would amount to say that in the last part of the trajectory the system initiates checking its performance by evaluating some sort of dynamic angular error, so that a coarse launch is refined in the second phase of rotation. It is also interesting that the first phase is not stereotyped: Gaze peak angular velocity increased significantly ($p < 0.001$) for increasing instruction angles, passing from 75 deg/s to 220 deg/s. This means that gaze rotation starts already roughly tailored to the to-be-covered angle. Again, this feature is found in reaching movements and head movements, where larger movements have higher peak velocities

(Zangemeister *et al.*, 1981a,b; Ghez *et al.*, 1997). The saccadic main sequence is a typical example of this behavior (Bahill *et al.*, 1979).

Another assumption regarding mental rotation is that its velocity is independent of the angle to be mentally covered. However, as the peak, also the mean angular velocity of the gaze increased significantly ($p < 0.001$) with the instruction angle, passing from 34 deg/s to 83 deg/s. Taken together, these features of gaze rotation contribute to regard visuo-spatial mental rotation as a kind of "mental reaching".

The heading of this section explicitly mentioned the evolution of mental rotation. So far, however, I have been rather cautious in directly linking gaze rotation to mental rotation, even if it is beyond doubt that eye movements in this task were controlled by a top-down process. At this point, the one-to-one relationship between gaze rotation and mental rotation is still a working hypothesis. It is now time to firm the working hypothesis up into a more confident statement. The claim is that gaze rotation faithfully reproduced the kinematics of visuo-spatial mental rotation. In what follows I will describe some experiments, together with their rationale, which support this claim.

Eye Movements Reproduce Faithfully the Kinematics of Motion Imagery

What makes studying mental rotation (and indeed any form of mental activity) problematic is that it cannot be directly measured. So we are faced with a seemingly circular problem. How can we know whether eye movements reflect a mental process if this process cannot be measured? The problem may be circumvented by making an analogy with a well-known procedure, namely a calibration procedure. The essence of any calibration is to assess the output of a system to known inputs in order to find out its general input–output relationship. As a second step, this knowledge is used to reconstruct the (unknown) input from a given output of the system, provided it falls within the established working range. It can be recognized at this point of our story that we are still lacking the first step needed to "calibrate" the oculomotor performance in response to known mental inputs. The rationale of the following experiments is to feed the oculomotor system with known mental inputs in order to see whether it is transparent to the inner process or whether distortion is introduced. In particular, as mental rotation is usually considered a constant velocity process, it is important to see whether a non-constant gaze rotation may result from a constant velocity mental input reproduced by the oculomotor circuitry in a distorted manner.

The critical point was how to induce a mental rotation process with known characteristics. To this aim, the experimental conditions have been slightly changed (Figure 7.7). Subjects were shown a motion template in the form of a spot rotating at constant velocity along the same target circle used in the previous experiments. After one cycle the spot disappeared. After a random pause, a portion of the circle was highlighted (rotation angle, that could be 40, 80 or 120°). The task was to mentally imagine the previously seen spot as rotating at the same velocity along the rotation angle and press a button upon the completion of the trajectory (motion imagery task). As can be seen, the task maintains the structure of the visuo-spatial mental rotation task, but with a

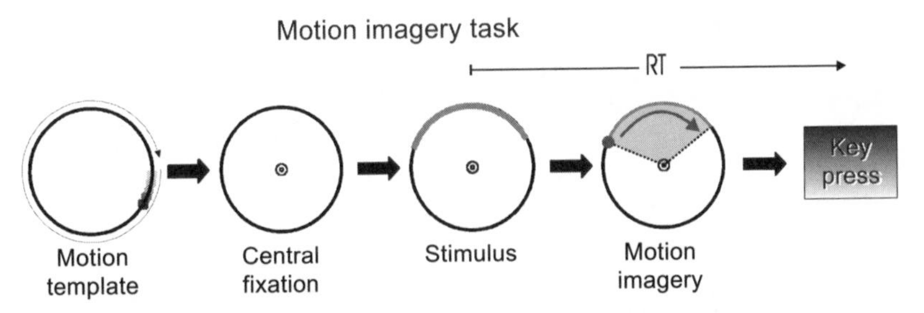

Figure 7.7: The motion imagery task. Subjects view a spot rotating at constant velocity along a target circle that disappears after one cycle. Then they fixate a central spot. After a random pause a portion of the circle is highlighted (rotation angle). Subjects have to mentally imagine the previously seen rotating spot as moving along the rotation angle in the clockwise direction, and press a key upon completion of the trajectory.

fundamental difference: Now subjects have to do explicitly in imagery what they were supposed to do implicitly in the visuo-spatial task according to the constant velocity hypothesis. The unknown mental rotation process has been substituted with an instructed motion imagery process. Still, it could be argued that there is no guarantee that imagery is really under control. However, if it can be shown that gaze rotation in this condition indeed occurs at constant velocity, then there is no reason to postulate that, in between a constant-velocity motion template and a constant-velocity gaze rotation, the intervening imagery process is not a constant-velocity process.

In Figure 7.8 are reported examples of raw sequences of saccades and the average trajectory for the three rotation angles employed in the motion imagery task. Recordings start at the presentation of the rotation angle. Ten subjects took part in this experiment and the experimental conditions were the same of the previous experiment on visuo-spatial mental rotation. The velocities of the motion templates (25, 40 and 55 deg/s, respectively for the three rotation angles) were quite low because it was important that motion could be easily imagined. The mean gaze angular velocity increased significantly ($p < 0.001$) with the rotation angle (16.9 deg/s ± 2.1 S.D., 34.8 deg/s ± 3.6 S.D. and 42.0 deg/s ± 6.4 S.D.), showing that the mean mental imagery speed could be controlled. For all rotation angles the mean gaze velocity was lower than that of the motion template, although only for the instruction angle of 40° the difference was statistically significant ($p < 0.01$). RTs were significantly higher ($p < 0.001$) for all rotation angles (2730 ms ± 305 S.D., 3072 ms ± 272 and 3547 ms ± 398 S.D. for the three rotation angles, respectively). These data show a tendency to underestimate the template velocity in imagery. Moreover, at variance with the visuo-spatial mental rotation task (Figure 7.9, left panel), here the angular progression of the gaze was clearly constant (Figure 7.9, middle panel), thus confirming that the experimental manipulations succeeded in inducing motion imagery at constant velocity.

Thanks to their small velocities, these stimuli were quite easy to imagine. However, a stronger test for the mental rotation hypothesis would require the velocity of the motion template to be as high as the mental rotation speed estimated from RTs in

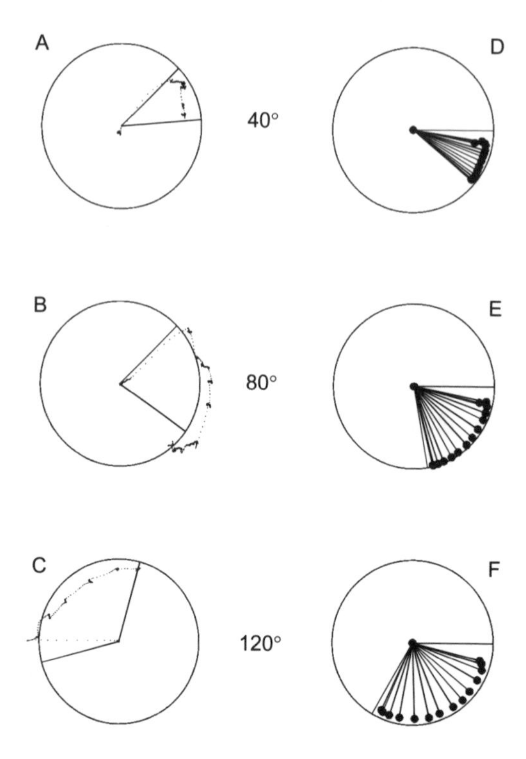

Figure 7.8: Spontaneous eye movements (**A–C**) and average gaze trajectories (**D–F**) in the motion imagery task, from stimulus onset to the button press. The gaze is initially in the central position. The gaze depicts the expected mental rotation very clearly.

the visuo-spatial task. At such high velocity it may happen that the saccadic system reaches some limits, thus bringing distortions into the traces. An experimental condition was added in which the constant velocity motion template had a fixed velocity of 250 deg/s, regardless the rotation angle (40, 80 and 120°). This corresponded rather strictly to the RT-based hypothesis of mental rotation. Preliminary results from four subjects indicated that RTs were much higher than the time taken by the template to cover the rotation angles (881 ms ± 219 S.D., 1128 ms ± 84 and 1259 ms ± 92 S.D. for the three rotation angles, respectively; $p < 0.001$ in all cases), suggesting that in imagery this high velocity had been particularly underestimated. Accordingly, gaze velocity was lower than expected on the basis of the motion template (78.4 deg/s ± 9.8 S.D., 116.8 deg/s ± 17.8 and 149.1 deg/s ± 19.6 S.D. for the three rotation angles, respectively; $p < 0.001$ in all cases). More importantly, in spite of the particularly demanding conditions of the task, the velocity profile of gaze rotation was constant again (Figure 7.9, right panel), thus further confirming that the visuo-spatial mental rotation task and the motion imagery tasks entail rotations with different time-courses. These kinematical differences might be due to differences in the underlying mechanisms, such as the presence, in mental rotation but not in motion imagery, of a final adjusting phase.

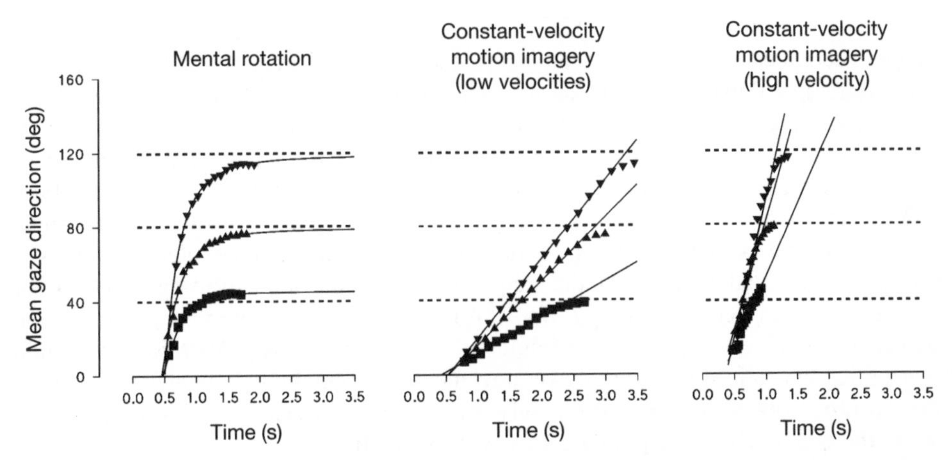

Figure 7.9: Comparison of the average gaze direction as a function of time in the visuo-spatial mental rotation task (left panel), in the constant-velocity motion imagery task with low-velocities motion templates (middle panel) and in the constant-velocity motion imagery task with a high-velocity motion template (right panel). At variance with the visuo-spatial mental rotation task, in the motion imagery task the gaze rotates at constant velocity. The continuous lines are the best fitting curves. To compute the linear fit (middle and right panels) the three final bins have not been considered. The dashed lines represent three instructions angles (mental rotation task) and the corresponding three rotation angles (motion imagery task).

Taken together, these results indicate that the gaze is able to reproduce accurately a constant velocity imagined rotation, both at high and low velocities, and lend credit to the hypothesis that, in the visuo-spatial mental rotation experiment too, gaze rotation was accurately tracing the kinematics of mental rotation.

About the Nature of Dynamic Mental Imagery

The question remained whether mental rotation is a continuous process or if it is of progressive but step-like nature. In the light of the strict association between saccades and attention (e.g., Chelazzi *et al.*, 1986; Shepard, *et al.*, 1986; Rizzolatti, *et al.*, 1994; Schneider & Deubel, 1995; Deubel & Schneider, 1996), it might be that visuo-spatial mental rotation is in fact a sequence of saccadic-like attentional shifts. The same question can be posed for motion imagery. Is motion represented in imagery as a sequence of discrete, saccadic-like shifts? A positive answer would amount to saying that we do not in fact imagine a moving dot but rather a succession of points in space.

However, many observations call for a different view. If we think at a bird in flight, we feel it as really moving, not just at different positions in space. Yet, despite compelling, our subjective experience of continuous motion during motion imagery is not by itself evidence that the core of the process is truly gradual. Evidence that motion imagery involves a continuous process comes from the finding that visual motion imagery activates cortical extrastriate areas, such as medial temporal (MT) and medial superior temporal (MST) cortex, known to be involved in optical flow processing, that is, in perceiving continuous motion (Goebel *et al.*, 1998). Indeed, a number of studies have shown that these neuronal structures are crucial for mental representation of continuous motion in the perception–cognition continuum (David & Senior, 2000; Kourtzi & Kanwisher, 2000). As far as visuo-spatial mental rotation is concerned, robust objective evidence pointing to the existence of a continuous internal process is the aforementioned rotation of the neuronal population vector during visuo-motor mental rotation (Georgopoulos *et al.,* 1989). In that case, a covert gradual rotation indeed occurred throughout the stimulus–response time lapse. Therefore, it would appear that the major limitation of eye movements in mimicking dynamic visuo-spatial mental processes is that in imagery only saccadic eye movements are released, even when the task requires imagining a slowly moving stimulus.

Yet, it may sound quite odd that the eyes cannot reproduce an imagined continuous motion by what would seem the most natural means, i.e., the smooth pursuit system. In fact, although the absence of smooth "pursuit" eye movements in imagery is in agreement with the well-known fact that these eye movements cannot be voluntarily initiated (Rashbass, 1961; Pola & Wyatt, 1991), firstly, the extrastriate areas MT/MST, which are upstream of the smooth pursuit circuitry (Lisberger *et al.*, 1987), are implicated in both motion perception and motion imagery (Goebel *et al.*, 1998) and, secondly, smooth eye movements can also be driven or modulated by non-retinal signals (Barr *et al.*, 1976; Steinbach, 1976; Whittaker & Eaholtz, 1982, Melvill-Jones & Berthoz, 1985; Barnes & Asselman, 1991; Barnes & Donelan, 1999; Barnes *et al.*, 2000). Thus, the mechanisms of inhibition of overt smooth "pursuit" eye movements during motion imagery remain to be clarified.

Concluding Remarks

In this chapter I have addressed the hypothesis that spontaneous eye movements can disclose the inner working of mental processes in different kinds of mental rotation tasks. Besides showing that visuo-spatial mental rotation has a behavioral correlate in the form of sequences of saccades often reproducing the supposed mental rotation trajectory, I have illustrated a number of quantitative aspects of the mean gaze rotation, particularly the kinematical aspects. Then I showed that rotary motion imagery tasks, in which mental rotation was explicitly dictated by instruction, resulted in very consistent eye movement patterns, whose kinematics was only marginally distorted compared to what was expected. The picture has emerged that the average, "platonic" gaze rotation indeed depicts the spatio-temporal angular evolution of the underlying "mental movements', in both the visuo-spatial mental rotation task and the motion imagery task. I also argued that the saccadic nature of eye movements in imagery simply reflects the constraints of the seemingly hard-wired working of the oculomotor circuitry, as, for unknown reasons, smooth eye movements in response to a smooth mental drive are normally inhibited.

This eye movement-based approach allows the testing of a number of hypotheses concerning limits and flexibility of motion imagery. Are we free to imagine any kind of motion? Can we imagine motions with various kinematics or even unnatural motions? If it is true that some ecologically-relevant constraints are embodied into our internal representations (Shepard, 1984), then it could be expected that natural motions have a somewhat privileged status in imagery.

A question that has been posed recently is whether we can freely imagine a particular kind of unnatural motion, namely, a motion that does not comply with the so-called two-thirds power law (Jeannerod, 1994; Actis Grosso *et al.*, 2001). This empirical law describes a regularity of certain human movements: The instantaneous tangential velocity of a movement such as writing, drawing or scribbling is a power function of its radius of curvature (Viviani & Terzuolo, 1982). Tracking with the hand stimuli whose motion does not obey the two-thirds power law is very difficult and error-laden (Viviani *et al.*, 1987, 1997). In addition, these stimuli determine strange percepts and look quite unnatural. For example, if a spot moving with non-biological velocity is tracing an elliptical trajectory, the perceived eccentricity of the ellipse is markedly biased, as if a veridical perception of form depended on the stimulus having an appropriate kinematics (Viviani & Stucchi, 1989). Similarly, a constant-velocity stimulus moving on a curvilinear path (that is, a stimulus violating the two-thirds power law) is not actually seen as having constant velocity (Viviani & Stucchi, 1992; de'Sperati & Stucchi, 1995). The question then arises whether we can freely imagine this unnatural movement that we are able neither to (re)produce nor to straightforwardly perceive. By taking advantage of the systematic sequences of spontaneous saccades during motion imagery and by considering that in smooth pursuit eye movements the saccadic component but not the smooth component is clearly modulated by the motion law of the target, even when it does not follow the two-thirds power law (de'Sperati & Viviani 1997), it can be ascertained whether, in mentally "tracking" such unnatural motions, the saccade sequences follow the two-thirds power law or are driven by an unconstrained motion imagery process.

Finally, "recording" the instantaneous spatio-temporal evolution of these "mental movements" allows the role of the brain structures implicated in visuo-spatial thinking to be investigated in greater detail. In particular, it would appear that methods with a high temporal resolution such as event-related potentials, magneto-encephalography and transcranial magnetic stimulation are well suited to be coupled with this eye-movement based approach in order to uncover the brain dynamics associated to these ongoing dynamic mental processes. Moreover, the ability to handle motion imagery experimentally may lead in the future to a better knowledge of the imagery components of those conditions involving a peculiar pattern of imaginative activity, both physiological, such as prolonged sensorial deprivation or altered consciousness states (e.g., hypnosis), and pathological, such as Alzheimer disease, hemineglect syndrome or drug-based hallucinatory states.

Acknowledgements

This work received financial support by M.I.U.R. (9905401355_003 — MM11338515_002).

References

Actis Grosso, R., de'Sperati, C., Stucchi, N., & Viviani, P. (2001). Visual extrapolation of biological motion. In: E. Sommerfeld, R. Kompass and T. Lachmann (eds), *Proceedings of the Seventeenth Annual Meeting of the International Society for Psychophysics* (pp. 261–266). Longerich: Pabst.

Bahill, A. T., Clark, M. R., & Stark, L. W. (1979). The main sequence, a tool for studying human eye movements. *Mathematical Bioscience*, *24*, 191–204.

Ballard, D. H., Hayhoe, M. M., & Pelz, J. B. (1995). Memory representations in natural tasks. *Journal of Cognitive Neuroscience*, *7*, 66–80.

Ballard, D. H., Hayhoe, M. M., Pook, P. K., & Rao, R. P. N. (1997). Deictic codes for the embodiment of cognition. *Behavioral and Brain Science*, *20*, 723–742.

Barnes, G. R., & Asselman, P. T. (1991). The mechanism of prediction in human smooth pursuit eye movements. *Journal of Physiology (London)*, *43*, 439–461.

Barnes, G. R., Barnes, D. M., & Chakraborti, S. R. (2000). Ocular pursuit responses to repeated, single-cycle sinusoids reveal behavior compatible with predictive pursuit. *Journal of Neurophysiology*, *84*, 2340–2355.

Barnes, G. R., & Donelan, A. S. (1999). The remembered pursuit task: Evidence for segregation of timing and velocity storage in oculomotor control. *Experimental Brain Research*, *129*, 57–67.

Barr, C. C., Schulteis, L. W., & Robinson, D. A. (1976). Voluntary, non-visual control of human vestibulo-ocular reflex. *Acta Otolaryngologica*, *81*, 365–375.

Brandt, S. A., & Stark, L. W. (1997). Spontaneous eye movements during visual imagery reflect the content of the visual scene. *Journal of Cognitive Neuroscience*, *9*, 27–38.

Carpenter, P. A., & Just, M. A. (1978). Eye fixations during mental rotation. In: J. Senders, R. Monty and D. Fisher (eds), *Eye Movements and Psychological Functions II*, Hillsdale: Erlbaum.

Chelazzi, L., Biscaldi, M., Corbetta, M., Peru, A., Tassinari, G., & Berlucchi, G. (1986). Oculomotor activity and visual spatial attention. *Behavioural Brain Research*, *71*, 81–88.

Cohen, M. S., Kosslyn, S. M., Breiter, H. C., DiGirolamo, G. J., Thompson, W. L., Anderson, A. K., Brrokheimer, S. Y., Rosen, B. R., & Belliveau, J. W. (1996). Changes in cortical activity during mental rotation. A mapping study using functional MRI. *Brain*, *119*, 89–100.

Cooper, L. A. (1976). Demonstration of a mental analog of an external rotation. *Perception and Psychophysics*, *19*, 296–302.

Coren, S. (1986). An efferent component in the visual perception of direction and extent. *Psychological Review*, *93*, 391–410.

Coren, S., & Girgus, J. S. (1978). Visual illusions. In: R. Held, H. Liebowitz and H. L. Teuber (eds), *Handbook of Sensory Physiology; Volume VIII* (pp. 549–568). New York: Springer.

David, A. S., & Senior, C. (2000). Implicit brain motion and the brain. *Trends in Cognitive Sciences*, *4*, 293–295.

de'Sperati, C. (1999). Saccades to mentally rotated targets. *Experimental Brain Research*, *126*, 563–577.

de'Sperati, C., & Stucchi, N. (1995). Visual tuning to kinematics of biological motion. *Experimental Brain Research, 105*, 254–260.

de'Sperati, C., & Stucchi, N. (1997). Recognising the motion of a graspable object is guided by handedness. *Neuroreport, 8*, 2761–2765.

de'Sperati, C., & Stucchi, N. (2000). Motor imagery and visual event recognition. *Experimental Brain Research, 133*, 273–278.

de'Sperati, C., & Viviani, P. (1997). The relationship between curvature and velocity in two-dimensional smooth pursuit eye movements. *Journal of Neuroscience, 17*, 3932–3945.

Deubel, H., & Schneider, W. X. (1996). Saccade target selection and object recognition: Evidence for a common attentional mechanism. *Vision Research, 36*, 1827–1837.

Ellis, S. R., & Stark, L. (1978). Eye movements during the viewing of Necker cubes. *Perception, 7*, 575–581.

Epelboim, J., & Suppes, P. (2001). A model of eye movements and visual working memory during problem solving in geometry. *Vision Research, 41*, 1561–74.

Finke, R. A. (1989). *Principles of mental imagery*. Cambridge, MIT Press.

Finke, R. A., & Shepard, R. N. (1986). Visual functions of mental imagery. In: K. R. Boff, L. Kaufman and J. P. Thomas (eds), *Handbook of Perception and Human Performance* (pp. 37.1–37.55). New York: Wiley.

Fischer, M. H., Deubel, H., Wohlschläger, A., & Schneider, W. X. (1999). Visuomotor mental rotation of saccade direction. *Experimental Brain Research, 127*, 224–232.

Ganis, G., Keenan, J. P., Kosslyn, S. M., & Pascual-Leone, A. (2000). Transcranial magnetic stimulation of primary motor cortex affects mental rotation. *Cerebral Cortex, 10*, 175–180.

Georgopoulos, A. P. (1990). Neural coding of the direction of reaching and a comparison with saccadic eye movements. *Cold Spring Harbor Symposium, LV*, 849–859.

Georgopoulos, A. P., Lurito, J. T., Petrides, M., Schwartz, A. B., & Massey J. T. (1989). Mental rotation of the neuronal population vector. *Science, 243*, 234–236.

Georgopoulos, A. P. and Massey, J. T. (1987). Cognitive spatial-motor processes. 1. The making of movements at various angles from a stimulus direction. *Experimental Brain Research, 65*, 361–370.

Ghez, C. Bermejo, R., Favilla, M., Ghilardi, M., & Gordon, J. (1997). Discrete and continuous planning of hand movements and isometric force trajectories. *Experimental Brain Research, 115*, 217–233.

Goebel, R. Khorram-Sefat, D., Muckli, L., & Singer, W. (1998). The constructive nature of vision: Direct evidence from functional magnetic resonance studies of apparent motion and motion imagery. *European Journal of Neuroscience, 10*, 1563–1573.

Hacisalihzade, S. S., Allen, J. S., & Stark, L. W. (1992). Visual perception and sequences of eye movements and fixations: A stochastic modelling approach. *IEEE Transactions System Man Cybernetics, 22*, 474–481.

Hallett, P. (1978). Primary and secondary saccades to goals defined by instructions. *Vision Research, 18*, 1279–1296.

Hebb, D. O. (1968). Concerning imagery. *Psychological Review, 75*, 466–477.

Jeannerod, M. (1988). *The Neural and Behavioural Organization of Goal-Directed Movements*. Oxford: Oxford Science Publication.

Jeannerod, M. (1994). The representing brain: Neural correlates of motor imagery and intention. *Behavioral and Brain Sciences, 17*, 187–245.

Klein, R. M. (2000). Inhibition of return. *Trends in Cognitive Sciences, 4*, 138–147.

Kosslyn, S. M. (1973). Scanning visual images: Some structural implications. *Cognitive Psychology, 8*, 441–480.

Kosslyn, S. M. (1980). *Image and Mind*. Cambridge and London: MIT Press.

Kosslyn, S. M. (1994). *Image and Brain*. Cambridge and London: Harvard University Press.

Kosslyn, S. M., Alpert, N. M., Thompson, W. L., Maljkovic, V., Weise, S. B., Chabris, S. E., Hamilton, S. L., Rauch, S. L., & Buonanno, F. S. (1993). Visual mental imagery activates topographically organized visual cortex: PET investigations. *Journal of Cognitive Neuroscience*, *5*, 263–287.

Kosslyn, S. M., Di Girolamo, Thompson, W. L., & Alpert, N. M. (1998). Mental rotation of objects vs. hands: Neural mechanisms revealed by positron emission tomography. *Psychophysiology*, *35*, 151–161.

Kosslyn, S. M., Ganis, G., & Thompson, W. L. (2001a). Neural foundations of imagery. *Nature Review Neuroscience*, *9*, 635–642.

Kosslyn, S. M., Thompson, W. L., Wraga, M., & Alpert, N. M. (2001b). Imaging rotation by endogenous versus exogenous forces: Distinct neural mechanism. *Neuroreport*, *12*, 2519–2525.

Kourtzi, Z., & Kanwisher, N. (2000). Implied motion activates extrastriate motion-processing areas. Response to David and Senior (2000). *Trends in Cognitive Sciences*, *4*, 295–296.

Kowler, E. (1995). Eye movements. In: S. M. Kosslyn and D. N. Osherson (eds), *Visual Cognition* (pp. 215–266). Cambridge and London: MIT Press.

Lisberger, S. G., Morris, E. J., & Tychsen, L. (1987). Visual motion processing and sensory-motor integration for smooth pursuit eye movements. *Annual Review of Neuroscience*, *10*, 97–129.

Liversedge, S. P., & Findlay, J. M. (2000). Saccadic eye movements and cognition. *Trends in Cognitive Sciences*, *4*, 6–14.

Melvill-Jones, G., & Berthoz, A. (1985). Mental control of the adaptive process. In A. Berthoz and G. Melvill-Jones (eds), *Adaptive Mechanisms in Gaze Control. Facts and Theories* (pp. 201–208). Amsterdam: Elsevier.

Neisser, U. (1976). *Cognition and Reality: Principles and Implications of Cognitive Psychology*. San Francisco: Freeman.

Noton D., & Stark L. (1971a). Scanpaths in eye movements during pattern perception. *Science*, *171*, 308–311.

Noton D., & Stark L. (1971b). Scanpaths in saccadic eye movements while viewing and recognizing patterns. *Vision Research*, *11*, 929–942.

Parsons, L. M. (1987). Imagined spatial transformations of one's hands and feet. *Cognitive Psychology*, *19*, 178–241.

Parsons, L. M. (1994). Temporal and kinematic properties of motor behavior reflected in mentally simulated action. *Journal of Experimental Psychology: Human Perception and Performance*, *20*, 709–730.

Parsons, L. M., Fox, P. T., Downs, J. H., Glass, T., Hirsch, T. B., Martin, C. C., Jerabek, P. A., & Lancaster, J. L. (1995). Use of implicit motor imagery for visual shape discrimination as revealed by PET. *Nature*, *375*, 54–58.

Pellizer, G., & Georgopoulos, A. P. (1993). Common processing constraints for visuomotor and visual mental rotations. *Experimental Brain Research*, *93*, 165–172.

Pola, J., & Wyatt, H. J. (1991). Smooth pursuit: response characteristics, stimuli and mechanisms. In J. R. Cronly-Dillon (ed.), *Vision and Visual Dysfunction: Vol. 8, Eye Movements* (pp. 138–156). London: Macmillan.

Rashbass, C. (1961). The relationship between saccadic and smooth tracking eye movements. *Journal of Physiology*, *159*, 338–362.

Rayner, K. (1998). Eye movements in reading and information processing: 20 years of research. *Psychological Bulletin*, *124*, 372–422.

Rizzolatti, G., Riggio, L., & Sheliga, B. (1994). Space and selective attention. In: C. Umiltà and M. Moscovitch (eds), *Attention and Performance XV* (pp. 231–265). Hillsdale: Erlbaum.

Schneider, W. X., & Deubel, H. (1995). Visual attention and saccadic eye movements: Evidence for obligatory and selective spatial coupling. In J. M. Findlay, R. W. Kentridge and R. Walker (eds), *Eye Movement Research: Mechanisms, Processes, and Applications* (pp. 317–324). Elsevier.

Scotto, M. A., Oliva, G. A., & Tuccio, M. T. (1990). Eye movements and reversal rates of ambiguous patterns. *Perceptual and Motor Skills*, *70*, 1059–1073.

Sekiyama, K. (1982). The kinesthetic aspects of mental representation in the identification of left and right hands. *Perception and Psychophysics*, *32*, 89–95.

Shepard, R. N. (1984). Ecological constraints on internal representation: Resonant kinematics of perceiving, imaging, thinking and dreaming. *Psychological Review*, *4*, 417–447.

Shepard, R. N., & Cooper, L. A. (1986). *Mental Images and Their Transformations*. Cambridge, London: MIT Press.

Shepard, S. M., & Metzler, J. (1971). Mental rotation of three-dimensional objects. *Science*, *171*, 710–703.

Sheperd, M., Findlay, J. M., & Hockey, R. .J. (1986). The relationship between eye movements and spatial attention. *Quarterly Journal of Experimental Psychology*, *38A*, 475–491.

Sirigu, A., & Duhamel, J. R. (2001). Motor and visual imagery as two complementary but neurally dissociable mental processes. *Journal of Cognitive Neuroscience*, *13*, 910–919.

Steinbach, M. (1976). Pursuing the perceptual rather than the retinal stimulus. *Vision Research*, *16*, 1371–1375.

Tanenhaus, M. K., Spivey-Knowlton, M. J., Eberhard, K. M., & Sedivy, J. E. (1995). Integration of visual and linguistic information in spoken language comprehension. *Science*, *268*, 632–634.

Viviani, P. (1990). Eye movements and visual search: cognitive, perceptual and motor control aspects. In E. Kowler (ed.), *Eye Movements and Their Role in Visual and Cognitive Processes* (pp. 353–394). Amsterdam, New York, Oxford: Elsevier.

Viviani, P., Campadelli, P., & Mounoud, P. (1987). Visuo-manual pursuit tracking of human two-dimensional movements. *Journal of Experimental Psychology: Human Perception and Performance*, *13*, 62–78.

Viviani, P., Redolfi, M., & Baud-Bovy, G. (1997). Perceiving and tracking kinesthetic stimuli: Further evidence of motor-perceptual interactions. *Journal of Experimental Psychology: Human Perception and Performance*, *23*, 1232–1252.

Viviani, P., & Stucchi, N. (1989). The effect of movement velocity on form perception: Geometric illusions in dynamic displays. *Perception and Psychophysics*, *46*, 266–274.

Viviani, P., & Stucchi, N. (1992). Biological movements look constant: Evidence of motor-perceptual interactions. *Journal of Experimental Psychology: Human Perception and Performance*, *18*, 603–623.

Viviani, P., & Terzuolo, C. (1982). Trajectory determines movement dynamics. *Neuroscience*, *7*, 431–437.

Wexler, M., Kosslyn, S. M., & Berthoz, A. (1998). Motor processes in mental rotation. *Cognition*, *68*, 77–94.

Whittaker, S. G., & Eaholtz, G. (1982). Learning pattern of eye motion for foveal pursuit. *Investigative Ophtalmology and Visual Sciences*, *23*, 393–397.

Wohlschläger, A., & Wohlschläger, A. (1998). Mental rotation and manual rotation. *Journal of Experimental Psychology: Human Perception and Performance*, *24*, 397–412.

Woodworth, R. S. (1899). The accuracy of voluntary movement. *Psychological Review*, *3*, Suppl 2.

Yarbus, A. L. (1967). *Eye Movements and Vision*. New York: Plenum.

Zangemeister, W., Lehman, S., & Stark, L. (1981a). Simulation of head movement trajectories: Model and fit to main sequence. *Biological Cybernetics*, *41*, 19–32.

Zangemeister, W., Lehman, S., & Stark, L. (1981b). Sensitivity analysis and optimization for a head movement model. *Biological Cybernetics*, *41*, 33–45.
Zelinsky, G. J., Rao, R. P. N., Hayhoe, M. M., & Ballard, D. H. (1997). Eye movements reveal the spatiotemporal dynamics of visual search. *Psychological Science*, *8*, 448–453.

Commentary on Section 1

Eye Movements and Visual Information Processing

John M. Findlay

The papers in this section explore the relations between eye movements and visual information processing. Recent years have seen increasing attempts to move forward from an earlier epoch when work both on the visual system and on the oculomotor system developed with little reference to the advances that were occurring elsewhere. Consideration of the links between the two areas raises a number of general questions. In order to set the context for discussion of the conference papers, the first part of this commentary gives a personal perspective on the two areas in order to provide a focus for the detailed discussion of the contributions.

Much early work in eye movements investigated the oculomotor system using very simple visual situations, often presenting observers with just a single "target" for the eye to track. The emphasis of work in the 1960s and 1970s was very much on the oculomotor system considered almost in isolation. The flavour of the period was reflected by work of individuals such as Robinson (e.g. 1975) and Stark (e.g. 1968) who were able, with considerable success, to develop ideas based on analogies with those used in the study of mechanical and electrical control systems. The self-evident scientific rigour of the approach, often backed with elaborate mathematical models, had considerable appeal although it may also have had a negative consequence in that the perception of eye movement research from people outside the area became that of a specialised esoteric corner. As general understanding of mental abilities developed, critics of the "systems" approach emerged (e.g. Steinman, 1986; Kowler, 1990) who stressed the fact that the systems approach did not allow ready extension to integrate cognitive factors and top-down control, essential for any understanding of how the eyes were controlled in normal vision.

The period around 1980 saw an exciting broadening of the scope of eye movement research to take this criticism on board. One important influence was the rapid progress made in the study of eye movements in text reading through the pioneering work of several groups (Rayner, 1975; McConkie & Rayner, 1975; Lévy-Schoen & O'Regan, 1980). The ECEM meetings series commenced in 1981. These meetings reflected the

The Mind's Eye: Cognitive and Applied Aspects of Eye Movement Research

ISBN: 0–444–51020–6

change of emphasis and it can also be argued that they continued to advance the change by providing a forum for eye movement workers of a variety of persuasions. To an extent, the trend can be detected earlier in the series of meetings organised by the US Air Force in the 1970s which acted in some ways as forerunner of the ECEM meetings with overlap occurring at the very first ECEM meeting (Groner *et al.*, 1983). The subsequent series of ECEM meetings have continued a rich tradition emphasizing how eye movements are employed in support of visual perception.

It is surprising therefore to note that work in oculomotor science still makes remarkably little contact with another rich tradition of work in vision and visual information processing. Much work in vision traces the way in which visual information received on the retina is transmitted and processed through subsequent stages of the visual pathways. Some interesting parallels can in fact be drawn between the history of eye movement studies and the historical development in this tradition. The spatial frequency approach of the 1970s and early 1980s (DeValois and DeValois, 1980) might be seen as having some analogy to the "systems" approach, albeit predicated on concepts from optical, rather than mechanical or electrical, engineering.

Much work in this tradition has been self-contained and it is still possible to find high quality texts on vision and visual perception that make almost no reference to the fact that the eye is mobile (Wandell, 1995; Regan, 2000) although others can be found that are more ready to cross the divide (the text by Palmer, 1999 can be commended as an instance). One reason for this state of affairs may relate to the widely held notion that the visual system operates as a self-contained module whose purpose is to allow "seeing". This view of vision emphasises the idea of processing carried out on the retinal image to achieve a well articulated mental representation of the visual world. Perhaps the most explicit advocate of this view was David Marr (e.g. 1982) and the view might be termed the Marrian approach. Its wide appeal seems in part due to our phenomenal experience that vision has a unified picture-like quality where we have direct awareness of our visual world.

From the Marrian viewpoint, eye movements have little role to play in visual perception since the visual process itself is held to have generated a rich mental representation. It is a very plausible further step to assume that some mental process can operate to select a specific part of this representation. An obvious extension of the Marrian viewpoint is the idea of covert attention, particularly as it is readily demonstrated that attention can be directed without moving the eyes. Work on covert visual attention accelerated considerably during the 1970s and 1980s, with pioneering contributions from Eriksen (e.g. Eriksen & Hoffman, 1972), Posner (e.g. Posner *et al.*, 1978; Posner, 1980) and Treisman (e.g. Treisman & Gelade, 1980). Although many alternative suggestions have been made to the "mental spotlight" analogy, the key notion about covert attention is that it operates by selection on some spatial basis at and around a defined location, generally away from the fovea. This idea of covert attention has also proved very attractive to oculomotor workers. Many theoretical accounts of oculomotor phenomena, particularly of eye scanning patterns, involve covert attention as a start point (Morrison, 1984; Henderson, 1992; Reichle *et al.*, 1998).

Covert attention might thus appear to provide a bridge between workers in vision and those working on eye movements. However it seems necessary to ask a few questions to test the strength of the foundations of this bridge. Can the idea of a mental spotlight be used without a commitment to the Marrian approach to vision? In the last decade, the dominance of the Marrian viewpoint has been challenged. Questions have been raised about both the claimed computational advantages (Ballard, 1991) and about the epistemological underpinning (Churchland *et al.*, 1994). Alternative, more heuristic and piecemeal accounts of vision have emerged (e.g. Milner & Goodale, 1995). The phenomenon of change blindness has cast doubt on the richness of visual representation. Can we use the term visual attention without some concomitant assumptions about its characteristics, possible at an implicit level? Any metaphor, such as that of a mental spotlight, comes with a variety of associations and connotations. William James famously wrote "everyone knows what attention is" and so it might appear given the large number of workers ready to draw upon the concept. Attentional explanations have proliferated with only occasional analysis of the axioms and frequently little distinction made between the concepts of attention as a selective process and attention as a limited resource (see Allport, 1993 for a careful critique of the multiple uses of the term).

To take a specific example, the well known feature integration theory of Treisman and Gelade (1980) explains visual search functions in conjunction search by assuming that search items are scanned element by element by a hypothetical attentional pointer. Such a pointer would appear to need the following properties. First, it can be moved quickly enough to scan displays rapidly (rates of 30 ms/item are frequently proposed). Second, it can select locations with no constraint on spatial resolution. More recent analyses have cast doubt on both these assumptions. Alternative explanations of visual search functions have been offered not requiring rapid attention deployment (Eckstein, 1998). It has been shown that there are severe limits to the spatial resolution capacity of covert attention (Intriligator & Cavanagh, 2001). Studies of visual search with free eye movements have resulted in a very different account of the search process (Motter & Belky, 1998a, 1998b; Findlay & Gilchrist, 1998, 2001).

Since visual attention is such a pervasive theme, this commentary chapter will proceed by examining in detail how the theme of attention is used in the various contributions in the section and will then attempt to draw out some important conclusions. Not all the contributions make equal use of the concept and other themes will appear in the course of the review.

The paper by Godijn and Theeuwes makes a good starting point because these authors explicitly explore the relationship between saccades and attention. Godijn and Theeuwes begin their contribution with a lucid overview of some mainstream ideas. In an information rich visual environment, selection is important — achieved by overt or covert attention. Overt attention is attention achieved by redirecting the eyes in saccadic movements. Covert attention is also related to selection, but in this case the selection is made by mental processes that do not involve moving the eyes. Although covert attention might operate in a great variety of ways, the one most emphasized is the ability to attend to a location in space other than the one to which the foveal axis is directed. This emphasis was set in the pioneering work of Michael Posner (1980) and has been maintained. Godijn and Theeuwes go on to describe the established division between endogenous and exogenous attentional drivers and give particular consideration to the way in which attention can be indexed operationally.

In the case of visual spatial attention, the standard approach has long been through the observable effects produced by covert attention to a location. Reaction times to a probe appearing at the attended location are faster than those to a probe at an equivalent unattended location. Visual discrimination at an attended location is superior to that at an unattended location. A further operational approach concerns temporal order judgements (TOJ), with probes in an attended location appearing to be registered earlier than those at unattended ones (Stelmach & Herdman, 1991; Shore *et al.*, 2001). However, Godijn and Theeuwes highlight a problem that emerges when these three measures are used to enquire whether attention "moves" to the target location of a endogenously generated saccade prior to the eye movement. A clear pre-saccadic discrimination advantage has been found by several workers. No effect at all is found when TOJ is used. Results with probe detection tasks are equivocal. To resolve this dilemma, Godijn and Theeuwes suggest that it might be necessary to distinguish two different types of covert attention, selective and preparatory.

Godijn and Theeuwes continue with results looking at whether attention is "moved" prior to exogenous saccades, making use of the ingenious oculomotor capture paradigm (Theeuwes *et al.,* 1998). Their careful analysis of this situation leads them to describe four properties of attention (p. 14 with deliberate alteration of the order and addition of italic emphases) (i) attention is captured by onsets *unless* attention is already engaged, (ii) attention is required at the saccade destination in order to program a saccade (iii) erroneous saccades to onsets are only perceived when attention is directed *for sufficient time* to a location, (iv) exogenous shifts are faster than endogenous ones. These properties are presented as assumptions, required to avoid abandoning the proposition that attention is invariably captured by visual onsets. While there is no doubt that these conclusions have been reached by careful and ingenious experimental work, the attentional properties have become rather complex and some way removed from the position that "everyone knows what attention is". There seems a considerable danger of restating results rather than explaining them.

Godijn and Theeuwes finally discuss a competitive integration model, having acknowledged similarities with the saccadic generation framework proposed by Findlay and Walker (1999) but also important differences. The most significant is the proposal that there are no separate "fixate" and "move" centres. It does indeed seem potentially more elegant to restrict the rigid WHERE/WHEN separation to stage 1 of the Findlay and Walker model (based on the brainstem pause and burst system) and envisage descending projections emerging from a common map at Stage 2 and higher stages, albeit with very different representation of fovea and non-foveal regions projecting to the brainstem gate.

Attention receives only one short paragraph (p. 17) of specific discussion in the competitive integration model, although in a broader sense the whole model concerns attention. Attention selects targets in the saccadic map *and* feeds back into early visual areas (in the manner of the VAM approach of Schneider (1995) to enhance visual discrimination. Although not particularly emphasized, Godijn and Theeuwes also bring in the *object-based* nature of that attentional selection process.

It is very interesting to note a significant change in emphasis here. Now, the selection of saccade targets has become primary and other attentional effects secondary (as explicitly noted on p. 19). This account of attention is in line with the pre-motor theory of attention. Although seen as radical when first posed (Rizzolatti *et al.*, 1987), support for the pre-motor theory has grown. The question of whether it is attention that should be seen as primary or whether alternatively, it is the selection of the next saccade target, is a watershed one which relates to the active vision message discussed at the end of this chapter. In favour of the argument that covert attention should have primacy is the demonstration that covert attention can occur without overt. Since in many visual situations, scanning eye movements are the norm, it might be particularly worthwhile to give careful consideration to situations when covert attention is used on its own. The chapter by Kean and Lambert, which also explores the relationship between covert and overt attention shifts, starts by giving consideration to the question just posed.

Kean and Lambert discuss naturally occurring situations in which it is advantageous to use covert attention rather than overtly moving the eyes and thereby gaining the advantage of higher foveal resolution. The most convincing examples concern social situations where overt shifts would provide information that an individual wishes to conceal. Deception (Kean and Lambert's example is of competitive sports such as soccer where it may be desirable to mislead an opponent), social hierarchies (where it may be rude, or even dangerous, to gaze at the individual to whom you are attending) form examples of such situations. Workers who assign primacy to covert attention may not all be entirely happy with the suggestion that they are investigating a phenomenon of social psychology but otherwise it seems incumbent on them to provide an alternative teleology of the phenomenon.

Kean and Lambert's account of attention makes use of the distinction between endogenous and exogeneous processes. However they emphasize the distinction, originating from Posner (1980), of endogenous processes being consciously directed and exogeneous ones free of conscious control. Although this distinction overlaps that of internally driven versus external driven process, the two are logically and, as their paper demonstrates, experimentally, separable. They discuss presumed neural substrates for these processes and argue that evidence supports the allocation of the superior colliculus (SC) to the exogenous and the dorsolateral prefrontal cortex (DLPFC) to the endogenous. However, it seems unlikely that a full account will be possible without referring to the posterior parietal cortex also.

Kean and Lambert review several recent experiments using eye movements as an indicator of *covert* attention. This measure has various advantages over the more frequently used manual response time measure although oculomotor preparation must always be occurring. The interest is in contingent attention, continuing Lambert's previous work over several years. Observers are trained to use an informative peripheral cue to generate a saccade either to the cue's location or to the opposite side (very like the classic anti-saccade task). This anti-task shows fast contingent dependent orienting *away* from the cue and is followed by inhibition of return. Kean and Lambert argue that this shows a process dependent upon implicit learning and trace this notion to William James, who used the term "derived cueing". The demonstration that this process can operate so rapidly is highly important, offering, amongst other things, a possible account of the otherwise puzzling short latency required to generate antisaccades.

Kean and Lambert emphasize the speed of this derived cueing process and argue that it is most appropriately described as exogenous (rapid time course, unconscious, implicit). They relate their finding to that of Kentridge *et al.* (1999) who point to a further dissociation between consciousness and attention cueing by showing that such cueing may occur in a blindsight patient in the absence of conscious awareness of the visual stimuli. The Findlay and Walker (1999) model envisaged that most of the higher processes controlling saccade guidance were "automatized". Kean and Lambert have provided an elegant demonstration of one way in which such automatized processes can be studied.

Hodgson and Golding are interested in a similar theme whereby an explicit saccadic response is made on the basis of a previously learned and automated association. Hodgson and Golding describe a number of tasks where this occurs. One is a version of the time-honoured Wisconsin card-sorting paradigm. Participants see a screen displaying cards each containing a symbol of a particular shape and colour. One card is designated the "stimulus" and the task is to make a manual response selecting one of three "response" cards. The card to be selected is the one that matches the response card on the current search dimension (colour or shape). The ability to switch from one search dimension to the other is a traditional measure of executive function, and is known to deteriorate in patients with frontal lobe damage. Hodgson and Golding show, by recording eye scans as well as manual responses, that in the early trials following a dimension switch, normal individuals will often make incorrect saccades to the pre-switch dimension, even though no incorrect manual response is made. Even when a correct saccade is made, analysis of saccade curvature demonstrates an effect of the previously learned association.

The phenomenon is suggestive of the role of implicit learning in saccadic goals as shown in the Kean and Lambert work discussed above and also the phenomenon of "priming of pop-out" (Maljkovic & Nakayama, 1994, 1996, 2000) which can affect saccades in visual search tasks (McPeek *et al.*, 1999). This work demonstrates elegantly that competition is a feature of saccadic target selection. In this case, the competition is between automatised internal processes and higher executive control, rather than top-down and bottom-up processes. Hodgson and Golding discuss "biased competition" race models that could provide a plausible neural implementation of this competition. They suggest that "rather than an all-controlling homunculus, the present studies suggest that executive control processes are a sleeping partner to more automatic mechanisms which take over once environment-behaviour associations have been learned". Hodgson and Golding point out that the selection between alternative stimulus-response mappings can be considered as a further manifestation of attentional selection. This is almost the only mention of the term attention within the paper and no attentional processes independent of the saccadic selection process are envisaged.

Shen, Reingold, Pomplun and Williams consider interaction between what is happening centrally and the selection of the next saccade target. Selection of the saccade target in a search task is of course dependent on processing in the visual periphery. Two approaches make contrasting predictions. According to the "attentional resource" approach: if more attention is required in the centre, less is "available" for periphery; hence selection will be *worse* with a difficult foveal task. The work of

Henderson and Ferreira (1990) is often cited in support of such a resource model. They studied a reading situation and showed that a difficult foveal word resulted in longer fixations on average on the subsequent word, interpreted as indicating less peripheral preview. A contrasting possibility, not making reference to any capacity limitation, starts from the assumption that any perceptual discrimination will improve if more time is available to process the information. This might lead to the prediction that a difficult task will result in longer fixations, and in consequence, *better* peripheral selection.

Shen, Reingold, Pomplun and Williams carry out a number of experiments, using a task closely based on that of Hooge and Erkelens (1999) with a display of a large number of Landolt ring items. Landolt rings are rings containing small gaps: the search task involves finding a ring with no gap. Ring thickness was manipulated as a variable which affects peripheral discriminability: subjects were instructed about the thickness of the search target. The scanning statistics in the Shen, Reingold, Pomplun and Williams experiments showed the effectiveness of the manipulations. Fixation durations increased as the target-distractor similarity is decreased and likewise search selectivity was shown, with about 70% of saccades directed to high similarity targets and fewer than 10% to low similarity targets. Critically, this proportion was unaffected by the difficulty of the discrimination on the fixation before the saccade, although as just noted this difficulty did affect the fixation duration.

Shen, Reingold, Pomplun and Williams relate this result to a similar one found by Findlay *et al.* (2001) in which it was shown that, in a colour-shape conjunction search task, the probability of a scanning saccade landing on target was independent of the duration of the previous fixation. It should be noted though that these results stand in contrast to findings showing that saccadic selectivity is affected by previous foveal events. Hooge and Erkelens (1999) found with a very similar search task that a manipulation, affecting fixation duration, also affected selectivity, although the relationship was a non-linear one, mainly manifest for very short fixations. One other significant factor might be that one comparison in the Hooge and Erkelens study was made over a factor (Fat C width) that was blocked. Thus the modification of fixation duration might have resulted from long-term strategic factors rather than the moment to moment variability tested by Shen, Reingold, Pomplun and Williams and Findlay *et al.* (2001).

How might the null result of Shen, Reingold, Pomplun and Williams and of Findlay *et al.* be interpreted? As noted above, both accounts involving perceptual discriminability predict that processing at fixation will affect selectivity, but in the opposite direction. It might be that both processes operate and the effects cancel out although such a proposal is lacking in scientific elegance. A further possibility might be suggested by the finding of McPeek *et al.* (2000). They used a simple three-item search task with a saccade contingent manipulation so that a switch could be made between distractor and target location at the time of the first saccade. The task elicited frequent occurrences of a sequence where an incorrect first saccade to a distractor was followed by a correct second saccade to the location of the target at the time of the first saccade. Crucially, these sequences were not affected by the contingent change in the location of the target unless the duration of the fixation following the first saccade was abnormally long (> 250 ms). In the vast majority of cases, the sequence was followed even when this no longer brought the eye to the target. The implication must be that, at least in the situation investigated, parallel processing of two saccades (termed "pipelined" processing by McPeek *et al.*) is generally occurring. If such pipelining is a general feature of saccadic programming, one consequence might be a less tight relationship between saccadic selection and events on the previous fixation.

Walker and Doyle's paper brings a reminder that sensory information is not exclusively visual. As well as competing visual targets for saccadic orienting, auditory and tactile stimulation can also create candidate targets. Walker and Doyle review a number of studies of human saccadic orienting to simple combinations of visual and auditory targets. A latency facilitation effect occurs if targets are spatially congruent and an amplitude compromise (similar to the visual global effect) is also found. For remote candidate targets, one or other is accurately selected and latency is increased. These effects parallel those found with two competing visual stimuli and Walker and Doyle point out the relevance of the research programme carried out by Barry Stein. Stein (Stein & Meredith, 1993; Stein & Wallace, 1996) has examined intersensory cells in cat superior colliculus and has found a correspondence between their properties in relation to multimode competition with the result shown in behavioural testing.

Walker and Doyle present an experiment in which the effects of auditory and tactile distractors on saccades to visual targets were examined. Particular interest centred around the detailed trajectory of saccades. A number of studies have confirmed and extended the demonstration by Sheliga *et al.* (1994) that saccade trajectories can systematically reflect underlying processing mechanisms. Walker and Doyle demonstrate that auditory and tactile distractors can produce systematic changes of curvature, although these effects are quite small and modulated by unexplained directional differences. The fact that trajectory changes appear to be compatible with a maintained saccade target, i.e. are self-compensating, has long been a challenge to theories of saccade generation (Jürgens *et al.*, 1981).

The concern in the paper of Engbert and Kliegl paper is the miniature microsaccades that can be detected during the process of fixation. A topic of intense interest in the 1950s and 1960s, their study has not been fashionable in more recent years. Kowler and Steinman (1979) probably contributed to the demise by arguing that microsaccades serve no useful purpose, although this claim was strongly opposed by Ditchburn (1980; see also Kowler & Steinman, 1980). Part of the basis of the claim was the finding (Steinman *et al.*, 1967) that the frequency of microsaccades can be drastically reduced by instructions without obviously deleterious consequences for vision. A modern interpretation of this finding might be that high-level control can influence the low-level balance between fixate and move systems (Findlay & Walker, 1999). Recent years have witnessed something of a revival of interest in these small movements (Krauzlis *et al.*, 1997; Martinez-Conde *et al.*, 2000).

Engbert and Kliegl report an observational study in which the pattern of microsaccades in each eye was recorded. They report a new phenomenon, that some microsaccades are monocular. This is quite surprising since early work (Krauskopf *et al.*, 1960; Riggs & Niehl, 1960) reported that microsaccades always showed conjugacy. Indeed, Krauskopf *et al.* reported no exception in 4000 cases examined. Engbert and Kliegl used an SMI recording system with a claimed resolution of 20 sec arc, although a computerised algorithm appears to have been necessary to extract some of the microsaccades from the background signal. It will be desirable to obtain confirmation of the phenomenon of monocular microsaccades from other laboratories. Nevertheless Engbert and Kliegl make a cogent case for their existence, on the basis that monocular movements show different population properties from the binocular ones. In particular binocular microsaccades show a dominant horizontal direction whereas monocular ones have a preferred vertical one.

Engbert and Kliegl investigate a further interesting phenomenon in connection with microsaccades. In a visual task involving fixation of a display that makes a sudden change, a strong suppression of microsaccades occurs shortly (100–200 ms) after the change occurs followed by a rebound period where microsaccades occur with a greater probability than the baseline one. The phenomenon had been observed previously but has received little attention. It appears analogous to the inhibitory processes hypothesized to account for the remote distractor effect of Walker *et al.* (1997) as discussed in Findlay and Walker (1999). Engbert and Kliegl find that binocular microsaccades are particular prone to the suppression effect.

Finally, the paper by de'Sperati returns to the issue of internal representations with some fascinating observations of eye movements from subjects carrying out tasks of dynamic mental imagery. The task used was a conceptually straightforward "mental rotation" task involving a point target, which might move in hypothesized ways to various locations around the circumference of a real and visible circle. In the first task, observers were asked to "mentally localise" the point on the viewed circle that was a specified angle (e.g. 100 deg) from a presented visible point. Observers were free to move their eyes and, although not required by the task, the eyes did frequently move. The movements were, as might be expected, exclusively saccadic movements. Individual instances were highly idiosyncratic but averaged records show a progression of average gaze location around the circle. This progression starts with a delay of around 300 ms, accelerates rapidly to a constant velocity of about 100 deg/sec and has a slower deceleration phase. One interpretation might be that a "mental" locus moved smoothly around the perimeter of the circle and this had some effect on the overt eye movement but was filtered through the saccadic machinery.

De'Sperati follows up this work by using a procedure in which a spot rotating at a fixed rate disappears and the observer is again instructed to continue tracking mentally. Again the average eye movements follow the designated trajectory closely. He describes this procedure as being "to feed the oculomotor system with known mental inputs in order to see whether it is transparent to the inner process or whether distortion is introduced". It might parenthetically be noted that Wolfe *et al.* (2000) employed a very similar logic to test whether a rapid item-by-item voluntary attention scan could occur. The singular failure of their manipulation did not affect their adherence to the position of Treisman and Gelade (1980) that such rapid rates of attentional scanning can occur in the absence of volition.

De'Sperati scarcely mentions the a-word directly, although his paper might well be seen as addressing attentional processes. However his cautious approach does very powerfully address the concerns noted at the beginning of this commentary. A final instance may make this clear. The averaged traces demonstrated by the experimental work reported in de'Sperati suggest that some attentional locus moves smoothly and steadily over the mental trajectory. Why then doesn't the eye follow such a smooth trajectory? It is known that smooth eye movements can be made in pursuit of a visible moving target. The traditional view that retinal movement is needed to generate smooth pursuit is not an absolute barrier. There are a number of instances, as de'Sperati notes, showing that the oculomotor system *does* have access to a mental signal for smooth visual motion in the absence of any equivalent retinal motion.

Such a paradox fits more easily with the piecemeal view of mental representation than with the Marrian view. Although visual attention might appear to provide a point of interaction between visual information processing and eye movements, it has been the theme of this chapter that considerable caution is needed. In particular the monolithic view of a single visual representation upon which visual attention can operate is increasingly suspect. Oculomotor workers have much to contribute to the study of visual information processing and visual attention provides common ground with workers in the visual system tradition. However oculomotor workers should not accept uncritically ideas of vision and visual attention that have arisen from this tradition. In contrast, their knowledge might encourage them to adopt the perspective of active vision (Hayhoe, 2000; Findlay & Gilchrist, 2001, in press) in which the movements of the eyes are regarded as a primary and basic feature of vision.

References

Allport, D. A. (1993). Attention and control: Have we been asking the wrong questions? A critical review of twenty-five years. In D. E. Meyer and S. Kornblum (eds), *Attention and Performance XIV* (pp. 183–218). Cambridge, MA: MIT Press.

Ballard, D. H. (1991). Animate vision. *Artificial Intelligence*, *48*, 57–86.

Churchland, P. S., Ramachandran, V. S., & Sejnowski, T. J. (1994). A critique of pure vision. In: C. Koch and J. L. Davis (eds), *Large Scale Neuronal Theories of the Brain* (pp. 23–60). Cambridge, MA: MIT Press.

DeValois, R., & DeValois, K. (1980). Spatial Vision. *Annual Review of Psychology*, *31*, 309–341.

Ditchburn, R. W. (1980). The function of small saccades. *Vision Research*, *20*, 271–272.

Eckstein, M. P. (1998). The lower visual search efficiency for conjunctions is due to noise and not serial attentional processing. *Psychological Science*, *9*, 111–118.

Eriksen, C. W., & Hoffman, J. E. (1972). Temporal and spatial characteristics of selective encoding from visual displays. *Perception and Psychophysics*, *12*, 201–204.

Findlay, J. M., Brown, V., & Gilchrist, I. D. (2001). Saccade target selection in visual search: The effect of information from the previous fixation. *Vision Research*, *41*, 87–95.

Findlay, J. M., & Gilchrist, I. D. (1998). Eye guidance and visual search. In: G. Underwood (ed.), *Eye Guidance in Reading and Scene Perception* (pp. 295–312). Amsterdam: Elsevier.

Findlay, J. M., & Gilchrist, I. D. (2001). Visual attention: The active vision perspective. In: M. Jenkin and L. R. Harris (eds) *Vision and Attention* (pp. 83–103). New York: Springer-Verlag.

Findlay, J. M., & Gilchrist, I. D. (in press). *Active Vision: The Psychology of Looking and Seeing.* Oxford University Press. Oxford Psychology Series.

Findlay, J. M., & Walker, R. (1999). A model of saccadic eye movement generation based on parallel processing and competitive inhibition. *Behavioral and Brain Sciences*, *22*, 661–721.

Groner, R., Menz, C., Fisher, D. F., & Monty, R. A. (1983). *Eye Movements and Psychological Functions: International Views*. Hillsdale NJ: Lawrence Erlbaum Associates.
Hayhoe, M. (2000). Vision using routines: A functional account of vision. *Visual Cognition, 7*, 43–64.
Henderson, J. M. (1992). Visual attention and eye movement control during reading and picture viewing. In: K. Rayner (ed.), *Eye Movements and Visual Cognition* (pp. 260–283). New York: Springer-Verlag.
Henderson, J. M., & Ferreira, F. (1990). Effects of foveal processing difficulty on the perceptual span in reading: Implications for attention and eye movement control. *Journal of Experimental Psychology: Learning, Memory and Cognition, 16*, 417–429.
Hooge, I. T. C., & Erkelens, C. J. (1999). Peripheral vision and oculomotor control during visual search. *Vision Research, 39*, 1567–1575.
Intriligator, J., & Cavanagh, P. (2001). The spatial resolution of visual attention. *Cognitive Psychology, 43*, 171–216.
Jürgens, R., Becker, W., & Kornhuber, H. H. (1981). Natural and drug-induced variations of velocity and duration of human saccadic eye movements: Evidence for a control of the neural pulse generator by local feedback. *Biological Cybernetics, 39*, 87–96.
Kentridge, R. W., Heywood, C. A., & Weiskrantz, L. (1999). Attention without awareness in blindsight. *Proceedings of the Royal Society of London, Series B, 266*, 1805–1811.
Kowler, E. (1990). The role of visual and cognitive processes in the control of eye movement. In: E. Kowler (ed.), *Eye Movements and Their Role in Visual and Cognitive Processes* (pp.1–70). Amsterdam: Elsevier/North-Holland.
Kowler, E., & Steinman, R. M. (1979). Miniature saccades: Eye movements that do not count. *Vision Research, 19*, 105–108.
Kowler, E., & Steinman, R. M. (1980). Small saccades serve no useful purpose: Reply to a letter by R. W. Ditchburn. *Vision Research, 20*, 273–276.
Krauskopf, J., Cornsweet, T. N., & Riggs, L. A. (1960). An analysis of eye movements during monocular and binocular fixation. *Journal of the Optical Society of America, 50*, 572–578.
Krauzlis, R. J., Basso, M. A., & Wurtz, R. H. (1997). Shared motor error for multiple eye movements. *Science, 276*, 1693–1695.
Lévy-Schoen, A., & O'Regan, J. K. (1980). The control of eye movements in reading. In: P. A. Kolers, M. E. Wrolstad and H. Bouma (eds), *Processing of Visible Language* (pp. 7–36). New York: Plenum Press.
Maljkovic, V., & Nakayama, K. (1994). Priming of pop-out 1. Role of features. *Memory & Cognition, 22*, 657–672.
Maljkovic, V., & Nakayama, K. (1996). Priming of pop-out 2. The role of position. *Perception and Psychophysics, 58*, 977–991.
Maljkovic, V., & Nakayama, K. (2000). Priming of pop-out 3. A short-term implicit memory system beneficial for rapid target selection. *Visual Cognition, 7*, 571–595.
Marr, D. (1982). *Vision*. San Francisco: W. H. Freeman.
Martinez-Conde, S., Macknik, S. L., & Hubel, D. H. (2000). Microsaccadic eye movements and firing of single cells in the striate cortex of macaque monkeys. *Nature Neuroscience, 3*, 251–258.
McConkie, G. W., & Rayner, K. (1975). The span of the effective stimulus during a fixation in reading. *Perception and Psychophysics, 17*, 578–586.
McPeek, R. M., Maljkovic, V., & Nakayama, K. (1999). Saccades require focal attention and are facilitated by a short-term memory system. *Vision Research, 39*, 1555–1566.
McPeek, R. M., Skavenski, A. A., & Nakayama, K. (2000). Concurrent processing of saccades in visual search. *Vision Research, 40*, 2499–2516.

Milner, D. A., & Goodale, M. A. (1995). *The Visual Brain in Action*. Oxford: Oxford University Press.

Morrison, R. E. (1984). Manipulation of stimulus onset delay in reading: Evidence for parallel programming of saccades. *Journal of Experimental Psychology, Human Perception and Performance*, *5*, 667–682.

Motter, B. C., & Belky, E. J. (1998a). The zone of focal attention during active visual search. *Vision Research*, *38*, 1007–1022.

Motter, B. C., & Belky, E. J. (1998b). The guidance of eye movements during active visual search. *Vision Research*, *38*, 1805–1818.

Palmer, S. E. (1999). *Vision Science: Photons to Phenomenology*. Cambridge MA: MIT Press.

Posner, M. I. (1980). Orienting of attention. *Quarterly Journal of Experimental Psychology*, *32*, 3–25.

Posner, M. I., Nissen, M. J., & Ogden, M. C. (1978). Attended and unattended processing modes: The role of set for spatial location. In: H. L. Pick and I. J. Saltzman (eds), *Modes of Perceiving and Processing Information* (pp. 137–157). Hillsdale NJ: Lawrence Erlbaum.

Rayner, K. (1975). The perceptual span and peripheral cues in reading. *Cognitive Psychology*, *7*, 65–81.

Regan, D. (2000). *Human Perception of Objects: Early Visual Processing of Spatial Form Defined by Luminance, Color, Texture, Motion, and Binocular Disparity*. Sunderland MA: Sinauer Associates.

Reichle, E. D., Pollatsek, A., Fisher, D. F., & Rayner, K. (1998). Toward a model of eye movement control in reading. *Psychological Review*, *105*, 125–147.

Riggs, L. A., & Niehl, E. W. (1960). Eye movements recorded during convergence and divergence. *Journal of the Optical Society of America*, *50*, 913–920.

Rizzolatti, G., Riggio, L., Dascola, I., & Umiltà, C. (1987). Reorienting attention across the horizontal and vertical meridians: Evidence in favor of a premotor theory of attention. *Neuropsychologia*, 25, 31–40.

Robinson, D. A. (1975). Oculomotor control signals. In: G. Lennestrand and P. Bach-y-Rita (eds), *Basic Mechanisms of Ocular Motility and their Clinical Applications* (pp. 337–374). Oxford: Pergamon Press.

Schneider, W. X. (1995). VAM: Neuro-cognitive model for visual attention control of segmentation, object recognition, and space-based motor action. *Visual Cognition*, *2*, 331–375.

Sheliga, B. M., Riggio, L., & Rizzolatti, G. (1994). Orienting of attention and eye movements. *Experimental Brain Research*, *98*, 507–522.

Shore, D. L., Spence, C., & Klein, R. M. (2001). Visual prior entry. *Psychological Science*, *12*, 205–212.

Stark, L. (1968). *Neurological Control Systems*. New York: Plenum Press.

Stein, B. E., & Meredith, M. A. (1993). *The Merging of the Senses*. Cambridge MA: MIT Press

Stein, B. E., & Wallace, M. T. (1996). Comparisons of cross-modality integration in midbrain and cortex. *Progress in Brain Research*, *112*, 289–299.

Steinman, R. M. (1986). The need for an eclectic, rather than a systems, approach, to the study of the primate oculomotor system. *Vision Research*, *26*, 101–112.

Steinman, R. M., Cunitz, R. J., Timberlake, G. T., & Herman, M. (1967). Voluntary control of microsaccades during maintained monocular fixation. *Science*, *155*, 1577–1579.

Stelmach, L. M., & Herdman, C. M. (1991). Directed attention and perception of temporal order. *Journal of Experimental Psychology, Human Perception and Performance*, *17*, 539–550.

Theeuwes, J., Kramer, A. F., Hahn, S., & Irwin, D. E. (1998). Our eyes do not always go where we want them to go. *Psychological Science*, *9*, 379–385.

Treisman, A., & Gelade, G. (1980). A feature integration theory of attention. *Cognitive Psychology*, *12*, 97–136.

Walker, R., Deubel, H., Schneider, W. X., & Findlay, J. M. (1997). The effect of remote distractors on saccade programming: Evidence for an extended fixation zone. *Journal of Neurophysiology*, *78*, 1108–1119.

Wandell, B. A. (1995). *Foundations of Vision.* Sunderland MA: Sinauer Associates.

Wolfe, J. M., Alvarez, G. A., & Horowitz, T. S. (2000). Attention is fast but volition is slow. *Nature*, *406*, 691.

Section 2

Eye Movements in Reading and Language Processing

Section Editor: Jukka Hyönä

Chapter 8

Where Do Chinese Readers Send Their Eyes?

Jie-Li Tsai and George W. McConkie

Most theories of eye movement control during reading assume that this control is word-based. Words are prominent perceptual units in most alphabetic writing systems. However, in Chinese text there is no perceptual indicator of where words begin and end; rather, the perceptually prominent units are characters. This chapter considers the question of whether the eyes are being sent to words or to characters when reading Chinese. Properties of both words and characters affect the likelihood of landing on a given character. Unlike alphabetic languages, there is no tendency for the eyes to land more frequently at the centers of Chinese words, but neither is there a tendency to land more frequently at the centers of characters. We argue against word-based control, and leave open the possibility of character-based control in spite of a lack of positive evidence.

Introduction

When reading a passage, our eyes move to different locations in the text in order to obtain the necessary information for comprehension. Eye behavior in reading is composed of successive fixations and saccadic movements. Several decades of studies have indicated that these movement patterns reflect, to some extent, the mental processes taking place (Just & Carpenter, 1980; Rayner, 1998). Although eye movement data are complex, at a basic level they result from just two types of decisions: *where* to move the eyes, and *when* to move them. This results in two basic measures, the duration of each eye fixation and the direction and length of each saccade. These are usually assumed to reflect cognitive and perceptual processes (Inhoff & Radach, 1998; Rayner & Pollatsek, 1981).

The Mind's Eye: Cognitive and Applied Aspects of Eye Movement Research

ISBN: 0–444–51020–6

The *where* decision in reading is generally considered to be made on a word-unit basis, selecting a word to send the eyes to next (McConkie, Kerr, Reddix & Zola, 1988; Reichle, Rayner & Pollatsek, 1999; Reilly & O'Regan, 1998). In most alphabetic writing systems, words are perceptually salient since they consist of closely-packed strings of letters separated by spaces. The most common location for the eyes to go in reading is near the center of the word, referred to as the Preferred Viewing Position (Rayner, 1979). One of the interesting characteristics of the Chinese writing system, however, is that there is no visual indicator of where words begin and end. Initial evidence suggests that there is not a Preferred Viewing Position at the centers of words in the reading of Chinese as there is with alphabetic languages (Yang & McConkie, 1999). The research reported in this chapter was an attempt to replicate this initial observation and to further explore the issue of where the readers' eyes tend to be sent as they read Chinese text, in particular, whether the Chinese readers' eyes are being sent to words or to characters.

The chapter begins by describing some features of the Chinese writing system that might relate to the decision of where to send the eyes during reading. It then summarizes some findings regarding the role of word units in selecting the next landing position when reading alphabetic text. Eye movement data from reading Chinese are then analyzed to examine conditions under which several factors affect where the eyes are sent. A final section discusses the implications for theories of eye movement control during reading.

Chinese Characters and Words

Modern written Chinese usually consists of horizontally-arrayed strings of characters, going from left to right (vertically-arrayed writing is also possible). Chinese characters can be regarded as the perceptual unit of the Chinese written system (Hoosain, 1991) because of its spatial structure and language function. Each character occupies a rectangular region of the same size, and characters are separated by space of an equal size, whether the characters are part of the same word or different words. The structure of a character can be further decomposed into component radicals, or even further into a series of individual strokes, for more detailed representations (DeFrancis, 1989; Hung & Tzeng, 1981). Since these radicals and sets of strokes are all contained within a compact area, separated by space from adjacent characters, they have the appearance of integrated, isolated visual objects. These characters vary greatly in their apparent complexity, which can be roughly indexed by the number of strokes required to write them, ranging from one to more than thirty. Another attribute of Chinese characters is that they are the written unit of the spoken language and usually map onto morphemes and syllables in the modern system. These attributes of characters suggest that the characters should be the basic unit of reading.

However, in the development of the Chinese language, there were not enough characters to represent the increasing number of new concepts. Instead of inventing new characters, many new words were created by compounding existing characters, representing the development of multi-syllabic words. As a result, some Chinese words are monomorphemic (one character) and others are polymorphemic (two or more characters, often referred to as compound words). According to the Chinese word corpus of Academia Sinica Taiwan (1998), only 9% of the words (types) found in continuous text are single characters while over 76% consist of two or three characters. That is, in a Chinese sentence, the great majority of the characters are actually constituents of compound words rather than being individual words. To make things more complex, many of these characters can stand alone as individual words when used in that manner. Furthermore, for compound words, the meanings of the constituent characters can be transparent or opaque to the word meaning. Transparent words are those whose meaning is quite obvious from the meanings of its constituent characters (e.g., the characters for "mind" and "logic" together mean "psychology"); opaque words are those for whom the meaning is not apparent from the meanings of the characters that make it up (e.g., the characters for "flower" and "growth" together mean "peanut"). Thus, while characters have meanings, their relationships to the meanings of words containing them are often not apparent.

The focus of the current study is on whether Chinese readers tend to send their eyes to words or characters. There are good arguments for either of these alternatives. Word meanings are, of course, more critical to the comprehension of the text than are character meanings. Alphabetic languages show a preferred viewing position on words and a refixation pattern that varies with the initial location of the eyes on the word, with fewer refixations if this fixation is toward the center of the word than if it is toward either end (O'Regan & Jacobs, 1992; McConkie, Kerr, Reddix, Zola & Jacobs, 1989). Words are identified faster if fixated at their center, from which O'Regan (1990) has argued that readers learn to send their eyes to this optimal location.

Of course, this raises the question of how the text string is segmented into words during reading. One possibility is that this segmentation can be accomplished visually from the overall configuration of compound words. Inhoff and Liu (1998) have found that Chinese readers, during their eye fixations, acquire information not only from the character to which the gaze is directed, but also from as much as two characters to the right. If this is the case, then at least at times it would be logically possible for the reader to perceive two-character words (the most common length in Chinese) to the right of the fixated character, which could then serve as the target for the next saccade. Although this may be possible for high frequency words, there are substantial difficulties in such a process. As noted above, most characters can also stand alone as words, or can combine with other characters to create additional words of varying lengths. Thus, the correct parsing of the characters into words is not immediately obvious and is fraught with the possibility of ambiguity and garden pathing (e.g., initially accepting one local parsing, only to find that further parsing becomes impossible, requiring a reparsing of the earlier area). Perceptually, words in a line of text still look like aligned characters rather than like distinctive wholes. Thus, the visual configuration of words is not as salient as that of characters as a basis for providing objects to which attention or saccades might be directed. The reading of alphabetic text is slowed when spaces between words are removed, and the preferred viewing locations are shifted to the beginnings of words (Rayner, Fischer & Pollatsek, 1998). This latter phenomenon may result from a tendency to send the eyes beyond the last parsed word (hence, to the beginning of the next word) when the boundaries of the next word cannot be located. Interestingly, adding spaces to Chinese text as a way of marking the locations of words does not increase reading speed (Liu, Yeh, Wang & Chang, 1974).

The alternative is that the eyes are being sent to selected characters. Chinese characters are perceptually distinct, and would be the natural units resulting from any preattentive parsing of the visual array. Thus, it seems quite likely that characters are the units by which Chinese text is initially attended, thereby serving as targets for saccades during reading. At the same time, if words are the primary units for developing textual meaning, guiding the eyes on the basis of characters would not seem to be an optimal strategy from a language processing perspective.

In summary, there are benefits and costs associated with each solution to the problem of selecting units to serve as saccade targets during reading of Chinese. It would be interesting to know whether the system selects a psycholinguistically-favored solution, the word, or a perceptually-favored solution, the character.

Sources of Evidence for Saccade Target Units

Three types of evidence can be used to help select among alternative possible units as the basis for oculomotor control in reading. The first concerns whether features of a given unit affect saccade decisions. In the present case, it is possible to examine whether the frequency of use of characters or of words influences the likelihood of the eyes being sent to selected characters. In English, studies have shown that higher frequency words are more likely to be skipped by the eyes during reading (Ehrlich & Rayner, 1981). In Chinese, the question is whether the frequencies of words or characters affect these choices.

The second source of evidence concerns the tendency of the eyes to go to the centers of words, the preferred viewing position (Rayner, 1979), in reading alphabetic languages. This tendency has been used to argue that saccades are being sent to words. It would be expected that there would be a similar Preferred Viewing Position Curve in Chinese reading, for the type of unit to which the eyes are being sent. Thus, if there were a tendency for the first fixation on characters or words to be near their center, this would serve as evidence for that unit serving as the saccade target.

In an earlier attempt to employ these first two methods, Yang and McConkie (1999) found no evidence for a preferred viewing position on either words or characters in Chinese, but at the same time observed word frequency effects on the likelihood of skipping characters and on the likelihood of refixating words.

The third source of evidence can come from a more fine-grained examination of where the eyes land in words and characters. McConkie *et al.* (1988) analyzed a large set of eye movement data from a normal reading task, examining the effects of three factors on the landing positions: word length, launch distance (distance from the location of prior fixation), and the duration of the prior fixation. There were several findings in the results. First, given a constant launch distance and word length, the landing sites in a word appeared to be normally distributed. These were called launch-site contingent landing position distributions. Second, the most frequently fixated location varied with the launch distance. When the launch site was further to the left of the word, the preferred landing site shifted to the left and the variance became larger. Third, the word length had very little effect on the means of the landing distributions or their variability. This finding implies a tendency to send the eyes to a "center of gravity" within the word.

McConkie *et al.* (1988) concluded that a functional target location, which lies near the center of a word, is the true target for saccades. However, the actual initial landing position within a word is determined by principles of oculomotor control. They further suggested that the preferred viewing position curve observed by Rayner (1979) is actually a weighted average across many launch-site contingent landing position distributions. The critical point for current purposes is that a more fine-grained examination of where the eyes tend to land during reading, and particularly of launch-site contingent landing position distributions, can give information about which units are being selected as saccade targets.

In the study described below, all three of these methods were employed in an attempt to determine whether the Chinese reader's eye movements are being sent to characters or to words.

Methodology

Eye movement data were collected from 18 college students in the National Yang-Ming University of Taiwan as they read four newspaper stories (3360 characters, total) selected from the Academia Sinica Balanced Corpus (1998). Only one line of text was shown at a time for each passage and the subject pressed a button to show the next line, which appeared at the same screen position as the previous line. Each line except for the last in the passage consisted of 25 characters. The size of a character shown on the screen was 24 × 24 pixels, with a space of 8 pixels between characters. The viewing distance is 70 centimeters and the width including a character and the space before it subtended 1 degree of visual angle.

There were forty-one lines of text in each passage and three comprehension questions were administered when subjects had finished reading each passage. The eye movements were recorded by an EYELINK eye-tracking system, sampling eye position at 250 samples/sec. Eleven percent of the fixations were excluded from the analysis, including fixations containing blinks and those that landed on the first or last two characters of each line of text. This yielded a total of 29,431 eye fixations.

Several character and word variables were used as predictors of where the eyes would land. The first was the visual complexity of characters, indexed by the number of strokes that it contained. The average number of strokes for these characters was 9.45. The proportions of stroke counts in the range of 1 to 10, 11 to 20, and above 20, were 63.31%, 34.66%, and 2.03%, respectively. The second variable was the frequency of each character in written Chinese, as estimated from the Academia Sinica Balanced Corpus (1998). The proportions of characters having frequencies in the range of less than 10^2 per 10 million characters, 10^2 to 10^3, 10^3 to 10^4, and greater than 10^4 per 10 million, were 0.23%, 4.25%, 27.96%, and 67.56%, respectively. A third variable was word frequency. There were 1795 words (tokens) in the reading material. The proportions of word frequencies in the range of less than 10^2 per million words, 10^2 to 10^3, 10^3 to 10^4, and larger than 10^4 per million words are 14.60%, 20.28%, 34.43%, and 30.70%, respectively. The fourth variable was word length. The proportions of words (tokens) with lengths of 1, 2, 3 and more than 3 characters, were 43.34%, 49.47%, 6.74%, and 0.45%, respectively.

Attractiveness of Character and Word Attributes

The first set of analyses investigated whether the character frequency or the frequency of the word containing the character is a better predictor of the frequency with which the eyes actually land on that character position, referred to as the "attractiveness" of that character. Separate analyses were conducted for characters lying 1, 2, 3, 4 or 5 positions to the right of the currently fixated character. A logistic regression model was used to estimate the odds ratios for the probability of the binary landing response (land or not land) being affected by character and word frequencies of the candidate character. The model can be summarized by the following conceptual expression:

$$\text{Landing Response} = f(\log(\text{character frequency}), \log(\text{word frequency}))$$

For the analyses, character and word frequencies were each transformed to the logarithm of the frequency count. The landing response was coded as 1 to indicate that the character position being examined was where the next fixation landed, or 0 otherwise. Five separate analyses were conducted, one each for characters lying different distances to the right of the currently fixated word. This allowed us to examine whether these variables affected the attractiveness of characters lying at different distances into the visual periphery. The data were excluded if the character at the target position was a single character word, because the correlation between character and word frequency is too high (about 0.97) for these cases. The attractiveness of a character is expected to be inversely related to the two predictors: higher frequency units should be less frequently fixated (Ehrlich & Rayner, 1981).

Table 8.1 presents the results of the five analyses, examining the frequencies of the eyes going to characters $N+1$ (the next character to the right of the currently fixated character), through $N+5$ (the fifth character to the right). The results indicate that both word frequency and character frequency affect the attractiveness of nearby characters while only character frequency affects more distant characters. A significant effect of word frequency is observed only for characters $N+1$ and $N+2$, while that for character frequency is observed for characters $N+1$ through $N+4$. Neither has an effect for character $N+5$. For those data sets where both word and letter frequency produce an effect ($N+1$ and $N+2$), we further conducted regression models for each single predictor. In the $N+1$ data set, the McFadden's-R^2s for the models with character frequency alone, word frequency alone, and both predictors were 0.0064, 0.0104, and 0.0116, respectively. In the $N+2$ data set, the McFadden's-R^2s were 0.0026, 0.0050, 0.0053,

Table 8.1: Logistic models for predicting the likelihood of the eyes' landing on different characters to the right of the currently fixated character (character *N*). Predictors include character and word frequencies.

		Character Position to the Right of the *N*th Fixation				
		***N* + 1**	***N* + 2**	***N* + 3**	***N* + 4**	***N* + 5**
Constant	Coefficient	–0.8687	–0.2469	–0.8162	–1.4040	–2.3296
	t-ratio	–5.2559	–1.9381	–5.9753	–8.2270	–10.4689
	p-value	(0.0000)	(0.0526)	(0.0000)	(0.0000)	(0.0000)
Character frequency	Coefficient	–0.1521	–0.0729	–0.0779	–0.1160	0.0066
	t-ratio	–3.4432	–2.1419	–2.1145	–2.4951	0.1105
	p-value	(0.0006)	(0.0322)	(0.0345)	(0.0126)	(0.9120)
Word frequency	Coefficient	–0.1471	–0.1001	0.0031	0.0358	–0.0064
	t-ratio	–7.3140	–6.4882	0.1790	1.6178	–0.2305
	p-value	(0.0000)	(0.0000)	(0.8579)	(0.1057)	(0.8177)
Chi-square (df = 2)		115.3404	82.6029	5.6098	6.3221	0.0533
p-value		(0.0000)	(0.0000)	(0.0605)	(0.0424)	(0.9737)
McFadden's-R^2		0.0116	0.0053	0.0004	0.0007	0.0000
Total *N*		12802	12637	12689	11767	10996
Landing response		1678	3834	3091	1688	984
No landing response		11124	8803	9598	10079	10012

respectively. These results indicate that word frequency made the greatest contribution to the attractiveness of the two character positions nearest the currently fixated character, though the amount of variance accounted for was quite low. Of course, only the effect of character frequency was significant for characters *N*+3 and *N*+4.

Thus, there were two main findings in the analyses of logistic regression models. First, the character frequency consistently influenced the landing response from the first to fourth character position to the right of the previous fixation. Second, the word frequency is more crucial than the character frequency when the predicted position is near to (i.e., within two character positions of) the previous fixation.

Effect of Character Complexity

The analyses just presented have shown that character frequency influences the attractiveness of characters in the competition for the next saccade. However, character frequency is known to be negatively correlated with character complexity, as indexed by the number of strokes it contains (Chan, 1982, cited by Hoosain, 1991). In the text used in the present study, the correlation between these variables is –0.43 (there is also a –0.18 correlation between the complexity of a character and the frequency of the word it is in). As Hoosain (1991) points out, this may be an example of a tendency to simplify characters that are written more often, similar to a tendency for more frequent words in English to be shorter (Zipf, 1949). In any case, since more complex characters are less well identified in the visual periphery (Yang, 1994), there is a need to position the eyes closer to them in order to identify them. As a result, character complexity may add to the attractiveness of characters in the competition for the next saccade. Alternatively, it is possible that, since less complex characters can be identified more peripherally, a reduction in attractiveness for already-identified characters could also lead to a tendency for reduced attractiveness for simpler characters. The question to be addressed here is whether this prediction of a relationship between character complexity and attractiveness is accurate, and, if so, whether character complexity might account for the relationship that was observed above between character frequency and character attractiveness.

Using the same analysis approach as in the previous section, character complexity was added together with character and word frequency as predictors of the landing response (i.e., likelihood of sending the eyes to that character) for characters at the *N*+1 to *N*+5 position to the right of the fixation. The results, presented in Table 8.2, show that when character complexity is added into the model, the character frequency has little or no influence on the landing response, except perhaps at position *N*+1.

Table 8.2: Logistic models for predicting the likelihood of the eyes' landing on different characters to the right of the currently fixated character (character *N*). Predictors include stroke count, character frequency and word frequency.

		Character Position to the Right of the *N*th Fixation				
		***N* + 1**	***N* + 2**	***N* + 3**	***N* + 4**	***N* + 5**
Constant	Coefficient	–1.2767	–0.5310	–1.2007	–1.5394	–2.3780
	t-ratio	–5.9387	–3.2614	–6.9007	–7.0051	–8.4240
	p-value	(0.0000)	(0.0011)	(0.0000)	(0.0000)	(0.0000)
Stroke counts	Coefficient	0.0189	0.0133	0.0182	0.0064	0.0023
	t-ratio	3.0344	2.8226	3.6282	0.9852	0.2795
	p-value	(0.0024)	(0.0048)	(0.0003)	(0.3245)	(0.7798)
Character frequency	Coefficient	–0.0974	–0.0340	–0.0259	–0.0978	0.0131
	t-ratio	–2.0243	–0.9212	–0.6512	–1.9492	0.2035
	p-value	(0.0429)	(0.3570)	(0.5149)	(0.0513)	(0.8388)
Word frequency	Coefficient	–0.1496	–0.1025	0.0002	0.0351	–0.0067
	t-ratio	–7.4165	–6.6234	0.0092	1.5821	–0.2412
	p-value	(0.0000)	(0.0000)	(0.9927)	(0.1136)	(0.8094)
Chi-square (df = 3)		124.4407	90.5415	18.6849	7.2892	0.1313
p-value		(0.0000)	(0.0000)	(0.0003)	(0.0632)	(0.9878)
McFadden's-R^2		0.0125	0.0058	0.0013	0.0008	0.0000

As before, word frequency has the main contribution to these models for characters *N*+1 and *N*+2 but character complexity still produces an effect for character *N*+3.

In summary, the results of these analyses have shown that both word-level and character-level factors have an influence on where the eyes are sent during reading. The cultural frequency of a word (i.e., the frequency with which it occurs in the language) affects the attractiveness of its component characters, but only for characters near the fixated character (i.e., the adjacent or next-to-adjacent character). Characters in higher frequency words have reduced attractiveness. These and more distant characters, up to four away from the current character, are affected by character complexity, which appears to be a primary basis for character frequency effects. Higher frequency and lower complexity reduce a character's attractiveness. This suggests that near-foveal words are being distinguished on some cognitive basis, since there are no strictly visual indicators available within the text string to isolate them, whereas complexity, a more perceptual factor, also influences character attractiveness both at these locations (less strongly than word frequency) and more peripherally where word frequency has no effect. Of course, there are many possible reasons for the word frequency influences, including facilitated perception and greater identification likelihood for higher-frequency words, differences in word predictability from the preceding context, and differences in predictability of the second character of a word from the first. Similarly, the character frequency/complexity effect may be due either to a general tendency for more complex characters to have greater attractiveness, or for less complex characters to be more likely to be identified at more peripheral locations, with identification leading to reduced attractiveness. Further research is needed to investigate the sources of these effects.

First Landing Position on a Character and a Word

As noted above, the existence of a preferred viewing position in English words has been used to argue for the word as being the unit to which saccades are sent. This suggests a second approach to determining whether saccades are being sent to characters or words in Chinese: we should observe a preferred viewing position, usually located near the center of the unit that is serving as the saccade target, though in studies of reading text without spaces it has been located nearer to the beginning of the word (Kajii, Nazir & Osaka, 2001; Rayner *et al.*, 1998). If the saccade target is a character, then we would expect to find a preferred viewing position within the space of the character, with fewer fixations in the space between characters; if it is a word then we would expect to find a preferred viewing position within the word, probably near its center even though that might be a space between the characters, and fewer fixations on the spaces between words.

In order to compare the Chinese preferred landing position data with that of English, a set of eye movement data from English readers was also analyzed, using a common spatial metric. As noted above, the width of each Chinese character including the space before it was one degree of visual angle. Similarly, each letter and space in English text that was previously read by a group of 66 adult English speakers, and used by

McConkie, *et al.* (1988) for the analysis of eye movements during reading, occupied 0.25 deg of visual angle. Thus, for purposes of comparison, each Chinese character, including the space before it, was divided horizontally into four units, with the width of each unit being 8 pixels and subtending about 0.25 degree of visual angle, equivalent to one letter position in the English text. The position of the space before a character was numbered as 0, and the three units in a character were numbered from 1 to 3 sequentially. For a Chinese two-character word, the position of the space before the first character was numbered as 0, and the remaining units were numbered from 1 to 7. In English text, the space before the word was numbered 0 and the letters were then numbered in order, going from left to right. Therefore, the width of a Chinese character was about the same as the length of a three-letter English word, and the width of a Chinese two-character word was about the same as a seven-letter English word.[1]

Four sets of data were used to examine the preferred landing positions in Chinese and English: the set of all first eye fixations on a 3-letter English word that did not have preceding or following punctuation (*N*=4165), all first eye fixations on seven-letter English words that did not have preceding or following punctuation (*N*=2281), all first eye fixations on a Chinese character (*N*=17,460), and all first eye fixations on a two-character Chinese word (*N*=9809). Fixations on the first and last two characters on each line were excluded in Chinese data, as were fixations on the initial and final words on a line in the English data, and all fixations preceded by a regressive saccade.

Figure 8.1 shows the proportions of initial fixations on a word that landed on each of the unit locations in each of these four sets of data. These are referred to as Preferred Viewing Position Curves. The Preferred Viewing Position Curves found with English data are in agreement with those in previous studies (McConkie *et al.*, 1988; Rayner, 1979). While the most common landing position in 3-letter words is at the end, the middle letter of 7-letter words is the most common landing position. However, the Chinese results are different from English. The Preferred Viewing Position Curves for both Chinese characters and words are flatter than those for corresponding-length English words, and neither showed evidence of a clear preferred viewing position. While Chinese readers show some tendency to fixate more on the center than the ends of the word, this difference is very small. In addition there is no tendency to have fewer fixations on the space between characters in a 2-character word (unit 4) as would be expected if saccades were being sent to characters. There is no evidence for an identifiable preferred landing position for either characters or words in the reading of Chinese.

The Influence of Launch Site

The work of McConkie *et al.* (1988) suggests two possible reasons for the failure to find a preferred landing position in the reading of Chinese. First, they point out that where the eyes land is not only a result of the word selected as the target for the saccade, but also of a type of error referred to as an oculomotor range effect (Kapoula, 1985; McConkie *et al.*, 1988), an error that varies with the distance of the launch site of the saccade from its target. There is a tendency to overshoot near objects and

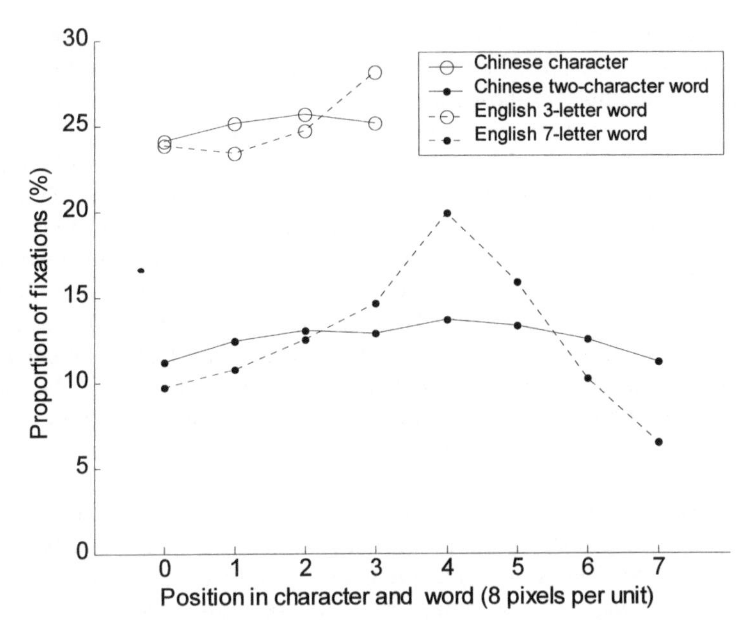

Figure 8.1: Preferred viewing position curves: Proportions of eye fixations that initially land at different positions within Chinese characters and two-character words, and within English three-letter and seven-letter words.

undershoot more distant ones. This is seen by plotting separate Preferred Viewing Position Curves for fixations following saccades launched from different distances from the words. These launch-site contingent landing position distributions are normally distributed, with a mean that varies linearly with launch site distance and a standard deviation that increases with that distance. Thus, the Preferred Viewing Position Curves presented in Figure 8.1 actually are a sum over a number of launch-site contingent landing position distributions on the target word. These individual distributions show a much smaller variance than the summed curve. Examining these launch-site contingent landing position distributions can reveal character or word-based properties of the distributions that are not seen in the summed curve. Second, McConkie *et al.* (1988) observed that these launch-site contingent landing position distributions, though having much less variance than the preferred viewing position curve, are still broad enough to be difficult to observe with shorter words. Thus, it would be quite possible to see these curves with English 7-letter words or Chinese 2-character words, but not necessarily with English 3-letter words or Chinese single characters which have only four data points.

In order to examine the launch-site contingent landing position distributions, the data used to plot Preferred Viewing Position Curves, reported in the prior section, were subdivided according to the launch site, or the distance of the prior fixation from the beginning of the word on which the eyes landed. These distributions for launch sites of –2 to –6 are shown in Figure 8.2 (a launch site of –2 is two one-fourth deg units or letter positions to the left of the space before the character or word on which the eyes landed). In addition, in those cases where the maximum was not at the first or last unit in the word or character, a normal curve was fit to the data.

Figure 8.2 indicates that, as McConkie *et al.* (1988) observed, English 7-letter words show well-defined, normally-distributed landing position distributions with means that move left as the launch site increases. These same distributions for 2-character Chinese words are very different, being much flatter than the curves for English words. They have the appearance, not of being the distinct, individual distributions observed in English reading data, but rather simply being segments of a much broader curve, namely, the total saccade length frequency distribution. Thus, there is no evidence that partitioning the data by launch site actually reduces the variance of the individual curves, relative to the total saccade length frequency distribution, an observation that fails to provide evidence for the kind of word-based control in the reading of Chinese that is found in English.

This conclusion is strengthened by additional information included in Figure 8.2. Crosses indicate predictions of the proportions of fixations that would be present in each unit bin if the data were being predicted directly by the full saccade length frequency distributions from the English and Chinese reading data. This prediction is made by assuming, for a given launch site, that the fixation location is at position 0, the first bin in the full saccade length frequency distribution, and identifying the units in that distribution that would be part of the space occupied by the word of interest. (In the case of a launch site of –2 with a 7-unit word, this space would include unit 2, which would be the space before the word, through unit 9, the position of the last letter of the word. The predicted number of fixations is generated from the fixation counts of unit 2 to unit 9 of the overall saccade length distribution.) The total number of fixations in this word space is then divided into the number of fixations in each of the eight unit bins of the word itself, giving proportions of fixations that are predicted to be in each of those bins in the launch-site contingent landing position distribution. This is essentially a prediction that the locations of words and characters in the text have no effect on where the eyes go. If the locations of words do play a role, the predictions from the saccade length distribution should depart from the launch-site contingent landing position distributions in a characteristic way, since the landing position distribution should be narrower (i.e., show less variance) than the predictions from the saccade length distribution. This would produce underpredictions in the center part of the saccade length distribution, and overpredictions toward the ends. This is the pattern that is clearly present in the English data in Figure 8.2, but not in the Chinese data. Rather, 2-character Chinese words show distributions that are accurately predicted from the saccade length distribution itself, with no apparent systematic deviations. Apparently, the Chinese reader's eyes are not being sent to word locations in the text string.

The landing position data for Chinese single characters and English 3-letter words are quite similar to each other and the predictions from the saccade length distribution are accurate for both. No systematic deviations are observed in either data set. Thus, there is no evidence that either 3-letter English words or single Chinese characters

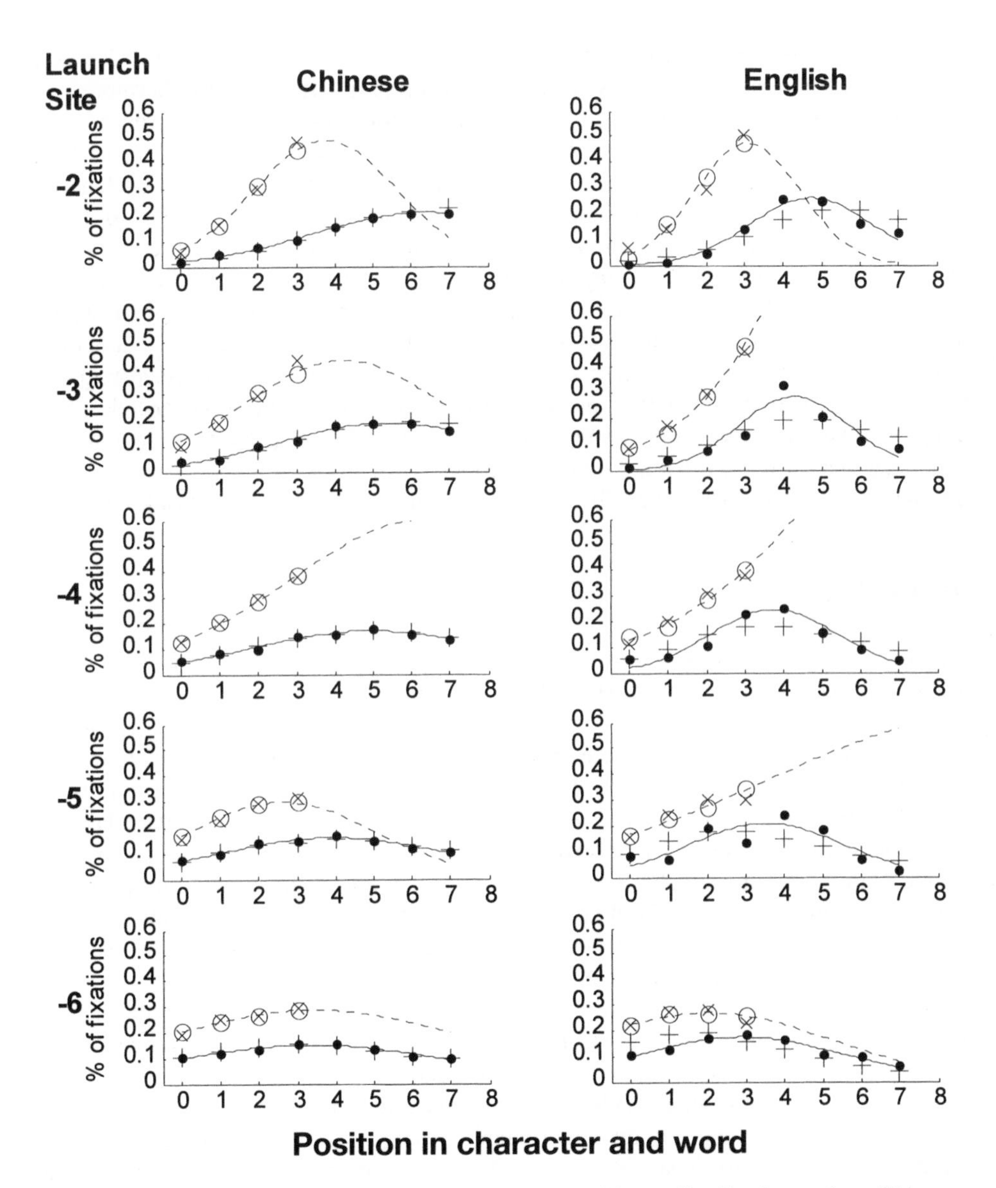

Figure 8.2: Launch-site contingent landing position distributions for Chinese characters (circles) and two-character words (dots) on the left, and for English words of corresponding length on the right. The curves are the best fitted normal curves for each distribution. For launch site –3, –4, –5 of English three-letter word, the fitting does not reach convergence. The crosses and Xs are the predicted proportions from the full saccade length distributions. See text for details.

serve as the targets for saccades during reading. However, we do not regard this as a strong test, since there are so few data points in these distributions. Furthermore, the launch-site contingent landing position distributions for 7-letter English words, and for words of other lengths as presented by McConkie *et al.* (1988), are much broader than the landing position distributions of 3-letter words. This means that if there are landing position distributions for individual Chinese characters that are similar to those observed in reading English words, the distributions for adjacent characters would overlap substantially, thus hiding their true shape. This issue must be addressed through the modeling of multiple overlapping distributions. At this point, then, we do not have evidence to establish the lack of character-based saccade targeting, but neither do we have sufficient evidence to support its existence. Pursuing this issue requires further investigation, but until proven otherwise, we will assume that saccade planning involves a competition among characters, and the eyes are then sent to the character that wins this competition, as proposed for objects in the visual field by Findlay and Walker (1999) and for words in reading by Yang and McConkie (2001).

Discussion

The goal of this study was to try to discover whether the character or the word is being chosen as the target for saccades made in reading Chinese. Three approaches were taken in studying this issue: identifying whether character or word properties affect where the eyes are being sent (here represented as the likelihood that the eyes will be sent to a particular character), whether Chinese readers show preferred viewing positions in characters or words, and whether they show launch-site contingent landing position distribution curves associated with words or characters that are similar to those found with words in English. Chinese readers do not appear to show either preferred viewing position curves or landing position distributions having characteristics that would suggest word-based saccade targeting. Neither was there supporting data for characters serving as saccade targets.

However, there are three reasons to keep the question open with respect to character-based saccade targeting. First, English words of the same length as the Chinese characters used in this study also failed to show these characteristics, possibly because such short words did not provide adequate data points to do the appropriate tests. Second, if Chinese characters do actually have launch-site contingent landing position distributions similar to those observed with English words, these distributions would extend quite far beyond the edges of the characters themselves, thus producing heavily overlapping distributions for adjacent characters. Such a situation could well produce the relatively flat preferred viewing position curves that were observed for Chinese characters and words. Third, since characters are made up of parts that differ in complexity and informativeness, this could produce some spread of the saccade target position in characters. This is a possibility that requires further investigation. Thus, it appears that there is reason not to reject the notion that the most perceptible objects in the Chinese writing system, the characters, may be the units that are competing to attract the eyes and that eventually serve as the objects to which the eyes are sent.

In contrast to the present results, Kajii *et al.* (2001) found that in reading Japanese the Kanji (Chinese) characters are more likely to be fixated and words made up of Kanji characters show a preferred viewing position, though it lay on the first character rather than at the center of the word. We suspect that these patterns result from the fact that Kanji characters are perceptually quite different from Kana characters, thus allowing Kanji words to be isolated perceptually in peripheral vision and to serve as saccade targets. No such perceptual basis exists for isolating words in Chinese text.

The alternative to word- or character-based guidance is to assume that the competition-based mechanism for the *where* decision in saccade control that is described by Findlay and Walker (1999), while appropriately describing saccade control in reading English in which words are well-marked perceptual objects, is not operating in the reading of Chinese. From this perspective the stimulus array is not treated as a collection of competing objects, but as an unsegregated field in which global effects operate to arrive at a decision concerning the distance and direction that the eyes will be sent on a given saccade. In this type of theory the saccade length frequency distribution could be weighted by characteristics of the characters. Our preference is to keep the assumption that eye movements in Chinese reading are being generated from character-based competition until evidence is produced that clearly contradicts this position. It is not clear, for example, how a mechanism in which global effects bias the overall saccade length frequency distribution could produce the word-based landing position distributions that are observed in the reading of alphabetic languages.

In spite of the fact that words are apparently not the units to which saccades are being sent in Chinese, their characteristics still affect the decision of where to send the eyes. Yang (1994) found that two-character Chinese words are less likely to be refixated following an initial fixation near their center than following an initial fixation farther from their center, as has been observed in the reading of English words (McConkie *et al.*, 1989; O'Regan & Jacobs, 1992). In the present study, the likelihood of the eyes going to character *N*+1 or *N*+2 (but not characters farther away) is affected by the cultural frequency of the word containing that character. It appears that the identification of words in Chinese can occur for characters within this region, and produce an influence on the selection of where to send the eyes.

One explanation of this influence is seen in theories of eye movement behavior that involve direct cognitive control, such as E-Z Reader (Reichle *et al.*, 1999; Pollatsek, Reichle & Rayner, this volume). This theory assumes a sequential consideration of the words during fixations, each being attended in turn. The fixated word (word *N*) is initially attended, and an evaluation is made of its frequency. This process takes longer for lower-frequency words, and its completion allows planning for a saccade to be initiated to the following word (word *N*+1). Once word *N* has been identified (which also takes less time for higher frequency words), attention is shifted to word *N*+1 and the cycle repeats. The completion of the frequency check on Word *N*+1, if accomplished early enough, cancels the saccade being planned and initiates a new plan to move the eyes to word *N*+2. Thus, frequency of a word affects the time it is attended in this word-by-word process, with a shorter attended time leading, on average, to a greater likelihood that the saccade plan for taking the eyes to that word will be cancelled and reducing the number of fixations on it.

Of course, Chinese writing adds another complexity to this theory: the necessity of first deciding what characters constitute a word. The reader must isolate a word before proceeding to estimate its frequency and then identify it. This theory also assumes that saccades are being sent to words, which, of course, was not supported by our analysis of landing position distributions in Chinese reading data. A modification to this theory could assume that the text is being attended character by character, with the character string being parsed into words in this process. Only when a word has been isolated does the process of estimating its frequency begin. In addition, it would have to be assumed that saccade targeting is either character based, or that the lack of perceptually distinct word units leads to some more global form of control. Neither of these seems to fit naturally with the current version of the E-Z Reader model.

An alternative theoretical position, competition/interaction theory proposed by Yang and McConkie (2001), assumes that the objects in the visual field compete to become the next saccade target, with a bias produced by strategy-based activation that favors objects to the right (in left-to-right writing systems). Assuming that characters serve as these competing objects when reading Chinese, various factors, such as the retinal location of a character and its complexity, can affect the attractiveness of characters in the competition. Of course, from this perspective the characters have their influence in parallel rather than being considered serially, as with E-Z Reader. But how might word frequency affect saccade targeting choices in this type of theory? The theory specifically excludes direct cognitive control except for cases of saccades following very long fixations on later fixations. The primary source of immediate, cognitively-based influence that is postulated for reading is an inhibition of the saccadic system when processing difficulty of some type occurs.

If we assume that lower-frequency words are more likely to produce processing difficulty with its resulting inhibition, then this is expected to happen more frequently when encountering lower frequency words. The effects of this inhibition includes the cancellation of some saccades so they occur only later, thus extending the fixation, eliminating the strategy-produced activation so that the rightward bias of the saccadic activity is reduced or eliminated, thus increasing the number of regressive saccades, and shortening the forward saccades. Since it appears that words are normally only recognized from characters at position *N*+2 or less, according to the region within which word frequency produces its effect on the likelihood of the eyes landing on selected characters, it must be words in this region that produce occasional processing difficulty. The mean forward saccade length in our data set is about 2.5 characters. Therefore, the shortening of saccades that is produced by inhibition would be expected to increase the number of fixations on characters less than 2.5 to the right; this includes characters *N*+1 and *N*+2, those on which lower word frequency was found to increase the number of fixations. Thus, from the perspective of the competition/interaction theory, it is not clear whether the increased frequency of fixating characters in lower frequency words is due to a direct increase in attractiveness for characters in such words, or to a stereotyped response to processing difficulty (Yang & McConkie, 2001). An alternative explanation, proposed by Hyönä and Pollatsek (2000) is that lower frequency words reduce the processing of more peripheral words, which shortens saccades.

While the present study does not succeed in reaching a firm conclusion about the nature of the unit to which the Chinese reader's saccades are directed, or, indeed, whether such a unit exists at all, it does rule out one reasonable candidate, and presents findings relevant to further explorations. It seems clear that, in reading Chinese, while words influence saccade decisions, they are not serving as targets for saccades. The above discussion also points out that this issue is of some interest to current theories of reading. It is clear that the study of Chinese reading is helping challenge, clarify and advance current theories of eye movement control during reading.

Acknowledgements

This research was conducted at National Yang-Ming University, Taipei, Taiwan, and supported by grants from Taiwan's National Science Council (NSC 89–2413-H-010–003 and NSC 90–2511-S-010–004) to Daisy Hung and Ovid Tzeng. This chapter was written while the first author was a Visiting Scholar at Beckman Institute, University of Illinois at Urbana-Champaign, supported by a Post-doctoral Scholar Grant from the National Science Council (NSC 90005P).

Notes

1 In dividing space in the Chinese text to correspond with letter-space units in the English text, we do not suggest the existence of any fundamental relationship between characters and letters. This is done simply as a means of examining whether landing position distributions exist in the eye movement patterns of Chinese readers over space ranges for which they have been shown by readers of English.

References

Academia Sinica balanced corpus. (Version 3) (1998). [CDROM]. Academia Sinica, Taipei, Taiwan.

Chan, M. Y. (1982). Statistics on the strokes of present-day Chinese script. *Chinese Linguistics, 1*, 299–305.

DeFrancis, J. (1989). *Visible Speech: The Diverse Oneness of Writing Systems*. Honolulu, HI, US: University of Hawaii Press.

Ehrlich, S. F., & Rayner, K. (1981). Contextual effects on word perception and eye movements during reading. *Journal of Verbal Learning & Verbal Behavior, 20*, 641–655.

Findlay, J. M., & Walker, R. (1999). A model of saccade generation based on parallel processing and competitive inhibition. *Behavioral & Brain Sciences, 22*, 661–721.

Hoosain, R. (1991). *Psycholinguistic Implications for Linguistic Relativity: A Case Study of Chinese*. Hillsdale, NJ, US: Lawrence Erlbaum Associates, Inc.

Hung, D. L., & Tzeng, O. J. (1981). Orthographic variations and visual information processing. *Psychological Bulletin, 90*, 377–414.

Hyönä, J., & Pollatsek, A. (2000). Processing of Finnish compound words in reading. In: A. Kennedy, R. Radach, D. Heller and J. Pynte (eds), *Reading as a Perceptual Process* (pp. 65–87). Oxford: Elsevier.

Inhoff, A. W., & Liu, W. (1998). The perceptual span and oculomotor activity during the reading of Chinese sentences. *Journal of Experimental Psychology: Human Perception & Performance, 24*, 20–34.

Inhoff, A. W., & Radach, R. (1998). Definition and computation of oculomotor measures in the study of cognitive processes. In: G. Underwood (ed.), *Eye Guidance in Reading and Scene Perception* (pp. 29–53). Oxford: Elsevier Science.

Just, M. A., & Carpenter, P. A. (1980). A theory of reading: From eye fixations to comprehension. *Psychological Review, 87*, 329–354.

Kapoula, Z. (1985). Evidence for a range effect in the saccadic system. *Vision Research, 25*, 1155–1157.

Kajii, N., Nazir, T. A., & Osaka, N. (2001). Eye movement control in reading unspaced text: The case of the Japanese script. *Vision Research, 41*, 2503–2510.

Liu, I. M., Yeh, J. S., Wang, L. H., & Chang, Y. K. (1974). Effects of arranging Chinese words as units on reading efficiency. *Chinese Journal of Psychology, 16*, 25–32.

McConkie, G. W., Kerr, P. W., Reddix, M. D., & Zola, D. (1988). Eye movement control during reading: I. The location of initial eye fixations on words. *Vision Research, 28*, 1107–1118.

McConkie, G. W., Kerr, P. W., Reddix, M. D., Zola, D., & Jacobs, A. M. (1989). Eye movement control during reading: II. Frequency of refixating a word. *Perception & Psychophysics, 46*, 245–253.

O'Regan, J. K. (1990). Eye movement and reading. In: E. Kowler (ed.), *Eye Movements and Their Role in Visual and Cognitive Processes* (pp. 395–453). Amsterdam: Elsevier.

O'Regan, J. K., & Jacobs, A. M. (1992). Optimal viewing position effect in word recognition: A challenge to current theory. *Journal of Experimental Psychology: Human Perception & Performance, 18*, 185–197.

Rayner, K. (1979). Eye guidance in reading: Fixation locations within words. *Perception, 8*, 21–30.

Rayner, K. (1998). Eye movements in reading and information processing: 20 years of research. *Psychological Bulletin, 124*, 372–422.

Rayner, K., Fischer, M. H., & Pollatsek, A. (1998). Unspaced text interferes with both word identification and eye movement control. *Vision Research, 38*, 1129–1144.

Rayner, K., & Pollatsek, A. (1981). Eye movement control during reading: Evidence for direct control. *Quarterly Journal of Experimental Psychology, 33A*, 351–373.

Reichle, E. D., Rayner, K., & Pollatsek, A. (1999). Eye movement control in reading: Accounting for initial fixation locations and refixations within the E-Z Reader model. *Vision Research, 39*, 4403–4411.

Reilly, R. G., & O'Regan, J. K. (1998). Eye movement control during reading: A simulation of some word-targeting strategies. *Vision Research, 38*, 303–317.

Yang, H. M. (1994). *Word Perception and Eye Movements in Chinese Reading*. Unpublished doctoral dissertation, University of Illinois at Urbana-Champaign, IL.

Yang, H. M., & McConkie, G. W. (1999). Reading Chinese: Some basic eye-movement characteristics. In J. Wang and A. W. Inhoff (eds), *Reading Chinese Script: A Cognitive Analysis* (pp. 207–222). Mahwah, NJ, US: Lawrence Erlbaum Associates, Inc., Publishers.

Yang, S. N., & McConkie, G. W. (2001). Eye movements during reading: A theory of saccade initiation times. *Vision Research, 41*, 3567–3585.

Zipf, G. K. (1949). *Human Behavior and the Principle of Least Effort*. Cambridge, MA: Addison-Wesley Press.

Chapter 9

The Perceptual Span During Music Reading

Elizabeth Gilman and Geoffrey Underwood

How much visual information do musicians require in order to read music? The following chapter examines this question, and presents an experiment in which a "moving window paradigm" is used to measure perceptual spans of musicians during a sight-reading task and a more complex transposition task. The data presented show that, even though eye-hand span varies with task difficulty and sight-reading skill, the perceptual spans of sight-readers remain relatively constant. This result contradicts Henderson and Ferreira's (1990) finding that increasing foveal load during text reading decreases perceptual span, and is likely to be a consequence of the time constraints imposed on the sight-reader.

Introduction

When musicians sight-read a piece of music, they are presented with previously unseen notation and asked to read and perform the music as accurately as possible. In this sense, sight-reading is very similar to reading a novel piece of text. In both cases, the symbols presented on the page must be translated into meaningful words or notes. Sloboda (1985) presents a highly detailed comparison of music and text reading, and suggests that music and language are very similar in terms of their structures and the way in which they are acquired. For example, both language and music can be described in terms of phonology, syntax and semantics. Therefore, a large portion of this chapter will include direct comparisons between music and text reading, discussing the benefits and limitations of borrowing experimental techniques and models of reading to explain music sight-reading.

During text reading, the reader fixates a word so that the word is projected onto the area of the retina with highest acuity (i.e. the fovea). Consequently, the symbols on the page can be perceived more easily and processed more efficiently. However, this

The Mind's Eye: Cognitive and Applied Aspects of Eye Movement Research

ISBN: 0–444–51020–6

does not mean that there is a direct association between the fixated word and the word being processed. In fact, a number of studies have provided evidence against the so-called "eye-mind assumption". For example, some words, particularly short words, are read without ever being fixated (e.g. Blanchard, Pollatsek & Rayner, 1989). Also, a certain amount of processing occurs parafoveally, which later facilitates processing when a word is subsequently fixated. For example, studies have shown that readers can process the first three letters of a parafoveal word, which is likely to initiate lexical access of that word (Rayner, Well, Pollatsek & Bertera, 1982; Lima & Inhoff, 1985; Lima, 1987; Inhoff, 1987, 1989a, 1989b; Inhoff, Bohemier & Briihl, 1992). Given the similarities between music and text reading, the same is likely to be true of music reading, and the available empirical data suggest that this is the case. For example, Kinsler and Carpenter (1995) measured musicians' eye-movements while they read and performed rhythm notation and report that the musicians did not read up to the end of the phrase. Rather, their ultimate fixation landed in the middle of the last measure and the final notes were read using peripheral vision.

Salis (1981) studied laterality effects using tachistoscopically presenting extrafoveal chords and dot patterns to good and poor music readers. The musicians' performance on the chord perception task was superior when chord patterns were presented in the right visual half-field whilst the reverse was true for dot patterns. Salis argues that the results provide evidence for left hemispheric processing of music material. However, the results could be interpreted as a direct effect of rightward reading habits. This argument is supported by studies of text reading, which have shown that the perceptual span extends towards the direction in which attention is normally directed (Pollatsek, Bolozky, Well & Rayner, 1981). For example, the perceptual span of readers of alphabetical orthographies extends approximately 14–15 characters to the right of fixation and only 3–4 characters to the left. Conversely, readers of languages printed from right to left show a left visual field superiority (Mishkin & Forgays, 1952; Orbach, 1952) and display evidence of perceptual spans that extend towards the left (Pollatsek *et al.*, 1981). Since music is comparable with alphabetical text in that it is read from left to right, it is likely that music readers have a rightward attentional bias for musical material.

At the very least, Salis' study provides evidence that musicians process extra-foveally presented music notation. However, it does not answer the question of the *amount* of parafoveal information used in a natural sight-reading task. Bean (1938) attempted to measure the "span of apprehension" of musicians by tachistoscopically presenting musical extracts and measuring the number of notes performed post-presentation. This, of course, produces an underestimate of the perceptual span since the task is mnemonic as well as perceptual. At best, this measure produces an estimate of the "note-identification" span, and like text-reading, full identification is not necessarily required in order to facilitate performance. For example, in text-reading, the "word identification span" is much smaller than the perceptual span, extending no more than 7–8 character spaces to the right of fixation in alphabetical orthographies (e.g. Rayner *et al.*, 1982). Studies have shown that readers process word length information (e.g. Ikeda & Saida, 1978; Morris, Rayner & Pollatsek, 1990; Rayner & Morris, 1992; Rayner, Sereno & Rayney, 1996), orthography (Rayner, Balota & Pollatsek, 1986), and phonology (Henderson, Dixon, Petersen, Twilley & Ferreira, 1995; Pollatsek, Lesch, Morris & Rayner, 1992) and even semantic information (Underwood, 1976; Underwood, Clews & Everatt, 1990) from parafoveal words, which can facilitate lexical access and/or guide eye-movements.

Other studies of music reading have measured musicians' eye-hand span (Goolsby, 1994a; 1994b; Sloboda, 1974, 1977; Weaver, 1943). The eye-hand span (EHS) is defined as being the distance between the musician's point of fixation and their point of performance. It is important to note that the EHS is different from the perceptual span. Whereas the EHS is a measure of how far musicians fixate ahead of their hands, the perceptual span is a measure of how much information the musician obtains around the point of fixation. Thus, by measuring EHS and perceptual span independently then combining the two measures, one can calculate the total amount of information extracted ahead of the point of performance.

Given the apparent similarities between music reading and text reading, it would be sensible to assume that these measurements can be obtained using techniques used in studies of text reading. Many studies of perceptual span in text reading, and indeed other domains such as visual search and scene perception, have implemented the "moving window paradigm" or "gaze-contingent window paradigm", developed by McConkie and Rayner (1975). This technique allows the experimenter to define an area around the observer's point of fixation (i.e. window), within which all stimulus information is retained whilst information outside the pre-defined area is occluded or degraded. By altering the shape and size of this window and making comparisons with a control condition where no window is presented, the experimenter can establish the size and shape of the observer's perceptual span. Variations on this technique, such as the moving mask technique (Rayner & Bertera, 1979) and boundary technique (Rayner, 1975) have been also employed to investigate the relative contributions of foveal and parafoveal information.

Truitt, Clifton, Pollatsek and Rayner (1997) (see also Rayner & Pollatsek, 1997) have already used this technique to investigate the perceptual span of pianists whilst sight-reading single line melodies. During this experiment, a control condition was compared with 2-beat, 4-beat and 6-beat window conditions. Sight-reading performance and eye-movements in the 2-beat window condition significantly differed from the same measures in the other three conditions, which did not differ from one another. Therefore, the experimenters conclude that the musicians' perceptual spans extend to approximately 3 beats (i.e. the fixated beat plus 2 beats to the right). However, the data provided no evidence to show that the perceptual span of skilled sight-readers differed from the perceptual span of less skilled sight-readers (i.e. there was no interaction between skill and window size). This is surprising given the effects of skill on perceptual span during text reading, reported by Rayner (1986) using the same technique. Skilled readers tend to have larger perceptual spans than those of less skilled readers, which may provide at least part of the explanation for their superior reading speed and efficiency. One reason why Truitt and her collaborators did not find such skill differences may have been a result of a small sample size (4 skilled vs. 4 less skilled). An alternative explanation is that the difference between the levels of skill in each group did not differ enough to flush out differences in perceptual span.

However, this is unlikely given that the number of years of piano experience ranged from two years to sixteen years. A more likely explanation is that the single-line melodies were over-simplistic and therefore insensitive to skill differences.

Therefore, the purpose of the experiment presented in this chapter was to (a) present more complex musical phrases with more than a single melody-line through a moving window, and (b) compare tasks of different levels of cognitive load. Given the effects of load on perceptual span reported by Henderson and Ferreira (1990), increasing the task demands should decrease perceptual span. The experiment also aimed to compare pianists of matched overall musical experience but varying sight-reading ability in order to attribute any differences in task performance to sight-reading skill alone.

Experiment

Method

The 40 grade 8 pianists who took part in the experiment were allocated to two sight-reading groups on the basis of a sight-reading pre-test. Pianists who scored higher than the median score were defined as "good sight-readers" (mean score = 91%) and those who scored below the median were defined as "poor sight-readers" (mean score = 67%). Before the experimental procedure began, the participant was seated 1 metre from the display monitor, and was asked to manoeuvre the keyboard so that (a) the keyboard could be played comfortably from where they were sitting and (b) the keys could be seen without the need for large head-movements. The participant was then asked to rest his or her head on a chin support (again to minimise unnecessary head-movement). An SMI head-mounted eye-tracker was then placed on the participant's head and calibrated, which involved the participant staring at a dot that appeared randomly at nine different points on the computer monitor.

The participant was then required to perform three different tasks, the order of which was randomly determined. For purposes of brevity, only the sight-reading task (medium load) and transposition task (high load) are described in this chapter. The third task, an error-detection task (low load) was included in the experiment because it required musicians to read the music without performing the notation on the keyboard. However, the nature of the task meant that the results were not comparable with the other two tasks because the patterns of fixations were strategic rather than typical of eye-movement behaviour during music reading.

In each task, 32 extracts from Bach chorales (as edited by Riemenschneider, 1941) were presented in a random order. All phrases were 3 bars long and included a 3-beat rest at the beginning of the first bar and a 1-beat rest at the end of the third bar. The entire stave was 743 pixels long and 105 pixels deep, which meant that the average bar width was approximately 248 pixels. The note heads were 8 pixels deep and 10 pixels wide (see Figure 9.1). Participants were seated 1 metre from the screen, and therefore there were approximately 65.7 pixels per degree of visual angle. Prior to the presentation of each phrase, the participant was asked to stare at a dot, which appeared where the phrase would start. The eye-tracker then corrected for any measurement error, in order to maintain a high level of calibration throughout the experiment. This inter-trial interval took approximately 3 seconds.

For the sight-reading task, the participant was asked to perform each phrase when it appeared on the screen at a comfortable tempo on a Casio 3800 keyboard. For the transposition task, the participant was presented with the same 32 phrases, and was asked to perform the phrase 2 semitones below the notes indicated in the score. In each task, participants viewed ¼ of the phrases through a 1-beat moving window, ¼ of the phrases through a 2-beat moving window, ¼ of the phrases through a 4-beat moving window, and ¼ of the phrases with no window. Inside the window area around the point of fixation, all stimulus information was retained, while information outside the window area was limited to a blank stave (see Figure 9.1). Since there was little variance in the horizontal distance between beats, the 1-beat window was always 64 pixels wide, the 2-beat window was 128 pixels wide, and the 4-beat window was 256 pixles wide. The window was offset so that information was available up to 10 pixels to the left of fixation, and therefore the currently fixated note was always visible. The height of the window was set to the height of the computer screen.

In order that each participant should see each phrase only once during each task, stimuli were rotated across four different groups of participants (i.e. group 1 participants saw group A stimuli through the 1-beat window, group 2 saw group B stimuli through the 1-beat window, etc.).

During the tasks, an SR Research Eyelink I eye-tracker was used to sample eye-movements every 4 ms (250 Hz). The position of the window was only updated when the position of the eye had moved more than 3 pixels to the right or to the left, so that the window would not be sensitive to microsaccades or drift. The eye-tracker and presentation of stimuli were both controlled by Compaq Deskpro computers, and stimuli were viewed on a 17 inch (43 cm) monitor (pixel resolution = 38 pixels per cm, screen

Figure 9.1: A diagram of a Bach chorale through a 2-beat moving window. The information inside the window is retained while the information outside the window is limited to a blank stave. In the diagram, the upper stave shows what the phrase looked like in the control condition, while the lower stave shows what the phrase looked like through the 2-beat window (the participant is fixating the first chord).

dimensions = 1024 × 768 pixels, refresh rate = 85 Hz). The maximum time taken to change the moving window display was approximately 22ms. Keyboard performance was recorded by one of the computers, which was connected to a Casio 3800 keyboard via a MIDI interface. At the beginning of the experiment, a time code was initiated so that eye-movement and MIDI data would be synchronised.

Results

Sight-reading All variables were analysed using mixed 2 × 4 (skill × window) analyses of variance and, apart from score and position of first error, were analysed using only data from trials in which participants made fewer than 30% errors (see *performance measures* for scoring algorithm). Consequently, three good sight-readers and seven poor sight-readers had to be excluded from the analysis because they did not perform enough trials to criterion level. Also, eye-movements were only included in the analysis if they occurred during keyboard performance (i.e. after the onset of the first note and before the onset of the last note). Finally, many of the variables had to be transformed using either a natural log transform (average fixation duration, number of fixations, performance time, note duration) or a cube root transform (keyboard number of fixations, keyboard fixation duration) to minimise errors associated with heterogeneous variances.

Eye-movement measures The significant main effect of window on fixation duration ($F3,84 = 22.323$; Mse = 0.007; $p < 0.0001$) revealed that fixations were significantly longer in the 1-beat condition (mean = 325 ms) than in the 2-beat (mean = 302 ms), 4-beat (mean = 286 ms) and nowin (mean = 276 ms) conditions ($p < 0.01$). Fixations were also longer in the 2-beat condition than in the nowin condition ($p < 0.001$). There was also a main effect of window on the number of fixations ($F3,84 = 20.554$; Mse = 0.013; $p < 0.0001$). Pianists made more fixations in the 1-beat condition (mean = 26.5) than in the 2-beat (mean = 22.4), 4-beat (mean = 21.0) and nowin (mean = 21.5) conditions ($p < 0.001$). Also, good sight-readers (mean = 20.5) made fewer fixations than poor sight-readers (mean = 26.0) ($F1,28 = 11.838$; Mse = 0.128; $p < 0.01$). The effect of window on saccades, which did not include saccades made to and from the keyboard, was also significant ($F3,84 = 12.742$; Mse = 12.751; $p < 0.0001$). Saccades were shorter in the 1-beat condition (mean = 56.4 pixels) than in the 2-beat (mean = 60.7 pixels), 4-beat (mean = 60.7 pixels) and nowin (mean = 61.4 pixels) conditions ($p < 0.001$).

The proportion of fixations made on different parts of the score and the corresponding mean fixation durations are shown in Table 9.1 below. The number of keyboard fixations and the total amount of time spent fixating the keyboard are also shown. The average amount of time spent viewing the keyboard is particularly low in the sight-reading task because there were many trials in which the keyboard viewing time was 0 ms (musicians did not need to look at the keyboard). For a fixation to be classified as on a note, the x location of the point of fixation had to be within 10 pixels from the note centre and the y location had to be within 8 pixels from the centre. For

Table 9.1: Proportions of fixations directed towards different areas of the score in the sight-reading task, and corresponding mean fixation durations. The table also shows the number of keyboard fixations and average keyboard viewing time.

	Spaces	**Notes**	**Tails**	**Signature**	**Barlines**		**Keyboard**
% fixations	30.80	57.53	6.14	4.39	1.14	Number of fixations	0.24
Mean duration (ms)	239.54	308.47	209.03	0.77	106.87	Total viewing time (ms)	92.80

fixations on tails, signatures and barlines, the fixation criterion was 5 pixels either side of the object. All other fixations were classified as fixations on spaces, and are therefore comparable with the "blanks" in the analysis performed by Truitt *et al.* Finally, for a fixation to be classified as a keyboard fixation, the y value of the fixation location had to be greater than 750.

A larger proportion of fixations were made on spaces in the 1-beat condition (mean = 33.3%) than in the 4-beat (mean = 29.5%) and nowin (mean = 29.5%) conditions ($F_{3,84} = 5.230$; Mse = 17.364; $p < 0.01$). Also, a smaller proportion of fixations were made on notes in the 1-beat condition (mean = 54.8%) than in the 2-beat (mean = 58.2%), 4-beat (mean = 59.0%) and nowin (mean = 58.1%) conditions ($F_{3,84} = 4.500$; Mse = 22.676; $p < 0.01$). A larger proportion of fixations were made on barlines in the 1-beat (mean = 1.5%) than in the 2-beat (mean = 0.9%) condition ($F_{3,84} = 3.019$; Mse = 0.574; $p < 0.05$), although this result is likely to be spurious since there were no differences between the 1-beat, 4-beat (mean = 1.1%) and nowin (mean = 1.1%) conditions. Finally, a larger proportion of fixations were made on the signature in the nowin condition (mean = 5.2%) than in both the 1-beat (mean = 4.1%) and 2-beat (mean = 4.1%) conditions ($F_{3,84} = 3.672$; Mse = 2.644; $p < 0.05$).

Eye-hand span The eye-hand span was calculated as the horizontal distance between the point of fixation and the point of performance (i.e. the difference between the x location of the centre of the performed note or chord and the x location of the point of fixation at the time of note or chord onset). If the onset of the note occurred during a saccade, then the x location of the next fixation was assumed to be the point of fixation. Also, if the musician made several attempts at a note or chord, then only the first attempt was taken into account. Good sight-readers (mean = 18 mm) displayed significantly larger EHSs than poor sight-readers (mean = 14 mm) ($F_{1,28} = 7.131$; Mse = 50.778; $p < 0.05$). When there was no moving window, good sight-readers' mean EHS was 19 mm, which approximates to 1 beat ahead. The poor sight-readers' mean EHS in the nowin condition was 15 mm, which approximates to ¾ beat. There was also a significant effect of window on EHS ($F_{3,84} = 5.670$; Mse = 11.102; $p < 0.01$). The EHS was significantly shorter in the 1-beat condition (mean = 14 mm) than in the 2-beat (mean = 16 mm), 4-beat (mean = 17 mm) and nowin (mean = 17 mm) conditions ($p < 0.05$).

Performance measures A "percentage correct" score was calculated as the number of notes performed at the correct pitch minus the number of additional notes or notes performed at the wrong pitch, divided by the number of notes in the phrase and multiplied by 100. As expected, good sight-readers (mean = 89.9%) performed more accurately than poor sight-readers (mean = 79.6%) ($F1,28 = 17.063$; $Mse = 185.050$; $p < 0.001$) and made their first error significantly later than poor sight-readers ($F1,28 = 4.349$; $Mse = 8.940$; $p < 0.05$). On average, good sight-readers made their first error on the 8.9th note played whereas poor sight-readers made their first error on the 7.8th note played. In addition, good sight-readers' performances (mean = 7.2s) were significantly shorter than poor sight-readers' performances (mean = 9.6s) ($F1,28 = 13.149$; $Mse = 0.175$; $p < 0.01$). There was also a main effect of window on performance accuracy ($F3, 84 = 10.814$; $Mse = 31.885$; $p < 0.0001$) and performance duration ($F3,84 = 30.160$; $Mse = 0.020$; $p < 0.0001$). Musicians performed less accurately in the 1-beat condition (mean = 80.6%) than in the 2-beat (mean = 86.3%), 4-beat (mean = 87.4%) and nowin (mean = 87.3%) conditions ($p < 0.01$) and performance durations were significantly longer in the 1-beat condition (mean = 10.3s) than in the 2-beat (mean = 8.2s), 4-beat (mean = 7.3s) and nowin (mean = 7.3s) conditions ($p < 0.001$). Also, performance durations were longer in the 2-beat condition than in the 4-beat and nowin conditions.

Transposition Data from the transposition task were analysed using the same methods as those employed for the sight-reading task, except that all variables, apart from score and position of first error, were analysed using only data from trials in which participants scored 60% or above. A slightly lower criterion level had to be used for the transposition task because the task was much more difficult and few participants regularly scored over 70%. Unfortunately, this still meant that only nine good sight-readers and five poor sight-readers were included in the analysis. Natural log transformations were used for average fixation duration, number of fixations, performance time, note duration, and % space fixations, and cube root transformations were used for keyboard number of fixations, keyboard fixation duration, % barline fixations, and % note stem fixations.

Eye-movement measures Good sight-readers (mean = 417 ms) displayed longer fixations than poor sight-readers (mean = 309 ms) ($F1,12 = 20.617$; $Mse = 0.053$; $p < 0.001$). Also, good sight-readers (mean = 31.9) used significantly fewer fixations than poor sight-readers (mean = 52.8) ($F1,12 = 15.319$; $Mse = 0.215$; $p < 0.01$). The significant effects of window on fixation duration ($F3, 36 = 10.439$; $Mse = 0.007$; $p < 0.0001$) and number of fixations ($F3,36 = 3.514$; $Mse = 0.017$; $p < 0.05$) revealed that fixation durations were longer in the 1-beat condition (mean = 417 ms) than in the 4-beat (mean = 361 ms) and nowin (mean = 349 ms) conditions ($p < 0.001$), and longer in the 2-beat condition (mean = 385 ms) than in the nowin condition ($p < 0.05$). Musicians also used more fixations in the 1-beat condition (mean = 43.4) than in the 2-beat condition (mean = 36.9).

The proportion of fixations made on different parts of the score and the corresponding mean fixation durations are shown in Table 9.2 below. The number of keyboard fixations and the total amount of time spent fixating the keyboard are also shown.

Table 9.2: Proportions of fixations directed towards different areas of the score in the transposition task, and corresponding mean fixation durations. The table also shows the number of keyboard fixations and average keyboard viewing time.

	Spaces	**Notes**	**Tails**	**Signature**	**Barlines**		**Keyboard**
% fixations	28.98	62.54	3.65	3.72	1.10	Number of fixations	2.64
Mean duration (ms)	275.97	405.82	298.55	398.27	108.39	Total viewing time (ms)	702.71

Good sight-readers (mean = 0.8) made fewer keyboard fixations than did poor sight-readers (mean = 6.0) ($F1,12 = 8.557$; Mse = 1.624; $p < 0.05$). Also, poor sight-readers (mean = 1009 ms) looked at the keyboard for longer than did good sight-readers (mean = 123 ms) ($F1,12 = 7.484$; Mse = 52.573; $p < 0.05$). No other effects reached significance.

Eye-hand span There was a significant effect of window ($F3,36 = 5.637$; Mse = 6.619; $p < 0.01$). The EHS was shorter in the 1-beat condition (mean = 9 mm) than in the 2-beat (mean = 12 mm), 4-beat (mean = 13 mm) and nowin (mean = 12 mm) conditions ($p < 0.05$). There was also a significant interaction between skill and window ($F3,36 = 3.614$; Mse = 6.619; $p < 0.05$). In the nowin condition, the EHS of good sight-readers (mean = 13 mm) was significantly larger than the EHS of poor sight-readers (mean = 9 mm) ($F1,48 = 5.286$; Mse = 11.554; $p < 0.05$). Also, only good sight-readers were affected by window size ($F3,36 = 9.471$; Mse = 6.619; $p < 0.001$). For good sight-readers, the EHS in the 1-beat condition (mean = 8 mm) was significantly smaller than the EHS in the 2-beat (mean = 13 mm), 4-beat (mean = 13 mm) and nowin (mean = 13 mm) conditions.

Performance measures There was a significant effect of skill ($F1,12 = 4.863$; Mse = 1221.883; $p < 0.05$), which revealed that good sight-readers (mean = 60.1%) performed more accurately than poor sight-readers (mean = 38.6%). Also, good sight-readers' performances (mean = 15.1 s) were significantly shorter than poor sight-readers' performances (mean = 22.0 s) ($F1,12 = 5.167$; Mse = 0.345; $p < 0.05$). Finally, the effect of window ($F3,36 = 8.597$; Mse = 0.017; $p < 0.001$) revealed that performances were longer in the 1-beat condition (mean = 20.9 s) than in the 2-beat (mean = 16.7 s), 4-beat (mean = 16.8 s) and nowin (mean = 16.0 s) conditions ($p < 0.01$).

Discussion

During this experiment, musicians were required to sight-read phrases written in harmony (i.e. the musicians were required to perform more than one note at a time).

Despite the inclusion of musical harmony, the study revealed similar results to those obtained by Truitt *et al.* who displayed much simpler single-line melodies. The largest window to differ from the control condition in the sight-reading task was the 2-beat window, which means that musicians process useful information from *more* than 2 beats. Therefore, the perceptual span (amount of useful information processed around the point of fixation) of musicians whilst sight-reading harmony is 3–4 beats (i.e. the fixated beat plus 2–3 beats to the right). It is impossible to conclude from this analysis whether the perceptual span is 3 or 4 beats, since a 3-beat window was not included in the study.

The similarity between this study and that of Truitt *et al.* is surprising given the findings of Henderson and Ferreira (1990) that increasing foveal load decreases perceptual span. The most likely explanation for this is that the music presented in this experiment did not comprise independent melody lines. Rather, the harmony consisted of a series of familiar musical units, or "chords" which could be recognised as patterns or "chunks". The idea of chunking was first introduced by Chase and Simon (1973) to explain grand-master chess players' extraordinary memory for board positions, and has been applied to a number of different domains including music (Waters, Underwood & Findlay, 1997; Waters, Townsend & Underwood, 1998). The process involves identifying familiar patterns when processing the constituent parts of these patterns in parallel. Interestingly, the process of chunking also applies to skilled text reading in that words are processed as whole units rather than as a series of individual letters. Therefore, in this experiment, reading chords was no more difficult than reading a single-line melody since the constituent notes were processed in parallel.

For the transposition task, the effects of window size on eye-movements were similar to those in the sight-reading task. However, the window had a reduced effect on task performance, although admittedly this could be a floor effect since most of the participants found this task extremely difficult. More importantly, in the transposition task, the window only affected *when* musicians moved their eyes (e.g. fixation duration) whereas in the sight-reading task, the window affected both *when* and *where* musicians moved their eyes (e.g. fixation locations, saccade length). While it is true that musicians made a larger percentage of note fixations in the transposition task than in the sight-reading task, the reader should not be mislead into thinking that the targets of saccades were more accurate in the former. Rather, musicians were simply more likely to make multiple fixations on chords in the transposition task since the task promoted a serial note-by-note strategy rather than chunking behaviour. As a result, musicians had a tendency to "edge" along the score rather than using peripheral information to plan saccade targets.

The perceptual span tells us how much information musicians use around the point of fixation, but how far did musicians fixate ahead of the point of performance? The eye-hand span of good sight-readers in the sight-reading task extended to about 1 beat in the control condition, which is less than the EHS reported for skilled sight-readers in the Truitt *et al.* study. Also, musicians' EHSs were much smaller in the transposition task compared with the sight-reading task. Therefore, unlike perceptual span, eye-hand span is largely affected by task demands and cognitive load. In addition, there is evidence to show that EHS, unlike perceptual span, varies with skill. It is possible that this is due to differences in performance tempo, since eye-voice span in text reading has been shown to increase with reading speed (Geyer, 1967). However, Truitt *et al.* controlled for tempo (although good sight-readers performed slightly faster) and still found that good sight-readers had larger eye-hand spans than poor sight-readers.

In the sight-reading task, the EHS of poor sight-readers extended to less than a beat and differed significantly from the 1 beat EHS of the good sight-readers. Therefore, combining eye-hand span with perceptual span, good sight-readers use information up to 1 bar (3–4 beats) ahead of the point of performance. Good sight-readers fixate one beat to the right of the point of performance and use information from at least 2 beats ahead of fixation. However, poor sight-readers use information up to 2 or possibly 3 beats ahead, since the EHS of poor sight-readers does not extend to a whole beat. In the transposition task, good sight-readers looked approximately ¾ beat ahead, while poor sight-readers looked approximately ½ beat ahead in the control condition. Therefore, both good and poor sight-readers used information from up to 2 or possibly 3 whole beats ahead of the point of performance in this task.

There were a number of notable differences between good and poor sight-readers for both the sight-reading and transposition tasks. Good sight-readers performed both tasks more accurately and rapidly, as one would predict, and tended to use fewer, shorter fixations. Furthermore, good and poor sight-readers did not differ in terms of the percentage of fixations directed towards notes, which means that overall, good sight-readers fixated fewer notes. This is consistent with the theory that expert sight-readers process note patterns rather than individual notes, thus processing more visual information within each fixation. Also, good sight-readers spent less time looking at the keyboard, which could be due to superior spatial representations of the keyboard layout or the use of more efficient finger positions that reduce the need to look at the keys (Parncutt, Sloboda, Clarke & Raekallio, 1997; Parncutt, Sloboda & Clarke, 1999).

Surprisingly, there were still no strong skill effects on perceptual span, despite the complexity of these tasks. The only interaction to reach significance was for EHS, which revealed that only good sight-readers were affected by the moving window. This supports the hypothesis that good sight-readers are more affected by preview than are poor sight-readers. However, there were no interaction effects on any other eye-movement measures. This indicates that the lack of skill effects in Truitt *et al.*'s study was not a result of over-simplistic musical materials. Rather, it appears that in any natural sight-reading task, good sight-readers use no more preview information than poor sight-readers. However, previous research has shown that in more artificial tasks, good sight-readers display larger perceptual spans (Waters *et al.*, 1998; Gilman, 2000). Therefore, the most likely conclusion is that good sight-readers can process musical material further into the periphery than poor sight-readers, but refrain from doing so in natural sight-reading tasks as a result of the strict time constraints imposed on the reader. During sight-reading, the rhythm and tempo of the music dictate the duration of each note. Therefore, looking too far ahead into the periphery would place high demands on working memory since any information processed must be stored up until the time that the notes should be performed. Good sight-readers already have an advantage in that they are fixating further ahead of the point of performance, and therefore, processing much further into the periphery would be of little benefit to processing efficiency and of great cost to working memory.

The results of this study generally lend support for applying research methods used for studying text reading to the domain of sight-reading. The moving window paradigm was adapted successfully in order to calculate the perceptual span of pianists whilst reading chordal harmony. One possible limitation of using this paradigm with music stimuli is that changes in the display are noticeable to the musicians, and consequently, it is difficult to determine whether effects of window size are due to limited parafoveal information or due to unusual reading conditions. However, display changes were noticeable in all three window conditions, and given that there were no differences between the 4-beat window and control conditions, the results cannot be entirely explained by display change effects. This problem has now been circumvented in studies of text-reading by presenting random letter strings rather than homogeneous strings of Xs outside the gaze-contingent window (see Rayner, 1998). A musical counterpart to random letter strings is difficult to create, however, because changes in pitch are more noticable than changes in letters. Consequently, display changes with random notes might be even more disruptive than display changes with blank staves. Future research in this area should be aimed at comparing alternative types of visual information outside the moving window area.

Another limitation is that this technique might not be so adaptable in the case of musical phrases in which musical patterns are less well defined spatially. For example, "contrapuntal" musical phrases comprise interweaving melodic lines rather than vertically arranged chords, and therefore the musician's perceptual span is likely to form a more irregular, unpredictable shape and size. Weaver (1943) shows that, when reading contrapuntal phrases, musicians move their eyes along one melodic pattern before moving to another, but unlike chords, it is hard to define what constitutes these melodic patterns and the point at which the musician will make transitions between them.

So far, the chapter has also proposed that the reading of text and music are highly comparable. For example, both text and music reading are performed more successfully by the identification and processing of "chunks" (i.e. words or chords rather than letters or individual notes). Furthermore, both skilled text reading and skilled music reading are affected by the familiarity and predictability of these chunks (Sloboda, 1976; Waters, Townsend & Underwood, 1998). The experiment itself has shown that in music reading, as well as text reading, notation is processed beyond the point of fixation and that the perceptual span extends in the direction in which attention is directed (although windows extending to the left of fixation were not investigated, the 4-beat window had no detrimental effect on performance or eye-movement behaviour even though information to the left of fixation was unavailable). This suggests that eye-movement models of reading, such as the E-Z Reader (Reichle, Pollatsek, Fisher & Rayner, 1998), might prove to be useful tools in predicting eye-movements during music sight-reading as well as text reading.

However, the findings of the experiment reported here have also outlined an important difference between music and text reading. The reason why skilled and less-skilled sight-readers did not differ in terms of perceptual span is likely to be linked to the fact that the sight-reader is under a strict time-constraint. Unlike silent reading, musical symbols dictate how long a note should be played for. In this sense, sight-reading is more similar to reading aloud, in which the rhythm of speech is important. For example, the duration of sounds can often change the inflection or the meaning of a word (e.g. did vs. deed). Kinsler and Carpenter (1995) have shown that eye-movements during sight-reading are tightly linked to the duration of notes, and consequently, any model of sight-reading must incorporate this relationship. Therefore, it is important to note that models of text reading are unlikely to account for all of the variance when used to predict eye-movements during music sight-reading. However, it is hoped that through a certain amount of modification, text reading models will prove to be invaluable tools in describing and understanding musicians' eye-movement behaviour.

Acknowledgement

We are grateful to the International Foundation for Music Research for supporting this work with research project RA136, and to Keith Rayner and George McConkie for comments on a previous draft of this report.

References

Bean, K. L. (1938). An experimental approach to the reading of music. *Psychological Monographs*, *50*, whole number 226.

Blanchard, H. E., Pollatsek, A., & Rayner, K. (1989). The acquisition of parafoveal word information in reading. *Perception & Psychophysics*, *46*, 85–94.

Chase, W. G., & Simon, H. A. (1973). Perception in chess. *Cognitive Psychology*, *4*, 55–81.

Geyer, J. J. (1967). Perceptual systems in reading: A temporal eye-voice span. *Dissertation Abstracts International*, *28*, 122–123.

Gilman, E. R. (2000). Towards an eye-movement model of sight-reading. Unpublished doctoral dissertation, University of Nottingham, Nottingham.

Goolsby, T. (1994a). Eye-movement in music reading — Effects of reading ability, notational complexity, and encounters. *Music Perception*, *12*, 77–96.

Goolsby, T. (1994b). Profiles of processing — Eye-movements during sight-reading. *Music Perception*, *12*, 97–123.

Henderson, J. M., Dixon, P., Petersen, A., Twilley, L. C., & Ferreira, F. (1995). Evidence for the use of phonological representation during transsaccadic word recognition. *Journal of Experimental Psychology: Human Perception & Performance*, *21*, 82–97.

Henderson, J. M., & Ferreira, F. (1990). Effects of foveal processing difficulty on the perceptual span in reading: Implications for attention and eye movement control. *Journal of Experimental Psychology: Learning, Memory, and Cognition*, *16*, 417–429.

Ikeda, M., & Saida, S. (1978). Span of recognition in reading. *Vision Research*, *18*, 83–88.

Inhoff, A. W. (1987). Parafoveal word perception during eye fixations in reading: Effects of visual salience and word structure. In: M. Coltheart (ed.), *Attention & Performance* (Vol. 12, pp. 403–420). London: Erlbaum.

Inhoff, A. W. (1989a). Lexical access during eye fixation in reading: Are word access codes used to integrate lexical information across interword fixations? *Journal of Memory & Language*, *28*, 444–461.

Inhoff, A. W. (1989b). Parafoveal processing of words and saccade computation during eye fixations in reading. *Journal of Experimental Psychology: Human Perception & Performance, 15*, 544–555.

Inhoff, A. W., Bohemier, G., & Briihl, D. (1992). Integrating text across fixations in reading and copytyping. In: K. Rayner (ed.), *Eye Movements and Visual Cognition: Scene Perception and Reading* (pp. 355–368). New York: Springer-Verlag.

Kinsler, V., & Carpenter, R. H. S. (1995). Saccadic eye-movements while reading music. *Vision Research, 35*, 1447–1458.

Lima, S. D. (1987). Morphological analysis in sentence reading. *Journal of Memory & Language, 26*, 84–99.

Lima, S. D., & Inhoff, A. W. (1985). Lexical access during eye fixations in reading: Effects of word-initial letter sequence. *Journal of Experimental Psychology: Human Perception & Performance, 11*, 272–285.

McConkie, G. W., & Rayner, K. (1975). The span of the effective stimulus during a fixation in reading. *Perception & Psychophysics, 17*, 578–586.

Mishkin, M., & Forgays, D. G. (1952). Word recognition as a function of retinal locus. *Journal of Experimental Psychology, 43*, 43–48.

Morris, R. K., Rayner, K., & Pollatsek, A. (1990). Eye movement guidance in reading: The role of parafoveal letter and space information. *Journal of Experimental Psychology: Human Perception & Performance, 16*, 268–281.

Orbach, J. (1952). Retinal locus as a factor in the recognition of visually presented words. *American Journal of Psychology, 65*, 555–562.

Parncutt, R., Sloboda, J. A., & Clarke, E. F. (1999). Interdependence of right and left hands in sight-reading, written and rehearsed fingerings of parallel melodic piano music. *Australian Journal of Psychology, 51*, 203–210.

Parncutt, R., Sloboda, J. A., Clarke, E. F., & Raekallio, M. (1997). An ergonomic model of keyboard fingering for melodic fragments. *Music Perception, 14*, 341–382.

Pollatsek, A., Bolozky, S., Well, A. D., & Rayner, K. (1981). Asymmetries in perceptual span for Israeli readers. *Brain & Language, 14*, 174–180.

Pollatsek, A., Lesch, M., Morris, R. K., & Rayner, K. (1992). Phonological codes are used in integrating information across saccades in word identification and reading. *Journal of Experimental Psychology: Human Perception & Performance, 18*, 148–162.

Rayner, K. (1975). The perceptual span and peripheral cues in reading. *Cognitive Psychology, 7*, 65–81.

Rayner, K. (1986). Eye movements and the perceptual span in beginning and skilled readers. *Journal of Experimental Child Psychology, 41*, 211–236.

Rayner, K. (1998). Eye movements in reading and information processing: 20 years of research. *Psychological Bulletin, 124*, 372–422.

Rayner, K., Balota, D. A., & Pollatsek, A. (1986). Against parafoveal semantic preprocessing during eye fixation in reading. *Canadian Journal of Psychology, 40*, 473–483.

Rayner, K., & Bertera, J. H. (1979). Reading without the fovea. *Science, 206*, 468–469.

Rayner, K., & Morris, R. K. (1992). Eye movement control in reading: Evidence against semantic preprocessing. *Journal of Experimental Psychology: Human Perception & Performance, 18*, 163–172.

Rayner, K., & Pollatsek, A. (1997). Eye movements, the eye-hand span, and the perceptual span during sight-reading of music. *Current Directions in Psychological Science, 6*, 49–53.

Rayner, K., Sereno, S. C., & Rayney, G. E. (1996). Eye movement control in reading: A comparison of two types of models. *Journal of Experimental Psychology: Human Perception & Performance, 22*, 1188–1200.

Rayner, K., Well, A. D., Pollatsek, A., & Bertera, J. H. (1982). The availability of useful information to the right of fixation in reading. *Perception & Psychophysics*, *31*, 537–550.
Reichle, E. D., Pollatsek, A., Fisher, D. L., & Rayner, K. (1998). Toward a model of eye movement control in reading. *Psychological Review*, *105*, 125–157.
Riemenschneider, A. (1941). *371 Harmonized Chorales by Johann Sebastian Bach*. New York: G. Schirmer, Inc.
Salis, D. L. (1981). Laterality effects with visual perception for musical chords and dot patterns. *Perception & Psychophysics*, *28*, 284–292.
Sloboda, J. A. (1974). The eye-hand span: An approach to the study of sight-reading. *Psychology of Music*, *2*, 4–10.
Sloboda, J. A. (1976). The effect of item position on the likelihood of identification by inference in prose reading and music reading. *Canadian Journal of Psychology*, *30*, 228–236.
Sloboda, J. A. (1977). Phrase units as determinants of visual processing in music reading. *British Journal of Psychology*, *68*, 117–124.
Sloboda, J. A. (1985). *The Musical Mind. The Cognitive Psychology of Music*. Oxford: Oxford University Press.
Truitt, F. E., Clifton, C., Pollatsek, A., & Rayner, K. (1997). The perceptual span and the eye-hand span in sight-reading music. *Visual Cognition*, *4*, 143–161.
Underwood, G. (1976). Semantic interference from unattended printed words. *British Journal of Psychology*, *67*, 327–338.
Underwood, G., Clews, S., & Everatt, J. (1990). How do readers know where to look next? Local information distributions influence eye fixations. *Quarterly Journal of Experimental Psychology*, *42A*, 39–65.
Underwood, N. R., & McConkie, G. W. (1985). Perceptual span for letter distinctions during reading. *Reading Research Quarterly*, *20*, 153–162.
Waters, A. J., Townsend, E., & Underwood, G. (1998). Expertise in musical sight-reading: A study of pianists. *British Journal of Psychology*, *89*, 123–149.
Waters, A. J., Underwood, G., & Findlay, J. M. (1997). Studying expertise in music reading: Use of a pattern-matching paradigm. *Perception & Psychophysics*, *59*, 477–488.
Weaver, H. E. (1943). A study of visual processes in reading differently constructed musical selections. *Psychological Monographs*, *55*, 1–30.

Chapter 10

The Reader's Spatial Code

Alan Kennedy, Rob Brooks, Lesley-Anne Flynn
and Carrie Prophet

This chapter reports the results of three studies each examining the way long saccades are used to reinspect elements of previously read text. In the first experiment participants showed evidence of accurately targeted reinspecting saccades but there was no suggestion that local punctuation in the form of a full stop acts to close off spatially coded information. The second experiment demonstrated that reinspecting saccades are coded with reference to a salient local "frame" bordering the text line. The third experiment confirmed the presence of large accurately located saccades, even when their presumed target was no longer visible.

Introduction

The starting point for this chapter is an experiment reported by Kennedy and Murray (1987). They examined eye movements as participants read displays like (1).

1. The novels in the library had started to go mouldy with the damp. novels

The defined target word (*novels*) was only displayed after the sentence had been read and the task was simply to indicate whether or not it had been present. Given that the sentences were short and not difficult to understand, reinspection would not be necessary for a correct response. However, participants sometimes launched remarkably accurate large saccades (e.g. 40–50 characters in extent) back into the sentence towards the defined target word. Average landing position was within a couple of letters of the centre of the defined target and saccade accuracy was unrelated to saccade extent, an outcome which seems difficult to square with the fact that the target was often in peripheral vision where its prior physical identification seems implausible, if not impossible. Kennedy and Murray concluded that "... readers must maintain, and use, a level of representation of text that involves the computation of spatial coordinates."

The Mind's Eye: Cognitive and Applied Aspects of Eye Movement Research

ISBN: 0–444–51020–6

In the intervening years there have been a number of attempts to elaborate and extend this "spatial coding" hypothesis. These have focused on the nature of the co-ordinate system (Baccino & Pynte, 1994; Kennedy & Baccino, 1994); the duration of the memorial record (Fischer, 1999); the role of spatialisation in syntactic parsing (Kennedy & Murray, 1984; Pynte, Kennedy, Murray & Courrieu, 1988); and the fact that low-level control over inter-word saccades in reading may differ between cases where the target is to a site which has already been inspected and cases where the target is novel (Radach & McConkie, 1998). In parallel with this work on eye movement control in reading there has been renewed interest in the role of spatially coded information in other tasks, for example, in object and picture processing (Brandt & Stark, 1997; Richardson & Spivey, 2000; Spivey & Geng, 2001).

To argue that spatialisation plays *some* role in reading skill may represent little more than an acknowledgement that reading poses unique spatio-temporal mapping problems. The temporal distribution of fixations is rarely isomorphic with the spatial distribution of information in the text, yet it is this latter which must be recovered for comprehension to occur. That is, the mental model derived from "John kissed Mary", must be independent of any particular sequence of fixations (and, in particular, the sequence "Mary", "kissed" and "John"). The reader must therefore either fixate text elements strictly in the order in which they are displayed or develop a spatial representation divorced from the temporal distribution of fixations (see Kennedy, 2000, for further discussion). Whether the spatially extended nature of text bestows processing advantages (Kennedy & Murray, 1984; Murray & Kennedy, 1988) or disadvantages (Cocklin, Ward, Chen & Juola, 1984; Juola, Ward & McNamara, 1982; Rubin & Turano, 1992) has, in fact, remained somewhat contentious. Even if it is concluded that some form of spatial coding is implicated in successful reading, we are largely ignorant as to its nature, its coordinate structure, its duration and its precise function. The metaphor of the page as a spatially-addressed "external memory" (Kennedy, 1983; O'Regan, 1992; Mackay, 1973) is appealing and has stimulated a great deal of experimental work, but in the majority of cases, the evidence recruited in its support has been indirect and very general in nature. Indeed, beyond the original study of Kennedy and Murray there is little evidence drawing directly on measured eye movements showing that readers know where on the page particular information is located and that they use this knowledge to direct reinspecting saccades.[1] The present chapter seeks to remedy this by reporting data from three experiments. The same experimental paradigm was employed in all three, involving the initial reading of a line of text followed by some form of decision (e.g. on the presence or absence of a target word). In each case, elaborate steps were taken to ensure that participants did not construe the task as *demanding* the execution of reinspecting saccades. In the first experiment, punctuation in the displayed sentence was manipulated to explore the influence of linguistic "closure" on the duration and accuracy of spatially coded information. In the second, the nature of the coordinate system was examined by embedding the displayed text in a salient local "frame" and manipulating its stability. In the third, the nature of the visibility of a defined target was manipulated. In all three experiments the over-riding objective was to estimate the frequency and accuracy of large reinspecting saccades.

Experiment 1

Punctuation as a Linguistic Control

Readers interpret punctuation signs as segmentation signals provided by the writer. In this respect, punctuation and inter-word spacing can be graded in terms of their ability to signal the closeness of adjacent text elements. For example, the full stop is ranked above the comma, which is itself a more powerful index of separation than the space (Fayol, 1997). Connectives of different kinds (e.g. "and", "after", "then") serve a parallel, if less well-defined, function in indicating to the reader the degree to which successive discourse elements should be bound together (Fayol, 1986). There have been very few eye movement studies looking directly at the function of punctuation in reading, although there is a growing awareness that punctuation marks have two consequences: (i) they increase the physical length of the word to which they are attached and this may influence low-level control over saccade programming; (ii) they convey linguistic information (primarily syntactic) which has at least the potential to affect saccade extent and/or the timing of fixations (Hill & Murray, 2000; Rayner, Kambe & Duffy, 2000).

Bringing these considerations to bear on an examination of long-range inter-word reinspections suggests that a profitable starting point would be the contrast shown in Figure 10.1 between the full stop, indicating closure of a sentence, and the connective "and", which invites the reader to provide thematic links between the two clauses. In the materials used in this experiment, the antecedent of the pronoun was ambiguous, increasing the possibility that readers may reinspect possible antecedents (e.g. "table" or "chair" in Figure 10.1) before making a decision. If the ability to launch accurate reinspecting saccades depends on text items being spatially indexed, the presence of the full stop takes on a special significance, since it raises the question as to whether linguistic closure reduces the accessibility of spatially indexed items in Clause 1.

Participants

Twenty-four English-speaking students of the University of Dundee volunteered to participate in the study. Participants were not paid to take part. All had normal or corrected-to-normal vision. Some had participated in other eye tracking experiments, but none were aware of the research hypotheses in the present study.

```
The table he found was broken.    The chair was also unstable.  Was it in good order?

The table he found was broken and the chair was also unstable.  Was it in good order?
```

Figure 10.1: Examples of materials used in Experiment 1.

Materials and Design

A set of 24 experimental sentences of the form illustrated in Figure 10.1 was created, each containing two clauses which could be separated by a full stop or concatenated with the conjunction "and". Associated with each line of text was a short question containing a pronoun, which could be related to either the first or second clause. As noted above, the antecedent of the pronoun was ambiguous, but the most plausible antecedent (in the judgement of the experimenters) was in the first clause for half of the items and the second clause for the other half. This method of construction was adopted deliberately to increase the likelihood of triggering reinspecting saccades, while avoiding participants seeing the *procedure* as one which demanded reinspections (that is, it was the text rather than the task which posed the problem in arriving at a decision). The experimental materials were interspersed with 40 fillers using a shuffling procedure, which ensured a random sequence of experimental items, and maximised spacing between their presentation. Filler items were of the same form and length and mixed ambiguous and unambiguous items. Two counterbalanced sets of items were constructed, assigning a random set of half of the experimental items to the full stop or "and" condition. Since it is well known that eye movement control is tightly coupled to physical properties of the text, spacing was matched in text containing the full stop, with additional spaces inserted so that the physical location of words on the display was identical in the two conditions (see Figure 10.1). A further set of six filler items was used for practice on the task. Factors in the design were Punctuation (full stop vs the word "and") and the location of the referent for the pronoun (Clause 1 or Clause 2). It should be borne in mind that since the categorisation of "target clause" was a matter of judgement by the experimenters, it might not have coincided with the participants' decisions.

Apparatus

The materials were presented in non-interlaced mode, in white-on-black polarity, on a monitor using a monochrome P4 phosphor running at 100 Hz frame rate controlled from a Control Systems Artist 1 graphics card mounted in an IBM compatible computer. At the viewing distance of 500 mm, one character subtended approximately 0.3 degrees of visual angle. Eye movements were recorded from the right eye using a Dr Bouis pupil-centre computation Oculometer interfaced to a 12-bit A-D device sampling X and Y position. This eye tracker has a resolution of better than 0.25 characters over the 60-character calibrated range (Beauvillain & Beauvillain, 1995). A dental wax bite bar and chin rest were used to minimise head movements. The eye movement recording system was calibrated prior to the presentation of each set of three experimental items. Viewing was binocular and horizontal eye position was recorded every two ms. Measures of fixation duration, fixation position, and intra- and inter-word saccade extent were computed off-line using statistical algorithms based on the effective resolution of the data for each individual participant with respect to the obtained noise in a given data set. The obtained resolution of the eye-tracking equipment was typically better than 0.5 characters.

Procedure

On arrival, participants were asked to read a set of printed instructions for the experiment. Each trial began with the display of a fixation marker in the form of an asterisk located two characters to the left of the initial letter of the sentence about to be displayed. Stable fixation of this for 500 ms led to display of the line of text. After reading this, participants pressed a right-hand button, which led to the display of a short question, located two spaces to the right of the displayed line, which remained on view (see Figure 10.1). Participants were instructed to respond "yes" or "no" to the question by pressing right-hand or left-hand buttons. It was explained that sometimes the correct answer would be unclear and in such cases they should respond with what they considered the best response. No feedback was given, and no reference was made explicitly or implicitly to reinspection of the sentences following the question display. Participants were initially given practice on the procedure and, once this had been mastered, the calibration technique was demonstrated. This required the fixation of five points distributed evenly across the horizontal axis of the screen at the point where the experimental items were to be displayed.

Results and Discussion

Reading time A preliminary analysis was carried out to check that the manipulation of punctuation produced measurable effects on reading performance. "Total pass" inspection time per word (Kennedy, Murray, Jennings & Reid, 1989) over the two clauses prior to presentation of the question was computed under the two presentation conditions and these data are shown in Figure 10.2.

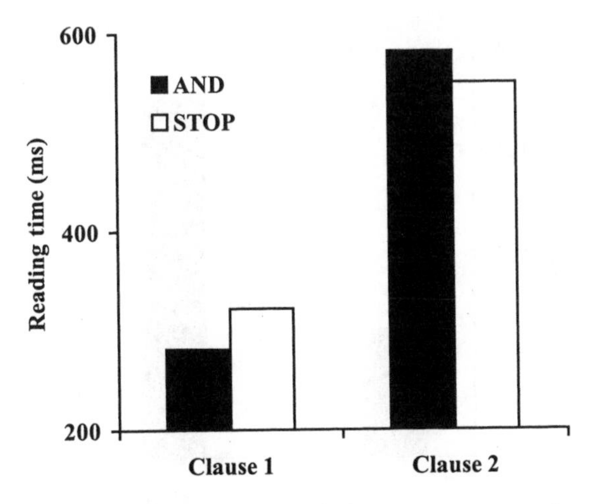

Figure 10.2: Total pass reading time per word (ms) in the two clauses of the experimental sentences in Experiment 1 as a function of the type of concatenation (full stop or the word "and").

Average processing time was much longer in the second clause than in the first (566 ms vs 302 ms), $F1$ (1, 22) = 75.3, $p < 0.001$, $F2$ (1, 22) = 142.3, $p < 0.001$. However, there was also a significant interaction between Clause and Punctuation, $F1$ (1, 22) = 8.64, $p < 0.01$, $F2$ (1, 22) = 7.58, $p < 0.01$. The presence of the full stop significantly lengthened reading time in Clause 1, $F1$ (1, 22) = 27.27, $p < 0.001$, $F2$ (1, 22) = 13.32, $p < 0.002$. The apparent trade-off into Clause 2 was not significant, $F1$ (1, 22) = 1.91, $F2$ (1, 22) = 2.43). There are several possible interpretations of this outcome, but consideration of these would involve issues other than those directly addressed in this chapter. The most plausible explanation is that "wrap-up" processes involving reinspections were triggered by the presence of the full stop.[2]

Probability of executing a large reinspecting saccade A large reinspecting saccade was defined as any left-going saccade more than 16 characters in extent leaving the zone defined by the boundaries of the displayed question and landing within five characters of the designated target. Of those left-going saccades leaving the region of the displayed question, 92% met this criterion. The relevant data are shown in Figure 10.3 and provide an answer to the primary question: the overall probability of making such saccades was 0.24. This was a much higher average rate than that reported by Kennedy and Murray (0.1), possibly because the "yes" decision in that study was very easy. Perhaps unsurprisingly, the probability of making a large reinspecting saccade to the second clause (0.36) was greater than to the first (0.13), regardless of the type of concatenation employed, F (1, 22) = 10.16, $p < 0.001$. Critically, however, the probabilities under the two types of concatenation (full stop vs "and") were virtually identical (0.240 and 0.245), and there was no hint of an interaction, all F values < 1. In summary, the incidence of this class of saccade was significantly above zero and

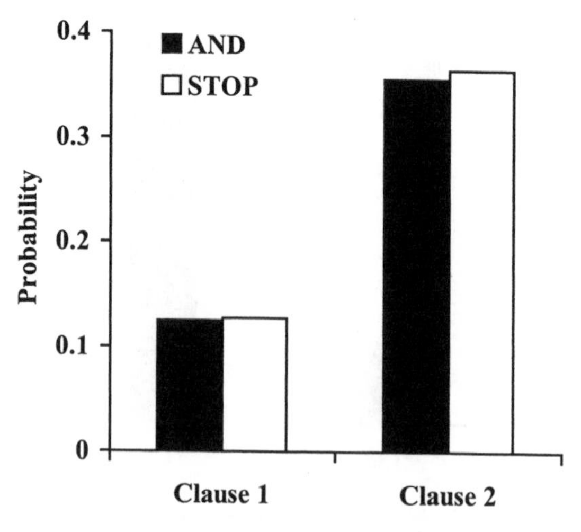

Figure 10.3: Probability of making a large reinspecting saccade as a function of target clause and type of concatenation (Experiment 1).

showed sensitivity to the distance of potential targets, but no differential effect of punctuation within the displayed line.

Saccade accuracy Measures of saccade accuracy were complicated by the fact that, for any item, although one target was more likely than the other, the experimental materials were ambiguous. As noted above, this problem was met by pre-defining a "best" target for each experimental item and only scoring reinspecting saccades, which fell within five characters of its boundary. Measures of signed and unsigned (i.e. absolute) error were derived with respect to the target centre[3] and are shown in Figure 10.4. Overall, large reinspecting saccades were very accurate, with an average signed error (taking the form of an undershoot) of less than one character, notwithstanding the fact that their average extent was 48.6 characters (i.e. approximately 16 degrees of visual angle). There was no effect at all of target distance ($F < 1$). That is, as predicted from the results of Kennedy and Murray (1987), there was no relationship between saccade extent and saccade accuracy. Furthermore, in parallel with the results on probability shown in Figure 10.3, the type of concatenation had no effect on average (signed) accuracy, $F < 1$.

Average absolute error (i.e. a measure of deviation from the target centre, regardless of whether the landing position was an overshoot or an undershoot) is possibly a more informative measure of accuracy. These data are also shown in Figure 10.4. On average, saccades fell within 2.7 characters of the target centre, a figure, which is remarkably close to, the 2.6 characters reported by Kennedy and Murray (1987) for left-going reinspections. Error did not vary as a function of target distance (Clause 1 vs Clause 2) or type of concatenation (all Fs < 1) and these factors did not interact ($F < 1$).

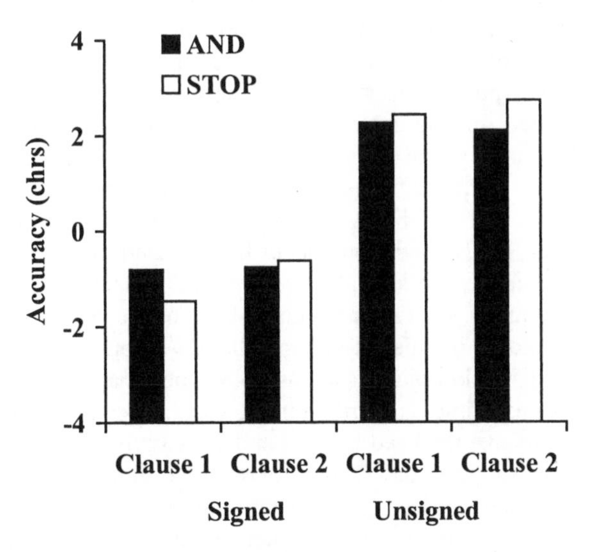

Figure 10.4: Signed and unsigned accuracy of large reinspecting saccades (Experiment 1). Data are plotted as a function of the target clause and type of concatenation. Negative values in the signed case are equivalent to an undershoot (i.e. a right-of-centre landing position).

In summary, this first experiment confirms that large saccades directed to sites in text, which have been previously inspected, are extremely accurate, typically landing within the boundary of their target. The two forms of concatenation influenced reading time and the distribution of eye movements within clauses, but had absolutely no influence at all on either the probability of launching large reinspecting saccades or on their accuracy.

We can conclude that the underlying memorial representation appears immune to local linguistic closure, even when its source, the full stop, is the most potent signal available. This proposition is, of course, restricted in the present case to consideration of a single horizontal line of text and to a single linguistic manipulation. With this caveat in mind, the outcome suggests that potential targets are coded less with respect to linguistic properties of the text than to physical properties of the displayed line. This proposition is examined more directly in Experiment 2, by manipulating the coordinate frame within which text is displayed.

Experiment 2

The Coordinate System: Shifting the Local Frame

A plausible interpretation of the results of Experiment 1 is that reinspecting saccades employ a form of local spatial code and that, for the experimental task employed, this may be defined by physical properties of the displayed line of text. In Experiment 2 we set out to examine this frame of reference in more detail. There have been numerous attempts to define the role of "frames of reference" in the control of eye movements and it would be hardly possible to summarise these here. Two broad research traditions may be identified. In one, stemming from the work of Marr (1982) the final "object-based" coordinate system is seen as the result of successive computations over largely "viewer-based" reference frames. In the other, eye movements are seen as coordinated entirely within a set of "object-centred" external reference frames (e.g. Ballard, Hayhoe, Pook & Rao, 1997). Both approaches have tended to ignore the limits set by neurophysiological evidence on spatial representation in saccade control (Findlay & Walker, 1999). But equally, the rather peculiar status of the "stimulus" in reading has been somewhat neglected in this on-going debate. Although displayed on various three-dimensional real-world objects such as books, newspapers and display screens, text is a two-dimensional stimulus. Further, in the vast majority of experimental studies measuring eye movements, the head is fixed and viewing is from a constant distance. These considerations simplify the search for the coordinate system in which text elements may be "spatially coded" (at least for simple laboratory procedures such as those considered here) but in spite of this fact the available evidence has produced strongly contrasting answers. On the one hand, success in tasks like "frame-stepped" reading (Bouma & de Voogd, 1974; Cocklin *et al.*, 1984) has tempted researchers to the conclusion that successive fixations need only be allocated retinal coordinates, although presumably involving at least some mechanism for combining the two retinal fields (Monk, 1985). On the other hand, the fact that the trade-off between landing and launch position, which obtains for progressive saccades is abolished in the case of reinspections, (Radach & McConkie, 1998) has been taken to suggest that readers do indeed know where already-inspected text elements are. But this is not to claim that such knowledge (i.e. the spatial coordinates of "word objects") is computed in a geocentric coordinate system. It seems much more probable that knowledge of where words lie on a page or screen will call on the representation of *relationships* between stimulus elements (what Wade & Swanston, 1996, call a "pattern-centric" code), in which salient local frames, such as that provided by the page boundary, may play a crucial role.

Experiment 2 set out to examine the role played by a local frame, using materials like those illustrated in Figure 10.5. The basic task — to indicate whether or not a target word had previously appeared — was identical to that employed by Kennedy and Murray (1987). However, although the displayed text was invariably stable with respect to the screen boundary (and the world at large), in one condition presentation of the defined target coincided with a shift in the boundary of a local frame. If potential target items are coded with reference to the rest of the visual world this manipulation should have little effect, since the text does not change position relative to the viewer. On the other hand, if spatial coding is carried out with reference to locally accessible frames, a more complex outcome is predicted. Relative to the most stable frame available to the viewer (the screen boundary) what changes in this task is the position of a local "frame marker" and the text itself can be defined as stable. However, relative to this local frame, the text may be perceived to shift (in this case to the right) in a way, which parallels induced movement.

Participants

Twenty-four English-speaking students of the University of Dundee volunteered to participate in the study. Participants were not paid to take part. All had normal or corrected-to-normal vision. Some had participated in other eye tracking experiments, but none were aware of the research hypotheses in the present study.

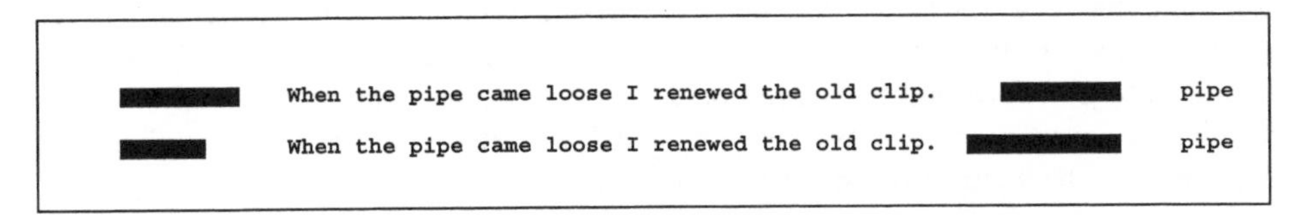

Figure 10.5: Example of the "frame shift" defined in Experiment 2. On presentation of the target word (*pipe*) a continuous block of characters defining a frame changed in size. Although the physical position of the sentence and target word did not alter, the displayed sentence can be described as shifting to the right relative to the boundary of the local frame. (Note the experimental sentences were longer than this example).

Materials and Design

A set of 48 experimental sentences, 61 characters in length, was constructed, together with a further set of six sentences used as practice. The materials were similar to those used by Kennedy and Murray (1987) with a single target word associated with each sentence (see Figure 10.5). For 24 of the sentences, targets were actually present in the sentence and for the remaining 24 they were a close synonym of a word in the sentence. In an equal number of sentences the defined target occurred either in the first half (defined as a "far" target) or in the second half (defined as a "near" target). Sentences were displayed on a single horizontal line between a prominent frame composed of a sequence of "characters" in which the complete array of 8×16 pixels defining a character was illuminated. This frame was separated from the displayed sentence by four character spaces at each end. The target word was displayed after a button press four characters beyond the right boundary of the frame. For half the materials, as the target word was displayed, the frame was changed in size by adding two characters on its right and removing two on its left (the procedure is illustrated in Figure 10.5). Two experimental lists were constructed, with the frame shift condition assigned to different half of the items in each list. Different random presentation sequences were used for each participant. The design thus had two within-subject factors: Target Distance (near vs far) and Frame Shift (stable vs unstable). Item List was treated as dummy factor in the analyses.

Procedure

The apparatus, data collection and calibration techniques were identical to those described for Experiment 1. Calibration took place after every three trials. On arrival, participants were asked to read a set of printed instructions for the experiment. Each trial began with the display of a fixation marker in the form of an asterisk presented for 500 ms and located two characters to the left of the initial letter of the sentence to be displayed. Stable fixation of this led to the sentence display. After reading the sentence a second fixation marker appeared on the right of the screen in the location to be occupied by the target word. Participants were instructed to fixate this and then press a button, at which point the target word was displayed. On a random half of the trials this button press also triggered a two-character frame shift. Participants were instructed to respond "yes" or "no" by pressing buttons to indicate whether the displayed target word had been presented in the sentence, which remained on view. No feedback was given and, as in Experiment 1, no reference was made, explicitly or implicitly, to the need to reinspect the sentence.

Results

The definitions of "large reinspecting saccade" and "target centre" were the same as those used in Experiment 1. Using the 16-character extent and five-character accuracy criteria, 95% of left-going saccades leaving the target word were defined as "large". For the purposes of analysis, only items demanding a "yes" response were scored. There was, therefore, no ambiguity as to the defined target of any large reinspecting saccade.

Probability of executing a large reinspecting saccade The probability of launching a large reinspecting saccade towards the defined target is shown in Figure 10.6. The overall probability of making such saccades, at around 0.2, was smaller than the average found in Experiment 1 (but, as noted above, the materials were probably easier). There was a pronounced effect of Target Distance, with saccades more likely when the target was relatively close (Near = 0.28, Far = 0.14), $F\ (1{,}20) = 9.35$, $p = 0.006$. As is clear from Figure 10.6, there was also a crossover interaction. Although this only approached significance, $F\ (1{,}20) = 3.80$, $p = 0.06$, it arose from a relatively strong tendency for more saccades to be launched to "near" targets when the frame shifted, $F\ (1{,}20) = 9.59$, $p = 0.005$. No other contrasts approached significance (including the apparent effect of Frame Shift for "far" targets), all $F < 1$.

Saccade accuracy As in Experiment 1, a landing position to the right of the target centre (i.e. an undershoot) was coded as negative. The relevant data are shown in Figure 10.7. There were too many missing cells to allow subject analyses, but a reasonable estimate of the reliability of the data could be arrived at by averaging across subjects for each item and computing the $F2$ statistic. There was a significant main effect of Frame Shift on signed error, $F2\ (1,\ 13) = 7.57$, $p = 0.02$, with a smaller average deviation from the target centre for items presented in the "frame shift" condition (0.13 chars) than in the "stable frame" condition (1.9 chars). Although at first sight it appears paradoxical that accuracy should be higher in the "frame shift" condition, this can be readily explained by noting that, for the experiment as a whole, it is the

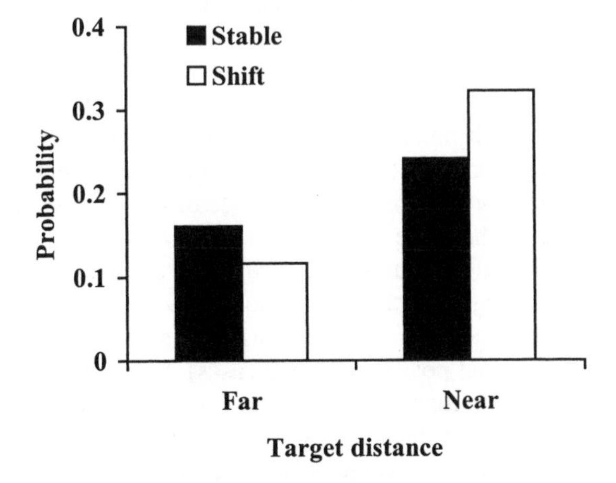

Figure 10.6: Probability of making a large reinspecting saccade as a function of target distance and frame stability (Experiment 2).

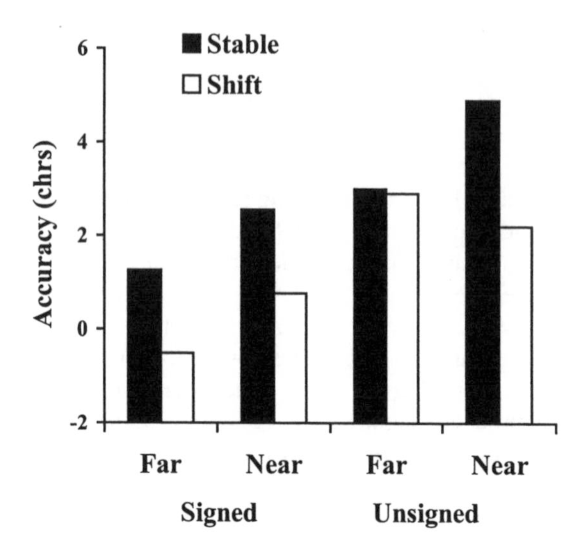

Figure 10.7: Absolute (signed) and relative (unsigned) accuracy of large reinspecting saccades (Experiment 2). Data are plotted as a function of the distance of the target and the stability of the frame. Negative values in the signed case are equivalent to an undershoot (i.e. a right-of-centre landing position).

"stable frame" condition, which represents baseline performance. That is, in the stimulus condition where nothing changes there is a tendency to overshoot the target centre by about two characters.[4]

Taking Figure 10.5 as an example, the outcome suggests that, in the "stable frame" condition, a saccade aimed at the target word "-pipe-" landed on the space before the word, whereas an equivalent saccade, launched under circumstances when the frame shifted two characters to the left, landed on the letter *i*, two characters further to the right. In other words, saccades appear to be computed relative to the frame and this has the effect of compensating almost exactly for what may be seen as an *apparent* shift to the right in text position. This outcome strongly supports the proposition that reinspecting saccades are spatially coded in terms of the most immediately accessible local frame, and not directly in terms of position defined, for example, with reference to the screen boundary or the more general visual environment.

Figure 10.7 also shows average unsigned error. Overall error in the "stable" condition, at 3.9 characters, was greater than that reported by Kennedy and Murray (1987), but nonetheless, the vast majority of large reinspecting saccades fixated the defined target (albeit not at an optimal position in some cases). Item analyses confirm a significant effect of Frame Shift, with a smaller average overshoot in the "shift" condition, $F2\ (1, 13) = 5.33$, $p = 0.04$, compensating for the apparent shift in text position. The more critical question is whether average error varied significantly with target distance and the results here clearly confirm the outcome of Experiment 1: there was no effect of the distance of the target on accuracy, and no interaction (Fs <1).

Experiment 3

Saccades to Non-visible Targets

The results of Experiments 1 and 2 offer reasonably convincing evidence that readers know where previously-presented text items are located, and make use of this spatially coded information to control aspects of re-reading. But we have not, as yet, considered one puzzling aspect of this outcome. None of the sentences presented were particularly difficult and all could surely have been held in memory, at least for the short interval following presentation. What purpose is served by re-examining particular words? Kennedy (1983) suggested that in the case of syntactically ambiguous material spatially indexed text elements represented an efficient way of "re-entering" words which had been the object of an inappropriate syntactic parse (see also, Pynte *et al.*, 1988). That is, since text is a permanent medium, so long as the reader knows where to look, the option to reinspect a physical token directly provides a more efficient route to reanalysis than reliance on a fading "echoic" memory. Oculomotor behaviour in this sense can be seen as somewhat similar to other pointing activities, attaching a spatial index to cognition.[5]

If information is spatially indexed in the way suggested by the outcomes of Experiments 1 and 2, it may be possible to dissociate the "pointing" and "reinspecting" aspects of eye movements. Typically, this has been examined using tasks where eye movements have been measured over the blank surface on which a visual array had previously been displayed (Christie & Just, 1975; Kennedy, 1982; Richardson & Spivey, 2000; Rothkopf, 1971; Spivey & Geng, 2001). Such eye movements do occur, but the accuracy of saccades directed to points in empty space has not been measured in a task using text. This is a significant omission, because it bears on the crucial question as to exactly what information is spatially coded. Three possibilities need to be considered. First, the target of reinspecting saccades over an otherwise blank surface could simply be a reflection of a general tendency to shift attention in one direction or another during problem-solving (Spivey & Geng, 2001; Weiner & Ehrlichman, 1976). In which case in the experimental task considered here the spatial distribution of "pointless reinspections" over a blank display would be effectively random. Second, saccades may be directed to the precise location held by a previously visible token (Kennedy, 1983; Kennedy & Baccino, 1994) in which case their "accuracy" should not differ from equivalent long-range saccades with a physically-present target. Finally, saccades may be launched to a location in space where some cognitive operation on a particular "target token" had taken place (Baccino & Pynte, 1994; Richardson & Spivey, 2000), in which case their "accuracy" might be expected to be less than when a target is physically present. Experiment 3 was conducted to distinguish between these alternatives.

The experiment used materials like those illustrated in Figure 10.8 in which display of a sentence was followed by a question containing either a pronominal or nominal anaphor. Two presentation modes were used: a "cumulative" mode in which the sentence remained visible after presentation of the question, and a "non-cumulative" mode, where presentation of the question removed the sentence display.

C: Mandy walked the dog and Martin kicked the ball.	Was he playing football?	[near yes]
NC:	Was he playing football?	[near yes]
C: Martin kicked the ball and Mandy walked the dog.	Was she playing football?	[near no]
NC:	Was she playing football?	[near no]

Figure 10.8: Examples of materials containing pronominal anaphors with "near" antecedents used in Experiment 3. There were an equal number with "far" antecedents. Each sentence contained either a near or a remote potential antecedent. The questions were not displayed until after a button-press indicating that the sentence had been read. Presentation was either "cumulative" [C] or "non-cumulative" [NC]. An equivalent number of sentences with "near" and "far" nominal antecedents was used [e.g. "after school Mary went to the library but Joe went fishing in the burn. Did Joe want a book?"].

Participants

Twenty-four English-speaking students of the University of Dundee volunteered to participate in the study. Participants were not paid to take part. All had normal or corrected-to-normal vision. Some had participated in other eye tracking experiments, but none were aware of the research hypotheses in the present study.

Materials and Design

A set of 96 experimental sentences like those illustrated in Figure 10.8 was constructed. A further 12 sentences were used as practice. Sentences varied in length, with an average of 60 characters. Half of the sentences were associated with a question containing a nominal anaphor and half with question containing a pronoun. Sentences were constructed such that the antecedent of the anaphor was present in either the first half (a "far" target) or the second half (a "near" target) of the sentence. Presentation mode (cumulative or non-cumulative) was blocked in the design, with half the participants receiving the conditions in one order and half in the other. Materials assigned to the two presentation modes were counterbalanced across the design. The experiment thus had factors of Target Distance, and Presentation Mode with Presentation Order treated as a dummy factor.

Procedure

The apparatus, data collection and calibration techniques were identical to those described for Experiment 1 and 2. On arrival, participants were asked to read a set of printed instructions for the experiment. Each trial began with the display of a fixation marker in the form of an asterisk presented for 500 ms and located two characters to the left of the initial letter of the sentence to be displayed. Stable fixation of this led to the sentence display. The designated question was not presented until participants pressed a button indicating that the sentence had been read. The question occupied the same physical location in both the cumulative and non-cumulative presentation modes, but in the non-cumulative mode the sentence was erased as the question display appeared.

Results

The definition of "large reinspecting saccade" and "target centre" were those adopted in the previous experiments and 82% of left-going saccades leaving the region of the displayed question met these criteria (but see note 5). In measures of accuracy, a landing to the right of centre (i.e. an undershoot) was coded negative.

Reading time To provide independent evidence that the manipulation of target distance had a measurable effect on reading behaviour, a preliminary analysis of overall reading time was carried out. Sentence-question pairs took longer to read when the target was "far" rather than "near" (6290 ms vs 6077 ms), $F\ (1,\ 22) = 6.75$, $p = 0.02$. There was no effect of Presentation Mode on overall reading time, $F < 1$. There were few errors (1.25 per subject) and no effect of either Presentation Mode or Target Distance on error rate (both $F < 1$) and no interaction ($F = 1.31$).

Probability of executing a large reinspecting saccade Figure 10.9 shows the probability of making a large reinspecting saccade in the various conditions of the design. Such saccades were much more likely in the cumulative presentation mode (0.50), when a target was physically present, than in the non-cumulative mode (0.13),

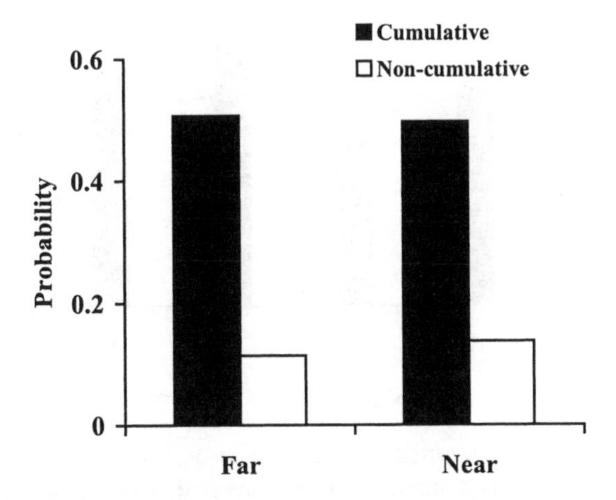

Figure 10.9: Probability of making a large reinspecting saccade as a function of target distance and presentation mode (Experiment 3).

F (1,22) = 46.00, $p < 0.001$. There was no effect of Target Distance, and no interaction, both $F < 1$. Although the probability of making a saccade at all in the non-cumulative mode was low, it was significantly greater than zero, F (1, 22) = 18.01, $p < 0.001$.

Saccade accuracy Measured accuracy under the cumulative presentation mode represents a reasonable estimate of baseline performance (see Figure 10.10). Across the experiment as a whole, in the measure of signed accuracy there was an overshoot of 4.7 characters in this condition. Participants were, therefore, considerably less accurate than in the previous two experiments. Significantly, however, there was no evidence that accuracy was worse in the non-cumulative presentation mode (it was, in fact, better, F (1,16) = 15.34, $p = 0.001$).[6] The measure of absolute (unsigned) accuracy suggests that performance was somewhat more variable in this experiment with a greater number of large reinspecting saccades landed outside the boundary of the target word. However, neither measure showed any effect of Target Distance and there was no interaction, all Fs < 1.

Saccade extent (characters) An analysis of saccade extent was carried out to test whether saccades in the non-cumulative presentation mode were qualitatively different from those resulting from a visible target. If saccades had no target in the non-cumulative condition, the effect of manipulated target distance should be restricted to the cumulative presentation mode, predicting a Target Distance x Presentation Mode interaction.

The relevant data are shown in Figure 10.11. The difference in extent appears slightly less marked in the non-cumulative presentation mode but analysis of these

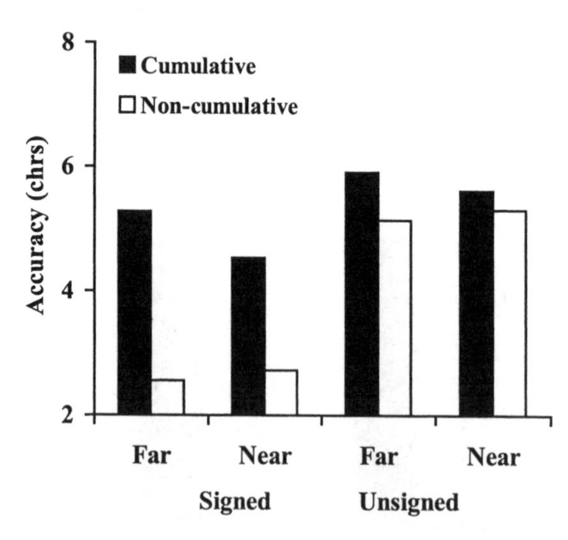

Figure 10.10: Signed and unsigned accuracy of large reinspecting saccades (Experiment 3). Data are plotted as a function of the target distance and presentation mode. Positive values in the signed case are equivalent to an overshoot (i.e. a left-of-centre landing position).

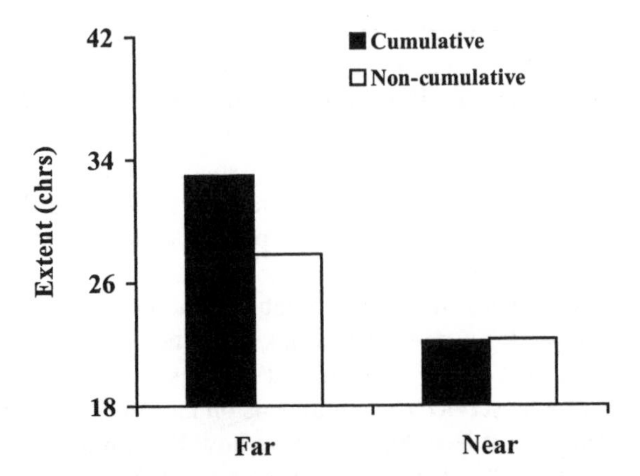

Figure 10.11: Saccade extent (chrs) of large reinspecting saccades as a function of target distance and presentation mode.

data showed no hint of a statistical interaction, $F\ (1,\ 22) = 1.19$, $p = 0.3$. It can be concluded that large reinspecting saccades found in the non-cumulative mode were sensitive to manipulated "target distance", even though the target was not visible. To confirm this, given the apparent trends in Figure 10.11, separate analyses were carried out for each presentation mode. There was a significant effect of Target Distance in both the cumulative, $F\ (1,\ 22) = 25.28$, $p < 0.001$, and non-cumulative modes, $F\ (1,\ 22) = 6.08$, $p = 0.02$.

Taken together, the results clearly do not arise as an artefact of some general increase in ocular motility associated with problem solving. Neither are they consistent with the related proposition that "pointless reinspections" occur to ill-defined regions in space where cognitive operations on a particular token had taken place.[7] Two outcomes point to an account in terms of operations over a spatially coded memory representation close to the physical layout of the text display. First, although rare, "accurate" saccades are launched to non-existent "targets"; second, manipulated target distance has virtually identical effects regardless of whether the target is visible or not.

Summary

These three experiments had a relatively modest objective: to demonstrate that readers can execute very accurate large reinspecting saccades to points in previously-inspected text located in peripheral vision, where physical identification of a potential target is impossible. The evidence from all three experiments supports this proposition and we will conclude with two brief comments on the wider implications of the data, one theoretical and one practical. Taken together, the experiments comment on the important theoretical notion that the printed page (or display screen) might function as an "external memory", to be inspected at will (what Mackay, 1973, called a "stable map").

There is substantial and convincing evidence that little detailed visual information survives the saccade and this has tempted some to the conclusion that our rich phenomenal experience is, in a sense, an illusion (O'Regan & Noë, 2001; but see Noë, Pessoa & Thomson, 2000). However, this is not to imply that perception lacks an underlying representational system, merely that its form (spatial pointers) is less obviously "visual" in nature. To gather information about the visual world demands that it is spatially tagged (Ballard, *et al.,* 1997) and the present data suggest that such tagging can be both precise and detailed.

The practical consequences of spatial coding in reading relate to the fact that the permanence of text provides a powerful means of reducing short-term memory load (and, incidentally, gives the reader a significant advantage over the hearer). Selective reinspection of a text in the service of comprehension is a much more efficient strategy than *ab-initio* re-reading. Little is known about how the ability to navigate through text in this way develops, but Kennedy and Murray (1988) looked at the different reading strategies deployed by children of above- and below-average reading skill, using a task very similar to that used in Experiment 3 here. The most striking feature of the results was that good readers were markedly more inclined to engage in selective reinspection, whereas poorer readers tended to "backtrack" or re-read. Comparison with a group of younger readers, constituting a "reading-age" control, strongly suggested that the more limited spatial knowledge of the poorer readers had played a causal role in their relative disability.

Acknowledgements

The authors are grateful to Robin Morris and an anonymous reviewer for helpful comments. This chapter was prepared while the first author was in receipt of an award from CNRS held in the Psychology Department, University Blaise Pascal, Clermont-Ferrand. Thanks to Michel Fayol and his colleagues for many hospitable discussions of the issues raised.

Notes

1 An exception is recent work by Vergilino and Beauvillain (2001), although this is restricted to consideration of two-word displays.
2 This interpretation was supported by an analysis of first-pass reading time (a measure excluding within-clause reinspections) which showed no significant effects of Punctuation.
3 In the case of targets with an even number of letters the "centre" was defined arbitrarily as the letter to the left of the physical bisection.
4 Interestingly, there was evidence of a range effect, with a smaller overshoot for more remote targets, $F2\ (1, 13) = 6.25$, $p = 0.03$.
5 This is not a novel observation. Lotze (1886) first proposed that the retina as a whole might be coded with respect to what he termed "Local Signs" (see also Ballard *et al.*, 1997)
6 This outcome can be related to the relatively inaccurate performance overall in the task which almost certainly arose because a proportion of saccades which met the "5-character" criterion with respect to the target boundary were actually intended for an adjacent word. It

appears that relatively fewer of these occurred in the non-cumulative presentation mode. A referee has correctly pointed out that a proportion of saccades intended to land on the verb might meet the 5-character criterion for the "actor" and could inflate undershoot scores in that condition. Why this might interact with presentation mode is unclear.

7 The data are also clearly inconsistent with the claim that gaze may be averted to improve recall by disengaging attention (Glenburg, Schroeder & Robertson, 1998).

References

Baccino, T., & Pynte, J. (1994). Spatial coding and discourse models during text comprehension. *Language and Cognitive Processes*, *9*, 143–155.

Ballard, D. H., Hayhoe, M. M., Pook, P. K., & Rao, R. P. N. (1997). Deictic codes for the embodiment of cognition. *Behavioral and Brain Sciences*, *20*, 723–767.

Beauvillain, C., & Beauvillain, P. (1995). Calibration of an eye movement system for use in reading. *Behavior Research Methods, Instruments and Computers*, *55*, 1–17.

Bouma, H., & de Voogd, A. H. (1974). On the control of eye saccades in reading. *Vision Research*, *14*, 273–284.

Brandt, S. A., & Stark, L. W. (1997). Spontaneous eye movements during visual imagery reflect the content of the visual scene. *Journal of Cognitive Neuroscience*, *9*, 27–38.

Cocklin, T. G., Ward, N. J., Chen, H. C., & Juola, J. F. (1984). Factors influencing readability of rapidly presented text segments. *Memory & Cognition*, *12*, 431–442.

Christie, J., & Just, M. A. (1975). Remembering the location and content of sentences in a prose passage. *Journal of Educational Psychology*, *68*, 702–710.

Fayol, M. (1997). On acquiring and using punctuation: A study of written French. In: J. Costermans and M. Fayol (eds), *Processing Interclausal Relationships* (pp. 157–178). New York: Erlbaum.

Findlay, J. M., & Walker, R. (1999). A model of saccade generation based on parallel processing and competitive inhibition. *Behavioral and Brain Sciences*, *22*, 661–674.

Fischer, M. (1999). Memory for word locations in reading. *Memory*, *7*, 79–116.

Glenburg, A. M., Schroeder, J. L., & Robertson, D. A. (1998). Averting the gaze disengages the environment and facilitiates remembering. *Memory & Cognition*, *26*, 651–658.

Hill, R. L., & Murray, W. S. (2000). Commas and spaces: Effects of punctuation on eye movements and sentence parsing. In: A. Kennedy, R. Radach, D. Heller and J. Pynte (eds), *Reading as a Perceptual Process* (pp. 565–589). Oxford: Elsevier.

Juola, J. F., Ward, N. J., & McNamarra, T. (1982). Visual search and reading of rapid serial presentations of letter strings, words and texts. *Journal of Experimental Psychology: General*, *111*, 208–227.

Kennedy, A. (1978). Eye movements and the integration of semantic information during reading. In: M. M. Gruneberg, R. N. Sykes and P. E. Morris (eds), *Practical Aspects of Memory*. London: Academic Press.

Kennedy, A. (1982). Eye movements and spatial coding in reading. *Psychological Research*, *44*, 313–322.

Kennedy, A. (1983). On looking into space: In: K. Rayner (ed.), *Eye Movements in Reading: Perceptual and Linguistic Processes* (pp. 237–250). New York: Academic Press.

Kennedy, A. (1992). The Spatial Coding Hypothesis. In: K. Rayner (ed.), *Eye Movements and Visual Cognition* (pp. 379–397). New York: Springer-Verlag.

Kennedy, A. (2000). Attention allocation in reading: Sequential or parallel? In: A. Kennedy, R. Radach, D. Heller and J. Pynte (eds), *Reading as a Perceptual Process* (pp. 193–220). Elsevier: Oxford.

Kennedy, A., & Baccino, T. (1994). The effects of screen refresh rate on editing operations using a computer mouse pointing device. *Quarterly Journal of Experimental Psychology, 48A*, 55–71.

Kennedy, A., & Murray, W. S. (1984). Inspection-times for words in syntactically ambiguous sentences under three presentation conditions. *Journal of Experimental Psychology: Human Perception and Performance, 10*, 833–849.

Kennedy, A., & Murray, W. S. (1987). Spatial coding and reading: Some comments on Monk (1985). *Quarterly Journal of Experimental Psychology, 39A*, 649–718.

Kennedy, A., Murray, W. S., Jennings, F., & Reid, C. (1989). Parsing complements: Comments on the generality of the principle of minimal attachment. *Language and Cognitive Processes, 4*, 51–76.

Lotze, H. (1886). *Outline of Psychology*. Boston: Ginn.

Mackay, D. M. (1973). Visual stability and voluntary eye movements. In: R. Jung (ed.), *Handbook of Sensory Physiology*, (Vol. 7). Berlin: Springer.

Marr, D. C. (1982). *Vision*. London: Freeman.

Monk, A. F. (1985). Theoretical Note: Coordinate systems in visual word recognition. *Quarterly Journal of Experimental Psychology, 37A*, 613–625.

Murray, W. S., & Kennedy, A. (1988). Spatial coding and the processing of anaphor by good and poor readers: Evidence from eye movement analyses. *Quarterly Journal of Experimental Psychology, 40A*, 693–718.

Noë, A., Pessoa, L., & Thomson, E. (2000). Beyond the grand illusion: What change blindness really teaches us about vision. *Visual Cognition, 7*, 93–106.

O'Regan, J. K. (1992). Solving the "real" mysteries of visual perception: The world as an outside memory. *Canadian Journal of Psychology, 46*, 461–488.

O'Regan, J. K., & Noë, A. (2001). A sensoriomotor account of vision and visual consciousness. *Behavioral and Brain Sciences, 24*.

Pynte, J., Kennedy, A., Murray, W. S., & Courrieu, P. (1988). The effects of spatialisation on the processing of ambiguous pronominal reference. In: G. Luer, U. Lass and J. Shallo-Hoffman (eds), *Eye Movement Research: Physiological and Psychological Aspects* (pp. 214–225). Gottingen: Hogrefe.

Radach, R., & McConkie, G. W. (1998). Determinants of fixation positions in words during reading. In: G. Underwood (ed.), *Eye Guidance in Reading and Scene Perception* (pp. 77–100). Oxford: Elsevier.

Rayner, K., Kambe, G. A., & Duffy, S. A. (2000). The effect of clause wrap-up on eye movements in reading. *Quarterly Journal of Experimental Psychology, 53A*, 1061–1080.

Richardson, D. C., & Spivey, M. J. (2000). Representation, space and Hollywood Squares: Looking at things that aren't there anymore. *Cognition, 76*, 269–295.

Rothkopf, E. Z. (1971). Incidental memory for location of information in text. *Journal of Verbal Learning and Verbal Behavior, 10*, 608–613.

Rubin, G. S., & Turano, K (1992). Reading without saccadic eye movements. *Vision Research, 32*, 895–902.

Spivey, M. J., & Geng, J. J. (2001). Oculomotor mechanisms activated by imagery and memory: Eye movements to absent objects. *Psychological Research, 65*, 235–241.

Vergilino, D., & Beauvillain, C. (2001). Reference frames in reading: Evidence from visually and memory-guided saccades. *Vision Research, 41*, 3547–3557.

Wade, N. J., & Swanston, M. T. (1996). A general model for the perception of space and motion. *Perception, 25*, 187–194.

Weiner, S. L., & Ehrlichman, M. (1976). Ocular motility and cognitive processes, *Cognition, 4*, 31–43.

Chapter 11

On the Processing of Meaning from Parafoveal Vision During Eye Fixations in Reading

Keith Rayner, Sarah J. White, Gretchen Kambe, Brett Miller and Simon P. Liversedge

Research dealing with parafoveal processing during eye fixations is reviewed. Four main topics are addressed: (1) parafoveal processing, (2) word skipping, (3) preview benefit effects, and (4) parafoveal-on-foveal effects. We argue that word skipping effects reflect the fact that a parafoveal word (word $n + 1$) has been identified on the fixation on word n. We also review evidence which strongly suggests that preview benefits during reading are not due to semantic processing of a parafoveal word. Finally, we review the more recent and more controversial research suggesting that the meaning of word $n + 1$ can influence the fixation time on word n, and argue that it is premature at this point to accept the validity of such findings with respect to normal reading. Implications of the research for serial attention shift models like the E–Z Reader model are also discussed.

Introduction

While a great deal has been learned about eye movements during reading over the past twenty-five to thirty years (Liversedge & Findlay, 2000; Rayner, 1978, 1998), there remain a number of unresolved issues (Starr & Rayner, 2001). In this chapter, we will focus our discussion on research related to one of these unresolved issues: parafoveal semantic processing of words. Research on this issue has apparently gained momentum because it has been assumed that if there were so-called parafoveal-on-foveal effects, or evidence that the meaning of the word to the right of fixation influences the duration of the fixation on the currently fixated word, it would be damaging to serial

The Mind's Eye: Cognitive and Applied Aspects of Eye Movement Research

ISBN: 0–444–51020–6

attention shift models such as the E-Z Reader model (Reichle, Pollatsek, Fisher & Rayner, 1998). We will return to this issue at the end of the chapter. However, before discussing the relevance of such research for the E-Z Reader model, we will first provide a general review of research on parafoveal processing and then discuss in turn (1) word skipping, (2) preview benefit effects, and (3) parafoveal-on-foveal effects.

We argue that readers can identify word $n + 1$ while fixating word n. When they do, its meaning becomes available and can influence fixation times on word n (and also word $n + 2$). Furthermore, if word $n + 1$ is identified, then the reader will skip that word. However, the primary argument we will make in this chapter is that if word $n + 1$ is not identified then its meaning does not become available and therefore cannot affect fixation time on word n. In such a situation, word $n + 1$ will typically not be skipped but must be fixated in order for word identification to occur. We argue that it would be premature at this point to accept evidence from studies that claim to show parafoveal-on-foveal effects as being very strong. Some of the evidence is based on tasks that may or may not easily generalize to reading and some of the evidence is inconsistent: while there are some studies showing parafoveal-on-foveal effects, there are also a number of studies that do not show such effects. While we will argue against the validity of some claims regarding parafoveal-on-foveal effects, we also readily agree that the case of word skipping according to our theoretical biases is prima facie evidence that the meaning of non-fixated words can be processed. Importantly, although word skipping shows that the meaning of parafoveal words can be processed, the characteristics of such words do not necessarily directly modulate the duration of the current fixation (parafoveal-on-foveal processing). To restate our general claim, we will argue that if a word is skipped, the meaning of that word was accessed on the current fixation (when the word in question was to the right of the currently fixated word). But, we will also argue that in the more frequent case when the word to the right of fixation is not skipped on the ensuing saccade that there is no solid support from the research data in reading tasks for (1) semantic preprocessing resulting in semantic preview benefits or (2) semantic preprocessing influencing the duration of the current fixation, that is parafoveal-on-foveal effects. This is not to say that such effects do not occur. Rather, our argument is that the evidence supporting such effects is questionable at this point in time.

Finally, before we turn to our review of parafoveal processing, we note that we are not arguing that no information is obtained from parafoveal words during reading. Indeed, in our review of preview benefit effects we will document that there are robust effects due to having a preview of a word before fixating on it. Furthermore, there is evidence for parafoveal-on-foveal effects due to unusual orthography at the beginning of the word to the right of fixation. Again, what we will be questioning is the extent to which the parafoveal-on-foveal effects are typical of normal reading.

Parafoveal Processing During Reading

It is clearly the case that foveal processing is critically important in reading. Indeed, it is generally agreed that the main purpose of eye movements during reading is to bring a region of text into foveal vision (the 2 degrees of central vision where acuity is highest) for detailed processing (Rayner, 1998). Readers need to get a good foveal glimpse of most of the words in the text for reading to proceed smoothly. On the other hand, reading on the basis of extra-foveal or parafoveal information is quite difficult, if not next to impossible (Rayner & Bertera, 1979; Rayner, Inhoff, Morrison, Slowiaczek & Bertera, 1981). Parafoveal vision is typically assumed to correspond to that part of the visual field falling from the end of the fovea out to about five degrees to the right of the fixation point, and likewise out to five degrees to the left of fixation. However, since information to the right of fixation is more important than information to the left for readers of English (Rayner, Well & Pollatsek, 1980), we will focus our discussion on parafoveal information to the right of fixation.

Although we have defined parafoveal vision in terms of degrees of visual angle, it is well known in reading that number of letters is the more appropriate metric to use than visual angle. This is because when the same text is read at different distances, even though the letters subtend different visual angles, the number of letters traversed by saccades is relatively invariant (Morrison & Rayner, 1981; O'Regan, 1983; O'Regan, Lévy-Schoen & Jacobs, 1983). Since for most normal sized text 3 or 4 letters equals one degree of visual angle, letters in parafoveal vision would typically extend from the 4th or 5th letter to the right of fixation out to the 15th or 20th letter to the right of fixation. Because acuity drops off steadily from the center of fixation, words presented in parafoveal vision are harder to accurately identify, and they take longer to identify, than words presented in foveal vision. Indeed, if a word of normal sized print is presented in parafoveal vision, it is identified more quickly and accurately when a saccade is made than when a saccade is not made (Jacobs, 1987; Rayner & Morrison, 1981). Thus, even though the program for planning and executing a saccade takes about 175–200 ms on average (Rayner, Slowiaczek, Clifton & Bertera, 1983), it is functional to move the eyes rather than hold them still.

The importance of parafoveal vision in reading was clearly demonstrated in the classic moving window studies (McConkie & Rayner, 1975; Rayner & Bertera, 1979). In these experiments, the eye-contingent display change technique was first introduced; readers' eye movements were monitored, and the amount of information that was available for processing on each fixation was controlled through the use of a moving window procedure (in which the appropriate text was available within the window and all letters outside of the window were perturbed in some way). These experiments, and many subsequent experiments (see Rayner, 1998 for an overview), demonstrated that when readers have the appropriate text information out to 14–15 letter spaces to the right of fixation (and information to the beginning of the currently fixated word or 3–4 letters to the left of fixation) that reading proceeds as if there was no window (i.e., the text was normal). So, clearly parafoveal information is being used in reading. Other experiments (see Rayner, Well, Pollatsek & Bertera, 1982) using the moving window paradigm demonstrated that if only the fixated word was available for processing within the window that reading rate was considerably slower than when more information was available. On the other hand, the reading rate was only about 20 words per minute slower than the full line condition when either readers had the currently fixated word and the word to the right of fixation available, or if they had the currently fixated word available as well as the first 3 letters of the next word (with the other letters replaced by visually similar letters). So, having only a single word slows reading, but having the fixated word and either the next word or the beginning letters of the next word (and the rest of the letters replaced with visually similar letters) available is almost as good as having the whole line. These results clearly demonstrate that both foveal and parafoveal information are important in reading.

The other thing that the original McConkie and Rayner (1975) study made apparent (also confirmed by a great deal of subsequent research, see Rayner, 1998) is that word length information (marked by the spaces between words) is critically important for programming saccades. Readers use the space information in parafoveal vision to program where their next saccade will go. It is also clear that some words in parafoveal vision can be identified without a direct fixation. Certainly, short words falling just to the right of fixation can be identified without direct fixation (Rayner, 1979; Rayner & McConkie, 1976). This leads us to the issue of word skipping, which we discuss in detail in the next section. Before moving to that topic, however, let us discuss the issue of parafoveal semantic preprocessing.

The idea of parafoveal semantic preprocessing was introduced by Underwood (1985) and referred to the notion that semantic processing of words to the right of fixation in the parafovea led to (1) faster recognition of that word on the next fixation (when the parafoveal word was directly fixated) and (2) rather intelligent guidance of the eyes to informative regions in text. We shall delay discussion of the first component of semantic preprocessing until the section on Preview Benefit. In the remainder of this section, we will briefly discuss the issue of eye guidance based on semantic preprocessing.

In a series of interesting experiments, Underwood and colleagues (Underwood, Bloomfield & Clews, 1988; Underwood, Clews & Everatt, 1990; Underwood, Clews & Wilkinson, 1989; Everatt & Underwood, 1992; Hyönä, Niemi & Underwood, 1989) examined the eyes' landing position in long words (10 or more letters) composed of informative and redundant halves. They reported that the eyes tend to initially move further into a word when the informative information is at the end of the word than at the beginning and suggested that semantic preprocessing of parafoveal words was responsible for the effect. However, the effect was sometimes small and sometimes non-significant. More importantly, in carefully controlled experiments, neither Rayner and Morris (1992) nor Hyönä (1995; see also Radach, Krummenacher, Heller & Hofmeister, 1995) were able to replicate this effect. They did demonstrate that readers quickly moved out of the beginning of a word when it was redundant (so that they could get to the more informative part of the word), but there was no evidence that the eyes initially moved further into words when the informative information was at the end of the word. In short, the data purporting to show semantic preprocessing in the parafovea do not provide compelling evidence to support the idea because of these failures to replicate the effect.

It is clearly the case that low level word length information acquired from parafoveal vision during reading is used in guiding eye movements to the next fixation location (Pollatsek & Rayner, 1982; Rayner, Fischer & Pollatsek, 1998). Furthermore, as shall be seen in the section on preview benefits, partial word information is obtained and used on the subsequent fixation to facilitate word recognition. But, our view is that readers typically don't acquire information regarding the meaning of parafoveal words. The obvious exception to this general statement is with respect to word skipping, to which we now turn.

Word Skipping During Reading

Whereas most of the words in a text are fixated, it is still the case that up to one third of the words are skipped. The most obvious case when words are skipped is when they are short (Blanchard, Pollatsek & Rayner, 1989; Brysbaert & Vitu, 1988; Rayner & McConkie, 1976; Rayner, Sereno & Raney, 1996). Since short words (under 3 letters long) are very frequent in English text, a lot of words are skipped simply because they are short. Whereas words that are 7–8 letters long are fixated most of the time in English, words that are under 3 letters long are skipped far more often than they are fixated. When two or three short words occur in succession in text, together they will typically only receive a single fixation. So, these words are apparently all identified on the same fixation. Likewise, when the word falling just to the right of the currently fixated word is short, it is typically also skipped and apparently identified on the fixation prior to the skip. However, it can't be the case that word length is the only factor that influences skipping. For example, O'Regan (1979, 1980) found that the function word **the** was skipped more frequently than 3-letter content words (see Gautier, O'Regan & LaGargasson, 2000 for more recent confirmation of this finding), presumably because it is more predictable, more visually familiar, and/or more frequent than the content words. Furthermore, Kennison and Gordon (1998) found differential skipping rates for different types of pronouns (of the same length), so perhaps the function that the pronoun serves is a factor. Thus, while it is clearly the case that word length exerts a very strong influence on skipping, it is also clear that other factors besides word length play a role.

Contextual constraint (or how predictable a word is from the prior context) also influences skipping behavior. This effect was first demonstrated by Ehrlich and Rayner (1981) who found that highly constrained words were skipped more frequently than unpredictable words. In subsequent work, Rayner and Well (1996) examined a range of predictability constraints so that target words were highly predictable (meaning that subjects could identify the target word in a cloze task 86% of the time), medium predictable (41% cloze accuracy), or low predictable (4% cloze accuracy). Table 11.1 shows example sentences and the basic results from the Rayner and Well study.

If we first consider the fixation time measures, we see that in first fixation duration (the duration of the first fixation on the target word independent of the number of fixations that were made) and gaze duration (the sum of all fixations on the target word before moving to another word), the high and medium predictability conditions were almost identical and both yielded shorter fixation times than the low predictability condition. However, when we consider the probability of fixating on the target word, we see that the medium and low predictability conditions did not differ from each other, and both were more likely to yield a fixation on the target word than the high predictable condition. Combined with results reported by Hyönä (1993), it is clear that only highly predictable words are skipped during reading (when the target words are not short).

Table 11.1: Example sentences and reading time measures on the target word (shown in bold) from Rayner and Well (1996).

High Pred: He mailed a letter without a **stamp** so it didn't arrive.
Medium Pred: Some of the ashes dropped on the **carpet** to her dismay.
Low Pred: They were startled by the sudden **voice** from the next room.

Constraint	**FFD**	**GAZE**	**FP**
High	239	261	0.78
Medium	240	261	0.88
Low	250	281	0.90

Note: FFD = first fixation duration (in ms), GAZE = gaze duration (in ms), and FP = Fixation probability.

Our view is that if a word is skipped, it was processed on the prior fixation. Pollatsek, Rayner and Balota (1986) reported that when a word is skipped, the duration of the prior fixation is inflated. Furthermore, there is some evidence suggesting that the duration of the fixation following a skip is also inflated (Reichle *et al.*, 1998). These results are a bit controversial at the moment since two analyses based on large corpora of data (Engbert, Longtin & Kliegl, 2002; Radach & Heller, 2000) did not yield strong evidence of inflated fixations prior to a skip. However, in a recent analysis, Rayner, Ashby, Pollatsek and Reichle (2003) again found an effect in which the duration of a fixation prior to skipping a predictable word was inflated by 23 ms. Longer fixation durations before skipping might reflect processing of the skipped word on the prior fixation or re-programming of the saccade target. It is difficult at this point to determine why some studies have reported inflated fixations and some have not. Inflated processing time prior to a skip is an important characteristic of the E-Z Reader model (Reichle *et al.*, 1998), and further research will need to determine more precisely the relationship between skips and fixations. It may be the case that not much processing cost due to skipping is inflicted on the processing system as a result of a skip. For the moment, however, and more central to the points addressed in this chapter, we continue to believe that skipping is prima facie evidence that the meaning of a parafoveal word can sometimes be processed on the current fixation.

Parafoveal Preview Benefits

One of the most robust findings in research on eye movements and reading is that a preview of the word to the right of fixation results in shorter fixations on that word when it is fixated on the next fixation. This was first demonstrated by Rayner (1975) and has been replicated many times (see Rayner, 1998). The size of the preview benefit is typically on the order of 20–50 ms. However, exactly how much of a preview benefit the reader obtains is influenced by characteristics of the text. For example, Balota, Pollatsek and Rayner (1985) found greater parafoveal preview benefit when the target word was predictable from the prior text, indicating that the extraction of parafoveal information is more efficient when aided by sentential context. Interestingly, they also found that a word is not nessarily accurately identified since nonwords that were visually similar to a target word were skipped some of the time. Furthermore, as a corollary to Balota *et al.*'s finding, it has also been found that when the difficulty associated with processing the fixated word is high that the extraction of parafoveal information decreases (Henderson & Ferreira, 1990; Kennison & Clifton, 1995; Rayner, 1986).

A number of studies have addressed the basis of preview benefit and the general conclusion is that benefit derives from abstract letter codes (McConkie & Zola, 1979; Rayner, McConkie & Zola, 1980), orthographic codes in the form of the beginning letters of words (Inhoff, 1989; Rayner *et al.*, 1982), and phonological codes (Henderson, Dixon, Petersen, Twilley & Ferreira, 1995; Pollatsek, Lesch, Morris & Rayner, 1992). Interestingly, and somewhat surprisingly, research aimed at determining the extent to which semantic codes contribute to the amount of preview benefit has not found evidence to support the importance of semantic codes.

Rayner, Balota and Pollatsek (1986) used the boundary paradigm introduced by Rayner (1975) to examine the extent to which semantic codes contributed to preview benefit. In their experiment, readers read a sentence such as "My younger brother has brilliantly composed a new song for the school play". When subjects began reading the sentence, either **song**, **tune**, **sorp**, or **door** occupied the target location. When the reader's eye movement crossed an invisible boundary location (the letter **w** in **new**), a display change occurred such that the preview stimulus changed to the target word **song**. Among the four previews, **song** is identical to the target word, **tune** is semantically related to the target word, **sorp** is a nonword that is orthographically and visually similar to the target word, and **door** is an unrelated word. In a standard priming experiment (in which the prime appeared for 200 ms in the center of vision followed by the target word in the same location), Rayner *et al.* found a significant priming effect for the related word (on the order of 20 ms). However, in the reading experiment (where the prime word appeared in the parafovea followed by the target word in the center of vision following an eye movement), when the same stimuli that produced a standard priming effect were used, there was no facilitation from the semantically related preview. On the other hand, the orthographically related preview did provide facilitation (see Table 11.2).

More recently, Altarriba, Kambe, Pollatsek and Rayner (2001) provided a stronger test of the extent to which semantic codes contribute to the amount of preview benefit. They had Spanish-English bilingual readers, who were equally fluent in each language, read sentences in which the preview was either identical to the target word (e.g. **sweet-sweet**), a cognate (which looked very similar to the target and meant the same thing as the target word e.g. **crema-cream**), a pseudo-cognate (which looked very similar to the target word, but did not mean the same thing e.g. **grasa-grass**), a non-cognate translation (which meant the same thing as the target word, but did not look like it e.g. **dulce-sweet**), or a control word (which was semantically and orthographically

Table 11.2: Reading time measures on the target word as a function of preview from Rayner, Balota and Pollatsek (1986).

Preview	FFD	GAZE
Identical (song-song)	214	246
Semantically related (tune-song)	230	286
Unrelated (door-song)	234	290
Visually similar nonword (sorp-song)	215	251

Note: FFD = first fixation duration (in ms), GAZE = gaze duration (in ms).

unrelated to the target word e.g. **torre-cream**). Note that the design of this study makes it possible to assess the extent to which semantic and orthographic codes contribute to preview benefit. That is, a comparison of the cognate, pseudo-cognate, and noncognate conditions provides more information than the Rayner *et al.* study was able to provide. Specifically, by using cognates and translations it was possible to more systematically vary the orthographic and semantic similarity between the target and preview. Table 11.3 shows the results of the study. Importantly, the non-cognate translation previews, which were semantically, but not orthographically, related to the target word, did not provide any preview benefit. The basic conclusion from the study is that orthographic codes yield preview benefit, but that semantic codes are not used in integrating information across saccades.

Other experiments (Lima, 1987; Kambe, 2003) have likewise provided no evidence that morphological codes are used in integrating information across saccades for readers of English. On the other hand, some recent studies (Deutsch, Frost, Pollatsek & Rayner, 2000; Deutsch, Frost, Peleg, Pollatsek & Rayner, 2003) have demonstrated parafoveal preview benefit effects in Hebrew due to morphology. In Hebrew, morphology tends to be more important than it is in English and the experiments by Deutsch and colleagues have demonstrated that a preview of the root morpheme

Table 11.3: Example stimuli and reading time measures on the target as a function of preview from Altarriba, Kambe, Pollatsek and Rayner (2001).

Condition	Preview	Target	FFD	GAZE
Identical	sweet	sweet	267	342
Cognate	crema	cream	270	344
Pseudo-cognate	grasa	grass	273	346
Non-cognate	dulce	sweet	287	367
Control	torre	cream	290	371

Note: FFD = first fixation duration (in ms), GAZE = gaze duration (in ms). The reading time measures show the combined results for Spanish and English targets.

facilitates processing of a target word. This finding is particularly interesting since the root morpheme is distributed throughout the word and not confined to the beginning of the word (where preview benefits in English are strongest). Future research will have to determine if such effects are apparent in other languages. However as noted earlier, with respect to English, the preview benefit (or priming effect) that is obtained is due to orthographic codes, abstract letter codes, and phonological codes. But, semantic information apparently does not contribute to the effect.

Parafoveal-on-foveal Effects?

There are two distinct types of parafoveal-on-foveal effects. Fixation durations on the foveal word n might be influenced by the characteristics of word $n + 1$ either when (1) word $n + 1$ is skipped or (2) when it is subsequently fixated. As discussed above, for the first such type of effect (word skipping) there is some suggestion that fixations on word n and word $n + 2$ are inflated when word $n + 1$ is skipped. Such effects could be explained by parallel processing of multiple words (Engbert *et al.*, 2002), or, as serial attention shift models (Reichle *et al.*, 1998) suggest, in terms of re-programming of saccades to word $n + 2$. In this section we will focus on parafoveal-on-foveal effects when both word n and word $n + 1$ are fixated. With respect to this type of parafoveal-on-foveal effect, Kennedy (1998, 2000), Kennedy, Pynte and Ducrot (2002), Murray (1998), Murray & Rowan (1998), Inhoff, Radach, Starr, and Greenberg (2000), Inhoff, Starr and Shindler (2000), and Underwood, Binns and Walker (2000) have all reported results in which some characteristic of the word to the right of fixation influenced how long readers looked at the fixated word. While such effects may be valid, we believe that at the moment there are at least three reasons why it might be premature to conclude that they demonstrate parafoveal-on-foveal effects during reading. First, we suspect that there is some question about the generalizability of the results. Second, studies examining the effect of the frequency of the word to the right of fixation have typically reported null effects. Third, not all studies show parafoveal-on-foveal effects. We will now discuss each of these points.

Generalizability

In the experiments reported by Kennedy and colleagues, while subjects are engaged in tasks that admittedly bear some similarities to reading, the fact remains that they are not reading. Specifically, in Kennedy and colleagues' studies the tasks usually require subjects to look at a fixation point and then words are presented to the right of fixation. The task is to determine if the words belong to a particular semantic category (e.g., clothing) or look or mean the same. In these studies, frequently a large number of variables are simultaneously orthogonally manipulated to produce results that are quite complex (and vary slightly from study to study). These experiments have produced parafoveal-on-foveal effects related to the frequency of the word to the right of fixation (though the effect is modulated by the informativeness of word $n + 1$ and the word length of word *n* and word *n* + 1) and orthographic properties of the letters at the beginning of the word to the right of fixation. While subjects do indeed fixate on the target words for times approximating reading fixations (though the mean gaze durations are usually somewhat longer than in reading), it is clearly the case that the subjects are not performing the majority of psycholinguistic processes that would normally occur during reading. To this extent, these tasks do not approximate reading. Instead, it seems to us that they are much more like a variant of a visual search task. While experiments using a visual search methodology have provided a substantial amount of data that are informative for saccade generation during the processing of visual arrays and scenes, it is not clear that they provide data relevant to the oculomotor control decisions relating to normal language processing. In support of this claim, Rayner and Fischer (1996) and Rayner and Raney (1996) demonstrated that when subjects search through text for a target word, the ubiquitous frequency effect wherein readers look longer at low frequency words than high frequency words (Inhoff & Rayner, 1986; Rayner & Duffy, 1986; see Rayner, 1998 for an overview) disappears. Thus, when normal language processing is not required in order to perform the task, very reliable phenomena occurring due to the linguistic characteristics of the text fail to occur. Note that although some artificial task studies of parafoveal-on-foveal effects have found frequency effects (Kennedy *et al.*, 2002), not all have, and such tasks simply do not demand many aspects of normal language processing (such as the effects in reading that are due to predictability and to sentence structure). Thus, we suspect that there are very good reasons to question the generalizability of these results to reading.

In the experiments by Murray (1998) and Murray and Rowan (1998), once again an artificial task that at best approximates reading is used. Specifically, subjects were presented with two sentences such as "The savages smacked the child" and "The uranium smacked the child" one below the other. Subjects were required to make a decision regarding whether the sentences were physically identical or not. When they were different, the two sentences differed by one word. Importantly, psycholinguistic processing is not required in order to make such a judgment. Indeed, one could argue that the task used by Murray is also some variant of a visual search task, or some form of visual discrimination task. Murray's key manipulation was that the sentences varied in the extent to which they were plausible or implausible. For example, the sentence "The uranium smacked the child" is obviously implausible because **uranium** is an inanimate object and therefore can't smack anything. Murray found that when readers fixated on word *n* (**uranium** in the example), if word *n* + 1 (**smacked**) resulted in an implausible reading, the fixation on word *n* was inflated even when word n+1 was subsequently fixated. This was especially true when the reader fixated near the end of word *n*. So, in contrast to many studies (see Garrod & Terras, 2000; Pickering & Traxler, 1998) in which effects of plausibility show up rather late in processing (often after the eyes have left the implausible region), Murray found effects of plausibility before the word which made the sentence implausible had been fixated. However, when we (Rayner & Miller) attempted to replicate this result in a natural reading situation in which readers were asked to read sentences and respond to comprehension questions, no evidence for a parafoveal-on-foveal effect was found. Indeed, the plausibility manipulation had very little effect on first pass reading time and was primarily limited to second pass reading times. Perhaps the fact that the primary task in Murray's experiment was to decide if the two sentences were physically identical led subjects to employ processes that are different from those they use when they read normally.

Unlike the studies we just discussed, two studies by Inhoff *et al.* (2000) did involve subjects in a reading task. Inhoff, Starr *et al.*'s experiment is very similar to other preview studies we described above in that a display change occurred when readers moved their eyes across an invisible boundary. Sentences such as the following were used: "He approached the yellow traffic light with some caution." When readers began reading the sentence, either the target word **light** was in the sentence or one of the following previews was present: **LIGHT** (an uppercase version preview of the target word), **qvtqp** (an orthographically illegal preview), or **smoke** (an unrelated preview word). When fixations on the target word following the display change were examined, the results were consistent with the results we mentioned earlier (the latter two conditions yielded longer fixation times on the word than the identical or uppercase condition). Thus, the results are consistent with orthographic/abstract letter code facilitation as the basis for the preview benefit. The other aspect of Inhoff, Starr *et al.*'s data which has been taken as support for parafoveal-on-foveal effects was that they also examined the fixation time on **traffic** as a function of the nature of the preview. While there was no effect due to the meaning of the word to the right of fixation (there was no difference between when **light** or **smoke** was the preview), the capital letter preview condition and the orthographically illegal preview condition both resulted in longer fixation times on **traffic**. Interestingly (see the next section), Inhoff, Starr, *et al.* reported a post-hoc analysis in which frequency was examined. They found no effect of the frequency of the word to the right of fixation on the current fixation time. Inhoff, Starr *et al.* therefore found no evidence that semantic processing of a word influences fixation durations on the previous word. However, they did find some evidence that unusual orthography at the beginning of word $n + 1$ resulted in longer fixations on word n. Underwood *et al.* (2000) also found that unusual orthography at the word beginning of word $n + 1$ can increase fixation durations on word n.

Inhoff, Radach *et al.* did find some evidence for semantic processing of the word to the right of fixation. They used sentences such as "Did you see the picture of her mother's mother at the meeting?" Reading times on **mother's** were compared when the following word was **mother** (identical), **father** (associated), or **garden** (unassociated). There were no text changes in this experiment. Inhoff, Radach *et al.* found that first fixation and gaze durations were shorter if the following word was either identical or associated to the fixated word, compared to when the following word was unassociated. Nevertheless, the fact that Inhoff, Starr *et al.* and the studies discussed below have not found evidence for semantic parafoveal-on-foveal effects indicates that the effects are not robust.

Furthermore, it is also interesting to note that the direction of parafoveal-on-foveal effects is not consistent. Sometimes the effect from the word to the right of fixation manifests itself in shorter fixations and other times it manifests itself in a longer fixation. For example, in a sentence reading experiment Underwood *et al.* (2000) found that fixations on the foveal word were longer when word $n + 1$ had an informative initial trigram. In contrast, in artificial task experiments by Kennedy (1998, 2000) fixations on the foveal word were shorter when word $n + 1$ had an informative initial trigram. While it might be possible to account for these differences (see Kennedy *et al.*, 2002), one might hope for more consistency.

While the studies by Kennedy, Murray and Inhoff *et al.* have been taken by some as evidence for parafoveal-on-foveal effects, it is important to note that there has been no consistent evidence that the meaning of the word to the right of fixation influences the fixation time on the current fixation. Indeed, Inhoff. Starr *et al.*'s primary finding (see also Underwood *et al.*) was that orthographic distinctiveness (either in the form of an unfamiliar uppercase string or an orthographically illegal string) of the word to the right of fixation can have some effect on the current fixation time. Such an effect has also been obtained by Kennedy and we shall see some further evidence of this effect below. But, the generalizability issue that we have raised (concerning the extent to which non-reading tasks generalize to reading) together with Inhoff, Starr *et al.*'s results are consistent with the idea that the meaning of the word to the right of fixation is not influencing the current fixation time.

Lack of Parafoveal-on-foveal Frequency Effects

As we noted above, there are very robust frequency effects for fixated words: low frequency words are reliably fixated longer than high frequency words (see Rayner, 1998 for a review). Some studies using artificial tasks have found effects of the frequency of word $n + 1$ on reading times on word n (Kennedy, 2000; Kennedy *et al.*, 2002) and others have not (Schroyens, Vitu, Brysbaert & d'Ydewalle, 1999). However, four different reading studies (Carpenter & Just, 1983; Henderson & Ferreira, 1993; Inhoff, *et al.*, 2000; Rayner *et al.*, 1998) have examined the effect of the frequency of the word to the right of fixation on the fixation time on the currently fixated word, and all of these studies have reported null effects. Namely, they found that the frequency of the word to the right of fixation did not influence fixation time on the fixated word. Such findings are clearly at odds with the view that semantic characteristics of the parafoveal word influence the fixation time on the currently fixated word. In order for the semantic characteristics of the parafoveal word to become available and influence foveal processing, readers must necessarily have identified the parafoveal word. Probably the most robust finding associated with word identification is the frequency effect. Consequently, we might therefore reasonably anticipate that the frequency of the parafoveal word would significantly modulate any semantic parafoveal-on-foveal effect that might occur. That is to say, if semantic parafoveal on foveal effects did occur, then one would expect longer fixations on the currently fixated word when the parafoveal word was low frequency and shorter fixations when the parafoveal word was high frequency. Rather, as we have noted, the fixation time is driven by the frequency of the currently fixated word, **not** the parafoveal word. It is also interesting to note that the effects reported Kennedy *et al.* (2002) are in the opposite direction with longer fixations on foveal words when $n + 1$ was high frequency (though they do provide arguments for why the direction of the effect might be different).

Lack of Parafoveal-on-foveal Effects

Although the topic of parafoveal-on-foveal effects has recently attracted a fair amount of attention, the issue was actually anticipated in the early Rayner (1975) study. Although this study is typically discussed in the context of preview benefit, it is the case that possible parafoveal-on-foveal effects were examined. Specifically, Rayner examined the duration of the last fixation prior to crossing the boundary and fixating on the target word. Furthermore, the duration of the last fixation was examined as a function of the launch site of the saccade that crossed the boundary. Figure 11.1 shows these data.

In Rayner's study, reader's read sentences such as "The soldiers guarded the palace throughout the day" with **palace** as the target word following the display change. The preview was either **palace**, a word that was orthographically similar to the target word (**police**), or nonwords that varied in their orthographic similarity to the target word (**pcluce**, **pyctce**, or **qcluec**). As is clear in Figure 11.1, there were no differences between any of the conditions when the saccade was launched from 4 or more character spaces from the beginning of the target word. When the eyes were 1 to 3 character spaces from the beginning of the target word, launch site fixations were longer for the nonword conditions than the two word conditions. Apparently then, the orthographic irregularity of the beginning of the nonwords was registered by the processing system resulting in a longer launch site fixation. Such a pattern was also observed in some of the studies we reviewed above. It should also be noted that in the original Rayner (1975) study, the boundary location was sometimes set partway into the target word. Thus, it was quite possible for readers to occasionally fixate on the preview word before it changed to the target word. In this case, the three nonword conditions resulted in fixations on the target word that averaged over 400 ms (in comparison to 218 ms for the two word conditions).

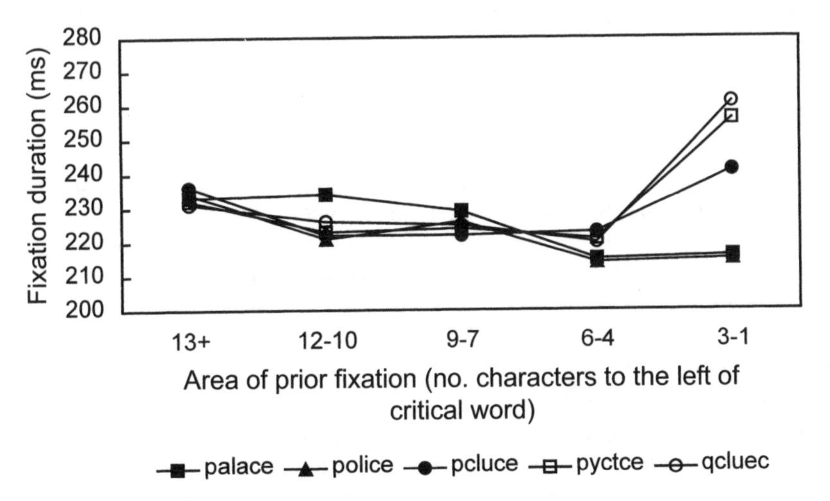

Figure 11.1: Mean fixation durations prior to fixating the critical word for each condition and for each area of prior fixation.

Three other studies that examined the duration of the fixation prior to fixating on a target word likewise revealed little difference as a function of the semantic information in the target word location. First, in the Rayner *et al.* (1986) study discussed earlier, examination of the last fixation duration and the gaze duration on the word fixated prior to crossing the boundary location and triggering the display change revealed no effects of the preview word (see Table 11.4). Second, examination of the last fixation prior to fixating on the target word in the Altarriba *et al.* (2001) study discussed earlier likewise revealed no effects of the preview word. Third, a recent study by White and Liversedge (2003) also found no effects on the launch site fixation.

White and Liversedge had readers read sentences in which a misspelling occurred at the beginning of a target word. A correctly spelled condition (e.g., **agricultural**) was compared to four misspelled conditions that created word beginnings with different degrees of orthographic irregularity. These ranged from initial trigrams that were pronounceable and high frequency (e.g., **acricultural**) to initial trigrams that never occur at the word beginning and which were unprounceable (e.g. **ngricultural**). There was no display change manipulation in this experiment so readers could directly fixate the misspelled word. Importantly, White and Liversedge did find that first

Table 11.4: Reading time measures on the word prior to the target as a function of preview from Rayner, Balota and Pollatsek (1986).

Preview	**Fixation *n*–1**	**GAZE**
Identical (song-song)	228	251
Semantically related (tune-song)	228	250
Unrelated (door-song)	222	251
Visually similar nonword (sorp-song)	219	248

Note: Fixation *n*–1 = duration of fixation prior to fixating the target (in ms), GAZE = gaze duration (in ms).

Table 11.5: Last fixation duration prior to fixating on the target word as a function of preview from Altarriba, Kambe, Pollatsek and Rayner (2001).

Preview	**Fixation *n*–1**
Identical (sweet-sweet)	265
Cognate (crema-cream)	266
Pseudo-cognate (grasa-grass)	270
Non-cognate (dulce-sweet)	269
Control (torre-cream)	268

Note: Fixation *n*–1 = duration of fixation prior to fixating the target (in ms). The results for Spanish and English targets have been combined.

fixation landing positions on the critical word were significantly nearer the word beginning if the word was misspelled compared to the correctly spelled condition. Therefore the misspellings were processed before the critical words were fixated. Consequently, if there are strong parafoveal-on-foveal effects there should be differences in the fixation durations prior to fixating the critical word. However, there were no effects of spelling on prior fixation durations (for word *n*–1), even for fixations very close to the left of the critical word. Figure 11.2 shows the fixation durations for each area of prior fixation and it is apparent that there is no consistent pattern in terms of the fixation duration as a function of the target condition. We also examined whether the last fixation was a refixation or a single fixation and found no difference in the data pattern between these two types of fixations. Likewise, there was no effect on refixation probability on word *n*–1 as a function of the target condition. Therefore White and Liversedge's results provide no support for parafoveal-on-foveal effects.

Finally, Kambe (2003) recently used the boundary paradigm to investigate the possibility of morphology contributing to preview benefit. She had readers read sentences in which target words were prefixed words (such as **preview**). Readers received either the identical word to the target word as a preview (**preview**), or just the prefix with the other letters replaced by similar letters (**preurcv**), or an unrelated string of letters (**qncurcv**). She, like Lima (1987), found no evidence for morphological priming effects.[1] However, we examined the launch site fixation duration in her study. Figure 11.3 shows these data. Once again, consistent with the Rayner (1975) study, we see that when the launch site was more than 4 letters from the beginning of the target word there was no difference between conditions. However, the orthographically illegal string at the beginning of the unrelated preview condition resulted in longer launch fixations when the eyes were within 3 characters of the beginning of the target word.

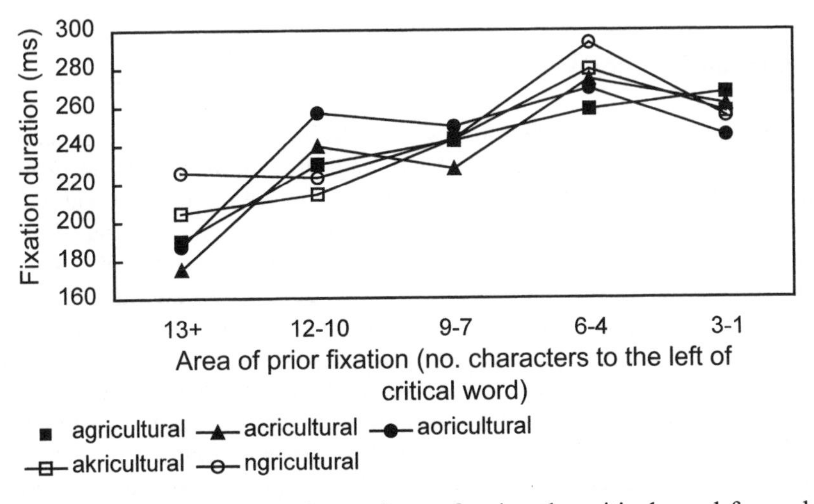

Figure 11.2: Mean fixation durations prior to fixating the critical word for each condition and for each area of prior fixation.

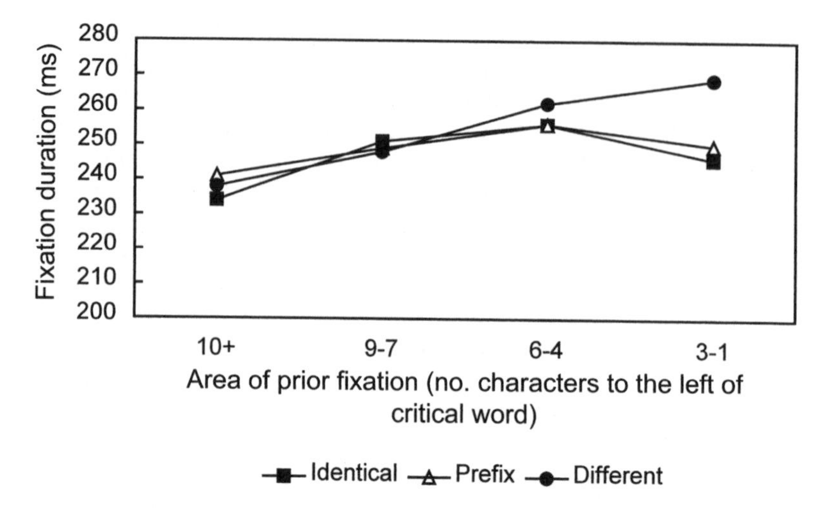

Figure 11.3: Mean fixation durations prior to fixating the critical word for each condition and for each area of prior fixation.

Taken together, the results of the studies we have discussed in this section provide rather clear evidence against parafoveal-on-foveal processing in terms of semantic effects. Some studies (Inhoff *et al.*, 2000; Kambe, 2003; Rayner, 1975) show effects due to unusual orthographic patterns at the beginning of a target region influencing the duration of the current fixation when the eyes are within 3 letter spaces of that region. However, other studies (White & Liversedge, 2003) do not even show an effect of unusual orthography at the beginning of a subsequent target string influencing the current fixation.

Summary and Conclusions

In this chapter, we have reviewed research dealing with semantic processing in parafoveal vision during reading. We have made five arguments based on the data reviewed. First, we argued that reading on the basis of parafoveal information is quite difficult (Rayner & Bertera, 1979; Rayner *et al.*, 1981). Second, we argued that word skipping is evidence that readers can sometimes obtain the meaning of a word to the right of fixation; our view is that when a word is identified to the right of fixation, it will typically be skipped. Third, we argued against the notion of semantic preprocessing wherein the eyes move further into words that have informative endings (Hyönä, 1995; Rayner & Morris, 1992). Fourth, we reviewed evidence that indicates that semantic codes are not the source of preview benefit effects in reading (Altarriba *et al.*, 2001; Rayner *et al.*, 1986); the source of the preview benefit appears to be some type of abstract letter code and/or phonological code. Finally, we have suggested that it would be premature at this point to assume that there are semantic parafoveal-on-foveal effects in reading. In this regard, we suggested that there may well be problems with the generalizability of some of the studies purporting to show such effects and have further suggested that in the context of reading per se, while there may be parafoveal-on-foveal effects due to unusual orthography at the beginning of the word to the right of fixation, there is not much convincing evidence for the effects being due to the meaning of the word to the right of fixation. Indeed, a number of studies have demonstrated that the frequency of the word to the right of fixation has no effect on the duration of the current fixation; however, the duration of the current fixation is strongly influenced by the frequency of the currently fixated word.

As we noted at the outset of this chapter, the main reason that parafoveal-on-foveal effects have attracted so much attention is because it has been assumed that such effects would be damaging to serial attention shift models like the E-Z Reader model. While it is the case that if the meaning of the word to the right of fixation exerted a strong effect on the current fixation duration it would be problematic for the model, it is also the case that parafoveal-on-foveal effects due to orthographic irregularity at the beginning of the parafoveal word are not necessarily a problem. Indeed, the most recent version of the model (Pollatsek, Reichle & Rayner, 2003, this volume; Reichle, Rayner & Pollatsek, 2002) has incorporated a mechanism that can accommodate such an effect. In the earliest versions of the model (Reichle *et al.*, 1998; Reichle, Rayner & Pollatsek, 1999), the eye movement program to move the eyes to the next unidentified word and the associated shift of attention to that word followed from the completion of a familiarity check and completion of lexical access, respectively. In the more recent version of the model, a pre-attention stage of processing has been incorporated. Thus, information about word length and letter information could be extracted prior to shifting attention. This contrasts with the earlier versions of the model in which no meaningful extraction of letter information occurred before an attention shift. The current version of the model was not necessarily constructed to account for orthographic irregularity effects, but it is not inconsistent with the model that unusual combinations of letters could be noticed by this pre-attentive processing and influence eye movement behavior. Thus, parafoveal-on-foveal effects due to unusual orthography at the beginning of a word would be independent of serial attention shifts associated with lexical access.

It should also be noted that even if it were the case that effects like those reported by Murray (1998) and Inhoff *et al.* (2000) proved to be reliable, the E-Z reader model would be able to partially account for the data. Specifically, if the effect was driven by fixations on the end of the word n, it could conceivably be the case that some of the time readers will undershoot their saccade and be intending to fixate (and attend to) word $n + 1$ while they're fixating near the end of word n. Such an undershoot explanation could conceivably account for effects such as those reported by Murray and Inhoff *et al.*

In conclusion, then, we accept that a word to the right of a fixated word can be identified and that when this happens the word is subsequently skipped. This usually happens for short words when the reader fixates very close to them. However, we feel that it would be very premature at this point to accept that there is any strong evidence obtained from a natural reading task to suggest that semantic information is obtained from the parafovea, except in those cases when the parafoveal word is subsequently skipped. It does appear that there is some evidence suggesting that orthographic irregularities at the beginning of the parafoveal word can influence the duration of the current fixation when the eyes are close (within three character spaces) to the beginning of that word. However, this finding is not problematic for the E-Z Reader model. Furthermore, we have argued that there is no convincing evidence for semantic preprocessing in terms of the eyes being drawn to the informative end of a parafoveal word. Finally, the basis for the robust parafoveal preview benefit obtained in numerous studies is not any type of semantic code, but rather abstract letter/phonological codes related primarily to the beginning letters of parafoveal words.[2] But, once again, as we noted at the outset of this concluding paragraph, it would be premature to argue that the meaning of the word to the right of fixation has any influence on the duration of the current fixation when that word is not skipped.

Acknowledgements

Preparation of this chapter was supported by Grant HD26765. It was written while the first author was the recipient of a Leverhulme Professorship at the University of Durham. We thank Raymond Bertram, Jukka Hyönä, Alan Kennedy, Alexander Pollatsek and Erik Reichle for their helpful comments on an earlier draft of this chapter.

Notes

1 Kambe (2003) also had a stem preview condition (**qscview**) in her experiment which is not discussed here. In another experiment, she made the prefix and stem previews visually distinct by replacing letters with **X**'s (**preXXXX**, **XXXview**). For more detail, see Kambe (2002).

2 As we noted earlier, in Hebrew there is evidence that morphological codes obtained from parafoveal words result in preview benefit (see Deutsch *et al.*, 2003).

References

Altarriba, J., Kambe, G., Pollatsek, A., & Rayner, K. (2001). Semantic codes are not used in integrating information across eye fixations in reading: Evidence from fluent Spanish English bilinguals. *Perception & Psychophysics*, *63*, 875–890.

Balota, D. A., Pollatsek, A., & Rayner, K. (1985). The interaction of contextual constraints and parafoveal visual information in reading. *Cognitive Psychology*, *17*, 364–390.

Blanchard, H. E., Pollatsek, A., & Rayner, K. (1989). The acquisition of parafoveal word information in reading. *Perception & Psychophysics*, *46*, 85–94.

Brysbaert, M., & Vitu, F. (1998). Word skipping: Implications for theories of eye movement control in reading. In: G. Underwood (ed.), *Eye Guidance in Reading and Scene Perception* (pp.125–148). Oxford, England: Elsevier.

Carpenter, P. A., & Just, M. A. (1983). What your eyes do while your mind is reading. In: K. Rayner (ed.), *Eye Movements in Reading: Perceptual and Language Processes* (pp. 275–307. New York: Academic Press.

Deutsch, A., Frost, R., Peleg, S., Pollatsek, A., & Rayner, K. (2003). Early morphological effects in reading: Evidence from parafoveal preview benefit in Hebrew. *Psychonomic Bulletin & Review*, in press.

Deutsch, A., Frost, R., Pollatsek, A., & Rayner, K. (2000). Early morphological effects in word recognition in Hebrew: Evidence from parafoveal preview benefit. *Language and Cognitive Processes*, *15*, 487–506.

Ehrlich, S. F., & Rayner, K. (1981). Contextual effects on word perception and eye movements during reading. *Journal of Verbal Learning and Verbal Behavior*, *20*, 641–655.

Engbert, R., Longtin, A., & Kliegl, R. (2002). A dynamical model of saccade generation in reading based on spatially distributed lexical processing. *Vision Research*, *42*, 621–636.

Everatt, J., & Underwood, G. (1992). Parafoveal guidance and priming effects during reading: A special case of the mind being ahead of the eyes. *Consciousness and Cognition*, 1, 186–197.

Garrod, S., & Terras, M. (2000). The contributions of lexical and situational knowledge to resolving discourse roles: Bonding and resolution. *Journal of Memory and Language*, *42*, 526–544.

Gautier, V., O'Regan, J. K., & LaGargasson, J. F. (2000). "The skipping" revisited in French: Programming saccades to skip the article "les". *Vision Research*, *40*, 2517–2531.

Henderson, J. M., Dixon, P., Peterson, A., Twilley, L. C., & Ferreira, F. (1995). Evidence for the use of phonological representations during transaccadic word recognition. *Journal of Experimental Psychology: Human Perception and Performance*, *21*, 82–97.

Henderson, J. M., & Ferreira, F. (1990). Effects of foveal processing difficulty on the perceptual span in reading: Implications for attention and eye movement control. *Journal of Experimental Psychology: Learning, Memory and Cognition*, *16*, 417–429.

Henderson, J. M., & Ferreira, F. (1993). Eye movement control during reading: Fixation measures reflect foveal but not parafoveal processing difficulty. *Canadian Journal of Experimental Psychology*, *47*, 201–221.

Hyönä, J. (1993). Effects of thematic and lexical priming on readers' eye movements. *Scandinavian Journal of Psychology*, *34*, 293–304.

Hyönä, J. (1995). Do irregular letter combinations attract readers' attention? Evidence from fixation locations in words. *Journal of Experimental Psychology: Human Perception and Performance*, *21*, 68–81.

Hyönä, J., Niemi, P., & Underwood, G. (1989). Reading long words embedded in sentences: Informativeness of word halves affects eye movements. *Journal of Experimental Psychology: Human Perception and Performance*, *15*, 142–152.

Inhoff, A. W. (1989). Parafoveal processing of words and saccade computation during eye fixations in reading. *Journal of Experimental Psychology: Human Perception and Performance*, *15*, 544–555.

Inhoff, A. W., Radach, R., Starr, M., & Greenberg, S. (2000). Allocation of visuo-spatial attention and saccade programming during reading. In: A. Kennedy, R. Radach, D. Heller and J. Pynte (eds), *Reading as a Perceptual Process* (pp. 221–246). Oxford: Elsevier.

Inhoff, A. W., & Rayner, K. (1986). Parafoveal word processing during eye fixations in reading: Effects of word frequency. *Perception & Psychophysics*, *40*, 431–439.

Inhoff, A. W., Starr, M., & Shindler, K. L. (2000). Is the processing of words during eye fixations in reading strictly serial? *Perception & Psychophysics*, *62*, 1474–1484.

Jacobs, A. M. (1987). On the role of blank spaces for eye-movement control in visual search. *Perception & Psychophysics*, *41*, 473–479.

Kambe, G. (2003). Parafoveal processing of morphologically complex (prefixed) words during eye fixations in reading: Evidence against morphological decomposition in reading. Submitted for publication.

Kennedy, A. (1998). The influence of parafoveal words on foveal inspection time: Evidence for a preprocessing trade off. In: G. Underwood (ed.), *Eye Guidance in Reading and Scene Perception* (pp.149–180). Oxford: Elsevier.

Kennedy, A. (2000). Parafoveal processing in word recognition. *Quarterly Journal of Experimental Psychology, 53A*, 429–455.

Kennedy, A., Pynte, J., & Ducrot, S. (2002). Parafoveal-on-foveal interactions in word recognition. *Quarterly Journal of Experimental Psychology, 55A,* 1307–1338.

Kennison, S. M., & Clifton, C. (1995). Determinants of parafoveal preview benefit in high and low working memory capacity readers: Implications for eye movement control. *Journal of Experimental Psychology: Learning Memory and Cognition, 21*, 68–81.

Kennison, S. M., & Gordon, P. C. (1998). Comprehending referential expressions during reading: Evidence from eye tracking. *Discourse Processes, 24*, 229–252.

Lima, S. D. (1987). Morphological analysis in sentence reading. *Journal of Memory and Language, 26*, 84–99.

Liversedge, S. P., & Findlay, J. M. (2000). Saccadic eye movements and cognition. *Trends in Cognitive Sciences, 4*, 6–14.

McConkie, G. W., & Rayner, K. (1975). The span of the effective stimulus during a fixation in reading. *Perception & Psychophysics, 17*, 578–586.

McConkie, G. W., & Zola, D. (1979). Is visual information integrated across successive fixations in reading. *Perception & Psychophysics, 25*, 221–224.

Morrison, R. E., & Rayner, K. (1981). Saccade size in reading depends upon character spaces and not visual angle. *Perception & Psychophysics, 30*, 395–396.

Murray, W. S. (1998). Parafoveal pragmatics. In: G. Underwood (ed.), *Eye Guidance in Reading and Scene Perception* (pp. 181–200). Oxford: Elsevier.

Murray, W. S., & Rowan, M. (1998). Early, mandatory, pragmatic processing. *Journal of Psycholinguistic Research, 27*, 1–22.

O'Regan, K. (1979). Saccade size control in reading: Evidence for the linguistic control hypothesis. *Perception & Psychophysics, 25*, 501–509.

O'Regan, K. (1980). The control of saccade size and fixation duration in reading: The limits of linguistic control. *Perception & Psychophysics, 28*, 112–117.

O'Regan, K. (1983). Elementary perceptual and eye movement control processes in reading. In: K. Rayner (ed.), *Eye Movements in Reading: Perceptual and Language Processes* (pp. 121–140). New York: Academic Press.

O'Regan, K., Lévy-Schoen, A., & Jacobs, A. M. (1983). The effect of visibility on eye movement parameters in reading. *Perception & Psychophysics, 34*, 457–464.

Pickering, M. J., & Traxler, M. J. (1998). Plausibility and recovery from garden paths: An eye-tracking study. *Journal of Experimental Psychology: Learning, Memory, and Cognition, 24*, 940–961.

Pollatsek, A., Lesch, M., Morris, R. K., & Rayner, K. (1992). Phonological codes are used in integrating information across saccades in word identification and reading. *Journal of Experimental Psychology: Human Perception and Performance, 18*, 148–162.

Pollatsek, A., & Rayner, K. (1982). Eye movement control in reading: The role of word boundaries. *Journal of Experimental Psychology: Human Perception and Performance, 8*, 817–833.

Pollatsek, A., Rayner, K., & Balota, D. A. (1986). Inferences about eye movement control from the perceptual span in reading. *Perception & Psychophysics, 40*, 123–130.

Pollatsek, A., Reichle, E. D., & Rayner, K. (2003). Modeling eye movements in reading: Extending the E-Z Reader model. In: J. Hyönä, R. Radach and H. Deubel (eds), *The Mind's Eye: Cognitive and Applied Aspects of Eye Movement Research* (pp. 361–390). Amsterdam: North-Holland.

Radach, R., & Heller, D. (2000). Relations between spatial and temporal aspects of eye movement control. In: A. Kennedy, R. Radach, D. Heller and J. Pynte (eds), *Reading as a Perceptual Process* (pp. 165–191). Oxford: Elsevier.

Radach, R., Krummenacher, J, Heller, D., & Hofmeister, J. (1995). Individual eye movement patterns in word recognition: Perceptual and linguistic factors. In: J. M. Findlay, R. Walker and R. W. Kentridge (eds), *Eye Movement Research: Mechanisms, Processes and Applications* (pp. 421–432). Amsterdam: North-Holland.

Rayner, K. (1975). The perceptual span and peripheral cues in reading. *Cognitive Psychology, 7*, 65–81.

Rayner, K. (1978). Eye movements in reading and information processing. *Psychological Bulletin, 85*, 618–660.

Rayner, K. (1979). Eye guidance in reading: fixation locations within words. *Perception, 8*, 21–30.

Rayner, K. (1986). Eye movements and the perceptual span in beginning and skilled readers. *Journal of Experimental Child Psychology, 41*, 211–236.

Rayner, K. (1998). Eye movements in reading and information processing: 20 years of research. *Psychological Bulletin, 124*, 372–422.

Rayner, K., Ashby, J., Pollatsek, A., & Reichle, E. D. (2003). The effects of frequency and predictability on eye fixations in reading: Implications for the E-Z Reader model. Submitted to publication.

Rayner, K., Balota, D. A., & Pollatsek, A. (1986). Against parafoveal semantic preprocessing during eye fixations in reading. *Canadian Journal of Psychology, 40*, 473–483.

Rayner, K., & Bertera, J. H. (1979). Reading without a fovea. *Science, 206*, 468–469.

Rayner, K., & Duffy, S. A. (1986). Lexical complexity and fixation times in reading: Effects of word frequency, verb complexity and lexical ambiguity. *Memory & Cognition, 14*, 191–201.

Rayner, K., & Fischer, M. H. (1996). Mindless reading revisited: Eye movements during reading and scanning are different. *Perception & Psychophysics, 58*, 734–747.

Rayner, K., Fischer, M. H., & Pollatsek, A. (1998). Unspaced text interferes with both word identification and eye movement control. *Vision Research, 38*, 1129–1144.

Rayner, K., Inhoff, A. W., Morrison, R. E., Slowiaczek, M. L., & Bertera, J. H. (1981). Masking of foveal and parafoveal vision during eye fixations in reading. *Journal of Experimental Psychology: Human Perception and Performance, 7*, 167–179.

Rayner, K., & McConkie, G. W. (1976). What guides a reader's eye movements? *Vision Research, 16*, 829–837.

Rayner, K., McConkie, G. W., & Zola, D. (1980). Integrating information across eye movements. *Cognitive Psychology, 12*, 206–226.

Rayner, K., & Morris, R. K. (1992). Eye movement control in reading: Evidence against semantic preprocessing. *Journal of Experimental Psychology: Human Perception and Performance, 18*, 163–172.

Rayner, K., & Morrison, R. M. (1981). Eye movements and identifying words in parafoveal vision. *Bulletin of the Psychonomic Society, 17*, 135–138.

Rayner, K., & Raney, G. E. (1996). Eye movement control in visual search: Effects of word frequency. *Psychonomic Bulletin & Review*, 3, 238–244.

Rayner, K., Sereno, S. C., & Raney, G. E. (1996). Eye movement control in reading: A comparison of two types of models. *Journal of Experimental Psychology: Human Perception and Performance, 22*, 1188–1200.

Rayner, K., Slowiaczek, M. L., Clifton, C., & Bertera, J. H. (1983). Latency of sequential eye movements: Implications for reading. *Journal of Experimental Psychology: Human Perception and Performance, 9*, 912–922.

Rayner, K., & Well, A. D. (1996). Effects of contextual constraint on eye movements in reading: A further examination. *Psychonomic Bulletin and Review, 3*, 504–509.

Rayner, K., Well, A. D., Pollatsek, A., & Bertera, J. H. (1982).The availability of useful information to the right of fixation in reading. *Perception & Psychophysics, 31*, 537–550.

Rayner, K., Well, A. D., & Pollatsek, A. (1980). Asymmetry of the effective visual field in reading. *Perception & Psychophysics, 27*, 537–544.

Reichle, E. D., Pollatsek, A., Fisher, D. L., & Rayner, K. (1998). Toward a model of eye movement control in reading. *Psychological Review, 105*, 125–157.

Reichle, E. D., Rayner, K., & Pollatsek, A. (1999). Eye movement control in reading: Accounting for initial fixation locations and refixations within the E-Z Reader model. *Vision Research, 39*, 4403–4411.

Reichle, E. D., Rayner, K., & Pollatsek, A. (2002) The E-Z reader model of eye movement control in reading: Comparisons to other models. *Brain and Behavioral Sciences*, in press.

Schroyens, W., Vitu, F., Brysbaert, M., & d'Ydewalle, G. (1999). Eye movement control during reading: Foveal load and parafoveal processing. *Quarterly Journal of Experimental Psychology, 52A*, 1021–1046.

Starr, M. S., & Rayner, K. (2001). Eye movements during reading: Some current controversies. *Trends in Cognitive Sciences, 5*, 156–163.

Underwood, G. (1985). Eye movements during the comprehension of written language. In: A. W. Ellis (Vol. ed.), *Progress in the Psychology of Language* (Vol. 2, pp. 45–71). London: Erlbaum.

Underwood, G., Binns, A., & Walker, S. (2000). Attentional demands on the processing of neighbouring words. In A. Kennedy, R. Radach, D. Heller and J. Pynte (eds), *Reading as a Perceptual Process* (pp. 247–268). Oxford: Elsevier.

Underwood, G., Bloomfield, R., & Clews, S. (1988). Information influences the pattern of eye fixations during sentence comprehension. *Perception, 17*, 267–278.

Underwood, G., Clews, S., & Everatt, J. (1990). How do readers know where to look next? Local information distributions influence eye fixations. *Quarterly Journal of Experimental Psychology, 42A*, 39–65.

Underwood, G., Clews, S., & Wilkinson, H. (1989). Eye fixations are influenced by the distribution of information within words. *Acta Psychologica, 72*, 263–280.

White, S. J., & Liversedge, S. P. (2003). Orthographic familiarity influences initial eye fixation positions in reading. *European Journal of Cognitive Psychology*, in press,.

Chapter 12

Bridging the Gap Between Old and New: Eye Movements and Vocabulary Acquisition in Reading

Robin K. Morris and Rihana S. Williams

This chapter reviews a series of eye movement studies that address the processes by which readers acquire new vocabulary knowledge during silent reading. Readers' initial encoding strategies, use of sentence context, generation of elaborative inferences, and memory for new word meaning are considered. The chapter provides a descriptive account of the process of establishing the meaning of a new word from a single encounter.

Introduction

Reading is a process of acquiring a meaningful message from written text. Yet, readers frequently encounter unfamiliar words in the course of their day- to-day reading. Every language contains thousands of rarely used words. New words and their meanings are constantly being added, either by borrowing from other languages or creating new words in response to technological and cultural change (Carroll, 1985; Clark, 1992; Lehrer, 1992).

Estimates of vocabulary growth suggest that one way or another we acquire the meaning of new words from reading. Estimates of the vocabulary size of high school graduates and college students range from 40,000 (Nagy, Anderson & Herman, 1987) to 100,000 words (Landauer, 1986; Miller, 1991, p. 138). Since these estimates are based on sampling from dictionaries which exclude a large number of names, stock phrases, and slang, true vocabulary size is almost certainly much larger, possibly by as much as 60% (Walker & Amsler, 1986).

Estimates of vocabulary size imply that people learn between 6 and 25 new words per day from age 2 to 20. It has been suggested that by high school, most of these

The Mind's Eye: Cognitive and Applied Aspects of Eye Movement Research

ISBN: 0–444–51020–6

new words are learned from text (Landauer & Dumais, 1997) and are learned without the help of formal definition, explicit explanation, or instruction (e.g., Nagy, Anderson & Herman, 1987; Nagy, Herman & Anderson, 1985; Stahl, 1991; Sternberg, 1987). Fraser (1999) provided self-report data that is consistent with this suggestion. She asked college-age readers to read an article, answer comprehension questions, and identify any unknown words that occurred in the text. During interviews that followed the reading task, readers reported that their preferred strategy for dealing with unknown words was to infer the meaning from context, rather than look in a dictionary or ignore the words.

There is an extensive literature demonstrating that readers successfully acquire knowledge of word meaning from silent reading (e.g. Carnine, Kameenui & Coyle, 1984; Daneman & Green, 1986; Fischer, 1994; Fraser, 1999; Herman, Anderson, Pearson & Nagy, 1987; Jenkins, Stein & Wysocki, 1984; Lampinen & Faries, 1994; Long & Shaw, 2000; McKeown, 1985; Nagy *et al.*, 1987; Nagy, Herman & Anderson, 1985; Mori & Nagy, 1999; Shefelbine, 1990; Shu, Anderson & Zhang, 1995; van Daalen-Kapteijns & Elshout-Mohr, 1981; van Daalen-Kapteijns, Elshout-Mohr & de Glopper, 2001; Werner & Kaplan, 1952) and there is evidence that this skill improves across the lifespan (Long & Shaw, 2000; Nagy *et al.*, 1987; Nagy & Scott, 1990; Werner & Kaplan, 1952). This body of evidence is quite compelling in its demonstration that readers can acquire word knowledge from text. However, the evidence exists primarily in the results of comprehension tests or vocabulary tests administered at varying points in time after the reading episode and thus, provides little insight into how skilled readers go about doing this.

This chapter explores the process of acquiring new word meaning from silent skilled reading. This exploration begins with an examination of readers' initial processing of a word that they have not seen before and lexical level factors that might influence the process at this stage. We then go on to consider the relationship between unfamiliar words and the context in which they occur and the role that sentence context may play in on-line development of word meaning. Finally, we consider issues related to the knowledge that the skilled reader takes away from a single encounter with a new word. Our goal is not to promote a particular model or theory of vocabulary acquisition in reading, but rather to provide a descriptive account of the processes involved in accomplishing this feat.

Lexical Level Effects on Vocabulary Acquisition in Reading

We know that readers are sensitive to differences in the degree of familiarity of the words that they know. For example, lexical familiarity, as assessed by printed word frequency (Francis & Kucera, 1982) influences a reader's initial processing time on a word, as measured by first fixation duration or gaze duration. Readers spend more time on low frequency than on high frequency words of equal length (e.g., Hyönä & Olson, 1995; Inhoff & Rayner, 1986; Rayner & Duffy, 1986; Rayner, Sereno & Raney, 1996; Schilling, Rayner & Chumbley, 1998; Vitu, O'Regan & Mittau, 1990). In addition to these initial processing time differences, the duration of the first fixation after the low frequency word is often inflated compared to the high frequency case (e.g., Rayner & Duffy, 1986). This is thought to reflect the processing of the low frequency word spilling over onto the next fixation. To what extent this is due to prolonged lexical processing and to what extent it is due to increased difficulty in integrating a low frequency word into the discourse representation remains an open question. However, correlational data reported in Rayner, Sereno, Morris, Schmauder and Clifton (1989) suggest that this is largely a text integration effect.

Subjective ratings of word familiarity also correspond to differences in word processing time. Gernsbacher (1984) demonstrated that lexical decision time on low frequency words was more accurately predicted using subjective ratings of familiarity than previous results that had relied solely on published estimates of word frequency as an index of lexical familiarity. Connine, Mullenix, Shernoff and Yelen (1990) replicated this result, even when words with extremely low familiarity ratings were eliminated from the experiment to ensure that participants knew the meanings of all of the words being tested. High frequency words were identified as words faster than low frequency words. In addition, low frequency words with higher ratings of familiarity were judged to be words faster and more accurately than their lower familiarity counterparts.

Williams and Morris (2003) explored readers' eye movement patterns in response to differences in both printed word frequency and subjective familiarity ratings. A set of very familiar-high frequency, somewhat familiar-high frequency, somewhat familiar-low frequency, and low familiar-low frequency nouns were embedded in sentence contexts such that neutral context preceded the target word and context consistent with the meaning followed the target word. All of the low frequency words were subjected to a vocabulary test to ensure that participants had some knowledge of the words' meanings. Readers spent less initial processing time on high frequency words than on low frequency words. When printed word frequency was held constant, differences in processing time due to differences in familiarity were only observed among low frequency words. That is, somewhat familiar-low frequency words were processed more quickly than low familiar-low frequency words, while familiarity differences within the high frequency category had no effect on initial processing. These findings are consistent with those reported previously in lexical decision tasks (e.g., Gernsbacher, 1984; Connine *et al.*, 1990) and suggest that subjective familiarity ratings are particularly useful when exploring readers' processing of less frequent words.

Given the extensive literature demonstrating that a reader's relative familiarity with a known word will influence initial processing time on that word, it seemed reasonable to begin our examination of readers' treatment of new words by comparing initial processing time on known words to processing time on novel words. Chaffin, Morris and Seely (2001) monitored readers' eye movements as they read sentence pairs like the following:

Joe picked up the (**guitar/zither/asdor**) and *began to strum a tune*.
He played the *instrument* to relax.

The first sentence contained a high familiar, a low familiar, or a novel word (in bold). Novel words were created for research purposes to assure that readers had no knowledge of their meaning prior to the experiments. All words were rated for subjective familiarity independent of the reading study (Chaffin, 1997). The target word was followed by context (italics) that allowed the reader to infer what kind of thing the target word referred to. This was followed by a second sentence in which the target word was referred to by a single related word (italics). Table 12.1 shows that readers spent more initial processing time on novel and low familiar words than they did on high familiar words.

Interestingly, the low familiar and novel words did not differ from each other on these measures. This was somewhat surprising given that eye movement measures have repeatedly been shown to reflect lexical processing difficulty. On the one hand, it suggests that readers were not altering their initial encoding strategies in the face of an unfamiliar word. But, this result may also be due in part to the way in which the novel and low familiar words were selected. The novel words conformed to standard orthographic properties of English and were no different in that respect from the low familiar words. While the subjective familiarity ratings for the low familiar words were higher than those obtained for the novel words, they were considerably lower than those used in studies that employed an a priori assessment of participants' knowledge of word meaning in addition to a rating of subjective familiarity (e.g., Connine *et al.*, 1990).

In a subsequent study (Williams & Morris, 2003), we applied more stringent criteria to the selection of the low familiar words to ensure that readers had some knowledge of the meaning of these words. Low familiar words were selected on the basis of printed word frequency, subjective familiarity rating, and performance on a vocabulary test. Only words that were low in printed word frequency, rated low in familiarity, and yielded greater than 75% accuracy on a multiple choice vocabulary test were included in our low familiar word category. This resulted in a group of low familiar words with familiarity ratings that roughly corresponded to the low familiar conditions of Connine *et al.* (1990) and that were not as low as the ratings included in Chaffin *et al.* (2001). Single sentence contexts were created such that either a high familiar, low familiar, or novel word could be inserted. Context prior to the target was neutral with respect to the meaning of the target word and the context following the target provided information about the target word meaning without making any explicit connection between target and context. For example:

Amie liked the (**scarf/tunic/nineer**) that was featured in the fashion show.

Table 12.1: Initial processing of the target word (Chaffin *et al.*, 2001).

Condition	First fixation (ms)	Gaze duration (ms)	Spill over (ms)
High familiar	247	302	246
Very low familiar	297	456	279
Novel	283	461	271

Table 12.2: Initial processing of the target word (Williams and Morris, 2002).

Condition	First fixation (ms)	Gaze duration (ms)	Spill over (ms)
High familiar	291	326	285
Low familiar	309	365	304
Novel	366	476	350

Under these conditions the effect of familiarity in the known word conditions observed in Chaffin *et al.* (2001) was replicated. That is, readers spent less time on high familiar words than on low familiar words. In contrast to the Chaffin *et al.* (2001) findings, Table 12.2 also shows that readers spent more processing time on the novel word than on a known but low familiar word at their initial encounter with the word. While these data revealed differences in processing time at initial encounter with novel words compared to known words, these times do not suggest that readers are invoking unique processing strategies at this point in time to deal with the unfamiliar word. However, we are aware of one other study (Just & Carpenter, 1980) that yields quite a different result. In the Just and Carpenter study skilled readers encountered words that they were unlikely to be familiar with, such as *thermoluminence*, in sentence contexts. Gaze durations on seven target words examined in this study ranged from 913–2431 ms. These are extremely long times compared to the times observed for the novel words in the studies that we have conducted. This could be due in part to the fact that these are extremely long words. But, given that these times are more than three times as long as the average gaze duration on a low frequency word they almost certainly reflect processing beyond the scope of what is typically considered lexical access.

Another possible explanation for the difference between the Just and Carpenter data and the data that we obtained comes from closer inspection into the information afforded the reader by morphological properties of the words. For example, although the reader may not be familiar with the word "thermoluminence", they may know that "thermo" is related to "heat" and "luminence" refers to "giving off light". The exaggerated processing time reported by Just and Carpenter may reflect that readers were dissecting the novel words into their morphological constituents, and in essence problem solving their way to a meaning for these words relying primarily on their lexical properties (see Sternberg & Powell, 1983 for a similar view).

Mori and Nagy (1999) provide more direct evidence of a relationship between morphological analysis and acquisition of word meaning. They showed that young readers make use of morphological information in deriving meanings for new words by demonstrating that young readers developed more precise meanings for words that were morphologically transparent than for words that were opaque (see also Bertram, Laine & Virkkala, 2000). This is not a dimension that we have considered in our own work. But, as the gaze duration data suggest, we would speculate that the impact of this sort of morphological problem solving is negligible in the studies we report given that most of our novel words were not transparent and did not contain familiar morphological roots.

The contrast between our findings and those of Just and Carpenter (1980) sets the stage for more detailed investigations into the lexical and sublexical properties of words and their respective influence on the early stages of word processing in reading. Morphological properties are not the only properties that warrant further investigation. There is also evidence that phonological information is available early in the processing of familiar words (see Morris & Folk, 2000 for a review) and there is indirect evidence that phonological information may influence the development of meaning for unknown words. For example, Chaffin (1997) showed participants a list of words and asked them to generate a semantically related word in response to each one. The erroneous responses to low familiar words were often sound-mediated responses. For example, persimmon was reported to be a spice, a response mediated by cinnamon (Chaffin, 1997). Further experiments examining the influence of word level features of the novel words clearly need to be done.

Sentence Context Effects on Vocabulary Acquisition

The previous section demonstrates that readers' processing time at their initial encounter with a word is influenced by word familiarity. We now turn to questions of how and when readers use the context in which the word occurs in order to determine word meaning.

It has been demonstrated that word meanings are more readily acquired from context than through definitional instruction (Miller & Gildea, 1987; Scott & Nagy, 1997; van Daalen-Kapteijns & Elshout-Mohr, 1981). Substituting new words for known words, for example learning that an argument can also be called an altercation, probably constitutes the simplest case of acquiring new vocabulary from context (Graves, 1987; Jenkins & Dixon, 1983; Nagy *et al.*, 1987). Nagy *et al.,* (1987) delineated three other distinct categories of new word reading encounters that the reader is likely to face: (a) The reader knows the concept by description and then encounters a one word label for it (e.g., The headdress that the March girls wear in *Little Women* is called a snood); (b) The reader does not know the concept but can learn it on the basis of prior knowledge and experience (e.g., based on your knowledge of horses and walking you could infer that canter is a type of stride); (c) The reader does not know the concept and would have to acquire new knowledge to understand it (this is often highly domain specific knowledge). Nagy *et al.* (1987) found that readers were successful at acquiring new vocabulary from reading alone in all but the final condition, even when the context did not explicitly state the relationship between novel and familiar words. All of the work from our laboratory involves substituting new words for known words in contexts that do not make that relationship explicit.

There is substantial evidence to indicate that initial processing time on a familiar word is affected by the context in which that word is read. For example, words that are predictable from the context are processed more rapidly than words that are not predictable (Ehrlich & Rayner, 1981; Inhoff, 1984; Rayner & Well, 1996; Schilling, Rayner & Chumbley, 1998; Zola, 1984). More importantly, words that are preceded by semantically related context are processed faster than words that are preceded by unrelated context, even when the word is not predictable and the context contains no strong semantic associates of the target word (Duffy, Henderson & Morris, 1989; Morris, 1994; Morris & Folk, 1998). These effects often emerge in first fixation duration, suggesting that this is a reflection of direct contextual influence on lexical access and not an effect of text integration processes feeding back to influence word processing. Finally, when readers encounter an ambiguous word that has two relatively equally likely meanings, processing time on the ambiguous word in neutral context is inflated relative to an unambiguous control word (e.g., Duffy, Morris & Rayner, 1988). However, this processing difficulty is eliminated when the disambiguating context precedes the ambiguous word, suggesting that context has boosted the activation of one meaning of the word making it more available for selection than the contextually unrelated meaning. Hence, it appears that readers can use context to facilitate recognition of familiar words and to differentiate between two known meanings of an ambiguous word.

There is also evidence that readers are able to discern relevant contextual information on-line as they are reading and differentially allocate processing time accordingly. This type of selective processing of relevant contextual information has been demonstrated in the amount of time readers spend on the context that follows an ambiguous word. Readers spend more time on disambiguating context following a balanced ambiguous word than an unambiguous control. But there is no difference in the amount of time spent on the context that follows the ambiguous word compared to an unambiguous word if the disambiguating information precedes the target word (e.g., Duffy *et al.*, 1988). How a reader allocates processing time across the text has been shown to depend on both the type of ambiguity and the type of information contained in the context (Seely, Morris & Schmauder, 2002). Seely *et al.* (2002) compared semantic and syntactic category ambiguity resolution. Participants read sentences like the following:

a. Obviously the forest **sheds** (cabins) were often maintained by the rangers.
b. Evidently the school **speakers** (desks) were never purchased at the book value.

The target word is in bold followed by the control word in parentheses. The next two words provide syntactic disambiguation and the rest of the sentence provides semantic disambiguation. Table 12.3 shows that when the critical word was ambiguous with respect to syntactic category assignment (sentence a), readers spent more time on the syntactically disambiguating context compared to a word that was ambiguous with respect to meaning within a single syntactic category. When the word had two or more possible meanings from a single syntactic category (sentence b), readers spent more time on semantically related context regions specifically, not on all contextual information that follows the word (Seely *et al.*, 2002). These results demonstrate that readers can differentiate more relevant from less relevant contextual information, as opposed to spending more processing time on any (or all) contextual information that follows the ambiguity. Given that readers are sensitive to relevant contextual information in the processing of familiar words, and that this sensitivity can be captured

Table 12.3: Lexical and syntactic category disambiguation (Seely *et al.*, 2002).

Condition	**Syntactic disambiguating region (ms)**	**Semantic disambiguating region (ms)**
N-V ambiguous	449	885
N-V control	378	924
N-N ambiguous	382	1026
N-N control	393	933

Note: N-N refers to words with multiple noun meanings and N-V refers to words with at least one noun and one semantically distinct verb meaning.

using eye movement measures gleaned from the records of silent readers, we can move on to explore questions regarding how readers use context to acquire the meaning of new words.

First, are readers able to recognize context that is informative with respect to the meaning of a new word, when they have little or no prior knowledge about the possible meanings of the word? Table 12.4 shows processing times in the informative context region that followed either a high familiar, low familiar, or novel word in sentences read by participants in the Williams and Morris (2003) study. Readers spent more total time in the informative context following a novel word than following a familiar word. They were also prompted to make more regressions out of the informative context region following novel words than familiar words. This pattern of results suggests that readers are using the contextual information provided to them in order to develop word meaning on-line as they read (see Chaffin *et al.*, 2001 for a similar pattern of results).

Alternatively, based on these data alone one could conclude that readers are not making any discriminations regarding the relevance of different aspects of the context. It could be that they would spend more time on the context that followed a novel word regardless of its relevance to deriving a meaning for the word. Chaffin *et al.* (2001) assessed this possibility by manipulating how informative the context was that

Table 12.4: Processing of the informative context (Williams and Morris, 2003).

Condition	**Total time** [a]	**Regressions out** [b]
High familiar	40	9
Low familiar	44	13
Novel	49	21

[a] Means are presented in ms per character.

[b] Regressions are computed as the proportion of looks out of or into a region after initially exiting from that region.

followed the novel word. By informative we refer to the degree to which the context constrains the possible meanings of the word. Readers were presented with sentence pairs like the following:

a. Joe picked up the **asdor** and *began to strum a tune.*
   He played the instrument to relax.
b. Joe picked up the **asdor** and *began to walk home.*
   He played the instrument to relax.

The target word is in bold and the context region is indicated in italics. In the more informative context condition (a) the context suggests that an asdor is some sort of musical instrument. In the less informative context condition (b), the context is more general. An asdor could be anything that a person could carry. The second sentence contained a word that was anaphoric to the target. It is apparent from the data patterns displayed in Table 12.5 that readers were able to differentiate between context that conveyed useful information for establishing a meaning for an unfamiliar word and context that did not. Readers spent more initial processing time and spent more time rereading the more informative context than the less informative context.

It is important to recognize that all contexts may not be equally effective in aiding the reader in the process of acquiring new word meaning (Beck, McKeown & Caslin, 1983). As Schatz and Baldwin (1986) aptly point out, an author's purpose is usually not to teach the meaning of a particular word, but rather to use that word to add information and/or constrain the meaning of the text. Thus, context cues are not necessarily reliable predictors of word meaning. Lampinen and Faries (1994) demonstrated that moderately constraining contexts, those that require the reader to generate some elaborative inference are more effective than either highly constraining contexts that make the connection between the new word and its meaning explicit, or minimally constraining contexts that do not clarify anything about the meaning. The sentences used by Chaffin *et al.* (2001) and by Williams and Morris (2003) meet the criteria laid out by Lampinen and Faries (1994) and the data suggest that readers are able to distinguish between the most informative contextual information and information that is less informative with respect to arriving at an appropriate interpretation for the unfamiliar word. Given that, what evidence is there that readers are actually connecting that contextual information with the target word?

There are two sources of evidence in our work that suggest that readers were in fact making connections between the informative context and the unfamiliar word. The

Table 12.5: Processing time on informative context in Sentence 1 (Chaffin *et al.*, 2001).

Condition	**Gaze duration (ms)**	**Total time (ms)**
More informative context	880	1203
Less informative context	718	1013

first comes from examining reading patterns in the target word and informative context regions. Table 12.6 shows that the informative context prompted readers to make more regressions when that context followed a novel word than when it followed a familiar word. In addition, readers were more likely to make regressions to the target word in the novel word condition than in any of the other three conditions. Finally, given that readers did go back to the target word, they spent more time there if the word was a novel word than if it was a known word.

The second line of evidence addresses the extent to which these reading patterns actually result in the readers' successful assignment of the intended meaning of the new word. The process of assigning meaning to new words in the context of silent reading is in some sense a process of generating an elaborative inference. Thus, in order to address this issue, we look to the existing eye movement literature on inference generation in discourse processing.

It is well established that readers routinely generate inferences as they are reading (see Graesser, Singer & Trabasso, 1994 for a review) and eye movement data have been informative in exposing this process (e.g., Calvo, Meseguer & Carreiras, 2001; Garrod, O'Brien, Morris & Rayner, 1990; Myers, Cook, Kambe, Mason & O'Brien, 2000; O'Brien, Raney, Albrecht & Rayner, 1997; O'Brien, Shank, Myers & Rayner, 1988; Poynor & Morris, 2003). O'Brien *et al.* (1988) demonstrated that readers generate elaborative inferences when the context is sufficiently constraining. Readers spent no more time on a target noun that referred to a previously inferred but not explicitly mentioned concept (e.g., the word knife following mention of a weapon) than to the same target word in a repeated condition (that is, knife following a previous mention of knife). This was taken as evidence that readers had inferred the concept knife prior to its appearance in print. Garrod *et al.* (1990) demonstrated that this was more likely to occur if the text established an anaphoric relation between the initial concept and the target noun than if there was no such relationship present.

Table 12.6: Reanalysis of target word and informative context (Williams & Morris, 2003).

Condition	**Context region**	**Target word**	**Target word**	**Context region**
	Regressions out [a]	**Regressions in** [a]	**Second pass time (ms)**	**Total time** [b]
High familiar	9	10	29	40
Low familiar	13	12	57	44
Novel	21	27	150	49

[a] Regressions are computed as the proportion of looks out of or into a region after initially exiting from that region.

[b] Means are presented in ms per character.

In Chaffin *et al.* (2001), readers encountered either a high familiar target word, a low familiar target word, or a novel target word in Sentence 1 of a two-sentence passage. Sentence 2 contained a word that was either a synonym or a category superordinate to the target word. This word was always placed in an anaphoric relation to the target word in Sentence 1. Using logic similar to that of Garrod *et al.* (1990), if readers are able to infer the meaning of the novel word as they read Sentence 1 then processing time on the associate in Sentence 2 should not vary as a function of target word familiarity. On the other hand, if there was not sufficient information available to infer a meaning from the context in the first sentence, then we might expect that readers would make that inference in response to encountering the related word in Sentence 2. In that case, we would expect that readers would spend more time on the associate in Sentence 2 following a novel word than following a known word.

Table 12.7 shows that there was no difference between familiar and novel word conditions in the amount of time readers spent on the related word in Sentence 2 when the context in Sentence 1 provided information about the novel word meaning, suggesting that readers had inferred the meaning of the target prior to encountering its referent in Sentence 2. In contrast, when the context in Sentence 1 did not provide the information necessary to infer the meaning of the novel word, readers spent more time on the related word in Sentence 2 and were more likely to engage in rereading upon encountering that word.

These results are similar to those obtained by Garrod *et al.* (1990) demonstrating that readers generate elaborative inferences about known words on-line. In our case, readers inferred a meaning for the novel word based on the context provided in Sentence 1 when that context was sufficiently informative. Thus, by the time they encountered the anaphor in the second sentence, its lexical entry was already activated and was processed as quickly following novel words as following known words. In contrast, when the context in Sentence 1 did not provide adequate information about the meaning of the word, readers spent more time on the associate in Sentence 2 and made more regressions. This additional processing effort suggests that under these circumstances readers are using the information in Sentence 2 to establish a meaning for the novel word. Thus, these results provide on-line evidence that readers have successfully inferred the intended meaning of the novel word prior to encountering the anaphor in the second sentence. But, do they retain knowledge of that new word meaning beyond the context of reading that sentence?

Table 12.7: Processing of the Sentence 2 definitional associate (Chaffin *et al.*, 2001).

Condition	**Gaze duration (ms)**	**Regressions out**
Familiar — more informative context	239	25
Novel — more informative context	233	24
Novel — less informative context	258	34

Memory for New Word Meaning

Williams and Morris (2003) provide more direct evidence regarding the meaning that readers have inferred and whether this new knowledge lasts beyond the reading session. In this experiment readers received a vocabulary test after the reading session was complete. The vocabulary test included all of the novel words that readers had seen in the reading portion of the experiment. Each novel word appeared with an answer choice that was closely related in meaning and an unrelated answer choice. Participants marked which answer was most closely related to the meaning of the word. Responses on this test were obtained after reading several intervening sentences, moving to another testing room, and reading a new set of instructions. On average the time elapsed between reading about a word and seeing the word as a test item was approximately 20–30 minutes. Readers were not told in advance that the experiment was related to vocabulary acquisition, nor were they told that there would be a test after the reading session. When performance on individual items was averaged across readers, performance was greater than chance for eight out of the 12 novel word items. This level of retention is consistent with studies that looked at retention for the meaning of novel words after one encounter in adolescent readers using a similar testing procedure (Nagy *et al.*, 1985). Of course, performance on the vocabulary test may have been prompted in part by the information provided at test. We do not dispute this. But it is important to note that there is nothing in the test to indicate to the participant which interpretation is correct unless they have some recollection of the information provided in the sentence reading. The answer choices in the vocabulary test were never words included in the sentence contexts so successful choices could not be made on the basis of surface level matching alone.

In an effort to gain more insight into the relationship between reading performance and successful acquisition of new word meaning we re-analyzed the gaze duration and second pass reading time data based on performance on the vocabulary test. Separate means were calculated for each of the eye movement measures for those items that were later answered correctly on the vocabulary test and directly compared to the means for items that were not answered correctly on the vocabulary test. When the reader initially encounters the novel word the context up to that point has provided very little information about what this word might mean. Spending additional time on the word at that point would do little to determine its meaning. Interestingly, readers spent less initial processing time (as measured by gaze duration) on novel words that they later answered correctly than on novel words that they missed on the vocabulary test. The context following the word provides information about what the word means. Readers devoted more second pass reading time to the novel words that they answered correctly than to the novel words that they answered incorrectly, further evidence that the second pass time on the novel word reflects the process of connecting the newly acquired contextual information to the unfamiliar word. The accuracy on the vocabulary test and the reading patterns combined provide further evidence that readers are not only generating the intended meanings for the novel words on-line, but also retaining these meanings for a period of time beyond the reading session.

Quality of the Meaning Representation

We have touched upon the issue of memory for new word meanings beyond the reading session in which they were introduced, but we have not begun to characterize the nature of those representations. Chaffin (1997) has suggested that readers are more likely to have what he terms definitional knowledge of word meaning. This sort of knowledge was reflected as synonyms or associates of the new word in a free association task. In this task setting participants were less likely to make responses that reflected thematic information such as information about events related to the new word. Whether this is an artifact of the testing procedure or a true reflection of the nature of meaning development has not yet been fully investigated.

Related to the issue of the content of the newly acquired lexical entry are issues related to the construction of the test. Our vocabulary test provided minimal information regarding the readers knowledge of the new words. Schwanenflugel, Stahl, and McFalls (1997) have successfully used sentence generation tasks to obtain more detailed information regarding the content of readers' meaning representations at the cost of obtaining slightly lower estimates of the number of new words retained. Studies by Shefelbine (1990) and van Daalen-Kapteijns *et al.* (2001) using a similar technique have demonstrated that readers are able to supply sensible inferences about the general categories that the novel words belong to and about additional properties the concept might have beyond those that were explicitly mentioned in the text.

The quality of the meaning representation as provided by the reader at test may also be influenced by individual differences in reasoning ability, working memory capacity, and vocabulary knowledge. Students with high reasoning ability as measured by the *Raven's Progressive Matrices Test* learned more about the meaning of low familiar words from reading a text than did students with low reasoning ability (Shefelbine, 1990). In addition, the ability to acquire meaning from context is positively correlated with the size of the readers' existing vocabulary. Adult readers with low working memory spans have been shown to extract less precise information about the meaning of novel words from context than high span readers (Daneman & Green, 1986). Perhaps the latter finding is related to the fact that readers with low working memory spans are less likely to generate elaborative inferences in general (Calvo, 2001). These are all issues deserving of further investigation.

Eye Movement Control

It is also interesting to consider the initial processing time data in light of current models of eye movement control in reading. Given the systematic relation between word frequency and initial processing time on a word, several prominent models of eye movement control in reading postulate that fixation time on a word is influenced primarily by the time required to access word meaning. That is, completion of lexical access is thought to be the trigger to move the eyes (e.g, Morrison, 1984). Yet we found no evidence that readers abandon their initial encoding strategies in the face of words that do not yet exist in their lexicon. If completion of lexical access were the trigger to move the eyes we might expect that readers would either spend an extremely long time on novel words or that there would be evidence of some sort of default mechanism that terminates processing in the absence of meaning. We saw no suggestion of this.

As with previous models, the E-Z Reader Model (see chapter by Pollatsek *et al.*, this volume) suggests that the command to move the eyes is triggered by lexical analysis of the currently fixated word. However, this model differs from past models in that it postulates a familiarity check prior to lexical access that allows eye movements to be initiated on the basis of the expected outcome of the lexical access process. For example, the orthographic and or phonological properties of a word might be sufficiently similar to other lexical entries to produce enough lexical activation to suggest that lexical access (that is, a match to a specific lexical entry) is imminent. Planning of the next eye movement would then be initiated based on crossing this threshold of lexical activation. On this account, eye movements for novel and low familiar words might be similar, if their orthographic and phonological features generate similar levels of lexical activity with similar rise times. Given similar activation arising from lexical features of the letter string, differences in initial processing time between the two word types (familiar and unfamiliar) would be influenced by differences in word level familiarity. Thus, in theory at least, it seems that the E-Z Reader Model could account for the initial processing time data that we have obtained.

Summary and Conclusions

The data presented indicate that readers do not abandon their usual encoding strategies when faced with a word that they don't know. Furthermore, we have provided evidence that readers are able to selectively attend to contextual information that is relevant to establishing a meaning for an unfamiliar word and that they infer the meaning of that word on-line. Vocabulary test data obtained after the reading session was terminated indicate that readers were successful at acquiring the intended meaning of the word and that this new knowledge persists beyond the reading episode. Finally, analysis of the eye movement data as a function of vocabulary test performance revealed a systematic relationship between reading patterns and memory for novel word meaning.

The extent to which the results of the eye movement studies converge with the existing literature on reading instruction and memory for new vocabulary is promising. In addition to furthering our understanding of vocabulary acquisition in silent reading, this work has provided new insights into other basic aspects of the reading process including issues related to eye movement control, word recognition, and discourse processing.

Acknowledgements

We wish to thank Jukka Hyönä, Manuel Calvo and Keith Rayner for their helpful comments on an earlier version of this chapter.

References

Beck, I. L., McKeown, M., & McCaslin, E. S. (1983). Vocabulary development: All contexts are not created equal. *The Elementary School Journal, 83*, 177–182.

Bertram, R., Laine, M., & Virkkala, M. M. (2000). The role of derivational morphology in vocabulary acquisition: Get by with a little help from my morpheme friends. *Scandinavian Journal of Psychology, 41*, 287–296.

Calvo, M. G. (2001). Working memory and inferences: Evidence from eye fixations during reading. *Memory, 9*, 365–381.

Calvo, M. G., Meseguer, E., & Carreiras, M. (2001). Inferences about predictable events: Eye movements during reading. *Psychological Research, 65*, 158–169.

Carnine, D., Kameenui, E. J., & Coyle, G. (1984). Utilization of contextual information in determining the meaning of unfamiliar words. *Reading Research Quarterly, 19*, 188–204.

Carroll, J. (1985). *What's in a Name.* New York: W. H. Freeman.

Chaffin, R. (1997). Associations to unfamiliar words: Learning the meanings of new words. *Memory & Cognition, 25*, 203–226.

Chaffin, R., Morris, R. K., & Seely, R. S. (2001). Learning new words in context: A study of eye movements. *Journal of Experimental Psychology: Learning, Memory, and Cognition, 27*, 225–235.

Clark, H. H. (1992). *Arenas of Language Use.* Chicago: University of Chicago Press.

Connine, C., Mullenix, J., Shernoff, E., & Yelen, J. (1990). Word familiarity and frequency in visual and auditory word recognition. *Journal of Experimental Psychology: Learning, Memory, and Cognition, 16*, 1084–1096.

Daneman, M., & Green, I. (1986). Individual differences in comprehending and producing words in context. *Journal of Memory and Language, 25*, 1–18.

Duffy, S. A., Henderson, J. M., & Morris, R. K. (1989). Semantic facilitation of lexical access during sentence processing. *Journal of Experimental Psychology: Learning, Memory, and Cognition, 15*, 791–801.

Duffy, S. A., Morris, R. K., & Rayner, K. (1988). Lexical ambiguity and fixation times in reading. *Journal of Memory and Language, 27*, 429–446.

Duffy, S. A., & Rayner, K. (1990). Eye movements and anaphor resolution: Effects of antecedent typicality and distance. *Language & Speech, 33*, 103–119.

Ehrlich, S., & Rayner, K. (1981). Contextual effects on word perception and eye movements during reading. *Journal of Verbal Learning and Verbal Behavior, 20*, 641–655.

Fischer, U. (1994). Learning words from context and dictionaries: An experimental comparison. *Applied Psycholinguistics, 15*, 551–574.

Francis, W., & Kucera, H. (1982). *Frequency Analysis of English Usage.* Boston: Houghton Mifflin Company.

Fraser, C. A. (1999). Lexical processing strategy use and vocabulary learning through reading. *SSLA, 21*, 225–241.

Garrod, S., O'Brien, E., Morris, R. K., Rayner, K., (1990). Elaborative inferencing as an active or passive process. *Journal of Experimental Psychology: Learning, Memory, and Cognition, 16*, 250–257.

Gernsbacher, M. (1984). Revolving 20 years of inconsistent interactions between lexical familiarity and orthography, concreteness, and, polysemy. *Journal of Experimental Psychology: General, 113*, 256–281.

Graesser, A. C., Singer, M., Trabasso, T. (1994). Constructing inferences during narrative text comprehension. *Psychological Review, 101*, 371–395.

Graves, M. F. (1987). The roles of instruction in fostering vocabulary development. In: M. G. McKeown and M. E. Curtis (eds), *The Nature of Vocabulary Acquisition* (pp. 165–184) Hillsdale, NJ: Lawrence Erlbaum Associates, Inc.

Herman, P. A., Anderson, R. C., Pearson, P. D., & Nagy, W. E. (1987). Incidental acquisition of word meaning from expositions with varied text features. *Reading Research Quarterly, 22*, 264–284.

Hyönä, J., & Olson, R. (1995). Eye fixation patterns among dyslexic and normal readers: Effects of word length and word frequency. *Journal of Experimental Psychology: Learning, Memory, and Cognition, 21*, 1430–1440.

Inhoff, A. (1984). Two stages of word processing during eye fixations in the reading of prose. *Journal of Verbal Learning and Verbal Behavior, 23*, 612–624.

Inhoff, A. W., & Rayner, K. (1986). Parafoveal word processing during eye fixations in reading: Effects of word frequency. *Perception and Psychophysics, 40*, 431–439.

Jenkins, J., & Dixon, R. (1983). Vocabulary learning. *Contemporary Educational Psychology, 8*, 237–260.

Jenkins, J., Stein, M. L., & Wysocki, K. (1984). Learning vocabulary through reading. *American Educational Research Journal, 21*, 767–787.

Just, M., & Carpenter, P. (1980). A theory of reading: From eye fixations to comprehension. *Psychological Review, 87*, 329–354.

Lampinen, J. M., & Faries, J. (1994). Levels of semantic constraint and learning novel words. In A. Ram and K. Eiselt (eds), *Proceedings of the 16th Annual Conference of Cognitive Science Society* (pp. 531–536). Hillsdale, NJ: Lawrence Erlbaum Associates, Inc.

Landauer, T. (1986). How much do people remember: Some estimates of the quantity of learned information in long-term memory. *Cognitive Science, 10*, 477–493.

Landauer, T., & Dumais, S. (1997). A solution to Plato's Problem: The latent semantic analysis theory of acquisition, induction, and representation of knowledge. *Psychological Review, 104*, 211–240.

Lehrer, A. (1992). Names and naming: Why we need fields and frames? In: A. Lehrer and E. Kittay (eds), *Frames, Fields, and Contrasts: New Essays in Semantic and Lexical Organization* (pp. 253–288). Hillsdale, NJ: Erlbaum.

Long, L. L., & Shaw, R. J. (2000). Adult age differences in vocabulary acquisition. *Educational Gerontology, 26*, 651–664.

McKeown, M. G. (1985). The acquisition of word meanings from context by children of high and low ability. *Reading Research Quarterly, 20*, 482–496.

Miller, G. (1991). *The Science of Words.* New York: Scientific American Library.

Miller, G., & Gildea, P. M. (1987). How children learn words. *Scientific American, 257*, 94–99.

Mori, Y., & Nagy, W. E. (1999). Integration of information from context and word elements in interpreting novel kanji compounds. *Reading Research Quarterly, 34*, 80–101.

Morris, R. K. (1994). Lexical and message-level sentence context effects on fixation times in reading. *Journal of Experimental Psychology: Learning, Memory, and Cognition, 20*, 92–103.

Morris, R. K., & Folk, J. R. (1998). Focus as a contextual priming mechanism in reading. *Memory & Cognition, 26*, 1313–1322.

Morris, R. K., & Folk, J. R. (2000). Phonology is used to access word meaning during silent reading: Evidence from lexical ambiguity resolution. In: A. Kennedy and R. Radach (eds), *Reading as a Perceptual Process* (pp. 427–446). Oxford: Elsevier Science.

Morrison, R. E. (1984). Manipulation of stimulus onset delay in reading: Evidence for parallel programming of saccades. *Journal of Experimental Psychology: Human Perception and Performance, 10*, 667–682.

Myers, J. L., Cook, A. E., Kambe, G., Mason, R. A., & O'Brien, E. J., (2000). Semantic and episodic effects on bridging inferences. *Discourse Processes, 23*, 1–24.

Nagy, W. E., Anderson, R. C., & Herman, P. A. (1987). Learning new words from context during normal reading. *American Educational Research Journal, 24*, 237–270.

Nagy, W. E., & Herman, P. A. (1987). Breadth and depth of vocabulary knowledge: Implications for acquisition and instruction. In: M. G. McKeown and M. E. Curtis (eds), *The Nature of Vocabulary Acquisition* (pp. 19–35). Hillsdale, NJ: Lawrence Erlbaum Associates, Inc.

Nagy, W., Herman, P., & Anderson, R. C. (1985). Learning new words from context. *Reading Research Quarterly, 19*, 304–330.

Nagy, W. E., & Scott, J. A. (1990). Word schemas: Expectations about the form and meanings of words. *Cognition and Instruction, 7*, 105–127.

O'Brien, E. J., Raney, G. E., Albrecht, J. E., & Rayner, K. (1997). Processes involved in the resolution of explicit anaphors. *Discourse Processes, 23*, 1–24.

O'Brien, E. J., Shank, D. M., Myers, J. L., & Rayner, K. (1988). Elaborative inferences during reading: Do they occur on-line? *Journal of Experimental Psychology, Learning, Memory, and Cognition, 14*, 410–420.

Poynor, D., & Morris, R. K. (2002). Inferred goals in narratives: Evidence from self-paced reading, recall and eye movements. Manuscript submitted for publication.

Rayner, K., & Duffy, S. A. (1986). Lexical complexity and fixation times: Effects of word frequency, verb complexity, and lexical ambiguity. *Memory & Cognition, 14*, 191–201.

Rayner, K., Sereno, S. Morris, R. K., Schmauder, A. R., & Clifton, C. (1989). Eye movements and on-line language comprehension processes. *Language and Cognitive Processes, 4*, SI21–SI49.

Rayner, K., Sereno, S., & Raney, G. (1996). Eye movement control in reading: A comparison of two types of models. *Journal of Experimental Psychology, Learning, Memory, and Cognition, 22*, 1188–1200.

Rayner, K., & Well, A. D. (1996). Effects of contextual constraint on eye movements in reading: A further examination. *Psychonomic Bulletin and Review, 3*, 504–509.

Reichle, E, Pollatsek, A., Fisher, D., & Rayner, K (1995). Toward a model of eye movement control in reading. *Psychological Review, 105*, 125–157.

Schatz, E. K., & Baldwin, R. S. (1986). Context clues are unreliable predictors of word meanings. *Reading Research Quarterly, 21*, 439–453.

Schilling, H., Rayner, K., & Chumbley, J. (1998). Comparing naming, lexical decision, and eye fixation times: Word frequency effects and individual differences. *Memory & Cognition, 26*, 1270–181.

Schwanenflugel, P. L., Stahl, S., & McFalls, E. L. (1997). Partial word knowledge and vocabulary growth during reading. *Journal of Literacy Research, 29*, 531–533.

Scott, J. A., & Nagy, W. E. (1997). Understanding the definitions of unfamiliar verbs. *Reading Research Quarterly, 32*, 184–200.

Seely, R. S., Morris, R. K., & Schmauder, A. R. (2002). The effect of syntactic category ambiguity on semantic resolution in reading. Manuscript submitted for publication.

Shefelbine, J. L. (1990). Student factors related to variability in learning word meanings from context. *Journal of Reading Behavior, 22*, 71–97.

Shu, H., Anderson, R. C., & Zhang, H. (1995). Incidental learning of word meanings while reading: A Chinese and American cross-cultural study. *Reading Research Quarterly, 30*, 76–95.

Stahl, S. A. (1991). Beyond the instrumentalist hypothesis: Some relationships between word meanings and comprehension. In: P. J. Schwanenflugel (ed.), *The Psychology of Word Meanings* (pp. 157–186). Hillsdale, NJ: Erlbaum.

Sternberg, R. J., & Powell, J. S. (1983). Comprehending verbal comprehension. *American Psychologist, 38*, 878–893

Sternberg, R. J. (1987). Most vocabulary acquisition is learned from context. In: M. G. McKeown and Curtis M. E (eds), *The Nature of Vocabulary Acquisition* (pp. 89–105). Hillsdale, NJ: Erlbaum.

van Daalen-Kapteijns, M. M., Elshout-Mohr, M., & de Glopper, K. (2001). Deriving the meaning of unknown words. *Language Learning, 51*, 145–181.

van Daalen-Kapteijns, M., Elshout-Mohr, M. (1981). The acquisition of word meanings as a cognitive learning process. *Journal of Verbal Learning and Verbal Behavior, 20*, 386–399.

Vitu, F., O'Regan, J. K., & Mittau, M. (1990). Optimal landing position in reading isolated and continuous text. *Perception & Psychophysics, 47*, 583–600.

Walker, D. E. & Amsler, R. A. (1986). The use of machine-readable dictionaries in sublanguage analysis. In: R. Grisham (eds), *Analyzing Languages in Restricted Domains. Sublanguage Description and Processing*. Hillsdale, NJ: Lawrence Erlbaum Associates.

Werner, H., & Kaplan, E. (1952). The acquisition of word meanings: A developmental study. *Monographs of the Society for Research in Child Development, 15*, 190–200.

Williams, R. S., & Morris, R. K. (2003). An eye movement analysis of vocabulary acquisition in skilled reading. Manuscript submitted for publication.

Zola, D. (1984). Redundancy and word perception during reading. *Perception & Psychophysics, 36*, 277–285.

Chapter 13

Application of Eye Tracking in Speech Production Research

Antje S. Meyer and Christian Dobel

During multiple object naming, the speakers' eye movements are closely linked to their speech planning: Speakers usually fixate upon new objects just before mentioning them. However, they sometimes skip "old" objects they have named before, especially if the referring expression is quite simple. The viewing time per object depends on the time required to fully plan the corresponding phrase. Finally, the available evidence suggests that lexical access to an object's name is unlikely to begin before fixation of the object. Because of these regularities, eye tracking is a promising tool for language production research.

Introduction

Eye monitoring has been a valuable research tool in psycholinguistics for many years, especially in studies of visual word recognition and reading (for reviews see Liversedge & Findlay, 2000; Rayner, 1998), but also in studies of auditory language comprehension (e.g., Dahan, Magnusen & Tanenhaus 2001; Tanenhaus, Spivey-Knowlton, Eberhard & Sedivy, 1995). Until now, eye tracking has not been much used in studies of language production. One reason for this may be a technical one: many eye tracking systems require the participants to sit absolutely still. This is often enforced by using a bite bar or chin rest, which interferes with articulation. Recently, however, eye tracking systems have been introduced that allow head and body movements. These systems can be used for many purposes, including language production research.

In this chapter, we will review studies of language production in which the speakers' eye movements were recorded. We will explain the assumptions underlying this work and we will show how results of eye movement analyses can supplement results of analyses of speech errors and onset latencies. Most of the reviewed studies concern the generation of simple utterances, typically noun phrase conjunctions such as "the cat and the chair", but in the final part of the chapter we will review studies targeting more complex utterances.

The Mind's Eye: Cognitive and Applied Aspects of Eye Movement Research

ISBN: 0–444–51020–6

Chronometric Studies of Word and Sentence Production

The production of single words can be studied in a number of different ways, for instance by analysing speech errors or momentary blockages of lexical retrieval processes (so-called tip-of-the-tongue states) in healthy speakers or patients diagnosed with various types of aphasia. In addition, one can analyse speech onset latencies in picture naming and other word production tasks (Bock, 1996). Latency analyses have been very fruitful in studies of single word production because a single word is likely to be fully planned before speech onset and the speech onset latency therefore reflects the difficulty of speech planning (but see Meyer, Roelofs and Meyer, in press). Based on a large body of evidence from different research paradigms, detailed models of single word access have been proposed (for reviews see Johnson, Paivio & Clark, 1996; Levelt, 1999; Rapp & Goldrick, 2000).

However, most of the time speakers do not produce isolated words but phrases or sentences. In order to do so, they must, of course, retrieve words from the mental lexicon as in single word production. One important research issue is how the retrieval processes for the words of a phrase or sentence are co-ordinated in time, whether, for instance, several words are processed in parallel, or whether words are retrieved in sequence. This issue is important in its own right, as any model of sentence production must state how different processing components are co-ordinated in time. In addition, the temporal co-ordination of word retrieval processes constrains syntactic and prosodic integration processes. In connected speech, words have to be selected such that they combine into syntactically and prosodically well-formed utterances. At any moment in time the speaker can only take the features of those concepts and words into account that have already been activated. Thus, the time course of the planning processes determines the information available to the syntactic and prosodic processors.

Complex sentences are rarely fully planned before speech onset. Instead speakers often only fully plan the first few words before they begin to speak and plan the rest as they talk. Experimental evidence supporting this view comes from studies in which speech onset latencies were measured for utterances varying in the complexity or difficulty of an utterance-initial or utterance-internal word or constituent. For instance, Smith and Wheeldon (1999) used a picture description task to elicit utterances that began with a complex phrase (e.g., "The dog and the kite move above the house") or ended with such a phrase ("The dog moves about the kite and the house"). They observed longer speech onset latencies for the former type of sentences, suggesting that speakers plan the first phase of an utterance more extensively than later phrases (for similar experiments see Ferreira & Swets, 2002; Wheeldon & Lahiri, 1997). The available findings from studies using this general research strategy suggest that the processing units at the highest planning level, the conceptual level, may encompass entire clauses, whereas at the lower, phonological, level the units often correspond to single phonological words (for a review see Ferreira & Swets, 2002).

Latency analyses permit researchers to classify speech planning processes as occurring before or after speech onset. In addition, it is sometimes possible to decide whether several processes running before speech onset do so in parallel or in sequence (e.g., Schriefers, de Ruiter & Steigerwald, 1999). However, speech onset latencies do not provide information about the relative timing of the planning processes occurring *after* speech onset. Some evidence about the timing of these processes can be gleaned from analyses of speech errors. For instance, the occurrence of word exchanges such as "can you shirt my irons?" suggests that the exchanging words were selected in parallel and uttered in the wrong order (e.g., Garrett, 1975). However, speech errors pose interpretative problems, as they do not constitute an on-line measure of processing and as they reflect, by definition, on derailed rather than intact sentence planning processes. In short, new research methods should be developed that permit researchers to study the co-ordination of speech planning processes before as well as after speech onset. As we hope to show, eye tracking may be such a method.

Eye Movements During the Production of Simple Noun Phrases

In this section we summarize a series of experiments in which we used eye tracking in order to study how lexical retrieval processes are co-ordinated when speakers name pairs of objects in noun phrase conjunctions such as "the cat and the chair". Figure 13.1 shows our working model of single word production (for details see Levelt, Roelofs & Meyer, 1999) and displays two hypotheses concerning the co-ordination of the planning processes for several words.

We display the main processing components involved in single word production in the columns of the figure. Object naming is broken down into a number of discrete processing components. First, there are visual-conceptual processes, which lead to the recognition of the object as an instance of a concept the speaker knows. Second, there are lexical access processes, which retrieve a name for the target concept. These processes are broken down into lemma selection (i.e., the selection of a syntactic word unit), morphological encoding processes (i.e., the selection and combination of the morphemes forming the word) and phonological encoding processes (i.e., the selection of the word's consonants and vowels, stress assignment and syllabification). Finally, there are post-lexical phonetic encoding processes, which take the syllabified phonological representation as their input and generate executable articulatory commands.

There are, of course, other models of object naming (e.g., Humphreys, Price & Riddoch, 1999; see Johnson *et al.*, 1996 for a review). All models assume that object naming involves a number of distinguishable processing steps. An important distinction they all make is between access to non-linguistic information about the properties and functions of objects and access to information about the object's name. According to some models (including our working model) these processes of object recognition and name retrieval constitute discrete, non-overlapping stages. According to other models (e.g., Humphreys *et al.*, 1999) the two sets of processes run in cascade, which implies that the non-linguistic processes need not be completed before the lexical retrieval processes begin. Some, but not all models assuming cascading processes allow for feedback from lower to higher processing levels (see Rapp & Goldrick, 2000 for further discussion).

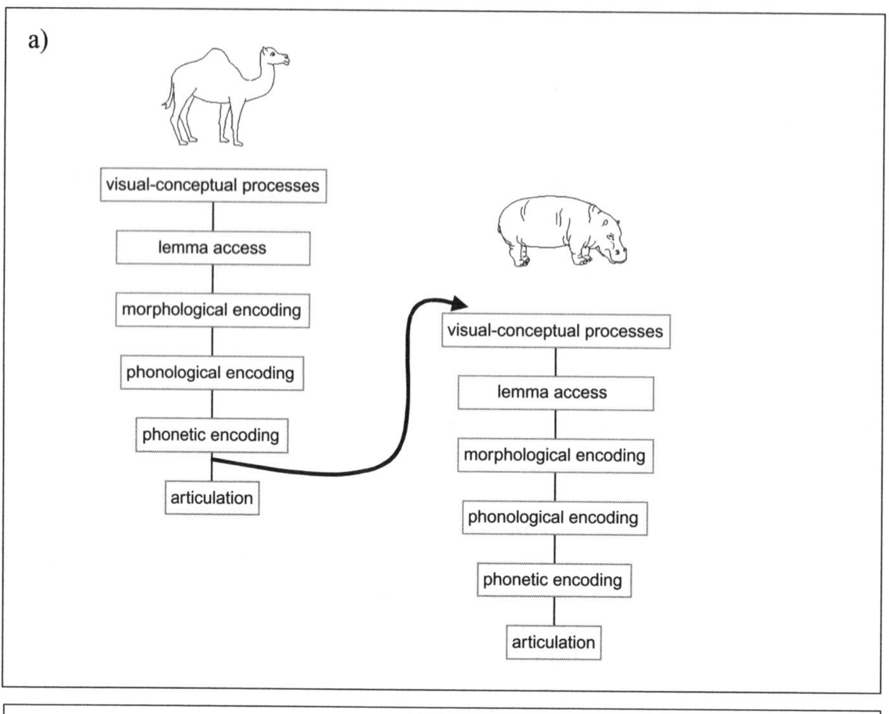

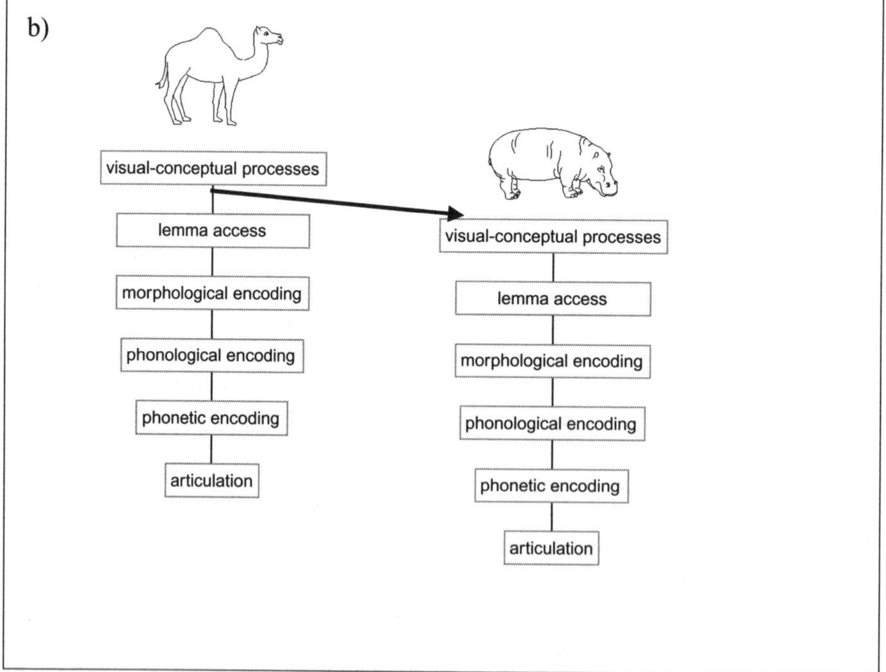

Figure 13.1: Temporal co-ordination of lexical retrieval processes according to the serial hypothesis (a) and the parallel hypothesis (b).

How could speakers co-ordinate these processes when they name two or more objects? Figure 13.1 illustrates two extreme hypotheses; intermediate hypotheses could be formulated as well. Figure 13.1a shows what we will call the serial hypothesis: Speakers focus on the first object until its name has been fully planned and then turn to the next object. In order to generate fluent speech, speakers will probably have to start processing a new object before initiating the first object's name, such that the time required for the articulation of the first object's name can be used to plan the second object's name (see Griffin, in press). According to this hypothesis, speakers initiate the processing of a new object as late as possible, thereby minimising the temporal overlap of the planning processes pertaining to different objects.

By contrast, according to an alternative hypothesis, called the parallel hypothesis hereafter and shown in Figure 13.1b, speakers initiate the processing of new objects as early as possible, thereby maximising the temporal overlap of the planning processes pertaining to different objects. The visual-conceptual processes leading to the identification of different objects probably have to be carried out in sequence (e.g., Treisman 1993), but the lexical retrieval processes for several words could, conceivably, run in parallel.

How can eye monitoring help us distinguish between these hypotheses? Evidently, we can determine whether, when and for how long speakers look at the objects they name. Whether this constitutes useful information depends on the existence of a systematic relationship between eye movements and speech planning. The main goal of a set of experiments conducted by Meyer, Sleiderink and Levelt (1998) was to determine whether such a link existed. The specific objectives were (i) to determine whether the objects were inspected in the same order as they were named and (ii) whether the difficulty of recognising and naming different objects would be reflected in the time speakers spent looking at them.

We asked speakers to name object pairs, as shown in Figure 13.2, starting with the left object. We recorded the speakers' speech onset latencies and their eye movements using an SMI EyeLink-Hispeed 2D eye tracking system. We found that the speakers almost always fixated upon each object they named in the order of mention. This is not particularly surprising. Since the distance between the objects on the displays was between 10 and 12 degrees of visual angle, the speakers had to fixate upon each of the objects in order to recognise them, and it certainly is a sensible strategy to inspect and name the objects in the same order. Concerning the timing of speech and gaze, we found that the eyes ran slightly ahead of the overt speech: Speakers looked at the left object for about 500 ms, then initiated the shift of gaze to the right object, and shortly afterwards began to say the first object's name.

Van der Meulen (2001) examined the gaze patterns of speakers naming sets of four objects. She found that, as long as speakers simply named the objects in a predetermined order, the alignment of gaze and speech was very regular. Speakers looked at the objects in the order of mention and their eye gaze usually ran ahead of the overt speech by about one object.

Having established that speakers looked at the objects in the order of mention, we examined which variables affected how long they would look at the objects. This time period, which we call the viewing time, was quantified as the period between the

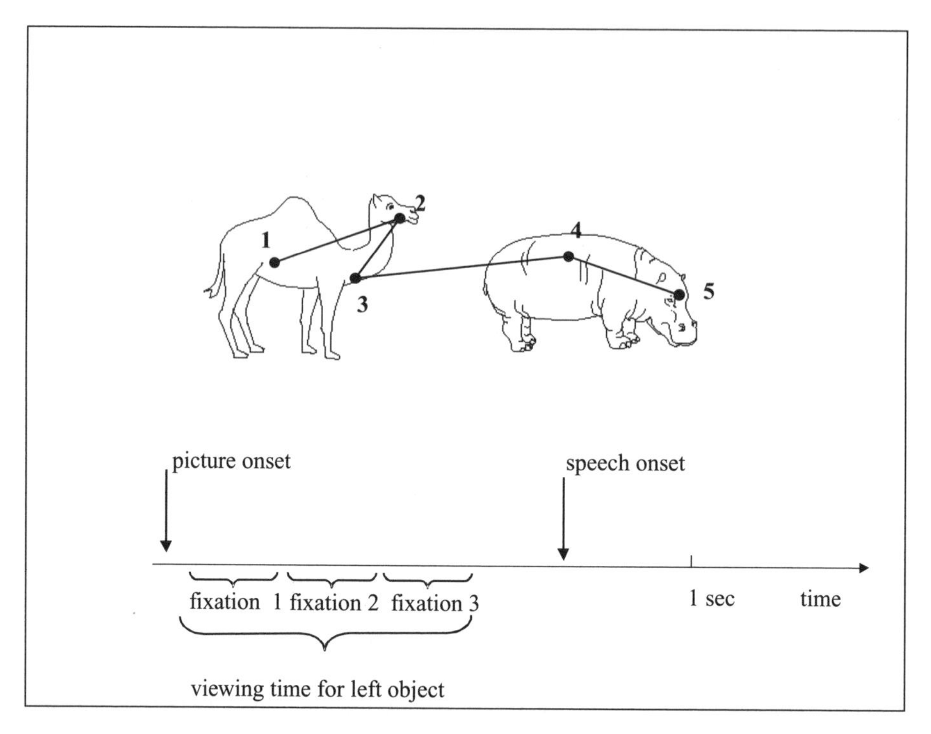

Figure 13.2: Typical scan path for two objects and viewing time for the left object.

beginning of the first fixation on an object and the end of the last fixation before the shift of gaze to the next object or, occasionally, to a location between objects.[1] The underlying logic in these experiments was as follows. We select a dependent variable that is known to affect a specific component of object naming and observe whether variations in this variable systematically affect the viewing time for the object. If they do, we can conclude that the shift of gaze from the target object to another object is initiated after the processing at the affected level has been completed. Thus, by observing which variables do or do not affect the viewing time for an object we can estimate which processes are carried out while an object is being looked at and which are carried out afterwards.

Following this general logic, Meyer *et al.* (1998) varied the ease of recognising target objects by presenting them as intact line drawings or in a degraded version, in which 50% of the contours were missing. As one would expect, speakers named the degraded drawings significantly more slowly than the intact ones. More importantly, the viewing times were also significantly longer (by 15 ms) for degraded than for intact objects. This shows that the speakers completed the visual-conceptual processes for the target object before the shift of gaze to the next object.

In the same experiment, we varied the frequency of the names of the objects, orthogonally to the ease of object recognition. Earlier studies (e.g., Jescheniak & Levelt,

1994) have shown that name frequency affects the retrieval of the morphological forms of words. We found, as expected, that the naming latencies were significantly shorter for objects with high frequency names than for objects with low frequency names. The viewing times were also significantly shorter (by 34 ms) for the objects with high frequency names. This suggests that the speakers only moved their eyes from the target to the next object after retrieving the morphological representation of the target object's name.

Recent findings from a similar experiment by Griffin (2001) corroborate our main conclusions. Griffin orthogonally varied the codability of the target objects and the frequency of their names. Codability is the ease of finding a suitable name for an object and has been shown in independent studies to affect the ease of lemma selection, the first step of lexical access. Griffin found that both codability and frequency systematically affected how long speakers looked at the objects. This study extends our findings by showing that there was a close relationship between viewing times and the ease of object naming not only for the first, but also for the second of three objects that were to be named. This is of great methodological interest because, as noted in the Introduction, the planning of utterance-internal parts of speech is difficult to study in reaction time experiments.

Meyer *et al.* (1998) and Griffin (2001) showed that speakers look at the objects they name at least until they had retrieved the morphological representation of their names. Meyer and van der Meulen (2000) examined whether the viewing times for objects would be sensitive to a phonological variable. Again, speakers were asked to name object pairs. At trial onset, they heard an auditory distractor word that was phonologically related to the first object's name or unrelated to it. For instance, speakers about to name a cat and bike would hear "cap" in the related condition or "fork" in the unrelated condition. As in earlier experiments (e.g., Meyer & Schriefers, 1991) the speech onset latencies were shorter after phonologically related than after unrelated distracters. The viewing time for the left object was shorter as well, by 50 ms, which shows that the speakers fixated upon the left object until they had retrieved the phonological form of its name.

Finally, several investigators have studied the viewing times for objects with long and shorter names. Zelinski and Murphy (2000) used a memory task, in which speakers first studied a display of four objects and, after a brief interval, saw one of the objects and had to indicate whether it had been part of the studied set. They found that during the study phase participants looked longer at objects with long than with shorter names (see also Noizet & Pynte, 1976). Meyer *et al.* (in press) asked speakers to name objects with monosyllabic and disyllabic names. They found that the naming latencies and viewing times were shorter for the items with short names than for those with longer names, provided that the items with long and short names were tested in separate blocks of trials. When the items were mixed, no length effects were found for latencies or viewing times. Meyer *et al.* argued that in mixed and pure blocks speakers used different criteria in deciding when to begin to speak. In the pure blocks the speakers fully prepared the phonological forms of the targets and retrieved the corresponding articulatory commands before beginning to speak and before initiating the shift of gaze to the next object. For monosyllabic targets they generated one syllable at the phonological and articulatory level; for disyllabic targets they generated two syllables at both levels. In mixed blocks, they still fully prepared the monosyllabic targets. By contrast, disyllabic targets were fully prepared at the phonological level, but not at the articulatory level. Instead articulation and the shift of gaze to the next object were initated as soon as the articulatory programme for the first syllable had been prepared. Computer simulations using the WEAVER++ model (Roelofs, 1977), which implements the model proposed by Levelt, Roelofs and Meyer, support this view. For the present purposes, the most important point is that parallel patterns of results were obtained for the latencies and viewing times: when a length effect was obtained for the latencies, there was an effect for the viewing times as well. When no length effect was observed for the latencies, there was no such effect for the viewing times either. This suggests that for simple naming the criteria governing the timing of the speech onset and of the shift of gaze to a new object are very similar. Speakers begin to speak and shift gaze to a new object when they have generated the complete phonological form of the first object's name and the articulatory commands for at least one syllable.

Eye Movements During the Production of Complex Noun Phrases

In the studies described so far, speakers referred to objects in simple noun phrases, such as "the cat". The viewing time for the first object to be named and the speech onset latency were highly correlated. Thus, it may seem that eye monitoring does not yield any information that could not be gained more easily from analyses of speech onset latencies. However, the relationship between viewing times and speech onset latencies changes dramatically when longer utterances are elicited. Meyer (submitted, see also Levelt & Meyer, 2000) showed Dutch speakers coloured pictures of objects appearing in different sizes. The speakers were either asked to name the objects in simple noun phrases (e.g., "de bal en de vork" — "the ball and the fork") or, in a different block of trials, to mention the size and colour of the objects as well, yielding more complex noun phrases ("de kleine groene bal en de vork" — "the little green ball and the fork"). As shown in Figure 13.3, the speech onset latencies for the two phrase types did not differ much. This was to be expected on the basis of earlier studies (e.g., Schriefers & Teruel, 1999) showing that speakers rarely fully plan complex noun phrases before speech onset. Speakers preparing to say "de kleine groene bal" probably only fully planned the determiner and the first adjective before speech onset, which did not take much longer than preparing the determiner and noun for "de bal". By contrast, the mean viewing time for the target objects was more than twice as long when complex than when simple noun phrases were required. Detailed analyses of the eye movements and speech output showed that for both phrase types, the shift of gaze to the next object was initiated about 250 ms before the noun onset.

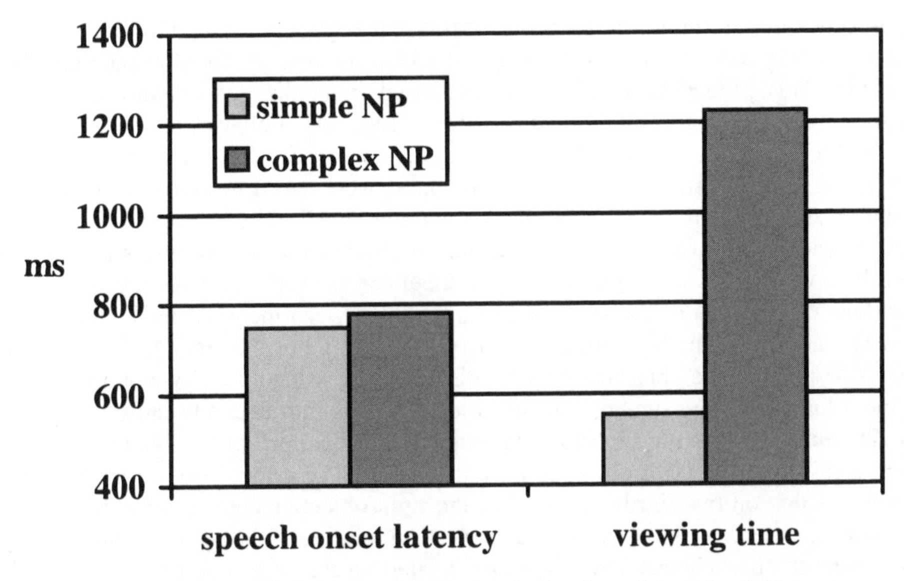

Figure 13.3: Mean speech onset latencies and viewing times (ms) for utterances beginning with simple and complex noun phrases.

In subsequent experiments, English speakers were asked to produce a variety of structures, including long and short noun phrases and relative clauses ("the ball, which is red, . . ."). In addition, Italian speakers were tested, who produced utterances such as "porta rossa" ("door red"), where the colour adjective follows the noun. The comparison between the results obtained for English and Dutch on the one hand and Italian on the other hand reveals whether the shift of gaze to the next object is time-locked to the last word of the phrase or to the noun. The results were clear: In all cases, the speakers' shifts of gaze occurred shortly before the onset of the phrase-final word. Thus, when speakers produce complex noun phrases, their eyes remain on the referent object until they have fully planned the phrase about the object and are about to initiate the phrase-final word.

Fixating versus Processing of Objects

We have shown that speakers naming several objects fixate upon them in the order of mention and that the time spent fixating upon an object is closely related to the time required to plan the utterance about the object. These results seem to support the serial processing hypothesis, according to which speakers co-ordinate the planning processes for different phrases such that the temporal overlap between planning processes is minimised. However, it is possible that objects that are visible extrafoveally begin to be processed before the eyes move there. Since extrafoveal information has been shown to be used in many other tasks (for an overview see Starr & Rayner, 2001), it may be exploited in multiple object naming as well.

In order to determine the extent of extraoveal object processing in our standard object naming task, we used a paradigm in which the display could change during a saccade. This procedure has been widely used in studies of reading and object processing (e.g., Henderson, 1992a,b, 1997; Henderson, Pollatsek & Rayner, 1987, 1989; McConkie & Rayner, 1975; Pollatsek, Rayner & Collins, 1984; Pollatsek, Rayner & Henderson, 1990). Again, participants were asked to name object pairs in noun phrase conjunctions, such as "the tree and the camel". The objects were displayed next to each other just above the horizontal midline of the screen, at a size of about five by five degrees of visual angle and about ten degrees apart (centre to centre), yielding a blank area of about five degrees width between them. At the bottom of the screen, one of two small symbols was displayed, a cross or the letter "x". The participants first named the objects and then indicated by pressing one of two buttons which symbol they saw. The symbol categorisation task was introduced because we needed to determine the viewing time for both objects and therefore had to induce a shift of gaze away from the right object to a new location. The critical feature of this paradigm was that on two thirds of the trials the right object changed during the saccade towards it, as soon as the eyes crossed the vertical midline of the screen. Consequently, the object the participants saw when they fixated on the right side of the screen (the target, which was to be named) could be different from the object they had seen extrafoveally before (the interloper). In the first experiment using this paradigm, we used interloper-target pairs that were identical or unrelated objects, or were different representations of the same object, e.g., drawings of different books or of different chairs.[2] The mean viewing time was shortest in the identical condition, intermediate in the conceptual condition, and longest in the unrelated condition. All three mean viewing times differed significantly from each other.

In next two experiments, the conceptually related condition was replaced by a semantically related condition, in which target and interloper were members of the same semantic category (e.g., lion and tiger or chair and lamp), or by a phonologically related condition, in which they had similar names (as in "cap" and "cat"). Again, the target viewing times were significantly shorter in the identical than in the unrelated condition. However, there was no difference between the unrelated and the semantically or the phonologically related condition.

The robust identity priming effect shows that the interloper was processed while the left object was being fixated upon. The conceptual priming effect (a picture of a chair priming recognition of a picture of a different chair) observed in the first experiment demonstrates that the processing of the extrafoveal object extended beyond the earliest steps of visual analysis. A semantic priming effect (with "tiger" priming "lion") would, depending on one's theory, be allocated at the conceptual or at the lexical level; but no such effect was observed here. A phonological priming effect (with "cap" priming "cat") would have to be a lexical effect, but no such effect was obtained either. Thus there is, from these data, no evidence that the retrieval of the name of the interloper began while the left object was being looked at. At present, we do not know whether the processing of the interloper was simply too slow to reach the lexical stage before the shift of gaze, or whether access to the name of an object is only triggered after the shift of gaze to the referent object has been initiated.

Pollatsek *et al.* (1984) obtained evidence for activation of the phonological form of the name of an extrafoveal object. However, there are many procedural differences between that study and ours. In the study by Pollatsek *et al.* the participants first passively viewed a fixation cross or picture and then moved their eyes to the interloper as soon as it appeared. During the saccade, the interloper was replaced by a target, which the participants named. Naming latency for the target was the dependent variable. In our experiments, the speakers named the left object, viewing time for the target was the dependent variable, and the proportion of related interloper-target pairs was much lower. Further research will have to determine which variables govern the extent of extrafoveal processing in object naming tasks. In Figure 13.1, we contrast two hypotheses concerning the temporal co-ordination of the recognition and name retrieval processes for two objects. As far as the speakers' eye movements are concerned, the serial hypothesis is supported. Speakers fixate upon the first object until they are ready to articulate its name and only then initiate a shift of gaze to the second object. However, the data also show that there is some overlap in the processing of the two objects: The visual and conceptual analysis of the second object appears to begin before fixation.

Repeated Reference

In all of the experiments discussed above, the speakers saw different objects on each trial. Van der Meulen (2001) and Van der Meulen, Meyer and Levelt (2001) studied what happened when speakers repeatedly referred to the same objects, as one often does in spontaneous speech. Van der Meulen (2001) asked speakers to describe object pairs in utterances such as "the green ball is next to the block" or "the ball next to the block is green". When speakers produced utterances of the first type, they fixated upon the left object for about 1400 ms and then turned to the right object. By contrast, when speakers produced utterances of the second type, they initially looked at the left object for about 800 ms, then turned to the right object just before naming it and finally looked at the left object again, about 600 ms before the onset of the adjective ("green" in the example). Thus, the object was revisited before its colour was named. We do not yet know whether the colour information was never submitted to working memory, or whether speakers did not find this representation reliable enough to use during colour naming. In any event, the findings are in line with the results of studies of action planning discussed in Hyönä *et al.* (2002) (see also Hayhoe, 2000). These studies showed that people usually look at objects they are about to use, that their eyes return to known objects in known locations when they use them again, and that they prefer basing actions on external visual information to using memory representations of the environment.

In a study by van der Meulen, Meyer and Levelt (2001) speakers saw pairs of displays on successive trials that shared the left object (see Figure 13.4). One group of speakers was asked to use noun phrases, such as "The angel is next to the camel. The angel is now next to the bed." The other group was asked to use a pronoun when an object was repeated, as in "The angel is next to the camel. It is now next to the bed." Before each pair of trials, a fixation point was presented in the middle of the screen. The trials of a pair were separated by brief blank intervals. The repeated objects

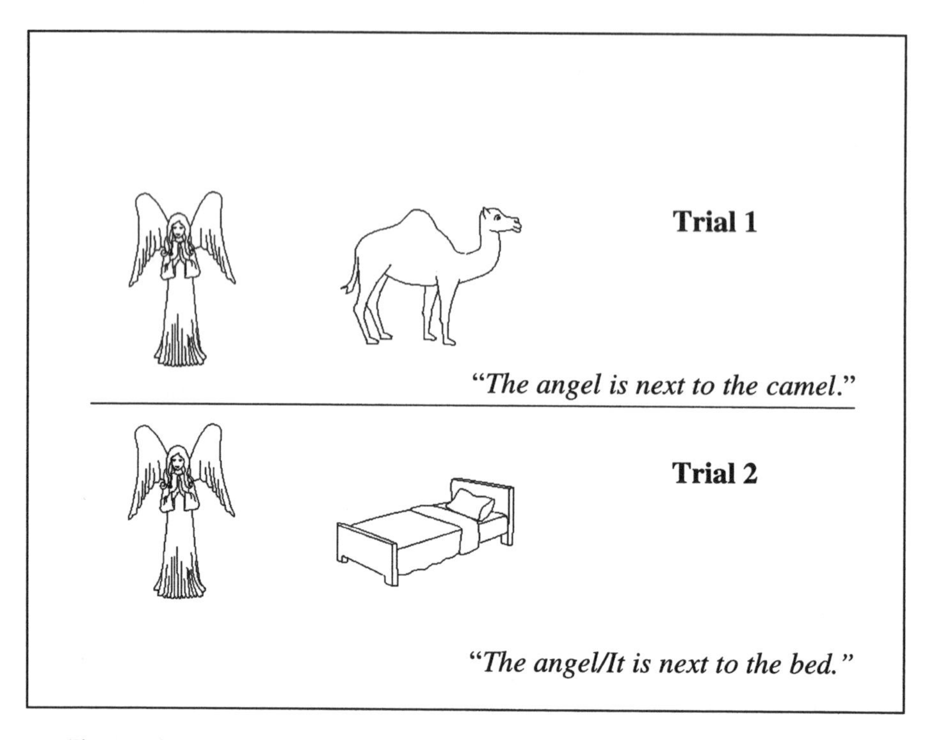

Figure 13.4: Displays used by van der Meulen, Meyer and Levelt (2000).

appeared in the same screen position on both trials of a pair. This experiment was carried out in Dutch, where the pronoun was "hij" ("it") on all trials, as well as in German, where speakers had to select one of three pronouns depending on the grammatical gender of the corresponding noun ("er", "sie" or "es"). In the German experiment, there were only three target objects in each block of trials, which differed in the grammatical gender of their names. Thus on each trial, the speakers in the noun phrase condition selected one of three noun phrases, and the speakers in the pronoun condition chose one of three pronouns.

The German and Dutch experiments yielded very similar results. Figure 13.5, for the German results, shows that the speakers were much more likely to look at the left object on the first trial of a pair than on the second trial, where the object was repeated. Interestingly, speakers were considerably less likely to look at the repeated left object when they produced a pronoun than when they produced a noun phrase, which may be due to the difference in frequency or length between the two types of expressions. Thus, speakers do not look at every object they refer to, as suggested by the earlier findings, but they may skip known objects, especially when they use very simple referring expressions.

How can the type of utterance being planned about an object affect whether the object will be looked at? The multiple object naming task bears some similarity to reading. Proficient readers do not fixate upon all words they read. Short, high frequency words and words that are highly predictable given the context are often skipped. Our

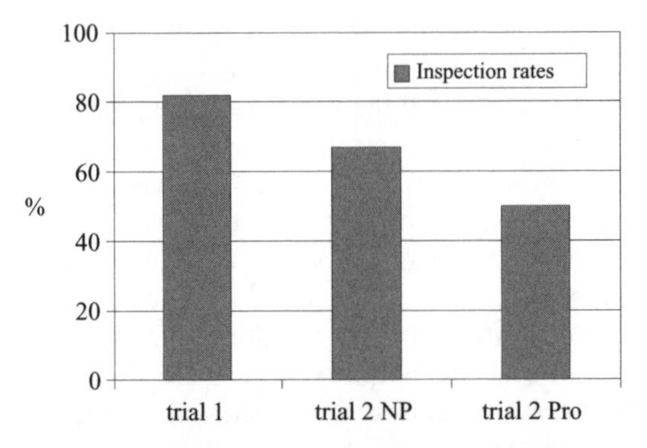

Figure 13.5: Mean inspection rates for the left object on the first trial of a pair and on the second trial in the noun phrase (NP) and pronoun (Pro) conditions.

preliminary account of the results obtained by van der Meulen *et al.* is inspired by the model of eye movement control during reading proposed by Reichle *et al.* (1998). We assume that speakers will, as default, plan an eye movement to each object to be named. However, if an object name becomes available fast enough, i.e. before the planning of the eye movement to the object has reached a ballistic phase during which it cannot be cancelled any more, the eye movement may be cancelled, and an eye movement to another object will be planned. An object name may be readily available because the speaker has named the object before and has stored a representation of the object or its name in working memory. In addition, object identification may be possible on the basis of extrafoveal information, as discussed above. In the experiment by van der Meulen *et al.* (2001) only three different left objects were used in each block of trials, each of which was named 12 times. Thus, their names should be highly available. This may explain why the inspection rate for the objects on the first trial of a pair was only 80% (compared to more than 90% in earlier experiments); on 20% of the trials the speakers apparently relied on peripheral information when naming the objects. On the second trial of a pair, the speakers probably remembered which object they had named less than ten seconds earlier. Interestingly, the inspection rate was higher when a noun phrase than when a pronoun was used, although the same noun phrase had been used on the preceding trial. This suggests that the speaker's working memory representation was a conceptual representation of the object rather than a presentation of the phrase used before. On the basis of this representation the pronoun could be generated more readily than the noun phrase, which accounts for the difference in the inspection rates. We are currently further exploring the validity of this proposal.

Eye Movements During Scene and Event Description

So far we have discussed experiments in which speakers were asked to name sets of objects in a predefined order. One may ask how much of the orderly co-ordination between eye gaze and speech will be maintained when speakers produce more complex utterances and when they decide themselves, as speakers usually do in spontaneous speech, what to name and in which order to do so.

Van der Meulen (2001) compared the speakers' gaze patterns when they described objects in a prescribed order using a prescribed sentence structure and when they determined the order of mention and syntactic structure themselves. On each trial, she presented a set of four objects. The objects were to be described in utterances such as "the cat is above the pen and the fork is above the watch" when all four objects were different, and in utterances such as "the cat and the fork are above a pen" when the bottom two objects were identical. In one condition, the materials were blocked such that throughout a block of trials the bottom objects were always identical, or they were always different. Therefore, the same syntactic structure could be used on all trials of a block. Here van der Meulen observed the tight co-ordination between eye gaze and speech described above: Speakers fixated upon each object just before naming it, about 500 to 600 ms before noun onset. In another condition, the two types of displays were mixed. Speakers therefore had to decide on each trial which of the two syntactic structures was appropriate. Here, the speakers were far more likely than in the blocked condition to inspect the bottom objects during the first 500 ms of the trial, probably in order to determine whether the objects were identical. Usually, the speakers looked at the bottom objects again just before naming them. In fact, the speakers' likelihood of inspecting one of the bottom objects during the 500 ms preceding the onset of the object's name was independent of whether or not they had looked at the object at trial onset. This suggests that the early inspection of the objects was functionally different from the later one: During the early inspection the speakers probably determined the identity of the objects and planned the utterance structure accordingly, whereas during the later inspection they retrieved the objects' names.

This suggestion is related to an earlier proposal by Griffin and Bock (2000, see also Griffin, 1998), who were, to our knowledge, the first to study the speakers' eye movements during the description of pictures of simple actions (e.g., of a man chasing a dog). They asked speakers to carry out several different tasks on the same set of pictures, among them a picture description task and a so-called patient detection task. In the latter task, the participants had to identify the event participant who was undergoing (rather than carrying out) the action. Griffin and Bock found that as early as about 300 ms after picture onset the likelihood of fixations on the agent and patient regions began to differ between the two tasks: In the description task speakers were far more likely to fixate upon the agent (who was mentioned first) than upon the patient, whereas the reverse held for the patient detection task. Griffin and Bock concluded that in both tasks viewers spent the first 300 ms or so in an effort to comprehend the event and to identify who is doing what to whom. Following this appraisal phase, the participants describing the displays fixated each event participant shortly before naming it, corroborating the results reported above. As Bock and Griffin point out, the results of their study support the distinction between two main phases of sentence generation postulated on independent grounds in many theories of language production (e.g., Levelt, 1989). The first phase is an appraisal or conceptualisation phase during which speakers determine the content of the utterance. The second phase is a formulation phase, during which they retrieve words and build syntactic structures expressing the utterance content.

We are currently further exploring the gaze patterns of speakers describing events and actions. In one experiment (Dobel, Meyer & Levelt, in preparation) Dutch speakers described three-participant events (see Figure 13.6) using double object constructions ("The cowboy gives the clown the hat") or prepositional structures ("The cowboy gives the hat to the clown"; we used a priming procedure to encourage speakers to use only these constructions). The agent always appeared on the left side of the screen.

Figure 13.7 displays the two major gaze paths the speakers used. The first path, shown in the right panel, led from the fixation cross at the top of the screen to the object region of the picture (the hat in the example), then to the agent (the cowboy),

Figure 13.6: Example for a three-participant event.

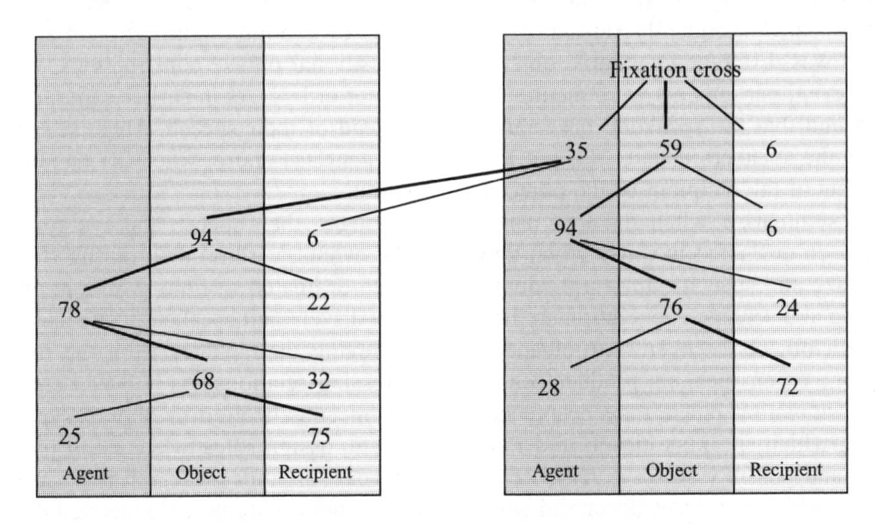

Figure 13.7: Main gaze paths (% fixations) for three-participant events.

back to the object and finally to the recipient (the clown). The second path, shown in the left panel, led from the fixation cross to the agent, then to the object, back to the agent, again to the object and finally to the recipient. Regardless of which scan path the speakers chose, they usually looked at the agent and the object before speech onset. By contrast, they rarely fixated upon the recipient before speech onset, and there was, contrary to our expectation, no systematic relationship between the likelihood of early (i.e. pre-speech onset) recipient fixations and the choice of syntactic structure.

The early attention given to the object of the action is surprising because the object was the only inanimate element in the pictures and would not be the first element to be fixated upon if speakers inspected the picture elements according to their left-to-right order on the display. However, the object appeared in the middle of the screen, directly below the fixation point. This location was a good vantage point to gain an overview of the scene. Moreover, a separate experiment, in which we asked participants to indicate where the action was depicted showed that the action region was the area around the inanimate object and the hands of the agent. Ferreira (2000) has argued on linguistic grounds that speakers must identify the action or event and select a verb before initiating a sentence. Thus, perhaps speakers fixated upon this region in order to encode the action and select the verb.

To test this hypothesis, we carried out a follow-up experiment, in which speakers either produced sentences to describe the actions or only named the event participants ("a cowboy, a hat, a clown"). We found that in the sentence condition the object region was significantly more often the first element to be fixated than in the list condition. This demonstrates that the early object fixations were not entirely stimulus-driven, but task-driven as well: One of the reasons why speakers frequently fixated upon the object region in the sentence probably was that they had to determine the event type and possibly the verb before speech onset.

As in our earlier experiments and as reported by Griffin and Bock (2000), we found that during speech most event participants were fixated upon shortly before being mentioned. Consequently, during speech, the order of fixating on the recipient and the object was different for double object and prepositional phrase constructions, where these event participants were mentioned in different orders. As in earlier experiments, we found that the speakers' gaze ran slightly ahead of their speech. Thus, at speech onset, the speakers usually did not look at the agent any more, who was mentioned first, but at the recipient or the object of the action.

To summarise, when speakers name sets of objects, their speech and eye gaze are tightly co-ordinated in time. When they describe events, we find that, during speech, eye gaze and speech output are equally tightly co-ordinated in time. Multiple object naming and sentence generation are both lexically driven, and speakers fixate upon the referent objects or event participants while retrieving their names from the mental lexicon. When speakers produce sentences expressing relationships between entities, this formulation phase may be preceded by a conceptualization or appraisal phrase during which speakers aim to understand the event, assign roles to the event participants and possibly select a verb. An important goal for future research is to specify exactly which processes occur and which types of representations are generated during the appraisal and formulation phase.

Conclusions

As noted in the Introduction, eye gaze has long been known to be systematically related to language processing in reading and speech comprehension. The eye tracking studies reviewed above showed that there is also a close link between eye gaze and language processing in the production of spoken language. Speakers naming several objects usually fixate upon each of them in the order of mention. The time speakers spend looking at objects they name is related to the time required for recognising the objects and for retrieving their names. Finally, when speakers describe slightly more complex displays and events and choose syntactic structures themselves, their gaze patterns become more complex as well but they are far from chaotic. The available data suggest that it may be possible to dissect the speaker's scan path into a section corresponding to an appraisal phase and a section corresponding to formulation processes.

Evidently, each of the studies reviewed above raises issues for further research. However, we have established systematic relationships between eye movements and speech planning, and therefore eye monitoring can now be used in a variety of ways to study language production. Researchers can analyse the speakers' gaze paths and determine which parts of a picture speakers inspect in which order. They can link this information to a record of the speech output and establish where the speaker is looking during the production of a critical word or phrase and which parts of the picture the speaker has inspected before. In addition, they can use the gaze-contingent display change methodology to alter pictures depending on the speakers' point of gaze. Finally, researchers can determine the viewing times for different regions of a picture in order to estimate how time consuming the processing of the region and the generation of the corresponding part of speech are.

How eye tracking is used will, of course, depend on the investigators' research interests. We have focused on the time course of lexical retrieval processes, and much more work needs to be done in this specific area. We have, for instance, only studied the simplest kinds of utterances. It will be interesting to see whether speakers engage in more extensive previews when they formulate more complex descriptions and whether this affects the speech-to-gaze alignment during speech. There are many other issues in language production research that can be addressed using eye tracking. One of them is how visual-conceptual information and linguistic constraints jointly determine the structure of utterances. One may, for instance, present one set of pictures to speakers of languages differing in word order rules and assess whether their gaze patterns differ only during the formulation phase, while event participants are being mentioned in different orders, or during the appraisal phase as well. Finally, eye tracking can be used to compare speech production across groups of speakers, for instance speakers using their first or second language, children or adults, young or older adults and healthy and brain-damaged speakers.

Acknowledgements

This research was supported by a Travel Grant of the Royal Society to the first author and by a grant of the Deutsche Forschungsgemeinschaft (DFG DO711/1–1) to the second author.

Notes

1 The viewing time for an object includes any intra-object saccades. By contrast, gaze duration, often used in studies of reading, is defined as the sum of fixations, excluding saccade durations.
2 The materials are available from the authors.

References

Bock, K. (1996). Language production: Methods and methodologies. *Psychonomic Bulletin & Review, 96,* 395–421.

Dahan, D., Magnuson, J. S., & Tanenhaus, M. K. (2001). Time course of frequency effects in spoken-word recognition: Evidence from eye movements. *Cognitive Psychology, 42,* 317–367.

Dobel, C., Meyer, A. S., & Levelt, W. J. M. (in preparation). Eye movements preceding and accompanying the production of syntactically complex sentences.

Ferreira, F. (2000). Syntax in language production: An approach using tree-adjoining grammars. In: L. Wheeldon (ed.), *Aspects of Language Production* (pp. 291–330). Howe: Psychology Press.

Ferreira, F., & Swets, B. (2002). How incremental is language production? Evidence from the production of utterances requiring the computation of arithmetic sums. *Journal of Memory and Language, 46,* 57–84.

Garrett, M. F. (1975). The analysis of sentence production. In: G. H. Bower (ed.), *The Psychology of Learning and Motivation: Vol 9.* (pp. 133–177). New York: Academic Press.

Griffin, Z. M. (1998). What the eye says about sentence planning. Unpublished doctoral dissertation. Urbana-Champaign: University of Illinois.

Griffin, Z. M. (2001). Gaze durations during speech reflect word selection and phonological encoding. *Cognition, 52,* B1–B14.

Griffin, Z. M. (in press). Pushing the limits of incremental word production. *Psychonamic Bulletin & Review.*

Griffin, Z. M., & Bock, K. (2000). What the eyes say about speaking. *Psychological Science, 11,* 274–279.

Hayhoe, M. (2000). Vision using routines: A functional account of vision. *Visual Cognition, 7,* 43–64.

Henderson, J. M. (1992a). Identifying objects across saccades: Effects of extrafoveal preview and flanker object context. *Journal of Experimental Psychology: Learning, Memory and Cognition, 18,* 521–530.

Henderson, J. M. (1992b). Object identification in context: The visual processing of natural scenes. *Canadian Journal of Psychology, 46*, 319–341.

Henderson, J. M. (1997). Transsaccadic memory and integration during real-world object perception. *Psychological Science, 8*, 51–55.

Henderson, J. M., Pollatsek, A., & Rayner, K. (1987). Effects of foveal priming and extrafoveal preview on object identification. *Journal of Experimental Psychology: Human Perception and Performance, 13,* 449–463.

Henderson, J. M., Pollatsek, A., & Rayner, K. (1989). Covert visual attention and extrafoveal information use during object identification. *Perception & Psychophysics, 45,* 196–208.

Humphreys, G. W., Price, C. J., & Riddoch, M. J. (1999). From objects to names: A cognitive neuroscience approach. *Psychological Research, 62,* 118–130.

Hyönä, J., Munoz, D. P., Heide, W., & Radach, R. (eds) (2002). *The Brain's Eye: Neurobiological and Clinical Aspects of Oculomotor Research.* Amsterdam: Elsevier.

Jescheniak, J. D., & Levelt, W. J. M. (1994). Word frequency effects in speech production: Retrieval of syntactic information and of phonological form. *Journal of Experimental Psychology: Learning, Memory, and Cognition, 20,* 824–843.

Johnson, C. J., Paivio, A., & Clark, J. M. (1996). Cognitive components of picture naming. *Psychological Bulletin, 120,* 113–139.

Levelt, W. J. M. (1989). *Speaking: From Intention to Articulation.* Cambridge: MIT Press.

Levelt, W. J. M. (1999). Models of word production. *TRENDS in Cognitive Sciences, 3,* 223–232.

Levelt, W. J. M., & Meyer, A. S. (2000). Word for word: Sequentiality in phrase generation. *European Journal of Cognitive Psychology, 12,* 433–452.

Levelt, W. J. M., Roelofs, A., & Meyer, A. S. (1999). A theory of lexical access in language production. *Behavioral and Brain Sciences, 22,* 1–38.

Liversedge, S. P., & Findlay, J. M. (2000). Saccadic eye movement and cognition. *TRENDS in Cognitive Sciences, 4,* 6–14.

McConkie, G. W., & Rayner, K. (1975). The span of effective stimulus during a fixation in reading. *Perception & Psychophysics, 17,* 578–586.

Meulen, F. F. van der (2001). *Moving Eyes and Naming Objects.* MPI series in Psycholinguistics. Nijmegen, The Netherlands.

Meulen, F. F. van der, Meyer, A. S., & Levelt, W. J. M. (2001). Eye movements during the production of nouns and pronouns. *Memory & Cognition, 29,* 512–521.

Meyer, A. S. (submitted). Eye movements during the production of long and short noun phrases.

Meyer, A. S., & Meulen, van der F. F. (2000). Phonological priming of picture viewing and picture naming. *Psychonomic Bulletin & Review, 7, 314–319.*

Meyer, A. S., Roelofs, A., & Levelt, W. J. M. (in press). Word length effects in language production. *Journal of Memory and Language.*

Meyer, A. S., & Schriefers, H. (1991). Phonological facilitation in picture-word-interference experiments: Effects of stimulus onset asynchrony and type of interfering stimuli. *Journal of Experimental Psychology: Learning, Memory, and Cognition, 17,* 1146–1160.

Meyer, A. S., Sleiderink, A. M., & Levelt, W. J. M. (1998). Viewing and naming objects: Eye movements during noun phrase production. *Cognition, 66,* B25–B33.

Noizet, G., & Pynte, J. (1976). Implicit labelling and readiness for pronunciation during the perceptual process. *Perception, 5,* 217–223.

Pollatsek, A., Rayner, K., & Collins, W. E. (1984). Integrating pictorial information across eye movements. *Journal of Experimental Psychology: General, 113,* 426–442.

Pollatsek, A., Rayner, K., & Henderson, J. M. (1990). Role of spatial location in integration of pictorial information across saccades. *Journal of Experimental Psychology: Human Perception and Performance, 16,* 199–210.

Rapp, B., & Goldrick, M. (2000). Discreteness and interactivity in spoken word production. *Psychological Review, 107,* 460–499.

Rayner, K. (1998). Eye movements in reading and information processing: 20 years of research. *Psychological Bulletin, 124,* 372–422.

Reichle, E. D., Pollatsek, A., Fisher, D. L., & Rayner, K (1998). Toward a model of eye movement control in reading. *Psychological Review, 105,* 125–157.

Roelofs, A. (1997). The WEAVER model of word-form encoding in speech production. *Cognition, 64,* 249–284.

Schriefers, H., de Ruiter, J. P., & Steigerwald, M. (1999). Parallelism in the production of noun phrases: Experiments and reaction time models. *Journal of Experimental Psychology: Language, Memory and Cognition, 25,* 702–720.

Schriefers, H. and Teruel, E. (1999). Phonological facilitation in the production of two-word utterances. *European Journal of Cognitive Psychology, 11,* 17–50.

Smith, M., & Wheeldon, L. (1999). High level processing scope in spoken sentence production. *Cognition, 73,* 205–246.

Starr, M. S., & Rayner, K. (2001). Eye movements during reading: Some current controversies. *TRENDS in Cognitive Sciences, 5,* 158–163.

Tanenhaus, M. K., Spivey-Knowlton, M. J., Eberhard, K. M., & Sedivy, J. C. (1995). Integration of visual and linguistic information in spoken language comprehension. *Science, 268,* 1632–1634.

Treisman, A. (1993). The perception of features and objects. In: A. Baddeley and L. Weiskrantz (eds), *Attention: Selection, Awareness and Control. A Tribute to Donald Broadbent.* Oxford: Clarendon Press.

Wheeldon, L., & Lahiri, A. (1997). Prosodic units in speech production. *Journal of Memory and Language, 37,* 356–381.

Zelinsky, G. J., & Murphy, G. L. (2000). Synchronizing visual and language processing: An effect of object name length on eye movements. *Psychological Science, 11,* 125–131.

Chapter 14

Eye Movements and Thematic Processing

Simon P. Liversedge

This chapter reviews studies investigating thematic processing during reading focusing on those that have used eye movement methodology. It is suggested that thematic processing represents an important aspect of comprehension that is an intermediary of relatively lower level psycholinguistic processes (such as lexical identification and parsing) and higher level discourse processing. It is also argued that *wh*-words induce thematic expectations and that thematic processing preferences for arguments influence eye movements very rapidly, whereas those for adjuncts have a delayed effect. The chapter concludes with a brief evaluation of the role of thematic processing in three theories of sentence processing.

Introduction

Thematic roles have played an important part in linguistic theory for almost 40 years. Early work by Gruber (1965), Fillmore (1968) and Jackendoff (1972) led to a substantial increase in interest in this linguistic approach to the characterisation of the semantic meaning of a sentence (e.g. Jackendoff, 1983, 1987; Dowty, 1991). Psycholinguistic interest in this approach has also developed steadily from early work such as that of Rayner, Carlson and Frazier (1983) and Carlson and Tanenhaus (1988) and Tanenhaus and Carlson (1989) through to more recent work (e.g. Altmann, 1999; Binder, Duffy & Rayner, 2001; Ferretti, McRae & Hatherall, 2001; Liversedge, Pickering, Branigan & Van Gompel, 1998b; Liversedge, Pickering, Clayes & Branigan, 2003; Mauner, Tanenhaus & Carlson, 1995; Mauner & Koenig, 1999; McRae, Ferretti & Amyote, 1997; Schütze & Gibson, 1999). In this chapter I will discuss what thematic roles are and what their role within language comprehension is, before providing a review of empirical research investigating thematic role assignment during sentence processing. I will place special emphasis on those studies that have employed eye movement monitoring techniques as would be appropriate for a volume of this nature. I will conclude with a discussion of the status of thematic roles in theories of sentence processing.

The Mind's Eye: Cognitive and Applied Aspects of Eye Movement Research

ISBN: 0–444–51020–6

When we read a sentence, we must necessarily identify the meaning and syntactic category of each of the words and then compute the syntactic structure of the sentence in order to form a representation of its semantic meaning (Rayner & Pollatsek, 1989). It is generally accepted that sentential interpretation occurs incrementally, roughly on a word-by-word basis (e.g. Marslen-Wilson, 1975; Pickering & Traxler, 1998). Importantly, whenever readers identify a word its meaning and syntactic category become available very rapidly. When the word is a verb, its subcategorisation frame and its thematic grid also become available. The subcategorisation frame specifies the syntactic categories of the arguments that the verb takes and the thematic grid specifies the meanings of those arguments. In addition to verbs, other constituents may also licence thematic roles (e.g. prepositions). Consider Example (1).

1. Mary put the food on the table.

When the verb *put* is read, its subcategorisation frame specifies that it takes three arguments and its thematic grid specifies that these arguments take the roles of Agent — the entity performing the action of the verb (*Mary*), Patient — the entity receiving the action of the verb (*the food*) and Location (*on the table*). Note that these arguments are an intrinsic part of the verb's meaning and consequently computing the thematic status of the different constituents of a sentence is central to the computation of the meaning of the sentence as a whole. In fact, Frazier (1987) has argued that thematic relations represent the interface between real world conceptual knowledge, the discourse representation and the syntactic processor. During sentence interpretation, the assignment of thematic roles can be looked upon as the stage in processing at which readers form their first shallow semantic representation of the sentence.

Experimental Investigations of Thematic Processing

Some early work that set out to provide an explicit account of the process of thematic assignment was that of Carlson and Tanenhaus (1988) and Tanenhaus and Carlson (1989). According to their view, all thematic roles associated with a verb are activated in parallel when the verb is identified. Thematic roles that are activated are assigned as soon as possible and any thematic roles that are not assigned to constituents are lodged in the discourse representation as unspecified entities or addresses. The account of thematic processing that Carlson and Tanenhaus put forward rests on three fundamental assumptions. First, the meaning of a verb is decomposable into two parts — the thematic roles and the verb's core meaning. Secondly, the way that thematic roles are assigned (i.e. to which particular constituents) has immediate consequences for the nature of the syntactic analysis that is assigned to the sentence. Finally, Carlson and Tanenhaus argue that not all thematic roles must be assigned in order for a coherent semantic representation of the sentence to be formed.

Tanenhaus, Burgess, Hudson D'Zmura and Carlson, (1987) tested these ideas in a whole sentence reading time study. They constructed context sentences like (2a) and (2b) and target sentences like (3).

2a. John had difficulty running fast to catch his plane.

2b. John had difficulty loading his car.

3. The suitcases were heavy.

Note that in the context sentence (2b) there is no object of the verb *load*, that is to say, the thing that John is loading is not specified. Consequently, according to Tanenhaus *et al.* the Location role associated with the verb *load* will remain empty and be lodged in the discourse representation. Importantly, in sentence (2a) all the empty thematic roles are filled. Thus, when readers process the second sentence (3) that contains a noun phrase that is a good candidate for filling the open thematic role in the discourse representation after reading sentence (2b) then they should assign the empty thematic role to the noun phrase. By contrast, such an assignment will not be possible when a sentence like (3) is read after a sentence like (2a). Consistent with these claims, Tanenhaus *et al.* found that participants took less time to read sentences like (3) and judged that they made sense more often after sentences like (2b) than after sentences like (2a). Although these results support the theoretical position outlined by Carlson and Tanenhaus (1988) and Tanenhaus and Carlson, (1989), it can be argued that the materials were not particularly well controlled across conditions. For example, there were significant differences in the linguistic content of context materials across conditions that might have induced processing biases other than the intended thematic manipulations. Furthermore, whole sentence reading times are not an ideal measure of the processes that occur during reading since they do not provide an on-line measure of when thematic variables first influenced processing. That is to say, they do not provide a moment-to-moment indication of the ease or difficulty readers are experiencing as each word of the sentence is processed.

Such information regarding the time course of processing (in this case thematic processing) is vital in two respects. First, it is very often important in reading research to establish when a linguistic variable first had an effect on processing. In order to determine the location, both temporally (i.e. at which fixation or fixations within the stream of fixations that comprise the eye movement record as a whole), and spatially (i.e. at which word, phrase or clause within the sentence) of when an effect first occurred it is necessary to use a methodology that provides excellent spatial and temporal resolution (Liversedge, Paterson & Pickering, 1998a). Secondly, time course information is essential if we are to relate the particular process that we are examining to other processes that occur during reading. Clearly, there are many processes associated with the comprehension of a written sentence (e.g. lexical identification, parsing, thematic role assignment, pronoun resolution, inferential processing, etc.), and these processes occur concurrently as we read. Also, some of these processes depend upon each other — for example, one process might necessarily have to be completed prior to the initiation of another (e.g. lexical identification of a word must occur prior to the incorporation of that word into a syntactic structure). By contrast, other processes may be independent of each other, in that one process may not require the output of another process prior to its initiation (e.g. thematic processing does not necessarily depend upon the completion of pronoun resolution). Information about time course of processing allows us to relate thematic processing to other psycholinguistic processes that occur when we read.

A measure that does provide adequate spatial and temporal resolution to investigate the time course of thematic influences and provides moment-to-moment information about the time course of processing is eye movement monitoring methodology. Eye-tracking methodology is generally regarded to be the most sensitive method of detecting early effects in reading (see Liversedge & Findlay, 2000; Rayner, Sereno, Morris, Schmauder & Clifton, 1989; Rayner, 1998). This methodology is non-invasive and simply requires participants to read sentences normally. There is no necessity for the participant to engage in additional tasks that may lead to task-specific strategies. It is for these reasons that the use of eye movement monitoring methodology is preferable to study thematic processing during reading.

A recent eye movement experiment investigating similar aspects of thematic role assignment similar to those carried out by Carlson and Tanenhaus (1988) and Tanenhaus and Carlson (1989) was conducted by Liversedge, Pickering, Branigan and Van Gompel (1998; see also Hanna, Spivey-Knowlton & Tanenhaus, 1996). We were interested in two aspects of thematic processing: First, whether the argument status of a prepositional phrase influenced the manner in which it was processed thematically; secondly, whether a context containing an interrogative *wh*-word could introduce an empty thematic role into the discourse representation that would influence thematic assignments to a subsequent prepositional phrase. Consider sentences (4) and (5).

4. The shrubs were planted by the apprentice.

5. The shrubs were planted by the greenhouse.

The argument status of a prepositional phrase dictates whether it is assigned a thematic role by the verb or not. In sentence (4) the *by*-phrase is an argument of the verb and since it specifies who performed the act of planting it is assigned an Agent role. In contrast, the *by*-phrase in (5) is not an argument of the verb, but is an adjunct. Adjuncts do not take their meaning from the verb in the same way that arguments do. They are not as intrinsically associated with the meaning of the verb. Furthermore, adjuncts are not assigned a thematic role by the verb but take a thematic role from the preposition. In our first experiment, we set out to investigate whether readers had a preference to initially process a thematically ambiguous *by*-phrase as an argument rather than an adjunct. We conducted an eye movement experiment using sentences like (4) and (5) in which the *by*-phrase was either an agentive argument or a locative adjunct. Importantly, we matched the disambiguating region of the *by*-phrase for length, frequency and plausibility across conditions to ensure that any differences in reading times were only due to the thematic status of the *by*-phrase. We split the sentence up into the following regions (indicated by slashes): *The shrubs/ were planted/ by the/ apprentice/ that morning.* We computed a number of reading time measures that provided information about exactly when and where disruption first occurred as the sentences were read. I will concentrate on three of these measures in this chapter. First pass reading times were defined as the sum of all the fixations in a region until the point of fixation left that region (zero first pass reading times were recorded if the region was skipped). Re-reading times for a region were defined as the sum of all the fixations made after a regression from that region until a fixation was made to the right of the region (note that although re-reading times are attributed to a region, the times themselves are comprised of fixations made to the left of that region). Finally, total reading times were defined as the sum of all fixations made in a region.

Our results showed that first pass reading times were shorter on the disambiguating noun of the *by*-phrase of argument sentences than of adjuncts (see Figure 14.1).

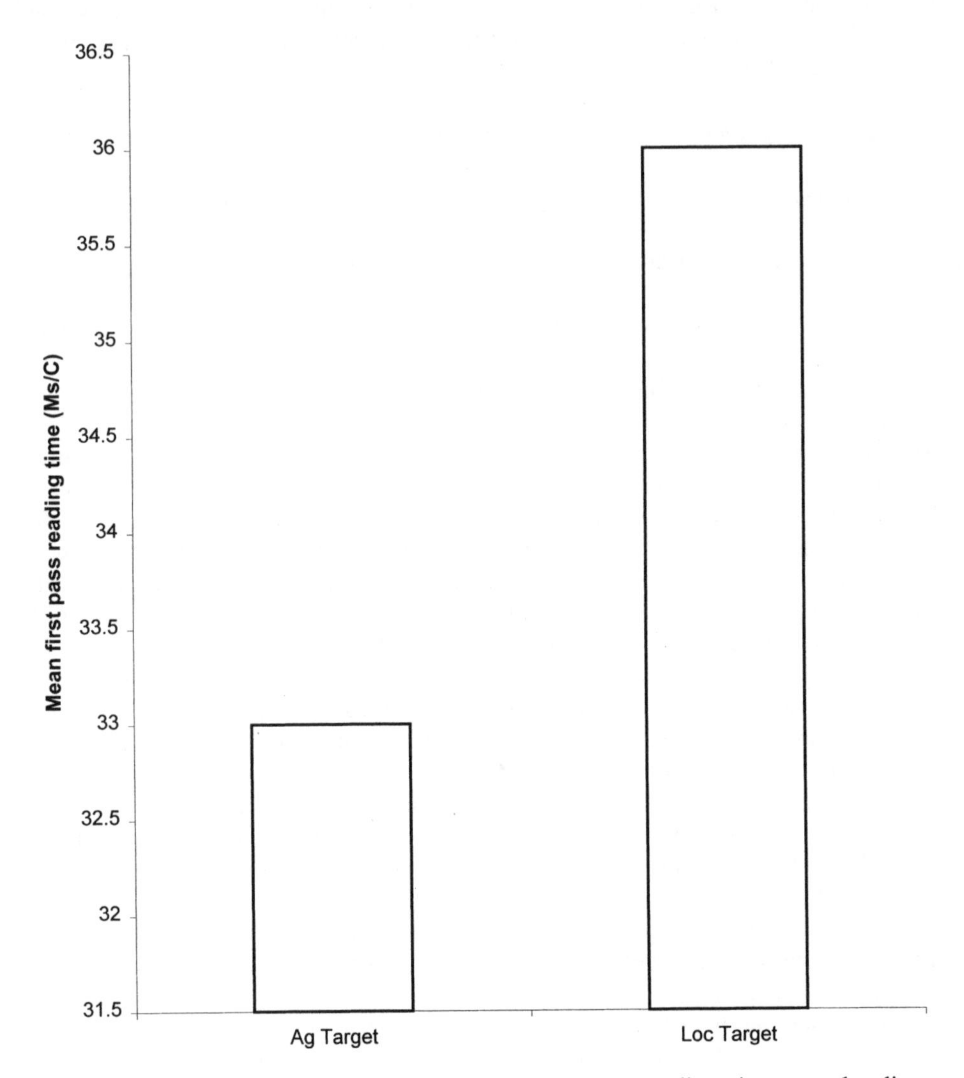

Figure 14.1: Liversedge *et al.* (1998b) Mean first pass reading times on the disambiguating region for agentive and locative target sentences presented in isolation.

We argued that this result indicates that readers have a preference to initially process ambiguous phrases as arguments over adjuncts. In particular, it suggested that the *by*-phrase was initially assigned an agent role by the verb. However, when the noun phrase was inconsistent with an agent role, readers were forced to treat the prepositional phrase as an adjunct and assign a location role licensed by the preposition *by*. Essentially, we claimed that readers had been thematically garden pathed when they read the *by*-phrase. Note also that since the disruption occurred during the first pass, the effects presumably reflect early thematic commitments.

In a second experiment we investigated whether the nature of a preceding context would modulate the argument preference we observed when sentences were processed in isolation. Context sentences were constructed that contained a *wh*-word (e.g. 6 and 7) which preceded the target sentences (e.g. 8 and 9).

6. The head gardener wondered who should plant the shrubs. (Agentive Context)

7. The head gardener wondered where to plant the shrubs. (Locative Context)

8. In fact, the shrubs were planted by the apprentice that morning. (Agentive Target)

9. In fact, the shrubs were planted by the greenhouse that morning. (Locative Target)

Following Carlson and Tanenhaus (1988) and Tanenhaus and Carlson (1989), we argued that such context sentences containing a *wh*-word should induce a thematic expectation. Note that example (6) indicates to readers that the planting of the shrubs should be performed by an as yet unspecified person. Similarly, example (7) indicates that planting of the shrubs should take place in an, as yet, unspecified location. We suggested that the *wh*-word might cause readers to lodge a semantically vacuous thematic role in the discourse representation in a similar way to that which occurs when a verb's arguments are not filled by the end of the sentence (Tanenhaus *et al.*, 1987). We argued that if this was the case, then when readers processed an agentive target sentence, regardless of whether the preceding context induced a locative or an agentive thematic expectation, the *by*-phrase in the agentive target sentences should be equally easy to read. For agentive target sentences an agent role should always be available from the thematic grid of the main verb of the target sentence and therefore no thematic misanalyses should occur. By contrast, when readers process a locative target sentence after a locative context, there should be no disruption when the *by*-phrase is read. The locative context should cause a semantically vacuous locative thematic role from the *wh*-word in the context to be lodged in the discourse representation. Thus, locative target sentences after a locative context should be as easy to process as agentive target sentences after either form of the context. By contrast, locative target sentences after an agentive context should cause readers processing difficulty when the *by*-phrase was encountered. The *wh*-word in the agentive context provides a semantically vacuous agent role in the discourse representation. Similarly, the main verb of the target sentence also provides an agent role. Thus, upon encountering the ambiguous *by*-phrase readers only have Agent roles available from the context and the thematic grid of the main verb and consequently should initially misassign an agent role to the locative *by*-phrase and be thematically garden pathed.

An alternative theoretical account to that offered by Liversedge *et al.* (1998b) is provided by the Constraint-based approach to language processing. Constraint-based theorists (e.g. MacDonald, 1994; MacDonald, Pearlmutter & Seidenberg, 1994a, 1994b; Spivey-Knowlton, Trueswell & Tanenhaus, 1993; Spivey-Knowlton & Sedivy, 1994; Tabossi, Spivey-Knowlton, McRae & Tanenhaus, 1994; Trueswell, Tanenhaus & Garnsey, 1994; Trueswell, Tanenhaus & Kello, 1993) claim that multiple sources of information (e.g. semantic, frequency and contextual information) constrain the parsing process. As each new word of the sentence is encountered, there is competition between alternative syntactic analyses favoured by the constraints imposed by the text thus far. Garden path effects occur when a constituent is encountered that conflicts with the previously favoured analysis. Importantly, in order to generate theoretical predictions advocates of the Constraint-based theory argue that measures of processing preferences can be obtained from "off-line" measures such as sentence fragment completions (e.g. Spivey-Knowlton & Sedivy, 1994). Their suggestion is that given a sentence fragment, then the proportion of fragment completions of one type of analysis or another reflects the strength of the constraints favouring each of those analyses at that point in the sentence. To test the predictions of the Constraint-based model Liversedge *et al.* (1998b) conducted a sentence completion experiment. We required participants to complete fragments like *In fact, the shrubs were planted by the . . .* when they followed either agentive or locative context sentences like (6) or like (7).

The results showed that the nature of the context dictated the types of completions participants produced. After a locative context, participants produced significantly more locative than agentive completions, and after an agentive context, participants produced significantly more agentive than locative completions. On the basis of the sentence completion results, Constraint-based theorists would predict a crossover pattern of data in the reading times — when the context and target sentences were congruent, reading times should be short whereas when the context and target sentences were incongruent reading times should be longer. These predictions contrast with those made by Liversedge *et al.* (1998b) as we argued that readers should only experience disruption to processing when reading locative target sentences after an agentive context.

We tested these predictions in a second eye movement experiment and found a pattern of first pass, re-reading and total reading times on the disambiguating *by*-phrase that was consistent with our own predictions rather than those of the Constraint-based theory. First pass reading times were longer for the *by*-phrase of locative sentences after agentive target sentences than the same region of the sentences in the other three conditions (see Figure 14.2). Furthermore, there were no reliable differences in first pass reading times for these other three conditions. Note again that these effects showed up early (during first pass) in the eye movement record. Our results provided evidence against the Constraint-based theory of sentence processing. They were, however, consistent with the predictions made by Liversedge *et al.* (1998b) that were based on the proposal by Carlson and Tanenhaus (1988) and Tanenhaus and Carlson (1989).

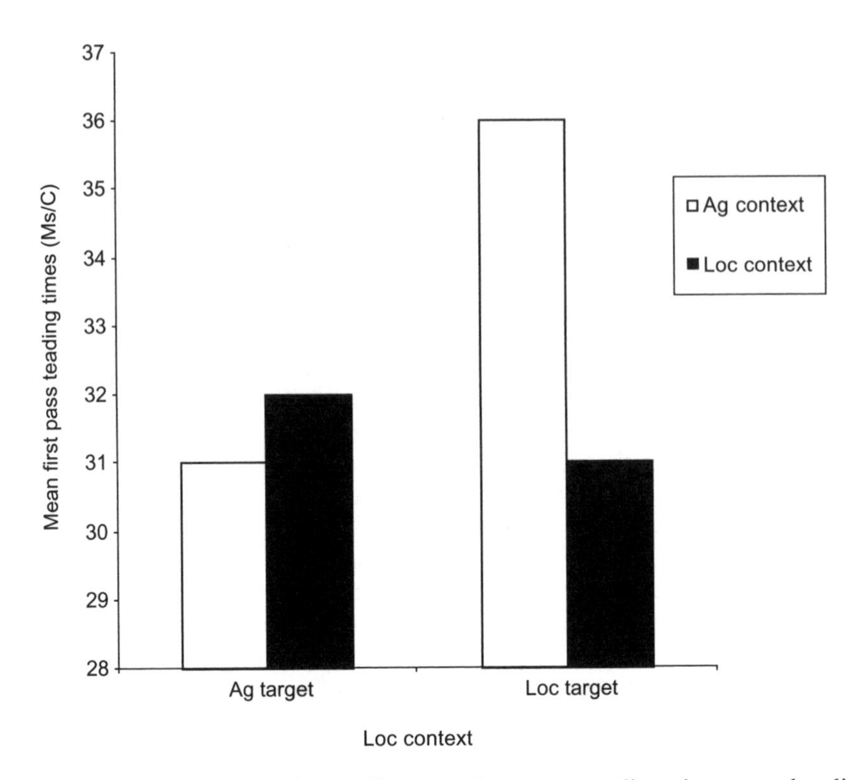

Figure 14.2: Liversedge *et al.* (1998) Mean first pass reading times on the disambiguating region for agentive and locative target sentences after agentive and locative context sentences.

In a more recent experiment, Liversedge, Pickering, Clayes and Branigan (2003) continued this line of research. Recall that in Liversedge *et al.* (1998b), we manipulated the argument status of the ambiguous prepositional phrase in the materials. The ambiguous *by*-phrase was either an argument of the verb in which case it was assigned an agent role, or it was an adjunct, in which case it was assigned a location role by the preposition. Note also that if the conclusions in the Liversedge *et al.* (1998b) study are correct, then for a prepositional phrase that is ambiguous between two alternative adjunct readings, it should be possible to induce initial thematic misanalyses for either of the two readings. Consider sentences (10) and (11) along with the account of thematic processing provided by Liversedge *et al.* (1998b).

10. In fact, the maid peeled the vegetables in the morning, with great care.

11. In fact, the maid peeled the vegetables in the kitchen, with great care.

Sentences (10) and (11) contain a prepositional phrase that is ambiguous between a temporal adjunct *in the morning* and a locative adjunct *in the kitchen*. For both types of sentence the ambiguous phrase is not assigned a thematic role by the verb, but instead by the preposition. We argued that through the use of preceding contexts containing *wh*-words it should be possible to induce a thematic expectation for either a locative or temporal reading. If readers process a context sentence like (13) containing the *wh*-word *where*, it should cause them to enter a semantically vacuous location role in their discourse representation.

12. The maid wondered when to peel the vegetables.

13. The maid wondered where to peel the vegetables.

Upon encountering a temporal target sentence like (10) or a locative target sentence like (11), readers should initially assign a location role to the phrase. For locative target sentences this assignment is correct and no processing difficulty should occur. However, for temporal target sentences the analysis is inappropriate and disruption to processing should occur when the misanalysis is detected and thematic re-assignment occurs. The converse should also be true for temporal context sentences like (12). After temporal context sentences, locative target sentences should produce processing difficulty, but temporal target sentences should not. Thus, for adjunct ambiguities in context we argued that reading times for the disambiguating region should show a crossover pattern. Another important aspect of the data was the time course of the effects. The effects that Liversedge *et al.* (1998b) observed for argument/adjunct ambiguities occurred during first pass on the disambiguating noun phrase of the prepositional phrase. However, as mentioned earlier, arguments take much of their meaning from the verb and are central to the core meaning of the sentence. Adjuncts, by contrast are not. Thus, it is quite possible that an inappropriate thematic assignment to an adjunct may not be detected as quickly since it is less central to sentential meaning than a verb argument. Thus, we anticipated that any disruption that did occur might appear in fixations made subsequent to fixations made during first sweep of the eyes through the sentence.

In addition to investigating the influence of context on adjunct processing, we set out to explore whether a semantically vacuous thematic role generated by a context containing a *wh*-word was associated with the particular lexical entry for a verb, or whether it was associated with a more general semantic representation of an event such as schemata (Rumelhart & Ortony, 1977), frames (Minsky, 1975) or scripts (Schank & Abelson, 1977). This question is currently receiving some attention within the literature.

Mauner and Koenig (1999) argue that the nature of thematic information about entities associated with the events described by a verb is specified as part of that particular verb's lexical representation, that is, as part of the verb's semantic argument structure (Fillmore, 1968). This view is consistent with the view of Carlson and Tanenhaus (1988) and Tanenhaus and Carlson (1989) outlined earlier. Importantly, according to this position, thematic roles associated with a verb are specific to that verb and are activated only when that particular verb is encountered.

Mauner and Koenig draw a distinction between the logical and linguistic requirement that verb arguments must be specified. They suggest that the logical requirement of the specification of an agent is stipulated by general conceptual knowledge. That is to say, the fact that cars normally require an agent to perform the action of starting them is something that we derive from our general knowledge of cars and how they function in the real world. The necessity of an agent is logically derived from general conceptual knowledge. By contrast, certain sentence forms linguistically require an agent whereas others do not. For example, the passive sentence (14) both logically and linguistically requires an agent whereas the intransitive sentence shown in (15) logically requires an agent, but does not linguistically require an agent.

14. The car's engine was started abruptly to test the adjustments that the mechanic had made.

15. The car's engine had started abruptly to test the adjustments that the mechanic had made.

Mauner and Koenig argued that if agent roles are derived from general conceptual knowledge, then the rationale clauses (*to test the adjustments*) should sound equally natural in both intransitive sentences like (15) and short passive sentences like (14) since both of these sentence forms logically entail an agent. By contrast, if the agent role is derived from the argument frame of the verb, then the rationale clause should sound more natural for the short passives than for the intransitive sentences since the short passives linguistically require an agent, whereas the intransitive sentences do not. Mauner and Koenig (2000) carried out several experiments using a word-by-word "stop making sense" methodology to test their claims. In one of their experiments they used sentences like (14 and 15). In support of their claims, Mauner and Koenig found that participants made more "NO" decisions and took longer to make decisions at the rationale clause for the intransitive sentences than for short passive sentences.

An alternative view to that of Mauner and Koenig is advocated by Ferretti, McRae and their colleagues (Ferretti, McRae & Hatherall, 2001; McRae, Ferretti & Amyote, 1997) who argue that thematic information associated with verb arguments is derived from situational knowledge stores such as schemata or scripts. Such information is stored as part of general world knowledge, and thus, a verb's argument structure gives immediate access to generalised schematic information about situations or events associated with verb meaning. In such a system, a verb's thematic roles are not shallow semantic characterisations of its arguments, but instead the verb's subcategorisation frame permits access to general world knowledge in order that specific likely entities associated with the action of the verb may be computed. For example, if the verb *arrest* is encountered then particular entities likely to perform that action (e.g. cop, policeman) will be activated as agents and particular entities likely to receive the action (e.g. crook, burglar) will be activated as patients. Essentially, Ferretti *et al.*'s view is that thematic roles are verb specific concepts available from general world knowledge, whereas the contrasting view of Mauner *et al.* (1995) is that thematic roles are very general categories that have a shallow semantic representation and are available from the lexical representation of the verb.

Ferretti *et al.* reported four experiments that they claim support their view. The first three experiments are lexical priming studies that indicate verbs can prime specific relevant agents, patients, and instruments (though interestingly not locations). These studies also show that verbs prime features of relevant agents. Finally, in an experiment in which auditory sentence fragments were used to prime relevant agents and patients, priming only occurred for agents and patients in an appropriate syntactic environment (e.g. *she was arrested by the* . . . primes *cop*, but not *crook* and *she arrested the* . . . primes *crook*, but not *cop*; see also Traxler, Foss, Seely, Kaup & Morris, 2000).

Although both the Mauner *et al.* and Ferretti *et al.* studies offer support for their respective viewpoints, both employed methodologies that some may argue do not necessarily reflect exclusively those processes involved in normal reading. In our experiment, we used eye movement monitoring methodology to examine this question. We prepared two versions of the context sentences containing the *wh*-word. In one version the main verb was the same as that in the target sentence and in the second version the verb in the context was different to that in the target sentence, but clearly instantiated the same scenario (e.g. *The maid wondered when/where to peel/prepare the vegetables. In fact, the maid peeled the vegetables in the morning/kitchen, with great care*). As with the earlier experiments, the critical regions were matched for length and frequency and the target sentences were matched for plausibility. We included a variety of connective phrases at the beginning of the target sentences (e.g. *in fact*, *surprisingly*, *actually*).[1]

We predicted that if a semantically vacuous thematic role generated by a *wh*-word in the context sentence was associated with particular lexical items then thematic congruency effects should occur when the verb was repeated between context and target sentence, but not when the verbs in the context and target sentence were different. Alternatively, we argued that if readers construct a discourse representation using real world knowledge accessed via a verb's subcategorisation frame, then perhaps thematic roles become associated with a more general representation of an event and we might therefore predict thematic congruency effects regardless of whether the verb is repeated or not.

As can clearly be seen from Figures 14.3 and 14.4, re-reading times and total reading times were longer for locative target sentences after temporal contexts than after locative contexts and longer for temporal target sentences after locative contexts than after temporal contexts. Thematic congruency effects occurred for both forms of the adjunct as was expected given the findings of Liversedge *et al.* (1998b). Interestingly, the effects did not occur during first pass as was the case for the *by*-phrase ambiguities. We argued that one reason that the congruency effects might have occurred downstream is because the materials in the Liversedge *et al.* (2000) experiment were ambiguous between two adjunct readings, whereas, those of Liversedge *et al.* (1998b) were ambiguous between an adjunct and an argument. Importantly, since arguments take their meaning from the verb they are more central to the core semantic meaning of a sentence and consequently, a misanalysis might be detected very quickly when an adjunct is initially misinterpreted as an argument.

Our repeated verb manipulation, however, produced less clear effects. When the verb was repeated there was a clear thematic congruency effect. There was a similar numerical difference when the verb in the context was different to that in the target sentence. Furthermore, the three-way interaction between verb type (repeated vs

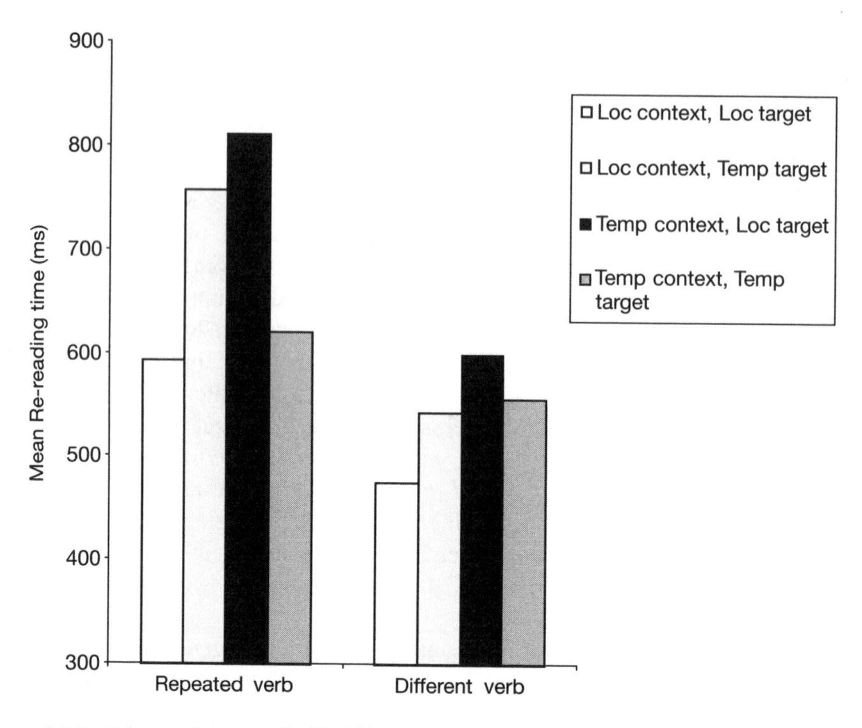

Figure 14.3: Liversedge *et al.* (2003) Mean re-reading times (ms) for temporal and locative target sentences after temporal and locative context sentences with and without a repeated verb.

different), context (locative vs temporal) and target sentence (locative vs temporal) was not reliable. Thus, a tentative conclusion is that the data are consistent with the hypothesis that readers very rapidly compute likely relevant entities that may fill thematic roles specified by a verb's subcategorisation frame.

To summarise, taken together the data from Liversedge *et al.* (1998b) and Liversedge *et al.* (2003) indicate that there is a preference to process ambiguous phrases as arguments over adjuncts. *Wh*-words induce thematic expectations by causing readers to lodge semantically vacuous thematic role in the discourse representation. Thus, through the use of context, it is possible to induce a thematic expectation in readers and cause them to misprocess a thematically ambiguous phrase. Importantly, it appears that it is possible to cause readers to initially misprocess an adjunct as an argument, but not to cause them to misprocess an argument as an adjunct. When readers initially misprocess an adjunct as an argument, detection of the misanalysis occurs very rapidly, whereas when readers misanalyse an adjunct they detect the misanalysis later during processing. Finally, it appears that semantically vacuous thematic roles are not associated with a particular verb, but instead are associated with a more general schematic representation of an event computed on the basis of real world conceptual knowledge (though this final conclusion should be treated with some caution).

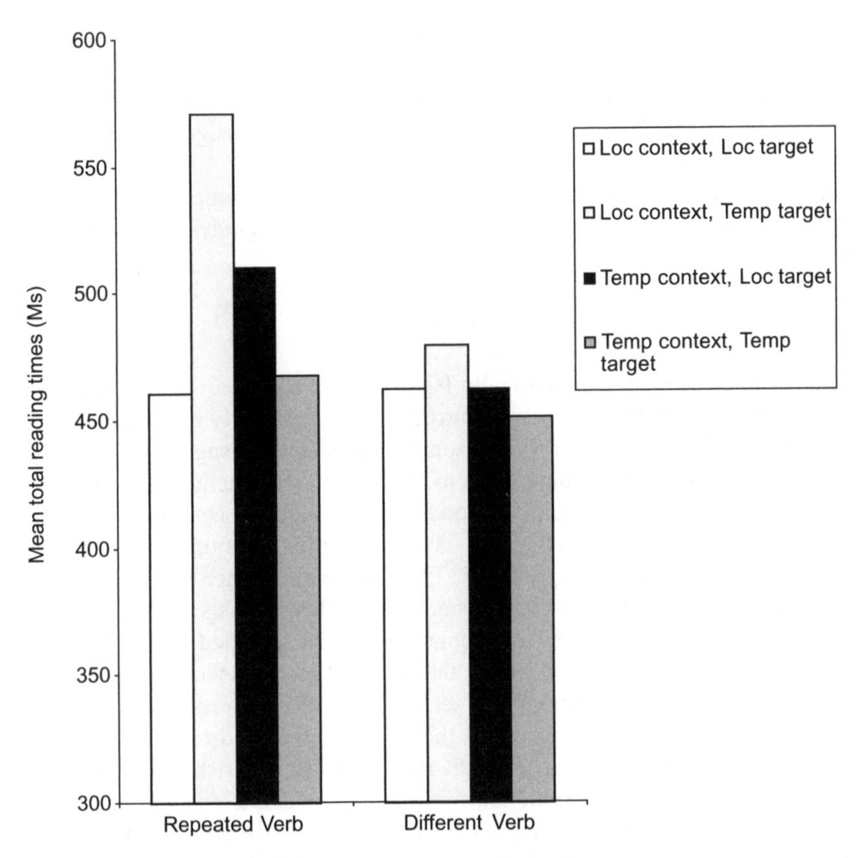

Figure 14.4: Liversedge *et al.* (2003) Mean total reading times (ms) for the spill-over region of temporal and locative target sentences after temporal and locative context sentences with and without a repeated verb.

Thematic Processing and Models of Sentence Processing

From the preceding discussion of the experimental work that has been carried out to investigate thematic processing, it should be clear that thematic role assignment is an important aspect of sentence comprehension that any comprehensive theory of sentence processing should specify. One of the first theories of sentence processing that explicitly outlined a role for thematic processing during sentence comprehension was the Garden Path theory (see Rayner *et al.*, 1983). According to the Garden Path theory (Frazier, 1979; Frazier & Rayner, 1982), syntactic processing proceeds according to two principles: Minimal Attachment and Late Closure. Minimal Attachment specifies that readers should initially adopt the syntactic analysis with the simplest phrase structure and Late Closure specifies that if a constituent is compatible with the phrase structure currently being constructed, then it will be incorporated into that structure. According to the Garden Path theory, these two rules are applied in an informationally encapsulated manner (Fodor, 1983). That is to say, the parser applies the rules without recourse to higher order conceptual information and conceptual knowledge cannot be used to initially guide the parser to assign one analysis rather than another. Consider sentence (16). According to Minimal Attachment, when such a sentence is processed readers will initially assign the simplest structure to the sentence and inappropriately interpret it to mean that the ginger hair was the instrument of the verb hit.

16. The woman hit the man with the ginger hair.

This analysis is perfectly grammatically legal and the only way in which readers can detect that an initial misanalysis has occurred is by semantically evaluating their initial analysis of the sentence. It is for this reason that Rayner *et al.* suggested a stage of thematic processing that occurs subsequent to the initial assignment of a syntactic analysis to the sentence. In the Garden Path model the thematic processor allows readers to re-analyse sentences for which there is a semantically more plausible analysis that is initially syntactically dispreferred. Thus, when a sentence like (16) is read, it is initially parsed according to Minimal Attachment. This analysis is then passed to the thematic processor that evaluates the thematic relations specified by the initial syntactic analysis against alternative possible thematic relations associated with a verb. If alternative thematic assignments offer a semantically more plausible analysis, then the initial syntactic analysis is rejected and the parser is required to undertake syntactic reanalysis. Thus, within the Garden Path model the stage of thematic processing is responsible for the semantic evaluation of the initial syntactic analysis of a sentence.

In relation to the current findings, the suggestion of a thematic processor that is designated as responsible for thematic role assignment is appealing. However, there are few details of exactly how the thematic processor functions. There is no explanatory mechanism for our finding that there is a thematic preference for arguments over adjuncts, nor that *wh*-words can induce a thematic expectation. Clearly the thematic processor of Rayner *et al.* (1983) would require substantial specification if it were to provide an adequate account of the current findings.

A second account of thematic effects during sentence processing is offered by the Constraint-based approach to sentence comprehension. Constraint-based theorists argue that different syntactic analyses compete against each other and readers adopt the analysis that is most strongly favoured by each of the linguistic constraints in the sentence at any particular point. In line with this view, thematic information is simply another linguistic constraint that impacts upon the process of competition. For example, within MacDonald's (1994) lexicalist Constraint-based framework, she argued that alternative argument structures associated with a verb may become partially activated as a function of both the frequency of the alternative structures and the thematic compatibility of that alternative with the other information contained in the sentence. According to her position, within a Constraint-based system thematic information stored with a verb's lexical entry will be used very rapidly to influence initial parsing decisions.

Although some may argue that thematic constraints are simply one more constraint that will influence competition between alternative analyses of a sentence, it is worth noting that the data obtained in the first of the two studies described here did not support a Constraint-based account. Recall that off-line sentence fragment completion data did not adequately predict the pattern of on-line reading times. Thus, while a Constraint-based account can, in principle, explain thematic differences, the off-line and on-line data obtained in the first of the two studies are problematic for the account. Further data that are relevant to this discussion are provided by Binder, Duffy and Rayner (2001). They showed that despite discourse and thematic information strongly biasing readers towards a relative clause analysis, it was not possible to induce a garden path effect in simple main verb constructions. Binder *et al.*'s study again indicates that although Constraint-based accounts should simply include thematic information as an additional constraint on the process of competition between alternative analyses, on-line data suggest that this does not necessarily happen.

Finally, the most recent theory that specifies a role for thematic processing is that of Townsend and Bever (2001). Townsend and Bever provide a comprehensive account of sentence processing in which thematic assignment is involved at a very early stage during comprehension. They argue for an analysis-by-synthesis model that has two stages. In the first stage associative habit based knowledge (termed pseudosyntax) is used to initially construct a basic representation of the meaning of the sentence. This initial representation of meaning is then input to a second synthetic stage of analysis in which abstract grammatical knowledge is used to construct a full formal syntactic representation of the sentence along with its corresponding meaning. Subsequent to the second stage of processing, there follows a checking procedure in which the full syntactic derivation is compared against a stored representation of the input sequence. If the formal syntactic derivation matches the input string, then comprehension proceeds, otherwise reanalysis is required.

For the purposes of the discussion here, the specific and complete details of the theory, along with its viability as an account of sentence comprehension are not of particular significance. What is perhaps more important is the fact that Townsend and Bever argue that the derivation of meaning at the initial associative stage is an approximation of the ultimate meaning of the sentence. The initial meaning of the sentence is obtained through the identification of low level phrases that are marked by function words and morphemes to which are applied sentential templates like NVN (corresponding to agent, action, patient). That is to say, at the very first stage of processing readers obtain a shallow semantic representation of the sentence through the use of templates that involve the assignment of thematic roles to low level constituents within the sentence. These canonical templates represent a resurrection of ideas first advocated by Bever and colleagues (Bever, 1970; see also Fodor, Bever & Garrett, 1974) who argued that they provide a powerful explanatory tool in terms of accounting for a substantial number of findings within the psycholinguistic literature, particularly work focusing on how readers process reduced relative clause sentences. Indeed, thematic preferences in line with an NVN template have recently been incorporated into explanations of the effects of focus operators on initial parsing decisions for reduced relative clause sentences (Paterson, Liversedge & Underwood, 1998;

Liversedge, Paterson & Clayes, 2002). Currently, there has been little experimental work to evaluate the theoretical claims of Townsend and Bever and it is not yet clear that their approach would provide an adequate account of the findings discussed in this chapter. However, what should be clear from the discussion of all three different theories of sentence processing is that the process of thematic role assignment deserves a prominent place in any comprehensive theory of sentence comprehension.

To summarise, this chapter represents a brief discussion of work investigating thematic processing. It provides a description of what thematic roles are and why they are a useful construct to allow psycholinguists to start to better understand the nature of the semantic interpretation of a sentence. Some of the experimental studies investigating thematic processing during sentence comprehension have been briefly reviewed. Particular attention has been paid to eye movement experiments since eye movement monitoring methodology provides an excellent tool for studying thematic processing during reading. Finally, three theoretical accounts of sentence processing that incorporate thematic role assignment as an important aspect of the comprehension process have been summarised.

Author Note

The research described in this chapter was supported by an Economic and Social Research Council Grant No R000 22 2647 to Simon Liversedge. I am grateful to Holly Branigan, Emma Clayes, Martin Pickering, and Roger van Gompel who were collaborators on the experiments discussed here and I would also like to thank Keith Rayner, Matt Traxler, Sarah White and an anonymous reviewer for helpful comments about this work.

Note

1 A full list of the materials may be obtained by contacting the author.

References

Altmann, G. T. M. (1999). Thematic role assignment in context. *Journal of Memory and Language*, *41*, 124–145.

Bever, T. G. (1970). The cognitive basis for linguistic structures. In: J. R. Hayes (ed.), *Cognition and the Development of Language* (pp. 279–362). Chichester: John Wiley & Sons.

Binder, K. S., Duffy, S. A., & Rayner, K. (2001). The effects of thematic fit and discourse context on syntactic ambiguity resolution. *Journal of Memory and Language*, *44*, 297–324.

Carlson, G., & Tanenhaus, M. K. (1988). Thematic roles and language comprehension. In: W. Wilkins (ed.), *Syntax and Semantics 21: Thematic Relations* (pp. 263–288). San Diego, CA: Academic Press.

Dowty, D. (1991). Thematic proto-roles and argument selection. *Language*, *67*, 547–619.

Ferretti, T. R., McRae, K., & Hatherall, A., (2001). Integrating verbs, situation schemas and thematic role concepts. *Journal of Memory and Language*, *44*, 516–547.

Fillmore, C. (1968). The case for case. In: E. Bach and R. T. Harms (eds), *Universals in Linguistic Theory* (pp. 1–90). New York: Holt, Rinehart & Winston.

Fodor, J. A. (1983). The modularity of mind. Cambridge, MA: MIT Press.
Fodor, J. A., Bever, T. G., & Garrett, M. F. (1974). *The Psychology of Language*. New York, McGraw-Hill.
Frazier, L. (1979). On comprehending sentences: Sentence parsing strategies. Ph.D. thesis, University of Connecticut. Indiana University Linguistics Club.
Frazier, L. (1987). Sentence processing: A tutorial review. In: M. Coltheart (ed.), *Attention and Performance XII* (pp. 559–586). Hillsdale, NJ: Erlbaum.
Frazier, L., & Rayner, K. (1982). Making and correcting errors during sentence comprehension: Eye movements in the study of structurally ambiguous sentences. *Cognitive Psychology*, *14*, 178–210.
Gruber, J. (1965). Studies in lexical relations. Unpublished Doctoral Dissertation, Massachusetts Institute of Technology, distributed by the Indiana University Linguistics Club.
Hannah, J. E., Spivey-Knowlton, M. J., & Tanenhaus, M. K. (1996). Integrating discourse and local constraints in resolving lexical thematic ambiguities. In: G. W. Cottrell (ed.), *Proceedings of the Sixteenth Annual Cognitive Science Society Meeting* (pp. 266–271).
Jackendoff, R. (1972). *Semantic Interpretation in Generative Grammar*. Cambridge, MA: MIT Press.
Jackendoff, R. (1983). *Semantics and Cognition*. Cambridge, MA: MIT Press.
Jackendoff, R. (1987). The status of thematic relations in linguistic theory. *Linguistic Inquiry*, *28*, 369–412.
Liversedge, S. P., & Findlay, J. M. (2000). Eye movements reflect cognitive processes. *Trends in Cognitive Sciences*, *4*, 6–14.
Liversedge, S. P., Paterson, K. B., & Clayes, E. L. (2002a). The influence of *only* on syntactic processing of "long" relative clause sentences. *Quarterly Journal of Experimental Psychology*, *55*, 225–240.
Liversedge, S. P., Paterson, K. B., & Pickering M. J., (1998a). Eye movements and measures of reading time. In: G. Underwood (ed.), *Eye Guidance in Reading and Scene Perception* (pp. 55–75). Oxford: Elsevier.
Liversedge, S. P., Pickering, M. J., Branigan, H. P., & Van Gompel, R. P. G. (1998b). Processing arguments and adjuncts in isolation and context: The case of *by*-phrase ambiguities in passives. *Journal of Experimental Psychology: Learning, Memory and Cognition*, *24*(2), 461–475.
Liversedge, S. P., Pickering, M. J., Clayes, E. L., & Branigan, H. P. (2002b). Thematic processing of adjuncts: Evidence from an eye tracking experiment. Manuscript under submission to *Psychological Bulletin and Review*.
MacDonald, M. C. (1994). Probabilistic constraints and syntactic ambiguity resolution. *Language and Cognitive Processes*, *9*, 157–201.
MacDonald, M. C., Pearlmutter, N. J., & Seidenberg, M. S. (1994a). The lexical nature of syntactic ambiguity resolution. *Psychological Review*, *101*, 676–701.
MacDonald, M. C., Pearlmutter, N. J., & Seidenberg, M. S. (1994b). Syntactic ambiguity resolution as lexical ambiguity resolution. In: C. Clifton Jr, L. Frazier and K. Rayner (eds), *Perspectives on Sentence Processing*. Hillsdale, NJ: Erlbaum.
Marslen-Wilson, W. D. (1975). Sentence perceptions as an interactive parallel process. *Science*, *189*, 226–228.
Mauner, G., & Koenig, J. (1999). Lexical encoding of event participation information. *Brain and Language*, *68*, 178–184.
Mauner, G., Tanenhaus, M. K., & Carlson, G. N. (1995). Implicit arguments in sentence processing. *Journal of Memory and Language*, *34*, 357–382.
McRae, K., Ferretti, T. R., & Amyote, L. (1997). Thematic roles as verb specific concepts. *Language and Cognitive Processes*, *12*, 137–176.
Minsky, M. (1975). A framework for representing knowledge. In: P. H. Winston (ed.), *The Psychology of Computer Vision* (pp. 211–277). New York: McGraw Hill.

Paterson, K. B., Liversedge, S. P., & Underwood, G. (1998). Quantificational constraints on parsing "short" relative clause sentences. *Quarterly Journal of Experimental Psychology, 52A*, 717–737.

Pearlmutter, N. J., & MacDonald, M. C. (1992). Plausibility and syntactic ambiguity resolution. In: *Proceedings of the 14th Annual Conference of the Cognitive Society* (pp. 498–503). Hillsdale, NJ: Erlbaum.

Pickering, M. J., & Traxler, M. J. (1998). Plausibility and recovery from garden paths: An eye tracking study. *Journal of Experimental Psychology: Learning, Memory and Cognition, 24*, 940–961.

Rayner, K., Carlson, M., & Frazier, L. (1983). The interaction of syntax and semantics during sentence processing: Eye movements in the analysis of semantically biased sentences. *Journal of Verbal Learning and Verbal Behavior, 22*, 358–374.

Rayner, K., & Pollatsek, A. (1989). *The Psychology of Reading*. Englewood Cliffs, NJ: Prentice Hall.

Rayner, K. Sereno, S. C., Morris, R. K., Schmauder, A. R., & Clifton, C. (1989). Eye movements and on-line comprehension processes. *Language and Cognitive Processes, 4*, 21–50.

Rumelhart, D. E., & Ortony, A. (1977). The representation of knowledge in memory. In: R. C. Anderson, R. J. Spiro and W. E. Montague (eds), *Schooling and the Acquisition of Knowledge* (pp. 99–135). Hillsdale, NJ: Erlbaum.

Schank, R. C., & Abelson, R. P. (1977). *Scripts, plans, goals and understanding: An inquiry into human knowledge structures*. Hillsdale, NJ: Erlbaum.

Schütze, C. T., & Gibson, E. (1999). Argumenthood and English prepositional phrase attachment. *Journal of Memory and Language, 40*, 409–431.

Spivey-Knowlton, M. J., & Sedivy, J. (1994). Resolving attachment ambiguities with multiple constraints. *Cognition, 55*, 227–267.

Spivey-Knowlton, M. J., Trueswell, J. C., & Tanenhaus, M. K. (1993). Context and syntactic ambiguity resolution. *Canadian Journal of Psychology, 47*, 276–309.

Tabossi, P., Spivey-Knowlton, M. J., McRae, K., & Tanenhaus, M. K. (1994). Semantic effects on syntactic ambiguity resolution: Evidence for a constraint-based resolution process. In: C. Umilta and M. Moscovitch (eds), *Attention and Performance XV* (pp. 589–616). Cambridge, MA: MIT Press.

Tanenhaus, M. K., Burgess, C., Hudson D'Zmura, S., & Carlson, G. (1987). Thematic roles in language processing. In: *Proceedings of the Ninth Annual Cognitive Science Society Meeting*, (pp. 587–596). Hillsdale, NJ: Erlbaum.

Tanenhaus, M. K., & Carlson, G. N. (1989). Lexical structures and language comprehension. In: W. D. Marslen-Wilson (ed.), *Lexical Representation and Process* (pp. 529–561). Cambridge, MA: MIT Press.

Townsend, D. J., & Bever, T. G. (2001). *Sentence Comprehension: The Integration of Habits and Rules*. Cambridge, MA: MIT Press.

Traxler, M. J., Foss, D. J., Seely, R. E., Kaup, B., & Morris, R. K. (2000). Priming in sentence processing: Intralexical spreading activation, schemas, and situation models. *Journal of Psycholinguistic Research, 29*, 581–595.

Trueswell, J. C., Tanenhaus, M. K., & Garnsey, S. M. (1994). Semantic influences on parsing: Use of thematic role information in syntactic disambiguation. *Journal of Memory and Language, 33*, 285–318.

Trueswell, J. C., Tanenhaus, M. K., & Kello, C. (1993). Verb-specific constraints in sentence processing: Separating effects of lexical preference from garden-paths. *Journal of Experimental Psychology: Language, Memory and Cognition, 19*, 528–553.

Chapter 15

On the Treatment of Saccades and Regressions in Eye Movement Measures of Reading Time

Wietske Vonk and Reinier Cozijn

This chapter deals with two issues concerning the calculation of measures of reading time based on eye movements: saccade duration inclusion and regression-contingent analysis. First, a claim is made to include the durations of saccades in accumulated measures of reading time because during a saccade language processing continues uninterruptedly. Second, the calculation of the first-pass processing time of a region should leave out cases in which a reader leaves the region with a regression to an earlier part of the text. A new measure is introduced, *forward reading time*, that, better than the *first-pass reading time,* reflects the first-pass processing of a region and that increases the probability of finding experimental evidence. The effects of the calculations are illustrated by analyses of an eye movement experiment on causal inferences.

Introduction

A major concern in any study using the eye movement recording technique to investigate reading behavior is the determination of an appropriate measure of reading time. The past twenty years of eye movement reading research, ever since Just and Carpenter asserted that there is a tight coupling of eye movements and cognitive processing (Just & Carpenter, 1980), the number of measures of reading time has grown and concern has risen to come to some sort of standardization (Inhoff & Radach, 1998; Murray, 2000). The present chapter deals with two issues pertaining to the question of standardization. The first issue is related to the treatment of saccade durations in the calculation of reading times and the second issue to the differential treatment of progressive and regressive reading behavior.

The Mind's Eye: Cognitive and Applied Aspects of Eye Movement Research

ISBN: 0-444-51020-6

The focus in this chapter is not on characteristics of eye movement measures in relation to perceptual processes of reading but on eye movement measures in relation to sentence (and discourse) processing. This is not to deny the importance of low-level, perceptual determinants of eye movement behavior during reading. It has been shown, indeed, that, for example, saccade launch sites, preferred viewing positions, parafoveal preview effects do influence where and when to move the eyes (see, e.g., Inhoff & Radach, 1998; Rayner, 1998; Vonk, Radach & Van Rijn, 2000). The concern here is, however, how eye movement data can be used to obtain an understanding of the difficulties readers encounter when reading sentences and text.

The relationship between eye movement behavior and language processing is not a simple one. The relationship is not as tight as Just and Carpenter (1980) have posited. They stated that all comprehension processes that operate on a word are started as soon as the word is viewed (immediacy assumption), and, secondly, that the eyes remain fixated on a word as long as the word is being processed (eye-mind assumption). Although there is indeed a relationship between the duration of a fixation on a word and the processing associated with that word, there are at least two problems with these assumptions. First, not all words in a text are fixated. In fact, a lot of words, mostly function words, are skipped. That is not to say that they are not processed. It has been shown that they are perceived and processed during the fixation preceding the skipped word (see Rayner & Pollatsek, 1989, for a review). This means that fixation duration does not necessarily reflect the processing of just the fixated word. A second problem is the existence of spill-over effects. It has been shown, for instance, that the prolonged processing time associated with fixating an infrequent word spills over to the next fixation (Rayner & Duffy, 1986). In short, although processing seems to be immediate, one should be cautious in the interpretation of what actually is being processed during the time spent in a region. Nevertheless, eye movements are a rich information source in the study of language comprehension processes.

An important advantage of the eye movement methodology is that eye movement patterns give insight into the temporal characteristics of the reading process. It allows for the separation of first viewing of a textual region from subsequent viewings of that region. This characteristic has helped in determining which strategies readers follow when encountering a difficulty in the text. Generally, readers follow one of three strategies to solve a problem while reading: They pause at the location where the difficulty arises, they regress to an earlier part of the text, or they postpone solving the problem and move on to the next part of the text hoping to find a solution there (cf. Frazier & Rayner, 1982; Liversedge, Paterson & Pickering, 1998). In order to capture reading behavior, several measures of reading time have been developed, and over time, the interest in attaining measurement standards in eye movement research is increased. We will discuss first some measures that are often reported in eye movement studies of language processing, namely first fixation duration, gaze duration or first-pass reading time, total reading time, and total-pass reading time.

First Fixation Duration

The duration of the first fixation on a word (or region) reflects the processing difficulty a reader experiences immediately on viewing that word. First fixation durations have been shown to reflect, for instance, word frequency effects (Rayner & Duffy, 1986; for an overview, see Rayner, 1998). Of course, this measure is of less use if the region is rather large. Larger regions require more fixations to be processed and the first fixation then is not indicative of the complete processing of that region.

Gaze Duration

The gaze duration is defined as the sum of the fixations on a word until the word is left in a backward or forward direction. The measure was first used by Just and Carpenter (1980) to indicate the reading time associated with the processing of a word. For a region, consisting of more than one word, the same measure appeared under different names, most notably as *first-pass reading time* (Inhoff & Radach, 1998; Liversedge *et al.* 1998; Rayner, 1998). However, a plea has been made to keep the name *gaze* for all of these measures, whether applied to one word or to regions containing more than one word (Hyönä, Lorch & Rinck, this volume; Murray, 2000). Counterintuitively, the measure that is calculated does not reflect the reader's "gaze" on a region but only the sum of the reader's fixations in a region leaving out the saccades. Whatever it is called, one should use one term for the measure whether it is applied to a region consisting of one word or more than one word.

Total Reading Time

The total reading time of a region is the sum of all fixations in a region, irrespective of first or subsequent passes through the sentence. The measure is sometimes assumed to reflect late effects of reading difficulty. It only is a worthwhile measure, if the effects are compared to effects obtained by measures of first-pass processing.

Total-pass Reading Time

To capture the reading time associated with regressive reading behavior, a measure has been developed that is calculated as the sum of the durations of fixations in a region, plus the fixations from that region to earlier parts of the text until the region is left in a forward direction (Duffy, Morris & Rayner, 1988). The measure was introduced as *total-pass reading time* (Kennedy, Murray, Jennings & Reid, 1989), but has been called *go-past reading time* (Clifton, Bock & Rado, 2000), *regression path duration* (Konieczny, Hemforth, Scheepers & Strube, 1995), and *cumulative region reading time* (Mitchell, Brysbaert, Grondelaers & Swanepoel, 2000) as well (see Murray, 2000, for an overview). The name *regression path duration* for this measure is actually a misnomer because it suggests that the measure represents only the regression part of the eye movement data associated with processing a region.

In a discussion on terminology, Murray (2000) has suggested to use the term *total-pass reading time* because this name best reflects the fact that the measure represents the complete processing of a region, that is, the inspection time of the region itself, plus the durations of fixations on earlier portions of the text that are spent in regressions from the region as a result of processing difficulties in the region.

Our interest lies in two aspects of the definition of measures of reading time based on eye movements. The first aspect relates to the method by which the measures of the time spent on a word or region are calculated. The standard method of calculating accumulated processing time is to sum fixation durations. The method is inspired by the fact that readers take in information during a fixation but not during the intermediate saccades. We will argue below that this argument falls short. The second aspect points to what happens when terminating the processing of a text region. Does the reader move on in the text to acquire new information or does he jump back to re-process old information? In most studies, the direction in which the reader continues is not taken into account in the measures that are used to estimate processing time. An exception was the work of Altmann, Garnham and Dennis (1992). They introduced a *regression-contingent analysis* to differentiate between region reading times prior to a regression to a previous region and those prior to a forward movement to the next region in the text. They compared these reading times to *gaze durations* and found similar results in the analyses of reading times prior to a forward movement but not in the analyses of reading times prior to a regression, which suggests that this aspect of eye movement based reading times should be taken into account.

In this chapter, we will discuss the issues of the calculation of accumulated region reading times and regression-contingent analyses. We will argue that saccade durations are part of the time spent in language processing and that, therefore, their durations should be included in accumulated measures of reading time. We will illustrate the effect of saccade inclusion by comparing analyses of *first-pass reading times,* in which saccade durations are not included, and first-pass reading times including saccade durations. We will show the benefit of a regression-contingent approach to the analyses of eye movement data. To that end we will introduce a new measure called *forward reading time* which represents the time spent in a region from the start of the first fixation in that region, provided that the reader enters that region for the first time, until the end of the last fixation in that region, including possible regressions within the region, before the region is left in a forward direction. We will illustrate our discussion with analyses of the data from an experiment that investigated the influence of the presence of the connective "because" on the processing of causal relation sentences. In this experiment, the same condition (sc. the presence of the connective) was expected to result in opposite effects in two different regions of the target sentence: on the one hand, in a decrease of processing time early in the sentence, and, on the other hand, in an increase in processing time late in the sentence. Therefore, the experiment is considered very suitable for demonstrating the effect of measures of reading time that differ in the way they are calculated.

The Experiment

The study presented here investigates the influence of a causal connective on the inferential processing of causal relations. Previous research has shown that readers spontaneously make causal inferences if they have prior knowledge of the situations or events that are causally related (Noordman & Vonk, 1992), but not if they lack such knowledge (Noordman, Vonk & Kempff, 1992). The causal relations used in these studies were linguistically signaled by the causal connective "because". The studies were not unequivocal with respect to the question what influence the connective had on making the causal inference. This question was investigated in our lab by Cozijn (1992). In a series of subject-paced, clause-by-clause reading experiments using a probe recognition task and a verification task, it was shown that a causal inference was made when the causal relation sentence was signaled by the causal connective "because", but not if the causal relation was not signaled. The causal relation sentences were embedded in short narrative texts on familiar topics. A sample text is presented in Table 15.1.

The sentence "he experienced a big delay, because there was a traffic jam on the highway" is justified by the inference that "a traffic jam causes delay". The connective "because" signals to the reader that the second clause of the sentence is to be understood as a cause for the consequence stated in the first clause. For familiar information, it triggers the memory retrieval process of the information by which this causal relation can be justified. The justifying inference contributes to the coherence of the discourse representation. In the course of the inferential justification process, which takes place at the end of processing the complete sentence, the first clause of the sentence is re-activated in memory. The results showed that this re-activation facilitated the recognition of a probe word taken from the first clause ("ochtend" in Table 15.1). The probe recognition times were faster if the connective was present than if it was absent. Similarly, the verification sentences, which contained the inferential information, were responded to faster if the connective was present than if was absent, indicating that the justifying inference was made during reading. However, the results of the probe recognition task and the verification task were not corroborated by the results of the clause reading times analyses. The clause reading times showed no effect of the presence of the connective, although the retrieval from memory of the justifying inferential information is assumed to be time-consuming and should show in an increase if the inference is made. How can this lack of an effect of the connective on the reading times be explained?

Noordman and Vonk (1997) distinguish a number of different functions that a connective serves. First, the connective suggests that an inference should be made. An inference leads to a longer reading time. This is the inference function of the connective "because". It is assumed that the inference is made after the complete sentence has been read, that is, at sentence wrap-up time (Just & Carpenter, 1980). A second function is that the connective indicates the way in which the clause has to be integrated with the previous context. The connective "because" indicates that the reader has to establish a causal coherence relation between the clauses. This is the integration function of the connective. There is evidence in the literature for this integration function

Table 15.1: Example of a text used in the experiments on the influence of the causal connective "because" on the processing of causal relations.

De heer Smit verliet rond half acht het huis.
Hij moest op zijn werk een belangrijke vergadering voorzitten.
Daarom was hij van plan om die morgen de papieren goed door te nemen.
Hij haalde zijn auto uit de garage en reed weg.
Op weg naar het werk had hij die **ochtend** een flinke vertraging[,/.]

Target clause with connective
omdat er een lange file was ontstaan op de snelweg.^**probe**

Target clause without connective
Er was een lange file ontstaan op de snelweg.^**probe**

Hij was blij dat hij wat eerder was vertrokken.
Hij hield er niet van om te laat te komen.

Verification statement
Een file op de snelweg leidt tot vertraging.

English translation:
Mister Smith left his house at about eight o'clock. At work, he had to chair an important board meeting. That is why he had planned to study the papers thoroughly. He got his car out of the garage and drove off. On his way to work that **morning**, he experienced a big delay[,/.]

Target clause with connective
because there was a traffic jam on the highway.^**probe**

Target clause without connective
There was a traffic jam on the highway.^**probe**

He was glad that he had left earlier. He hated to be late.

Verification statement
A traffic jam on the highway causes delay.

(Haberlandt, 1982; Millis, Golding & Barker, 1995; Millis & Just, 1994). For instance, Haberlandt (1982) investigated the role of sentence–initial adversative and causal connectives and found a facilitating effect of these connectives on the reading of the immediately following words. The effect did not occur on the reading of the last words of the sentences. He suggested that "reading comprehension is facilitated when the reader's expectations are guided by the presence of a surface marker which explicates the semantic relationship between adjacent sentences" (Haberlandt, 1982: 243).

In the subject-paced reading experiments mentioned above, the reading times were measured clause-by-clause. The reading times, therefore, reflected the processing of complete clauses. The fact that no increase in reading times was found can be explained by assuming, as indicated above, that the connective not only triggers an inference that leads to an increase in processing time of the clause (at the end of the clause), but also facilitates integration that leads to a shorter processing time of the clause (right after the connective).

This hypothesis was investigated in our lab (Cozijn, 2000) in an eye movement experiment because this technique allows for an inspection of reading times on smaller textual units than the clause. In the experiment, the same condition is expected to result in opposite effects in two different regions of the target sentence: on the one hand in a decrease of processing time early in the sentence, and on the other hand in an increase in processing time late in the sentence. This eye movement experiment we consider, therefore, very suitable to demonstrate the effect of measures that differ in the way they are calculated.

Method

Forty students of Nijmegen University were paid to participate in the experiment. They all had normal, uncorrected vision. Participants read the texts with both eyes but only the movements of the right eye were measured. Horizontal and vertical eye movements were sampled at a rate of 200 Hz (5 ms per measurement) by an Amtech ET3 infrared pupil reflectance eye tracker (Katz, Müller & Helmle, 1987). The eye tracker has a spatial resolution of 5 to 10 minutes of arc, that is, approximately 0.25 degrees of visual angle. Participants were restricted in their head movements by the use of a chin rest, a forehead rest, and a bite bar. The distance between the participant's eye and the monitor screen was 59 cm. Each character on the screen subtended approximately 0.28 degrees of visual angle. The texts were presented in a black non-proportional font (Courier New, 12 pt) on a light gray background.

The same texts were used as in the subject-paced reading experiments. There were 24 experimental texts and 24 filler texts that contained connectives other than "because". There were two factors in the experiment: familiarity of the causal relation and presence of the connective. Since the manipulation of familiarity turned out not to interact with the presence of the connective it is left out in the presentation. Each participant saw each experimental text in one condition, with or without the causal connective.

For the analyses, the target sentences were divided into regions of one or more words. The regions of interest here are the middle and the final region of the target clause, regions 5 and 6 in Table 15.2 (regions are indicated by numbered slashes).

The size of the middle region (region 5) was determined by the consideration that it was the smallest region consisting of the same words, albeit in a different word order, in the conditions with and without the causal connective ("een lange file was ontstaan" vs. "was een lange file ontstaan"; see the target clauses in Table 15.1) that contained the causal information by which the inference could be made. The final region (region 6) consisted of a prepositional phrase that did not bear on the causal relation, allowing the wrap-up process to be measured after the causal information has been presented. It was made sure that the regions of interest were always presented at

Table 15.2: Screen layout of the target sentence and analysis regions, indicated by numbered slashes.

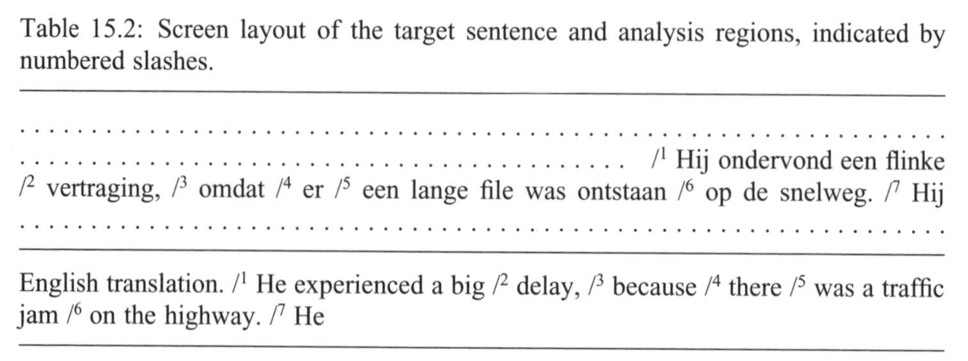

. . . .
. . . . /[1] Hij ondervond een flinke
/[2] vertraging, /[3] omdat /[4] er /[5] een lange file was ontstaan /[6] op de snelweg. /[7] Hij
. . . .

English translation. /[1] He experienced a big /[2] delay, /[3] because /[4] there /[5] was a traffic jam /[6] on the highway. /[7] He

Note: The numbered slashes were not presented; they are for the convenience of the reader only.

the center of the screen and that they were preceded and followed by other regions (regions 2 and 7, respectively) on the same line, thus avoiding that return-sweeps would affect the reading times.

The reading experiment started with a twelve-point calibration routine, by which the correspondence between eye movements and screen coordinates was established. Each experimental text was preceded by an asterisk, indicating the start of the text. The participant had to fixate the asterisk and press a button to start reading the next text. The participants were instructed to read the text thoroughly but quickly, in order to understand it, and to press the button as soon as they had finished reading. Immediately after reading a text, they had to judge a verification statement, that was either true or false with respect to the content of the text, urging them to read well. Before the presentation of the next text, a re-calibration routine was started. It consisted of four calibration points presented in the center of the screen. The experiment consisted of two blocks of 24 texts. Each block was preceded and followed by a calibration. Between the blocks there was a break, allowing the participant to take a rest. The total duration of the experiment was one hour.

Results

The reading times were analyzed in two analyses of variance, one with participants as random variable (*F1*) and one with items as random variable (*F2*). The participants analysis contained participant group as a between factor and the items analysis contained item group as a between factor. These factors were entered in the analysis to reduce error variance, as suggested by Pollatsek and Well (1995). Both types of analyses contained presence of the connective "because" as a within-factor.

Reading times belonging to items on which the participants had made a verification error, as well as outliers exceeding 2.0 Sd. from the participant and the item means were excluded from the analyses. Apart from these excluded observations, there were missing data as a result of skipping or blinks. In total, this resulted in 93.0% valid data in the analysis of the middle region and 91.7% in the analysis of the final region. The analyses were performed on the *first-pass reading times*, the first-pass reading times including saccade durations, and the *forward reading times*. In order to facilitate the comparison of the results of the measures used in the present study, the means and statistics have been gathered in two comprehensive tables, Tables 15.3 and 15.4, respectively. First, the results of the standardly applied *first-pass reading times* are presented. As indicated above, *first-pass reading time* is defined as the sum of the fixation durations in a region until the region is left in a backward or forward direction. The mean *first-pass reading times* are listed in Table 15.3, section a., and the related statistics are presented in Table 15.4, column a.

The results of the middle region of the *first-pass reading times* showed an effect of presence of the connective. The reading times were shorter if the connective was present than if it was absent. For the final region, an opposite effect of the connective was found, albeit in the analysis by participants only. The mean reading times were longer if the connective was present than if it was absent. The analysis of the two regions combined showed an interaction between region and presence of the connective, and here also only in the analysis by participants.[1]

The results of the analyses indicated that the presence of the connective led to a decrease in reading times of the words immediately following the connective and to an increase in reading times of the final words of the sentence. This counteracting effect explains why no increase in reading times was found in the self-paced reading experiments mentioned earlier where clause reading times were obtained. An analysis of the mean clause reading times of the target clause (not including the connective) in the eye movement experiment also showed no effect of the presence of the

Table 15.3: Overview of reading time measures. Means (ms) of the measures of first-pass reading times, reading times including saccades, and forward reading times in two critical regions (middle and final) as a function of the presence of the connective "because".

	Connective		
Region	**Present**	**Absent**	**Difference**
a. First-pass reading times (not including within-saccades)			
Middle	501	549	–48
Final	378	359	19
b. Reading times including within-saccades			
Middle	548	602	–54
Final	403	382	21
c. Forward reading times (including within-saccades)			
Middle	605	677	–72
Final	431	395	36

Table 15.4: Overview of statistics of the analyses of the reading time measures. Statistics of the measures of first-pass reading times, reading times including saccades, and forward reading times in two critical regions (middle and final) as a function of the presence of the connective "because", and the interaction between critical region and the presence of the connective.

Effect	**First-pass reading times (not including within-saccades)** a.	**Reading times including within-saccades** b.	**Forward reading times (including within-saccades)** c.
Connective, middle region			
*F*1	5.53	5.74	10.24
MSE	8209	10370	10060
p	0.024	0.022	0.003
*F*2	4.17	4.12	11.80
MSE	6870	8963	7043
p	0.054	0.054	0.003
Connective, final region			
*F*1	5.29	6.01	11.25
MSE	1253	1485	2326
p	0.027	0.019	0.002
*F*2	1.27	1.43	3.13
MSE	3121	3813	4867
p	0.273	0.245	0.092
Interaction region × connective			
*F*1	11.69	11.97	16.15
MSE	3708	4783	7217
p	0.002	0.001	0.000
*F*2	3.42	3.55	9.31
MSE	7889	10098	9101
p	0.079	0.074	0.006

Note: Number of degrees of freedom: $F1(1,36)$; $F2(1,20)$.

connective (all *F*s < 1). The means were 1268 and 1256 milliseconds for the conditions with and without connective, respectively. Apparently, the two effects canceled each other out.

Saccade Durations and Measures of Reading Time

The *first-pass reading times* are based on the sums of the durations of the *fixations* in a region. The question is whether this measure adequately reflects the processing time associated with the regions. The regions consist of several words and, as a rule, attract more than one fixation. Fixations are separated by saccades and the decision to sum only the fixation durations is implicitly based on the assumption that saccade durations do not contribute to comprehension time.

That saccades do not contribute to comprehension time might be concluded from the fact that when the eyes move, no information is taken in, a phenomenon called saccadic suppression (Matin, 1974). Furthermore, saccade length and, consequently, saccade durations are generally assumed not to be under the control of higher-level cognitive processes, but to be determined solely by low-level perceptual processes. However, saccadic suppression and low-level control do not imply that higher-level cognitive processes do not continue during a saccade. In fact, it is rather unlikely that all cognitive processing is disrupted during a saccade. As Irwin (1998: 2) puts it:

> [. . .] the cumulative effect of cognitive suppression during saccades could be quite large; if one assumes that the average person makes 2–3 saccades each second [. . .] then cognition would be suppressed for a total of 60–90 minutes a day!

Research on cognitive suppression during saccades has shown that only cognitive processing that shares processing resources with saccadic programming, saccade execution, or visual processing is affected (see Irwin, 1998, for an overview). Irwin investigated whether lexical processing is affected as well but found no influence of the making of saccades on word recognition and word identification processes. These lexical processes continue uninterruptedly.

Since lexical processing apparently is unaffected by saccades, there is no reason to assume that higher-level language processes, such as sentence and discourse processes, are affected. The conclusion, therefore, is that if language comprehension processes continue during a saccade, the duration of a saccade that connects two consecutive fixations in a region should be added to the comprehension time associated with that region.[2]

If saccade durations are included in the calculation of reading times, the question is to what extent they influence the outcome of the experiment. Table 15.3 (section b.) shows the mean reading times with the inclusion of the durations of *within-saccades*, that is, saccades that connect the consecutive fixations in a region.

The results (see Table 15.4, column b.) are similar to the results of the analyses of the *first-pass reading times* based on summing fixations only. In the middle region, there was an effect of the presence of the connective. The reading times were shorter if the connective was present than if it was absent. Conversely, in the final region, the reading times were longer if the connective was present compared to if it was absent, albeit in the by-participants analysis only. The combined analyses of the two regions showed an interaction effect of region with presence of the connective in the analysis by participants, and a trend in the analysis by items.

Although the inclusion of within-saccade durations did not substantially change the outcome of the experiment, the differences between the means and the resulting F-values were somewhat larger in the analysis of the reading times including the within-saccade durations (Table 15.3, section a.; Table 15.4, column a.) than in the analysis of the *first-pass reading times* (Table 15.3, section b.; Table 15.4, column b.).

The influence of within-saccade durations on reading time is directly related to their number of occurrences. As can be seen, at first glance, in Table 15.5 (section a., numbers in superscript) it is obvious that the number of within-saccades is not distributed equally over conditions: There were more within-saccades in the middle region when the connective was absent than when it was present, and more within-saccades in the final region when the connective was present than when it was absent. Table 15.5 (section a.) also shows the mean summed within-saccade durations for the two critical regions of the target clause.

Separate analyses of the mean summed within-saccade durations revealed similar effects as in the *first-pass reading times* analyses. The analyses of the middle region showed an effect of the presence of the connective: $F1(1,36) = 5.06$, $MSE = 111$, $p < 0.05$; $F2(1,20) = 4.80$, $MSE = 115$, $p < 0.05$. The mean summed within-saccade

Table 15.5: Overview of measurement related data. Means of summed within-saccade durations (ms), regressions (percentage of valid data), and means of reading times prior to a regression in two critical regions (middle and final) as a function of the presence of the connective "because" (absolute numbers in superscript).

	Connective	
Region	**Present**	**Absent**
a. Mean summed within-saccade durations (ms)		
Middle	58.7^{645}	64.0^{738}
Final	40.9^{364}	37.4^{327}
b. Regressions (% of valid data)		
Middle	13.8^{62}	15.6^{70}
Final	19.0^{84}	13.7^{60}
c. Mean reading times (including within-saccades) prior to a regression (ms)		
Middle	284^{62}	238^{70}
Final	295^{84}	317^{60}

durations were shorter if the connective was present than if it was absent. For the final region, an opposite effect was found in the analysis by participants: $F1(1,36) = 4.98$, $MSE = 50$, $p < 0.05$; $F2(1,20) = 2.66$, $MSE = 31$, $p = 0.12$. The analyses of the two regions combined showed an interaction effect between region and the presence of the connective: $F1(1,36) = 9.20$, $MSE = 84$, $p < 0.01$; $F2(1,20) = 5.52$, $MSE = 96$, $p < 0.05$.

The separate analyses of the summed within-saccade durations showed significant results for the influence of the presence of the connective. This effect is attributable to the number of saccades and not to their duration. In fact, the duration of the saccades decreased with an increase in the number of saccades. The overall mean within-saccade durations, which were 32.49, 31.81, and 30.57 milliseconds for cases of one, two, or three and more within-saccades, respectively, showed a linear trend: $F(1,1265) = 9.23$, $MSE = 58$, $p < 0.01$. It seems simply to be the case that when there are many fixations in a region the distances between the fixations are shorter than when there are few fixations in a region. The shorter distances lead to shorter saccade durations.

The analysis of mean summed within-saccade durations, presented above, is not a plea to perform such analyses separately from analyses of summed fixation durations. They are presented simply to show that, with a sufficient number of within-saccades, within-saccade durations contribute to the effects of the experimental manipulation. Therefore, on theoretical as well as empirical grounds, within-saccade durations should be included in measures of reading time.

First-pass Reading Times and Regressions

The *first-pass reading time* measure, as indicated above, is calculated as the sum of the fixations on a region *until the region is left in a backward or forward direction.* It implies that the nature of the consecutive observation is of no consequence to its measurement. In other words, it does not matter whether the next fixation is situated in the following or in the preceding part of the text. This, however, seems questionable. Altmann *et al.* (1992) introduced a *regression-contingent* analysis of *first-pass reading times*. In this analysis, reading times prior to a regression to an earlier part of the text are analyzed separately from reading times prior to a forward saccade to the next region of the text. Their data show that reading times prior to a regression are shorter than those prior to a forward movement. An explanation for this difference might be that the decision to jump back in the text in order to resolve a comprehension problem occurs before the processing of that region is completed. The completion then takes place during the re-reading of an earlier part of the text. If, on the other hand, the reader concludes the processing of a region (and solves the problem) while fixating that region, reading times on that region will be longer. Although regressions normally are attributed to comprehension difficulties (see, e.g., Rayner, 1998; Rayner & Pollatsek, 1989), it is not the case that if a comprehension difficulty is encountered, a regression is always made (Rayner & Sereno, 1994). Frazier and Rayner (1982) observed three distinct patterns of eye movements when readers encountered a parsing problem during comprehension: (1) they continued reading in a forward direction but with long fixation durations; (2) they made a regression to the region of the text where the problem could be solved; and (3) they continued reading until the end of the sentence after which they regressed to the beginning of the sentence in order to re-read it. The conclusion to be drawn from Altmann *et al.*'s (1992) and Frazier and Rayner's (1982) studies is that one should not calculate *first-pass reading times* without taking regressive behavior into account. If readers encounter a comprehension problem in a region and follow reading strategy (1), their reading times of that region will be longer, and if they follow reading strategy (2), their reading times will be shorter than if no comprehension problem occurred. Not taking regressive behavior into account might even lead to the paradoxical situation that if regressions are frequent, the *first-pass reading times* are shorter in the difficult condition than in the easy condition, which goes against the prediction of the effect of comprehension difficulty on reading times.

In the Altmann *et al.* (1992), Frazier and Rayner (1982), and Rayner and Sereno (1994) studies, mentioned above, large numbers of regressions were obtained (e.g., up to 76% in Altmann *et al.*, 1992). These studies were set up to investigate garden-pathing. In the present experiment, there were 15% regressions on average, which is the number that is found in normal reading circumstances (Rayner, 1998). Nevertheless, the experiment did compare the processing of an easy condition with a more difficult condition, and one would expect this to show in the number of regressions in these conditions as well.

Crucial to the determination of the influence of regressive behavior on the *first-pass reading times* is how many regressions occur and how they are distributed over conditions. Table 15.5, section b., shows the percentages of the regressions of valid data for the two regions in the conditions with and without the connective "because". In the middle region, it seemed that there were more regressions when the connective was absent than when it was present, but the distribution did not deviate from chance: $\chi^2 = 0.62$, df = 1, $p = 0.432$. In the final region, the distribution of regressions showed a trend toward more regressions when the connective was present than when it was absent: $\chi^2 = 3.79$, df = 1, $p = 0.052$. Indeed, although the number of regressions was small, the data indicated that more regressions occurred in the more difficult conditions than in the easier conditions.

What determines the influence of regressive behavior on the *first-pass reading times* is how soon after entering a region a regression is made. Table 15.6 allows for a closer inspection of the regressions in the experiment. It shows the distribution of fixations in the two critical regions of the causal relation sentence as a function of the presence of the connective "because" and of whether the region was left in a forward direction (no regression) or a backward direction (regression).

Of all cases where a regression occurred, most were made immediately after the first fixation. The percentage of regressions that were made immediately after the first fixation in the middle and the final regions, with and without connective were 79.0% (49 out of 62 cases), 87.1% (61 out of 70), 67.9% (57 out of 84), and 61.7% (37 out of 60), respectively (see Table 15.6).

It can be expected that the effects of the unequal distribution of regressions over conditions has consequences for the *first-pass reading times*. Table 15.5, section c., shows the reading times including saccade durations for cases where a regression occurred. The mean reading times prior to a regression show a pattern that is exactly opposite to the pattern obtained for the *first-pass reading times* (cf. Table 15.3, section a.). The reading times in the difficult conditions (middle region without and final region with connective) are shorter than in the easy conditions.[3] However, as explained earlier, this paradoxical finding is completely understandable. If on encountering a difficulty in a region a regression is made, the reading time of the region will not reflect the complete processing of the region and the reading time will be short.

Interestingly, we also found the *first fixation durations* that were immediately followed by a regression to be smaller than those prior to a forward saccade (see Table 15.7). Note, however, that the *first fixation durations* contribute to a very small extent to the effect of the regressions on *first-pass reading times*.

Table 15.6: Frequency of occurrence as a function of number of fixations, and regression and presence of the connective "because" in two critical regions (middle and final).

			Number of fixations									
Region	**Regression**	**Connective**	**1**	**2**	**3**	**4**	**5**	**6**	**7**	**8**	**9**	**Σ**
Middle	No	Present	40	176	111	39	10	8	2	1	1	388
		Absent	10	135	145	51	22	5	3	2		373
	Yes	Present	49	8	2	1	2					62
		Absent	61	6	1	1	1					70
Final	No	Present	109	185	53	9	1	1				358
		Absent	143	185	40	8	1	1				378
	Yes	Present	57	19	6	2						84
		Absent	37	18	4	1						60

Table 15.7: Mean first fixation durations (ms) in two critical regions (middle and final) as a function of regression immediately after first fixation (absolute numbers in superscript).

	Regression after first fixation	
Region	**No**	**Yes**
Middle	218^{761}	189^{110}
Final	223^{736}	202^{94}

In short, the results of the reading times prior to a regression make clear that the inclusion of these cases in the measure of *first-pass reading time* cause an underestimation of the effect of the experimental manipulation. Therefore, more appropriate measures have to be devised to indicate first-pass processing of a region. If readers encounter a problem during reading and pause to solve the problem before they move on in the text, the proper way to capture their behavior is to base the reading times on only those cases in which the region is left in a forward direction. The measure that results from this method we call *forward reading time*. It is defined as the time spent in a region from the start of the first fixation in that region, provided that the reader enters that region for the first time, until the end of the last fixation in that region, including possible regressions within the region, before the region is left in a forward direction. In this measure, the reading times prior to a regression are excluded. It is clear from the data presented in Table 15.5, section c., that the differences between the conditions will be larger in the *forward reading times* than in the *first-pass reading times* (see Table 15.3, sections c. and b., respectively, for a proper comparison, including within-saccade durations). Including the reading times prior to a regression away from the target region, which show a pattern that is opposite to that of the *forward reading times,* diminishes the differences in the *first-pass reading times*.

The mean *forward reading times* for the two critical regions of the causal relation sentence were analyzed analogously to the analyses reported earlier. Table 15.4, column c., presents the statistics. There was an effect of the presence of the connective in the middle region. The middle region was processed faster if the connective was present than if it was absent. In the final region, the presence of the connective led to longer forward reading times. Finally, the interaction between region and presence of the connective was significant.

Again, the results are in line with those of the *first-pass reading times*. However, the difference between the means is larger in the *forward reading times* than in the *first-pass reading times* without and including saccades (compare Table 15.3, sections a. and c., and sections b. and c., respectively), and the effect size is larger, notably in the $F2$-analyses of the main effect of the presence of the connective in the middle and the final region and those of the interactions between region and the presence of the connective (compare Table 15.4, columns a. and c., and columns b. and c.). We believe that the *forward reading times* give a more accurate view of the processing time associated with a region than the *first-pass reading times* do. They are not contaminated with a decrease in reading time caused by the inclusion of cases in which the reader leaves the region with a regression. In this sense, they reflect the processing of a region when readers stay in a region to resolve a reading problem and move on when they are finished.

Note that the *forward reading times* are longer than the *first-pass reading times*. (Table 15.3, sections c. and a.). This is not only due to the inclusion of within-saccade durations but also to the exclusion of cases prior to a regression, as explained above. Concerning the exclusion of cases prior to a regression, a similar finding has been reported by Altmann *et al.* (1992). Rayner and Sereno (1994), on the other hand, did not find longer reading times for their forward reading times (no saccade durations included). However, they defined forward reading times over parts of the sentence preceding and including the region of concern. This means that cases were excluded in which readers had made a regression earlier in the sentence. If these cases are eliminated, it is not possible to tell whether the reading times will increase or decrease. In fact, analyses of these extended forward reading times in studies in our lab, conducted on data reported in Mak, Vonk and Schriefers (submitted) support this contention.

Discussion

Two issues related to eye movement measures were addressed: the inclusion of saccade durations in accumulated measures and the regression-contingent analysis of *first-pass reading times*.

The Inclusion of Saccade Durations

There are good theoretical arguments to include within-saccade durations in the calculation of accumulated measures of processing time based on eye movements. However, it is striking to note that the research community still refrains from treating them accordingly. Even those who are explicitly concerned with the methodology of eye movement reading research ignore or dismiss the issue (Inhoff & Radach, 1998; Liversedge *et al.*, 1998; Murray, 2000; Rayner, 1998). The general argument is that their contribution to reading time is negligible. For instance, Inhoff and Radach (1998: 41) state that "it appears unlikely that the inclusion of movement time [. . .] will lead to a notable change", and Murray (2000: 657) suggests that "it will probably make relatively little difference in the majority of instances". Rayner (1998: 378) recognizes the relevance of the issue: "On the other hand, when regions larger than a single word represent the unit of analysis, effects of saccade duration are larger and hence are more likely to aid in revealing differences between conditions". However, he did not explicitly indicate that they therefore should be included. Our argument, of course, is that, however small their influence, they *should* be included in measures of reading time as soon as fixation durations are accumulated, i.e., if more than one fixation is contained in the measure, because they contribute to language processing time as well as fixations do. From our discussion, it should be clear that the influence of the inclusion of within-saccade durations to reading time is attributable to their number of occurrence. Two factors have an impact on their frequency of occurrence. First, their number is directly related to the size of the region to which the measurement of the reading time is applied. This means that the contribution of the inclusion of within-saccade durations to the measurement of the overall processing time of a region will increase with the size of the region to which the measurement is applied. Second, and more importantly, the contribution of the inclusion of within-saccade durations will increase with processing difficulty because the number of fixations in a region will increase if the reader runs into trouble.

A question that has not been addressed is how to treat saccades that connect regions. Should they be included in the reading time of the region from which they originate or to which they are directed? This is not an easy issue. Saccade length is to a great extent determined by saccade launch position and the length of the targeted word. The longer the targeted word is, the longer the saccade to that word will be (O'Regan, 1990). Furthermore, saccade length is modulated to a certain degree by preview effects of the targeted word, for instance, by the informativeness of word beginnings. Whether or not these effects are orthographic or reflect higher-level processing is a matter of debate (Everatt & Underwood, 1992; Hyönä, 1995; Hyönä, Niemi & Underwood, 1989; Rayner & Morris, 1992; Underwood, Clews & Everatt, 1990; Underwood, Hyönä & Niemi, 1987; Vonk, Radach & Van Rijn, 2000). Apparently, during a fixation, parafoveal information is extracted from the visual field in order to determine the size of the next saccade. If it is assumed that this means that the reader has already started processing the targeted region, it would make sense to attribute any processing occurring during the saccade to that region. However, if durations of saccades toward a region are to be included in the processing time associated with that region, word length should be controlled for in the launch region as well as in the target region. In general, one should not include saccades between regions in the calculation of reading times of regions from which they are launched (but see Inhoff & Radach, 1998).

Regression-contingent Analysis of First-pass Reading Time

The present study has demonstrated the necessity to distinguish two components in the calculation of *first-pass reading times* of a region. The first component is related to cases where the reader leaves the region in a forward direction, and the second to cases where the reader regresses from the region. Neglecting this distinction results in *first-pass reading times* that underestimate the processing time associated with a region. The reason is that reading times in a region are shorter if a regression is made than if the region is left in a forward direction. The measure associated with the progressive reading behavior is coined *forward reading time.* It is the time spent in a region from the start of the first fixation in that region, provided that the reader enters that region for the first time, until the end of the last fixation in that region before the region is left in a forward direction. The measure includes within-saccade durations and regression durations *within the region* as contrasted to accumulated progressive fixation duration (cf. Frazier & Rayner, 1982). The measure reflects the processing time of a region associated with solving a reading problem immediately while pausing at the region where the problem occurred, and increases the probability of finding experimental evidence for differences between conditions compared to *first-pass reading times*.

To investigate the processing time of a region when a reader regresses, one can resort to *total-pass reading time*. It is the time spent in a region, *plus* the time spent in regressing to earlier parts of the text until the region is left in a forward direction (note that saccade durations are included). This measure reflects the processing time of a region associated with solving a reading problem either by staying in that region or by re-reading earlier portions of the text.

Both measures can be used simultaneously to capture the two types of reading behavior. However, by doing so, one risks the inflation of Type I error rates (Liversedge *et al.*, 1998) because the measures are not statistically independent. There are several possibilities to avoid this dependency. Liversedge *et al.* (1998) proposed the *re-reading time*, which is the *total-pass reading time* excluding the reading time of the region from which a regression has been made (see also Hyönä *et al.*, this volume). Alternatively, one could use a restricted version of the *total-pass reading time* that only contains the cases in which a regression occurred. Finally, Konieczny (1996) suggested the calculation of the *load contribution*, that is, the *total reading time* in a region within the regression path of another region. Of course, these measures are to be used in combination with other measures, such as *forward reading time*, to obtain a complete picture of the eye movement patterns associated with processing a region of the text.

Crucial to the meaningfulness of these measures in a particular experiment is how many regressions actually occur. If there are many regressions, measures like *re-reading time* and *load contribution* make sense. If there are not so many regressions, as in the experiment discussed here, these measures are less usable.

Conclusion

It should be clear from our exposition that the issue of standardization of measures in eye movement reading research (Inhoff & Radach, 1998; Liversedge *et al.*, 1998; Murray, 2000) cannot easily be resolved. Much depends on what kind of reading behavior the experimental setting elicits. We consider it mandatory to always report the *percentages of regressions*, overall and per condition. The overall percentage of regressions determines the applicability of duration measures, whereas the percentages per condition determine their conclusiveness. If the number of regressions differs between conditions, *first-pass reading time* is not an appropriate measure. We suggest using *forward reading time* as a measure of "first pass" reading. We support the idea to always report *total-pass reading time* (Murray, 2000). Dependent on the number of regressions other more specific regression based measures, including load measures, should be applied. Finally, we make a plea for including within-saccade durations in all accumulated measures of reading time, irrespective of the size of the region to which they are applied.

Acknowledgements

We like to thank Leo Noordman as well as the editor Jukka Hyönä and the reviewers Keith Rayner and Mike Rinck for valuable comments on an earlier version of this chapter. We also like to express special thanks to Pim Mak for reanalyzing some data sets with respect to extended forward reading times.

Notes

1 As a control, analyses were performed on the reading times of regions before the target clause. As it should, they showed no effects of the presence of the connective.

2 It seems reasonable that small blinks within a region should be treated similarly.
3 Analyses of variance were performed on the mean reading times prior to a regression, but no effects of the presence of the connective were obtained. This is most probably due to the small number of cases.

References

Altmann, G. T., Garnham, A., & Dennis, Y. (1992). Avoiding the garden path: Eye movements in context. *Journal of Memory and Language*, *31*, 685–712.

Clifton, C., Jr., Bock, J., & Rado, J. (2000). Effects of the focus particle *only* and intrinsic contrast on comprehension of reduced relative clauses. In: A. Kennedy, R. Radach, D. Heller and J. Pynte (eds), *Reading as a Perceptual Process* (pp. 591–619). Oxford: Elsevier.

Cozijn, R. (1992). Inferential processes during the construction of a coherent text representation. In: M. van Oostendorp (ed.), *Proceedings of the CLS Ph.D. Conference 1992* (pp. 1–14). Nijmegen/Tilburg, The Netherlands: Center for Language Studies.

Cozijn, R. (2000). Integration and inference in understanding causal sentences. Unpublished doctoral dissertation, University of Tilburg, Tilburg, The Netherlands.

Duffy, S. A., Morris, R. K., & Rayner, K. (1988). Lexical ambiguity and fixation times in reading. *Journal of Memory and Language*, *27*, 429–446.

Everatt, J., & Underwood, G. (1992). Parafoveal guidance and priming effects during reading: A special case of the mind being ahead of the eyes. *Consciousness and Cognition*, *1*, 186–197.

Frazier, L., & Rayner, K. (1982). Making and correcting errors during sentence comprehension: Eye movements in the analysis of structurally ambiguous sentences. *Cognitive Psychology*, *14*, 178–210.

Haberlandt, K. F. (1982). Reader expectations in text comprehension. In: J. F. Le Ny and W. Kintsch (eds), *Language and Comprehension* (pp. 239–249). New York: North-Holland.

Hyönä, J. (1995). Do irregular letter combinations attract readers' attention? Evidence from fixation locations in words. *Journal of Experimental Psychology: Human Perception and Performance*, *21*, 68–81.

Hyönä, J., Niemi, P., & Underwood, G. (1989). Reading long words embedded in sentences: Informativeness of word parts affects eye movements. *Journal of Experimental Psychology: Human Perception and Performance*, *15*, 142–152.

Inhoff, A. W., & Radach, R. (1998). Definition and computation of oculomotor measures in the study of cognitive processes. In: G. Underwood (ed.), *Eye Guidance in Reading and Scene Perception* (pp. 29–54). Oxford: Elsevier.

Irwin, D. E. (1998). Lexical processing during saccadic eye movements. *Cognitive Psychology*, *36*, 1–27.

Just, M. A., & Carpenter, P. A. (1980). A theory of reading: From eye fixations to comprehension. *Psychological Review*, *87*, 329–354.

Katz, B., Müller, K., & Helmle, H. (1987). Binocular eye movement recording with CCD arrays. *Neuro-Ophthalmology*, *7*, 81–91.

Kennedy, A., Murray, W. S., Jennings, F., & Reid, C. (1989). Parsing complements: Comments on the generality of the principle of minimal attachment. *Language and Cognitive Processes*, *4*, 51–76.

Konieczny, L. (1996). Human sentence processing: A semantic-oriented parsing approach. Doctoral dissertation, IIG-Reports 3/96, Universität Freiburg, Institut für Informatik und Gesellschaft.

Konieczny, L., Hemforth, B., Scheepers, C., & Strube, G. (1995). PP-attachment in German: Results from eye movement studies. In: J. M. Findlay, R. Walker and R. W. Kentridge (eds),

Eye Movement Research: Mechanisms, Processes, and Applications (pp. 405–520). New York: North-Holland.

Liversedge, S. P., Paterson, K. B., & Pickering, M. J. (1998). Eye movements and measures of reading time. In: G. Underwood (ed.), *Eye Guidance in Reading and Scene Perception* (pp. 55–76). Oxford: Elsevier.

Mak, W. M., Vonk, W., & Schriefers, H. (submitted). When rocks crush hikers: Processing relative clauses in Dutch.

Matin, E. (1974). Saccadic suppression: A review. *Psychological Bulletin*, *81*, 899–917.

Millis, K. K., Golding, J. M., & Barker, G. (1995). Causal connectives increase inference generation. *Discourse Processes*, *20*, 29–49.

Millis, K. K., & Just, M. A. (1994). The influence of connectives on sentence comprehension. *Journal of Memory and Language*, *33*, 128–147.

Mitchell, D. C., Brysbaert, M., Grondelaers, S., & Swanepoel, P. (2000). Modifier attachment in Dutch: Testing aspects of construal theory. In: A. Kennedy, R. Radach, D. Heller and J. Pynte (eds), *Reading as a Perceptual Process* (pp. 493–516). Oxford: Elsevier.

Murray, W. (2000). Sentence processing: Issues and measures. In: A. Kennedy, R. Radach, D. Heller and J. Pynte (eds), *Reading as a Perceptual Process* (pp. 649–664). Oxford: Elsevier.

Noordman, L. G. M., & Vonk, W. (1992). Reader's knowledge and the control of inferences in reading. *Language and Cognitive Processes*, *7*, 373–391.

Noordman, L. G. M., & Vonk, W. (1997). The different functions of a conjunction in constructing a representation of the discourse. In: J. Costermans and M. Fayol (eds), *Processing Interclausal Relationships: Studies in the Production and Comprehension of Text* (pp. 75–93). Mahwah, NJ: Erlbaum.

Noordman, L. G. M., Vonk, W., & Kempff, H. J. (1992). Causal inferences during the reading of expository texts. *Journal of Memory and Language*, *13*, 573–590.

O'Regan, J. K. (1990). Eye movements and reading. In: E. Kowler (ed.), *Eye Movements and their Role in Visual and Cognitive Processes* (pp. 395–453). Amsterdam: Elsevier.

Pollatsek, A., & Well, A. D. (1995). On the use of counterbalanced designs in cognitive research: A suggestion for a better and more powerful analysis. *Journal of Experimental Psychology: Learning, Memory, and Cognition*, *21*, 785–794.

Rayner, K. (1998). Eye movements in reading and information processing: 20 years of research. *Psychological Bulletin*, *124*, 372–422.

Rayner, K., & Duffy, S. A. (1986). Lexical complexity and fixation times in reading: Effects of word frequency, verb complexity, and lexical ambiguity. *Memory & Cognition*, *14*, 191–201.

Rayner, K., & Morris, R. K. (1992). Eye movement control in reading: Evidence against semantic preprocessing. *Journal of Experimental Psychology: Human Perception and Performance*, *18*, 163–172.

Rayner, K., & Pollatsek, A. (1989). *The Psychology of Reading*. Englewood Cliffs, NJ: Prentice Hall.

Rayner, K., & Sereno, S. C. (1994). Regressive eye movements and sentence parsing: On the use of regression-contingent analyses. *Memory & Cognition*, *22*, 281–285.

Underwood, G., Clews, S., & Everatt, J. (1990). How do readers know where to look next? Local information distributions influence eye fixations. *Quarterly Journal of Experimental Psychology*, *42A*, 39–65.

Underwood, G., Hyönä, J., & Niemi, P. (1987). Scanning patterns on individual words during the comprehension of sentences. In: J. K. O'Regan and A. Levy-Schoen (eds), *Eye Movements: From Physiology to Cognition* (pp. 467–477). Amsterdam: North-Holland.

Vonk, W., Radach, R., & van Rijn, H. (2000). Eye guidance and the saliency of word beginnings in reading text. In: A. Kennedy, R. Radach, D. Heller and J. Pynte (eds), *Reading as a Perceptual Process* (pp. 269–299). Oxford: Elsevier.

Chapter 16

Eye Movement Measures to Study Global Text Processing

Jukka Hyönä, Robert F. Lorch Jr and Mike Rinck

In this chapter, we demonstrate the usefulness of the eye tracking method in studying global text processing. By "global text processing," we refer to processes responsible for the integration of information from sentences that are not adjacent in the text. Potential eye movement measures indexing global text processing are discussed using as examples the processing of topic-introducing sentences and the processing of inconsistencies. In addition to the existing measures of regional gaze duration and lookback fixation time, we advocate new measures that may be applied to the study of global text processing. These include a new extended first-pass fixation time measure that allows lookbacks to previous text regions without necessarily terminating the first-pass reading, and first-pass rereading time that sums up all the reinspective fixations made during first-pass reading. We also demonstrate the potential usefulness of analyzing the origin and destination of eye movement sequences, such as lookback sequences.

Introduction

The goal of the present chapter is to consider the applicability of the eye tracking method to study global text processing. The starting point is that eye tracking has been successfully adopted to the study of basic reading processes and to that of syntactic parsing, but there are surprisingly few studies where eye tracking is employed to examine global text processing (but see Blanchard & Iran-Nejad, 1987; Hyönä, 1995; Hyönä, Lorch & Kaakinen, 2002; Kaakinen, Hyönä & Keenan, 2002; Vauras, Hyönä & Niemi, 1992). There may be several reasons for this state of affairs, but we focus on one likely reason; namely the apparent lack of consensus on the measures to be used to tap into global text processing. In what follows, we will consider the

The Mind's Eye: Cognitive and Applied Aspects of Eye Movement Research

ISBN: 0–444–51020–6

applicability of conventional eye movement measures to the study of global text processing, and we will also suggest new, potentially useful measures.

Global Text Processing

We begin by defining what we mean by global text processing. Global text processes are those processes that identify and represent relationships between pieces of text information that span relatively long distances in a text. Our definition excludes mental processes related to building coherence between consecutive sentences and focuses on processes that link together information from sentences that are not adjacent in the text.

Global text processing probably becomes more concrete by the following examples. The first type of global text processing pertains to cases where successful comprehension requires a reinstatement in working memory of information mentioned in the preceding text. For example, in expository text, it is common that the writer probes the reader to reactivate relevant prior information ("As mentioned earlier . . .", "Recall that . . ."). This may be achieved mentally, but it may also involve overt behavior, such as rereading the to-be-reinstated information. An example of a reinstatement process related to the comprehension of narratives pertains to activating in working memory the goal of the main character's current actions. It has been shown that information about the goal is pertinent to the interpretation of the various actions carried out by the protagonist (van den Broek & Lorch, 1993). When the relevant goal information is not active in the reader's working memory, it may be retrieved by rereading the text where this information is provided. Yet another example of this type of global text processing concerns the comprehension of inconsistencies in meaning between two text segments. Inconsistencies must be resolved if the reader wishes to build an internally coherent representation of the text. Resolving inconsistencies may require reprocessing of text segments that do not cohere with each other. Later in the chapter we will discuss in more detail how this type of global text processing could be studied with eye tracking.

A second category of global text processing that is discussed more thoroughly below has to do with processes pertaining to the comprehension of multiple-topic expository texts. A typical expository text is constructed around a global topic that is developed in a hierarchically related set of topics and subtopics. Thus, comprehension of such a text requires, in part, that readers represent the text's topics and subtopics and their relationships. When a topic is identified, it, in turn, serves as a context for comprehending subsequent information that elaborates upon it.

Standard Eye Movement Measures

Eye tracking methods have been used most extensively to study lexical access and syntactic parsing. The measures that have been developed provide the researcher with information about the time course of processing. Thus, the measures can be divided into those that index immediate effects and those that index more delayed effects of processing. In the study of lexical access (for more, see Inhoff & Radach, 1998), immediate effects have usually been studied using the *duration of the first fixation* on the critical word as the primary measure (see Table 16.4). The most widely used measure, *gaze duration* is also assumed to reflect relatively immediate effects in processing. Gaze duration is computed as the sum of all individual fixations landing on the critical word before exiting it. When there is only one fixation on the word, gaze duration equals first fixation duration (see Table 16.4). When refixations occur, gaze duration indexes less immediate effects than first fixation duration. For example, Hyönä and Pollatsek (1998; Pollatsek, Hyönä & Bertram, 2000) used these measures to study the identification of long compound words when they were embedded in sentences. They found that both the frequency of compound word constituents (first and second) as well as the whole-word frequency affected gaze duration, but only the frequency of the first constituent produced an immediate effect as reflected in first fixation duration.

For delayed effects, the following measures are widely used: the duration of first fixation after leaving the target word, the duration of regressions back to the target word, and the total fixation time computed as the sum of gaze duration and regression time (see Table 16.4). The common denominator for these measures is that they reflect events in the reader's eye behavior after the critical word is once exited. Total fixation time is a composite measure indexing both immediate and delayed effects. An example of a study showing a delayed effect in the absence of an immediate effect is that of Bertram, Hyönä and Laine (2000). These researchers sought evidence for the view that words are accessed via their constituent morphemes. They observed that for inflected words the frequency of the word's stem produced no immediate effects (i.e., no effects on gaze duration or first fixation duration), but a reliable lagged effect reflected in gaze duration for the $N+1$ word.

As one moves from the study of lexical processing to syntactic processing, the potential units of analysis increase both in number and size. Whereas the word is the natural unit of analysis in lexical processing, there are four relevant levels of processing in the study of syntactic processing: (a) the word at which a parsing choice is expected to be made or a syntactic ambiguity to reveal itself, (b) the phrase, (c) the clause, and (d) even the whole sentence. Related to the increase in the number and size of potentially interesting units of analysis, the mental processing associated with syntactic processes is more complex and varied than the mental processing associated with lexical processing. Thus, syntactic effects on eye movements are correspondingly more complex than lexical effects on eye movements.

Studies of syntactic parsing have used a multitude of processing measures. Immediate and delayed effects in processing have been studied using first-pass and second-pass measures for the critical sentence regions. First-pass fixation time for a text region, which can be a word, a phrase, or a clause, is defined analogously to gaze duration as the summed fixation time on a text region before exiting it (either right or left). Murray (2000) proposes that it should be called gaze duration instead of first-pass fixation time, as it has been typically referred to in the syntactic parsing literature. Murray argues that separate names imply that these two measures would differ in some ways, although their underlying principle is exactly the same. Although we agree with this argument, we use a slightly different term, *regional gaze duration,* to avoid confusion with the standard gaze duration related to single words (see Table 16.4). The second-pass fixation time that has been suggested to measure more delayed effects in parsing (e.g., a reanalysis of the syntactic structure) is analogous to the regression time measure used in the word identification studies (see Table 16.4). As the name indicates, the measure sums up fixations that return to a text region after it has been fixated at least once (i.e., during the first-pass reading).

In recent parsing studies, a measure often called regression path reading time has gained increased popularity (see e.g., Konieczny, 1996; Liversedge, Paterson & Pickering, 1998). This measure was invented to capture difficulty effects in parsing that were obscured in the standard measures (for more details, see Liversedge *et al.*, 1998). For example, when encountering a text region that entails syntactic ambiguity, readers may quickly go back to previous parts of the sentence to start over the parse in order to resolve the ambiguity (see also Vonk & Cozijn, this volume). This type of behavior would be reflected in shorter regional gaze durations for the region revealing the ambiguity — an effect opposite to what would be predicted. On the other hand, the predicted disruption in processing is shown in the reinspective fixations landing on earlier sentence regions. The regression path measure is designed to index exactly this kind of processing difficulty effect. To compute the regression path reading time, the researcher sums up all the reinspective fixations that occur once the target region is reached and before exiting the target region to the right (see Table 16.4). The distinctive feature of regression path reading time is that it sums up temporally contiguous fixations regardless of their spatial locations (as long as they are directed to previous sentence regions with respect to the target region). An alternative to this procedure is to cluster the fixations constituting the regression path on the basis of their spatial location, in which case they would be defined as second-pass fixations for the sentence regions they land on. Liversedge *et al.* (1998) demonstrate that this alternative to the regression path reading time may substantially weaken the chances to detect a disruption effect.

Within the regression path, it is possible to further separate out fixations that are directed from the target region to earlier sentence regions from fixations that land on the target region during its first-pass reading (i.e., prior to backtracking). These reinspective fixations constitute a measure called rereading time (Liversedge *et al.*, 1998) or first-pass regression time (Van Gompel, Pickering & Traxler, 2001). An example of a study in which all the measures mentioned above were utilized is that of Van Gompel *et al.* (2001). The study established no effect of attachment ambiguity in the first-pass fixation time, but reliable effects in the regression path reading time and the rereading time. These results were taken as evidence for the view that the studied type of syntactic ambiguity primarily increases the likelihood of computing a reanalysis of the syntactic structure.

Could the standard eye movement measures summarized above be applied to the study of global text processing? It is perhaps quite obvious that sentence- or clause-level measures of eye behavior will be more relevant than word-level measures. This is because the expected effects are rarely confined to individual words but to larger text regions. Thus, the measures of regional gaze duration and second pass fixation time as well as regression path reading time developed in the syntactic parsing research may be readily applicable to the study of global text processing. On the other hand, it is possible that even larger units of analysis may need to be defined (e.g., an entire paragraph or a subsection). Moreover, some types of global processing may appear with notable delays, which may pose a need for new measures so that all aspects of global processing would become apparent from the eye tracking records. In what follows, we present a more detailed description of two types of global text processing, one related to the processing of a text's topic structure and another related to resolving textual inconsistencies. For each case, we discuss how eye tracking can be applied to tap the processes and what measures, existing or new, will be needed to conduct a thorough analysis of the eye behavior in question.

Case Study 1: Processing Topic-introducing Sentences

Sentences that introduce new discourse topics in an expository text place heavy processing demands on readers. Examine the excerpt from a descriptive expository text presented in Table 16.1. This excerpt consists of the first four (abbreviated) paragraphs from a text that compares and contrasts two fictional countries with respect to several attributes. The topic sentences of the third and fourth paragraph have been singled out for consideration ("$Topic_1$" and "$Topic_2$," respectively), along with the final sentence of the third paragraph ("End_1"). These examples illustrate two important functions of topic sentences with respect to the global structure of a text. First, a topic sentence is an explicit statement of a macroproposition; that is, it is a superordinate statement that integrates several of the statements in the paragraph or subsection in

Table 16.1: Excerpt from an expository text discussing characteristics of two fictional countries.

Although the countries of Culatta and Morinthia share a border, they differ in many different ways. In this article, we will discuss what makes each country interesting and unique.

The population of Morinthia is comprised primarily of the descendants of European immigrants. Most of the settlers came to the country in the 18th century. They consisted mainly of poor families who were willing to risk the unknowns of a new and unsettled country for the opportunity to own land and build their futures.

The people of Culatta have a fascinating history. ($Topic_1$) Physically and culturally, they are quite distinct from the people of the bordering countries. They are descendants of a race and culture that has existed in the country for at least 12,000 years. **Scientists hypothesize that the race originated in Asia. (End_1)**

The geography of Morinthia is striking in its contrasts. ($Topic_2$) The western border of the country consists of a rugged mountain range that plunges to the sea. East of the mountains is a vast tropical jungle that dominates the central part of the country. . . .

---

which it is placed. Second, when it is placed in a paragraph-initial position, a topic sentence serves to introduce a new discourse topic.

Consider the topic-introducing function of (many) topic sentences. An adequate understanding of an expository text requires that readers represent the major topics of the text and their relationships (Gernsbacher, 1990; Kieras, 1981; Lorch, Lorch & Matthews, 1985). As explicit statements of discourse topics, topic sentences are particularly relevant to the processing of the topic structure of a text (Kieras, 1980; Lorch *et al.*, 1985). At the time a topic sentence is first encountered during reading, it should entail relatively high processing demands. Indeed, topic sentences are processed more slowly than sentences that elaborate established discourse topics (Hyönä, 1994; Lorch *et al.*, 1985). In addition, a topic sentence is processed more quickly when the discourse topic it introduces can be related to the immediately preceding discourse topic than when it cannot (Lorch *et al.*, 1985; Lorch, Lorch & Mogan, 1987). These results are consistent with the hypothesis that readers update a representation of the text's topic structure each time they encounter a topic-introducing sentence and that the updating includes a computation of the relationship of the new topic to previously established topics. Additional support for this hypothesis is provided by the finding that rereading a text results in greater facilitation of processing of topic sentences than non-topic sentences (Hyönä, 1995). The selective speed up of rereading of topic sentences is presumably because readers do not need to construct a topic structure representation during the second reading of a text.

In addition to its topic-introducing function, a topic sentence integrates much of the information in the paragraph or subsection that it dominates. In fact, this is the defining characteristic of a topic sentence. As an important context for the interpretation and integration of subsequent information, readers might be expected to refer back to topic sentences with some regularity. For example, if readers encounter difficulty understanding a statement in a paragraph, they might attempt to relate the statement back to the topic sentence. Even in the absence of comprehension difficulties, readers might systematically refer back to topic sentences as part of a metacomprehension strategy to check understanding while reading. A particularly likely place to see evidence of such processing is the end of a paragraph (see End_1 in Table 16.1). The white space at the end of a paragraph and the corresponding preview of the indentation of the following line of text both predict a possible upcoming change of topic. Readers who systematically review their understanding as they read might be expected to show evidence of such a strategy at the ends of paragraphs. In fact, at least some readers use a strategy of looking back from "end" sentences to topic sentences and headings (Hyönä *et al.*, 2002).

Most of our knowledge of how readers process information about discourse topics while reading is based on experiments using single-sentence presentation of text. This is a serious limitation because it omits text layout information that is likely to be an important influence on the online processing of information relevant to the text's topic structure (Hyönä *et al.*, 2002). Further, the experimental procedure prevents readers from looking back to previous text and forces them to rely on memory for prior text. These correlated characteristics of the procedure are likely to distort our understanding of the nature of readers' strategies for processing topic-relevant statements while reading. To consider how the use of eye tracking measures might augment our understanding of topic processing, we first elaborate the types of processes that might be involved in topic processing.

Topic Processing as Structure Building

Gernsbacher (1990) has proposed that shifts at all levels of structure in a discourse entail special processing efforts on the part of the reader. Applying her ideas specifically to the processing of topic structure, the following processing is hypothesized to occur when readers encounter a topic-introducing sentence. First, a topic sentence corresponds to a shift in the discourse structure from an old discourse topic to a new one. The reader responds to this shift by terminating processing of the current discourse topic and suppressing it (i.e., causing it to become less accessible). Further, a new structure is initiated to represent the new topic. We think of this shift to a new structure as involving moving from a subnetwork representing the previous section of the text, to finding the appropriate location in the text's hierarchical structure to start a new branch. Once this location is established (i.e., the relationship of the new topic to the existing topic structure representation is computed), it serves as the foundation for mapping new information. The reader continues to the next sentence and attaches the information in that sentence onto the new branch of the text representation. As long as the text continues to elaborate the new topic, information is mapped into the subnetwork corresponding to that topic.

Topic sentences play a critical role in coordinating the online processing of the text with the construction of the subnetwork representation associated with a topic. Encountering a topic-introducing sentence while reading triggers the reader to begin construction of a new subnetwork, as already described. Further, the topic sentence is an important point of access to the text representation. A reader who wishes to review information associated with the topic, or who needs to access specific information stored about the topic, must either rely on memory or must look back to the appropriate part of the text. If the reader consults memory, an effective strategy is to access the text representation by locating the relevant topic in the topic structure representation (i.e., locate the entry point to the appropriate subnetwork). If the reader wishes to search the text, an effective strategy is to find the topic-introducing sentence and search forward from that point.

Potential "Areas of Interest" in Tracking Topic Processing

The theoretical framework just described suggests at least four types of sentences that should be informative with respect to the nature of readers' attempts to process topic structure information during reading. In addition to topic sentences themselves, sentences that begin a new paragraph and sentences that conclude paragraphs and/or subsections may play a special role in a topic structure processing strategy. Of course, the initial sentence of a paragraph often is a topic sentence. Because this correlation is presumably high and experienced readers are aware of it, paragraph-initial sentences may receive special attention from readers independently of whether they introduce a new topic. Finally, sentences that refer back to a previous topic at some point after discussion of the topic has concluded are also potentially informative about the nature of readers' representation and processing of a text's topic structure. We will restrict our discussion to topic sentences in considering how eye tracking measures may help us to understand topic processing during reading.

Suppose that readers perform the mental operations implied by the structure building framework (Gernsbacher, 1990) when they realize that the discourse topic has shifted. That is, they suppress the previous text topic, shift to the new topic, and initiate construction of a new subnetwork. Assuming that those operations require time and that they are initiated relatively close in time to when pertinent information is encountered during reading (Just & Carpenter, 1980), evidence of such processing should appear in the eye movement record sometime during reading of a topic-introducing sentence. But what unit of analysis is most appropriate to the detection of such effects? This is ultimately an empirical question that may not have a single, simple answer. In fact, the earliest possible effect of a topic shift might actually be before any part of the topic sentence has been fixated. Consider the sentence labeled "Topic_2" in Table 16.1. Readers could conceivably shift from the old topic during the course of reading End_1 on the assumption that each paragraph discusses a new topic (although such a strategy seems risky). This might appear as an increase in gaze duration on the final word and/or final verb phrase of End_1.

Even if they show some anticipation of a change of topics, readers cannot shift to a new topic until it is identified. If a single word fully communicates the new topic, then the earliest possible point at which this information becomes available is the first content word of the topic sentence. If the topic is communicated by a phrase as in the text illustrated in Table 16.1, then the initial noun phrase of the topic sentence is the earliest point at which the new topic may be identified. Thus, it is theoretically possible that topic processing might be initiated at the first new noun or the initial noun phrase containing a new noun. However, it is possible that readers do not commit themselves to updating their topic structure representations until the clause or sentence is completed. In that case, effects reflecting topic processing may not appear until the end of the clause or sentence. If so, defining the unit of analysis as the clause or sentence may be informative, although analyzing the final word or phrase of the topic sentence may catch "wrap up" effects. The situation may be even more complicated than this. Consider the possibility that readers are conservative about committing to a change of topics until the new topic is confirmed by the following sentence. This hypothesis stretches the immediacy principle uncomfortably (Murray, 2000), but researchers have observed such "spillover effects" in studies of readers' processing of global coherence relations (Myers, O'Brien, Albrecht & Mason, 1994). In short, analyses at the various grain levels that have been investigated in other studies are also likely to be useful in the application of eye tracking methods to the study of topic processing while reading. Further, we would like to hold open the possibility that still larger units of analysis may be relevant to the study of topic processing. Given that topics dominate paragraphs and even larger sections of text, these larger units of text organization may be useful units of analysis in the study of online topic processing. For example, if a statement refers to information established earlier in a text, a reader might search backwards through the text for the relevant information. Such a search process might be organized according to text layout information such as paragraph indentation or headings within the text (Klusewitz & Lorch, 2000).

Potential Measures of Topic Processing

Again, it is largely an empirical question as to what processing measures may be most informative in the study of topic processing. Nevertheless, we might anticipate that topic processing lags somewhat behind the theoretically earliest points at which such processing might be initiated. If that turns out to be the case, then measures of initial processing (e.g., regional gaze duration) will not reveal topic processing effects. Further, it is likely to be the case that the most useful units of analysis will be units larger than single words. In that event, much of the conventional arsenal of eye movement measures (e.g., gaze duration on individual words, probability of fixation, probability of regressing from or to a given word) will not be applicable (Inhoff & Radach, 1998; Rayner, 1998; Rayner, Raney & Pollatsek, 1995). Rather, the most useful measures will be of two sorts. First, conventional measures of the processing of phrases and sentences such as regional gaze duration, total reading time, and the probability of looking back from or to a phrase or sentence (see Table 16.4; and Rayner, 1998; Rayner *et al.*, 1995). Second, measures that attempt to analyze how readers search for information relevant to understanding a statement that they are processing (Liversedge *et al.*, 1998; Murray, 2000).

If the structure building framework provides a reasonable approximation to the (topic) processing involved in understanding a topic sentence, there are many alternative ways in which such processing might be manifested in eye movement records. One possibility is that readers can rely on memory to access the information needed to comprehend a topic sentence (at least some readers some of the time); therefore, they can complete all necessary processing of the topic sentence without looking back to preceding sentences or forward to subsequent sentences in the text. In this situation, the more conventional eye tracking measures should be sensitive to the nature of the readers' topic processing. Consider one hypothetical scenario for the example of Topic_1 in Table 16.1.

For Topic_1, the information that the discourse topic has changed is first available when the word "Culatta" is fixated because it is at this point that it is clear that the country under discussion has changed. Readers might demonstrate sensitivity to the change of topic in several ways. If readers respond immediately to the topic switch, then their initial fixation duration on "Culatta" should be slow (relative to an appropriate control). If their response is a little delayed, gaze duration and total time on "Culatta" might be relatively long. The pattern of regressions within a topic sentence may also provide useful information about when readers commit to a change of discourse topic. Specifically, regressions may be initiated from "Culatta" back to the start of the sentence if readers respond relatively immediately to a potential topic change; alternatively, readers may often launch regressions from the end of a sentence that introduces a topic change (Hyönä, 1995).

If readers must identify where to attach the new topic in their text representation, this computation probably involves retrieving the preceding text topic (i.e., population of Morinthia) and determining that the two topics should attach to the same superordinate (i.e., the general topic is the people of the two countries). When this computation occurs during the course of reading the topic sentence is an open question. It might occur before a forward fixation from the word "Culatta"; it might occur in the course of sentence wrap-up; it might be delayed still further. This uncertainty might be resolved by comparing the processing profile of $Topic_1$ with the processing profile of a topic sentence that introduces a topic whose location in the text structure is less easily computed (e.g., $Topic_2$ in Table 1), (cf., Lorch *et al.*, 1985, 1987).

It is entirely possible that readers will not rely exclusively on their memories to do all necessary processing of topic sentences. Rather, they may search the text for relevant information (Liversedge *et al.*, 1998; Murray, 2000). In that event, some conventional measures may not be very meaningful. For example, if readers often interrupt processing of a topic sentence before reaching its end, it is not clear how to interpret a measure like regional gaze duration or even total fixation time (Liversedge *et al.*, 1998). To elaborate, suppose that in the course of reading $Topic_2$, a reader begins looking back to the previous text after reading the initial noun phrase but before any fixations occur on the verb phrase. There are at least two measurement problems here. First, by its conventional definition, the look away from the sentence terminates regional gaze duration despite the fact that the reader has not completed processing of the sentence. If regional gaze duration is intended as a measure of the initial processing of a sentence, then an appropriate measure must attempt to aggregate the initial fixations on all parts of a sentence. That is, when the reader returns to a sentence that was incompletely processed before a look away, first-pass reading on the remainder of the sentence should be added to first-pass reading on the initial part of the sentence (see Hyönä & Juntunen, submitted, for an example of such an algorithm). We will call this measure *extended first-pass fixation time* (see Table 16.4).

The second measurement problem concerns the lookbacks themselves. We define "lookbacks" as all fixations that occur on text that is prior to the most recently-fixated sentence. Thus, lookbacks include both regressive fixations and forward fixations. Assuming that the lookbacks are triggered by an attempt to process the information in the topic sentence, analysis of the lookbacks is critical to understanding the reader's processing of the topic sentence. To begin with, it is informative if a lookback (or series of lookbacks) is initiated after processing the noun phrase of $Topic_2$. In the structure building framework, this action might be interpreted as a demonstration that the reader has recognized the change of topic and is attempting to compute the location of the new topic (i.e., geography of Morinthia) in the text representation. In this example, readers may well search back to the paragraph about the population of Morinthia because that was the last mention of a common superordinate. If the readers' memory for the location of that paragraph and its topic sentence is less than perfect, the search back may involve several fixations on intervening material or a jump back to the beginning of the text followed by a forward search. In a text with headings, there might be a systematic search utilizing the headings to identify potentially relevant information about the relationship of the new topic to previous text topics. In short, it is plausible to imagine multiple fixations on prior statements that vary in their relevance to the processing that the reader is attempting to complete. In this circumstance, the landing position of the initial lookback is unlikely to correspond to the goal of the search. Rather, a more thorough analysis of the pattern of lookbacks would be in order.

If processing of a topic sentence is interrupted by a series of lookbacks to prior text, several measures may be jointly informative. One potentially useful measure is the time between the initiation of a lookback sequence and the first forward fixation past the location from which the sequence was initiated, or "rereading time" (Liversedge *et al.*, 1998; see Table 16.4). Rereading time provides information about the amount (or difficulty) of processing necessary at some point in time, but it is not informative about the nature of that processing. The sequence of fixations along with information about the relative duration of fixations may begin to reveal what a reader is trying to compute. The sequence of fixations alone may not be very informative because some fixation locations may not provide the reader with useful data; rather, they may just be the consequence of less than perfect memory for the location of potentially useful text information. When a lookback does land the eyes on a relevant text location, however, the reader will presumably spend more time inspecting that phrase or sentence. Thus, a record of the total lookback time on each sentence (or smaller unit of analysis) during a lookback sequence provides information about what information the reader found most useful. Putting together the information about where the lookback sequence originated, how the prior text was searched, and what information received the most attention during the sequence should provide important information about what computation the reader was attempting to perform and how the computation was accomplished.

Finally, as an index of the relative difficulty of understanding a particular sentence, it may be useful to sum the duration of all rereading times within the sentence before a forward fixation is made to the next sentence. The assumption here is that if readers are having difficulty understanding a sentence, they will reprocess it before continuing to the next sentence. Thus, the extended first-pass rereading time on a sentence provides a measure of the extent of difficulty in comprehending the sentence (Hyönä *et al.*, 2002). Of course, the assumption underlying the use of this measure may be wrong. Perhaps there are times when a reader attempts to resolve a comprehension problem by reading ahead. In such cases, a comprehension difficulty may manifest as looks back to the problematic sentence shortly after reading ahead.

Case Study 2: Processing Inconsistencies

Inconsistencies and the Updating of Situation Models

What does it mean to "understand" a text? Recent theories of text comprehension agree on at least a partial answer: deep comprehension of a text involves the creation and updating of situation models (e.g., Gernsbacher, 1990; Kintsch, 1988, 1998; van Dijk

& Kintsch, 1983; Zwaan, Langston & Graesser, 1995; Zwaan & Radvansky, 1998). A situation model is a representation of the situation described by the text, rather than a representation of the text itself. Situation models are multidimensional; they contain information about the causal, temporal, and spatial relations of the situation. In case of narrative texts, the represented dimensions also include the protagonists' goals, traits, beliefs, and emotions. Moreover, a situation model needs to be consistent, that is, it has to represent a state of affairs that is possible in the world described by the text (this does not exclude states that are only possible in fictional worlds such as fairy tales).

Zwaan and Radvansky (1998) proposed a detailed account of how readers construct situation models during text comprehension. In their general processing framework, they distinguish between three types of situation models: the current, the integrated, and the complete model. They also distinguish four types of processes operating on these models: the construction, updating, and retrieval of situation models, and the foregrounding of specific situation model elements. According to their framework, readers construct a current model of the situation described by a sentence or clause. As they continue reading, each sentence yields a new current model. The information from all sentences read so far is being integrated into a single model, which is therefore called the integrated model. The process of incorporating a new sentence into the integrated model is called updating of the model. During reading and updating, readers may focus more on some types of information than on others, a process called foregrounding. When all sentences have been read, the integrated model is stored in long-term memory as the complete model. Later, the complete model or elements of it may be retrieved from memory in an attempt to remember what has been read (see Zwaan & Radvansky, 1998, for details of these processes).

Inconsistencies affect this chain of processes by interfering with updating of the situation model: If a sentence contradicts earlier ones, it becomes difficult or even impossible to update the current model and to create a consistent integrated model. Inconsistencies have been used in many studies to explore the different types of information represented in situation models. The general argument in these studies is the following: If readers represent a certain type of information in the situation model, they should exhibit comprehension difficulties upon encountering a sentence that is inconsistent with regard to this information. For instance, the current model might represent the protagonist of a narrative as *located inside* a building. If the protagonist's spatial location is indeed represented in the situation model, readers should find it difficult to integrate a sentence stating that the protagonist *went inside*. In fact, inconsistencies such as this one were used to demonstrate the representation of spatial information in situation models (e.g., O'Brien & Albrecht, 1992; de Vega, 1995). Other studies employing this paradigm revealed the representation of emotional information (e.g., Gernsbacher, Goldsmith & Robertson, 1992) character information (Albrecht & O'Brien, 1993; de Vega, Diaz & Leon, 1997), and temporal information (Rinck, Hähnel & Becker, 2001) in situation models created from narratives. In addition, inconsistencies contained in expository texts yielded similar comprehension difficulties (Lorch, Shannon, Lemberger, Ritchey & Johnston, 2000).

Most of the studies employing this *inconsistency paradigm* have not used eye tracking to study the processing of inconsistencies. Instead, reading times of single sentences were recorded. This may be illustrated by the following sample text taken from Rinck *et al.* (2001).

1. Today, Mark and Claudia would finally meet again.
2a. Mark's train arrived at Dresden Central Station 20 minutes *after* Claudia's train.
2b. Mark's train arrived at Dresden Central Station 20 minutes *before* Claudia's train.
3. Mark was very excited when his train stopped at the station on time.
4. He tried to think of what he should say when he met her.
5. Many people were crowding on the platform.
6. Claudia was already waiting for him when he got off the train with his huge bag.
7. They both were so happy.

Participants read this text (without the italics) one sentence at a time in a self-paced manner, and reading time of each sentence was recorded. Each participant read either Sentence 2a or Sentence 2b. Sentence 6 is consistent with Sentence 2a, but inconsistent with Sentence 2b: Claudia cannot be waiting for Mark, if he has arrived before her. Therefore, participants who have read 2b should find it more difficult to integrate Sentence 6 into their current situation model than participants who have read 2a. The critical dependent variable was reading time of the sixth sentence, and indeed, a strong *inconsistency effect* occurred: Reading time of Sentence 6 was reliably longer in the inconsistent version of the text (see Rinck *et al.*, 2001, for details).

There are a number of reasons why it may be advantageous to record eye movements within the inconsistency paradigm. First and foremost, eye tracking measures may yield additional information about the processing of inconsistencies; information not available from the "sentence-by-sentence reading" version of the paradigm. Just as topic processing, the processing of inconsistencies may be distorted by the presentation of single sentences because it prevents readers from looking back to previous sentences and forces them to rely on memory instead. Second, many studies have used the inconsistency paradigm, yielding large and reliable inconsistency effects on reading times for many different dimensions of situation models. Thus, we are dealing with an effect which is general as well as reliable. Third, the experimental manipulations employed in the inconsistency paradigm may be neatly controlled, excluding confounds of experimental conditions and materials. Thus, for each text it is possible to compare an inconsistent version to a consistent one, while the two versions are almost identical in wording. Finally, the most critical locations relevant for eye tracking measures are fairly obvious, namely the two sentences that contain the potentially conflicting pieces of information (Sentence 2 and Sentence 6 in the example above).

Potential "Areas of Interest" in Tracking Inconsistency Processing

Most of the texts employed to study inconsistencies were structured very similarly: A critical sentence appearing early in the text (e.g., the second sentence in the example above) explicitly states a critical piece of information. A number of sentences (usually at least three of them) follow, which are consistent with both versions of the critical sentence and with following the target sentence. The target sentence is presented next (e.g., as the sixth sentence in the example above). This sentence is either consistent with all previous information or it explicitly contradicts the information contained in the earlier critical sentence. Thus, the critical sentence and the target sentence are most crucial for processing of the inconsistency. For all participants, comprehension of the text should proceed smoothly up to the target sentence. As soon as the target sentence is presented, however, processing should differ. While readers of the consistent target sentence should not exhibit any comprehension problems, readers of the inconsistent sentence should find it difficult or even impossible to integrate the sentence into the current situation model. If they notice the inconsistency (not all of them do, see Rinck *et al.*, 2001, 2002), they should engage in repair processes in order to resolve the apparent inconsistency. Although the nature of these repair processes has not been investigated in detail, a number of possibilities seem plausible: First, readers might go on reading, hoping to find a resolution in the following sentences. Second, they might doubt the validity of their current situation model, wondering whether they misunderstood earlier parts of the text (e.g., the critical sentence). Third, they might disbelieve the target sentence rather than the current model, wondering whether the target sentence contains a mistake. Finally, they might try to solve the comprehension problem by elaborating possible resolutions for the apparent contradiction. Using sentence reading times, it is impossible to tell which of these repair processes is occurring because all of them are compatible with the increase in reading time observed for inconsistent target sentences compared to consistent ones.

Existing and Potential Measures of Inconsistency Processing

From the possible repair processes just outlined, a number of promising eye tracking measures may be derived. Using these, the goal is to tell how processing of inconsistent statements differs from processing of consistent ones, and what kind of repair processes are likely to occur during the processing of inconsistencies. In order to do so, Rinck *et al.* (2002) as well as Lorch and Lemberger (2001) recorded eye movements during the reading of narratives and expository texts, respectively. In both studies, complete texts were presented instead of single sentences, and the participants were free to look back and forth during reading. Also, sentences rather than words or phrases were used as the unit of analysis, similar to the suggestions made above for the study of topic processing. Some of the measures suggested in this chapter — although not all of them — may be illustrated by the experiments reported by Rinck *et al.* (2002). In this study, individual eye fixations within a single sentence were ignored, but individual fixation durations were summed up to yield regional gaze duration for each sentence. Eye movements from a sentence to another sentence presented earlier were analyzed as *lookbacks*, whereas *regressions*, i.e., backward eye movements within sentences, were not separately analyzed. Within these restrictions, Rinck *et al.* (2002) computed the following measures: *Regional gaze duration* of a sentence was defined in the standard way as consisting of all fixations within the sentence during the first reading of it, before moving on or moving back to a different sentence. The summed duration of lookback fixations comprised the *second-pass fixation time*. Finally, *total text fixation time* was computed by summing up all fixation times registered for the text (see Table 16.4).

Using these measures, it was possible to identify eye movement patterns associated with the processing of inconsistencies. For instance, with regard to temporal inconsistencies, Rinck *et al.* (2002) found that an inconsistent sentence does not seem to cause readers to fixate this sentence for a longer time due to the increase in processing difficulty. Thus, regional gaze durations and second-pass fixation times of the critical target sentences did not mirror the sentence reading times observed in earlier experiments. Rather, the inconsistency caused readers to look for information that might be used to check and resolve the inconsistency. As we hypothesized above for topic processing, it was also found for inconsistency processing that lookbacks did not always land directly on the critical second sentence. However, most of the lookbacks were indeed aimed at the second sentence where readers would find the conflicting piece of information. Thus, more lookbacks to the second sentence and longer second-pass fixation times of this sentence were observed in the inconsistent condition compared to the consistent one. Longer second-pass fixation times were also observed for the final sentence of the text, indicating that readers also checked this sentence for an explanation of the inconsistency.

Rinck *et al.* (2002) concluded that the inconsistency effect observed with single-sentence presentations may lack ecological validity: no increase in regional gaze duration of the target sentence was observed, indicating that the increase in sentence reading time is an artifact of the single-sentence presentation. If readers are able to see the complete text, they do not stay with the sentence containing the inconsistent information. Instead, they look back to related information earlier in the text or move on to the next sentence searching for an explanation. The single-sentence presentation does not allow for this type of visual search; it forces readers to perform a memory search instead. It seems that the commonly used sentence-by-sentence reading paradigm reliably indicates that inconsistencies do cause *some kind* of additional processing. Measures of eye movement patterns, however, may supply a much better picture of *what* these additional processes may be.

In addition to the fairly standard measures employed by Rinck *et al.* (2002), other measures may prove to be useful as well. In general, these measures are similar to those suggested for topic processing. There we suggested a measure of *extended first-pass fixation time* for which the first-pass reading is *not* terminated by a lookback sequence to a previous sentence provided that (a) a lookback sequence is initiated well before completing reading the sentence and (b) that the reader returns to the sentence to read the remaining part of it (see Table 16.4). Using this new measure, an effect of inconsistency may be established also for the target sentence, as the effect may be obscured in the regional gaze duration measure due to its premature termination by a lookback sequence initiated in the middle of the sentence. An apparent weakness in the new measure is that it obscures the information about first-pass reading consisting of two temporarily separate fixation sequences intervened by a lookback sequence. This weakness could be remedied by tagging these trials as such, which would then make possible their exclusion or inclusion in the analyses (see also Vonk & Cozijn, this volume). Moreover, *total lookback time* may be recorded for the text preceding the target sentence. This measure would yield the total time invested to solve the textual inconsistency. Thus, it is similar to rereading time and regression path measures used in the study of syntactic parsing (Liversedge *et al.*, 1988; Van Gompel *et al.*, 2001).

With texts consisting of a fairly limited number of sentences, it is also feasible to record all between-sentences movements and store their frequencies in an *eye movement matrix*, which shows how often readers moved their eyes from any possible starting sentence to any possible destination sentence (see Table 16.4), similar to the transition matrix suggested by Ponsoda, Scott and Findlay (1995). Again, within-sentence eye movements would be ignored in the eye movement matrix. For instance, looking back from the second sentence to the first sentence and making several fixations on the first sentence would be counted as one movement sequence. A sample matrix of this kind is shown in Table 16.2. This matrix contains hypothetical data that might be collected during reading of consistent texts made up of seven sentences: There are only a few lookbacks, they are distributed fairly evenly across all sentences, and most eye movements lead from one sentence to the following one. Moreover, the matrix contains a pattern of lookbacks that Rinck *et al.* (2002) did indeed observe: many lookbacks lead from the final sentence to the first sentence because participants tended to reread every text, in order to prepare for a comprehension question following each text.

In comparison, Table 16.3 contains hypothetical eye movement sequences that one might expect to observe during reading of inconsistent texts. In this matrix, there are more lookbacks, and particularly more lookbacks from the sixth and seventh sentence towards the critical second sentence. These eye movement matrices are particularly informative when used to compare two experimental conditions contained in short texts. With more conditions and more sentences, the matrices may become too many and too large, the number of movement sequences in each cell too low, and the observed patterns too complicated to allow statistically significant conclusions. Moreover, instead of simply counting the number of eye movement sequences (or their percentages) in each cell, it might be informative to include the *fixation durations* following

Table 16.2: Hypothetical eye movement matrix for consistent texts.

Starting sentence	**Destination sentence**						
	1	**2**	**3**	**4**	**5**	**6**	**7**
1	–	89	4	2	5	3	2
2	5	–	93	1	7	0	1
3	9	4	–	91	1	2	4
4	3	9	2	–	88	4	3
5	2	6	4	3	–	95	2
6	8	5	6	8	6	–	83
7	35	8	7	6	9	7	–

Table 16.3: Hypothetical eye movement matrix for inconsistent texts.

Starting sentence	Destination sentence						
	1	**2**	**3**	**4**	**5**	**6**	**7**
1	–	79	14	12	10	8	2
2	15	–	83	9	12	29	6
3	19	14	–	81	11	7	9
4	13	19	22	–	78	11	8
5	12	16	14	13	–	85	5
6	18	45	16	18	16	–	93
7	29	38	17	16	19	17	–

the movements as well. As argued above for topic processing, it may be important to distinguish which sentences are processed from how long they are being processed. In any case, the matrices may be most useful in generating hypotheses regarding typical eye movement patterns associated with different experimental conditions, if precise hypotheses have not been derived a priori from a theoretical basis. In doing so, later analyses may be restricted to the most important cells of the matrix, for instance, those involving the second and the sixth sentence of consistent versus inconsistent texts.

For the processing of inconsistencies, it is also possible to derive hypotheses regarding the most frequent *eye movement scan paths* (see Table 16.4). For consistent texts, these paths should follow the linear order of sentences "1–2–3–4–5–6–7", possibly repeated once in preparation for a comprehension question. For inconsistent texts, paths including lookbacks to the second sentence should be more frequent, for instance "1–2–3–4–5–6–2" or "1–2–3–4–5–6–7–2". This pattern of results was indeed observed by Rinck *et al.* (2002), although the analyses were not powerful enough to yield statistically significant differences. With more powerful designs, it may be possible to show that some prototypical paths occur more often in one experimental condition than in another. These analyses would add further information to the results obtainable with eye movement matrices.

Finally, all eye movement measures of inconsistency processing discussed so far are based on sentences as the unit of analysis. This is convenient and useful, but sometimes it may be more appropriate to study smaller units of inconsistent texts. For instance, the target sentence "*Claudia was already waiting for him when he got off the train with his huge bag*" shown above could be rewritten to "*When he got off the train, Claudia was already waiting for him*". For this sentence, only the second clause is consistent or inconsistent with earlier information. Therefore, regional gaze durations and second-pass fixation times might be restricted to the second clause instead of the complete sentence. An even narrower focus may be applied to texts that have been used to investigate emotional inconsistencies. With single-sentence presentation, emotional inconsistencies have yielded extremely large inconsistency effects. The materials used in these studies do not contain any explicit statement of the protagonist's emotional reactions. Instead, readers have to infer the reactions from the contents of the story, which they reliably do (e.g., Gernsbacher *et al.*, 1992). For instance, a story may describe how the protagonist commits a crime, for which a close friend is blamed and punished. The target sentence then states that the protagonist "felt extremely guilty" versus "felt extremely proud". Thus, the consistent target sentence and the inconsistent one differ by only one word, rendering fixations of this target word and eye movements originating from it particularly interesting. On the other hand, there is no explicit statement that is being contradicted by the target word. Therefore, it will be particularly challenging to identify the eye movement patterns associated with processing of emotional inconsistencies.

Summary

In the present chapter, our aims have been twofold. First, we have tried to demonstrate the potential applicability of eye tracking to study global text processing. Despite its apparent usefulness, the method has been applied only infrequently for this purpose. One reason for the paucity of such studies is probably the fact that in the existing literature, no standards have been developed for specific measures to be used. Thus, our second goal in the present chapter has been to discuss the potential usefulness of existing eye movement measures and to suggest new measures capable of reflecting different aspects of global text processing. The most important measures discussed in this chapter are shown in Table 16.4. To sum up our discussion, we list in the following the measures that we believe hold the most promise for the study of global text processing. The default unit of analysis is here assumed to be the sentence, although the measures may also be applied to other text units.

Existing Measures

1. ***Regional gaze duration***: Summed duration of fixations made on a text region before exiting it.
2. ***Lookback fixation time*** (or second-pass fixation time): Summed duration of fixations that return to a text region after its first-pass reading.

New Measures

3. ***Extended first-pass fixation time***: It differs from the conventional regional gaze duration in allowing fixations to previous text regions in the middle of first-pass reading without terminating the first-pass reading, as long as the reader returns to the target region to complete reading it. This would mean that the reader makes a series of fixations that land further into the sentence than any of the fixations prior to backtracking (see Hyönä & Juntunen, submitted).

Table 16.4: Eye movement measures described in this chapter.

Measure	**Standard or common unit**	**Timing of effects**	**Definition**
First fixation duration	Word	Immediate	Duration of first fixation on the target word
Gaze duration	Word	Immediate	Summed duration of all fixations on the target word before exiting it
First fixation duration after leaving	Word	Delayed	Duration of first fixation after leaving the target word
Regression	Word	Delayed	Fixation of previously processed target word, usually associated with "backward" eye movement
Regression time	Word	Delayed	Duration of all regressions back to the target word
Total fixation time	Word	Delayed	Sum of gaze duration and regression time
Regional gaze duration (First-pass fixation time)	Region (word, phrase, clause, sentence)	Immediate	Summed duration of all fixations on the target region before exiting it
Lookback fixation time (Second-pass fixation time)	Region	Delayed	Duration of all regressions back to the target region
Regression path reading time	Region	Delayed	Summed duration of all reinspective fixations before exiting target region to the right
First-pass rereading time	Sentence	Delayed	Summed duration of all reinspective fixations on the target sentence during its first-pass reading
Lookback	Sentence	Delayed	Any fixation on text prior to the most recently fixated target sentence, including backward and forward fixations as long as they do not return to the target sentence
Lookback time	Sentence	Delayed	Duration of lookbacks

Table 16.4: continued.

Measure	**Standard or common unit**	**Timing of effects**	**Definition**
Extended first-pass fixation time	Sentence	Immediate and delayed	Sum of first-pass fixation time and additional fixation times on target sentence, if (a) lookbacks occur before completing the target sentence and (b) eyes return to remaining part of sentence before fixating later sentences
Total text fixation time	Sentence	Immediate and delayed	Sum of all fixations on complete text
Eye movement matrix	Sentence	Immediate and delayed	Contingency table containing the frequencies or durations of all between-sentence movements, from any starting sentence to any destination sentence
Scan path sequence	Sentence	Immediate and delayed	Frequency of a particular sequence, in which the sentences are fixated.

Note: The terms "time" and "duration" have been used synonymously in the literature.

4. ***First-pass rereading time***: Summed duration of all reinspective fixations landing on the target region during its first-pass reading. By definition, reinspections are fixations that land on a subregion (i.e., word) that has already been fixated. A reinspective cycle is initiated by a regression, but subsequent fixations may be either regressive or progressive (see Hyönä *et al.*, 2002).
5. ***Eye movement matrices***: This measure is applied to the analysis of eye movement sequences between text regions. For example, going back from region *N* to region *N*–3 and making a series on second-pass fixations on region *N*–3 is considered a lookback sequence, whose origin is region *N* and destination region *N*–3. Both the frequency of between-region movements (see Tables 16.2 and 16.3) and the summed duration of fixations constituting the movements may be analyzed.

References

Albrecht, J. E., & O'Brien, E. J. (1993). Updating a mental model: Maintaining both local and global coherence. *Journal of Experimental Psychology: Learning, Memory, and Cognition, 19,* 1061–1070.

Bertram, R., Hyönä, J., & Laine, M. (2000). The role of context in morphological processing: Evidence from Finnish. *Language and Cognitive Processes, 15,* 367–388.

Blanchard, H.E, & Iran-Nejad, A. (1987). Comprehension processes and eye movement patterns in the reading of surprise-ending stories. *Discourse Processes, 10,* 127–138.

de Vega, M. (1995). Backward updating of mental models during continuous reading of narratives. *Journal of Experimental Psychology: Learning, Memory, and Cognition, 21,* 373–385.

de Vega, M., Diaz, J. M., & Leon, I. (1997). To know or not to know: Comprehending protagonists' beliefs and their emotional consequences. *Discourse Processes, 23,* 169–192.

Gernsbacher, M. A. (1990). *Language Comprehension as Structure Building.* Hillsdale, NJ: Erlbaum.

Gernsbacher, M. A., Goldsmith, H. H., & Robertson, R. R. W. (1992). Do readers mentally represent characters' emotional states? *Cognition and Emotion, 6,* 89–111.

Hyönä, J. (1994). Processing of topic shifts by adults and children. *Reading Research Quarterly, 29,* 76–90.

Hyönä, J. (1995). An eye movement analysis of topic-shift effect during repeated reading. *Journal of Experimental Psychology: Learning, Memory, and Cognition, 21,* 1365–1373.

Hyönä, J., & Juntunen, M. Irrelevant speech effects on sentence processing during reading. Submitted for publication.

Hyönä, J., Lorch Jr, R. F., & Kaakinen, J. (2002). Individual differences in reading to summarize expository text: Evidence from eye fixation patterns. *Journal of Educational Psychology, 94,* 44–55.

Hyönä, J., & Pollatsek, A. (1998). Reading Finnish compound words: Eye fixations are affected by component morphemes. *Journal of Experimental Psychology: Human Perception and Performance, 24,* 1612–1627.

Inhoff, A. W., & Radach, R. (1998). Definition and computation of oculomotor measures in the study of cognitive processes. In: G. Underwood (ed.), *Eye Guidance in Reading and Scene Perception* (pp. 29–53). Oxford: Elsevier.

Just, M. A., & Carpenter, P. A. (1980). A theory of reading: From eye fixations to comprehension. *Psychological Review, 87,* 29–54.

Kaakinen, J., Hyönä, J., & Keenan, J. M. (2002). Perspective effects on on-line text processing. *Discourse Processes, 33,* 159–173.

Kieras, D. E. (1980). Initial mention as a signal to thematic content in technical passages. *Memory & Cognition, 8,* 345–353.

Kieras, D. E. (1981). Component processes in the comprehension of simple prose. *Journal of Verbal Learning and Verbal Behavior, 20,* 1–23.

Kintsch, W. (1988). The role of knowledge in discourse comprehension: A construction-integration model. *Psychological Review, 95,* 163–182.

Kintsch, W. (1998). *Comprehension.* New York: Cambridge University Press.

Klusewitz, M. A., & Lorch Jr, R. F. (2000). Effects of headings and familiarity with a text on strategies for searching text. *Memory & Cognition, 28,* 667–676.

Konieczny, L. (1996). *Human sentence processing: A semantics-oriented parsing approach.* Freiburg: University of Freiburg, IIG-Berichte 3.

Liversedge, S. P., Paterson, K. B., & Pickering, M. J. (1998). Eye movements and measures of reading time. In: G. Underwood (ed.), *Eye Guidance in Reading and Scene Perception* (pp. 55–76). Oxford: Elsevier.

Lorch Jr, R. F., & Lemberger, C. (2001). Detecting inconsistencies while reading. Poster presented at the 11th European Conference on Eye Movements, Turku, Finland.

Lorch Jr, R. F., Lorch E. P., & Matthews, P. D. (1985). On-line processing of the topic structure of a text. *Journal of Memory and Language, 24,* 350–362.

Lorch Jr, R. F., Lorch E. P., & Mogan, A. M. (1987). Task effects and individual differences in on-line processing of the topic structure of a text. *Discourse Processes, 10,* 63–80.

Lorch Jr, R. F., Shannon, L., Lemberger, C., Ritchey, K., & Johnston, G. S. (2000). Accessing information during reading. Paper presented at the 41st meeting of the Psychonomic Society, St. Louis.

Murray, W. S. (2000). Sentence processing: Issues and measures. In: A. Kennedy, R. Radach, D. Heller and J. Pynte (eds), *Reading as a Perceptual Process* (pp. 649–664). Oxford: Elsevier.

Myers, J. L., & O'Brien, E. J. (1998). Accessing the discourse representation during reading. *Discourse Processes, 26,* 131–157.

Myers, J. L., O'Brien, E. J., Albrecht, J. E., & Mason, R. A. (1994). Maintaining global coherence during reading. *Journal of Experimental Psychology: Learning, Memory, and Cognition, 20,* 876–886.

O'Brien, E. J., & Albrecht, J. E. (1992). Comprehension strategies in the development of a mental model. *Journal of Experimental Psychology: Learning, Memory, and Cognition, 18,* 777–784.

O'Brien, E. J., & Myers, J. L. (1987). The role of causal connections in the retrieval of text. *Memory & Cognition, 15,* 419–427.

Pollatsek, A. Hyönä, J., & Bertram, R. (2000). The role of morphological constituents in reading Finnish compound words. *Journal of Experimental Psychology: Human Perception and Performance, 26,* 820–833.

Ponsoda, V., Scott, D., & Findlay, J. M. (1995). A probability vector and transition matrix analysis of eye movements during visual search. *Acta Psychologica, 88,* 167–185.

Rayner, K. (1998). Eye movements in reading and information processing: 20 years of research. *Psychological Bulletin, 124,* 372–422.

Rayner, K., Raney, G. E., & Pollatsek, A. (1995). Eye movements and discourse processing. In: R. F. Lorch Jr and E. J. O'Brien (eds), *Sources of Coherence in Reading* (pp. 9–36). Hillsdale, NJ: Erlbaum.

Rinck, M., Gámez, E., Díaz, J. M., & de Vega, M. (2002). *Processing of Temporal Information: Evidence from Eye Movements.* Manuscript submitted.

Rinck, M., Hähnel, A., & Becker, G. (2001). Using temporal information to construct, update, and retrieve situation models of narratives. *Journal of Experimental Psychology: Learning, Memory, and Cognition, 27,* 67–80.

van den Broek, P., & Lorch Jr, R. F. (1993). Network representations of causal relations in memory for narrative texts: Evidence from primed recognition. *Discourse Processes, 16,* 75–98.

van Dijk, T. A., & Kintsch, W. (1983). *Strategies of Discourse Comprehension.* New York: Academic Press.

Van Gompel, R. P. G., Pickering, M. J., & Traxler, W. J. (2001). Reanalysis in sentence processing: Evidence against current constraint-based and two-stage models. *Journal of Memory and Language, 45,* 225–258.

Vauras, M., Hyönä, J., & Niemi, P. (1992). Comprehending coherent and incoherent texts: Evidence from eye movement patterns and recall performance. *Journal of Research in Reading, 15,* 39–54.

Zwaan, R. A., Langston, M. C., & Graesser, A. C. (1995). The construction of situation models in narrative comprehension: An event-indexing model. *Psychological Science, 6,* 292–297.

Zwaan, R. A., & Radvansky, G. A. (1998). Situation models in language comprehension and memory. *Psychological Bulletin, 123,* 162–185.

Commentary on Section 2

Advancing the Methodological Middle Ground

Albrecht W. Inhoff and Ulrich Weger

We argue that the oculomotor approach occupies a methodological middle ground on which a relatively high degree of experimental control is achieved under relatively natural task conditions. The complexity of the oculomotor response creates measurement issues and measurement opportunities, as a multitude of potential measures can often be construed. We advocate the computation of multiple measures in which different types of temporal and spatial contiguity are used for the principled computation of task-specific measures. These measures need to be empirically grounded until standards are established that should be anchored in a theory of oculomotor control.

Introduction

The chapters on empirical studies of reading and language production demonstrate the effective use of eye movement-based methods in the study of a strikingly broad range of domains. The chapters examine eye movement guidance during the reading of Chinese text (Tsai & McConkie, this volume), perceptual processes underlying music reading (Gilman & Underwood, this volume), trans-saccadic visual word recognition (Rayner, White, Kambe, Miller & Liversedge, this volume), the specification of novel word meaning (Morris & Williams, this volume), spatial memory for word locations (Kennedy, Brooks, Flynn & Prophet, this volume), thematic and global text processing (Hyönä, Lorch & Rick, this volume; Liversedge, this volume), and cognitive processes underlying speech production (Meyer & Dobel, this volume).

In spite of their heterogeneity, these chapters also reveal remarkable commonalities. All examine performance under relatively complex and relatively natural task conditions, and the experimenter obtains a continuous record of on-line information use while the task is performed. Together, the chapters also reveal two methodological

The Mind's Eye: Cognitive and Applied Aspects of Eye Movement Research

ISBN: 0–444–51020–6

trends: the development of new paradigms, and the evolution of new oculomotor measures. In the following, we will use these commonalities and trends to discuss what we will call a "methodological middle ground".

A Methodological Middle Ground

One methodological approach by which our knowledge of cognitive processes has advanced employs the use of highly controlled experimental situations in which subjects are exposed to relatively simple stimuli. Like a chemist, who seeks to remove all impurities from a to-be-examined substance, the experimenter seeks to create impoverished and highly controlled experimental stimuli so that pure effects of information usage can be distilled.

Critically, this approach also requires that the experimenter removes all impurities from the performance measure which generally consists of an overt response that is extraneous to the experimental task. The ideal overt response is simple, stereotypic, and completely prescribed. It should be simple (and singular) to minimize variability in response planning and response execution; it should be stereotypic, so that the same set of motor commands and movement parameters is used throughout a sequence of trials, and it should be prescribed, meaning that the subject knows exactly whether, how, and when to execute the to-be-measured response, so that the linkage between the experimental manipulation and the response is self-evident.

Although this approach has a long tradition, going back to Wundt and the structuralists, and although it has yielded valuable insights into the constituents of complex performance, it has found its detractors. Impoverished laboratory conditions may lack ecological validity and can induce situation-specific processing strategies. That is, experimental conditions may induce subjects to use information that is not sought under typical viewing conditions or, conversely, experimental conditions could discourage the use of information that is sought under more natural conditions. Recently, O'Regan and Noe (in press) have argued that cognitive and perceptual processes are designed to determine and exploit sensori-motor contingencies. If this were the case, then contingency-deprived laboratory conditions with highly impoverished stimuli and responses are, by definition, ill suited for the study of complex cognitive processes.

At the other methodological extreme, tight experimental control is abandoned and the experimenter exposes the subject to natural variation and records concomitant responses. In one study of this type, Dishard and Land (1998) examined the use of visual cues by novice and expert drivers while they steered vehicles around curves on rural roads. Eye movements, which are part and parcel of the driving task, were measured and used to index drivers' use of visual cues. The results were intriguing: experts sought information from a relatively distant tangent point, to predict curvature, and from closer to the car's current position, to maintain the position in the lane. Novice drivers used more variable and less effective information acquisition strategies.

Although such field studies go beyond laboratory studies, in that they not only show that information *can* be used but also that it *is* used, they are ill suited for some purposes. Selected natural variation is often quite limited, and critical variables are often subject to confounds, making this approach ill suited for theory testing. Moreover, responses are unconstrained. They can be complex and exceedingly variable which makes it difficult to isolate those aspects of the response pattern that are influenced by the examined natural variation. Consequently, evidence from field studies often needs to be accompanied by laboratory testing. In Dishard and Land's case, this involved driving in an off-road simulator.

The chapters of the current section reveal yet another approach to the study of cognitive processes. This approach combines key elements of both experimental extremes, thus occupying what we call "a methodological middle ground". On this middle ground, the subject deals with a relatively complex and often quite natural visual environment, and the experimenter maintains a relatively high level of experimental control. In the current chapters, experimental stimuli consisted of "natural" lines of notes in the study on music reading, of lines of text in studies examining word recognition, saccade programming, memory for word locations, and text comprehension, and stimuli consisted of relatively complex multi-object displays in the experiments on speech production.

Nevertheless, this experimental middle ground offers considerable experimental control. The experimenter defines the experimental task and maintains full control over the properties and composition of the visual environment. For instance, in the chapters on music and text reading, the experimenter determines the musical, linguistic and spatial properties of to-be-studied items, and in Meyer and Dobel's (this volume) study of speech production, the experimenter defines the experimental task and maintains full control over picture content.

Critically, complex relevant responses can be executed on this middle ground. Even the execution of a single saccade involves the specification of multiple parameters, movement direction and of movement size (landing position), and a subsequent fixation can, consequently, follow saccades of a wide range of sizes and directions. Generally, the subject decides not only when and how to move the eyes but also how often to move them. In spite of the potential complexity of the response, relevant components can be discerned, as the experimenter defines the area (or areas) of interest and retains control over aspects of the movement pattern, e.g., the location of movement onset and offset.

Yet, some experimental control is lost on the middle ground. The onset and offset of the processing of an item is often difficult to establish. When subjects can determine whether, when, where, and how often to look, a particular item within the complex display may not be selected for processing. It may not be looked at and if it is, the experimenter has little control over its temporal visibility and over its exact projection on the retina. When multiple fixations are executed, the experimenter loses some control over the order of information acquisition.

Furthermore, measurement issues arise at various levels (see Inhoff & Radach, 1998). Only a subset of a sequence of executed responses may pertain to the experimental manipulation and, when this is the case, an "effective" subset of eye movements must be selected. Various subsets can be defined, and a plethora of possible oculomotor measures can be devised especially when the stimulus environment and the corresponding eye movement pattern are relatively complex.

Some of the loss of experimental control can be recouped. Several of the current chapters use eye movement contingent display changes to control the spatial visibility of information and the time course of information acquisition. Other chapters offer novel principled solutions for the computation of more effective oculomotor measures.

Expanding Experimental Control

Post-hoc analyses can be conducted to examine effects that are difficult to control. For instance, effects of viewing location, or of the number of critical item fixations, or of a particular fixation sequence, can be determined by examining effects of an experimental manipulation as a function of (initial) landing position, of the number of fixations on the item, or of the order of fixations with which the item was examined.

Eye movement contingent display techniques offer another more direct means of spatial and temporal control by yoking a specific spatial or temporal manipulation of the visual stimulus to a particular oculomotor event. In McConkie and Rayner's (1975) moving window technique, the experimenter defines a spatial window within which useful visual information is available and outside of which some or all useful information is denied. The window moves in synchrony with the eyes, so that the eyes assume a dual role: they implement the spatial window manipulation and they provide an on-line measure of window-size and window-type effects. Gilman and Underwood (current volume) use this technique to show that skilled pianists seek useful visual information from up to two beats ahead of a current fixation.

Other eye movement contingent display methods have been devised to study the time course and nature of visual information use. Rayner's (1975) boundary technique involves the change of information at a critical location when the eyes cross a predetermined spatial location. The relationship between the information available at the critical location before and after boundary crossing is manipulated, and the effect of this manipulation on critical area viewing is measured. Boundary studies have shown that readers obtain several types of useful information from a (parafoveally visible) word before it is fixated (see Rayner, 1998, for a review), the key issue being whether useful semantic information can be obtained from the parafovea prior to a word's fixation. Rayner *et al.* (this volume), conclude that this is not the case.

Meyer and Dobel (this volume) use a variant of the boundary technique to examine speech production in serial object naming. Interestingly, their results revealed benefits from token previews, e.g., parafoveal preview of one type of chair facilitated the naming of a different type of chair when it was subsequently fixated. Yet, the meaning of the parafoveal object may not have been activated, as preview of a semantically related object, e.g., a table, did not yield a benefit, which is consistent with Rayner *et al.*'s key contention.

Eye movement contingent display changes can also be used to control the temporal availability of specific types of visual — and nonvisual — information. Sereno and Rayner (1992) used a boundary technique to mask a target word within a sentence with noninformative letters until the eyes moved across a pre-target boundary. This initiated two contingent display changes: an immediate change, that replaced the letter string with a prime, and a subsequent change after a pre-determined prime duration, that replaced the prime with the target (which then remained visible during the remainder of the fixation).

Recently, we (Inhoff, Connine & Radach, in press) used a novel eye movement contingent technique to manipulate nonvisual information. In our technique, a movement of the eyes across a predefined spatial sentence boundary initiated the presentation of a spoken word, and effects of the spoken word on sentence reading were examined. In one instantiation, readers heard a spoken word that was identical to, phonologically similar, or phonologically dissimilar to a visual target word upon its fixation. Eye movements during target and post-target reading revealed robust interference from a similar or dissimilar spoken word during target reading. Moreover, similar spoken words — but not dissimilar spoken words — hampered post-target reading presumably because the similar spoken word interfered with the visual target's representation in (phonological) working memory.

Other eye movement contingent techniques can be devised to bring the rigor of traditional experimental paradigms to the methodological middle ground. Often results of the two approaches converge. Consistent to the results of "impoverished" single word classification studies, the fast-priming technique has shown that readers can use phonological information prior to semantic information during word recognition. The fast-priming technique further showed that readers do so under relatively natural task conditions.[1]

The Selection and Aggregation of Oculomotor Measures

When a relatively complex visual display is examined with a sequence of eye movements, as occurs on the methodological middle ground, the experimenter must select one or several "pertinent" oculomotor responses to index the effect(s) of the experimental manipulation. Although all principled selections use the spatial location of a fixation and often its temporal contiguity to determine its eligibility, selection issues arise because different spatial and/or spatio-temporal selection criteria can often be adopted. When more than one eligible eye movement occurs, the experimenter also needs to determine whether and how the sequence of eye movements is aggregated into a composite measure. Ideally, a processing hypotheses should define to-be-measured spatial unit(s), e.g., a word, phrase, sentence, or supra-sentence unit in studies of reading, and the selection and aggregation of eye movements to index the cognitive processing of these units should be prescribed by a theory of oculomotor control that specifies how a particular type of process can be expressed in a sequence of eye movements. Since no prescriptive standards have been established, the selection and aggregation of oculomotor events is largely determined by the experimenter.

Selection and aggregation can be straightforward. Kennedy *et al.* (this volume) were interested in readers' knowledge of the spatial location of previously identified words. Hence, they measured individual saccades (regressions) to a previously inspected critical word location to index a reader's memory for that location. To determine whether readers of Chinese directed the eyes to character or word units, Tsai and McConkie (this volume) measured the landing locations of forward directed saccades.

Selection and aggregation issues are more prominent when effects of the experimental manipulation can be distributed over multiple eye movements and when processing time is of primary interest, as is the case in most of the studies of this section. The eligibility of an oculomotor event for selection is generally determined by its spatial relationship to a pertinent location or area. Typically, this comprises the area of a spatially defined unit, e.g. a word, phrase, or picture segment. Eligibility can also be determined by spatial referent points, e.g., fixations to the right or left of a critical location. Aggregation is generally accomplished by one of two principles, temporal contiguity and/or spatial patterning. By temporal contiguity, we mean that a successive sequence of eligible oculomotor events is included in the measure until the spatial eligibility criteria are no longer met. That is, an oculomotor event is included in the measure only if the immediately preceding oculomotor event is also included. By spatial patterning we mean that an oculomotor event is included only when it has particular directional properties.

In the experiments on vocabulary acquisition (Morris & Williams, this volume), an oculomotor event is considered eligible when it falls on a particular area, the manipulated target unit or the next word in the text, or when it has a particular spatial reference point, i.e., when it departs to prior text from the target word, and when it returns to the target unit after post-target text has been read. A nonaggregated and an aggregated viewing duration measure is computed for target and post-target units. The nonaggregated measure consists of the duration of the first fixation on the word and the aggregated, temporally contiguous measure consists of the cumulated viewing durations of all fixations on the word until another word is fixated (this measure is also referred to as gaze duration or as first pass reading time). Together the different measures provide a coherent and compelling account of information acquisition when an unknown "word" is encountered.

Yet, this relatively comprehensive set of oculomotor measures is by no means exhaustive. Other oculomotor events might be considered eligible, e.g., the saccade to the target unit and/or to the post-target unit. Furthermore, the pertinent area could be expanded and other principled means of aggregation could be devised. Computations of first fixation durations and gaze durations generally do not distinguish between instances in which a word is read only once and instances in which it is subsequently reread. Yet, a case could be made that first fixations and gaze durations should yield a more accurate word recognition measure when only those instances are considered in which *all* fixations on the target and post-target word were temporally contiguous, i.e., when no subsequent re-reading occurred. Furthermore, it might well be possible that the gaze measure could be further improved if the spatial pattern of the saccade sequence was considered. (see Vonk & Cozjin, this volume).

The comparison of effects for different types of aggregation is critical to Liversedge's (this volume) studies that seek to determine processing preferences for thematic assignments. A noncontinguous measure, total viewing duration, that aggregates first pass viewing duration on a pertinent segment of text and the time spent re-fixating it after nontarget text has been viewed, is contrasted with a temporally contiguous measure, first pass viewing durations. Differential effects of thematic assignment on the two measures are used to discern processing preferences. Similar experimental effects for first pass and total viewing durations are considered the hallmark of preferred assignments; the signature of a nonpreferred assignment, by contrast, is a selective increase for total viewing durations over first pass viewing durations, as the change from the preferred to the nonpreferred assignment tends to be accompanied by a subsequent (discontinuous) reinspection of the pertinent segment.

Vonk and Cozjin's (this volume) and Hyönä *et al.*'s (this volume) chapters explicitly deal with the aggregation issue. Vonk & Cozjin's analyses show that first pass viewing durations are slightly more sensitive to the experimental manipulation when the time spent moving the eyes is included in the measure than when it is excluded (as is commonly the case). More important, they also show that consideration of the spatial patterning of a single saccade during first pass direction, the saccade that leaves the target area, can substantially improve the quality of the first pass measure. For the reported materials, effect sizes were larger and more robust when only those trials were included in the computation of first pass reading in which the pertinent area was exited with a forward directed saccade.

Hyönä *et al.* advocate three novel types of aggregations for the study of global text processing. A noncontinuous first pass fixation time measure that is similar to extant contiguous first pass measures in that it aggregates temporally contiguous fixation on a pertinent area until the eyes exit the area. It is also dissimilar to extant first pass measures in that aggregates all first pass viewing times on different segments of the pertinent area even when the viewing of different segments of the area is disrupted by the viewing of nonpertinent text. The second measure, the time spent re-reading different parts of a pertinent area after other text has been inspected, also involves aggregation of temporally noncontiguous fixations. Their third measure, origin and destination of eye movement sequences, is a novel temporally continuous measure that cumulates the time spent going back from one region to another region.

The chapters of the current section report almost half a dozen different types of first pass viewing duration measures, and a virtual plethora of other aggregated and nonaggregated measures is conceivable. What measure — or set of measures — should be computed? Ideally, a theory of oculomotor control should explicate when and how effects of a particular experimental manipulation are expressed during a sequence of eye movements. Experimental measures also need to be empirically grounded, Vonk and Cozjin's (this volume) work being an example of this type of work, and theoretical specification and empirical grounding should mutually reinforce each other. Clearly, the choice of a particular eligibility criterion and aggregation type should not simply be determined by the resulting effect size.

As part of this endeavor, we propose, first, that at least one theoretically motivated standard measure be adopted, second, that other principled types of selection and aggregation be used to compute additional oculomotor measures and, third, that effect sizes be compared across measures. Adoption of a standard measure is desirable so that oculomotor effects can be compared across studies. In view of their theoretical anchoring and prominence, we suggest that gaze durations should generally assume the role of measurement standard, perhaps in conjunction with a movement pattern criterion, until theoretical advances recommend a different standard. Computation of additional principled measures should be used to fine-tune the oculomotor measure to task demands and to experimental manipulations, and the discovery of a novel effective measure should be as newsworthy as the establishment of an experimental effect. Rather than being a disadvantage, different and novel means of selection and aggregation provide unique measurement opportunities that dramatically increase the analytic power of the methodological middle ground. Comparisons of effect sizes for different types of principled selection and aggregation may not only establish an experimental effect but also reveal its time course and its fine-grained nature.

Over the past quarter of a century, the oculomotor approach has rapidly evolved and expanded (see Rayner 1978, 1998 for reviews). Theoretical and methodological advancements continue to emerge that should add prominence to what we have called a methodological middle ground.

Acknowledgements

This work was supported by NSF grant BCS0002024. We would like to thank Brianna Eiter for her helpful comments on the paper. Correspondence should be addressed to Albrecht Inhoff, Department of Psychology, State University of New York at Binghamton, Binghamton, NY, 13902–6000, USA, or per email to inhoff@binghamton.edu.

Note

1 Display change techniques require careful control conditions, especially when the visual display change is implemented after fixation onset. These display changes appear to lead to a time-locked inhibition of a programmed saccade. Without proper controls, the oculomotor inhibition may be mistaken for an interruption of linguistic processes (Reingold & Stampe, 2000).

References

Dishard, D. C., & Land M. F. (1998). The development of eye movement strategies of learner drivers. In: G. Underwood (ed.), *Eye Guidance in Reading and Scene Perception* (pp. 419–430). Oxford: Elsevier Science Ltd.

Inhoff, A. W., & Radach, R. (1999). Definition and computation of oculomotor measures in the study of cognitive processes. In G. Underwood (ed.), *Eye Guidance in Reading and Scene Perception* (pp. 29–54). Oxford Elsevier Science Ltd.

Inhoff, A. W., Connine, C., & Radach, R. (in press). A contingent speech technique in eye movement research on reading. *Behavior Research Methods Instruments & Computers*.

McConkie, G. W., & Rayner, K. (1975). The span of the effective stimulus during a fixation in reading. *Perception & Psychophysics, 17,* 578–586.

O'Regan, J. K., & Noe, A. (2001). A sensori-motor account of vision and consciousness. *Brain and Behavioral Sciences, 24,* 930–1031.

Rayner, K. (1975). The perceptual span and peripheral cues in reading. *Cognitive Psychology, 7,* 65–81.

Rayner, K. (1978). Eye movements in reading and information processing. *Psychological Bulletin, 85,* 618–660.

Rayner, K. (1998). Eye movements in reading and information processing: 20 years of research. *Psychological Bulletin, 124,* 372–422.

Reingold, E. M., & Stampe, D. M. (2000). Saccadic inhibition and gaze contingent research paradigms. In: A. Kennedy, R. Radach, D. Heller and J. Pynte (eds), *Reading as a Perceptual Process* (pp. 119–145). Oxford: Elsevier Science Ltd.

Sereno, S. C., & Rayner, K. (1992). Fast priming during eye fixations in reading. *Journal of Experimental Psychology: Human Perception and Performance, 18,* 173–184.

Section 3

Computational Models of Eye Movement Control in Reading

Section Editor: Ralph Radach

Chapter 17

Using the Saccadic Inhibition Paradigm to Investigate Saccadic Control in Reading

Eyal M. Reingold and Dave M. Stampe

In two experiments, during reading, 33 ms flickers occurred to the left or the right of the point of gaze at random intervals. We documented a decrease in saccadic frequency following the onset of this task-irrelevant flicker. This effect was referred to as saccadic inhibition. It was found that the saccadic inhibition effect differed depending on whether the flickers were congruent or incongruent with the direction of the next saccade. Specifically, a large flicker produced a stronger inhibition in the congruent than the incongruent condition, whereas the reverse was true for a small flicker. The implications of these findings for models of saccadic control in reading are discussed.

Introduction

Models of saccadic control in reading differ dramatically with respect to the hypothesized role of visuo-spatial attention in the programming and execution of eye movements (see Rayner, 1998 for a review). The attentional guidance model postulates tight coupling between attention and saccadic control in reading. An early version of this model, proposed by Morrison (1984; see also Just & Carpenter, 1980), argued that an attention shift in the direction of the next saccade occurs prior to its execution. Specifically, this model assumes that attention is initially centered on the foveated word (word *N*) during fixation. Following lexical encoding of the fixated word, attention covertly shifts in the direction of reading and a saccade aimed at fixating the newly attended word (word $N+1$) is programmed. If parafoveal lexical encoding of word $N+1$ is completed prior to the execution of the next saccade, attention further shifts to word $N+2$ and a new saccade aimed at fixating word $N+2$ is programmed. In such

The Mind's Eye: Cognitive and Applied Aspects of Eye Movement Research

ISBN: 0–444–51020–6

instances either word $N+1$ is skipped (i.e., the N to $N+1$ saccade is canceled) or, if the point of no return is exceeded, word $N+1$ is only briefly fixated on route to fixating word $N+2$ (see Rayner & Pollatsek, 1989 for a review and discussion). Morrison's (1984) model was very influential and several modified versions aimed at extending it were proposed (e.g., Henderson, 1992; Henderson & Ferreira, 1990; Kennison & Clifton, 1995; Pollatsek & Rayner, 1990; Rayner & Pollatsek, 1989; Reichle, Pollatsek, Fisher & Rayner, 1998). In addition, alternative models were formulated, which assumed that nonlexical, low-level information determines saccadic control in reading (e.g., Kowler & Anton, 1987; McConkie, Kerr, Reddix & Zola, 1988; McConkie, Kerr, Reddix, Zola & Jacobs, 1989; O'Regan, 1990, 1992). Such models argued that the influence of higher-level cognitive or attentional processes on saccadic control in reading is very limited. Currently, the issue of the relationship between attention and saccadic control in reading remains controversial (e.g., Brysbaert & Vitu, 1998; Deubel, O'Regan & Radach, 2000; Kliegl & Engbert, 2003; McConkie & Yang, 2003; Pollatsek, Reichle & Rayner, 2003; Rayner, 1998; Reilly & Radach, 2003; Vitu, O'Regan, Inhoff & Topolski, 1995; Vitu & O'Regan, 1995).

One clear prediction that can be derived from the attentional guidance model is the occurrence of perceptual enhancement in the direction of the next saccade, due to preallocation of attention to the target of the saccade. In order to empirically test this prediction, Fischer (1999) introduced the dynamic orienting paradigm. During reading, a probe (an asterisk appearing just above a line of text) was presented with a delay of either 25 ms (early probe conditions) or 170 ms (late probe conditions) following the onset of a randomly selected fixation (see Panel A of Figure 17.1). Participants were required to indicate probe detection with a speeded manual response. Reading comprehension and probe detection were designated as primary and secondary tasks respectively in this dual task paradigm. The spatial location of the probe was either 5 or 10 characters to the left (incongruent conditions) or the right (congruent conditions) of the participants' gaze position, or directly above the fixated character. Based on the attentional guidance model, Fischer (1999) predicted that if the probe is presented late in the fixation following the preallocation of attention to the next saccadic target (words $N+1$ or $N+2$), faster probe detection reaction times (RTs) should be obtained when the probe was displayed in a location congruent with the direction of reading than when it was presented in the opposite direction. If, however, the probe is presented early in the fixation, before attention is preallocated, no such difference should be seen. This pattern was successfully demonstrated in a visual search condition in which participants searched for target letters embedded in reading-like displays of horizontal letter strings. In contrast, the predicted pattern of results was not obtained in either a reading condition (i.e., where participants read for comprehension) or a "mindless reading" condition (i.e., where all letters in the text were replaced by the letter "z" and participants pretended to read these z-strings; see Rayner & Fischer, 1996; Vitu *et al.*, 1995). In addition, across all three primary tasks (visual search, reading, and mindless reading) and for both the early and late probe conditions it was demonstrated that probes interfered with eye movements producing longer fixations and shorter saccades. Furthermore, this interference was attenuated when the probe was presented near the location of the next saccadic target (i.e., in the congruent conditions) as compared to

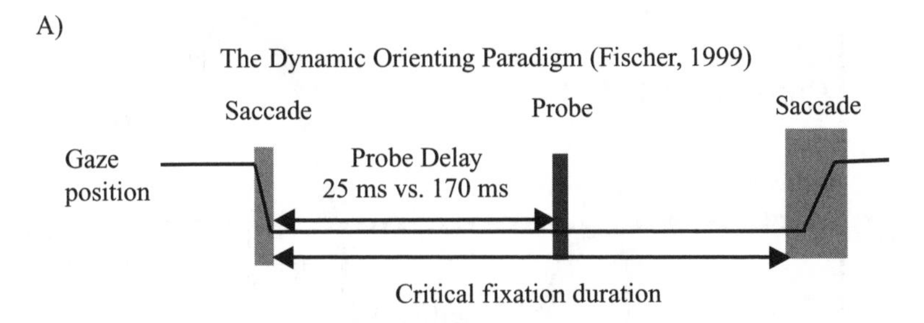

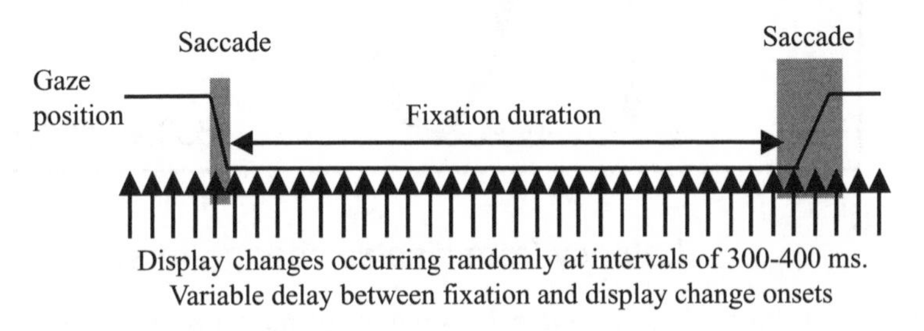

Figure 17.1: Panel A illustrates the timing of probe presentation in the dynamic orienting paradigm (Fischer, 1999) and Panel B illustrates the display change timing in the saccadic inhibition paradigm (Reingold & Stampe, 1997, 2000, 2002, in press; Stampe & Reingold, 2002) (see text for details).

when the probe was presented in the opposite hemifield (i.e., in the incongruent conditions). Importantly, if this congruency effect (i.e., the smaller interference in congruent than incongruent conditions) was due to the preallocation of attention to the next saccadic target it should have been demonstrated in late probes but not in early probe conditions. However, the size of the congruency effect was fairly comparable across the late and early probe conditions. Thus, probe detection performance and oculomotor behaviour documented by Fischer (1999) did not provide support for the predictions of the attentional guidance model.

In contrast to the findings reported by Fischer (1999), Reingold and Stampe (in press) documented perceptual enhancement in the direction of the next saccade in reading and the magnitude of this enhancement was shown to be influenced by higher level cognitive or attentional task demands. This study employed the saccadic inhibition paradigm (Reingold & Stampe, 1997, 2000, 2002, in press; Stampe & Reingold, 2002). In this paradigm, while participants were reading for comprehension, task-irrelevant transient display changes (e.g., a part of the text screen was replaced by a black screen for 33 ms resulting in the subjective experience of a flicker) occurred at intervals that varied randomly between 300 ms to 400 ms. (see Panel B of Figure 17.1). Figure 17.2 illustrates a typical histogram of saccadic frequency following the onset of the display change. Specifically, for the first 50 ms following the display change,

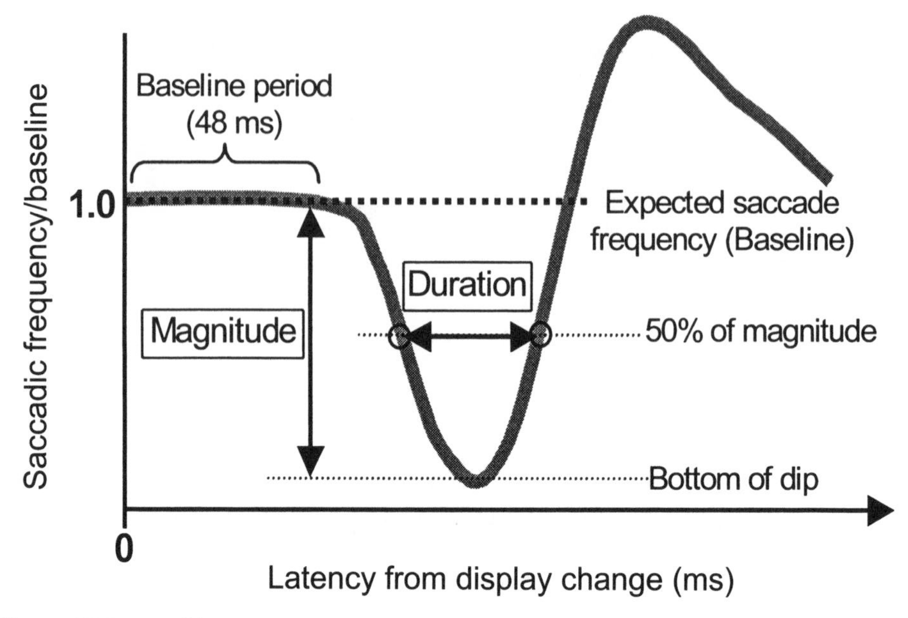

Figure 17.2: An illustration of a typical histogram of normalized saccadic frequency following a display change in the saccadic inhibition paradigm (Reingold & Stampe, 1997, 2000, 2002, in press; Stampe & Reingold, 2002). The saccadic inhibitions measures used in Experiments 1 and 2 are also illustrated (see text for details).

the proportion of saccades remained constant. Approximately 60 to 70 ms following the onset of the display change, the proportion of saccades decreased below this initial level forming a dip that reflects saccadic inhibition. Following the dip, an increase above the initial level of saccadic frequency occurred, forming a peak, which likely reflects the recovery from inhibition. Finally, following the peak, the proportion of saccades returned to initial levels. Reingold and Stampe (in press) devised and validated a variety of quantative measures of the strength and latency of the saccadic inhibition effect. Two of the measures that are the most relevant for the present investigation are illustrated in Figure 17.2. The *magnitude* of saccadic inhibition was defined as the proportion of saccades inhibited when inhibition was at its maximum (i.e., at the centre of the dip) and the *duration* measure was defined as the temporal interval during which inhibition was greater than or equal to 50% of its magnitude.

Reingold and Stampe (in press) documented that the saccadic inhibition effect varied as a function of the saliency of the visual event, with more salient display changes producing stronger and more sustained inhibition. Most importantly, saccadic inhibition was also shown to be sensitive to higher-level cognitive or attentional factors. Specifically, in a reading condition, the magnitude of saccadic inhibition was 22.8% stronger and its duration was 22.7% longer when the display change (a flicker displayed within the boundaries of a gaze contingent, 10° square window; see Panel A of Figure 17.3) location was congruent with the direction of the saccade, relative to

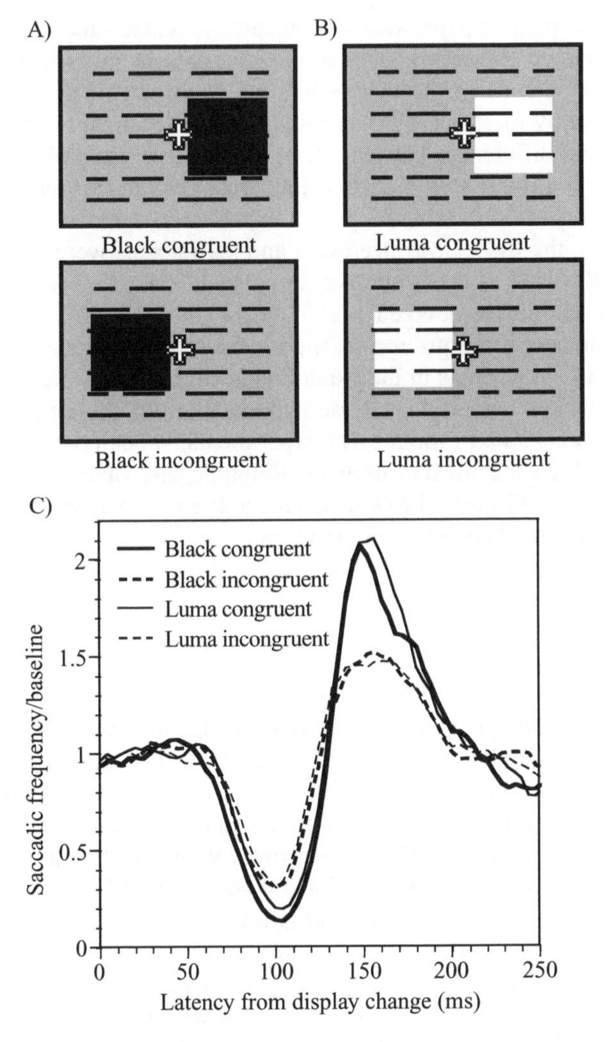

Figure 17.3: Results of Experiment 1. Panel A shows the black congruent flicker (top) and the black incongruent flicker (bottom); (the cross represents point of gaze in each image). Panel B shows the luma congruent flicker (top) and the luma incongruent flicker (bottom); (the cross represents point of gaze in each image). Panel C plots the histogram of normalized saccadic frequency following the flicker onset in the black congruent, black incongruent, luma congruent, and luma incongruent conditions (see text for details).

when the flicker was incongruent with the direction of the saccade. In contrast, the corresponding congruency effects in a mindless reading condition were only 6.5% and 11.7% for the magnitude and duration measures respectively.

The goal of the present paper was to replicate and extend the findings reported by Reingold and Stampe (in press). In addition, the present investigation attempted to empirically identify some of the factors that may account for the dissimilar patterns of results obtained by Fischer (1999) and Reingold and Stampe (in press). As follows, there are important differences between these studies that may hold some explanatory value. In Fischer's (1999) study, participants were required to respond to the probe (i.e., a dual task condition), whereas in the experiments conducted by Reingold and Stampe (in press), the flicker was irrelevant and participants were instructed to ignore it. In addition, the size of the probe was very small (an asterisk appearing above a letter), whereas the flicker covered a 10° square. Furthermore, in Fischer's (1999) study, the appearance of the probe was time-locked to the beginning of fixation (see Panel A Figure 17.1), whereas in the experiments conducted by Reingold and Stampe (in press), the flicker occurred at a random delay from the beginning of fixation (see Panel B Figure 17.1). In the present experiments, we demonstrate that the difference in size between the flickers used by Reingold and Stampe (in press) and the probes employed by Fischer (1999) account at least in part for the differences in the pattern of findings obtained across studies.

Experiment 1

This experiment was designed to replicate the basic congruency effect reported by Reingold and Stampe (in press). The display change used was a flicker that was displayed within the boundaries of a gaze contingent, 10° square window that was displaced such that its nearest edge was 1° to the left or right of the point of gaze. For left to right saccades in English (i.e., forward saccades), the flicker to the left is incongruent and the flicker to the right is congruent with the direction of the saccade. According to the prediction of the attentional guidance model, a congruent flicker will be perceptually enhanced, whereas an incongruent flicker will be attenuated, leading to stronger inhibition in the congruent condition relative to the incongruent condition. This is the pattern demonstrated by Reingold and Stampe (in press). In the present experiment another variable was added to the design. Specifically, participants were reading black text presented on a gray background and the flicker was created by changing the background into black (as was the case in the experiments conducted by Reingold & Stampe, in press; see Panel A of Figure 17.3) or by changing the background into white (see Panel B of Figure 17.3). These black versus luma flicker conditions differed in that the text disappeared in the former, but not in the latter, condition. By comparing the congruency effect across these conditions it would be possible to determine the influence of the disappearance of the text, if any, on the size of this effect.

Method

Participants A group of 10 participants were tested. All participants had normal or corrected to normal vision, and were paid $10.00 per hour for their participation.

Apparatus The SR Research Ltd EyeLink eye tracking system used in this research has a sampling rate of 250 Hz (4 ms temporal resolution). The EyeLink system uses an Ethernet link between the eye tracker and display computers, which supplies real-time gaze position data. The on-line saccade detector of the eye tracker was set to detect saccades with an amplitude of 0.5° or greater, using an acceleration threshold of 9500°/sec^2 and a velocity threshold of 30°/sec. Participants viewed a 17″ (43 cm) ViewSonic 17PS monitor from a distance of 60 cm, which subtended a visual angle of 30° horizontally and 22.5° vertically. The display was generated using an S3 VGA card, and the frame rate was 120 Hz.

During the flicker, a gaze contingent window was used to limit the transient image to a 10° square area that was displaced such that its nearest edge was 1° to the left (incongruent) or right (congruent) of the point of gaze (see Panels A and B of Figure 17.3). As participants moved their eyes, their gaze position on the display was computed and used to set the region in which the transient image would be displayed during the next flicker. The average delay between an eye movement and the update of the gaze-contingent window was 14 ms.

Materials and randomization Participants read a short story for comprehension. The text was presented in black on a gray background. Anti-aliased, proportional spaced text was used with an average of three characters per degree of visual angle and an average of 10 lines of text per page. Each trial displayed one page of text in the story, with pages presented in the same order to all participants. Participants read a total of 100 pages of text, with 50 pages randomly assigned to each of the two flicker conditions (black, luma). The pairing of pages to conditions was determined randomly for each participant, constrained to allow no more than three contiguous trials of the same condition. For each of these flicker conditions the direction of the flicker (i.e., to the left or the right of current gaze position) varied randomly across consecutive display changes. Display changes lasted for 33 ms and occurred at inter-flicker intervals that varied randomly between 300 to 400 ms.

Procedure A 9-point calibration was performed at the start of the experiment followed by a 9-point calibration accuracy test. Calibration was repeated if the error at any point was more than 1°, or if the average error for all points was greater than 0.5°. Participants were instructed to read the text for comprehension. They were told that they would be asked questions about the content of the story when they finished reading. On average participants answered over 95% of these questions accurately, indicating that they complied with the instructions and were not simply scanning the text. After reading each page, participants pressed a button to end the trial and proceed to the next page of the story.

Data analysis The timing of the display changes was recorded along with eye movement data for later analysis. Eye tracker data files were processed to produce histograms of saccade frequency as a function of latency from the display change. Only forward saccades were included in the analysis. A separate histogram was compiled for each participant and condition, and analyzed to produce the magnitude duration measures of the evoked saccadic inhibition (see Figure 17.2). Forward saccades following a flicker to the left were classified as part of the incongruent conditions and forward saccades following a flicker to the right were classified as part of the congruent conditions.

Results and Discussion

Panel C of Figure 17.3 presents the normalized saccadic frequency histograms for all four experimental conditions, and the derived saccadic inhibition measures (magnitude, duration) are shown in Table 17.1. For each of these dependent variables, a 2 × 2 within participants ANOVA, which crossed flicker congruency (congruent versus incongruent) by flicker type (black versus luma), was performed. Most importantly, as can be seen in Figure 17.3, when the direction of the saccade is congruent with the flicker, a stronger and more sustained inhibition results, and this congruency effect is quite similar across the black and luma flicker conditions. Consistent with this observation, the main effect of congruency was significant (magnitude: $F(1, 9) = 95.15$, $p < 0.001$; duration: $F(1, 9) = 11.86$, $p < 0.01$) and the interaction between congruency and flicker type was not significant for both the magnitude and duration measures (both Fs < 1). In addition, the black flicker produced stronger and more sustained inhibition than the luma flicker (magnitude: $F(1, 9) = 11.34$, $p < 0.01$; duration: $F(1, 9) = 16.93$, $p < 0.01$), probably due to the greater saliency of the former relative to the latter display change (see Reingold & Stampe, in press; Stampe & Reingold, 2002). Thus, the present experiment replicated the congruency effect reported by Reingold and Stampe, (in press) and demonstrated that the disappearance of the text has no influence on the size of this effect.

Table 17.1: Means and standard errors of the magnitude and the duration of saccadic inhibition across conditions in Experiments 1 and 2.

Experiment	**Condition**	**Saccadic Inhibition**	
		Magnitude (Proportion)	**Duration (ms)**
Experiment 1	Black congruent	0.856 (0.024)	49.2 (2.4)
	Black incongruent	0.719 (0.040)	43.2 (2.7)
	Luma congruent	0.796 (0.031)	44.6 (2.4)
	Luma incongruent	0.689 (0.030)	37.9 (1.9)
Experiment 2	Small congruent	0.497 (0.054)	33.4 (2.6)
	Small incongruent	0.632 (0.037)	35.2 (2.8)
	Large congruent	0.886 (0.017)	49.2 (2.6)
	Large incongruent	0.759 (0.025)	44.0 (2.0)

Experiment 2

As mentioned earlier, Fischer (1999) reported that the probes used in his paradigm interfered with eye movements producing longer fixations and shorter saccades and that this interference was greater when the probe was incongruent with the direction of the next saccade as compared to when the probe was congruent with the direction of the next saccade. In contrast, the congruency effect documented by Reingold and Stampe (in press), and replicated in Experiment 1, reflects stronger interference in the congruent than the incongruent condition. The present experiment was designed to investigate if the difference between the large flicker and the small probe may account for the dissimilar patterns of results obtained by Fischer (1999) and Reingold and Stampe (in press). In this experiment two flicker conditions were used: a 10° square flicker identical to the black flicker condition used in Experiment 1 and employed by Reingold and Stampe (in press), and a 1° square flicker presented centered 4° to the left or the right of the current gaze position (see Panels A and B of Figure 17.4).

Method

General A group of 10 participants who had not taken part in Experiment 1 was tested. All participants had normal or corrected to normal vision, and were paid $10.00 per hour for their participation.

Participants read a total of 100 pages of text, with 50 pages randomly assigned to each of the two flicker conditions (10° square, 1° square). The pairing of pages to conditions was determined randomly for each participant, constrained to allow no more than three contiguous trials of the same condition. For each of these flicker conditions the direction of the flicker (i.e., to the left or the right of current gaze position) varied randomly across consecutive display changes. All other aspects of the design, procedure, and data analysis were identical to Experiment 1.

Results and Discussion

Panel C of Figure 17.4 presents the normalized saccadic frequency histograms for all four experimental conditions, and the derived saccadic inhibition measures (magnitude, duration) are shown in Table 17.1. For each of these dependent variables, a 2×2 within participants ANOVA, which crossed flicker congruency (congruent versus incongruent) by flicker size (large: 10° square, small: 1° square), was performed. Most importantly, the interaction between congruency and flicker size was significant for both the magnitude ($F(1,9) = 36.29$, $p < 0.001$) and the duration ($F(1,9) = 6.73$, $p < 0.05$) measures. As can be seen in Figure 17.4, a dramatic reversal occurred in the direction of the congruency effect as a function of flicker size. Specifically, the large flicker condition replicated the results of Experiment 1 producing a stronger and more sustained inhibition in the congruent than in the incongruent condition (magnitude: $t(9) = 6.37$, $p < 0.001$; duration: $t(9) = 4.21$, $p < 0.01$). In contrast, for the small flicker

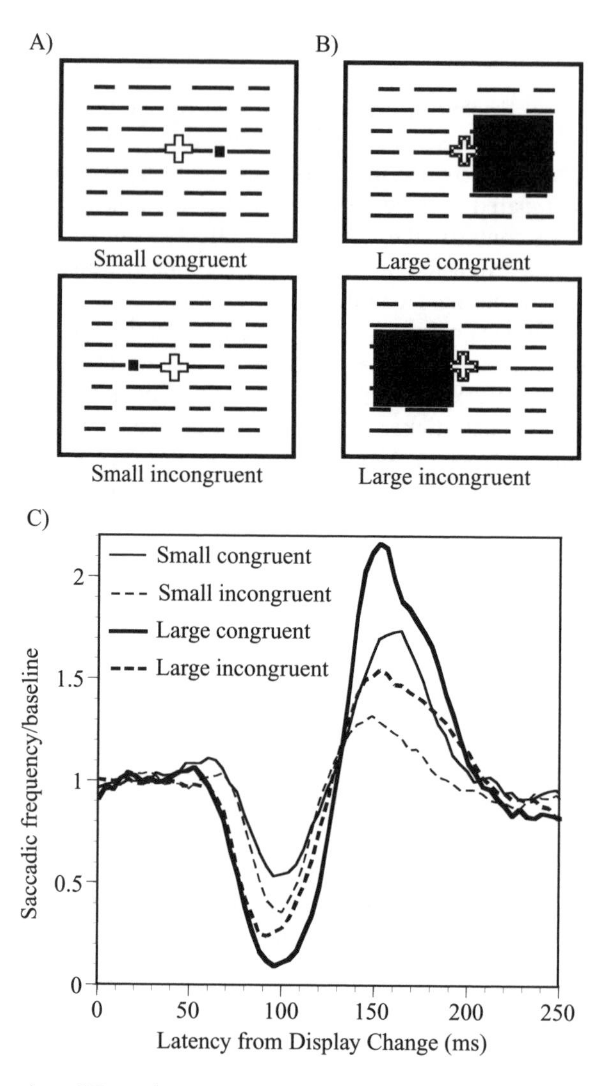

Figure 17.4: Results of Experiment 2. Panel A shows the small congruent flicker (top) and the small incongruent flicker (bottom); (the cross represents point of gaze in each image). Panel B shows the large congruent flicker (top) and the large incongruent flicker (bottom); (the cross represents point of gaze in each image). Panel C plots the histogram of normalized saccadic frequency following the flicker onset in the small congruent, small incongruent, large congruent, and large incongruent conditions (see text for details).

the incongruent condition produced a stronger inhibition than the congruent condition, and the congruency effect for duration was not significant (magnitude: $t(9) = 2.68$, $p < 0.05$; duration: $t < 1$). Thus, the results obtained with the large flicker replicate the pattern reported by Reingold and Stampe (in press), whereas the results obtained with the small flicker are consistent with the findings demonstrated by Fischer (1999).

General Discussion

The present findings replicated the congruency effect reported by Reingold and Stampe (in press). This effect is consistent with the prediction of the attentional guidance model of saccadic control in reading, which assumes that prior to the execution of saccades, attention is covertly preallocated to the saccadic target (e.g., Henderson, 1992; Henderson & Ferreira, 1990; Kennison & Clifton, 1995; Morrison, 1984; Pollatsek & Rayner, 1990; Rayner & Pollatsek, 1989; Reichle *et al.*, 1998). Furthermore, these results are consistent with a growing body of research that suggests that whereas attention can be shifted covertly in the absence of eye movements, eye movements are preceded by an attentional shift to the saccadic target (e.g., Deubel & Schneider, 1996; Henderson, 1993; Hoffman, 1998; Hoffman & Subramaniam, 1995; Kowler, Anderson, Dosher & Blaser, 1995; Rafal, Calabresi, Brennan & Sciolto, 1989; Rayner, McConkie & Ehrlich, 1978; Remington, 1980; Schneider & Deubel, 1995; Shepherd, Findlay & Hockey, 1986, but see Stelmach, Campsall & Herdman, 1997 for evidence against this preallocation hypothesis). This view concerning the link between the oculomotor and attentional systems is also supported by neurophysiological data (e.g., Goldberg & Wurtz, 1972; Kustov & Robinson, 1996; Mohler & Wurtz, 1976; Wurtz & Mohler, 1976) and work with neuropsychological populations such as neglect patients (e.g., Johnston & Diller, 1986; Walker & Young, 1996).

The present experiments extended the prior findings reported by Reingold and Stampe (in press) in two important ways. First, Experiment 1 clearly demonstrated that the disappearance of the text had no influence on the size of the congruency effect. Second, the results of Experiment 2 indicated that a crucial variable that mediated the direction of the congruency effect is the size of the sudden-onset (i.e., display change). Specifically, a large flicker produced a stronger inhibition in the congruent condition than in the incongruent condition, whereas the reverse was true for a small flicker. Further studies are required in order to investigate whether, as suggested here, the influence of flicker size on the congruency effect may account for the dissimilar patterns of results obtained by Fischer (1999) and Reingold and Stampe (in press). In addition, although empirically the effect of flicker size on the congruency effect has been convincingly demonstrated, currently, we can only speculate about the theoretical explanation for this finding. Reingold and Stampe (in press) interpreted the stronger inhibition in the congruent than the incongruent condition as resulting from a perceptual enhancement of the large flicker due to attentional preallocation in the direction of the next saccade. Fischer (1999) provided two potential explanations for the stronger interference in the incongruent than the congruent condition produced by the presentation of the small probe. The first explanation was based on the idea that the sudden-onset probe redirects attention away from the intended saccadic target and that the cost of this attentional shift is greater in the incongruent condition due to the greater mismatch between the locations of the probe and the saccadic target. However, Fischer (1999) acknowledged that rather than attention, the effect of the probe on eye movements might be mediated by a low-level oculomotor effect known as the remote distractor effect. This effect signifies the slowing of saccadic reaction time (SRT), which occurs when a saccadic target and a distractor stimulus are presented simultaneously at different locations in the visual field. Importantly, SRTs are slower when the distractor is presented in the opposite hemifield than the target (this is the case in the incongruent probe conditions) than when the distractor is presented near the saccadic target (this is the case in the congruent probe conditions). Thus, an important goal for future research would be to attempt and disentangle these alternative explanations. Finally, the present study illustrates the potential contributions of the saccadic inhibition paradigm to the study of higher-level cognitive or attentional factors that underlie reading.

Acknowledgement

Preparation of this paper was supported by a grant to Eyal Reingold from the Natural Science and Engineering Research Council of Canada. We wish to thank Elizabeth Bosman and Colleen Ray for their helpful comments on an earlier version of this paper.
Correspondence should be addressed to Eyal Reingold, University of Toronto, Department of Psychology, 100 St. George Street, Toronto, Ontario, Canada, M5S 3G3. Electronic mail may be sent to: reingold@psych.utoronto.ca.

References

Brysbaert, M., & Vitu, F. (1998). Word skipping: Implications for theories of eye movement control in reading. In: G. Underwood (ed.), *Eye Guidance in Reading and Scene Preparation* (pp. 125–148). Oxford: Elsevier.

Deubel, H., O'Regan, J. K., & Radach, R. (2000). Commentary on section 2. Attention, information processing and eye movement control. In: A. Kennedy, R. Radach, D. Heller and J. Pynte (eds), *Reading as a Perceptual Process* (pp. 355–374). Amsterdam: Elsevier

Deubel, H., & Schneider, W. X. (1996). Saccade target selection and object recognition: Evidence for a common attentional mechanism. *Vision Research, 6*, 1827–1837.

Deubel, H. M., & Stampe, D. M. (2000). Commentary on Section 2. Attention, information processing and eye movement control. In: A. Kennedy, R. Radach, D. Heller and J. Pynte (eds), *Reading as a Perceptual Process* (pp. 355–374). Oxford: Elsevier.

Fischer, M. H. (1999). An investigation of attention allocation during sequential eye movement tasks. *Quarterly Journal of Experimental Psychology, 52A*, 649–677.

Goldberg, M. E., & Wurtz, R. H. (1972). Activity of superior colliculus in behaving monkey: I. Visual receptive fields of single neurons. *Journal of Neurophysiology, 35*, 542–559.

Henderson, J. M. (1992). Visual attention and eye movement control during reading and picture viewing. In: K. Rayner (ed.), *Eye Movements and Visual Cognition: Scene Perception and Reading* (pp. 260–283). New York: Springer-Verlag.

Henderson, J. M. (1993). Visual attention and saccadic eye movements. In: G. d'Ydewalle and J. Van Rensbergen (eds), *Perception and Cognition: Advances in Eye Movement Research* (pp. 37–50). Amsterdam: North-Holland/Elsevier.

Henderson, J. M., & Ferreira, F. (1990). Effects of foveal processing difficulty on the perceptual span in reading: Implications for attention and eye movement control. *Journal of Experimental Psychology: Human Perception and Performance, 16*, 417–429.

Hoffman, J. E. (1998). Visual attention and eye movements. In: H. Pasher (ed.), *Attention* (pp. 119–153). East Sussex, UK: Psychology Press.

Hoffman, J. E., & Subramaniam, B. (1995). The role of visual attention in saccadic eye movements. *Perception & Psychophysics, 57*, 787–795.

Johnston, C. W., & Diller, L. (1986). Exploratory eye movements and visual hemi-neglect. *Journal of Clinical and Experimental Neuropsychology, 8*, 93–101.

Just, M. A., & Carpenter, P. A. (1980). A theory of reading: From eye fixations to comprehension. *Psychological Review, 87*, 329–354.

Kennison, S. M., & Clifton, C. (1995). Determinants of parafoveal preview benefit in high and low working memory capacity readers: Implications for eye movement control. *Journal of Experimental Psychology: Learning, Memory, and Cognition, 21*, 68–81.

Kliegl, R., & Engbert, R. (2003). SWIFT explorations. In: J. Hyönä, R. Radach and H. Deubel (eds), *The Mind's Eye: Cognitive and Applied Aspects of Eye Movement Research* (pp. 391–411). Amsterdam: North-Holland.

Kowler, E., Anderson, E., Dosher, B., & Blaser, E. (1995). The role of attention in the programming of saccades. *Vision Research, 35*, 1897–1916.

Kowler, E., & Anton, S. (1987). Reading twisted text: Implications for the role of saccades. *Vision Research, 27*, 45–60.

Kustov, A. A., & Robinson D. L. (1996) Shared neural control of attentional shifts and eye movements. *Nature, 384*, 74–77.

McConkie, G. W., Kerr, P. W., Reddix, M. D., & Zola, D. (1988). Eye movement control during reading: I. The location of initial eye fixation in words. *Vision Research, 28*, 1107–1118.

McConkie, G. W., Kerr, P. W., Reddix, M. D., Zola, D., & Jacobs, A. M. (1989). Eye movement control during reading: II. Frequency of refixating a word. *Perception & Psychophysics, 46*, 245–253.

McConkie, G. W., & Yang, S.-N. (2003). How cognition affects eye movements during reading. In: J. Hyönä, R. Radach and H. Deubel (eds), *The Mind's Eye: Cognitive and Applied Aspects of Eye Movement Research* (pp. 413–427). Amsterdam: North-Holland.

Mohler, C. W., & Wurtz, R. (1976). Organization of monkey superior colliculus: Intermediate layer cells discharging before eye movements. *Journal of Neurophysiology, 39*, 722–744.

Morrison, R. E. (1984). Manipulation of stimulus onset delay in reading: Evidence for parallel programming of saccades. *Journal of Experimental Psychology: Human Perception and Performance, 10*, 667–682.

O'Regan, J. K. (1990). Eye movements and reading. In: E. Kowler (ed.), *Eye Movements and their Role in Visual and Cognitive Processes* (pp. 395–453). Amsterdam: Elsevier.

O'Regan, J. K. (1992). Optimal viewing position in words and the strategy-tactics theory of eye movements in reading. In: K. Rayner (ed.), *Eye Movements and Visual Cognition: Scene Perception and Reading* (pp. 333–354). New York: Springer-Verlag.

Pollatsek, A., & Rayner, D. (1990). Eye movements and lexical access in reading. In: D. A. Balota, G. B. Flores d'Arcais and K. Rayner (eds), *Comprehension Processes in Reading* (pp. 143–164). Hillsdale, NJ: Erlbaum.

Pollatsek, A., Reichle, E. D., & Rayner, K. (2003). Modeling eye movements in reading: Extensions of the E-Z reader model. In: J. Hyönä, R. Radach and H. Deubel (eds), *The Mind's Eye: Cognitive and Applied Aspects of Eye Movement Research* (pp. 361–390). Amsterdam: North-Holland.

Rafal, R. D., Calabresi, P. A., Brennan, C. W., & Sciolto, T. K. (1989). Saccade preparation inhibits reorienting to recently attended locations. *Journal of Experimental Psychology: Human Perception and Performance, 15*, 673–685.

Rayner, K. (1998). Eye movements in reading and information processing: 20 years of research. *Psychological Bulletin, 124*, 372–422.

Rayner, K., & Fischer, M. H. (1996). Mindless reading revisited: Eye movements during reading and scanning are different. *Perception & Psychophysics, 58*, 734–747.

Rayner, K., McConkie, G. W., & Ehrlich, S. F. (1978). Eye movements and integrating information across fixations. *Journal of Experimental Psychology: Human Perception and Performance, 4*, 529–544.

Rayner, K., & Pollatsek, N. (1989). *The Psychology of Reading*. Englewood Cliffs, NJ: Prentice Hall.

Reichle, E. D., Pollatsek, A., Fisher, D. L., & Rayner, K. (1998). Toward a model of eye movement control in reading. *Psychological Review, 105*, 125–157.

Reilly, R. G., & Radach, R. (2003). Foundations of an interactive activation model of eye movement control in reading. In: J. Hyönä, R. Radach and H. Deubel (eds), *The Mind's Eye: Cognitive and Applied Aspects of Eye Movement Research* (pp. 429–455). Amsterdam: North-Holland.

Reingold, E. M., & Stampe, D. M. (1997). Transient saccadic inhibition in reading. Paper presented at the 9th European Conference on Eye Movements, Ulm, Germany.

Reingold, E. M., & Stampe, D. M. (2000). Saccadic inhibition and gaze contingent research paradigms. In: A. Kennedy, R. Radach, D. Heller and J. Pynte (eds), *Reading as a Perceptual Process* (pp. 119–145). Oxford: Elsevier.

Reingold, E. M., & Stampe D. M. (2002). Saccadic inhibition in voluntary and reflexive saccades. *Journal of Cognitive Neuroscience, 14*, 371–388.

Reingold, E. M., & Stampe D. M. (in press). Saccadic inhibition in reading. *Journal of Experimental Psychology: Human Perception and Performance.*

Remington, R. W. (1980). Attention and saccadic eye movements. *Journal of Experimental Psychology: Human Perception and Performance, 6*, 226–244.

Schneider, W. X., & Deubel, H. (1995). Visual attention and saccadic eye movements: Evidence for obligatory and selective spatial coupling. In: J. M. Findlay, R. Walker and R. W. Kentridge (eds), *Eye Movement Research: Mechanisms, Processes and Applications* (pp. 317–324). Amsterdam: North-Holland/Elsevier.

Shepherd, M., Findlay, J. M., & Hockey, R. J. (1986). The relationship between eye movements and spatial attention. *The Quarterly Journal of Experimental Psychology, 38A*, 475–491.

Stampe, D. M., & Reingold, E. M. (2002). Influence of stimulus characteristics on the latency of saccadic inhibition. In: J. Hyönä, D. Munoz, W. Heide and R. Radach (eds), *The Brain's Eye: Neurobiological and Clinical Aspects of Oculomotor Research* (pp. 73–87). Amsterdam: Elsevier.

Stelmach, L. B., Campsall, J. M., & Herdman, C. M. (1997). Attention and ocular movements. *Journal of Experimental Psychology: Human Perception and Performance, 23*, 823–844.

Vitu, F., & O'Regan, J. K. (1995). A challenge to current theories of eye movements in reading. In: J. M. Findlay, R. Walker and R. W. Kentridge (eds), *Eye Movement Research: Mechanisms, Processes, and Applications* (pp. 381–393). New York: Elsevier.

Vitu, F., O'Regan, J. K., Inhoff, A. W., & Topolski, R. (1995). Mindless reading: Eye movement characteristics are similar in scanning letter strings and reading text. *Perception & Psychophysics, 57*, 352–364.

Walker, R., & Young, A. W. (1996). Object-based neglect: An investigation of the contributions of eye movements and perceptual completion. *Cortex, 32*, 279–295.

Wurtz, R., & Mohler, C. W. (1976). Organization of monkey superior colliculus: Enhanced visual response of superficial layer cells. *Journal of Neurophysiology, 39*, 745–765.

Chapter 18

Modeling Eye Movements in Reading: Extensions of the E-Z Reader Model

Alexander Pollatsek, Erik D. Reichle and Keith Rayner

The evolution of our attempts to model eye movements while people read text is described. All of our models have the core assumption that the primary "engine" that drives the eyes forward in text is the serial identification of words in text. However, the relationship between the identification of a word and the forward movement of attention and the program to fixate the subsequent word is not completely straightforward. The chapter is divided into four major sections. In the first, we describe the "design principles" that guide our modeling. In the second, we summarize our prior published modeling work (Reichle, Pollatsek, Rayner & Fisher, 1998; Reichle, Rayner & Pollatsek, 2002). The third section describes our recent efforts to modify our model, both to improve the fit to a corpus of data in English and to make the assumptions more psychologically reasonable. The final section expands our modeling efforts to explain the fixation patterns on Finnish compound words in text, allowing the model to be responsive to constituents of words as well as whole words.

Introduction

Silent reading is one of the most important skills humans master, and it is certainly one of the most frequent complex cognitive acts most adults perform. One can attempt to understand this skill in many different ways; however, a very powerful tool for understanding reading is examining the pattern of eye movements while people silently read text for meaning (Rayner, 1998). This claim is largely non-controversial now, but not too many years ago, it was believed that little could be learned from eye movements in reading because it was hypothesized that the cognitive acts in reading were too slow to be responsive to a system that was generating eye movements 3–4 times

The Mind's Eye: Cognitive and Applied Aspects of Eye Movement Research

ISBN: 0–444–51020–6

a second (Kolers, 1976). However, it is now clear that the moment-to-moment pattern of eye movements is responsive to aspects of the meaning of the text, such as the frequency of a word in the language, and therefore cognitive acts, such as recognizing a word, do influence the behavior of the eyes (at least some of the time) in a moment-to-moment fashion (Rayner, 1998).

We will discuss below some of the major phenomena of interest in reading, but first we would like to discuss how one comes to understand these phenomena. Being able to explain the complete pattern of eye movements (both where the eyes land on each fixation and the duration of each fixation) in a complex task like silent reading is clearly a difficult, if not impossible, task. Thus, we have to accept provisional understanding of these phenomena. Part of the process of understanding the cognitive processes in reading and how they control the pattern of eye movements is conducting experiments to determine which factors in the text the eyes are sensitive to. As indicated above, the length of time one fixates a word in text is influenced by the frequency of that word (even when the length of the word is controlled). Other experiments have been conducted that indicate that factors at very different levels of processing, ranging from the type of font to syntactic ambiguity to discourse level factors also affect the process of reading (and the pattern of eye movements).

As a result of these experiments, we now have a wide range of phenomena that we know influence eye movements. However, knowing that these variables influence eye movements — somehow — is clearly only a partial (and fairly unsatisfactory) solution to the problem of understanding eye movements in reading and the cognitive processes behind them. One not only wants to know whether some variable affects reading, but exactly how it is influencing both the cognitive processes involved in understanding the text and the pattern of eye movements that are at least partially driven by these cognitive processes. As a result, people have theorized about eye movements in reading for over a hundred years. However, it is only in the last few years that people have attempted quantitative models that have provided some initial attempts to study the reading process with some degree of precision. Although there were earlier attempts to model how language processing affects eye movements during reading (e.g., Just & Carpenter, 1980), these models generally lacked quantitative precision and/or made assumptions that were at odds with existing data.

We will later make more detailed comments about the benefits of such modeling efforts, but an important one is that quantitative modeling provides a more rigorous test of ideas. For example, consider trying to understand how the frequency of a word influences fixation time on that word. Qualitatively, this seems reasonable and not very remarkable. However, to model this phenomenon quantitatively, one has to explain how such influence can really go on in real time (i.e., whether one can make reasonable assumptions about the time to access a word in the mental lexicon, eye movement programming times, etc., and still be able to account for this phenomenon). In fact, such modeling is possible, but not without some effort, and this effort indicates that there are many non-trivial constraints on the system generating the pattern of eye movements.

A major decision in any modeling enterprise is defining the scope of the data one is attempting to model (see Reichle & Rayner, 2002; Reichle, Rayner & Pollatsek, 2002). Models of reading have run the gamut, from those that have attempted to model the relation between the entire cognitive process of reading to the pattern of eye movements (e.g., Just & Carpenter, 1980) to those that attempted to model eye movements in reading largely from fairly low-level visual factors such as the acuity of the letters (Clark & O'Regan, 1999; Legge, Klitz & Tjan, 1997). Our modeling work to date has taken a middle ground and focused almost exclusively on relating word identification to the pattern of eye movements and accounting for "higher-level" processing only through its influence on the speed of word identification. As a result, our modeling efforts fall quite short of explaining eye movements in reading, as there are clearly "higher order" influences on eye movements (e.g., syntactic "garden path" effects and discourse level effects). Nonetheless, we feel that our modeling has allowed us to understand the control of the eyes during reading more deeply, and furthermore, that there is still a lot more to be learned by pursuing modeling at this level. In the next section, we briefly describe what we think are the salient data in reading that any quantitative model of reading should need to explain. We will also describe what we think is known about saccadic eye movements that constrain how cognitive processes can control eye movements in reading. This will help us to argue why we have chosen this scope of explanation and modeling strategy.

In this chapter, we will focus exclusively on our own modelling efforts. Since we published the first version of the E-Z Reader model (Reichle *et al.*, 1998) a number of other models have appeared. Some of these models are quite complementary to the E-Z Reader model and others make very different assumptions (while still trying to account for some of the same data). The chapters in this volume by Reilly and Radach, McConkie and Yang, and Kliegl *et al.* provide details of some of the most prominent of these competing models. Elsewhere, we (Reichle *et al.*, 2002) have provided overviews of these alternative models and have provided an account of how well we believe these models fare in comparison to the E-Z Reader model. In this chapter, however, we will not discuss the alternative models at all, as our goals in the present chapter are to provide (a) a clear overview of the rationale for basic E-Z Reader model, (b) some recent extensions of the model, and, (c) in the process, give the reader a sense of how our model is a useful heuristic device as well as a formal model.

What Do We Know about Cognitive Processes in Reading?

Word Identification

Certainly, one of the most studied topics in cognitive psychology is word identification. Word identification has been extensively studied both in isolation and in the context of reading text. Many variables have been shown to influence word identification in isolation using techniques such as lexical decision and naming time: for example, the frequency of the word, the length of the word, whether it is "regular" in terms of spelling to sound correspondance, the influence of "neighbors" (i.e., similarly spelled words), and whether "parts" of a word (e.g., morphemes) have a role in word identification. In fact, most of these variables have been shown to influence fixation times on words in text as well (see Rayner, 1998). In addition, it is clear that there are "top-down" influences on word identification, most notably the predictability of a word in the sentence context (usually assessed by getting "off-line" estimates of the probability of predicting a word when given the text in a sentence prior to the word of interest). Moreover, in addition to the many experiments indicating that presenting a semantically or associatively related word prior to a target word facilitates processing of the word in isolation, there are experiments that indicate that such semantic "priming" can occur when people are reading text (Carroll & Slowiaczek, 1986; Morris, 1994).

As a result of these findings, there have been many models proposed for "lexical access", either using the metaphor of a mental dictionary that is "looked up" by the incoming stimulus (e.g., Coltheart, Rastle, Perry, Langdon & Ziegler, 2001) or more complex distributed systems (e.g., Seidenberg & McClelland, 1989), where there is no necessary orthographic dictionary unit for a word. Although the metaphors differ for these models, they generally converge on similar solutions to the problem of word identification. The models almost all use letter detectors, letter cluster detectors, and similar intermediaries between the visual stimulus and word identification, and almost all are systems in which these units are activated in parallel, with the system converging on a "solution" to the identity of the word. The "top-down" influences are usually handled by positing excitation flowing to a dictionary unit from semantically or associatively related dictionary units. (Parallel distributed processing, or PDP, models generally assume that these units are distributed across a collection of processing elements and that patterns of activity across these elements — rather than individual elements — represent individual words.) Many of these models (both those that employ lexical units and those that have distributed representations) also employ lateral inhibition among various representations at the same level to help the system converge on a solution faster.

In addition, there are experiments that help to trace the time course of word encoding during reading and how they relate to eye position and eye movements. Most important are the *moving window* (McConkie & Rayner, 1975; Rayner & Bertera, 1979) and *boundary* (Rayner, 1975) experiments in which display changes are made contingent on the location of fixation. In moving window experiments, a "window" of normal text is presented around the fixation point and "garbage" (such as random letters) is presented outside of this window. The moving window experiments have established that word processing does not proceed far in advance of fixating a word, as people can read normally (i.e., with normal speed and comprehension) when the window includes only the fixated word and the two words to the right of it. In fact, reading speed is only reduced about 10% when the window includes only the fixated word and the word to the right of it (Rayner, Well, Pollatsek & Bertera, 1982). Thus, processing of a word appears to begin at most two words in advance of fixating it and usually only when the word to the left of it is being fixated.

In moving window experiments, there is a display change after virtually every saccade. In contrast, in boundary experiments, there is only one display change — when a single invisible boundary is crossed. These experiments thus allow one to be more diagnostic about what information is being extracted from the text prior to fixating it that facilitates processing when a word is fixated. In these experiments, the typical manipulation involves giving various *previews* of a target word prior to fixating it (e.g., a visually similar preview, a phonologically similar preview, or a synonym) and then assessing the speed of processing of the target word with either a completely dissimilar preview or the same word as the target word as a baseline. The findings are that orthographically similar previews and homophone previews provide significant benefit, whereas synonyms provide no benefit.[1] These findings converge on the conclusion that significant processing is done on a word before it is fixated, but that little or no activation of meaning occurs until it is fixated. However, as many words in text are skipped, it is clear that the meaning of some words is extracted before they are fixated. Thus, it appears unless the meaning is fully extracted from a word that appears to the right of fixation in text, the partial activation of its meaning is not used on subsequent fixations.

Recently, there has been considerable discussion (Inhoff, Starr & Shindler, 2000; Kennedy, 1998, 2000; Murray & Rowan, 1998) about whether the meaning of the word to the right of fixation can influence the duration of the current fixation. We won't discuss this in any detail here, as the chapter by Rayner, White, Kambe, Miller and Liversedge in the present volume contains an extended discussion of this issue. We will simply comment that we agree with Rayner *et al.* that there is no convincing evidence that the word to the right of fixation influences the current fixation duration when people are engaged in normal reading. However, there is some evidence suggesting that an orthographically illegal string of letters at the beginning of the word to the right of fixation can influence the current fixation when the reader fixates very close to the beginning of the parafoveal word.

As a result of the amount of data collected and the modeling efforts directed at understanding how printed words are identified, one can comfortably say that the word recognition process is "sort of understood". On some level, it appears to be a "look-up" process, and one has a pretty good idea of the type of process, its time course, and many of the variables that influence its time course.

Syntax, Parsing and Understanding the Meaning of Discourse

In contrast, it is far less clear how to represent either what the reader has understood, either in terms of the syntactic structure of a sentence or the "meaning" of the discourse or the process by which the reader has arrived at this understanding (see Rayner & Clifton, 2002). For example, although something like phrase-structure grammar is probably a good approximation to a reader's understanding of relatively simple syntactic structures, it is far from clear how to represent more complex syntactic structures. Moreover, while there are some interesting general principles about how sentences should be parsed, there are few detailed theories about the moment to moment way in which these trees are constructed. Understanding how the meaning of prose is understood (through theories such as "story grammars") is even sketchier, as there is far less constraint on meaning than on syntax.

As a result, there is little, at present, that is known about these levels of processing that is likely to be particularly helpful in guiding a detailed quantitative model of reading. Most of the phenomena that have been studied in these domains tend to be manipulations that place the reader "in trouble". For example, in the syntactic domain there is the work of Frazier and Rayner (1982) and many subsequent researchers with syntactic "garden-path" sentences. This work employs sentences that have temporary ambiguities such that these sentences are likely to be initially misparsed, and furthermore have specific locations where this misparsing is likely to be realized by the reader. The experiments demonstrate that the reader does tend to realize the misparsing fairly close to the earliest point where the problem could be identified and immediately engages in some sort of "clean-up" operation (e.g., fixating for a long time in one location or regressing back into the earlier part of the sentence). Other experiments involving anaphor are similar in that they employ pronouns that either have ambiguous or very distant antecedents and examine what happens when the reader encounters the pronoun (Ehrlich & Rayner, 1983). Finally, experiments that manipulate the ease of understanding the meaning of a "story" often involve examining what happens when the simplest construal of the story is not realized and a re-evaluation is needed (Myers & O'Brien, 1998).

From all of this research, it is not at all clear that either syntactic or higher-level semantic processing is affecting the course of eye movements across the text when they are running smoothly. Instead, these higher-order processes may only affect the eye movement process when they are not running smoothly and a recomputation is needed. As a result, we believe it is a tenable hypothesis that the normal movement of the eyes through text is directed by the identification of individual words and that higher-order processes intervene only when some processing difficulty at these higher levels is encountered and needs to be straightened out before more words can usefully be encoded. That is, we are not denying that these higher-order processes exist or are important in understanding the meaning of the text. Instead, we think they usually exist in the background, lagging behind the process of word encoding and are usually invisible in the process of word encoding. A major reason for this is that these processes usually require knowing the meaning and syntactic properties of individual lexical items and thus need to lag behind lexical encoding; as a result, waiting for successful completion of these stages before moving the eyes forward is likely to produce a reading process much slower than skilled reading. Thus, the optimal solution for skilled reading may be to move forward when words are identified (or almost identified in our model) and hope that the higher-order processes constructing syntactic and semantic structures will be able to keep up. Such a process may have only a small cost in repairing difficulties when they occur; moreover, these difficulties may not occur too often in felicitiously written prose.

As a result, another justification for our scope of modeling is that the system of encoding individual words may represent a close approximation to a "module" in understanding reading. That is, there may be a self-contained system driving the eyes forward that relies only on this information, and higher-order influences may represent a quite different (and more complex) system that (a) may only intervene occasionally and (b) may represent a qualitatively different type of control on eye movements in reading. Consistent with this view, we typically model only reading protocols that contain no interword regressions in the region of text that we are interested in. Of course, this working hypothesis may be wrong and it may be that "higher-order" processes intervene in reading continually. However, we know of no data that force that conclusion, and we think our approach is a fruitful way of starting to understand the reading process and the pattern of eye movements during that process. Moreover, as indicated earlier, we think this limited complexity of our model makes it far easier to understand the relationship between the model and the reading process, and therefore be able to assign "blame" sensibly when the model does not account for some aspect of the data.

Three Modeling Attempts

The above indicates the general scope and aims of our modeling efforts. In this section, we discuss three different modeling attempts. The first is intended as "background", as it summarizes our initial presentation of the E-Z Reader model of eye movements in reading English (Reichle *et al.*, 1998). The second is a description of our new version of the model, which we feel is conceptually better and also gives a somewhat improved fit to the same data. The third is a first attempt to extend the model to account for the processing of more complex words. In this case, it is an attempt to fit the fixation patterns and durations on Finnish compound words encountered in text.

Our Prior Model

As the discussion of the prior section indicates, we attempted to model a corpus of data (Schilling, Rayner & Chumbley, 1998) using the basic assumption that encoding words is the driving force behind moving the eyes forward in the text. In addition, as indicated earlier, because regressions back to a prior word in the text are assumed to be due to a different process, we eliminated data from the corpus in which there were regressions in the sentence. Furthermore, we included only single-line sentences in our analysis, because we thought that the process of making a large return sweep (and the reader's eyes probably only landing approximately where they were programmed to go) was too complex to deal with at this stage. In our paper, we presented some successive approximations to a relatively complete model. It is this last model (Model 5 in Reichle *et al.*, 1998) that we will outline below.

Cognitive processing assumptions The basic "engine" of our model is quite simple and was drawn from an earlier (non-quantitive) model of Morrison (1984): the reader encodes words in the text sequentially and, when various stages of word processing are completed, programs an eye movement and shifts attention to the next word. The only aspect of the model that is not completely straightforward is that the signal to program a saccade to the next word is different from the signal to shift attention to that word. In contrast, in some theories of spatial attention, the act of shifting attention is the signal to move the eyes. However, such an assumption would not allow one to explain several phenomena observed in reading, such as "spillover effects" — that the frequency of one word will often influence fixation times on the next word or two. As a result, our modeling proposes two stages of word processing, the *familiarity check* stage and the *lexical completion* stage. Completion of the familiarity check stage results in a program to fixate the next word in the text (i.e., the word to the right in English). Completion of the (later) lexical completion stage signals a shift of attention to the next word. The way we currently conceptualize these stages is that the former stage is the point at which the orthographic entry of the word (and possibly the phonological entry of the word) is excited to a sufficient degree that it is "safe" to assume that the word will be processed successfully, but the latter stage is when the orthographic, phonological, and semantic processing of the word is complete and thus attention can move to the next word with no confusions about word order and no (or minimal) crosstalk between the codes extracted from the two words. It is also worth noting that we assumed, in our original modeling, that the starting time for processing of a word was when attention shifted to that word (i.e., no lexical processing of a word occurred until attention was directed at it).

More specifically, each of these processes is assumed to be a function both of the log of the frequency of $word_n$ in the language, $\ln(freq_n)$, and the predictability of $word_n$ from the prior text in the discourse, p_n. The equations that we found most satisfactory for this purpose were the following.

$$t(f_n) = \{f_b - [f_m * \ln(freq_n)]\} * (1 - \theta * p_n) * \epsilon_1^x \quad (1)$$

$$t(lc_n) = \Delta * \{f_b - [f_m * \ln(freq_n)]\} * (1 - p_n) * \epsilon_2^x \quad (2)$$

Note that the second equation is for the *additional* time to complete lexical access after the familiarity check stage has completed. Also note that both equations are linear functions of the log of word frequency (with f_b and f_m, the free parameters of this linear function) and both are also functions of the word's predictability from the prior text. However, they are somewhat different functions of the two variables. We assume that the later stages of word processing are more influenced by the context (i.e., predictability) as indicated by the θ parameter in Equation 1 that attenuates the influence of predictability. However, note that the slope and intercept parameters in the two equations, f_b and f_m, are the same and that, as indicated above, the frequency values were obtained from norms and the predictability values were collected from norming studies we conducted (see below). Also note that the last term in the equations adjusts the speed of processing of a word by the distance the word is from the fixation point. However, our adjustment here was relatively crude, as the distance, x, is measured in words ($x = 0$ when $word_n$ is fixated, 1 when $word_{n-1}$ is fixated, etc.) The free ϵ parameters modulate this distance effect. The reason we used two different ϵ parameters here is that the familiarity check process was assumed to be cruder than lexical access, so that partial identification of letters and/or letter features is likely to be more sufficient for a familiarity judgment, than for full lexical access.

Before moving on to outline the rest of the machinery of the model, a few comments are in order. First, we are definitely not committed to these two variables, log frequency and predictability, as being the only, or even the best, predictors of the time to identify a word in text. In the first place, even if the number of times a reader saw a word in text was a perfect predictor of the time to access a word, it is clear that the standard Francis and Kučera norms (and most other frequency norms) are taken from a corpus that is likely to be unrepresentative of what most readers have encountered. (The Francis and Kučera norms are also drawn from a relatively small corpus.) Moreover, it is likely that many other variables could also influence the time to encode a word, such as the concreteness of the word, the part of speech of the word, the frequency that it is encountered in spoken language, and the age of acquisition. Similarly, the predictability norms that we are using are also crude estimates. They are obtained when the reader has no visual information about the target word, but has a large amount of time to use all the words in the sentence prior to the target word to "guess" what the next word will be. First, in real reading, the interaction between context and uptake of visual information is likely to be more interactive, with the partial visual information extracted helping to guide the influence of the context. Second, in reading text, the time course is reduced quite a bit, so that the predictability given several seconds to think about it may be only a crude estimate of how context can influence "on line" processing.

In sum, we are not using these measures because we believe they are the total explanation of word processing in reading. Instead, we are using them because they have both been shown to produce significant effects in reading when holding other variables constant. That is, even if they are flawed estimates, they are clearly related to important processing factors (i.e., how often the reader has seen the word before and how much top-down influence there is on the word). In fact, as later sections will indicate, we will also need to implicitly introduce word length as a separate variable and, when modeling compound words, we will need to incorporate encoding of word constituents as part of the word recognition process.

Eye control assumptions The above indicated how we viewed the word identification system to be hooked up to the attentional and eye movement systems. Below we spell out the details of how we assumed the eye movement system to work. The key assumption we made was that eye movements can be programmed in parallel, such that a new command can go out to the eye movement system regardless of whether a command has already been sent out and, furthermore, there is no queue. That is, all the eye movement programs go into a "buffer" storage and each are executed depending solely on when they reached the buffer (i.e., one program doesn't have to wait for another to complete before it starts to be executed). However, we do not assume that the commands go into the buffer and are independently executed. A key assumption of the model is that later commands can, in some cases, cancel earlier eye movement commands.

This cancellation assumption was inspired by work by Becker and Jürgens (1979), which employed a much simpler task: fixating a moving point of light which moved in two discrete steps, first to location 1 and then to location 2. Their key manipulation was varying the delay between the the first movement and the second. They found that when the delay was large, people executed saccades to location 1 and then to location 2. (The fixation between the two saccades was often short.) In contrast, when the delay was small, people only executed a saccade to location 2. (Moreover, they found that for intermediate delay intervals, people often made a saccade to a location intermediate between the two locations.) These data suggest (a) that eye movements to the two locations are programmed in parallel and (b) that the later program can either cancel or interact with the earlier program.

In our modeling, we did not try to incorporate any assumptions consistent with their finding that saccades were sometimes programmed to intermediate locations. Instead, to keep the model simple, we merely incorporated assumptions that were consistent with the data that indicated that execution of eye movement programs could be cancelled. As will be seen below, these assumptions are crucial in explaining how readers skip words. The way we modeled this was to assume two discrete states for an eye movement program that has been initiated: an initial labile state and a subsequent non-labile state. As long as the eye movement program is in the labile state, it is canceled by any subsequent eye movement program that is initiated, but once the program goes into the non-labile state, it can't be cancelled and will be executed when that state is terminated.

The cancellation of programs in the labile stage accounts for word skipping. That is, if while word_n is fixated, a program is sent out to fixate word_{n+1} followed quickly by a program to fixate word_{n+2}, then the program to fixate word_{n+1} will be cancelled by the program to fixate word_{n+2} and word_{n+1} will be skipped. When will this happen? Most often when word_{n+1} can be processed quickly: when it is high frequency and/or predictable from prior context. We should note that two qualitative phenomena naturally "fall out" of our model. First, the model predicts that when word_{n+1} is skipped, fixation times on word_n are longer than when word_{n+1} is not skipped. This is because skipping involves cancelling an earlier program with a later program. In fact, we obtained this pattern in two studies (Pollatsek, Rayner & Balota, 1986; Schilling, *et al.*, 1998, reported in Reichle *et al.*, 1998). However, Radach and Heller (2000) reported a contradictory finding (slightly shorter fixations preceding skips). Limited space precludes a full discussion of the issue, but unfortunately, all their analyses are flawed to some extent because they rely on correlational analyses, so that one is comparing fixations and skips on different words. However, in the Pollatsek *et al.* (1986) study, the analyses were limited to a small number of target words of approximately controlled frequency and length in fairly similar sentence frames. In contrast, it appears that in the Radach and Heller (2000) analysis, a large number of (fairly uncontrolled words) were used in an extended passage of text.[2] Second, fixation times on word_{n+2} are also lengthened by skipping word_{n+1}. This is accounted for by our assumptions above about word processing being influenced by the eccentricity of the word: when word_{n+1} is skipped, the reader is viewing word_{n+2} from a less advantageous place prior to fixating it than when word_{n+1} is not skipped.

The model, as outlined so far, can successfully predict the effects of frequency and predictability both on the *gaze duration* on a word (i.e., the total time spent fixating a word before it is exited by a fixation to the right) and the probability it is fixated. Each of these gaze durations, however, would be a single fixation on a word because the model, as outlined so far, has no mechanism for predicting refixations. Thus, it would predict that each word is either skipped or fixated exactly once. As a result, unless one wants to posit that refixations only occur due to errors in programming the target of a saccade, an additional mechanism is needed to predict refixations. In our initial modeling, the mechanism we posited for refixations was the following: immediately when a word is fixated, a program is produced to refixate the word (with the target at the word center). The motivation for this was that the middle of a word is not necessarily where the eyes initially land (due to programming errors), and one doesn't want to get stuck in a bad place. Although there is always a second fixation on a word programmed when the eyes land on a word, the second fixation often does not occur because the saccade to refixate the word can be canceled by the the program to fixate the next word. This cancellation occurs just as with between-word saccades (as this is all going on in the motor system which doesn't know one type of saccade from another). Qualitatively, this set of assumptions produces the correct pattern of data — that more frequent and more predictable words will be refixated less often. This is because when a word is easier to process (i.e., is more frequent or is more predictable), the program to fixate the next word will occur earlier and thus be more likely to cancel the program to refixate.

This model was applied, as indicated above, to the Schilling *et al.* (1998) corpus of data. The frequency of each word was obtained and predictability values for each word in the corpus were also obtained. Our modeling was at two levels. First we tried to predict the differences observed on target words that were either low-frequency (mean = 2 per million) or high-frequency (mean = 141 per million), and second, we tried to predict the entire pattern of fixation durations over the whole corpus of text. We will concentrate on the latter simulations here. When we applied the model to this corpus of data with the model outlined above (Model 5 in Reichle *et al.*), we obtained quite a good fit of the data. Moreover, almost all the parameter values of the model seemed reasonable (see Table 18.1). In order to simplify presentation of the simulations, we collapsed observed and predicted values over 5 frequency classes of words. As seen in Table 18.1, the model predicted the gaze durations on words and the probability of skipping quite well. It was a bit less successful in predicting the pattern for the duration of first fixations on a word and the probability of skipping. (Note that all fixation durations reported here are conditional on a word being fixated; that is, a skip of a word is counted as missing data rather than as a zero duration fixation.)

Before going on to describe how we have modified the model in ways that we think are both more psychologically plausible and will account for the data better, we need to fill in one detail that was passed over in our initial modeling efforts: exactly where the eyes land. In the modeling efforts above, we assumed that a word was fixated and that the saccade landed on the word. Aside from the issue of whether the targeted word is always the fixated word, a more complete model of eye movements in reading has to specify exactly where the eyes land. To fill in this gap, we made our assumptions about where saccades land more precisely, largely guided by an elegant analysis of landing positions in reading conducted by McConkie and his colleagues (McConkie, Kerr, Reddix & Zola, 1988; McConkie, Zola, Grimes, Kerr, Bryant & Wolff, 1991). Briefly, their data are fit by a model that posits that the middle of a word is the targeted location, but that there is both systematic error and random error influencing where the saccade actually lands. The systematic error is linear, with a bias to

Table 18.1: Observed and predicted values of gaze durations, probability of skipping, making a single fixation, and making two fixations for E-Z Reader 5 for five frequency classes of English words. (Predictions are from E-Z Reader 5.)

Freq. class	Mean freq.	Gaze duration		First fixation duration		Single fixation duration	
		Observed	Predicted	Observed	Predicted	Observed	Predicted
1	3	293	291	248	251	265	274
2	45	272	271	234	253	249	263
3	347	256	257	228	246	243	252
4	4889	234	226	223	223	235	224
5	40700	214	211	208	210	216	210

Freq. class	Probability of skipping		Probability of making a single fixation		Probability of making two fixations	
	Observed	Predicted	Observed	Predicted	Observed	Predicted
1	0.10	0.09	0.68	0.73	0.20	0.17
2	0.13	0.16	0.70	0.76	0.16	0.07
3	0.22	0.27	0.68	0.68	0.10	0.04
4	0.55	0.49	0.44	0.50	0.02	0.01
5	0.67	0.68	0.32	0.32	0.01	0.00

Notes:
The best fitting parameters for E-Z Reader 5 are: $f_b = 195$ ms, $f_m = 17$ ms; $\Delta = 0.70$, $t(m) = 150$ ms; $t(M) = 50$ ms; $\epsilon_1 = 1.25$; $\epsilon_2 = 1.75$; $\theta = 0.5$. The within-word motor programming parameters, $t(r)$ and $t(R)$, are assumed to have the same values as $t(m)$ and $t(M)$, respectively. Root mean squared deviation = 0.198.

overshoot for short saccade programs and a bias to undershoot for longer saccade programs. Moreover, there is random error, which increases as both the length of the intended saccade increases and as the fixation duration on the launch site decreases.

When we grafted this on to our original model of reading, we also obtained quite good fits to the pattern of landing positions on a word (Reichle *et al.*, 1999). This version of the model, E-Z Reader 6, was successful in that the model generated Gaussian landing site distibutions that were centered on the words. Also consistent with the McConkie *et al.* results, the variability of this predicted landing site distributions increased with increasing saccade length and as the durations of the launch site fixations decreased. Finally, an interesting emergent property of the model was discovered; namely, that the probability of making a refixation was minimal on short words and on words following initial fixations near the optimal viewing position. Both phenomena have been reported in the literature (Rayner, Sereno & Raney, 1996) and emerge from the model by virtue of the fact that the familiarity check is more likely to complete — and hence cancel the program to refixate — following fixations near the optimal viewing position (due to visual acuity constraints). Given our space limitations, we won't present the details of these simulations, but instead move on to our current modeling efforts.

An Improved Model of Reading in English

As indicated above, the model predicted both the gaze durations and the probability of skipping words quite well. However, it succeeded somewhat less well at predicting the durations of individual fixations on a word. Most notably, the model tended to predict first fixation durations that were not related to frequency in a monotonic fashion. The reason for this discrepancy between the observed and predicted data is that, in the model, the duration of the first fixation is determined by the outcome of a "race" between two processes: the familiarity check and the labile program to make a refixation. This means that, with low-frequency words, lexical processing is slow enough that the labile program to refixate almost always completes before the familiarity check. Thus the rapid programming and execution of these refixations causes the first fixations on low-frequency words to be quite short in duration. The net result, then, is the attenuation of — or, in the case of the lowest-frequency words, a complete elimination of — the word frequency effect on first fixation duration.[3]

This suggests that we may essentially have the right mechanism for fixating the next word, but not exactly the right mechanism for refixating a word. There were two other aspects of the model, however, that appeared to be in need of modification. The first was conceptual. Our model assumed that no processing of an upcoming word occurred prior to shifting attention to it. Among other things, this seems a bit incongruous, as some visual information needs to be obtained from the word in order for an eye movement to be programmed to its center. Although one could posit that only this very low-level information could be processed pre-attentively (but no letter information could be processed), we felt this was an unnatural division of attentive and pre-attentive processing. As a result, in our new modeling efforts (Reichle, Rayner & Pollatsek, 2002), we introduced what amounts to a third stage of word identification before the familiarity check stage, the pre-attentive stage. As this is intake of relatively raw featural information, the amount of processing in this stage is not influenced by the frequency of the word or its predictability. It is merely a function of physical characteristics of the word and how far the word is from fixation. The amount of information extracted during this preattentive stage (in term of ms of visual processing time, *vpt*) is given by Equation 3.

$$vpt = t \,/\, \epsilon_1^{\{\Sigma_i |\text{letter } i \text{ — fixation}| \,/\, N\}} \qquad (3)$$

In Equation 3, t represents time (in ms), N is the number of letters in the word being processed, and ϵ_1 is a free parameter that modulates how the average spatial disparity between the word's letters and the fixation location (i.e., the fovea) affect the amount of visual processing done from that location. The time needed to complete the initial visual processing of a word thus increases as the distance between a word's center and the current fixation location increases. Also note that this time increases with word length because the individual letters of longer words will (on average) be further away from the point of fixation than will the letters of shorter words. Consequently, our addition of this assumption allows the model to handle word length effects; that is, with all else being equal, longer words will take longer to identify than shorter words.[4]

This change in our word processing assumptions was necessary because our initial modeling did not take word length into account. Although this probably would not appear as a glaring problem in fitting a large corpus of data — as word length and word frequency are highly correlated — we viewed this as a serious conceptual problem. In the limit, if a word is so long that either its beginning or end letters are outside the window from which such meaningful information can be extracted, the word can not be processed on a single fixation, unless one wants to introduce a significant guessing mechanism into the model. As a result, we also modified our assumptions about word processing, making the initial stage of lexical processing (i.e., the familiarity check) more difficult, the further each individual letter was from fixation (see Equation 4). As a result of these assumptions, if the centers of two words are equally far from fixation and are equated on frequency and predictability, then the longer one will take longer to process. However, we should make clear that our equations do not explicitly posit that processing is slower for longer words; that characteristic falls out of our eccentricity assumptions. We think it is likely that other visual factors come into play as well. For example, there is some evidence (Bouma, 1973) that there is lateral inhibition between letters, such that end letters are processed faster than interior letters (as the space next to them reduces the lateral inhibition). However, to keep things simple, we did not introduce any assumptions about lateral inhibition. Note that Equation 4 is similar to Equation 3 with respect to how the average disparity between a word's letters and the center of visual processing affect the processing rate.

$$t(f_n) = \{f_b - [f_m * \ln(freq_n)]\} * (1 - \theta * p_n) * \epsilon_2^{\{\Sigma_i |\text{letter } i - \text{fixation}| / N\}} \tag{4}$$

As in Equation 1, f_b, f_m, and θ are free parameters that determine how the familiarity check time is modulated by each word's frequency and predictability. As in Equation 3, N is the number of letters in the word and ϵ_2 is a free parameter that determines the degree to which acuity limitations (and hence word length) affect the rate of processing. Because the last stage of word identification (i.e., the completion of lexical access) is less constrained by the raw visual information (but more affected by predictability), we deemed it reasonable that the processing rate for this later stage of processing should not be modulated by visual acuity. Thus, in E-Z Reader 7, visual acuity delimits both the rate of pre-attentive visual processing and the familiarity check, but not the completion of lexical access.

The above equations indicate how our word processing assumptions have been modified. However, they don't address the issue of how words are refixated. One problem that occurred to us is that our modeling efforts tended to predict that the first of two fixations can never be really short, as they are programmed when the person lands on a word. Moreover, when we examined the Finnish compound word data that we discuss in the next section, it appeared that we would never be able to predict mean first fixation durations as short as they were. Thus, it appears that the reader does not necessarily wait until fixating a word to program a refixation. We also noticed that there were data (not from reading, but from a simple oculomotor task involving judgments on two multimorphemic words) that indicated that two fixations on a word appeared to be programmed in parallel at the same time (Vergilino & Beauvillain, 2000).

Accordingly, we changed the basic mechanism for refixating a word. Instead of assuming that a refixation is automatically programmed when the word is landed on, we posited that readers program a second fixation to land on $word_{n+1}$ at the same time that they program the first one. Due to random variability in the duration of processes, however, they get executed at different times, and due to random variability in the location of where the saccade actually lands, the two saccades will usually land in different locations. Moreover, other saccade programs can cancel either or both of these programs if they are still in the labile stage. However, because they are programmed simultaneously, they do not cancel each other. Both are posited to land in the middle of the word. (It should be noted that the error associated with the second saccade depends upon the distance between its target — not where it is initiated — and the location of the first fixation.) Thus, in E-Z Reader 7, the primary saccade from one word to the next is programmed in parallel with a saccade that, if completed, will result in a refixation. The probability with which this second (refixation) saccade will be programmed is determined by two variables: the length of the up-coming word and the word's predictability, p_n. The probability of programming a refixation, *rp*, is given by Equation 5, where α is a free parameter that scales the degree to which word length and predictability affect the probability of refixating. Equation 5 is appealing because of its simplicity and because one might reasonably argue that long words and/or words that are unpredictable should be the recipients of multiple refixations as such words should be the most difficult to identify. Furthermore, both types of information (i.e., word length and predictability) become available very rapidly. Word length can be ascertained through low-spatial frequency processing in the visual periphery, and word predictability can be determined prior to being fixated via higher-level linguistic processing. Strictly speaking, one might expect that only a refixation on the most predictable word would be affected. However, one can view predictability norms as reflecting stochastic processes; that is, the word, which averaged over people's ratings, has a predictability norm of 0.5 may not be the word that half the people would predict in that situation.

$$rp = (1 - p_n) * \alpha * \text{length} \tag{5}$$

In addition to these changes to the structure of the model that we used to fit the data in Table 18.1, we also explicitly modeled the variability in where saccades land by incorporating the assumptions about both systematic and random error in where a saccade goes to (all saccades are assumed programmed to the center of a word). These assumptions imply that the "wrong" word may be fixated not infrequently. However, this is no problem for the model; the processing assumptions don't require that the saccade be accurate. The currently attended word is processed and its processing is determined by how far the letters in it are from fixation. The less accurate the saccade, the slower the processing, but that is the only consequence of the error in programming. Of course, this can sometimes (approximately 12% of simulation trials using the Schilling *et al.* sentences) result in interword regressions. This occurs whenever a saccade from the end of a word that is intended to produce a refixation overshoots its target (the word's center) and thereby results in the eyes landing on the preceding word. In such cases, the simulation trial is simply discarded so as to exclude regresssions. (As stated earlier, our model is not intended to explain interword regressions because we believe that, in many cases, these occur whenever high-level processing falters.)

We incorporated all these changes into E-Z Reader 7 and obtained an even better fit to the Schilling *et al.* data (see Table 18.2). Not only were the fits quantitatively better, they were also qualitatively better. Most noticeably, the pattern of predicted first fixation durations looked more like the observed values, as there is now a monotonic increase of first fixation duration with increasing word frequency. This is due to the fact that refixations are no longer determined by the outcome of a race between a labile program to make an intra-word saccade and the completion of the familiarity check; instead, the program to make an intra-word saccade is programmed in parallel to the program that initially moves the eyes to the word. As already mentioned, this change in the model was motivated by data indicating that — at least under some circumstances — two saccadic programs can be generated at the same time (Vergilino & Beauvillain, 2000).

We also think that many of the processing time assumptions in E-Z Reader 7 are more reasonable, largely because of the addition of the pre-processing stage. For example, in E-Z Reader 5, the processing time assumed for many of the function words in frequency class 1 was assumed to be unnaturally short (especially as the signal needs to travel from the eye to the cortex). In contrast, the assumed processing times for words in E-Z Reader 7 all seem quite reasonable and consistent with what is known about word processing and the visual system (e.g., the minimal time needed to fully identify the most frequent word in English, "the," when it is completely predictable is approximately 142 ms). As a result, E-Z Reader 7 is consistent with the known physiological constraints on the rate of visual processing, such as the fact that it takes approximately 90 ms for visual information to propagate along the optic pathway from the retina to the visual cortex. Similarly, the evidence from physiological studies (e.g., ERP) place the best estimates for the minimal amount of time needed for word identification in the range of 125–180 ms (Sereno, Rayner & Posner, 1998). Finally, our estimate of 245 ms for the mean time needed to program a saccade (including visual input time) is also more in line with previous estimates. For example, in a task that merely required subject to look at newly appearing visual targets, Becker and Jürgens (1979, p. 980) estimate the mean time to program and then initiate saccades at 285 ms. Likewise, in a task that required subjects to fixate sequentially, Rayner, Slowiaczek, Clifton and Bertera (1983) reported mean saccadic latencies in several of

Table 18.2: Observed and predicted fixation durations and probabilities for five frequency classes of English words. (Predictions are from E-Z Reader 7.)

		Gaze duration		**First fixation duration**		**Single fixation duration**	
Freq. class	**Mean freq.**	**Observed**	**Predicted**	**Observed**	**Predicted**	**Observed**	**Predicted**
1	3	293	287	248	248	265	265
2	45	272	274	234	242	249	257
3	347	256	255	228	238	243	245
4	4889	234	225	223	221	235	222
5	40700	214	208	208	207	216	207

	Probability of skipping		**Probability of making a single fixation**		**Probability of making two fixations**	
Freq. class	**Observed**	**Predicted**	**Observed**	**Predicted**	**Observed**	**Predicted**
1	0.10	0.12	0.68	0.73	0.20	0.14
2	0.13	0.16	0.70	0.72	0.16	0.11
3	0.22	0.24	0.68	0.69	0.10	0.06
4	0.55	0.50	0.44	0.48	0.02	0.01
5	0.67	0.66	0.32	0.35	0.01	0.00

Notes:
The best fitting parameters for E-Z Reader 7 are: $f_b = 160$ ms; $f_m = 5$ ms; $\Delta = 0.4$; $\theta = 0.5$; $t(m) = 170$ ms; $t(M) = 70$ ms; $\epsilon = 1.08$; $\lambda = 0.05$; $\psi_b = 7$; $\beta_b = 0.85$; $\beta_m = 0.11$; $\Omega_b = 7.3$; $\Omega_m = 4.5$. Root mean squared deviation = 0.119.

their conditions well in excess of 200 ms. Thus, E-Z Reader 7 indicates that word identification can in fact be the signal to move the eyes without having to posit unreasonably short saccadic latencies.

Finally, it is worth emphasizing that all of the above modifications were made without changing the model's basic assumptions (e.g., serial shifts of attention), nor by compromising other aspects of the model's performance. As reported elsewhere (Reichle *et al.*, 2002), E-Z Reader 7 actually does a better job predicting all of the various eye-movements-in-reading phenomena (e.g., refixation probabilities, parafoveal preview effects, etc.) than do previous versions of the model. Admittedly, E-Z Reader 7 has some additional free parameters, but these have not been introduced for mere "curve fitting" convenience; they have been introduced to capture real phenomena in visual processing, such as greater difficulty in processing individual letters, the further they are from fixation. As a result, we think that E-Z Reader 7 is a very good approximation to how the eyes move forward in the silent reading of English text.

Modeling the Processing of Finnish Compound Words

All the above modeling assumes that words and letters are the only relevant components in word identification, and that stages in the identification of the entire word — the familiarity check on the whole word and the lexical completion of the whole word — are the only signals sent out by the language processing system to the eye movement and attentional systems in the process of silent reading. However, there are now several studies that indicate that the gaze duration on a word and other measures of fixation time and location are affected by subword units, notably constituents and morphemes (e.g., Niswander, Pollatsek & Rayner, 2000). Perhaps the most extensive set of eye movement data on the processing of polymorphemic words is a series of studies on 12-letter Finnish compound words by Hyönä and Pollatsek (1998) and Pollatsek, Hyönä and Bertram (2000). In Experiment 2 of Pollatsek *et al.* (2000), the whole-word frequency of a two constituent compound word was shown to have a significant effect on gaze duration and other measures. However, the frequency of both the first constituent (Hyönä & Pollatsek, 1998, Experiment 2) and the second constituent (Pollatsek *et al.*, 2000, Experiment 1) also had a large effect on various fixation duration measures. Moreover, the length of the first constituent had a significant effect on where people refixated on the word. These experiments thus clearly indicate that processing of word parts is influencing the eye movement system.

The obvious question, then, is whether a model such as E-Z Reader 7 can be adapted to fit the data from these experiments or whether the model has to be fundamentally reconceptualized to account for these data. Qualitatively, the Finnish compound word data appear to be consistent with a "dual process" view of processing complex words, in which there are two "routes" to accessing the meaning of the word: (a) a compositional route, in which the constituents are accessed and then combined; and (b) a direct route, in which the entire word form is accessed as a whole. Moreover, as indicated above, the data suggest that both "routes" have access to the eye movement system. The process of quantitatively modeling the data makes clear, however, that all sorts of decisions have to be made about both routes — both how they work and relate to each other and how they influence the eye movement system. This necessary increase in precision allows us to draw some conclusions that would be very difficult to make without such modeling. Specifically, as we shall see, the modeling exercise casts doubt on whether the simplest (and most common) version of such a dual route model is an adequate explanation for the processing of polymorphemic words in reading.

One decision that has to be made is whether the two constituents of the word are accessed in parallel or in series.[5] We thought that, consistent with our view of how words in text are processed, it was more parsimonious to assume that the constituents were accessed in series (from left to right). One also has to decide how the two routes relate to each other. Do they massively interact or are they parallel independent routes to the meaning of the compound word? Again, in the spirit of parsimony, we decided to try the simplest option first: a parallel "race" model, in which access of the word's meaning would be accomplished by whichever route finished first. (Further processing by the route that loses the race is assumed either to be terminated or irrelevant to the language processing and eye movement systems.) This summarizes the basic structure of the assumed cognitive processing. However, we now need to fill in details about both (a) how long processing is supposed to take place as a function of word characteristics and constituent characteristics and (b) how word processing and constituent processing communicate with the eye movement system.

Simulation results Our first simulation used the standard version of E-Z Reader 7 (referred to below as version 7A) which, as indicated above, assumed a parallel "horse-race" model for these compound words. Note that, as with our basic Model 7, we assumed there is a preprocessing stage on the second constituent occurring in parallel with attentional processing of the first constituent. We assumed that the times for the familiarity check and lexical access stages for either of the constituents or the whole word are given by its frequency in the language and the eccentricity of the letters as given above in Model 7.

If the familiarity check on the word completes before the familiarity check on the first constituent, then the oculomotor system starts programming a saccade to the next word, and lexical processing on the compound word (both the whole word and its first constituent) continues. The completion of lexical access of the whole word then causes attention to shift to the next word. Thus, when the whole-word lookup process beats the first constituent lookup process, the current model is exactly the same as previous versions of E-Z Reader. However, when the familiarity check completes on the first constituent before it completes on the whole word, the oculomotor system begins programming a saccade to the second constituent (rather than to the next word). If lexical processing on the first constituent then finishes, attention shifts to the second constituent and lexical processing of that unit is initiated.[6] Subsequently, the familiarity check on the second constituent may complete, causing the oculomotor system to program a saccade to the next word. (Alternatively, if the whole-word familiarity check completes, this will cancel any pending labile saccadic programs — including those that would otherwise move the eyes to the second constituent — and will instead initiate a program to move the eyes to the next word.) Thus, in cases where the complex word is identified through constituent composition, each constituent is functionally equivalent to a smaller word.

The model, as described thus far, was used to simulate the Hyönä and Pollatsek (1998) and Pollatsek *et al.* (2000) experiments. In these experiments, as with the Schilling *et al.* corpus, people read single sentences, and there was no return sweep prior to encountering the target word region. Also (similar to Schilling *et al.*), the design was a counterbalanced one in which either a matched high- or low-frequency version of a compound word appeared in a sentence frame for a given subject. In each of the simulations reported below, we first computed the mean normative frequencies and lengths of the compound words and their constituents. (The mean frequencies are reported at the bottom of Table 18.3.) The compound words (i.e., targets) were then inserted into sentence frames constructed from the Schilling *et al.* (1998) corpus; by inserting the targets into the sixth word position of the Schilling *et al.* sentences, we ensured that the words immediately preceding and following the targets would vary with respect to all of the characteristics that are known to affect eye movements (e.g., frequency, length, etc.). This would help ensure that our simulation results were not dependent on a particular type of frame. However, they are inaccurate to the extent that these frames are different from the frames actually used. Each simulation was completed using 1,000 statistical subjects per experimental condition. Finally, unless otherwise noted, the simulations were also completed using the same free parameter values that have been used in our simulations of the Schilling *et al.* sentences.

Somewhat unexpectedly, the results of our first simulation showed absolutely no effect of the frequency of either constituent, although they predicted a 17 ms effect for word frequency. For example, in our simulation of Experiment 1 of Pollatsek *et al.* (2000), the low- and high-frequency second constituents produced exactly the same gaze durations (mean = 390 ms), whereas the observed values were 626 ms and 531 ms, respectively. These results indicate that such a model is quite inadequate as an explanation of compound word processing, unless quite different assumptions are made about the speed of processing these word and component entities than are made about the processing of standard words.[7]

To deal with this problem, we abandoned the "horse-race" model and assumed that the identification of compound words proceeds only through morphemic composition.[8] Thus, in the second incarnation of the model (E-Z Reader 7B), the compound words are only identified via serially processing the two constituents of each target word. On some level, this modification makes the model simpler, as compound words can only be identified through morphemic composition, and it can also be viewed as being an endorsement of the basic claim that attention is allocated serially during reading (see Footnote 6). That is, if our model is correct, then attention shifts from one linguistic unit (word or word constituent) to the next as the individual units are identified.

As one might expect, E-Z Reader 7B predicted sizeable constituent frequency effects. For example, it predicted gaze durations of 516 and 483 ms for target words having low- and high-frequency first constituents in Hyönä and Pollatsek (1998) and gaze durations of 515 and 492 ms for targets having low- and high-frequency second constituents, in Pollatsek *et al.*'s (2000) first experiment. Despite this improvement in the model's performance, however, the overall gaze durations were still much too short (e.g., predicted values of 515 and 492 ms vs. observed values of 626 and 521 ms) and the predicted constituent frequency effects were too small. Furthermore, as might be expected, the removal of the model's whole-word lookup route completely eliminated the effects of whole-word frequency. Thus, by removing whole-word lookup to handle one aspect of the data (constituent frequency effects), we changed the model so that it can no longer handle another aspect of the data (whole word frequency effects).

This problem led us to think more deeply about whether our basic assumptions were plausible. It occurred to us that looking up the constituents was not tantamount to looking up the compound word. That is, even for putatively transparent compounds, such as "cowboy", given that one knows the meaning of "cow" and "boy", one still does not know what the compound word means (e.g., is a "cowboy" a calf?). Thus, it appears that some sort of composition stage is needed in the processing of the compound word as well.[9] If one adds such a "constituent gluing" stage to our model, and furthermore makes the reasonable assumption that this "glue time" is sensitive to the frequency of the whole word, then one has a chance to account for both constituent and whole word frequency effects with the model. Accordingly, we added this stage (which starts after the lexical identification of the second constituent), and assumed that the "glue" time, *gt*, is a linear function of whole-word log frequency, where γ_b and γ_m are free parameters that determine how *gt* (in ms) is related to word frequency. In the simulation results presented below (in Table 18.3), the values of γ_b and γ_m were set equal to 115 ms and 70 ms, respectively. (The value of *gt* was set equal to zero in those cases where *t* would otherwise be less than zero.)

$$gt = \gamma_b - \gamma_m * \ln(\text{frequency}) \qquad (6)$$

The addition of morphemic "glue" time, however, only produces whole-word frequency effects in the eye movement record if the signal to move from the compound word to the next is the completion of this process and not the completion of the familiarity check on the second morpheme. This being the case, we modified the model so that only the completion of the combinatory process (following the completion of lexical access of the second morpheme) signals the oculomotor system to move the eyes to the next word. Given our assumption that the "identification" of compound words occurs only after the individual constituents have been both identified and combined, we reasoned that the completion of this process would play the same functional role as the familiarity check on single-constituent words. In addition, we also modified the model so that the completion of the familiarity check on the second morpheme would cause the eyes to move to yet another location within the compound word. We reasoned that this last change would increase the overall number of fixations, which in turn should both increase the gaze durations and decrease the durations of the individual (i.e., first and second) fixations.

Model 7C now successfully predicted both constituent and whole-word frequency effects. However, the predicted constituent frequency effects were much too small. One possible reason for this shortcoming is the fact that not only were the low frequency constituents low frequency, they were unique. That is, the compound word one of the constituents was in was the only compound word in the language that had this constituent in that position. This suggests that the overall frequency of the low frequency constituents may be overestimating how rapidly they are processed in this unusual context. To correct this problem, we added a modest amount of time (15 ms) to both processing stages (i.e., the familiarity check and the completion of lexical access) for the low-frequency constituents of our target words. As expected, this change enhanced the size of the predicted constituent frequency effects. In the simulation of Hyönä and Pollatsek's (1998) second experiment, for instance, Model 7D predicted a 66 ms gaze duration difference between targets having low- and high-frequency first constituents, which was close to the observed difference of 87 ms (see Table 18.3). One other aspect of the simulation is noteworthy. The individual constituent frequencies are — as observed — having their effects at different points in time. In our simulation of Hyönä and Pollatsek's (1998) second experiment, the frequency of the first constituent produced a frequency effect on both the first and second fixation durations (14 and 17 ms frequency effects, respectively). In contrast, our simulation of Pollatsek *et al.*'s (2000) first experiment successfully predicted that

Table 18.3: E-Z Reader 7D simulation results.

Hyönä & Pollatsek (1998) Experiment 2	Observed			Predicted		
	LF[1] First constit.	**HF[1] First constit.**	**Diff.**	**LF First constit.**	**LF[1] First constit.**	**Diff.**
Gaze dur. (ms)	522	435	87**	610	544	66
1st fix. dur. (ms)	197	188	9**	249	238	11
2nd fix. dur. (ms)	197	188	9*	293	276	17
Prob. of 1st fix.	1	1	0	1	1	0
Prob. of 2nd fix.	0.87	0.85	0.02	0.88	0.84	0.04
Prob. of 3rd fix.	0.45	0.30	0.15**	0.31	0.24	0.07

Pollatsek *et al.* (2001) Experiment 1	Observed			Predicted		
	LF[2] Second constit.	**HF[2] Second constit.**	**Diff.**	**LF[2] Second constit.**	**HF[2] Second constit.**	**Diff.**
Gaze dur. (ms)	626	531	95**	675	638	37
1st fix. dur. (ms)	215	214	1	236	242	–6
2nd fix. dur. (ms)	212	204	8*	262	255	7
Prob. of 1st fix.	0.99	1	–0.01	1	1	0
Prob. of 2nd fix.	0.91	0.88	0.03	0.95	0.93	0.02
Prob. of 3rd fix.	0.58	0.44	0.14*	0.62	0.55	0.07

Pollatsek *et al.* (2001) Experiment 2	Observed			Predicted		
	LF[3] Word	**HF[3] Word**	**Diff.**	**LF[3] Word**	**HF[3] Word**	**Diff.**
Gaze dur. (ms)	427	345	82**	618	535	83
1st fix. dur. (ms)	200	195	5*	250	249	1
2nd fix. dur. (ms)	180	164	16**	261	283	–22
Prob. of 1st fix.	0.97	0.97	0	1	1	0
Prob. of 2nd fix.	0.70	0.65	0.05**	0.89	0.80	0.09
Prob. of 3rd fix.	0.32	0.15	0.17**	0.47	0.19	0.28

Notes:

1. LF = 9 per million; HF = 551 per million
2. LF = 2 per million; HF = 359 per million
3. LF = 1 per million; HF = 41 per million
4. ** indicates differences that were statistically reliable.
5. * indicates differences that were marginally reliable.

the frequency of the second constituent did not affect the first fixation duration but instead only would affect the durations of subsequent fixations.

Although Model 7D predicts all the gaze duration effects and the effects of the constituent frequencies on the pattern of individual fixations quite well, there appear to be two major discrepancies between the model's predictions and the data. The first is that, in all of the simulations mentioned thus far, the model consistently predicted individual fixation durations that were too long. In the observed data, for instance, the mean first fixation durations were between 189 and 215 ms, whereas in the simulations using model 7D these same means were between 232 ms and 252 ms.

We attempted to remedy this problem by adjusting three of the free parameter values: First, we decreased the time needed to complete the labile stage of saccadic programming from 170 ms to 150 ms. Our logic for doing this was as follows: By decreasing the time needed to complete the labile stage of saccadic programming, we would also decrease the durations of individual fixations because the length of the latter are determined by the amount of time needed to move the eyes from one location to another. Second, we increased the value of the parameter that modulates the degree to which the length of an upcoming saccade target (word or second morpheme) influences automatic refixations. By doing this, we reasoned that we would increase the probability with which long saccade targets would be the recipients of automatic refixations, and would thereby both increase the overall proportion of refixations and decrease the durations of individual fixations (as automatic refixations are short). Finally, we decreased the preferred saccade length (i.e., the saccade length for which the oculomotor system tends to neither over- or undershoot its intended target) from seven to 3.5 character spaces for intraword saccades. Again, we reasoned that this change would increase the number of refixations to the extent that intraword saccades would be less likely to overshoot their targets — the words' second morphemes. These three changes (which we will refer to as model 7E), though admittedly ad hoc, are nevertheless consistent with a rather simple notion that Finnish readers are more prone (than are English readers) to trade accuracy for speed by making several rapid, short saccades within long words. These assumptions were used in the simulations presented in Table 18.4.

Overall, the results predicted by the model are quite similar to those observed across the experiments. As can be seen in Table 18.4, our modifications of model 7D had the intended results: The overall proportion of second and third fixations increased, causing the durations of individual fixations to decrease. Moreover, these changes did not adversely affect the model's overall performance; both constituent frequency and whole-word frequency effects are evident, these effects are approximately the correct magnitude, and the frequency of the first constituent has an earlier effect than does the frequency of the second constituent.

There are two main discrepancies between model 7E and the data. The first is that the model did not predict any effect of whole-word frequency on the second fixation duration. We think there is a solution to the problem, but it would require collecting additional data. That is, it is plausible that whole-word frequency exerts its influence not only by increasing the speed of the glue time (which mainly affects refixation probabilities), but also by making the second constituent more predictable given the first

Table 18.4: E-Z Reader 7E simulation results.

Hyönä & Pollatsek (1998) Experiment 2	**Observed**			**Predicted**		
	LF[1] First constit.	**HF[1] First constit.**	**Diff.**	**LF[1] First constit.**	**HF[1] First constit.**	**Diff.**
Gaze dur. (ms)	522	435	87**	567	490	77
1st fix. dur. (ms)	197	188	9**	194	175	19
2nd fix. dur. (ms)	197	188	9*	211	201	10
Prob. of 1st fix.	1	1	0	1	1	0
Prob. of 2nd fix.	0.87	0.85	0.02	0.99	0.97	0.02
Prob. of 3rd fix.	0.45	0.30	0.15**	0.61	0.46	0.15

Pollatsek *et al.* (2001) Experiment 1	**Observed**			**Predicted**		
	LF[2] Second constit.	**HF[2] Second constit.**	**Diff.**	**LF[2] Second constit.**	**HF[2] Second constit.**	**Diff.**
Gaze dur. (ms)	626	531	95**	625	564	61
1st fix. dur. (ms)	215	214	1	178	180	–2
2nd fix. dur. (ms)	212	204	8*	192	176	16
Prob. of 1st fix.	0.99	1	–0.01	1	1	0
Prob. of 2nd fix.	0.91	0.88	0.03	1	0.99	0.01
Prob. of 3rd fix.	0.58	0.44	0.14*	0.91	0.84	0.07

Pollatsek *et al.* (2001) Experiment 2	**Observed**			**Predicted**		
	LF[3] Word	**HF[3] Word**	**Diff.**	**LF[3] Word**	**HF[3] Word**	**Diff.**
Gaze dur. (ms)	427	345	82**	593	537	56
1st fix. dur. (ms)	200	195	5*	188	189	–1
2nd fix. dur. (ms)	180	164	16**	192	223	–31
Prob. of 1st fix.	0.97	0.97	0	1	1	0
Prob. of 2nd fix.	0.70	0.65	0.05**	0.99	0.97	0.02
Prob. of 3rd fix.	0.32	0.15	0.17**	0.81	0.46	0.35

Notes:
1. LF = 9 per million; HF = 551 per million
2. LF = 2 per million; HF = 359 per million
3. LF = 1 per million; HF = 41 per million
4. ** indicates differences that were statistically reliable.
5. * indicates differences that were marginally reliable.

constituent (all else being equal). However, to access this adequately, we would need to get predictability data. The second problem is that the predicted refixation probabilities are too high (although the differences between the high- and low-frequency conditions are predicted reasonably well). This may also indicate that we've over-relied on the automatic refixation mechanism in model 7E. That is, a mechanism that relies on predictability will influence both fixation durations and the probability of making a refixation, whereas the glue-time mechanism (occurring very late in the process) is primarily affecting the number of fixations (and is creating too many of them).

We think this modeling exercise was successful, as we have demonstrated that one can predict the general pattern of eye movements that are observed when the morphemic characteristics of compound words are manipulated. Clearly, one needs additional assumptions and free parameters, as one is trying to model new phenomena that clearly draw on additional processes (processing of subword linguistic units). We also think this demonstration is informative because it suggested, contrary to our initial intuitions, that a parallel "race" model of morphemic processing does not appear to be able to handle the observed data for these long compound words. Our simulation results instead suggest that one is likely to be able to explain both constituent and whole-word frequency effects only by assuming a strictly compositional model of morphemic processing. Thus, word constituents are, to a large extent, treated like smaller words (which are not separated by spaces) by those systems that guide the eyes during reading and are processed sequentially. However, we are skeptical that this strictly serial model would generalize to processing of other polymorphemic words in English, such as inflected or derived words, where one of the morphemes (the prefix or suffix) is often only 1–3 letters.

Finally, although we have so far focused exclusively on predicting fixation durations and fixation probabilities in the Finnish corpus, it is worth noting that the model also did a fairly good job of predicting the locations of individual fixations. For example, in their first experiment, Hyönä and Pollatsek (1998) manipulated the length of the first constituent while holding word length, constituent frequency, and whole-word frequency constant. Although the length of the first constituent did not affect where the eyes first landed within the compound words, subsequent fixations were reliably affected by first constituent length. For instance, when the first constituent was short (mean length = 4 letters), the second fixation landed (on average) 8.2 letters into the targets, whereas when the first constituent was long (mean length = 8.5 letters), the second fixation landed 8.7 letters into the targets. The model predicted second fixation locations that were in close agreement to the observed values: With short and long first constituents, the mean locations of the second fixations were 7.2 and 7.9 letters, respectively. In both cases, the second fixation landed one half of a character space further into the word when the first morpheme was long. The complete set of observed and predicted fixation locations for Hyönä and Pollatsek's first experiment are presented in Table 18.5. At least on first pass, the model predicts both the absolute and relative locations of individual fixations in the data being simulated. That is, successive fixations move the eyes further into the compound words, and there is a disparity between how far the eyes move into the words with the eyes moving further into the words containing long first morphemes. Finally, note that the absolute fixation locations are in fairly close agreement to the observed locations; in all cases, the predicted locations are within one character space of the observed locations.

Table 18.5: Observed and predicted fixation locations.

	Observed (Hyönä & Pollatsek, 1998, Experiment 1)			Predicted (E-Z Reader 7E)		
Fixations	**Long (8.5 character spaces) 1st constit.**	**Short (4 character spaces) 1st constit.**	**Difference**	**Long (8.5 character spaces) 1st constit.**	**Short (4 character spaces) 1st constit.**	**Difference**
1st	4.6	4.6	0	4.2	4.1	0.1
2nd	8.7	8.2	0.5*	7.9	7.2	0.5
3rd	9.0	8.8	0.2*	9.3	8.1	1.2

Notes: 1. * indicates differences that were statistically reliable.

Final Comments

Perhaps, in conclusion, we should "step back" a bit from the details of the modeling and again take a look at the purpose of our modeling enterprise. Our goal was to model eye movements in text with a model of suitable complexity such that one should be able: (a) to quantitatively fit a reasonably interesting and complex data set, but (b) to know whether something is basically wrong with one's assumptions and either modify them accordingly or conclude that the model is fundamentally inadequate. We think our latest simulations in English (shown in Table 18.2) indicate that the basic model we have posited is certainly a good approximation to the truth of how word frequency affects eye movement behavior in reading.

In contrast, there are clearly unresolved questions about the adequacy of our attempts to model Finnish compound words. First, we needed to put in a relatively ad-hoc assumption to explain the sizes of the constituent frequency effects observed in these experiments. However, it should be noted that in some more recent experiments by Pollatsek & Hyönä (2001) in which the first constituent frequency was varied in a less extreme way, the first constituent frequency effect was only about 30 ms, and it is likely that the model could predict the size of this effect without resorting to ad-hoc assumptions. A second possible problem is that we had to introduce a new assumption about the whole-word composition process taking over and determining whether a fixation would go to the next word. That is, it prevented successful completion of processing of the second constituent from allowing the eyes to move on. This is somewhat contrary to the spirit of the initial E-Z Reader model because this composition process is not necessarily dependent on continued visual processing of the letters. This suggests that we should look for an alternative explanation for whole-word frequency effects. Perhaps some version of "predictability" (conditional frequency of the second constituent given the first constituent) may work.

Perhaps a final comment is in order about our abrupt dismissal of the "horse-race" model for compound word processing. Although, we have not had the time to try all plausible assumptions of processing times for the whole words and for the constituents to be able to rule out any possible such model, we think it is extremely unlikely that any such model one could predict both sizeable contituent frequency effects and word frequency effects. What might work is a dual-route interactive model, where the whole-word processes "feed into" the constituent processing, and vice versa. In fact, our current model 7E, in which there is a constituent glueing stage after the constituent parts are identified, could be viewed as a simple and special case of such an interactive model. The problem with such an interactive parallel model, however, is that it is close to being a "Turing machine" (i.e., some version could probably predict any pattern of data, whether observed or not).

Regardless of the final answers to these questions, however, we think that our initial modeling efforts here have been extremely valuable. That is, when one views the pattern of data of these Finnish compound words experiments, the "obvious" explanation (qualitatively) is an independent dual-route "race" type of model. (Similar qualitative models have been proposed in other domains of word processing where there are compositional and "whole-word" effects.) What quantitative modeling adds is the precision to determine whether such hypotheses really can account for the data. And as we have argued, in this case, it's not at all clear that any such model can.

Notes

1 A recent experiment (Altarriba, Kambe, Pollatsek & Rayner, 2001) similarly found that when English–Spanish bilinguals read text in either language that a preview that was a translation of the target word provided no significant benefit over an unrelated control preview.

2 The problem with a long (and varied) text is that it could vary in difficulty quite substantially from passage to passage, which could also change global reading strategies, such as people requiring higher thresholds of activation for word identification in difficult passages. Such changes in strategies would induce a positive correlation (i.e., higher thresholds would mean less skipping and longer fixation durations).

3 There's a change in the model, different from the one described in the current section, that is closer to the original model and that may fix this problem. That is, in the Becker and Jürgens (1979) experiments, there was also evidence for "compromise saccades" (i.e., intermediate in position between the initial and second saccade targets) when the interval between the two stimuli was intermediate between both saccades being executed and a single one executed to the second target. Thus, a later saccade program intended for the next word (triggered by the familiarity stage) can interrupt and delay the saccade, but then produce an intermediate saccade that is still on the first word. As a result, first fixation durations on a word may indeed be sensitive to word frequency (and related variables) with a model of refixations closer in spirit to the original model than the model outlined in the current section.

4 Strictly speaking, Equations 3 and 4 produce word length effects (holding the eccentricity of the center of the word constant) only if the word straddles the fixation point. We used the arithmetic mean of the absolute distances in these formulas because of computational simplicity. However, if this were changed to some other combination rule (e.g., the geometric mean), then the equation would predict word length effects in all cases.

5 Finnish is a highly inflected language, so that a few of the constituents of these Finnish compound words were polymorphemic, containing a root morpheme and an inflectional ending. However, we did not try to model at the morphemic level in this analysis; we treated a constituent as a whole whose ease of recognition was given by the frequency of the constituent.

6 This assumption is seemingly at odds with the core assumption of E-Z Reader regarding the serial allocation of attention; that is, if attention shifts from one word to the next in a strictly serial manner, then how can it be both fixed on a whole word unit (i.e., a compound word) and simultaneously moved from the word's first constituent to its second? This inconsistency provides one logical basis for our later conclusions that attention is initially restricted to the first constituent, and that complex words are identified solely through morphemic composition. Under this type of strong serial-attention assumption, individual constituents are functionally equivalent to words that are not separated by spaces; attention shifts from one linguistic unit (word or constituent) to the next as each new unit is identified.

7 The race model might even be in worse shape if component processing is slowed down further, relative to whole-word processing, by an initial "parsing" stage needed to break down the compound into its components.

8 Given our comments above justifying the sequential nature of the component processing, it seemed somewhat inconsistent to assume that these long compounds could be processed as a whole — especially to assume that the whole word could be processed in parallel starting at the initial fixation. One might be able to modify the "direct look-up" route to make it more plausible by starting it after a couple of fixations have rendered most of the letters visible. However, we suspect that such a model might be conceptually fairly similar to the model we end up with.

9 This would be true even in a race model, unless the compositional stage were irrelevant. That is, when the compositional stage "wins", one still needs this assembly process.

References

Altarriba, J., Kambe, G., Pollatsek, A., & Rayner, K. (2001). Semantic codes are not used in integrating information across eye fixations: Evidence from fluent Spanish-English bilinguals. *Perception and Psychophysics*, *63*, 875–891.

Becker, W., & Jürgens, R. (1979). An analysis of the saccadic system by means of double step stimuli. *Vision Research*, *19*, 967–983.

Bouma, H. (1973). Visual interference in the parafoveal recognition of initial and final letters of words. *Vision Research*, 1*3*, 767–782.

Carroll, P., & Slowiaczek, M. L. (1986). Constraints on semantic priming in reading: A fixation time analysis. *Memory & Cognition*, *14*, 509–522.

Clark, J. J., & O'Regan, J. K. (1999). Word ambiguity and the optimal viewing position in reading. *Vision Research*, *39*, 843–857.

Coltheart, M., Rastle, K., Perry, C., Langdon, R., & Ziegler, J. (2001). DRC: A dual route cascaded model of visual word recognition and reading aloud. *Psychological Review*, *108*, 204–256.

Ehrlich, K., & Rayner K. (1983) Pronoun assignment and semantic integration during reading: Eye movements and immediacy of processing. *Journal of Verbal Learning and Verbal Behavior*, *22*, 75–87.

Frazier, L., & Rayner, K. (1982). Making and correcting errors during sentence comprehension: Eye movements in the analysis of structurally ambiguous sentences. *Cognitive Psychology, 14*, 178–210.

Hyönä, J., & Pollatsek, A. (1998). Reading Finnish compound words: Eye fixations are affected by component morphemes. *Journal of Experimental Psychology: Human Perception and Performance, 24*, 1612–1627.

Inhoff, A. W., Starr, M., & Shindler, K. L. (2000). Is the processing of words during eye fixations in reading strictly serial? *Perception & Psychophysics, 62*, 1474–1484.

Just, M. A., & Carpenter, P. A. (1980). A theory of reading: From eye fixations to comprehension. *Psychological Review, 87*, 329–354.

Kennedy, A. (1998). The influence of parafoveal words on foveal inspection time: Evidence for a processing trade-off. In: G. Underwood (ed.), *Eye Guidance in Reading and Scene Perception* (pp. 149–179). Oxford: Elsevier.

Kennedy, A. (2000). Parafoveal processing in word recognition. *Quarterly Journal of Experimental Psychology, 53A*, 429–455.

Kolers, P. (1972). Experiments in reading. *Scientific American, 227*, 84–91.

Legge, G. E., Klitz, T. S., & Tjan, B. S. (1997). Mr Chips: An ideal-observer model of reading. *Psychological Review, 104*, 524–553.

McConkie, G. W., Kerr, P. W., Reddix, M. D., & Zola, D. (1988). Eye movement control during reading: I. The location of initial eye fixations on words. *Vision Research, 28*, 1107–1118.

McConkie, G. W., & Rayner, K. (1975). The span of the effective stimulus during a fixation in reading. *Perception & Psychophysics, 17*, 578–586.

McConkie, G. W., Zola, D., Grimes, J., Kerr, P. W., Bryant, N. R., & Wolff, P. M. (1991). Children's eye movements during reading. In: J. F. Stein (ed.), *Vision and Visual Dyslexia* (pp. 251–262). London: Macmillan Press.

Morris, R. K. (1994). Lexical and message-level sentence context effects on fixation times in reading. *Journal of Experimental Psychology: Learning, Memory, and Cognition, 20*, 92–103.

Morrison, R. E. (1984). Manipulation of stimulus onset delay in reading: Evidence for parallel programming of saccades. *Journal of Experimental Psychology: Human Perception and Performance, 10*, 667–682.

Murray, W. S., & Rowan, M. (1998). Early, mandatory, pragmatic processing. *Journal of Psycholingusitic Research, 27*, 1–22.

Myers, J. L., & O'Brien, E. J. (1998). Accessing the discourse representation while reading. *Discourse Processes, 26*, 131–157.

Niswander, E, Pollatsek, A. & Rayner, K. (2000). The processing of derived and inflected suffixed words during reading. *Language and Cognitive Processes, 15*, 389–420.

Pollatsek, A. & Hyönä, J. (2001). The role of morphological constituents in reading Finnish compound words. *Journal of Experimental Psychology: Human Perception and Performance, 26*, 820–833.

Pollatsek, A., Hyönä, J., & Bertram, R. (2001). The roles of first morpheme frequency and semantic transparency in processing Finnish compound words in reading. Paper presented at the morphology workshop, Nijmegen, Holland. (June).

Pollatsek, A., Rayner, K., & Balota, D. A. (1986). Inferences about eye movement control from the perceptual span in reading. *Perception & Psychophysics, 40*, 123–130.

Radach, R., & Heller, D. (2000) Relations between spatial and temporal aspects of eye movement control. In: A. Kennedy, R. Radach, D. Heller and J. Pynte (eds), *Reading as a Perceptual Process* (pp. 165–192). Oxford: Elsevier.

Rayner, K. (1975). The perceptual span and peripheral cues in reading. *Cognitive Psychology, 7*, 65–81.

Rayner, K. (1998). Eye movements in reading and information processing: 20 years of research. *Psychological Bulletin, 124*, 372–422.

Rayner, K., & Bertera, J. H. (1979). Reading without a fovea. *Science, 206*, 468–469.

Rayner, K., & Clifton, C. Jr. (2002). Language comprehension. In: D. L. Medin (Vol. ed.) *Stevens' Handbook of Experimental Psychology: Volume X* (pp. 261–316). New York: Wiley.

Rayner, K., Sereno, S. C., & Raney, G. E. (1996). Eye movement control in reading: A comparison of two types of models. *Journal of Experimental Psychology: Human Perception and Performance, 22*, 1188–1200.

Rayner, K., Slowiaczek, M. L., Clifton, C., & Bertera, J. H. (1983). Latency of sequential eye movements: Implications for reading. *Journal of Experimental Psychology: Human Perception and Performance, 9*, 912–922.

Rayner, K., Well, A. D., Pollatsek, A., & Bertera, J. H. (1982) The availability of useful information to the right of fixation during reading. *Perception and Psychophysics, 31*, 537–550.

Reichle, E. D., Pollatsek, A., Fisher, D. L., & Rayner, K. (1998). Toward a model of eye movement control in reading. *Psychological Review, 105*, 125–157.

Reichle, E. D., & Rayner, K. (2002). Cognitive processing and models of reading. In: G. K. Hung and K. J. Ciuffreda (eds), *Models of the Visual System*. New York: Plenum.

Reichle, E. D., Rayner, K., & Pollatsek, A. (1999). Eye movement control in reading: Accounting for initial fixation locations and refixations within the E-Z reader model. *Vision Research, 39*, 4403–4411.

Reichle, E. D., Rayner, K., & Pollatsek, A. (2002). The E-Z Reader model of eye movement control in reading: Comparisons to other models. *Behavioral and Brain Sciences* (in press).

Schilling, H. E. H., Rayner, K., & Chumbley, J. I. (1998). Comparing naming, lexical decision, and eye fixation times: Word frequency effects and individual differences. *Memory & Cognition, 26*, 1270–1281.

Seidenberg, M. S., & McClelland, J. L. (1989). A distributed, developmental model of word recognition and naming. *Psychological Review, 96*, 523–568.

Sereno, S. C., Rayner, K., & Posner, M. I. (1998). Establishing a time-line of processing during reading: Evidence from eye movements and event-related potentials. *NeuroReport, 9*, 2195–2200.

Vergilino, D., & Beauvillain, C. (2000). The planning of refixation saccades in reading. *Vision Research, 40*, 3527–3538.

Chapter 19

SWIFT Explorations

Reinhold Kliegl and Ralf Engbert

SWIFT is a computational model of eye guidance in reading. It assumes (1) spatially distributed lexical processing, (2) a separation of saccade timing from saccade target selection, and (3) autonomous and parallel generation of saccades with inhibition by foveal targets. The model accounts for fixation probabilities as well as various measures of inspection time in their relation to lexical processing difficulty. We illustrate the dynamics associated with saccade generation and inhibition by foveal targets. In addition, we generate predictions for an experiment involving gaze-contingent display change.

Introduction

Is it sufficient to assume that reading involves sequential shifts of attention from one word to the next or are several words within the perceptual span processed in parallel? There is still a very productive controversy surrounding this issue summarized recently by Starr and Rayner (2001). These authors concluded that

> One potential solution would be to abandon the serial framework of attention models of eye-movement control and replace it with a parallel mechanism. . . . Words would thus be processed in parallel, although the processing of information would be most accurate at the center of the attentional distribution. . . . However, such a model seems rather complicated and would be difficult to implement in a computational model. Thus a challenge for proponents of a parallel mechanism of attention during reading is to delineate the parameters of such a framework (Starr & Rayner, 2001, p. 162).

We proposed a model that fits this description (Engbert, Longtin & Kliegl, 2002). The model is based on three principles: spatially distributed lexical processing, a partial

The Mind's Eye: Cognitive and Applied Aspects of Eye Movement Research

ISBN: 0–444–51020–6

separation of saccade timing from saccade target selection, and autonomous Saccade generation With Inhibition by Foveal Targets. From the last principle we also derived an acronym for the model (i.e., SWIFT). In the following we present a synopsis of the model components presenting in greater detail our assumptions about foveal inhibition and the dynamics of saccade generation and saccade cancellation. In addition, we evaluate the model with respect to predictions for an experiment with gaze-contingent display changes. With these explorations of the SWIFT model, we aim at a greater transparency of its core principles and demonstrate the utility of such a computational model for accounts of extant data as well as the prediction of novel aspects of eye guidance in reading.

The SWIFT Model

Lexical Processing

Figure 19.1 provides an overview of the model components. The "Lexical processing" box encapsulates how words are processed relative to the current eye position. We call this processing "foveal lexical activity." In contrast to other computational models such as E-Z Reader (Reichle, Pollatsek, Fisher & Rayner, 1998; Reichle, Rayner & Pollatsek, 1999) or the model proposed by Engbert and Kliegl (2001) we assume that the perceptual span encompasses four words, namely the word currently fixated as well as the one to the left and the two words to the right. The (normalized) processing rate depends on fixation location: It is largest for the fixated word [parameter estimate: $\lambda(0) = 0.798$] and considerably smaller for the left and right neighbor [$\lambda(1) = \lambda(-1) = 0.077$] and even smaller for the second word to the right [$\lambda(2) = 0.048$]. As the sum of λs was fixed at 1.0, we used only two degrees of freedom for the three parameter estimates. Details about parameter estimation will be presented in the "Model evaluation" section; the model was estimated with a total of 11 free parameters (see Table 19.1).

A word is processed as soon as it is within the perceptual span leading to a dynamic change in what we call lexical activity associated with this word. We assume that there is a maximum of lexical activity with each word depending on its frequency and its predictability from the prior sentence context, using the specification proposed by Reichle *et al.* (1998), that is $L_n = (1-p_n)(\alpha - \beta \log f_n)$, where p_n represents the predictability of $word_n$ and f_n the printed frequency of $word_n$, α and β are model parameters which were estimated as 148.5 and 5.71, respectively. Thus, the maximum of lexical activity will range between 148.5 (=α) for an unpredictable, very-low-frequency word and 0 for a perfectly predictable word. Over time lexical activity increases from zero for unprocessed words to the maximum lexical activity and then decreases back to zero. The processing time associated with the increase in lexical activity is called lexical preprocessing; the processing time required for the return to the zero-baseline is called lexical completion — again in rough analogy to a two-level processing introduced by Reichle *et al.* (1998). Obviously, preprocessing and completion could be conceived as two independent processes. Rather than estimating a single processing rate

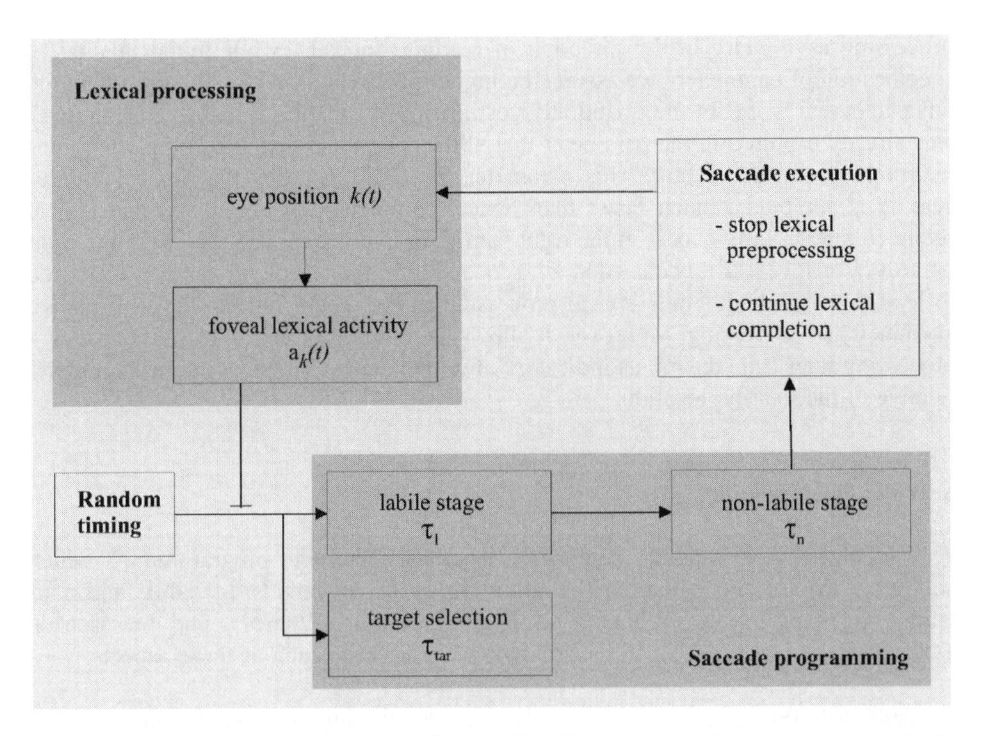

Figure 19.1: Schematic diagram of SWIFT. The main subsystems are saccade programming and lexical processing. These two subsystems are coupled via a foveally-inhibited random timing system and a saccade execution system which moves the eyes during saccades (from Engbert *et al.*, 2002).

and preprocessing factor, we could estimate two separate rates for these processes with the first process "running" from zero to maximum lexical activity and the second process from maximum to zero lexical activity. The remaining link between these two processes is that they use the same maximum of lexical activity for a given word as a stopping value for the process. As an analogy of this conceptualization it may be useful to interpret the maximum lexical activity of a word as an indicator of its "attractiveness" to the mind. In this sense, unpredictable or low-frequency words are more interesting to the mind and it takes longer for this interest to subside than in the case of predictable or high-frequency words. Note that the current formula based on printed frequency and predictability serves only as a descriptive interface to the general difficulty of the words in the perceptual span. Ideally, such lexical activity values should be provided by a theoretical account of sentence processing. Predictability and printed frequency are but convenient proxies of theoretically unspecified contributions of syntactic, semantic, and pragmatic sources of variance for the processing difficulty associated with a given word in its sentence context.

Processing rates for left and right parafoveal word could be constrained to the same estimate without loss of fit (i.e., 0.077). However, it is well-known that there is a processing asymmetry in the direction of reading. Indeed, in our initial qualitative specification of parameters we assumed a higher processing rate for the right than the left parafoveal word. In the quantitative estimation it turned out that this asymmetry was already captured in the parameter that allows preprocessing and completion rates to differ by constant factor. This parameter ($f = 62.5$) suggests that lexical preprocessing is completed much faster than lexical completion. It turned out that lexical preprocessing usually occurs in the right parafovea and lexical completion in the right parafovea or fovea. If it occurred at all, processing in the left parafovea was restricted to lexical completion. Thus, lexical processing in the model reflects the well-known asymmetry of processing. Moreover, it allows us to interpret word-position dependent processing rates (i.e., the λs) as indicators of retinal acuity which would be symmetric relative to the fixation location.

Saccade Programming

The second major model component specifies "Saccade programming" which comprises (a) saccade initiation, (b) inhibition by foveal targets, (c) labile and non-labile stages of timing (When?), (d) target selection (Where?), and (e) saccade execution (see Figure 19.1). In the following we describe each of these aspects.

Saccade initiation In the SWIFT model saccade initiation occurs autonomously (see "Random timing" box in Figure 19.1), an idea already implemented in our earlier sequential-attention shift model (Engbert & Kliegl, 2001) after a random interval generated by a timer (t_S). Assuming a gamma distribution, t_S was estimated with a mean of 187.1 ms and a relative standard deviation of 0.239 of the mean (i.e., 44.7 ms). (A single value for relative standard deviations was estimated for saccade initition times and other saccade-related timing distributions, see next paragraph). The assumption of autonomous saccade initiation is fundamentally different from E-Z Reader and its predecessor models (e.g., Morrison, 1984) where saccade initiation is strictly coupled to some aspect of lexical processing.

Inhibition by foveal targets The intuition guiding the assumption of autonomous saccade initiation is that during reading we initiate saccade programs according to some preferred mean rate. However, we want to allow for some influence of lexical processes. Specifically, we assume that high lexical activity delays the saccade initiation. In other words, if there is a chance that comprehension may lag behind the autonomous generation of saccades and, consequently, comprehension and eye position threaten to desynchronize, then saccade initiation can be postponed. In Figure 19.1, this intervention is represented by the inhibitory link from the foveal lexical activity to the link between "Random timing" and "Saccade programming". We assume an additive contribution of foveal lexical activity [$ha_K(t)$] to the random interval generated by the timer (t_S). Thus, a new saccade program is started after the time interval $t' = t_S + ha_K(t)$. For a word of maximum difficulty ($\alpha = 148.5$, $p = 0$), no parafoveal preprocessing and purely foveal processing the maximum "inhibition time"

$ha_K(t)$ amounts to 181 ms for the current set of parameter estimates [$\lambda(0) = 0.798$; $h = 50.3$] (see Appendix A). Thus, the next saccade will be initiated at the latest 181 ms after the value drawn from the distribution of saccade initiation times. (The calculation in the appendix also shows that the maximum inhibition time does not depend on the precise value of h as long as h is sufficiently large; for an infinite value of h, the maximum inhibition time would be 186 ms. Thus, in principle this free parameter might not be necessary in future model versions.) In general, however, the delay will be much smaller because of lower lexical activity, parafoveal preprocessing, asynchrony of maximum lexical activity, and determination of inhibition time during foveal processing. Indeed, the amount of fixation time due to foveal inhibition amounted to less than 15% for low-frequency words in the simulation. Nevertheless, foveal inhibition was necessary to explain the dependency of first fixation duration on word frequency. We will discuss a similar proposal by Yang and McConkie (2001; McConkie & Yang, this volume) in the final section of this chapter.

Labile and nonlabile stages of saccade program ("when?") Once a saccade is initiated, that is once a saccade program is started, we assume two stages, a labile and a non-labile stage. Assuming gamma distributions, labile and non-labile times were estimated to last on average 128.6 (SD = 30.7 ms) and 41.6 ms (SD = 9.9 ms), respectively. The model assumes that a saccade can be cancelled and saccade targets can be modified during the labile phase. A distinction between labile and non-labile stage is implemented in E-Z Reader as well (Reichle *et al.*, 1998). However, E-Z Reader does not allow direct target modification. Rather a saccade program is always initiated to the next word at the completion of the first stage of lexical processing (i.e., the familiarity check); target modification can occur indirectly through the cancelation of a saccade during the labile stage. In the SWIFT model, a new saccade can be initiated during the preparation of an older one. Such interactions of a new saccade program with an older one will be the topic of the next section.

Target selection ("where?") The distinction between "when?" and "where?" is motivated by neurophysiological results (e.g., Carpenter, 2000; Findlay & Walker, 1999; see Reilly & Radach, this volume, for another implementation of this distinction in a computational model of reading). Saccadic target selection (where to move next) is specified as largely independent of saccade timing (when to move). Target selection was estimated to occur after 87% of the labile phase, that is on average after 112.1 ms in the current implementation. This dependence could be relaxed substantially in future versions of the model. This target of the next saccade, one of the words within the current perceptual window, is stochastically determined according to the values of current lexical activities. Thus, the word with the largest current lexical activity is the most likely target; perfectly predictable words with a lexical activity of zero will be skipped. The differences in processing rates associated with fixation position and the differences in lexical preprocessing and completion rates generate a "bow wave" of lexical activity pulling the eye in direction of reading across the sentence. The reason is that the currently fixated word is processed at the highest rate (see above) and is likely to be already in the stage of lexical completion with a continuous decrease in lexical activity at the time of target selection. Words to the right of the fixated word are processed much slower than the fixated word and are likely to be still in the stage of lexical preprocessing with a continuous increase in lexical activity (or perhaps even "trail" the fixated word in the stage of lexical completion). Consequently, at the time of target selection the lexical activity is likely to be lower for the currently fixated words than its right neighbor. Therefore, the latter is more likely to be selected as the next saccade target.

Consequences of target selection mechanism The stochastics of the selection process as well as differences between maximum lexical activities will lead to refixations of the current word and refixations to the previous word. In addition, there is the possibility that the eye moves on before a word (i.e., the word to the left of the fixated one) is completely processed with the residual lexical activity remaining at a constant level as long as the word is outside the perceptual span. These words will remain potential targets for saccades with a probability derived from their residual lexical activities. As the residual lexical activity is typically rather low, the selection is likely to occur late when the eye has moved towards the end of the sentence. We call such regressions to words to the left of the current perceptual span "long regressions". Their empirical plausibility (i.e., the accuracy of regressions to previous words in a text) has been demonstrated by Kennedy and Murray (1987).

Saccade execution The end of the non-labile stage of saccade programming triggers "Saccade execution" (see Figure 19.1). At this time, lexical preprocessing is suspended but lexical completion continues. The rationale for this distinction is that lexical preprocessing requires perceptual input which is suppressed or strongly attenuated during saccade execution. Lexical completion should not be affected by saccadic suppression. In the simulation, times for saccade executions were fixed with a mean of 25 ms and a standard deviation of 8.3 ms assuming a gamma distribution of latencies. The saccade execution shifts the position of the eye which in turn leads to a change in the foveal and non-foveal lexical activities.

Dynamics of Saccade Generation

Most of the model assumptions are fairly straightforward but there are some intricacies associated with the initiation of saccade programs. In particular, due to autonomous saccade generation it is possible, that a new saccade program is started while the previous saccade program is still "under construction." The effects of such parallel saccade programs on the resulting fixation durations depend on the stage of completion of the first saccade program (completed, labile stage, non-labile stage, or execution stage). As described in the last section, lexical processing has an effect on the initiation of the next saccade program solely via the inhibition by foveal lexical activity. Lexical activity can only delay the initiation of a saccade program, it cannot start or cancel a saccade program. Thus, if the currently fixated word is very difficult and if there is a high lexical activity when the initiation time for the next saccade is sampled, then the gamma-distributed latency will be extended accordingly. Therefore, for the following illustration the effects of lexical activity of foveal inhibition can be subsumed under the initialization time. Figure 19.2 illustrates a hypothetical sequence of saccade programs assuming for now a deterministic scheme with labile stages lasting 125 ms, non-labile stages 50 ms, and execution times 25 ms. We also assume that targets are selected 100 ms after saccade initiation. Now we turn to a description of the various cases of overlap between saccade programs.

Non-overlapping Saccades

When a saccade program (SP) is started, the time for the initiation of the next saccade is determined as well. The initial saccade program SP0 is started with an initialization of $t = 0$ (see Figure 19.2). At the beginning of SP0 the initialization time for saccade program SP1 is sampled according to a gamma distribution, the dashed horizontal line indicates that SP1 is to start at $t = 250$ ms. According to the above specifications, for the first fixation duration F0 we simply compute the sum of labile and non-labile stages which amounts to 175 ms. The target for SP0 was specified after 100 ms during the labile stage. The first fixation (i.e., the nonlabile saccade program stage) is terminated with the execution of the saccade lasting 25 ms. Thus, the next fixation F1 starts at 200 ms. Note that there are still 50 ms of processing (250 ms — 200 ms) before SP1 is initiated. Thus, the inital saccade program does not overlap with the first one; SP1 is started during F1 at absolute time $t = 250$ ms. At this point in time a latency for SP2 is chosen with a value of 225 ms. Consequently, there is again enough time for SP1 to complete without interference from SP2: Its labile phase will end at absolute time 375 ms; the non-labile phase at absolute time 425 ms (amounting to a F1 duration of 225 ms = 50 ms + 175 ms); target selection occurred at absolute time 350 ms. The execution of the second saccade lasts from 425 ms to 450 ms. This illustration shows that there is no simple link between fixation duration and the time intervals at which saccade programs are initiated because the difference between the time for the initiation of the next saccade program and the time required for carrying out the current saccade program is allocated to the next fixation duration.

Saccade Initiation During Labile Stage of Current Saccade Program

At absolute time 475 ms, saccade program SP2 is started and the next saccade program SP3 is to be initiated with a latency of 75 ms. The initiation of SP3 falls in the labile stage of SP2. In this case, SP2 is simply canceled and replaced with SP3 (i.e., the duration in the labile stage is reset to zero; any target would be de-selected as well). Obviously such a cancelation extends the duration of current fixation by the amount of time already spent in the labile stage of canceled saccade program (see F2). Note also that cancelation of a saccade program implies that only one saccade program is active.

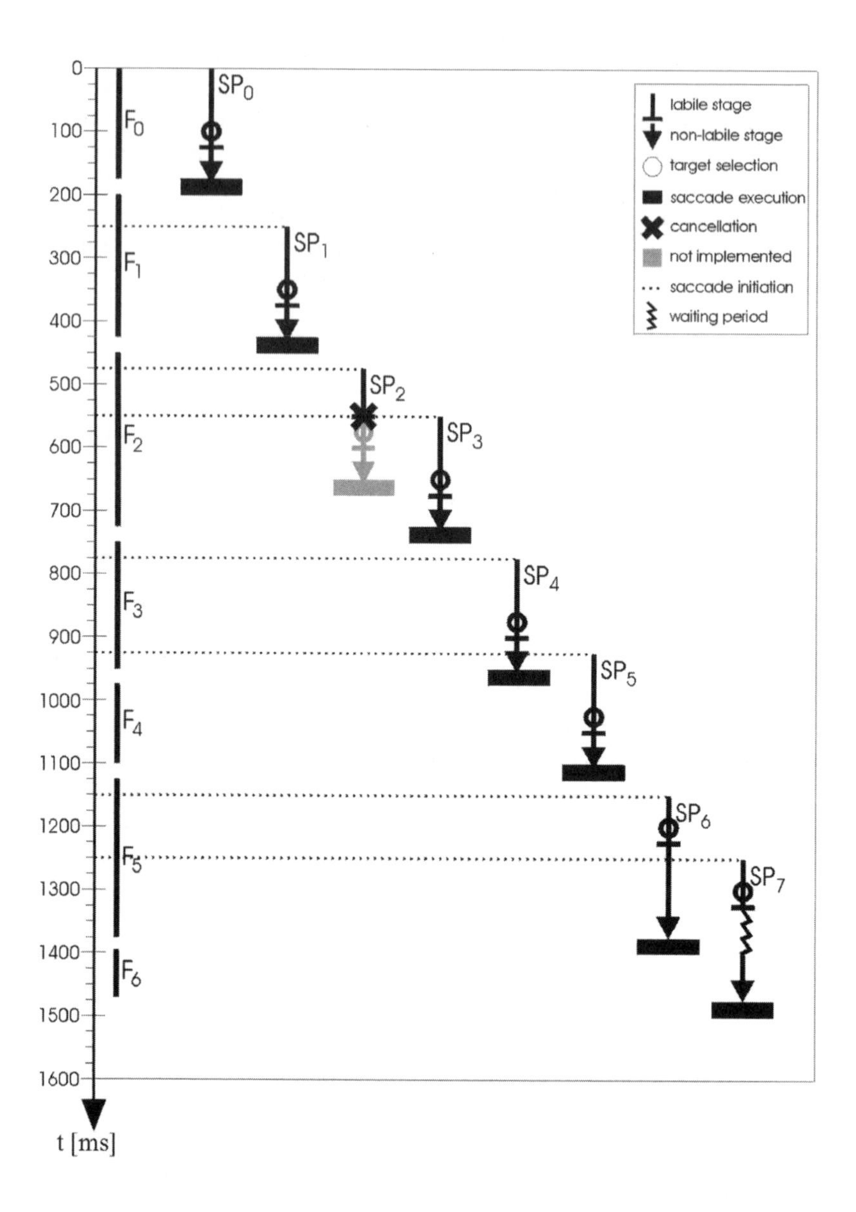

Figure 19.2: Dynamics of saccade generation in SWIFT. When a saccade program (SP) is started, the time for the initiation of the next saccade is determined as well. This will lead to non-overlapping (SP0 and SP1) and various cases of overlapping saccade programs. Initiation of the second program during the labile stage of the first one leads to cancellation of the first program (SP2 and SP3). Initiation during the non-labile stage does not interfere (SP4 and SP5) but the two programs can not be in the non-labile stage at the same time (SP6 and SP7). In this case the second program has to wait. Fixation durations are shown next to time axis (F0 to F6).

Saccade Initiation During Non-labile Stage of Current Saccade Program

The case of a saccade initiation during the non-labile stage of a saccade program is illustrated for SP4 and SP5. As implied by the name of the stage, there is no consequence for a saccade program in the non-labile stage. The program will run its course and land at the target word selected earlier. There is also no consequence for the initiation of the new saccade program. Thus, in this case there are two saccade programs running in parallel without interference. The same logic applies if the initiation of a new saccade program occurs during the execution of the current saccade.

Two Saccade Programs in Non-labile or Execution Stages

Durations of labile and non-labile stages were specified as stochastic variables. Consequently, the second saccade program might enter the execution stage while the first saccade is being executed. Obviously the eyes cannot go two targets at the same time. In this case the later saccade program simply has to wait for the current one to be completed. This case is illustrated with SP6 and SP7. Note that this case can lead to very short fixations because the eye is in rest only for the duration of the nonlabile stage of SP7.

Model Evaluation

We used the corpus of Schilling, Rayner and Chumbley (1998) to fit the model and estimate parameters. The corpus comprises 48 sentences with a total of 536 words. Frequency and predictability values (collected by Reichle *et al.*, 1998) are available for each word. Moreover, Schilling *et al.* reported statistics of gaze, first, and single fixation duration as well as the probabilities of single, double, and zero fixations (i.e., skipping). Trials with regressions were discarded from the analysis. The same corpus was used by Reichle *et al.* (1998) and Engbert and Kliegl (2001). For each sentence we obtained 100 simulations. The model was fitted with 11 free parameters. These parameters, the best fitting values, and the associated standard deviations are listed in Table 19.1.

The panels of Figure 19.3 represent three simulations of the same sentence using the parameter estimates of Table 19.1. The sentence read was: "Mark told Jane that he would meet her after baseball practice." Time runs from left to right. The solid black line indicates the position of the eye at each point in time. Vertical gray lines indicated saccades; the vertical dashed line marks the end of reading. The thin lines above the solid eye-position line show the time course of lexical activity for each word. Finally, the solid black segments on the time axes indicate delays of saccade program initiation due to foveal inhibition. Obviously, aside from differences in total reading time between simulations, the three examples illustrate the high degree of complexity that follows from the theoretical principles and the inherent dynamics of the model and the large variance of eye movement traces that result from them. The same principles and dynamics,

Table 19.1: Model parameter estimates.

	Parameter	Symbol	Value	SD
Lexical parameters	Difficulty, intercept	α	148.5	3.6
	Difficulty, slope	β	5.71	0.29
Processing rate	Foveal	λ (0)	0.798	0.017
	Parafoveal	λ (1, −1)	0.077	0.017
	Parafoveal	λ (2)	0.048	0.017
	Preprocessing	f	62.5	5.8
Saccade parameters	Random timing (ms)	t_S	187.1	2.6
	Labile stage (ms)	τ_l	128.6	3.2
	Nonlabile stage (ms)	τ_n	41.6	4.7
	SD (relative to mean)	ρ	0.239	0.021
	Target selection (% labile stage)	τ_{tar}	0.872	0.056
	Inhibition factor	h	50.3	14.1

Notes: Standard deviations (SD) were based on five runs of parameter estimation by genetic algorithm (see Engbert *et al.*, 2002, Appendix A). Processing rates were estimated with the constraint: $\lambda(0) + 2\lambda(1, -1) + \lambda(2) = 1$, yielding a total of 11 free parameters for the model.

however, also strongly constrain the type of paths the eye can take through the sentence (Engbert, Longtin & Kliegl, in press). Indeed, such statistics may prove very useful for comparing computational models with each other and with human data.

The traces also illustrate links between lexical activity and various types of eye movements. The top panel represents a trace consisting solely of forward moves. At the beginning the eye is on the first word and the model starts processing the first three words as indicated in the increase and subsequent decrease of lexical activity visualized for each word. The rate of change is highest for the fixated word. After about 500 ms the eye moves to the second word. The decision point to select this word as the next target occured around 100 ms into the first fixation. Note that at this time lexical activity of the fixated word was probably already smaller than that of the second and the third word. The height of lexical activity indicates the probability with which a word is selected as the next target. Therefore, chances were high for the selection of the second or third word with the second word "winning" in this case. Note that in the second and third trace for this sentence (middle and bottom panels of Figure 19.3) the third word was selected as the target with the second word being skipped in the process. Words are also skipped if they are already processed completely in the parafovea of earlier fixations. Examples of such skips are the words "that", "her" and "practice" in the first trace.

The second and third trace illustrate two different regressions to the skipped word "told". In the second trace (middle panel of Figure 19.3), "Janet" has the highest lexical activity among the four words in the perceptual span (i.e., "told Janet that he") during the second fixation and was selected as the target for the third saccade, leading to a refixation on "Janet" indicated with a circle on the trace. Obviously, such refixations are more likely for low-frequency, unpredictable words. The same mechanism will also yield regressions back to earlier words. For example, during the third fixation (i.e., the refixation of "Janet") "told" was selected as the next saccade target, leading to a regression back to this word.

We already encountered regressive movements of the eye to an earlier word within the perceptual span. A word will also be the target of a long regressive movement if its processing was not completed while it was in the perceptual span. For example, the word "told" in the bottom panel of Figure 19.3 was left in such an unfinished state because first "Janet" and then "he" were selected as targets causing "told" to fall to the left of the perceptual span. In the model the residual lexical activity of this word will stay at its last value in the perceptual span until it is selected again as a target. Any word with residual lexical activtiy will compete for target selection irrespective of whether or not it still is in the perceptual span. Typically such a residual lexical activity is low due to earlier processing and therefore the chances of being selected as a target are small as long as there are unprocessed words to the right. However, the predictability of words increases with serial word position and, consequently, maximum lexical activity will decrease across a sentence. Thus, as the eye approaches the end of the sentence chances increase again that words with residual lexical activity will be selected. In the third trace (bottom panel of Figure 19.3) "told" was selected as a target when the eye fixated "would." The following words had been processed completely; therefore reading resumes at the next word with lexical activity outside the perceptual span which is "after" in this case. (Incidentally, "after" was first processed during the fixation on "meet". Due to the subsequent regression, processing ceased while lexical activity was fairly high.)

There is good evidence that such long regressions are typically very precise, suggesting that reading sets up a spatial representation of word locations (Kennedy, 2000). Experimental evidence for long regressions due to residual lexical activity would constitute very strong support for the model because, traditionally, such long regressions have been attributed only to high-level sentence parsing problems, such as revisions of an interpretation in a garden-path sentence. Predictions relating to this distinction remain to be tested in experiments but at this point they illustrate the innovative potential of a dynamic model.

From such simulations we can compile summary statistics of various measures of inspection time as well as various measures of fixation probability and compare them with experimental results. These comparisons are displayed in the two panels of Figure 19.4 as a function of logarithmic word frequency. In general, the model reproduces the qualitative patterns very well. Most notable are the increase in skipping probability and the drop in single-fixation probability for high-frequency words. Also the decrease in inspection time measures as a function of logarithmic word frequency is reproduced very well. In the current implementation the height of lexical activity at the time of target selection is used as weight for determining the next saccadic goal. Explorations of the model showed that data could also be fitted if height determined the next target (i.e., a "winner-take-all" rule). Thus, the stochastic selection of target selection was not very critical. However, substantially lower parafoveal than foveal processing rates were critical for obtaining adequate model fits.

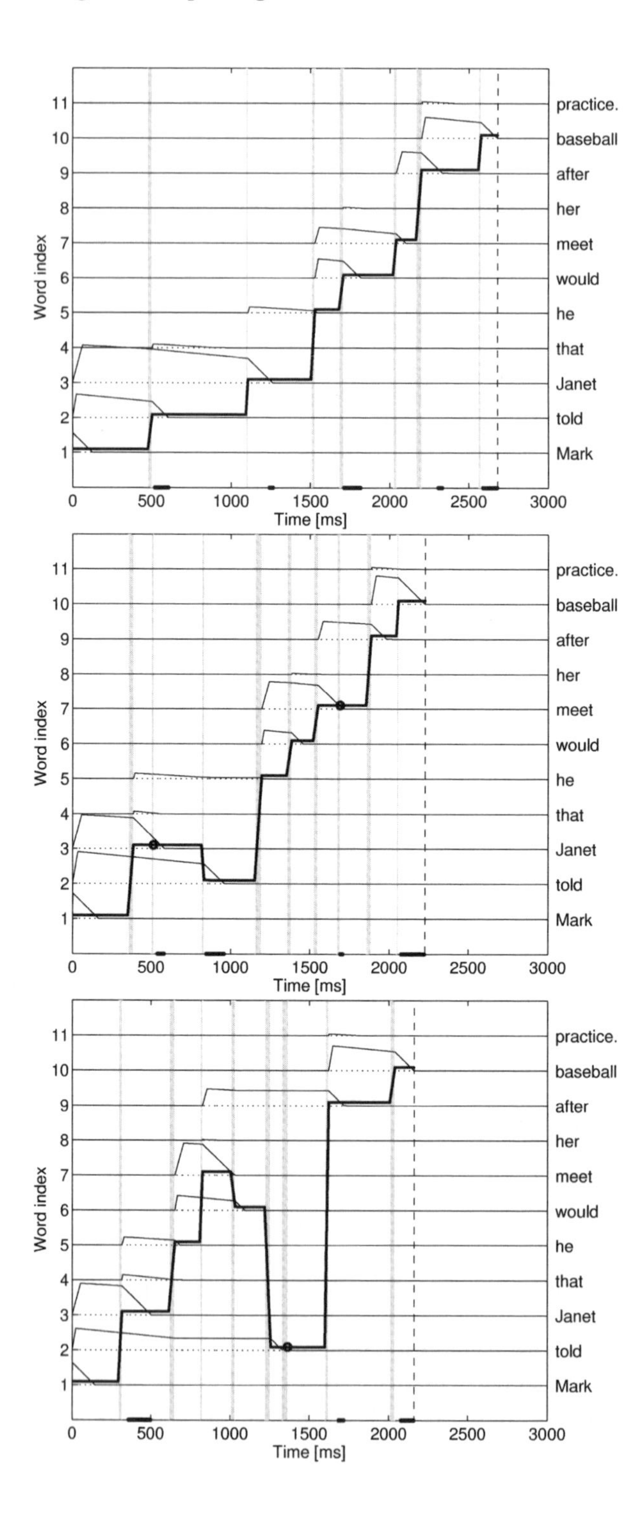
Word index
Time [ms]
0
500
1000
1500
2000
2500
3000
1
2
3
4
5
6
7
8
9
10
11
practice.
baseball
after
her
meet
would
he
that
Janet
told
Mark

Simulation of Eye-contingent Display Changes

In this section, we demonstrate that SWIFT can be used to predict several measures of a typical reading experiment. Properties of the perceptual span in reading are fundamental to the model. To derive the model we used psychologically plausible assumptions about the perceptual window. To test whether these assumptions lead to realistic behavioral patterns, we investigate predictions that can be obtained from numerical simulations with respect to an experimental manipulation of the display during reading (Binder, Pollatsek & Rayner, 1999; we restrict our analysis to the preview condition). In an eye-contingent display change experiment, a target word in a sentence was changed during preview. As soon as the eyes entered the target region, the preview word was replaced by the target word. The display change was performed during the saccade to the target region and the subjects tested were unaware of this manipulation.

Some results of the Binder *et al.* (1999) experiment are summarized in Table 19.2. If the preview was not changed, that is if an identical base word appeared at the target location, a skipping probability of 0.30 resulted. If a different word was used during preview, that is if the word was replaced with a different word during the approaching saccade, skipping probability decreased to 0.165. Also there was an increase of first fixation durations and regression probability for the changed word. These results highlight the importance of parafoveal lexical processing during reading.

In our numerical simulations, we used a sentence of the corpus by Schilling, Rayner and Chumble (1998) to demonstrate the effect of reduced preview in our model. Obviously, we only aimed for a qualitative reproduction of key results because of differences in the sentence material and experimental set-up. To this end, we reset the lexical activity of the target word to zero as soon as a saccade to the target region occurs (see Figure 19.5). As a result, we find a comparable reduction in skipping probability from 0.47 to 0.15. Furthermore, like Binder *et al.* (1999) we observe an increase in the first fixation durations in the changed compared to the identical preview condition (Table 19.2). Finally, to demonstrate the extraction of information to the left of the fixation point, we analyzed the probability of regressions to the target word. Our theoretical expectation was that the reset of lexical activity will lead to an increase of the number of regression. For the target word used here, the probability of regressing was 0.042 without preview manipulation. The increase of the regression probability to 0.07 as a consequence of the reset of lexical activity is much smaller than the increase observed in the experiment by Binder *et al.* (1999). This discrepancy, however, may be explained by the fact that, in the model, regressions are completely caused by

---

Figure 19.3: Trajectories for the same sentence from three simulation runs of SWIFT with parameter estimates of Table 19.1. Lexical activities (thin lines) are plotted over time together with the eye position (bold line). The execution of saccades is indicated by the shaded vertical regions. The beginning of a refixation is indicated by a circle. Foveal inhibition (delay of initiation of saccade program) is marked by the bold segments on the time axis. Sentence and data on word frequencies and predictability were taken from the Schilling *et al.* (1998) experiment.

incomplete lexical processing. Different sources of regressions, as reviewed for example by Rayner (1998), are beyond the scope of the current version.

In summary, the model can be used to predict typical measures used for the analysis of eye movement experiments with gaze-contingent display changes. The influence of preview manipulations and the mechanism of extraction of information to the left of the fixated word are qualitatively in good agreement with experimental results. These results underline the model's psychological plausibility and how properties of the perceptual window can be thought to influence the dynamics of eye movements.

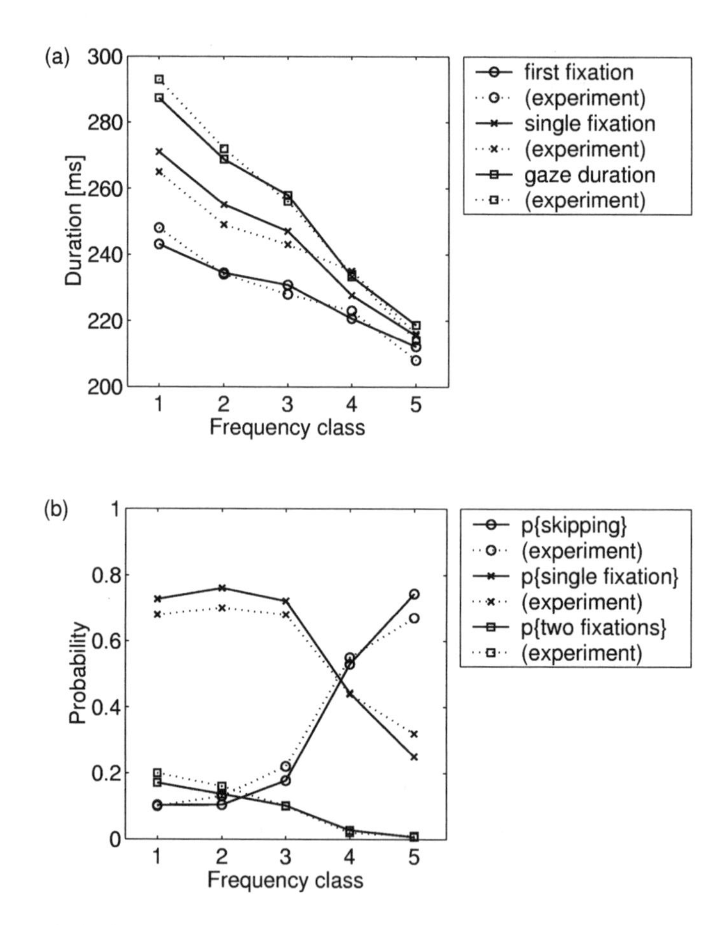

Figure 19.4: Statistical evaluation of SWIFT performance; experimental data are from Schilling *et al.* (1998). (a) First fixation duration, gaze duration, and single fixation duration as a function of word frequency class (averaged over 1000 statistical realizations from SWIFT simulations, i.e. 1000 simulations of the model over the same corpus of sentences but with different pseudo-random numbers). (b) Probabilities for word skipping, performing a single fixation, and making two fixations (computed from the same runs as in (a)) as a function of word frequency class (from Engbert *et al.*, 2002).

Table 19.2: Gaze-contingent display changes: a comparison of SWIFT and Binder *et al.* (1999).

	SWIFT		**Binder *et al.* (1999)**	
Reading measure/Preview	**Identical**	**Different**	**Identical**	**Different**
Probability of skipping preview	0.470	0.150	0.300	0.165
First fixation on target (ms)	187	201	228	246
Probability of regressing to target	0.042	0.070	0.080	0.220

Note: Values for different preview of Binder *et al.* (1999) represent mean of related and unrelated preview changes.

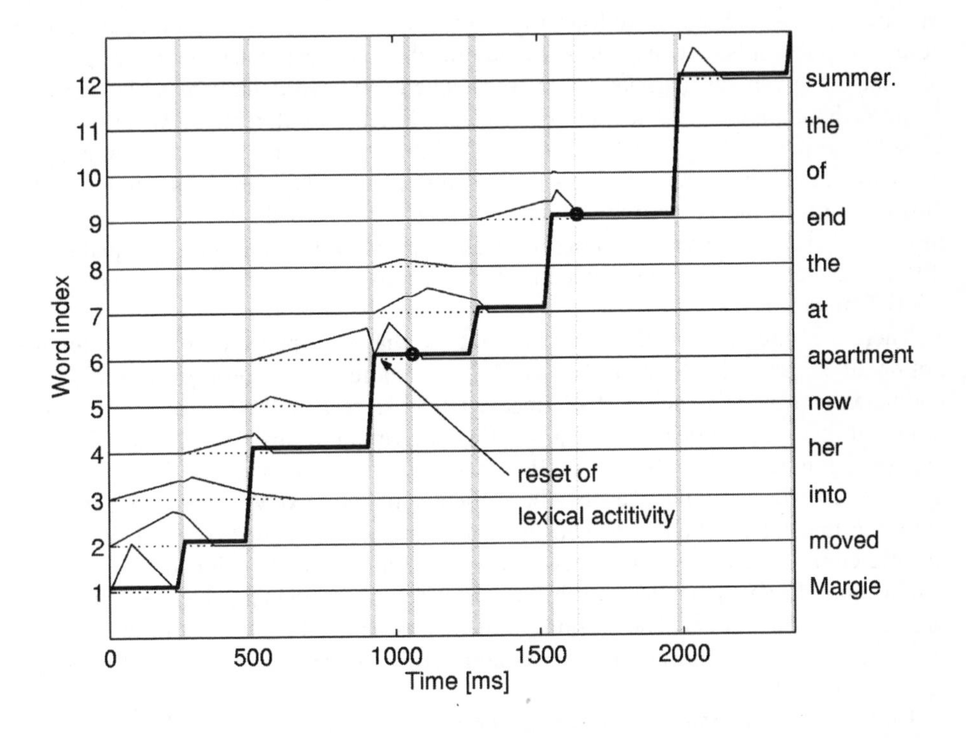

Figure 19.5: Numerical simulation of the preview experiment by Binder *et al.* (1999). A preview display was changed during the first saccade entering the target region ($word_6$). In our simulations, the lexical activity of the target word was reset to zero, when the first saccade to the target region was performed. In the example trajectory shown here, this occurs at time $t = 930$ ms. As a result, the fixation duration of $word_6$ increases. Also the word is skipped less often and is more likely to attract a regressive saccade.

Comparisons and Perspectives

Comparison with E-Z Reader

Computational model of eye movements in reading hold much promise for providing unifying accounts of rich and diverse sets of experimental results. Recent years have witnessed the emergence of a few attempts in this direction (Engbert & Kliegl, 2001; Reichle *et al.*, 1998, 1999; Reilly & Radach, this volume). The SWIFT model was developed as an alternative to the E-Z Reader (Reichle *et al.*, 1998), which at the time, in our opinion, was the most advanced computational model in the domain of attentional and ocular control during reading. Consequently, central assumptions guiding the design of E-Z Reader were adopted for SWIFT, such as two stages of lexical access (including the formula for combining word frequency and predictability) and the distinction between labile and nonlabile stages of saccade programs terminating with saccade execution. We also followed Reichle *et al.*'s (1998) lead to simulate reading at the level of words, not characters, which probably underestimates the role of perceptual factors given the correlation of word length and word frequency, but greatly reduced the complexity of model construction. Finally, as Reichle *et al.* (1998), we used the corpus of data from Schilling *et al.* (1998) to evaluate the model. In preliminary model comparisons, goodness of fit statistics were typically quite comparable. However, given the large architectural differences between the models, quantitative model fits may not be very meaningful. Rather we want to point to a few qualitative differences between the models which we considered to be critical in the design of the SWIFT model.

There are three pieces of empirical evidence that may prove problematic for the notion of sequential attention shifts (SAS) in reading and, consequently, also for computational models such as E-Z Reader (Reichle *et al.*, 1998) or our own earlier model of this type (Engbert & Kliegl, 2001) subscribing to this assumption (for references see Engbert *et al.*, 2002, p. 622f.): First, there is some evidence that processing of a fixated word is influenced by the difficulty of the next word. Second, there is evidence that information is picked up to the left of the fixated word. Third, there is little empirical support for longer fixation durations prior to skipped words — an implication of sequential attention shift models. The experimental evidence for the first problem is controversial. The second problem was handled at the cost of an increase in model complexity in more recent versions of the model (see Pollatsek, Reichle & Rayner, this volume).

However, in our opinion, there is no easy solution to the third problem: In SAS models, word skipping requires that a saccade targeting the next word is cancelled and re-programmed to the following word. Saccade cancellation necessarily increases the fixation duration prior to the saccade. Indeed both E-Z Reader and our own previous SAS model exhibited a strong effect of this sort (173 ms in E-Z Reader 5; 75 ms in Engbert & Kliegl, 2001). The three empirical studies cited by Reichle *et al.* (1998) in support of longer fixation durations prior to skipped words reported the following effects: 3 to 7 ms (Hogaboam, 1982, Tables 18.1, 18.3), 21 ms (Pollatsek, Rayner &

Balota, 1986, p. 126), and 38 ms (Reichle *et al.*, 1998, p. 147). Given that the effect can not be secured empirically in large data sets (e.g., McConkie, Kerr & Dyre, 1994; Radach & Heller, 2000), the SAS assumption should be reconsidered and SAS models minimally require architectural revision. One solution might be to postpone target selection, similar to the partial separation of saccade initiation and target selection in the SWIFT model. Accordingly, for SWIFT simulations we observed a range of 10 to 21-ms increases of fixation durations prior to skipped words which appears to be in agreement with the empirical data. Incidentally, in our opinion, it is to the credit of E-Z Reader and other SAS models that they can be seriously challenged with experimental evidence.

A general problem of E-Z Reader relates to the order-of-processing methodology. Even slight modifications of the model typically increase the number of states to be considered and seriously limit the complexity of the dynamics that can be covered in the model. Given that SWIFT was designed with at least some of these problems in mind, it is not surprising that it is not affected by these problems. And we should point out that SWIFT does not (yet) account for the full scope of the Schilling *et al.* (1998) corpus, most notably we did not fit distributions of fixation durations as function of logarithmic word frequency.

Comparison with Competition–Inhibition Theory

Yang and McConkie (2001, also McConkie & Yang, this volume) proposed a Competition–Inhibition theory which is very specific with respect to the timing of saccades, that is the "when?" component of saccade programs. Although the theory is not yet implemented as a computational model, there are two central assumptions of the SWIFT model that are conceptually very much in agreement with the Competition–Inhibition theory: autonomous timing of saccades and inhibition of foveal targets due to lexical factors. First, autonomous timing of saccades represents active search for new information as well as predictions and expectations about where relevant information is to be found. This is very different from models assuming that the completion of lexical or cognitive processes triggers the initiation of new saccades; the familiarity check in the E-Z Reader model is such an example (Reichle *et al.,* 1998). The proposal that the eye is acting on expectations that are corrected by sensory feedback if necessary (i.e., the "motor prediction" perspective of Wolpert & Flanagan, 2000), is also in line with more general theories of the relation of eye movements and complex actions such as driving or cricket batting (Land & Furneaux, 1997; Land & McLeod, 2000). It seems plausible that eye movements in reading will eventually be understood as a special case of a more general theory of eye movements and action.

The second notion common to SWIFT and Competition–Inhibition theory is that lexical factors affect saccadic latency (and indirectly fixation duration) via an inhibition by foveal targets. The assumption is that the process of autonomous saccade generation can be delayed by lexical difficulty. In the current implementation of the SWIFT model, the sampled time interval for initiating the next saccade program increases with the lexical difficulty of the fixated word. Obviously, this prolongs the fixation on the current word — up to a maximum of an additional 181 ms. Yang and McConkie's (2001) inhibition process is very sophisticated. They distinguish between three types of saccades (early saccades initiated about 100 — 125 ms after fixation; normal saccades initiated after 175 — 200 ms; and late saccades initiated after 225 ms). Their most relevant result for the current discussion is that display changes of text content (word to nonword) affected only the initiation of late saccades. If there are qualitatively different saccade types and if lexical factors influence only the late ones, then SWIFT will have to be changed to accommodate this high degree of specificity, for example, by linking lexical processing with the labile phase of saccade generation conditional on some minimum amount of processing of the foveal word.

The Next Version of SWIFT

We opted for some glaring simplifications in the evaluation and specification of the SWIFT model. They were motivated by keeping model complexity down and model comparability high. There are at least three necessary extensions. The first two extensions are to increase the scope of the data base that should be accounted for without major revisions of the current model. Specifically, as mentioned earlier, unlike E-Z Reader the SWIFT model does not account for distributions of fixation durations as a function of logarithmic word frequency. Moreover, although SWIFT uses a single mechanism to generate all types of within-line eye movements (i.e., word-to-word, skippings, refixations, and regressions), we have not modeled regression probabilities, perhaps also based on a distinction between regressions within the perceptual span and long regressions. The reason for this omission was that the Schilling *et al.* (1998) corpus had removed sentences with regressions prior to the analysis. Fitting regression probabilities requires a new corpus of eye movement data.

The more serious extension of the SWIFT model concerns a switch from word-based to letter-based processing. The model should reproduce typical landing-position probabilities as a function of word length and saccadic launch distance (for a review of the relevant literature we refer to Radach and Heller, 2000). Moreover, the relation between fixation positions and fixation durations has been of lasting concern in eye movement research. Specifically, there is evidence for two independent effects of fixation position on fixation duration: (1) Fixation durations are longer for fixations in the center of words, (irrespective of whether they are single or first fixation durations (Vitu, McConkie, Kerr & O'Regan, 2001) and (2) fixation durations increase with the launch distance of the last saccade (Radach & Heller, 2000; Vitu *et al.*, 2001). Finally, in this context it may be useful (or even necessary) to allow for a dynamic adjustment of the letter-based perceptual span in response to lexical difficulty. Obviously, such an extension requires a data base that includes information about landing positions of fixations at the letter level in addition to the statistics that were reported in Figure 19.4. Such an increase in data base is necessary to constrain the model parameter space. If successful, such an extension would provide a very desirable modeling framework for the joint consideration of oculomotor, perceptual and low-level cognitive control issues.

Acknowledgments

This work was supported by Deutsche Forschungsgemeinschaft (DFG grants KL 955/3–1, 3–2, 3–3). A SWIFT applet and source codes of the model can be found at: http://www.agnld.uni-potsdam.de/~ralf/swift/. We thank André Longtin, Ralph Radach, Ronan Reilly, and an anonymous reviewer for constructive comments.
Address for correspondence: Reinhold Kliegl, Department of Psychology, University of Potsdam, PO Box 601553, 14415 Potsdam, Germany. E-mail: kliegl@rz.uni-potsdam.de (Reinhold Kliegl), engbert@rz.uni-potsdam.de (Ralf Engbert).

References

Binder, K. S., Pollatsek, A., & Rayner, K. (1999). Extraction of information to the left of the fixated word in reading. *Journal of Experimental Psychology: Human Perception and Performance, 25,* 1162–1172.

Carpenter, R. H. S. (2000). The neural control of looking. *Current Biology, 10,* R291–R293.

Engbert, R., & Kliegl, R. (2001). Mathematical models of eye movements in reading: A possible role for autonomous saccades. *Biological Cybernetics, 85,* 77–87.

Engbert, R., Longtin, A., & Kliegl, R. (2002). A dynamical model of saccade generation in reading based on spatially distributed lexical processing. *Vision Research, 42,* 621–636.

Engbert, R., Longtin, A., & Kliegl, R. (in press). Complexity of eye movements in reading. *International Journal of Bifurcation and Chaos.*

Findlay, J. M., & Walker, R. (1999). A model of saccade generation based on parallel processing and competitive inhibition. *Behavioral and Brain Sciences, 22,* 661–721.

Hogaboam, T. W. (1983). Reading patterns in eye movement data. In: K. Rayner (ed.), *Eye Movements in Reading* (pp. 309–332). New York: Academic Press.

Inhoff, A. W., Radach, R., Starr, M., & Greenberg, S. (2000). Allocation of oculo-spatial attention and saccade programming during reading. In: A. Kennedy, R. Radach, D. Heller and J. Pynte (eds), *Reading as a Perceptual Process* (pp. 221–246). Amsterdam: Elsevier.

Kennedy, A. (2000). Attention allocation in reading: Sequential or parallel? In: A. Kennedy, R. Radach, D. Heller and J. Pynte (eds), *Reading as a Perceptual Process* (pp. 193–220). Amsterdam: Elsevier.

Kennedy, A., & Murray, W. S. (1987). Spatial coding and reading: Some comments on Monk (1985). *Quarterly Journal of Experimental Psychology, 39A,* 649–718.

Land, M. F., & Furneaux, S. (1997). The knowledge base of the oculomotor system. *Philosophical Transactions of the Royal Society London, B352,* 1231–1239.

Land, M. F., & McLeod, P. (2000). From eye movements to actions: How batsmen hit the ball. *Nature Neuroscience, 3,* 1340–1345.

McConkie, G. W., Kerr, P. W., & Dyre, B. P. (1994). What are "normal" eye movments during reading: Toward a mathematical description. In: J. Ygge and G. Lennestrand (eds), *Eye Movements in Reading* (pp. 315–348). Oxford: Elsevier.

Morrison, R. E. (1984). Manipulations of stimulus onset delay in reading: Evidence for parallel programming of saccades. *Journal of Experimental Psychology: Human Perception and Performance, 10,* 667–682.

Pollatsek, A., Rayner, K., & Balota, D. A. (1986). Inferences about eye movement control from the perceptual span in reading. *Perception & Psychophysics, 40,* 123–130.

Radach, R., & Heller, D. (2000). Spatial and temporal aspects of eye movement control. In: A. Kennedy, R. Radach, D. Heller and J. Pynte (eds), *Reading as a Perceptual Process* (pp. 165–191). Oxford: Elsevier.
Rayner, K. (1998). Eye movements in reading and information processing: 20 years of research. *Psychological Bulletin, 124,* 372–422.
Reichle, E. D., Pollatsek, A., Fisher, D. L., & Rayner, K. (1998). Toward a model of eye movement control in reading. *Psychological Review, 105,* 125–157.
Reichle, E. D., Rayner, K., & Pollatsek, A. (1999). Eye movement control in reading: Accounting for initial fixation locations and refixations within the E-Z Reader model. *Vision Research, 39,* 4403–4411.
Schilling, H. E. H., Rayner, K., & Chumbley, J. I. (1998). Comparing naming, lexical decision, and eye fixation times: Word frequency effects and individual differences. *Memory & Cognition, 26,* 1270–1281.
Starr, M.S., & Rayner, K. (2001). Eye movements during reading: some current controversies. *Trends in Cognitive Science, 5,* 156–163.
Vitu, F., McConkie, G.W., Kerr, P., & O'Regan, J.K. (2001). Fixation location effects on fixation durations during reading: An inverted optimal viewing position effect. *Vision Research, 41,* 3513–3533.
Yang, S.-N., & McConkie, G. W. (2001). Eye movements during reading: A theory of saccade initiation times. *Vision Research, 41,* 3567–3585.
Wolpert, D. M., & Flanagan, J. R. (2000). Motor prediction. *Current Biology, 11,* R729–R732.

Appendix A:

Analytical Calculation of the Theoretical Maximum of Inhibition Time

In the SWIFT model, the time between two subsequent decisions to start a saccade program is given by a random time interval t_s and an additive contribution of foveal inhibition $h \cdot a_k(t)$. Let us denote the time of end of the last saccade (or, equivalently, the start of the current fixation) by t'. The next command to start a saccade program is generated at time

$$t' = t_s + h \cdot a_k(t) \tag{1}$$

The theoretical maximum of the contribution of the inhibition process, i.e. max $\{h \cdot a_k(t)\}$ can be calculated. For simplicity, we can choose $t = t'$. The inhibition mechanism reaches its maximum under three conditions:

1. The random component is — by chance — zero: $t_s = 0$.
2. There has been no preprocessing of the foveal word: $a_k(0) = 0$, i.e. the lexical activity of the foveal word is zero at the start of the fixation. Since lexical preprocessing time is short compared to lexical completion, however, we assume that $a_k(0) = L_k$ to further simplify our calculations.
3. The foveal word has a very low frequency, which implies that its lexical difficulty is $L_k = \alpha$.

Since the foveal word is lexically processed with rate $\lambda(0)$, its lexical activity decreases linearly according to the relation

$$a_k(t') = \alpha - \lambda(0) \cdot t' \tag{2}$$

Putting together this equation with Equation 1 with $t_s = 0$, i.e. $t' = h \cdot a_k(t')$, we obtain the relation $t'/h = \alpha - \lambda(0) \cdot t'$, which can be rearranged to the final equation for the maximum of the inhibition time

$$t' = \frac{\alpha}{\lambda(0) + \frac{1}{h}} \tag{3}$$

We interpret this result by discussing the two limiting cases of a vanishing or infinite inhibition parameter h:

$$h \to 0: \quad t' = 0$$

$$h \to +\infty: \quad t' = \frac{\alpha}{\lambda(0)} \tag{4}$$

In the first case, without inhibition, the maximum of t' is obviously zero. In the second, and more interesting case, even an arbitrary large inhibition parameter leads to a finite contribution of $t' = \alpha/\lambda(0) = 186$ ms (for the estimated value $h = 50.3$, we obtain a slightly lower value $t' = 181$ ms). The small increase of 5 ms between the cases $h = 50.3$ and $h \to +\infty$ also explains the large standard deviation estimated for the inhibition factor h (see Table 19.1).

Chapter 20

How Cognition Affects Eye Movements During Reading

George W. McConkie and Shun-nan Yang

Based on recent data (Yang & McConkie, 2001), we argue against direct cognitive control of most saccades made during reading. Rather, cognitive influences appear to be more indirect. We propose that saccades are executed at random times, with a pattern over the fixation period that is described by a hazard curve that is low initially, rises quickly to an asymptote and then typically drops more slowly or remains fairly constant. Cognition influences saccade onset times (which produce the fixation durations) in three ways. First, processing difficulties can inhibit and delay saccades. Second, parameters can be adjusted in response to text or task conditions. Third, given enough time, cognition can control saccades more directly. The latter appears to occur quite infrequently. Regressions are also not necessarily cognitively controlled.

Introduction

Decades of research using eye movement data to study reading have established the fact that eye behavior is being influenced in real time by the cognitive processes taking place. Fixations tend to be longer on average, and more likely to be followed by a refixation, when the eyes are directed at (or were just previously directed at) less frequent words (Rayner & Raney, 1996; Schilling, Rayner & Chumbley, 1998), words containing spelling errors (Zola, 1984), more ambiguous words (Kennedy & Murray, 1984; Rayner & Duffy, 1987), inappropriate words given the interpretation of the previous context (Balota, Pollatsek & Rayner, 1985; Balota & Rayner, 1983; Ehrlich & Rayner, 1981), and word locations requiring more processing (O'Regan, 1989; Rayner, 1979; Underwood, Clews & Everatt, 1990). The eyes are less likely to be sent to highly predictable words (Kerr, 1992; Brysbaert & Vitu, 1998; Ehrlich & Rayner, 1981).

The Mind's Eye: Cognitive and Applied Aspects of Eye Movement Research

ISBN: 0–444–51020–6

Given these consistently-observed relationships, it is reasonable to assume that there is a direct and tight relationship between cognition and eye movement control: that is, that the *when* and *where* decisions are the direct result of the cognitive processes taking place, in order to provide the language processes with the visual information needed for their activities. It is most commonly assumed that a saccade is enabled when some critical cognitive event occurs, and that the eyes are then moved to the location currently being attended, presumably because that is the location from which visual information is currently needed. Theories have differed on the nature of the cognitive saccade triggering event that is postulated, including inability to make a critical language decision because of inadequate visual information (McConkie, 1979), completion of processing permitted by the currently-fixated word (Just & Carpenter, 1980), completion of word identification (Morrison, 1984) and completion of a frequency assessment of the currently-attended word (Reichle, Pollatsek, Fisher & Rayner, 1998; Pollatsek, Reichle & Rayner, this volume). We will refer to this class of positions collectively as *direct cognitive control theories*.

Yang and McConkie (2001) have argued against this basic assumption of the direct cognitive control theories on the basis of the effects of non-textual patterns on the frequency distributions of saccade onset times during reading. These investigators occasionally replaced the text with other patterns, including strings of random letters and Xs, for a single fixation, with both the appearance and disappearance of the abnormal patterns occurring during saccades so the stimulus changes themselves would not be detected. They found that strings of random letters and Xs which preserve the word spacing pattern produce very little effect on the distribution of saccade onset times for the first 175–200 ms, and apparently have no effect on the onset time of most saccades. Thus, a majority of saccades occur at the same time whether the current stimulus is normal text or random letters, a fact that is difficult to reconcile with a theory that assumes that most saccades are triggered by some cognitive event in language processing. A similar result was found with pseudowords, in the absence of display changes (McConkie, Reddix & Zola, 1992)

To account for these and other results, Yang and McConkie proposed an alternative theory of eye movement control in reading based on the theorizing of Findlay and Walker (1999) and recent neurophysiological investigations of eye movement control. Designated the Competition/Interaction (C/I) theory, it assumes that both the timing and target selection in making eye movements result from competition processes between move and fixate centers, as described by Findlay and Walker (1999). The concept of fixate and move centers is based on distinctive functional roles of various neural regions related to oculomotor processes; the functional division within the superior colliculus has been suggested to contribute extensively and perhaps exclusively to the process of regulating fixate and move activity. Generally speaking, it is proposed that activation of the fixate center inhibits saccadic activity, and activation of the move center increases saccadic activity, increasing the likelihood of an object (in reading, a word) in the activated region winning the competition to become the target of the next saccade (Reilly & Radach, this volume).

The C/I theory proposes that saccade preparation can begin as soon as a fixation begins, even before the visual input provided by the text becomes available to the oculomotor system. In addition, the asymmetry of saccade direction is also linked to the different levels of movement activity at the regions representing rightward and leftward eye movements. According to this theory, saccades are assumed to be triggered by an oculomotor strategy that has been developed in the process of learning to read (Levy-Shoen, 1981) rather than directly by cognitive events themselves. This strategy consists of adding activation to the region of the move center that corresponds to the near periphery in the right visual field (assuming a left-to-right writing system), which adds to the activation resulting from visual stimulation. This additional activation means that, in most cases, a word in the near right visual field will win the competition and become the target of the next saccade, and eye movements will tend to show a rightward flow along the line.

Variance in the onset times of these saccades is assumed to reflect the variance in competition resolution time, which is affected very little by language processing under normal conditions (Kliegl & Engbert, this volume). This has been supported by studies of pseudoreading (Vitu, O'Regan, Inhoff & Topolski, 1995; Yang, 2002; cf., Rayner and Fischer, 1996) that have shown frequency distributions of fixation duration and saccade length, in the absence of any influence from higher-level psycholinguistic processes, that are quite similar to those observed in normal reading.

A particularly useful form of data representation for examining the dynamics of saccadic activity over time during a fixation is the hazard curve (Yang & McConkie, 2001; Yang, 2002), which is an alternative way to represent the same information that is contained in a frequency distribution of fixation durations (or saccade latencies). A percentage frequency distribution is produced by dividing the number of cases in each time bin (here, 25 ms bins) by the total number of cases. A hazard curve is produced by dividing the number of cases in each time bin by the number of cases still surviving (that is, for which no saccade has yet occurred) at the beginning of that time bin. Thus, the hazard level at any point in time is the proportion of surviving fixations that end in a saccade during that time period. The curve indicates changes over time, through the period of the fixation, in the level of saccadic activity taking place.

There are two disadvantages to this type of data representation. First, since each successive proportion is based on a smaller number of surviving fixations, hazard curves become increasingly unstable at later time intervals. Second, when data are pooled across subjects to create such a curve, the data at later intervals come increasingly from those subjects who make the longest fixations, indicated by the lowest hazard levels. This necessarily happens because subjects who make shorter fixations produce most of their saccades at earlier periods, with few or no fixations remaining until the longer intervals. This has the effect of causing an artificial reduction in hazard levels at later time intervals. Thus, hazard levels at the later intervals must be interpreted with caution.

Figure 20.1 shows a typical fixation duration hazard curve (upper panel) for a group of skilled readers, together with the fixation duration frequency curve (lower panel) for the same data. Three time periods can be seen in the hazard curve: an early period during which saccadic activity is very low and rising slowly, a middle period during which the saccadic activity is rising quickly, and a late period in which saccadic activity stabilizes. McConkie and Dyre (2000) represented this curve with four

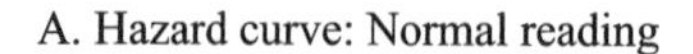

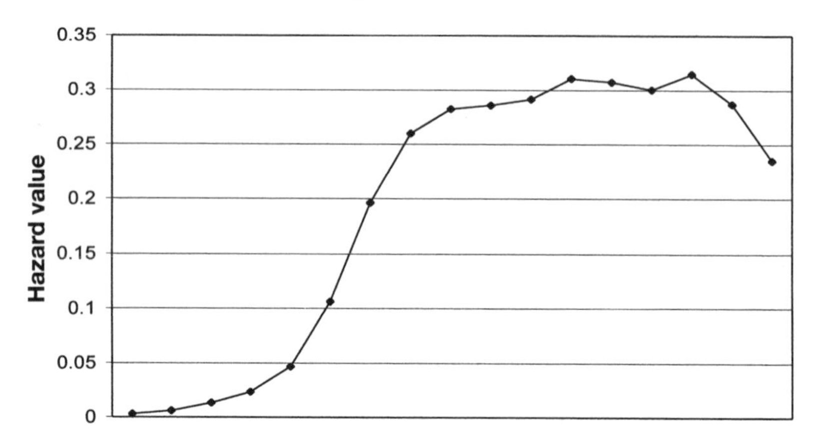

B. Frequency distribution: Normal reading

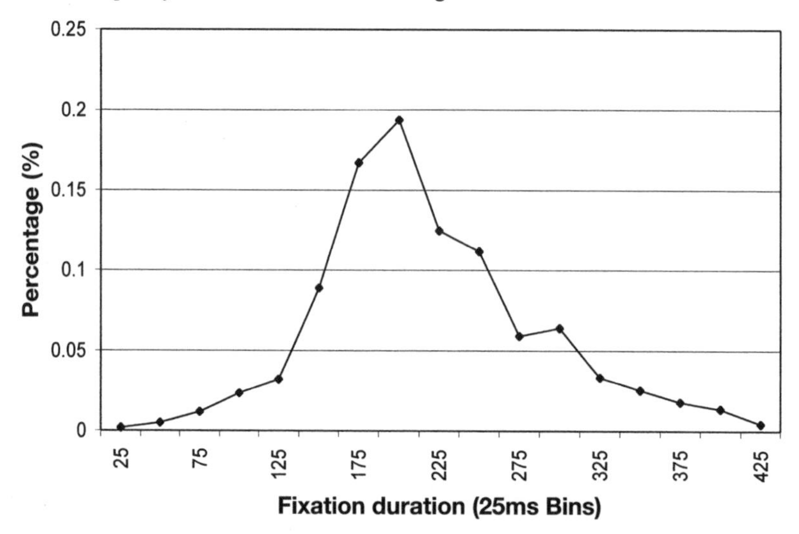

Figure 20.1: Fixation duration hazard function (upper panel, A) and frequency distribution (lower panel, B) in reading. The hazard values are based on both forward and regressive saccades at each time interval. The data are combined from 37 subjects, with a total of 1621 cases.

parameters, one representing the slope of the linear part of the curve in the early period, two representing the onset and rate of the rise in the middle period, and one representing the asymptotic level.

The C/I theory proposes that at or near the end of each saccade the saccadic system is briefly inhibited, which reduces the likelihood of a new saccade occurring immediately after the previous one. Thus, there is a low level of activity at the beginning of the new fixation. The lifting of that inhibition allows a rapid rise in saccadic activity, leading to an asymptote when the inhibition has fully dissipated.

The above principles describe a basic mechanism that repetitively produces saccades without direct cognitive control, generating them at relatively random times determined by the process of saccade preparation itself. However, there are three ways in which cognitive activities can influence the saccadic activity: through processing-based inhibition, through parameter adjustment, and through late-acting cognitive control. We will describe each of these briefly.

Processing-based Inhibition

Yang and McConkie (2001) reported that when an abnormal, text-like pattern is encountered during a single fixation, the hazard level begins to drop below that of the control condition (that is, the level of saccadic activity declines) at a time that depends on the nature of the alternative pattern. Thus, some saccades that would have occurred are cancelled and occur only later (McConkie *et al.*, 1992). They attributed this to an inhibition of the saccadic system, perhaps through the activation of oculomotor regions that generate fixation-related activity. Figure 20.2 shows the hazard curve for saccade onset times (or fixation durations) in the control condition (normal text present during the fixation) and in a nonword condition (strings of random letters present during the fixation). The inhibition of saccades (reduction in saccadic frequency) is reflected in the nonword condition hazard levels becoming lower than those of the control condition beginning in the 200 ms interval; beginning about 275 ms the non-word condition hazard level rises as the cancelled saccades begin to occur, including many that are regressive. A similar, but less dramatic, phenomenon occurs when the eyes fall upon pseudowords in the text with no display changes (McConkie *et al.*, 1992; Yang, 2002).

We propose that these non-text letter strings or pseudowords create processing problems at some level of text processing, and that the brain centers being disturbed send a distress signal that results in inhibition in the saccadic system. If this inhibition occurs before the system is committed to making the saccade (referred to as the point of no return, McConkie *et al.*, 1985, or as the beginning of the non-labile period, Reichle, *et al.*, 1998), then the saccade is likely to be cancelled and followed by the programming of another saccade. Whether a saccade is actually canceled by the inhibition or not depends primarily on the outcome of a race between the normal strategy-produced saccade signal and the processing-based inhibition (when it occurs) that results from processing difficulties. We assume that the saccade delay produced by this inhibition is not finely regulated by the needs of the cognitive system; rather it is a discrete response to the fact that a problem occurred. The fact that more severe problems result in increasingly longer mean fixation durations is not the result of regulating the saccade delay according to the processing requirements, but to the fact that the earlier and more severe the processing difficulty, the larger the proportion of saccades that are canceled and thus delayed by the inhibition. A proposal similar to this has been described in the SWIFT model (Engbert, Longtin & Kliegl, 2002; Kliegl & Engbert, this volume),

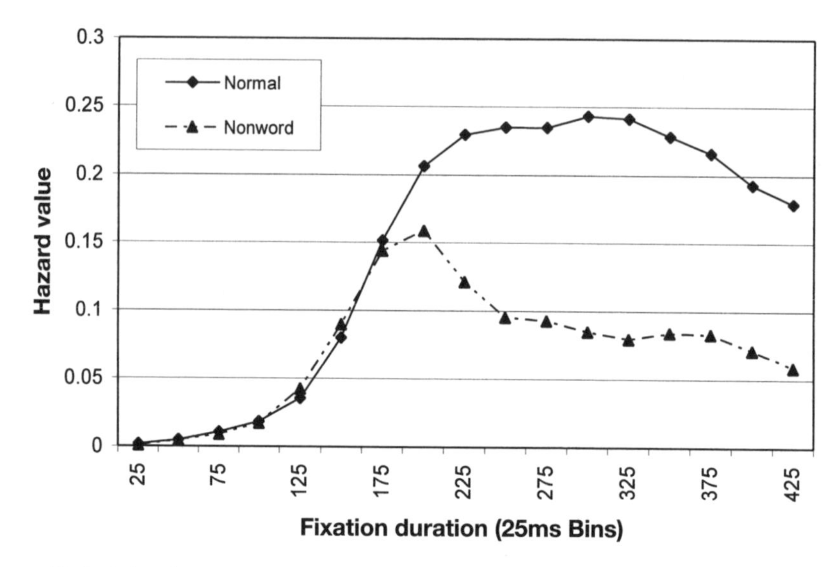

Figure 20.2: Fixation duration hazard function for fixations preceding forward saccades. Normal text condition data come from critical fixations on which the original text was present. Random letter condition data come from cases in which all words are replaced with random letters, with spaces and punctuation being preserved, for a single fixation. Data are combined from 37 subjects, with a total of 1273 and 657 cases in the two conditions.

although they regard the size of the effect of inhibition on mean fixation duration as being a function of word frequency, which shifts saccade latency distributions. Our view is that the size of this effect is determined by the proportion of saccades that are affected by the inhibition.

Yang & McConkie (2001) observed that there appear to be three categories of saccades: early saccades that occur regardless of the nature of the stimulus, even in the presence of a blank screen; normal saccades (the majority of saccades) that occur in the presence of text-like stimulus patterns (Xs or random letter strings that preserve the spacing between the words) but whose onset time is not directly related to the language processing taking place (that is, they are unaffected by whether the stimulus consists of real text or strings of non-words); and late saccades consisting of the delayed saccades that occur after an earlier saccade was canceled. The effect of processing-based inhibition is primarily to reduce the number of normal saccades and thus increase the number of late saccades that occur, many of which are regressions. The severity of the processing difficulty and how early in the fixation the difficulty occurs determine the number of normal saccades that are canceled and, through that, the amount of change in the mean fixation duration in that condition, as compared to the control condition. This is proposed as the basis for the increases in mean fixation duration that are observed when processing difficulties are encountered, rather than there being a graded increase in fixation duration resulting from the need for more processing time, as Reichle *et al.* (1998) and Engbert *et al.* (2002) suggest. Notice that this implies a discrete basis for control of the durations of individual fixations, where, in most cases, either the normal saccade occurs or it is canceled and another saccade is generated later, rather than a fine-grained adjustment of fixation time based on the current language processing needs.

The method used by Yang and McConkie (2001) was to replace the entire page of text with an alternative page during occasional selected saccades, and then to return the normal text to the screen during the following saccade, so the abnormal pattern was occasionally present for a single fixation during ongoing reading. Since the stimulus changes occurred while the eyes were in a saccadic movement, the subjects did not directly detect the change itself, and the observed response changes are the result of the abnormal stimulus pattern. In this study, and in follow-up studies that are yet unreported, the onset time of the inhibition varies with the nature of the abnormal stimulus pattern, depending on the processing level it is designed to disrupt. The pattern of saccadic inhibition observed is different from that reported by Reingold and Stampe (in press; this volume) from display changes that occur during eye fixations when the stimulus motion accompanying the change is directly perceived. In this case the saccadic inhibition occurs with a much shorter latency (about 80 ms) and with a briefer inhibition period.

Parametric Adjustment

We assume that the oculomotor control system is a flexible system that can be tuned to optimize performance in supporting the reading activity. We further assume that this tuning process is accomplished through changes in neural functioning that can be represented as the adjusting of parameters in a mathematical description. An example of this type of adjustment can be seen by comparing the hazard curves of fixation durations for the control conditions in two recent studies we have conducted. One study (Yang & McConkie, 2001) is described above: during occasional, randomly-selected fixations, separated by at least nine saccades, the text was replaced by alternate patterns (random letter strings, strings of Xs, strings of dashes or a blank page). The stimulus change occurred during one saccade and the text returned during the next, so the alternate stimulus pattern was present for a single fixation. Subjects were asked to ignore the alternative stimuli and read as normally as possible. The second study was similar but had only two types of alternate stimuli: random letters and strings of Ss, both with the spaces between words preserved. Subjects were instructed to ignore one of these patterns, but to stop their eyes and hold their fixation when the second occurred.

Each study included a control condition in which saccades were selected in the same way but no change was made in the text for that fixation. Thus, the actual stimulus pattern on most fixations, including those in the control condition, were identical for the two studies as was the primary task of reading and comprehending the text; the only difference was in the nature of the alternate stimuli that occasionally occurred, and the reader's secondary task. Subjects were drawn from the same subject pool. Any difference in the data between the two studies, then, is attributable to differences in the reading strategies elicited in the two studies by task and experimental fixation stimulus differences.

Figure 20.3 shows the hazard curves for saccade onset times (or fixation durations) for the critical fixations in these two control conditions. The curves are quite similar during the first period, and begin to rise rapidly at about the same time. The primary difference is in the asymptote level and, associated with that, the rate at which the curves rise to reach this asymptote. We have observed that this asymptote level, r, represented by McConkie and Dyre's (2000) fourth parameter, often shows substantial differences among conditions in other data sets we have compared, and in differences among subjects. A lower asymptote essentially increases the average time before the following saccade occurs, thus producing longer fixations on average. A flat hazard curve is characteristic of a Poisson, or random waiting time, distribution; thus, a reasonable interpretation is that during the third period most saccades are being generated at random times, but with the average rate being controlled by the state of some aspect of the system, represented by this parameter. An increased general level of activation of Findlay and Walker's (1999) fixate center, for example, could inhibit saccadic activity somewhat, increasing the time before saccades tend to be initiated, thus slowing reading and providing longer average times for processing to occur during fixations.

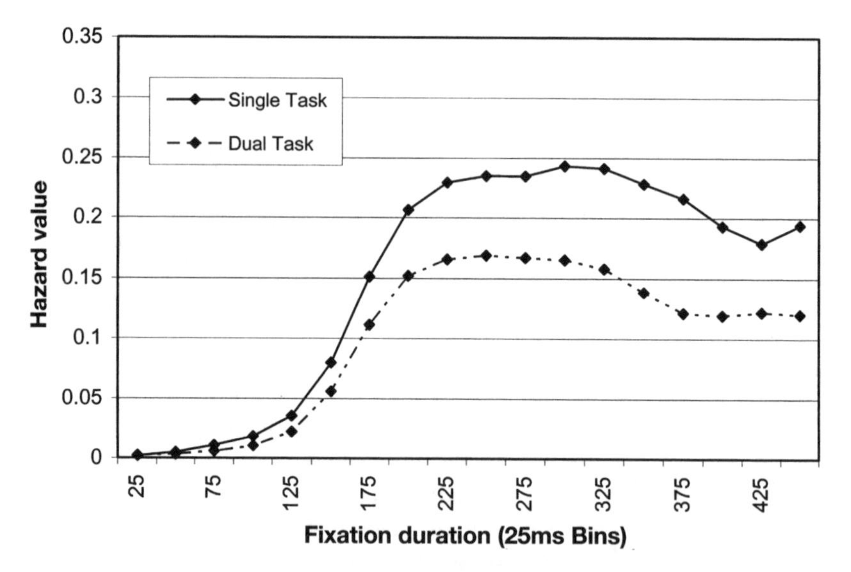

Figure 20.3: Fixation duration hazard functions preceding forward saccades. The single-task function results from a condition in which subjects read text for comprehension, ignoring occasional, single-fixation presentations of abnormal text patterns. The dual-task function results from a condition in which subjects ignored one abnormal, single-fixation text pattern (random letters or all Ss) but tried to stop their reading when the other pattern appeared. Both curves are from the control condition for which no altered text patterns were present. Data are combined from 35 subjects for each curve, with totals of 1037 and 685 cases.

The hazard curve asymptote parameter appears to be particularly responsive to varied reading task conditions. However, we assume that there must be other parameters, as well, that are responsive to (or settable by) aspects of the cognitive or motivational state of the reader, adjusting the general characteristics of the saccadic activity so that it is appropriate to the needs of the language processing taking place. This proposal raises a number of issues requiring further research: the development of an inventory of these parameters, identification of the conditions under which they are adjusted, examination of the behavioral effects of their adjustment and how quickly the adjustments occur, understanding the implications of parameter value differences among individual readers, and discovering the neurological bases for these differences.

Late-acting Cognitive Control

People clearly can exercise a great degree of cognitive control over their eye behavior when they wish to. This cognitive control is not complete; Kramer, Cassavaugh, Irwin, Peterson and Hahn (2001) report that the sudden onset of a stimulus object can draw observers' eyes even when they are trying to avoid directing their gaze to the object. But we do have the ability to direct our gaze to a specified word and to maintain the gaze for an extended period. Recognizing this ability appears to open the door to accepting the basic assumption of cognitive control theories; the system can simply direct the gaze toward a word until some cognitive event is reached and then purposively shift the gaze to a different word. Our claim is that, while this type of cognitive control is possible, the system does not normally work this way during reading because this type of control is too slow and is quite taxing on the cognitive system.

In order to obtain an initial estimate of the time required to exert specific cognitive control during reading, we conducted a study (Yang & McConkie, in preparation; Yang, 2002) in which, as people read, the text was occasionally replaced for a single fixation by strings of a single letter which served as a request for a particular response: Rs indicated that subjects should move their eyes regressively (somewhat like an anti-saccade task, Hallett, 1978; Everling & Fischer, 1998), and Fs indicated that they should move their eyes progressively (somewhat like a pro-saccade task, Hallett & Adams, 1980). The control condition (no changes) provides a baseline. Figure 20.4 shows the hazard curves for fixation durations for these conditions, plotting forward and regressive hazard curves separately. Summing forward and regressive hazard curves produces the type of general hazard curve observed in earlier figures. Separating the data in this manner allows us to observe the points in time at which there is an increase in regressions in the R condition, compared to the other conditions, and an increase in forward saccades in the F condition, indicating compliance with the instructions.

An examination of Figure 20.4 (a) indicates that both experimental conditions show a similar deviation from the control condition at about 200 ms. This early inhibition onset time is similar to that observed in Yang and McConkie (2001) when the text was replaced by segmented strings of Xs with instructions to ignore the alternative stimuli. Thus, this is a generalized effect of processing-based inhibition that results rather automatically from encountering non-textual patterns. The actual initiation of cognitive

a)

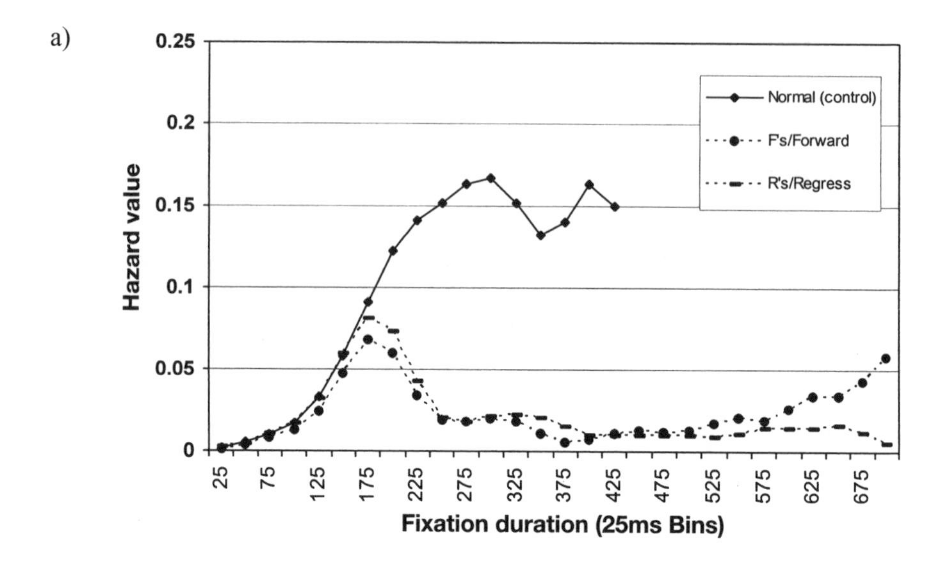

b)

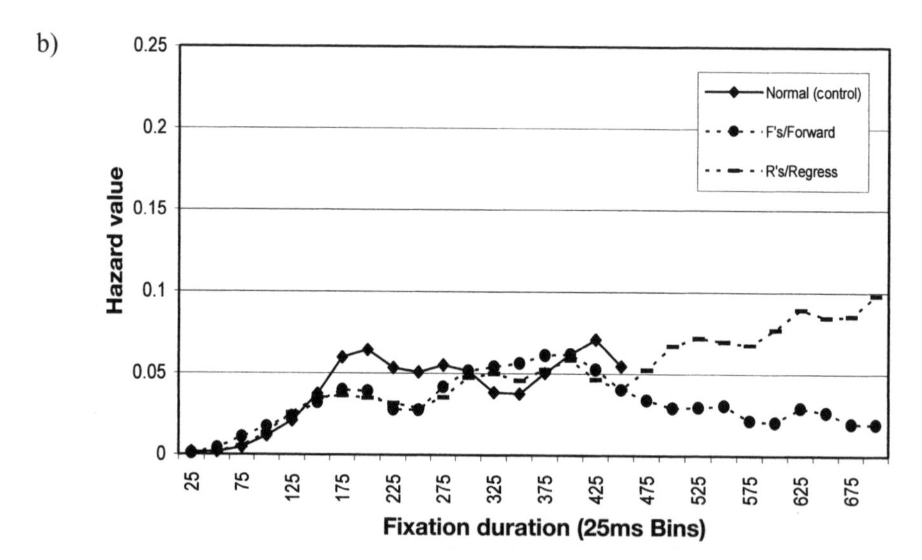

Figure 20.4: Fixation duration hazard functions preceding forward and regressive saccades. Normal text condition data come from critical fixations on which the original text was present. Purposive movement condition data come from fixations on which the text was replaced by strings of Fs or Rs, with the subject asked to make a forward saccade when Fs appear, and a regressive saccade when Rs appear. An increase in the F condition hazard function in (a), relative to the other conditions, indicates the occurrence of purposive saccade control, as does an increase in the R condition in (b). Data are combined from 35 subjects for each curve, with totals of 685, 481, and 333 cases for forward saccades, and with totals of 309, 434, and 539 cases for regressive saccades.

control, indicated by an increase in the number of saccades in the requested direction, appears much later: Figure 20.4b shows an increase in regressive saccades in the R condition beginning at about 475 ms after the onset of the fixation, and Figure 20.4a shows a lesser rise in forward saccades in the F condition beginning even later. Fixations that terminated prior to this time (94% of all fixations in the control condition) ended before such direct cognitive control was possible. Thus, we propose that, while it is possible for a reader to exercise direct cognitive control over the timing and direction of saccades, this only happens in cases with very long fixations, after the normal saccade has been cancelled due to processing difficulty. This type of direct cognitive control, based on currently-available visual information, is not the normal means by which eye behavior is produced during reading. It is possible that in some cases an intentional saccade can be planned over two or more fixations (for example, see McConkie, Underwood, Wolverton & Zola, 1988), but this requires further investigation.

It should be noted that fixations in tasks in which appropriate eye movement behavior is less predictable, such as during picture viewing, tend to be longer on average than fixations in reading. For example, fixations in the 500 ms range are quite rare during reading but not uncommon in picture viewing data. This additional time may allow direct cognitive control to come into play more frequently in these situations. Also, the latency of direct cognitive control may be reduced through extensive practice, resulting in shortening of response time; the degree to which cognitive control can be hastened remains a matter for further investigation.

Regressive Saccades

There has been much speculation about the role of regressive saccades in reading. They are often interpreted as resulting from direct cognitive control: the eyes are being sent back to a word because of failure to identify it or to deal with processing difficulties involving that word (Radach & McConkie, 1998; Rayner & Sereno, 1994). However, Yang and McConkie (2001) found that, although abnormal stimulus patterns greatly increased the number of regressions made, with some conditions producing 2.5 times as many regressions as other conditions, the frequency distributions of saccade lengths did not differ among these conditions. For example, many new regressions occurred when the computer screen was blanked during the fixation, as compared to the control condition with normal text, but these additional saccades still had a saccade length distribution similar to that of the control condition, even with this drastic change in the stimulus pattern. Other studies have also noted the lack of effect of experimental conditions on the lengths of regressive saccades during reading (Vitu *et al.*, 1998; Radach & McConkie, 1998). Thus, it appears that there is little support for the notion that typical regressive eye movements during reading are taking the eyes to specific text locations requiring further consideration. This supports the above argument that many, and perhaps most, regressive saccades are not generated by the purposive control of eye movements. C/I theory includes the assumption that when processing difficulties occur and inhibition of the saccadic system results, this disrupts the strategy-based activation process, thus reducing or eliminating the rightward bias in saccadic activity. This allows more regressions to occur. Yang and McConkie (2001) observed an increase to 50% or even more of the saccades being regressive in some conditions. This can be cognitively functional, of course, preventing the eyes from proceeding to new text when a processing problem occurs, and thus providing more time for trying to resolve it. But it does not appear to be the result of direct cognitive control. We presume that this is because, as indicated above, direct cognitive control can occur only after fairly long delays, usually resulting in an intervening saccade being initiated on some other basis in the meantime. In the F and R study described above, it was seldom that subjects who were trying to initiate regressions on signal were able to actually produce a regression as the immediately following saccade.

Delayed Effects

It is important to note that the analyses reported here have all focused exclusively on the effects that stimulus patterns (and their presumed processing effects) have on the onset time, direction and length of the saccade that ends the fixation on which the pattern was perceived. This is referred to as Saccade $N+1$, assuming that Saccade N was the saccade on which the initial display change occurred that brought the abnormal stimulus pattern onto the screen. Changes in the time and properties of Saccade $N+1$ are referred to as "immediate effects." There are also "delayed effects" in which the time and properties of Saccade $N+2$ are affected by the abnormal stimulus patterns (Rayner, 1998). We have not yet studied the cognitive influences on these saccades in detail. However, it is clear that delayed effects can occur in the absence of immediate effects during reading (McConkie *et al.*, 1988). Also, delayed effects provide much more time in which cognitive influences could develop to affect the properties of saccade $N+2$. Thus, it would not be surprising to find a larger role of cognition in delayed eye movement control in reading. However, it is still necessary to document such influences, and to examine the means by which they occur, as we have tried to do for immediate effects. A general implication of the research described above is that effects that may at first appear to result from direct cognitive control can actually be produced by non-cognitive or indirect cognitive influences on eye behavior. Whether this is also true of delayed effects, and whether the study of these effects will reveal additional types of cognitive influences in eye behavior during reading, remains to be determined.

Since information that produces delayed effects on Saccade $N+2$ was acquired during the prior fixation, that ending in Saccade $N+1$, it is important to note that these effects are likely to be interpreted as "peripheral preview" effects. Thus, it is not clear whether a consistent distinction can be maintained between delayed effects and effects resulting from a peripheral preview of the upcoming text.

Summary

In this paper, we have argued against theories that attribute most eye movement behavior during reading to direct cognitive control. We have described a recently proposed theory in which eye behavior is controlled in a much less cognitive manner through a rather automatic system of activation and inhibition within the saccadic system. This raises the question of how cognition, which is known to affect eye behavior, exerts its influence in such a system. Three ways are suggested: inhibition of saccadic activity when processing difficulties are encountered, adjustment of parameters in the saccadic control system, and, in some cases, direct cognitive control, though this tends to have a long latency. Thus, the proposal is that eye behavior in reading is usually produced on a non-cognitive basis but is influenced in indirect ways by certain cognitive activities that are taking place in reading. A basic question is whether reading eye behavior can be accounted for in general with these types of low-level principles, and whether and when it is necessary to employ higher-level, more cognitively-related principles.

References

Balota, D. A., Pollatsek, A., & Rayner, K. (1985). The interaction of contextual constraints and parafoveal visual information in reading. *Cognitive Psychology, 17*(3): 364–390.

Balota, D. A., & Rayner, K. (1983). Parafoveal visual information and semantic contextual constraints. *Journal of Experimental Psychology: Human Perception & Performance, 9*(5): 726–738.

Brysbaert, M., & Vitu, F. (1998). Word skipping: Implications for theories of eye movement control in reading. In: G. Underwood (ed.), *Eye Guidance in Reading and Scene Perception.* (pp. 125–147). Oxford: Elsevier.

Ehrlich, S. F., & Rayner, K. (1981). Contextual effects on word perception and eye movements during reading. *Journal of Verbal Learning & Verbal Behavior, 20*(6): 641–655.

Engbert, R., Longtin, A., & Kliegl, R. (2002). A dynamical model of saccade generation in reading based on spatially distributed lexical processing. *Vision Research, 42*, 621–636.

Everling, S., & Fischer, B. (1998). The antisaccade: A review of basic research and clinical studies. *Neuropsychologia, 36*(9): 885–899.

Fadden, S. (1995). Eye movement measures and reading: Toward an understanding of the relationship between eye movements and cognitive processing. Master Thesis at University of Illinois at Urbana-Champaign. Champaign, Illinois.

Findlay, J. M., & Walker, R. (1999). A model of saccade generation based on parallel processing and competitive inhibition. *Behavioral & Brain Science, 22*(4): 661–721.

Hallett, P. E. (1978). Primary and secondary saccades to goals defined by instructions. *Vision Research, 18*(10): 1279–1296.

Hallett, P. E., & Adams, B. D. (1980). The predictability of saccadic latency in a novel voluntary oculomotor task. *Vision Research, 23*, 329–339.

Just, M. A., & Carpenter, P. A. (1980). A theory of reading: From eye fixations to comprehension. *Psychological Review, 87*(4): 329–354.

Kennedy, A., & Murray, W. S. (1984). Inspection times for words in syntactically ambiguous sentences under three presentation conditions. *Journal of Experimental Psychology: Human Perception & Performance, 10*(6): 833–849.

Kerr, S. (1992). Eye movement control during reading: the selection of where to send the eyes (word skipping). *Dissertation Abstracts International,* Vol. 53(10-A), April 1993, 3485.

Kramer, A. F., Cassavaugh, N. D., Irwin, D. E., Peterson, M. S., & Hahn, S. (2001). Influence of single and multiple onset distractors on visual search for singleton targets. *Perception & Psychophysics, 63*(6): 952–968.

Levy-Schoen, A. (1981). Flexible and/or rigid control of oculomotor scanning behavior. In: D. F. Fisher, R. A. Monty and J. W. Senders (eds), *Eye movements: Cognition and Visual Perception* (pp. 299–314). Hillsdale, NY: Erlbaum.

Logan, G. D., & Cowan, W. B. (1984). On the ability to inhibit thought and action: A theory of an act of control. *Psychological Review, 91*(3): 295–327.

Logan, G. D., & Irwin, D. E. (2000). Don't look! Don't touch! Inhibitory control of eye and hand movements. *Psychonomic Bulletin & Review, 7*(1): 107–112.

McConkie, G. W. (1979). On the role and control of eye movements in reading. In: P. A. Kolers, M. Wrolstad and H. Bouma (eds), *Processing of Visible Language I* (pp. 37–48). New York: Plenum Press.

McConkie, G. W., & Dyre, B. P. (2000). Eye fixation durations in reading: Models of frequency distributions. In: R. Radach, A. Kennedy, D. Heller and J. Pynte (ed.), *Reading as a Perceptual Process* (pp. 683–700). Oxford: Elsevier.

McConkie, G. W., Reddix, M. R., & Zola, D. (1992). Perception and cognition in reading: Where is the meeting point? In: K. Rayner (ed.), *Eye Movement and Visual Cognition* (pp. 293–303). New York: Springer-Verlag.

McConkie, G. W., Underwood, N. R., Zola, D., & Wolverton, G. S. (1985). Some temporal characteristics of processing during reading. *Journal of Experimental Psychology: Human Perception & Performance, 11*(2): 168–186.

Morrison, R. E. (1984). Manipulation of stimulus onset delay in reading: Evidence for parallel programming of saccades. *Journal of Experimental Psychology: Human Perception & Performance, 10*(5): 667–682.

O'Regan, J. K. (1989). Visual acuity, lexical structure, and eye movements in word recognition. In: B. G. Elsendoorn and H. Bouma (eds), *Working Models of Human Perception.* (pp. 261–292). London, England: Academic Press.

Radach, R., & McConkie, G. W. (1998). Determinants of fixation positions in words during reading. In: G. Underwood (ed.), *Eye Guidance in Reading and Scene Perception.* (pp. 77–100). Amsterdam: Elsevier.

Rayner, K. (1979). Eye guidance in reading: Fixation locations within words. *Perception, 8*(1): 21–30.

Rayner, K. (1998). Eye movements in reading and information processing: 20 years of research. *Psychological Bulletin, 124*(3): 372–422.

Rayner, K., & Duffy, S. A. (1987). Eye movements and lexical ambiguity. In: J. K. O'Regan and A. Levy-Schoen (eds), *Eye Movements: From Physiology to Cognition* (pp. 521–529). Amsterdam: Elsevier Science.

Rayner, K., & Fischer, M. H. (1996). Mindless reading revisited: Eye movements during reading and scanning are different. *Perception & Psychophysics, 58*(5): 734–747.

Rayner, K., & Raney, G. E. (1996). Eye movement control in reading and visual search: Effects of word frequency. *Psychonomic Bulletin & Review, 3*(2): 245–248.

Rayner, K., & Sereno, S. C. (1994). Regressive eye movements and sentence parsing: On the use of regression-contingent analyses. *Memory & Cognition, 22*(3): 281–285.

Reichle, E. D., Pollatsek, A., Fisher, D. L., & Rayner, K. (1998). Toward a model of eye movement control in reading. *Psychological Review, 105*(1): 125–157.

Reingold. E. M., & Stampe, D. M. (in press). Saccadic inhibition in reading. *Journal of Experimental Psychology: Human Perception and Performance.*

Schilling, H. E. H., Rayner, K., & Chumbley, J. I. (1998). Comparing naming, lexical decision, and eye fixation times: Word frequency effects and individual differences. *Memory & Cognition, 26*(6): 1270–1281.

Underwood, G., Clews, S., & Everatt, J. (1990). How do readers know where to look next? Local information distributions influence eye fixations. *Quarterly Journal of Experimental Psychology, 42A*, 39–65.

Vitu, F., McConkie, G. W., & Zola, D. (1998). About regressive saccades in reading and their relation to word identification. In: G. Underwood (ed.), *Eye Guidance in Reading and Scene Perception* (pp. 101–124). Amsterdam: Elsevier.

Vitu, F., O'Regan, J. K., Inhoff, A. W., & Topolski, R. (1995). Mindless reading: Eye-movement characteristics are similar in scanning letter strings and reading texts. *Perception & Psychophysics, 57*(3): 352–364.

Yang, S.-N. (2002). Inhibitory control of saccadic eye movements in reading: A neurophysiologically based interaction-competition theory of saccade programming. unpublished doctoral dissertation. University of Illinois at Urbana-Champaign.

Yang, S.-N., & McConkie, G. W. (2001). Eye movements during reading: a theory of saccade initiation times. *Vision Research, 41*, 3567–3585.

Yang, S.-N., & McConkie, G. W. (in preparation). Oculomotor-based and purposive manipulation of saccade preparation: An inhibition-based explanation of saccade control during reading.

Zola, D. (1984). Redundancy and word perception during reading. *Perception & Psychophysics, 36*, 277–284.

Chapter 21

Foundations of an Interactive Activation Model of Eye Movement Control in Reading

Ronan G. Reilly and Ralph Radach

This chapter describes an interactive activation model of eye movement control in reading, which we refer to as "Glenmore", that can account within one mechanism for preview and spillover effects, and for regressions, progressions, and refixations. The model decouples the decision about when to move the eyes from the word recognition process. The time course of activity in a "fixate centre" determines the triggering of a saccade. The other main feature of the model is the use of a saliency map that acts as an arena for the interplay of bottom-up visual features of the text, and top-down lexical features. These factors combine to create a pattern of activation that selects one word as the saccade target. Even within the relatively simple framework proposed here, a coherent account has been provided for a range of eye movement control phenomena that have hitherto proved problematic to reconcile.

Theoretical Background and an Outline of the Model

Current Models of Eye Movement Control in Reading

In recent years, research on eye movements in reading has made substantial progress. A key new development in the field is the emergence of computational models of eye movement control during reading (see Kennedy, Radach, Heller & Pynte, 2000 and Reichle, Rayner & Pollatsek, in press, for detailed discussions). The modelling principles and the algorithms that these models implement reflect the theoretical viewpoints of their authors. For example, in the influential E-Z Reader model (Reichle, Pollatsek, Fisher & Rayner, 1998) sequential lexical processing is suggested to be the

The Mind's Eye: Cognitive and Applied Aspects of Eye Movement Research

ISBN: 0–444–51020–6

obligatory trigger for the generation of all eye movements made in normal reading. In contrast, Reilly & O'Regan (1998), following the theoretical framework developed by O'Regan (1990), attempted to demonstrate that a good account for the positioning of fixations in reading can be achieved by using a set of rather dumb oculomotor heuristics. We believe that both of these positions have their merits and can account for important aspects of eye behaviour during reading. On the other hand, both approaches have also serious limitations.

It is quite clear that a pure visuomotor account as proposed by Reilly and O'Regan (1998) is not sufficient to explain many phenomena that are apparent in human reading behaviour. This has been acknowledged already by O'Regan, Vitu, Radach and Kerr (1994) who suggested that visuomotor and cognitive theories of eye movement control in reading will need to be combined: "The resulting intermediate theory contains both an underlying scanning strategy that can manifest itself even in the absence of linguistic material to process; it contains a modulator influence of linguistic processing . . . Future work will show how exactly these components must be combined" (p. 345).

Thus, the question of interest is not whether eye movements are determined by visuomotor factors or linguistic processing, but to what degree these two types of factors are involved and how they interact. Taking the likelihood of fixating a word as an example, Brysbeart and Vitu (1998) have shown in an elegant meta-analysis that most of the variance in "word skipping" is accounted for by word length.[1] At the same time there are significant influences from two cognitive factors, word frequency and contextual predictability. A similar example is the analysis of refixations by McConkie, Kerr, Reddix, Zola and Jacobs (1989), who first showed that the likelihood of immediately refixating a word based on initial fixation position can be expressed as a u-shaped (quadratic) function. Importantly, only the vertical shift parameter of the refixation function varies with word frequency, while the slope of the function is determined by a visual factor, the eccentricity of the initial fixation position relative to the word centre. Analyses of this type appear to suggest that eye viewing behaviour is co-determined by low-level visual and higher-level cognitive factors, possibly in terms of low-level default routines that are affected substantially by cognitive modulation.

The critique of models that are based on sequential lexical processing and discrete shifts of attention has taken different directions. From a computational point of view, Engbert and Kliegl (2001) have shown that about the same fit of empirical reading data can be achieved using a model that includes simpler principles of operation. Their model differs from E-Z Reader in at least two important respects: the two phases of the word recognition process (familiarity check and lexical access) are replaced with an all-or-none process, and sometimes "autonomous" saccades are triggered independent of lexical processing.

A second line of criticism is based on empirical work suggesting that word processing during reading may be spatially distributed rather than confined to one word at a time as suggested by sequential attention shift models. These observations usually take the form of demonstrating that characteristics of a parafoveal word can influence the duration of a fixation on the currently fixated foveal word (e.g., Kennedy, 1998; Kennedy *et al.,* 2002). Inhoff, *et al.* (2000) have shown that information is regularly being acquired from positions left of the currently fixated word. Moreover, a recent experiment by Inhoff, Lemmer, Eiter and Radach (2001) suggests that information from a parafoveal word can be acquired very early during a reading fixation. Although these results may not yet be conclusive, several sources of evidence point to the possibility that linguistic processing during reading can operate on (at least) two words in parallel.

This view is corroborated by another line of criticism dealing with the time line of processing events. As discussed by Deubel, O'Regan and Radach (2000) and Radach, Inhoff and Heller (2002), basic oculomotor research using the double step paradigm indicates that the (re-)programming of a saccade must be initiated at least 70 ms to 90 ms before the end of the current fixation. This is similar to the suggestion by McConkie, Underwood, Zola and Wolverton (1985) of an interval of 80–100 ms before saccade onset as the deadline for stimulus influences during a fixation. Sereno, Rayner and Posner (1998) report ERP results in a word recognition paradigm indicating that, starting at 132 ms after stimulus onset, word frequency differences can be observed. Since the lexical familiarity stage in the E-Z Reader Model is sensitive to word frequency effects, this phase of word processing cannot take place much earlier than the 132 ms estimate. Taken together, the conclusion is that the effective time window for direct (immediate) linguistic influences during a fixation of 250 or 275 ms is very limited. Looking at the scenario that sequential attention models suggest to account for "word skipping" (see e.g., Pollatsek, Reichle & Rayner, this volume), it is not clear how operations like lexical access on the origin word, a shift of attention to the skipped word and a familiarity check on the skipped word could all be completed in sequential way within this brief time interval.

One way to accommodate the problems discussed above is to relax assumptions concerning the allocation of attention, allowing more than one word to be lexically processed concurrently. Following LaBerge and Brown (1989), Inhoff *et al.* (2000) suggested replacing the idea of a sequentially moving attentional spotlight by an attentional gradient around the point of fixation that can include several words. A similar route has been taken by Engbert, Longtin and Kliegl (2002) who developed a computational model of eye movement control in reading that incorporates the notion of spatially distributed lexical processing. Other core principles of their model are an autonomous timing of saccades that tend to be generated at a preferred rate and the inhibition of saccade initiation by foveal lexical processing. On the other hand, Engbert *et al.* (2002) have maintained some key features of the original Easy Reader model, for example, a division of lexical processing into an early versus late stage, and a distinction between a labile and a non-labile phase of saccade programming (see Kliegl & Engbert, this volume, for a discussion).

An Outline of Our Theory and Model

The model that will be described in this chapter represents a more radical departure from the sequential attention shift conception. While our theory has some apparent similarities with the SWIFT model and the competition-inhibition model by Yang and McConkie (2001), there are also important differences.

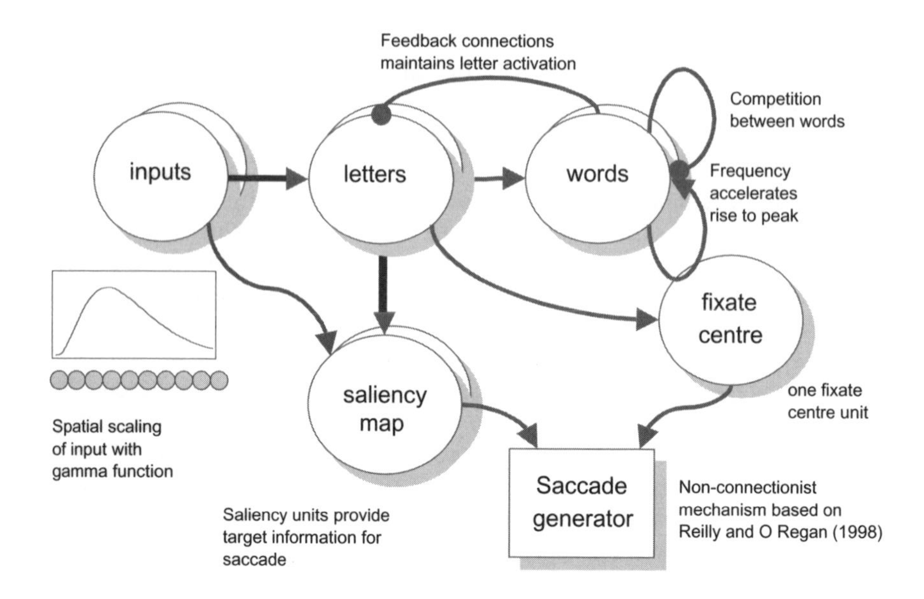

Figure 21.1: Model overview.

This figure represents the main components of the Glenmore model. The circles represent connectionist components, the rectangle a non-connectionist component. Connections with filled circles represent negative connections, those with arrows positive ones. Note that the negative connection from words to letters act to maintain activity in the letter units, when those units have a cumulative Gaussian transfer function. This is because the negative top-down values will impede the rate at which the activation values saturate as a function of their bottom-up inputs.

Our theory and its first computational implementation, the "Glenmore" model,[2] are closely related to the general theory of saccade generation developed by Findlay and Walker (1999). They proposed that saccade target selection is accomplished via parallel processing and competitive inhibition within a two-dimensional salience map. The triggering of a saccade is controlled by a fixate centre that is sensitive to input from several routes of cognitive processing. Once a saccade is triggered, it will go to the target that has emerged as a winner in the saliency competition. Radach (1999) has proposed that the saliency map mechanism could provide an elegant basis for a theory of saccade generation in reading, avoiding many complexities inherent to the class of sequential attention models. In particular, we deviate from the view that the word-by-word sequence within a line of text provides a chain of default saccade targets and that the observable eye movement behaviour is generated via frequent cancellation and reprogramming of default saccades.

In most cognitive activities that require systematic visual scanning, every saccade is directed to a specific target object. In the case of reading this selection takes the form of deciding which of the words within the current "perceptual span" should be the target for the next saccade. In the vast majority of cases, this will include the word currently fixated, the preceding word, and the two or three following words. The most important low-level sources of influence on this decision are the length and the eccentricity of words located around the current point of fixation (Kerr, 1992; McConkie, Kerr & Dyre, 1994). Following Findlay and Walker, we assume that potential target words are represented (perhaps as low spatial frequency objects) on a salience map and, depending on the particular visual configuration, their salience values form a preference list of potential targets.

We further assume that at the beginning of each fixation low level visual information, coded as a saliency vector, is available that allows for the triggering of a saccade without any cognitive influence.[3] Over time, the saliency values representing potential targets will change in response to information about ongoing linguistic processing. This simple principle can be illustrated using the following example: Suppose that a reader is fixated on a letter in the right half of a word of medium length, the next word ($N+1$) is short and the word $N+2$ again of medium size. Given this visual configuration, it is likely that word $N+2$ will have the highest initial salience value and will be the target for the next interword saccade. If however, word $N+1$ turns out to be difficult to process, its saliency value may rise quickly and it will become a more attractive target. As will be discussed later, in this scenario there is competition between words for limited processing resources, opening a route to explain fovea-on-parafovea and parafovea-on-fovea effects. The above example shows also that within a spatial saliency framework the notion of "word skipping" becomes meaningless, as there is no default saccade program to $N+1$ that needs to be cancelled and reprogrammed (see Brysbaert & Vitu, 1998, for a similar idea).

In addition to the saliency map representation, the Glenmore model includes a visual input module, a word processing module, a fixate centre and a saccade generator, producing the actual saccadic movement (see Figure 21.1). A visual input vector codes the visual configuration around the current fixation position. The computation of initial saliency values is based on a letter processing function (McConkie & Zola, 1987) and effectively accounts for effects of word length and eccentricity. Visual information is transferred to the saliency map representation and to a linguistic processing module that implements processing on the letter and word level within an interactive activation (IA) framework (Grainger & Jacobs, 1998). From the linguistic processing module information about ongoing processing is transferred in two directions. The vector of letter unit activation is transmitted to the saliency map, where it is used to update continuously the saliency values of potential saccade targets. At the same time, feedback on the general level of excitation in the word processing network is sent to the fixate centre. The triggering of a saccade is based on activity in a fixate module that operates in relation to the dynamics of spatial saliency. Over the course of a fixation, activity in the fixate centre will tend to fall, a process that has a random component (similar to autonomous saccade triggering in SWIFT or the random waiting time component in Yang & McConkie's model) and a non-spatial processing component. The saccade will be executed by the saccade generator module after a latency period and will always be directed to the target with maximum saliency. The saccade generator implements the front end behaviour of the eyes as described by McConkie, Kerr,

Reddix and Zola (1988) and implemented in Reilly and O'Regan (1998) and Reichle, Rayner and Pollatsek (1999).

We believe that one important feature of our spatial saliency theory of eye movement control in reading is its neuroscientific plausibility. The general architecture is in harmony with neurobiological constraints and information processing principles suggested by oculomotor research (see e.g., Wurtz, 1996; Carpenter, 2000; Munoz, 2002). An important feature of the Glenmore model is that it operates on the level of individual letters *and* words both in terms of visual and linguistic processing and eye movement control. The inclusion of a realistic interactive activation network of letter and word processing is motivated by the fact that IA models have proven especially useful for capturing the time course of parallel activation and competitive inhibition between processing units on a hierarchy of levels. We see the letter and word processing part of the model as a step towards the necessary integration of modeling efforts in the neighbouring domains of word recognition and continuous reading (Grainger, 2000; Jacobs, 2000).

It is also worth pointing out that the spatial saliency theory of eye movement control in reading does not refer to the concept of "visual attention" in any way. We agree with Findlay and Walker (1999) who noted that not much is gained by assuming that "attention" is disengaged, moved and re-allocated as a function of saliency and that it is these process which, in turn, trigger saccade programming. From an epistemic point of view it appears that proposing that "attention" moves from word to word is only shifting the problem of explanation to another level: if attention is supposed to be an entity that moves somewhat independently from eye movements, this *movement* will not only need to be triggered. It will also need to be programmed, it will need to have a targeting mechanism, a latency and a duration, requiring a machinery of "attention generation" to co-exist concurrently with that of saccade generation.

Before the details and dynamics of the model are described in more detail, we will discuss some general issues related to computational modelling in the area of eye movement control in reading.

Computational Modelling

As mentioned in the introduction, there has been a significant increase in the use of computational modelling techniques to explore in a more rigorous fashion various theories of eye movement control in reading. We believe that this has helped to clarify some important theoretical issues and to eliminate some proposals that proved less viable when put under computational scrutiny. The traditional version of the Morrison (1984) model comes to mind here. Its assumptions regarding the interplay of the time course of saccadic programming and lexical identification proved unsupportable when implemented computationally (Reilly & O'Regan, 1998), which may have contributed to motivating the revised version of Morrison's model embodied in E-Z Reader (Reichle *et al.*, 1998).

Computational modelling serves to clarify theories, but cannot of itself resolve conflicts between them. Guidelines or meta-principles of computational modelling are required that will allow, among other things, the ready comparison of one model with another. A significant weakness of the current state of the field is in the area of model comparison, and the lack of an agreed methodology for doing so. Current computational models are usually compared on the basis of how well they fit a given set of data and how parsimoniously they do so. Authors usually describe the success of their model in terms of reproducing many of the global features of eye movements in reading — fixation and gaze durations, word skipping, refixations, etc., all as a function of word processing difficulty. However, when it comes to comparing the performance of models of differing complexity, with, for example fewer free parameters, we see the limitations of this approach to model comparison. There is a need here for a more in-depth comparison than the admittedly important one of comparing how well the models fit the data. At the very least, one should describe how many parameters are used in the model, how they were estimated, and how the numbers compare to the competing model. While an appeal to parsimony may not be entirely appropriate for models that ultimately rest upon a biological foundation (one suspects that evolution is not always transparently parsimonious), some comparison of numbers of parameters and their motivation would be helpful. Even here, the issue of what parameters are relevant to a comparison is moot. For example, if we compare E-Z Reader (Reichle *et al.*, 1998) to a connectionist model (e.g. Reilly, 1993a,b), do we count the modifiable weights of a neural network as free parameters? While providing answers to these questions is beyond the scope of the present discussion, these are issues that need to be tackled, particularly with the proliferation of computational models in the area.

The Glenmore Model

Background

The brain is a computational device best formalised by the *differential* rather than the *propositional* calculus. Dynamical systems theory is increasingly exploited as a means of understanding brain function both at a neural and cognitive level. In cognitive science, a field which has been traditionally dominated by a paradigm in which cognition is taken to be the manipulation of internal symbols, limitations in the symbol-based approach have become increasingly apparent. Researchers throughout cognitive science have been casting around for alternative theoretical frameworks. One of the most productive of these is the dynamical systems framework, according to which cognitive processes are behavioural patterns of non-linear dynamical systems and are best studied using the mathematics of dynamical modelling and dynamical systems theory (Port & Van Gelder, 1995; Kelso, 1995).

A dynamical systems approach has the potential to cast new light on an old issue in the eye movement control literature: What is the role of linguistic factors in the moment-to-moment control of eye movements in reading? Rather than facilitating all-or-none style explanations, a dynamical systems account can potentially accommodate explanations that argue for the interplay, over time, of linguistic and oculomotor factors. Such accounts might have oculomotor facts playing a dominant role early in the processing of the visual input, with lexical and linguistic factors entering the picture further into the processing time line.

One of the goals of Glenmore is to explore this class of dynamical model, one that allows the interplay of factors from multiple levels of representation. The most appropriate class of modelling frameworks for this approach would seem to be connectionist models, and specifically interactive activation (IA) models (McClelland & Rumelhart, 1981; Rumelhart & McClelland, 1982). A typical IA model comprises a set of interconnected neuron-like units. Activity is transmitted through the network over weighted connections. The units comprising the network implement a transfer function that combines the unit's inputs and generates an output based on these combined inputs. The nature of the transfer function varies from model to model, but the general design philosophy is to keep the operation it implements relatively simple. So, a typical transfer function might take the weighted sum of inputs and perform some sort of normalising operation using, for example, the sigmoid function. The network "computes" by circulating activation (i.e., real valued numbers) throughout the network until some stopping criterion has been reached. This might be the achievement of a threshold, or the stabilising of levels of activity.

Architecture

The connectionist architecture of the Glenmore model is relatively simple, comprising input units, letter units, saliency units, and word units (see Figure 21.1). In addition, there is a "fixate centre" unit that controls the decision when to execute a new saccade. This decision is based on the general level of activity in the letter units, once activation in the fixate centre falls below a given threshold, a saccade is targeted to the most salient word "blob" on the saliency map.

As already mentioned, each class of unit has an associated transfer function that determines what kind of output it generates from its inputs, and how this changes over time. The model uses two transfer functions: Gaussian and sigmoid. The Gaussian transfer function allows the unit to accumulate input, such that the output of the unit rises and decays over time. The precise rate of change is determined by the shape of the distribution, which in turn is determined by the two parameters of the Gaussian (mean and standard deviation). These parameters are fixed for the model described here ($m = 50$, $sd = 0.3m$), such that the output of the function is 1.0 for $m = 50$. The sigmoid transfer function operates in a similar way to the Gaussian, except that its output does not decay over time.

The only parameters free to vary are the weights connecting the units, and even here all weights of the same type (e.g., letter-to-word weights) are given the same value. These variable model parameters are selected by a parameter fitting process based on the Alopex learning algorithm (Unnikrishnan & Venugopal, 1994). More details of the parameter fitting will be given in the section below.

A more detailed picture of the model is given in Figure 21.2. The inputs from the fixation, in the form of a 30 spaces wide "perceptual span" vector of 1s and 0s

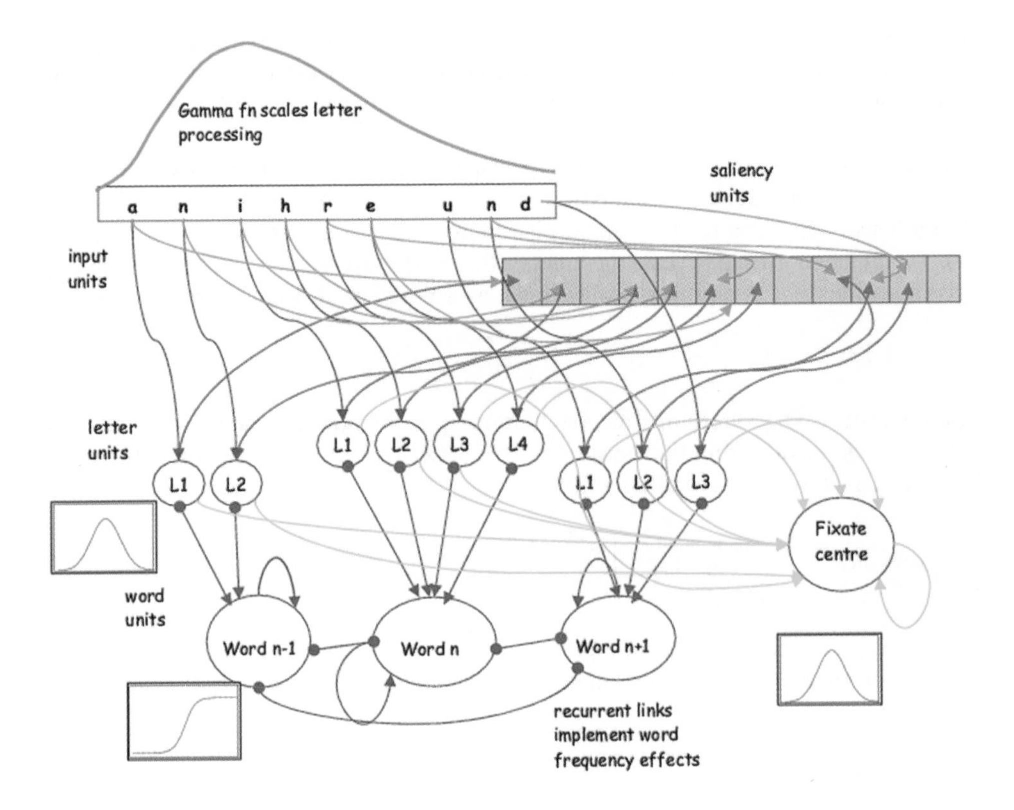

Figure 21.2: Model detail.

This figure is a schematic representation of the Glenmore's internal connectivity. The transfer functions of the relevant units are represented graphically in boxes adjacent to them. Activity in the 30 input units, scaled by the gamma function to represent variability in spatial resolution, is propagated to the saliency map (the "where" pathway) and to the letter units (the "what" pathway). Activity in the letter units feeds forward to the word units, which in turn feed activation back to the letter units. Note that the feedback from words to letters is negatively weighted, so that letters receiving a large amount feedback have their activation maintained for longer. The recurrent connections on the word units are used to implement word frequency effects. The more familiar or frequent the word, the more rapidly its activation will rise and decay. The fixate centre unit takes input from the letter units. When that activity falls below a certain threshold, a saccade is triggered to the word with the largest peak in the saliency map. There is also a stochastic component to this process that will cause the activity of the fixate centre to decay at varying times.

(indicating the presence or absence of a letter) are fed forward to the letter units. The letter units form the nexus of the processing network, connecting to the word and saliency units. The word units then act as a source of top-down support for the letter units, augmenting the letter activations. The saliency units preserve the spatial representation (or map) of the input, and are the representational structure used in the selection of a saccade target. A saccade is triggered when activation in the fixate centre (FC) unit falls below an adjustable threshold. FC activity is a function of the global level of activity of the letter units. The threshold of the fixate centre can be adjusted on the basis of strategic factors such as reading task and difficulty of the material. Variations in this threshold thus permit the early or late triggering of saccades.

Input units The architecture assumes a visual field of 30 character spaces, with the fovea at position 11. An asymmetric perceptual span is implemented by the probability density function of the gamma distribution centred on the fovea (see Equation 4). The function is used to scale the inputs, where the presence of a character is initially given a value of 1.0, and then scaled down as a function of distance from the fovea.

$$i(x) = \Gamma(x,3.5,4.0) \tag{1}$$

Letter units The letter units receive bottom-up activation from the input units, and top-down activation from the word units. The letter unit transfer function is the probability density of a Gaussian distribution.

$$g(x;m,sd) = \frac{1}{\sqrt{2\pi \cdot sd}} \exp - \left[\frac{(x-m)^2}{2 \cdot sd^2} \right] \tag{2}$$

Where x is the accumulated net input to the unit, $m = 50$, and $sd = 0.3m$. Note that a given x_i at time $t+1$ is the accumulated weighted sum of the inputs to the unit calculated as:

$$x_{i,t,+1} = x_{i,t} + \Sigma_j w_{ij}\, o_{j,t+1} \tag{3}$$

where w_{ij} is a weight connecting unit i to unit j, and $o_{j,t+1}$ is the output from unit j at time $t + 1$. At present, the current model indicates the presence or absence of a letter in the visual field through the activation of a letter unit. The letter unit is connected to its appropriate word unit. The determination of what letter unit is connected to what word units is done *a priori* by the model. The current focus of the model is saccade target selection, and the development of a more complex word recognition module is planned as a further extension to the model.

Saliency units The saliency units receive activation from both input and letter units. The input from the letter units represents crosstalk between the "what" and "where" processing pathways, and provides a direct top-down "cognitive" contribution to the evolution of the saliency values for specific regions of the visual field. The saliency unit transfer function is the probability density of the Gaussian distribution with the same parameters as that of the letter units. The units accumulate activity over time and reach a peak of activation after 50 time steps, corresponding roughly to the eye-brain transmission lag (McConkie, 1983; Sereno *et al.*, 1998).

The role played by the saliency map in the model is to support the target selection process, whereby the word "blob" with the highest activity at a certain point in the processing of a fixation acts as the target for the next saccade (Findlay & Walker, 1999). This can potentially be the currently fixated word, the preceding word, and one or other of the succeeding words. Once the blob with the highest level of activity is selected, a saccade generator module is used to execute a saccade in a way that implements the metrical properties of saccade amplitudes in reading described by McConkie *et al.*, (1988; see below).

Word units Word units receive inputs from their respective letter units, and in turn send activation back to these letter units. There are seven word units, since this is the maximum number of words that were found to be contained in the visual field of width 30 in the text used in the experiments. These units use the following sigmoid transfer function:

$$s(x) = \frac{1}{1 + \exp\left[-\frac{(x-m)}{8}\right]} \quad (4)$$

where x is the accumulated net input to the unit and m is, again, 50. This function outputs in the range 0 through 1, with 0.5 for an input of m. The divisor 8 is used to linearise somewhat the S-shaped sigmoid function.

The net input, x, to this equation is slightly more complex than for other units.

$$x_{i,t+1} = x_{i,t} + \frac{\Sigma_j w_{ij} L_{j,t+1}}{n} + \Sigma_k w_{ik} W^r{}_{k,t+1} - \Sigma_m w_{im} W^o{}_{m,t+1} \quad (5)$$

The terms W^r and W^o denote recurrent inputs and inputs from other words, respectively. Note that the letter input is averaged over word length n, so that word length does not affect the rate of activation accumulation, just the average activity of the component letters.[4] The value of the self-recurrent connections is a function of the word's frequency. The higher the frequency the more activation the word receives, and the more rapidly its output peaks. The specific values of the connections are determined by the parameter search mechanism. Note again that the same connection value is used to for each connection type.

The use of word frequency is the only high-level factor that comes into play in this instance of the model. There is scope, however, for using the threshold adjustment of the Fixate Centre unit to implement a strategic modulation of the reading process.

In addition to the self-recurrent connections, the word units are also connected to neighbouring word units with inhibitory connections. This implements a competition for word processing resources. Within this framework, words can be processed in parallel from a given fixation, but one word will tend to dominate at a given point in time. In this way, we have a mechanism for reconciling the different strands of evidence that, on the one hand, appear to suggest a sequential left-to-right processing of words, and on the other, the simultaneous processing of more than one word in a given fixation (see Inhoff & Weger, this volume, for a discussion).

Determining saccade metrics From the data of McConkie *et al.* (1988), we have a good idea of the end-point behaviour of the eye in reading. The intended target of a saccade appears to be the centre of the word, and this is attained with varying accuracy as a function of the eye's launch distance from the target.

McConkie *et al.* demonstrated that the distributions of landing sites on a word tended to be Gaussian in shape. The centre of these distributions and their standard deviations appeared to be determined primarily by oculomotor factors. They found a general tendency for the eye to land around the centre of the word, and a leftward shift of distribution means with the increase in launch distance. They proposed that the pattern of landing site distributions can possibly be accounted for by five principles: (1) The centre of the word is the functional target of a saccade; (2) a systematic range error causes the eye to be increasingly deviated from this target as a linear function of distance from the launch site; (3) this range error is somewhat less, the longer the eye spends at the launch site;[5] (4) there is a random, Gaussian-shaped distribution of landing sites around the target location; and (5) the spread of this distribution increases as a function of launch distance. These five principles can be summarised in three equations. The first is a linear equation (see Equation 6) describing how the mean landing site (m) on a word deviates as a function of launch distance (d). Note that both m and d are defined to be zero at the centre of the targetted word. In the case of a four-letter word, this would be half way between the second and third letter positions.

$$m = 3.3 + 0.49d \tag{6}$$

The second is a cubic equation (see 7) describing the spread of landing positions around m.

$$sd = 1.318 + 0.000518\, d^3 \tag{7}$$

The third is a Gaussian equation (see 8) accounting for the random distribution of landing sites, and for which m and sd are the parameters.

$$f(x;m,sd) = \frac{1}{\sqrt{2\pi \cdot sd}} \exp\left[-\frac{(x-m)^2}{2 \cdot sd^2}\right] \tag{8}$$

In the present version of the model, once the activation of the FC unit falls below threshold, these equations are used to determine the amplitude of a saccade aimed at the centre of the word with the highest saliency.

Fixate centre unit The FC is a single unit with a recurrent connection and connections from the letter units. It implements a Gaussian transfer function identical to that used in the letter and saliency units, so that recurrent inputs and inputs from the letter units gradually cause the unit output to rise to a peak and fall away. There is also a stochastic element to the behaviour of the FC unit. While for all other Gaussian equations we use a value of 50 as their mean, the FC unit Gaussian's mean can vary random between 50 ± 10. This effectively speeds up the rise and fall of activity in the FC Gaussian, which permits us to model relatively brief fixations that can occur on a word.

Once the FC output drops below a certain threshold, a saccade is executed to the word location in the saliency map that has the highest saliency value.

Dynamics

The dynamics of the model are typical of the broad class of "interactive activation" models. At the start of a fixation, the 30 element input vector of units is activated with values that are a function of whether there is a character present in a specific location or not, and the eccentricity of that location. As mentioned above, a gamma function is used to weight the inputs.

Once there is a pattern of activity on the input units, the network connections are dynamically configured to ensure that the appropriate letter units connect to the appropriate word units, and vice-versa. Obviously, this is not meant to be analogous to any biological process, and is used here as a computational convenience to reduce the size of the network needed to run the simulation. The default state of the network is for every letter unit to be connected by bi-directional connections to every word unit. The process of configuration that occurs at the beginning of each fixation eliminates spurious connections. Note that the activation values of word and letter units are carried over from one fixation to the next. By this mechanism, spill-over and preview effects are implemented.

With the network configuration complete, the input activation is fed forward to a set of letter units and a set of saliency units, each of which comprises 30 units. There are one-to-one connections from the input units to the letter and saliency units. There are also feedback connections from the word units to the letter units. Because of the use of the Gaussian PDF as the transfer function for the letter and saliency units, the activity of the letter units reaches a peak after a number of cycles of activation. This has been set at 50 cycles, so that the letter unit representing input from the fovea of the visual field will peak after 50 time steps, and will then start to decline. The further one moves away from the fovea, however, the more slowly the level of activation accumulates. The activation of non-foveal letters will reach the same peak value, but will take an increasingly larger number of time steps the further one moves from the fovea.

The letter units receive top-down input from the word units whose level of activity is a function of the *average* letter input from the letter units, and the frequency of occurrence of that word in the language. The more frequent is the word, the more rapidly it is activated, and the more rapidly it asymptotes to an output value of 1.0. Frequency effects are implemented by a positive self-recurrent connection that is proportional to the frequency of the word. Thus the activation levels of high-frequency words rise more rapidly than lower frequency words, but this is also a function of the activity of the letter units, which in turn is a function of the eccentricity of the letters in the visual field. Consequently, visually eccentric high frequency words will be more rapidly identified (i.e., their activity will peak earlier) than their low frequency counterparts.

During the processing of words in a fixation, there is competition between words, mediated by inhibitory connections between word units. Once a word has peaked, it ceases to compete, leaving the way open for other words to complete their processing. In this way, several words can be processed simultaneously, but usually one word takes the lion's share of processing resources.

Activation from the letter units is also sent to the saliency units where it combines with the activation from the input units. Again the transfer function is a Gaussian PDF, which models in one function the accumulation of activation and its decay over time. Areas of high-activation peak and decay more rapidly. Given the varying resolution of the input units (modelled by the gamma function), saliency units receiving foveal inputs will peak and decay more rapidly than other units. So after a certain number of iterations, the saliency values will drop in the foveal regions of the saliency units, implementing a form of "inhibition of return" (Gibson & Egeth, 1994).

Activity from the letter units is passed to the fixation centre unit. This unit acts as a spatially undiscriminating summary of saliency unit activity. So it will, in principle, trigger a refixation, regressive or progressive saccade, depending on where the saliency maximum is located.

Parameter Fitting

In order to determine an appropriate set of weights for the various unit connections in the simulation, a parameter search algorithm was devised. The algorithm used was based on the Alopex neural network-learning algorithm (Unnikrishnan & Venugopal, 1994). Instead of using an *error gradient* to guide the changes in weight (or parameter) values, Alopex uses *local correlations* between changes in individual weights and changes in a global error measure. The algorithm does not make any assumptions about the transfer functions of individual units, and does not explicitly depend on the functional form of the error measure. This makes it ideal for parameter fitting, where we might want to combine a number of factors into a complex cost function that we wish to minimise. Thus, we can ensure the algorithm selects a set of parameters that satisfies the global temporal and spatial characteristics of reading, specifically fixation duration and saccade length.

The algorithm is initially *stochastic* in its search, and uses a "temperature" parameter in a manner similar to that in simulated annealing (Kirkpatrick, Gelatt & Vecchi, 1983) to gradually make the search more deterministic as the algorithm converges on a desirable set of parameters.[6]

The parameter search involves making small perturbations (e.g., ±0.01) to the parameters based on whether the previous change resulted in a reduction of the cost function. The parameter $w_{ij}(n)$ in Equation 9 below, refers to the connection between units i and j at time n. This parameter is perturbed by $\pm\delta$, where δ is a constant value.

$$w_{ij}(n) = w_{ij}(n-1) + \delta_{\iota\vartheta}(n) \tag{9}$$

$$\delta_{ij}(n) = \begin{cases} -\delta \text{ with probability } p_{ij}(n) \\ +\delta \text{ with probability } 1-p_{ij} \end{cases} \tag{10}$$

However the changes depend probabilistically on the cost function value, as can be seen from Equation 11.

$$p_{ij}(n) = \frac{1}{1 + \exp\left(\delta \dfrac{\Delta E(n)}{T(n)}\right)} \tag{11}$$

This probability is modulated by a temperature variable (Equation 12), which is derived from the overall cost function value. As this value decreases, the selection of parameter change moves from being stochastic to deterministic.

$$T(n) = \frac{\delta}{N} \sum_{n'=n-N}^{n-1} |\Delta E(n')| \quad \text{if } n \text{ is a multiple of } N,$$
$$T(n) = T(n-1) \quad \text{otherwise} \tag{12}$$

where δ is the constant parameter change (usually around 0.01), and $\Delta E = E(n-1) - E(n-2)$ is the change in error between the two previous iterations.

The Alopex algorithm shows reasonable convergence, but is not as efficient as a gradient descent based algorithm. Nonetheless, the flexibility it provides in the specification of arbitrarily complex cost functions is worth the slower convergence. This is especially true when the number of parameters to be estimated is relatively small.

In the case of the model described in this chapter, there were nine free parameters to be estimated, furthermore, the sign of seven of these was constrained to be positive, and the maximum absolute value permitted was 10.0. The parameters were the following connections: input-to-letter, input-to-saliency, letter-to-word, letter-to-saliency, word-to-letter, word-to-word, word-to-fixate centre, fixate centre-to-fixate centre. All except word-to-fixate centre were constrained to be positive in sign.

The text used for the parameter fitting was a 2500 word German text on the topic of the Inuit. The cost function used was:

$$Cost = (n-1)^2 + \left(\frac{1}{m} - 1\right)^2 + ((p-11)-8)^2 + (t-0.25)^2 \tag{13}$$

where n is the number of peaks (at a minimum for just one peak), m is the maximum value of the peak (at a minimum when 1.0), p is the location of the maximum peak (at a minimum for 8 characters to the right of the fovea), and t is the threshold for the FC unit (the desired threshold is 0.25).

Evaluation

The model can deal with a variety of low-level reading phenomena in an integrated and parsimonious manner. In this section, we will demonstrate the operating principles of the model and how they can account for such phenomena as preview effects, spillover effects, refixations, regressions, and word skipping. The advantage of this model over others is, we believe, the variety of phenomena that can be accounted for with letter level accuracy within a rather simple computational framework.

Figure 21.3 illustrates how the word units, saliency units, and fixate centre interact on a given fixation. Once the activity in the fixate centre falls below a certain threshold, a saccade is triggered to the word with the highest saliency value, irrespective of where it is. In this case, the word with the highest saliency is $n + 1$, resulting in a progressive interword saccade. Also illustrated in Figure 21.3 is the mechanism mediating preview effects. These effects arise through the activation value of the new word with

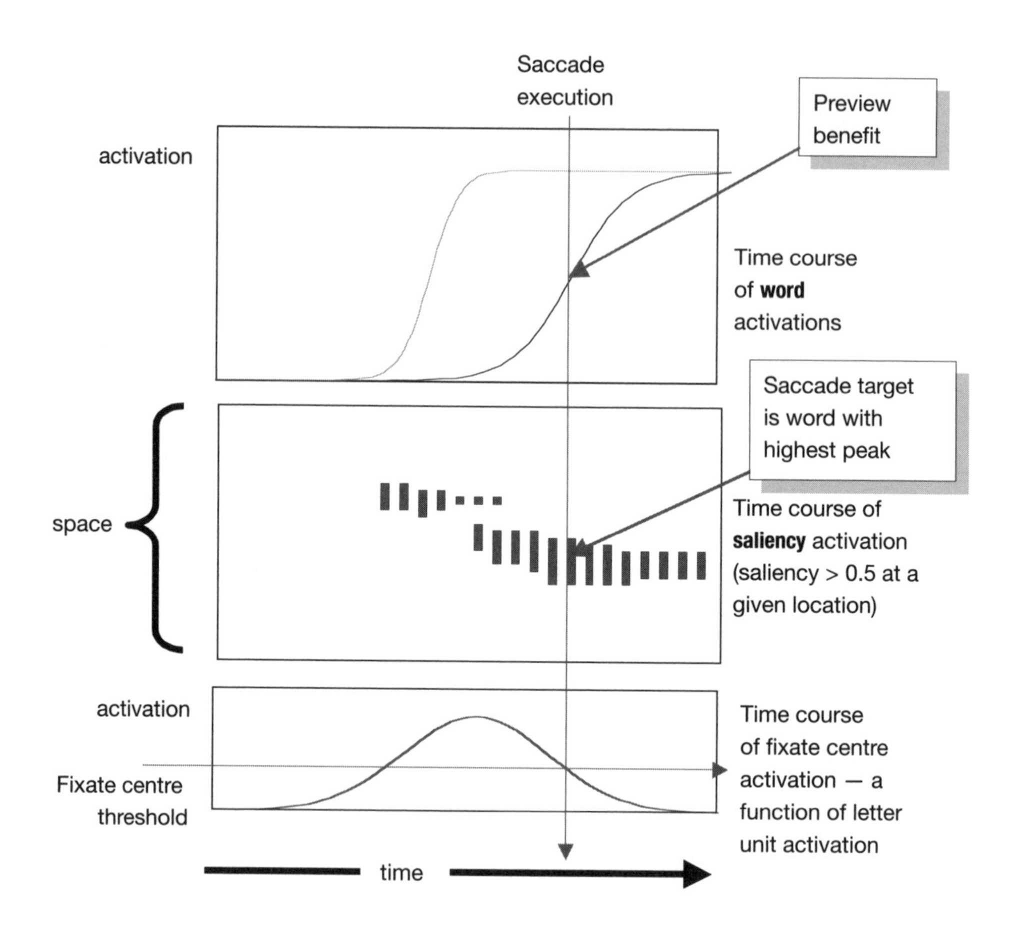

Figure 21.3: Time course of processing.

This figure represents the time course of processing in a number of the components of the network (the letter units are not shown to reduce the complexity of the figure). The top panel represents the time course of activation of two word units. The activation of word units is carried over from fixation to fixation. The vertical line down through the panels indicates the time at which a saccade was triggered. At that time, the word represented by the lighter line has asymptoted and the activation value of the second word has started to rise. This level of activation represents the preview benefit for that word when it comes to the start of the next fixation. The second panel is a spatio-temporal representation of the activity levels across the saliency map. The bars indicate regions in saliency map that have activation values greater than 0.5 at a given time step. When a saccade is triggered, the word with the highest saliency peak is the saccade target. The bottom panel is a representation of the time course of activity of the fixate centre. When this falls below an adjustable threshold, a saccade is triggered. The activity of the fixate centre is a function of the activity of the letter units.

the rising level of activity is being carried over to the next fixation. Note that preview does not result from the disengagement and re-engagement of an attentional mechanism. Rather we propose a continuous mechanism that dynamically modulates the processing load across the words in the fovea and immediate neighbourhood as a function of their relative difficulty. Thus, more than one word can be processed at a given time, though there is competition for lexical processing resources between words. Note that once a word reaches its asymptotic value, it no longer competes with other words.

The model is capable of accounting for the modulation of preview benefit by the difficulty (in this case, low frequency) of the currently fixated word (see Figure 21.4). By modelling the processing of the words as a continuous asymptotic process, we can account for the dynamic interplay between the processing of word n and word $n + 1$. In Figure 21.4, the frequency of word n is varied. Where word n is a high frequency word, the level of word activity rapidly rises to a peak, thus removing it from competition with word $n + 1$, and allowing it, in turn, to be processed. When word n is a low frequency word, less progress is made in processing word $n + 1$, thus reducing any preview benefit for it.

Spillover effects from the processing of the previous word on the currently fixated word can be accounted for in precisely the same way. In Figure 21.5, we see that the processing of the low-frequency word has not asymptoted prior to the fixation. Recall that the trigger for executing a saccade is a drop in the level of activity of the fixate centre below a certain threshold. It bears no *direct* relationship to the successful, or otherwise, processing of the currently fixated word. If a high frequency word precedes the current fixation, there is less likely to be processing of that word continuing into the next fixation.

Refixations

In Figure 21.6, we show the performance of the time course of activation over the letter field when an eight-letter word is fixated for the first time at its last letter. Note that there has been no preceding fixation of this word or the one preceding it. The situation is equivalent to there having been a long saccade from the right to this point in the text. There is a build up of activation in two competing word targets on the saliency map, with the currently fixated word (*beweisen*) marginally winning the competition. This word is then selected as the target for the next saccade.

Regressions

In Figure 21.7, we have a similar graph showing the build up of activation taking place over the location of the preceding word. Again, there was no fixation on the currently fixated or preceding word prior to this one. The fixation position is at the beginning of an eight letter high-frequency word. Consequently it drops in saliency fairly rapidly, leaving the preceding word the most salient target. If a saccade is triggered at this point, the result would be a regression to the preceding word.

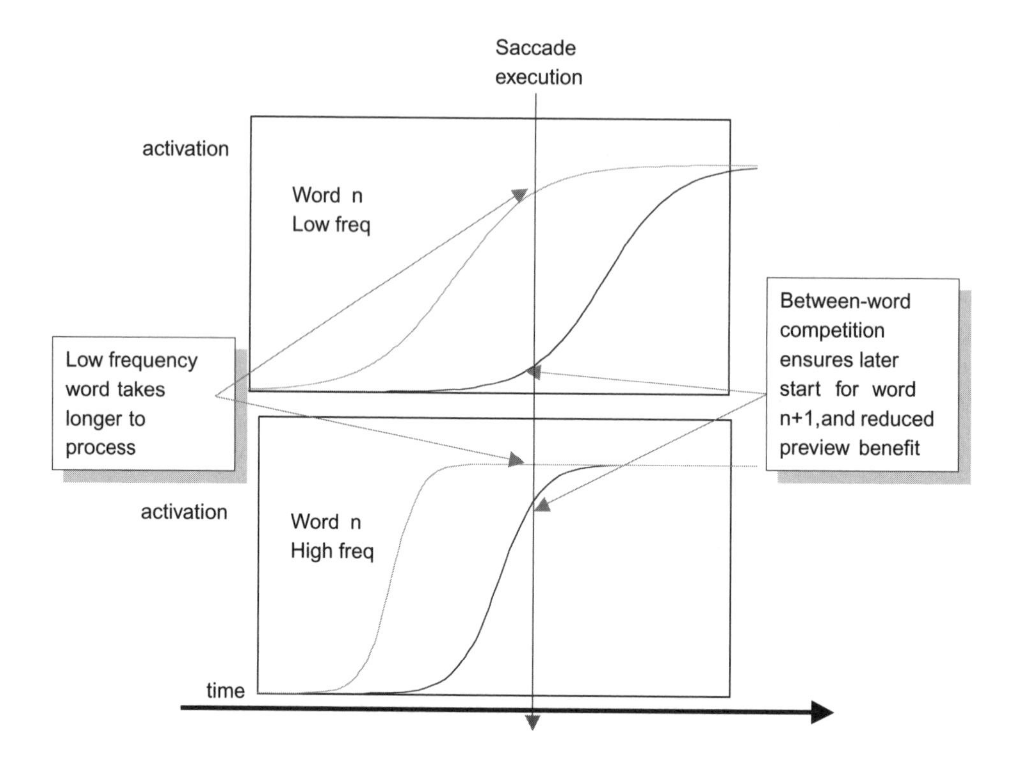

Figure 21.4: Preview modulation by frequency of fixated word.

This figure represents schematically the interaction between word frequency and peripheral preview. The time course of word activation for two pairs of words is represented, where word n in each pair (lighter curve) is either a low frequency word (top panel) or a high-frequency word (bottom panel). Because there is competition between active word units (until they reach asymptote), if word n remains active longer it will limit the processing of word $n + 1$ (darker curve). This is represented by differences in the rise time of the word $n + 1$, as a function of the frequency of word n.

Likelihood of Fixating a Word ("Word Skipping")

Figure 21.8a and b illustrates how the model can account for the likelihood of fixating a word as a function of its frequency. As mentioned above, we do not believe that there is any default tendency to aim a saccade at each word on a line of text; therefore we put the traditional term "word skipping" in quotation marks. Note that in this simulation example, each of the graphs shows the second of two fixations on the word or within the vector, the preceding fixation having been on the word to left of the current fixation (fixation locations are indicated by arrows). In Figure 21.8a, we can see that the high frequency word in the right parafovea does not receive a fixation, whereas in 21.8b the low frequency word gets fixated. This effect is achieved by the slower accumulation of activation for the low frequency word, and the subsequent maintenance of activity in the saliency map in the region of the low frequency word.

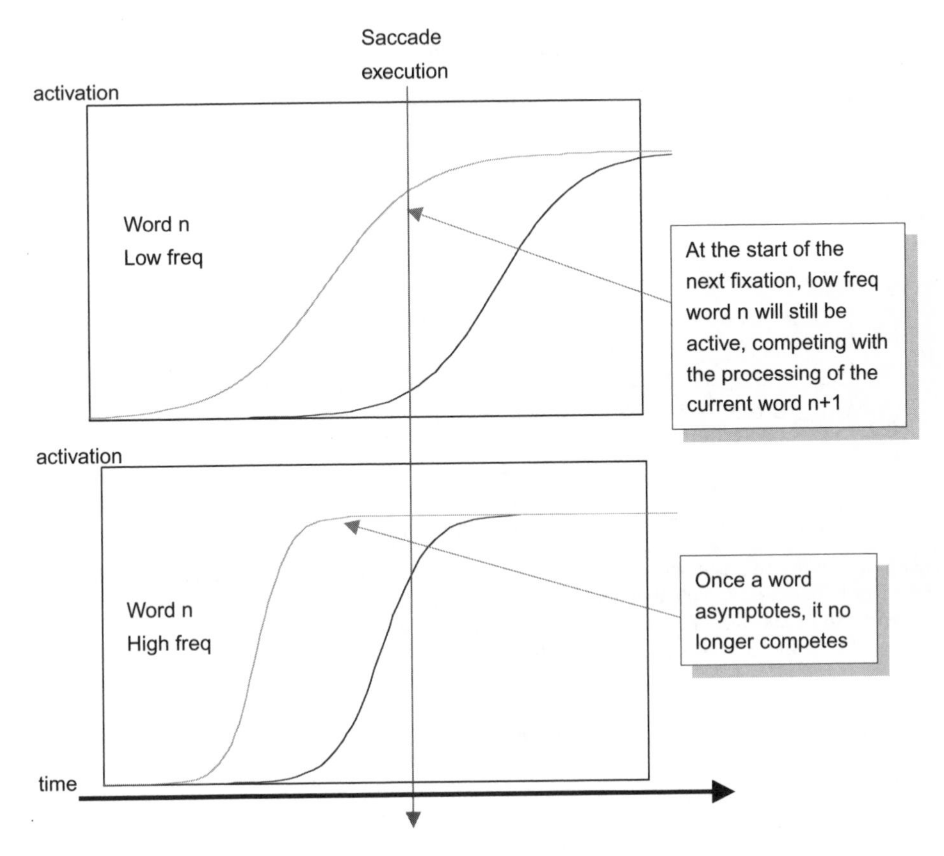

Figure 21.5: Spillover effects from preceding word.

This figure schematically represents the production of spillover effects. The time course of word activation for two pairs of words is represented, where word *n* in each pair is either a low frequency word (top panel) or a high-frequency word (bottom panel). As illustrated in the top panel, if a saccade occurs before word *n* reaches asymptote, its activation will carry over to the next fixation, competing for processing in the succeeding fixation. This will not happen where word *n* is a high frequency word (bottom panel).

Conclusion

The Glenmore model has been described in outline, and some qualitative and quantitative simulation results have been presented. We have proposed a radical departure from the more traditional attention shift models of eye movement control reading. Our model does not use a notion of attention shift, but rather permits the processing of words in parallel, with a limited amount of competition between words in a given fixation. The model decouples the decision about when to move the eyes from the word recognition process, but allows for a substantial influence of linguistic processing on the movement decision. More specifically, the time course of activity in a "fixate centre"

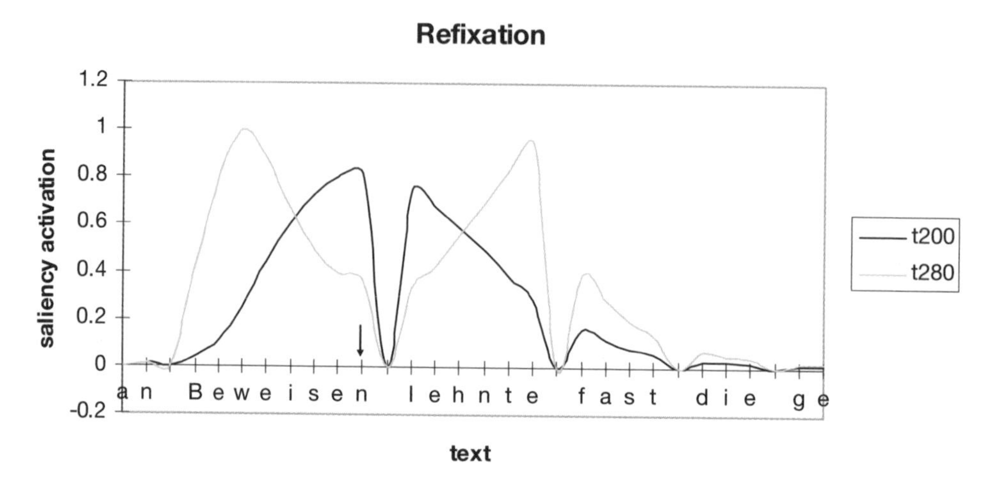

Figure 21.6: Modelling a refixation.

This figure shows two time-slices of activation in the saliency map for a fixation on the last letter of the word "Beweisen". Note that neither the current nor previous word had previously been fixated. The activation peak over the current word is marginally ahead of word $n + 1$. A saccade triggered at either 200 or 280 simulated msecs would, therefore, result in a refixation of the current word.

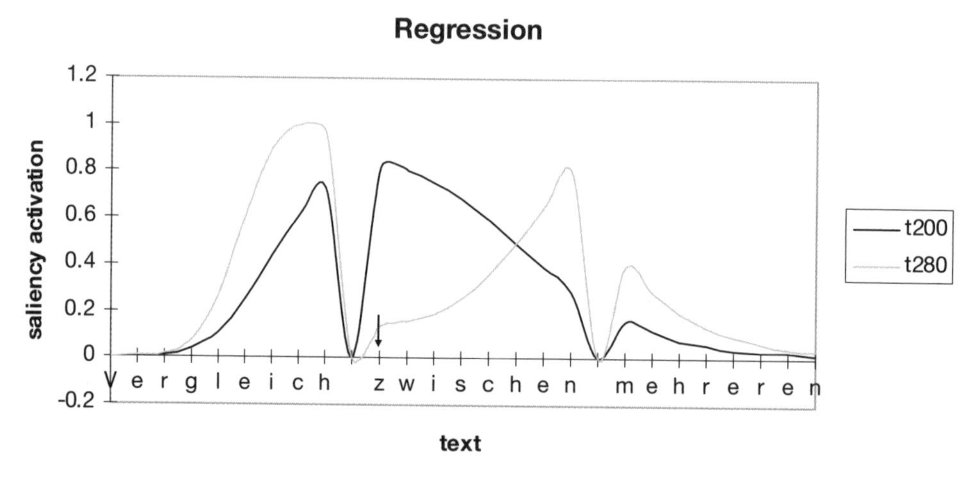

Figure 21.7: Modelling a regressive fixation.

This figure shows two time-slices of activation in the saliency map for a fixation on the first letter of the word "zwischen." As with Figure 21.6, neither the current nor previous word had previously been fixated. The activation peak over the preceding word results in a regressive fixation to word $n - 1$.

a)

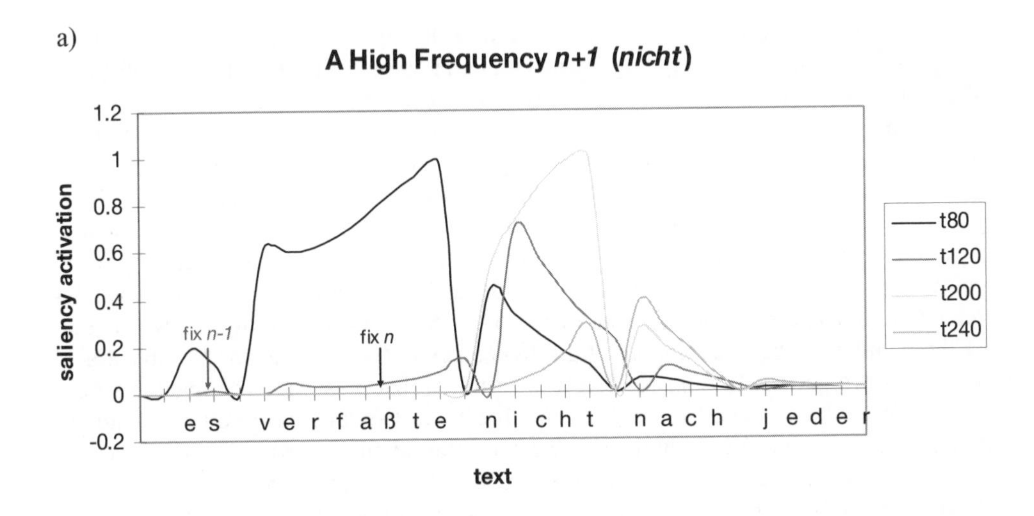

b)

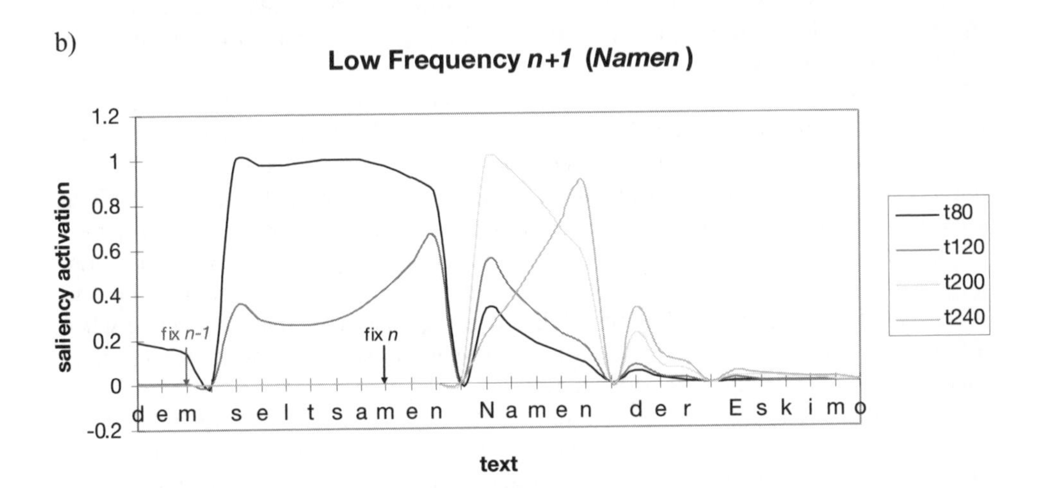

Figure 21.8: Frequency effects on word fixation likelihood.

This figure shows the effect of the frequency of word $n + 1$ on the likelihood of it being fixated. The black arrow indicates the current fixation. The arrow labelled "fix n" indicates the current fixation. Note that for both figures the preceding fixation was on the last letter of the preceding word, indicated by the "fix n −1" arrow. Because "nicht" is a higher frequency word than "Namen", and because saliency is influenced by word frequency, the saliency location of the more frequent word declines more rapidly than the less frequent word. A saccade triggered at 240 simulated msecs to the word location with the highest peak will result in a higher frequency $n + 1$ not being targeted for a fixation (a), but with the opposite the case for a lower frequency $n + 1$ (b).

determines the triggering of a saccade in a way that is co-determined by a random component and ongoing processing on the letter and word level. The other main feature of the model is the use of a saliency map that acts as an arena for the interplay of bottom-up visual features of the text, and top-down lexical features. These factors combine to create a pattern of activation that selects one word as the saccade target.

As we have emphasized in the introductory section, our theoretical approach is closely related to the more general theoretical framework developed by Findlay and Walker (1999). While our model is in principal quite similar to their conception, it is much more precise with respect to a number of reading-specific mechanisms. It is less detailed in other respects, and in some ways it departs from specific suggestions made by these authors. The most important difference lies in the mechanisms and dynamics of saccade triggering. In both models it is assumed that a saccade is triggered when activity in the fixate centre falls below the threshold. However, according to Findlay & Walker, increased activity in the move centre (the saliency map) promotes a decline of activity in the fixate centre via reciprocal inhibitory connections. In addition, a decrease in activity in the fixate centre can also be triggered "via descending influences" from higher processing centres (p. 663). However, their ideas on the nature of these higher order influences remain vague, and need to be specified in a model on a task as specific as reading.

Why is there no reciprocal connection between the saliency map and the fixate centre in the Glenmore model? Findlay and Walker's theory is concerned with explaining the latency of a saccade that is generated in a way that can account for a number of experimental effects like the gap effect or the remote distractor effect that are all variations of a dramatically changing visual input. In the context of explaining the latency of a single saccade under such conditions, competition between potential targets on the visual level plays a key role. There is a lively discussion on how to model these influences up to the possibility that such saccades may be elicited when a threshold level of activation is surpassed in a spatially selective move system rather than a fixate centre (Dorris & Munoz, 1999).

In reading, the situation is quite different in that the visual environment is stable and consists of well defined elements in a highly structured spatial arrangement. Hence, there is no active competition of targets on the level of visual processing other than what is captured with assigning initial saliency values to the elements in the perceptual span vector. The rest of the saliency dynamics comes entirely from processing on the letter and word level. It would therefore not add to the explanatory power of the model to include an extra connection from the saliency vector to the fixate centre. The optimal way to model the substantial influence of linguistic processing on fixation duration in reading appears to be via the implemented transfer of integrated activity from the letter processing vector to the fixate centre. This mechanism has the advantage of accounting in a parsimonious way for effects on *both* the orthographic and lexical processing level on the duration of fixations.

Within the modelling framework described in this chapter, it is possible to account within one mechanism for preview and spillover effects, and for regressions, progressions, and refixations. Even with this simple model, a range of phenomena that have hitherto proved problematic to account for, have been accommodated in a way that appears to be in harmony with neorophysiological constraints to the degree that such constraints can be made explicit for the task of reading. How does the Glenmore model compare to the similar approaches that have recently been put forward by Yang and McConkie (2001) and by Engbert *et al.* (2002; see also the chapters by McConkie & Yang, and by Kliegl & Engbert in this volume).

The interaction/competition theory by Yang and McConkie (2001) is also based the theoretical framework suggested by Findlay and Walker (1999). The theory is very explicit about the specific mechanisms of saccade triggering and the resulting distributions of fixation durations which they primarily attribute to non-cognitive factors. In comparison the Glenmore model, the I/C theory is less specific about the spatiotemporal dynamics of saliency within the perceptual span. It also does not attempt to model letter level and word level processing in any explicit way. On the other hand, it makes relatively explicit claims about ways in which cognition can influence saccade initiation times. There are three possibilities for such influences, referred to as processing-based inhibition, parametric adjustment and late-acting cognitive control. The first two of these mechanisms could be accounted for within the Glenmore model. Yang and McConkie (2001) presented text with occasional non-text letter strings or pseudowords and observed that saccades normally triggered after fixation durations of more than about 200 ms were substantially delayed. They propose "that these non-text letter strings or pseudowords create processing problems at some level of text processing, and that the brain centers being disturbed send a distress signal that results in inhibition in the saccadic system" (McConkie & Yang, this volume). In the Glenmore model, entering a pseudoword into the letter processing vector would make this string an attractive target for fixation (via transfer of processing activation to the saliency map), with associated long-lasting activation of the fixate centre, resulting in delayed saccade onsets. The second mechanism for cognitive influences proposed in the I/C theory, parametric adjustment, is similar to accounting for strategic influences on reading behaviour that we have briefly discussed earlier in this chapter. It can be accounted for in the Glenmore model by adjusting the critical threshold in the fixate centre. In sum, we see the Interaction/Competition theory in a complementary relation to our own approach, the major difference being our less extreme viewpoint concerning the relative importance of visual vs. cognitive factors.

As mentioned in the introductory section, the Glenmore model has striking similarities with the SWIFT-model proposed by Engbert *et al.* (2002) (see Kliegl & Engbert, this volume for the latest version). Kliegl & Engbert emphasise that their model is based on three principles: a partial separation of saccade timing from saccade target selection, spatially distributed lexical processing, and autonomous saccade generation with inhibition by foveal word processing. One major difference is that in Glenmore linguistic processing is implemented at the letter and word level and that the influence of processing on the timing of saccade triggering is in terms of the integrated processing activity within the perceptual span rather than exclusive processing of the foveal word. Moreover, in SWIFT the decision which word should be the target of the next fixation is stochastically determined as a function of lexical activity, with the word having the largest current lexical activity being the most likely target. In contrast, the way of modelling a hierarchy of potential saccade targets in Glenmore is via combined visual information and processing dynamics in a spatial saliency map. The SWIFT model is much more specific with respect to the time line of saccade programming, considering in detail various possibilities of temporal overlap in the processing of successive saccades. This is an issue that needs to be reconsidered for Glenmore, including the possibility that our assumptions regarding a simple relation between saccade triggering and execution were too naïve.

Seen in conjunction, theories and models like I/C, SWIFT and Glenmore represent a class of approaches that may differ greatly in detail but appear to share a common philosophy. However, as documented in the chapter by Pollatsek, Reichle & Rayner in this volume, there are also good arguments in favour of their view that reading saccades are triggered on the basis of sequential lexical processing. This range of opinion is likely to generate lively theoretical discussions and to provoke empirical studies to test predictions associated with various models. This may contribute to replacing the low level vs. high level controversies that have dominated the last decade with a more productive debate on *how* visuomotor and cognitive elements combine to determine the movements of our eyes while we read.

Notes

1 It could be argued that word length carries linguistic information, for example, in terms of reducing the number of word candidates activated in parafoveal preprocessing. However, as Inhoff, Radach, Eiter and Juhasz (in press) have recently shown, this appears not to be the case.

2 The name Glenmore originates from the location of a remote cottage in the south west coast of Ireland, where the foundations of our modelling work were laid during a one-week retreat there.

3 This idea is related to the notion of a visual scanning routine as proposed by Levy-Schoen (1981). She suggested that "The routine does not determine absolute saccade length but rather criteria according to which saccade length will be programmed so that the eyes move in a way relevant to the task" (p. 301). It is assumed that readers learn scanning patterns individually as a way to navigate through configurations of low spatial frequency word objects. During this process, a reader will develop routines that, on average, provide optimal information acquisition for letter and word processing. Our implementation represents a first approximation to explicitly modelling this mechanism.

4 This is empirically justified by the aforementioned finding by Inhoff *et al.* (in press) that parafoveal word length information is not used for lexical processing.

5 This suggestion rested on a trend in the data of McConkie *et al.* (1988) that was not statistically tested. It was not replicated in the data of McConkie, Grimes, Kerr and Zola (1990) and a detailed analysis of a large corpus of individual reading data also did not find evidence in favour of this idea (Radach & Heller, 2000). We therefore decided not to include it in our specification of saccade metrics.

6 The temperature term in simulated annealing has the effect of increasing the stochasticity of the parameter search when the error is high. It's analogous to the process used in steel manufacture of heating and slow cooling of the metal to encourage the formation of more stable crystalline structure and thus increase the metal's strength.

References

Blanchard, H. E. (1985). A comparison of some processing time measures based on eye movements. *Acta Psychologica, 58,* 1–15.

Brysbaert, M., & Vitu, F. (1998). Word skipping: Implications for theories of eye movement control in reading. In: G. Underwood (ed.), *Eye Guidance in Reading and Scene Perception* (pp. 125–148). Amsterdam: Elsevier.

CELEX German Database. Release D25. Computer software. Nijmegen: Centre for Lexical Information, 1995.

Deubel, H., O'Regan, K., & Radach, R. (2000). Attention, information processing and eye movement control. In: A. Kennedy, R. Radach, D. Heller and J. Pynte (eds), *Reading as a Perceptual Process* (pp. 355–374). Oxford: Elsevier.

Dorris, M. C., & Munoz, D. P. (1999). The underrated role of the "move system" in determining saccade latency. *Behavioral and Brain Sciences 22*(4): 681–682.

Engbert, R., & Kliegl, R. (2001). Mathematical models of eye movements in reading: A possible role for autonomous saccades. *Biological Cybernetics, 85,* 77–87.

Engbert, R., Longtin, A., & Kliegl, R. (2002). A dynamical model of saccade generation in reading based on spatially distributed lexical processing. *Vision Research, 42,* 621–636.

Findlay, J. M., & Walker, R. (1999). A model of saccade generation based on parallel processing and competitive inhibition. *Behavioral & Brain Sciences 22*(4): 661–721.

Gibson, B. S., & Egeth, H. E. (1994). Inhibition of return to object-based and environment-based locations. *Perception & Psychophysics 55*(3): 323–339.

Grainger, J. (2000). From print to meaning via words? In: A. Kennedy, R. Radach, D. Heller and J. Pynte (eds), *Reading as a Perceptual Process* (pp. 147–161). Oxford: Elsevier.

Grainger, J., & Jacobs, A. M. (1998). On localist connectionism and psychological science. In: J. Grainger and A. M. Jacobs (eds). *Localist Connectionistic Approaches to Human Cognition* (pp. 1–38). Mahwah, NJ: Erlbaum.

Inhoff, A.W., Lemmer, S., Eiter, B., & Radach, R. (2001). Saccade programming and parafoveal information use in reading: More evidence against attention-shift models. Poster, *11th European Conference on Eye Movements*, Turku, Finland.

Inhoff, A. W., Radach, R., Starr, M., & Greenberg, S. (2000). Allocation of visuo-spatial attention and saccade programming in reading. In: A. Kennedy, R. Radach, D. Heller and J. Pynte (eds), *Reading as a Visual Process* (pp. 221–246). Oxford: Elsevier.

Inhoff, A., Radach, R., Eiter, B., & Juhasz, B. (in press). Parafoveal processing: Distinct subsystems for spatial and linguistic information. *Quarterly Journal of Experimental Psychology*.

Jacobs, A. M. (2000). Five questions about cognitive models and some answers from three models of reading. In: In A. Kennedy, R. Radach, D. Heller and J. Pynte (eds), *Reading as a Perceptual Process* (pp. 721–732). . Oxford: Elsevier.

Just, M. A., & Carpenter, P. A. (1987). *The Psychology of Reading and Language Comprehension*. Boston: Allyn and Bacon.

Kelso, J. A. S. (1995). *Dynamic Patterns: The Self-Organization of Brain and Behavior*. Cambridge, MA: MIT Press.

Kennedy, A. (1998). The influence of parafoveal words on foveal inspection time: Evidence for a processing tradeoff. In: G. Underwood (ed.), *Eye Guidance in Reading and Scene Perception* (pp. 149–180). Oxford: Elsevier.

Kennedy, A., Radach, R., Heller, D., & Pynte, J. (eds) (2000). *Reading as a Perceptual Process*. Oxford, Elsevier.

Kennedy, A., Pynte, J., & Ducrot, S. (2002). Parafoveal-on-foveal interactions in word recognition. *Quarterly Journal of Experimental Psychology, 55A*, 1307–1337.

Kerr, P. W. (1992). Eye movement control during reading: The selection of where to send the eyes. Unpublished Doctoral Dissertation, University of Illinois at Urbana-Champaign.

Kirkpatrick, S., Gelatt Jr, C. D., Vecchi, M. P. (1983). Optimization by simulated annealing. *Science, 220,* 671–680.

LaBerge, D., & Brown, V. (1989). Theory of attentional operations in shape identification. *Psychological Review, 96,* 101–124.

McClelland J. L., & Rumelhart D. E. (1981). An interactive activation model of context effects in letter perception: Part 1. An account of basic findings. *Psychological Review, 88,* 375–407.

McConkie, G. W. (1983). Eye movements and perception during reading. In: K. Rayner (ed.), *Eye Movements in Reading. Perceptual and Language Processes* (pp. 65–96). New York: Academic Press.

McConkie, G. W., Grimes, J. M., Kerr, P. W., & Zola, D. (1990). Children's eye movements during reading. In: J. F. Stein (ed.), *Vision and Visual Dyslexia* (pp. 000–000). MacMillan.

McConkie, G. W., Kerr, P. W., & Dyre, B. P. (1994). What are "normal" eye movements during reading: Toward a mathematical description. In: J. Ygge and G. Lennerstrand (eds), *Eye Movements in Reading* (pp. 315–328). Oxford: Pergamon.

McConkie, G. W., Kerr, P. W., Reddix, M. D., & Zola, D. (1988). Eye movement control during reading: I. The location of initial eye fixations on words. *Vision Research, 28,* 1107–1118.

McConkie, G. W., Kerr, P. W., Reddix, M. D., Zola, D., & Jacobs, A. M. (1989). Eye movement control during reading: II. Frequency of refixating a word. *Perception and Psychophysics, 46,* 245–253.

McConkie, G. W., & Rayner, K. (1975). The span of the effective stimulus during a fixation in reading. *Perception and Psychophysics, 17,* 578–586.

McConkie, G. W., Underwood, N. R., Zola, D., & Wolverton, G. S. (1985). Some temporal characteristics of processing during reading. *Journal of Experimental Psychology: Human Perception and Performance, 11*, 168–186.

McConkie, G. W., & Zola, D. (1987). Visual attention during eye fixations while reading. In: M. Coltheart (ed.), *Attention and Performance* (Vol. 12, pp. 385–401). London, NJ: Erlbaum.

Morrison, R. E. (1984). Manipulation of stimulus onset delay in reading: Evidence for parallel programming of saccades. *Journal of Experimental Psychology: Human Perception and Performance, 10,* 667–682.

O'Regan, J. K., Vitu, F., Radach, R., & Kerr, P. (1994). Effects of local processing and oculomotor factors in eye movement guidance in reading. In: J. Ygge and G. Lennerstrand (eds), *Eye Movements in Reading* (pp. 329–348). New York: Pergamon.

Port, R. F., & Van Gelder, T. (1995). *Mind as Motion: Explorations in the Dynamics of Cognition.* Cambridge, MA: MIT Press.

Radach, R. (1999). Top-down influences on saccade generation in cognitive tasks. *Behavioral and Brain Sciences 22*(4): 697–698.

Radach, R., & Heller, D. (2000). Relations between spatial and temporal aspects of eye movement control. In: A. Kennedy, R. Radach, D. Heller and J. Pynte (eds), *Reading as a Perceptual Process* (pp. 165–191). Oxford: Elsevier.

Radach, R., Inhoff, A. W., & Heller, D. (2002). The role of attention and spatial selection in fluent reading. In: E. Witruk, A. D. Friederici and T. Lachmann (eds), *Basic Function of Language, Reading, and Reading Disability* (pp. 137–154). Boston: Kluwer.

Reichle, E. D., Pollatsek, A., Fisher, D. L., & Rayner, K. (1998). Toward a model of eye movement control in reading. *Psychological Review 105*(1): 125–157.

Reichle, E. D., Rayner, K., & Pollatsek, A. (1999). Eye movement control in reading: Accounting for initial fixation locations and refixations within the E-Z Reader model. *Vision Research, 39,* 4403–4411.

Reichle, E., Rayner, K., & Pollatsek, A. (in press). Comparing the E-Z reader model to other models of eye movement control in reading. *Behavioral and Brain Sciences.*

Reilly, R., (1993a). A connectionist attentional shift model of eye movement control in reading. In: Proceedings of the fifteenth annual conference of the Cognitive Science Society, University of Colorado at Boulder, pp. 860–865.

Reilly, R. (1993b). A connectionist framework for modeling eye-movement control in reading. In: G. d'Ydewalle and J. Van Rensbergen (eds), *Perception and Cognition: Advances in Eye-Movement Research* (pp. 193–212). Amsterdam: Elsevier.

Reilly, R. G., & O'Regan, J. K. (1998). Eye movement control during reading: A simulation of some word-targeting strategies. *Vision Research, 38,* 303–317.

Rumelhart, D. E., & McClelland, J. L. (1982). An interactive activation model of context effects in letter perception: Part 2. The contextual enhancement effect and some tests and extensions of the model. *Psychological Review, 89,* 60–94.

Sereno, S. C., Rayner, K., & Posner, M. I. (1998). Establishing a timeline of processing during reading: Evidence from eye movements and event-related potentials. *NeuroReport, 9,* 2195–2200.

Unnikrishnan, K. P., & Venugopal, K. P. (1994). Alopex: A correlation-based learning algorithm for feedforward and recurrent neural networks. *Neural Computation, 6,* 469–490.

Wurtz, R. H. (1996). Vision of the control of movement. *Investigative Ophthalmology & Visual Science 37*(11): 2131–2145.

Yang, S.-N., & McConkie, G. W. (2001). Eye movements during reading: A theory of saccade initiation times. *Vision Research, 41,* 3567–3585.

Commentary on Section 3

Moving Eyes and Reading Words: How Can a Computational Model Combine the Two?

Jonathan Grainger

Historically, since the boom of reading research in the 1970s, work using eye movement techniques, and work on isolated word reading have formed two quite independent and mutually impermeable sectors of research. This is quite a natural evolution given that it was not at all obvious, at that time at least, that eye movement recordings could provide any information relevant to the study of isolated word recognition. During this period, referred to as the third era of eye movement research by Rayner (1998), research into eye movements in reading best interacted with psycholinguistic research at the level of sentence processing. For example, eye movement recordings were used to examine how the syntactic processor handles ambiguous situations such as those created by reduced relative clauses in English (e.g., "the (famous) horse walked past the barn fell").[1] There are obviously some examples of fruitful interactions across research on visual word recognition and research on eye movements in reading, but these are more the exception than the rule. The point here is that the field has, I believe, suffered from a lack of communication across these sub-disciplines. I will now discuss some of the possible reasons for this lack of communication before going on to discuss the more recent, and very encouraging, interaction that appears to be developing. Several of the chapters in this section nicely illustrate this trend.

The interaction between eye-movement research in reading and research on visual word recognition has probably been hampered by several factors, some of which will be discussed here. First, the vast majority of research on visual word recognition has focused on the processing performed "in a single glance" in relatively short, often monosyllabic words (research on morphological processing is an exception here). Second, while many researchers interested in visual word recognition tended to see eye movements as a cumbersome tool that had yet to prove its utility to the field, some eye movement researchers interested in single word recognition tended to see the use of eye movement technology as a superior, more ecological tool that would

The Mind's Eye: Cognitive and Applied Aspects of Eye Movement Research

ISBN: 0–444–51020–6

eventually replace "artificial" laboratory tasks such as lexical decision. Third, some people apparently believed that there was nothing more to learn about visual word recognition and that the interesting questions for reading research were to do with comprehension at the sentence level and beyond.[2] Last but not least, before the early 1980s it was not at all clear that lexical processing had any influence on how the eyes moved through printed text. It is this last point that has generated some debate in the area of eye movement research over the last 20 years, and many of the chapters in this section pick up on this on-going controversy.

In this commentary, I will examine the critical role played by computational modelling in developing a healthy cross-fertilisation of visual word recognition research and research on eye movement control in reading. The different contributions to the present section will serve as a basis for this analysis. In particular, the way the different models solve some critical issues in current reading research will be used to illustrate the importance of computational modelling for this field. Finally, I will briefly address the thorny problem of model comparison and evaluation, before examining possible future directions for research in this field.

Computational Modelling and Functional Overlap

Today, the general field of reading research is experiencing an optimistic move toward a possible reconciliation between the sub-disciplines of visual word recognition and eye-movement control in reading. A number of papers over the last ten years have systematically compared results obtained with eye movement measurements and results obtained from the same experimental manipulation in visual word recognition paradigms (e.g., Folk & Morris, 1995; Grainger, O'Regan, Jacobs & Segui, 1992; Inhoff, Briihl & Schwartz, 1996; Inhoff & Topolski, 1994; Perea & Pollatsek, 1998; Schilling, Rayner & Chumbley, 1998). Schilling *et al.* (1998), for example, manipulated the printed frequency of word targets and measured subjects' performance to these words in the standard word recognition tasks of lexical decision and naming. They compared these data to the results obtained using eye movement recordings from the same subjects reading the same set of words embedded in sentence contexts. The good correlation between eye movement measurements and both lexical decision and naming latencies (better for the latter) led the authors to conclude that "both the naming and lexical decision tasks yield data concerning word recognition processes that are consistent with effects found during silent reading". Although this is admittedly an important empirical observation, caution must be exercised when generating model-free conclusions from such data. When two different tasks produce the same pattern of effects, this does not necessarily imply that the two task are sensitive to a common underlying mechanism that is responsible for generating the effects. Schilling *et al.* report a significant correlation across lexical decision and naming latencies (by subjects and by items) in their experiment, while Carreiras, Perea and Grainger (1987) failed to find a significant correlation across these two tasks. Effects of a given variable (e.g., orthographic neighbourhood density) have often been found to be facilitatory in both naming and lexical decision (e.g., Andrews, 1989), yet very different mechanisms may be responsible for this facilitatory effect in the two tasks (Grainger & Jacobs, 1996). Modelling functional overlap is one remedy for this problem, and its prerequisite is the development of computational models. These have been available to visual word recognition researchers since the early 1980s (McClelland & Rumelhart, 1981; Paap, Newsome, McDonald & Schvaneveldt, 1982), and are now available to the field of eye movement control in reading.

Related to the question of modelling functional overlap, I recently suggested (Grainger, 2000) that one means of providing a better integration of these two areas of research would be to abandon the principle that eye movement recordings somehow provide a privileged, more ecological, instrument for measuring reading performance. Certainly at the level of single words, eye movement measurements could be considered as another experimental paradigm at the same level as the classic paradigms of visual word recognition. The critical point here is that, just like you need an appropriate model of the lexical decision task in order to successfully link lexical decision data to a model of word reading, you also need an appropriate model of eye movement control in reading in order to successfully link eye movement data to the same model of word reading. So, one not only needs a *model of the process* that is being investigated (word reading, for example), but one also needs a *model of the task* that is being used to investigate the process. With respect to investigations of single word reading, models of eye movement control therefore play a similar role as models of other laboratory tasks such as lexical decision and word naming. Finally, in order to maximise the amount of constraint provided by data obtained with a given technique, the model of the task must be as highly specified as possible, and expressed at least with the same degree of precision as the model of the process that the task is being used to investigate. Jacobs and Grainger (1994; see also Grainger & Jacobs, 1996) discussed the notion of functional overlap as an essential ingredient of a multi-task approach to understanding visual word recognition. It is the combination of the following factors that provides the critical constraints for model development: (1) empirical data obtained using different tasks, (2) a computational model of the process that one wants to understand, (3) computational models of the tasks used to investigate the given process, and (4) specification of what is task-specific in each of these task-models.

Many of the chapters in this section are a perfect illustration that this level of theorising has been achieved, and that we have now entered a new exciting era (the fourth era)[3] of research on eye movements in reading, where computational models of eye movement control and computational models of lexical processing can mutually constrain each other. I might immediately add that the benefits are clearly for both communities of researchers, since a model of isolated word recognition could not be considered complete without demonstrating how this model functions in a normal reading situation. A significant first step in this direction was made with the publication of the E-Z Reader model of eye movement control in reading (Reichle, Pollatsek, Fisher & Rayner, 1998; see also Pollatsek, Reichle & Rayner, this volume). The fact that lexical processing plays a central role in eye movement control in the E-Z Reader model shows very clearly how the two subfields can now fruitfully interact. However, the chapters by McConkie and Yang, and Reilly and Radach argue convincingly, I believe, against a strong version of the lexical control hypothesis as implemented in the E-Z Reader model (i.e., that each saccade is triggered by ongoing lexical processing). I will examine the arguments proposed in the different chapters of the present section for and against lexical control, but let me first say that a strong version of this hypothesis is not a necessary condition for fruitful interaction among word recognition and eye movement researchers.

In reading the chapters of the section on computational models of eye movement control in reading, one is impressed by the consensus around what are the critical phenomena to be explained. One of the strengths of the modelling efforts presented in this section is the general agreement as to what constitutes a reasonable data-base at present. The scope of the models is deliberately limited for purposes of tractability, and the to-be-modelled phenomena are, in the great majority, well replicated, unquestionable facts about how the eyes move during reading. This is an excellent starting point for any modelling enterprise, and this situation is only possible today thanks to the many years of effort from dedicated researchers in the field (see Rayner, 1998, for a review). As mentioned above, there appears to be a general consensus as to what are the critical phenomena that require explaining. However, two very central questions are still open to debate, as reflected by the varying positions adopted by the contributors to this section. These concern two of the three issues[4] recently summarised by Starr and Rayner (2001): (1) to what extent do higher-level cognitive (linguistic) processes influence eye movement behaviour? and (2) is word recognition in reading strictly serial (one word at a time)? I will examine the different response to these two questions provided by the chapters in this section, in an attempt to converge on a tentative synthesis.

Cognitive Control of Eye Movements in Reading

The original E-Z Reader model, as well as later versions (Pollatsek *et al.*, this volume), is characterized by the way lexical processing is used as the major component of the complex system that guides the eyes through text. To someone, like myself, interested in visual word recognition, this appears to be a very reasonable proposal. Given that the minimal goal of reading behaviour is to extract meaning from individual words, to be combined to allow interpretation of higher level structures such as phrases and sentences, then it seems reasonable to assume that the process of individual word recognition should influence reading behaviour. As such, the E-Z Reader model endorses a rather extreme form of the linguistic control hypothesis, or what McConkie and Yang refer to as *direct* cognitive control. The E-Z reader model is certainly the first computational model of eye-movement control in reading that makes use of a quite complex processing machinery to determine when a saccade will be generated. More specifically, lexical processing on the currently fixated word is evaluated online. When an orthographic representation (and possibly a phonological representation) for a word is sufficiently activated (but not yet identified), then the lexical processor "decides" that the word will be processed correctly and that it is now safe to move the eyes onto the next word. At first sight there appears to be a very interesting parallel between the familiarity check mechanism used in the E-Z Reader model and the mechanism used by several researchers to describe the process of lexical decision (e.g., Balota & Chumbley, 1984; Grainger & Jacobs, 1996). However, although a familiarity check makes perfect sense for an information processing device that has to perform lexical decision (where familiar stimuli have to be distinguished from unfamiliar stimuli), there is, on the contrary, no clear theoretical justification for its use in a model of eye movement control in reading.

Kliegl and Engbert's chapter presents the latest version of the SWIFT model and evaluates this model relative to its competitors using the empirical data of Schilling *et al.* (1998) and parafoveal preview effects obtained using eye-contingent display changes (Binder, Pollatsek & Rayner, 1999). Although there are many similarities between the SWIFT model and E-Z Reader, there are some fundamental differences, one at the level of cognitive control, and the other in terms of serial versus parallel processing. Concerning the issue of cognitive control, Kliegl and Engert abandon the familiarity check mechanism that the E-Z Reader model uses for saccade initiation. Instead, in the SWIFT model saccade initiation occurs autonomously. However, lexical activity can influence this autonomous saccade generation programme by delaying saccade initiation. The lexical activity associated with the currently fixated word modifies the timing interval that is sampled for determining the initiation of the next saccade.

McConkie and Yang's chapter begins with the presentation of some critical data against direct cognitive control theories of eye movement control in reading. Yang and McConkie (2001) demonstrated that strings of random letters and Xs which preserve the word spacing pattern had hardly any effect on the distribution of onset times of most saccades. McConkie and Yang conclude that saccade initiation times depend very little on whether the currently fixated stimulus is a word or random letters. This is not what one would expect if some measure of lexical processing was determining when a saccade will be triggered. The Interaction/Competition (I/C) theory discussed by McConkie and Yang adopts the mechanism described by Findlay and Walker (1999) whereby saccade timing and target selection in eye movement control result from competitive processes between so-called move and fixate centres. This mechanism is adapted to the specific case of reading and provides a system for saccade timing that is not directly influenced by cognitive events. Similarly to the SWIFT model, this saccade timing mechanism can be indirectly influenced by cognitive activity via inhibitory control. The basic idea is that when some processing difficulty is encountered, then a kind of "distress signal" is triggered that results in inhibition in the saccadic system. This processing-based inhibition influences fixation and gaze durations primarily by reducing the number of normal saccades and increasing the number of late saccades. It is interesting to note that a similar proposal had already been made by McConkie (1979) who claimed that the eyes are moved in response to unsuccessful processing of parafoveal words: "Assume that at some point during a fixation, as the attended region is shifted along the line of text visual information is sought from a region too far from central vision to readily supply the visual detail needed for identification . . . Assume that seeking visual detail from a retinal region that is not readily available causes the saccadic system to initiate a saccadic eye movement . . . This is assumed to be the primary basis for saccadic eye movement control in reading" (p. 43). Just as in the I/C theory, the system is hypothesized to respond to the fact that something "does not compute", rather than responding to successful processing as in the E-Z Reader model.[5]

Reilly and Radach present a computational model of eye movement control in reading (Glenmore) that has many points in common with McConkie and Yang's I/C theory. Both of these approaches appear to be heavily influenced by (i) Findlay and Walker's (1999) general theory of saccade generation; and (ii) by the earlier work of the "Groupe Regard" in Paris, and most notably the visual scanning routine proposed by Lévy-Schoen (1981), and O'Regan's (1990) strategy and tactics theory. Findlay and Walker's theory provides a specific mechanism (the saliency map) for implementing a "dumb" oculomotor strategy in a model of eye movement control in reading. However, Reilly and Radach go one very significant step further than their mentors. They describe a specific mechanism for how lexical processing influences oculomotor heuristics.

Coming back to the issue at stake here (cognitive control of eye movements in reading), Reilly and Radach hit the proverbial nail on the head when they say that ". . . the question of interest in not whether eye movements are determined by visuo-motor factors or linguistic processing, but to what degree these two types of factors are involved and how they interact". The relative involvement of these two factors and their interaction can only be appropriately described within the framework of a computational model, as very nicely illustrated by several of the chapters of this section. So the debate around the issue of cognitive control of eye movements (see the chapters of this section, and Deubel, O'Regan & Radach, 2000; Rayner, 1998) could well be abandoned. One does not need to know whether or not linguistic factors are the *main* driving force behind eye movements in reading in order to make progress in this area. Adopting a much more practical stance, one simply needs to specify when and how linguistic factors can intervene and influence the observed pattern of eye movements. This amounts to describing the functional overlap between the system that controls eye movements in reading and the system that recognizes printed words.

Serial vs. Parallel Processing of Words

The second main issue addressed by all the contributions to this section concerns whether or not serial attention shifts precede eye movements during reading, thus dictating a strictly serial processing of words. Reingold and Stampe's chapter is centred on this particular question. These authors present a new phenomenon related to saccadic control in reading, referred to as saccadic inhibition, that is taken as evidence in favour of attentional guidance models. In a generic attentional guidance model (as originally proposed by Morrison, 1984), an attention shift occurs to the next word to be fixated before the saccade to that word is executed. So, when reading normal text, attention skips along from word to word just slightly ahead of the eyes. According to Reingold and Stampe, this predicts that one ought to be able to observe some form of perceptual enhancement in the direction of the next saccade, due to preallocation of attention to that location. By creating screen flicker either to the right (congruent) or to the left (incongruent) side of a currently fixated point, these authors observed a stronger effect of a large flicker manipulation in the congruent condition. They interpret this finding as resulting from a perceptual enhancement of the large flicker due to attentional preallocation in the direction of the next saccade.

The E-Z Reader model described by Pollatsek *et al.* adheres to the basic principles of attentional guidance models. The familiarity check mechanism (based on partial lexical processing) governs saccade generation, and attention is allocated to the newly targeted word once lexical processing is complete. The timing associated with saccade programming allows attention to be shifted to the next word before a saccade is actually executed. However, apart from the E-Z Reader model, all the other models described in this section have abandoned a strictly serial allocation of attention. Each of the relevant chapters summarizes the key results that have led the authors to abandon this hypothesis. Since this is one of the critical differences between SWIFT and the E-Z Reader model, Kliegl and Engbert provide a good summary of the empirical findings. They describe three results that are difficult to handle by a serial attention shift mechanism. These are: (1) parafoveal on foveal influences; (2) influence of information extracted to the left of the fixated word; and (3) variation in fixation durations as a function of whether the next word is skipped or not. For Kliegl and Engbert, although there is still some doubt as to the first two effects, it is the third type of evidence that is critically damaging.[6] According to serial attention shift models, word skipping arises when a saccade to the next word is cancelled and re-programmed to the following word. This necessarily causes an increase in the fixation duration prior to the re-programmed saccade. However, this clear prediction has not been upheld in some empirical analyses, such that the issue remains under debate (McConkie, Kerr & Dyre, 1994; Radach & Heller, 2000; see Kliegl & Engbert, this volume, for a discussion).

These considerations have led Kliegl and Engbert, and Reilly and Radach to abandon the idea of a serial attention shift mechanism in favour of spatially distributed processing around fixation in the form of an attentional or processing gradient. This takes the specific form of a saliency map in the Glenmore model. In this particular approach, there is no deliberate shifting of attention during reading, just a gradient of bottom-up activation spreading right and left of fixation that can be modulated by higher-level factors (Findlay & Walker, 1999). This activation gradient or saliency map provides a mechanism for explaining parafoveal influences on foveal target processing. Its precise instantiation in the Glenmore model raises the very interesting issue of how different words at different retinal locations can be processed at the same time. I will come back to this point.

Which Model is Best?

The apparent success of the computational models described in this section to account for a given target set of data, raises the obvious question as to how one might be able to decide which one does the job best. Jacobs (2000) has already tackled this question and provided some answers for computational modellers of eye movement control in reading. The chapters in the present section provide some admirable applications of basic principles in the development and testing of computational models. As already noted by Jacobs (2000), the E-Z Reader "suite" (continued in the Pollatsek *et al.*

chapter) is a nice illustration of the practice of "nested modelling", where the old version of the model is embedded in the new version, thus preserving the core principles of the original model. One can also applaud the testing strategy applied by Kliegl and Engbert where two models of similar structure (E-Z Reader and SWIFT) are evaluated relative to fits with the same set of data, thus allowing strong scientific inference (Estes, 1975).

However, rather than getting bogged down in the complex intricacies of model comparisons (number of free parameters, comparing free parameters to weight strengths, number of units and layers of algorithmic models, etc., etc.), here I would like to make a more general comment on *modelling style*. There are many different styles of computational modelling in cognitive science, and the present contributions illustrate part of the spectrum. There are some more mathematical-style models (E-Z Reader and SWIFT) using closed-form expressions, and one example of an algorithmic model of the interactive-activation family (Glenmore). In the more general area of reading research, currently dominated by research on single word recognition, the vast majority of models are of three basic kinds: verbal-boxological (pre-quantitative), mathematical, and algorithmic connectionist models (as opposed to algorithmic symbolic models, although Coltheart, Rastle, Perry & Ziegler's, 2001, dual-route model is an example of a hybrid symbolic-connectionist algorithmic model). There are two types of algorithmic connectionist models of the reading process: localist connectionist models (e.g., McClelland & Rumelhart, 1981, for printed word perception), and PDP models with distributed representations (e.g., Seidenberg & McClelland, 1989, for reading aloud printed words).

In the preface to their volume on localist connectionism, Grainger and Jacobs (1998) summarized the advantages of the localist approach as opposed to a more distributed approach (typically involving the use of the backpropagation learning algorithm) under the headings: continuity, transparency, and unification. In terms of continuity, Grainger and Jacobs (1998; see also Page, 2000) have argued that algorithmic models of the localist connectionist variety are the most apt at preserving an understandable link with less formal verbal-boxological theorizing. More specifically, the architecture of the processing system can be quite accurately described in a boxological model (e.g., Grainger & Ferrand, 1994), before being implemented in an algorithmic model (e.g., Jacobs, Rey, Ziegler & Grainger, 1998). The study of eye-movement control in reading has benefited from a good deal of informal, pre-quantitative theorizing (e.g., Morrison's, 1984, attention-guidance model, and O'Regan's, 1990, strategy-tactics theory) that serves as the basic ground matter for developing computational models. This is true for most areas of cognitive science, where the normal course of events is to first sketch a verbal theory, which may or may not be implementable as a mathematical model (depending on the complexity and the precision of the verbal theory), and then develop a computational model on the basis of prior theorizing.

The tight link between localist connectionist models and their verbal-boxological predecessors, allows these models to provide a (more) transparent mapping between model structure and model behaviour. In some PDP models with distributed representations the model is as complex as the phenomena it seeks to explain. This led Forster (1994) to wonder about the utility of such models. He discusses this point using his next-door neighbour analogy: "Suppose I discover that my next-door neighbor can correctly predict the outcome of every word recognition experiment that I do. This would be a surprising discovery, certainly, but it would not have any scientific utility at all until it was discovered how she was able to do it. I could scarcely publish my next-door neighbor as a theory without having explicated the reasoning involved (p. 1295)". It is the "black box" aspect of fully distributed models that hinders explanation.

Grainger and Jacobs (1998) argued that localist connectionist models provide a unifying account of human cognition. Page's (2000) "localist manifesto" is another such appeal for a unified account of human information processing that adopts a localist connectionist stance. Page (2000) documents the long-standing role played by Stephen Grossberg as a key figure in this area of computational modelling. Grossberg's modelling work is characterised by the application of a small set of computational principles (e.g., adaptive filter, lateral inhibition, 2/3 matching rule, masking fields) that are applied to the explanation of a very wide range of human information processing, from low-level vision (e.g., Grossberg, 1999) to higher-level cognitive processes such as word recognition and recall (Grossberg & Stone, 1986). I suspect that progress in the development of computational models of eye movement control during reading will show, once again, the superiority of localist connectionism as a general approach to the study of human cognition. The Glenmore model is a good example of how such an approach facilitates integration across neighbouring fields of research (a theory of eye movement control, and a model of visual word recognition).

Finally, all computational modellers must ask the question: *what has been gained from the modelling enterprise*? The critical gain must be expressed in terms of improved understanding of the phenomena to be explained, in terms of improved understanding of the processes under study, and in terms of the generation of new predictions for further experimentation. Converting a verbal statement to a mathematical equation is generally accepted as an improvement in science (e.g., Estes, 1975), but the gain in understanding can be quite minimal. For example, a mathematical model of visual word recognition that predicts word recognition time as a function of word frequency and the number and frequency of all orthographic neighbours of the target word (using Luce's , 1959, choice rule, for example), would not be doing much more than re-stating the empirical evidence that word frequency and orthographic neighbourhood density and frequency influence visual word recognition (e.g., Luce & Pisoni's (1998) neighbourhood activation model of spoken word recognition). The critical gain only becomes obvious when the computational model proves its utility (1) as an explanatory device and (2) as a heuristic for generating scientific research. The "winning" model is typically one for which a general consensus builds up around these last two points. A single critical test between two models is rarely a deciding factor in this field of research.

Reading Words and Moving Eyes: Future Developments

It appears that the recent development of quantitative models of eye movement control in reading has opened the door for more intense and fruitful interaction with the field of printed word perception. The importance of such an interaction has become more obvious given the dominant role played by lexical processing as a precise implementation of the (direct or indirect) "cognitive control" aspect of these models. After reading the chapters in this section, it seems that one key question that has guided recent research in this field, and should continue to guide future research, could be formulated as follows: Given initial oculomotor and linguistic constraints, how would an information processing device proceed in order to extract visual information from text with the aim to translate it into meaning? Some tentative answers to this question have been formulated in the different chapters of this section, in the form of explicit descriptions of the information processing that is performed (i.e., as computational models). However, this general question or research goal can be usefully broken up into smaller questions that could serve to guide future research. These "sub-goals" reflect the fact that, although some very simple oculomotor strategies appear very capable of providing a good first response to the general question, several of the chapters in this section argue that the timing and targeting of eye-movements also depends to some extent on lexical processing. It is therefore important to be able to specify (1) how oculomotor and perceptual constraints affect eye movement control in a "reading-like" situation; (2) how linguistic constraints affect visual word recognition in the absence of eye movements; and (3) how these two situations can combine to generate the phenomena observed in eye movement investigations of reading. For each of these points we need to specify the critical empirical observations, and define the potentially viable explanations in terms of computationally explicit mechanisms.

Thus, concerning the first point, we need to specify the basic phenomena associated with eye movements in a reading-like situation but in the absence of linguistic input (e.g., Vitu, O'Regan, Inhoff & Topolski, 1995; Yang & McConkie, 2001). Here the goal is to determine the basic oculomotor constraints that govern eye movements before any lexical influences have operated. This will enable us to define the functional overlap between a general model of eye movement control (e.g., Findlay & Walker, 1999) and a model of eye movement control in reading. Pursuing the development of "non-linguistic" models of eye movement control in reading will help define the limits of such an approach, and therefore facilitate the integration phase described in point three.

Concerning the second point, we need to specify the basic phenomena associated with visual word recognition that a model of reading should attempt to capture. Here, the goal is to describe the constraints on lexical processing that apply independently of whether or not eye movements are required to enable this processing. Jacobs and Grainger (1994) proposed a list of basic phenomena observed with standard visual word recognition tasks. The list included the word frequency effect, the word superiority effect, the regularity/consistency effect, and the effects of orthographic neighbourhood, which appeared a reasonable list at that time for evaluating different models of visual word recognition. The field has evolved since then, and researchers working with isolated word recognition techniques still need to establish a solid set of replicable phenomena that reflect the core mechanisms of printed word perception. For example, phenomena observed in the word naming task but not in other word recognition tasks (e.g., the regularity/consistency effect) are likely to reflect the operation of mechanisms involved in generating an articulatory output, and are probably not very good candidates for providing constraints on models of silent reading.[7] Once again, applying the principle of modelling functional overlap is critical here. For a given effect observed in a given task it is vital to know whether the machinery generating the effect will overlap with the lexical processes implicated in eye movement control during reading.

This leads us to the final integration phase, point three. Here we raise the question that most of the contributions to the section on computational modelling have attempted to address. Namely, how oculomotor constraints combine with lexical constraints to produce the patterns of eye movements that are recorded in empirical investigations of eye movement control in reading. However, all the models described in this section take the performance of skilled adult readers as the target to be explained by the different models. Developmental investigations of eye movement control in reading will become increasingly important for the field. Observations of how the oculomotor system learns to adapt to different types of linguistic stimuli (isolated words, sentences, texts) in the process of learning to read will provide significant additional constraints on computational models of eye movement control in reading. Localist connectionist learning models, such as adaptive resonance theory (Grossberg, 1980) may prove useful in establishing the link between algorithmic models of adult performance and an account of how oculomotor and attentional control develop during the process of learning to read.

Finally, research on printed word perception now needs to make the complementary step in the other direction, reflecting on how issues of eye-movement control can affect the way we think about basic issues in visual word recognition. This step has already been partly made in research considering how initial fixation location can affect the way we process printed words (O'Regan & Jacobs, 1992; Nazir, Heller & Sussmann, 1992; Clark & O'Regan, 1998; Stevens & Grainger, 2003).The concept of spatially distributed processing and the saliency map described by Reilly and Radach automatically spark the debate as to how parallel processing of words can operate. Clearly, if two words that receive input from different spatial locations can compete within a single lateral-inhibitory word identification network, then we are led to make some very clear predictions relative to paradigms involving simultaneous presentation of two or more printed words. This opens up a whole new area of investigation, even for those researchers who remain stubbornly attached to word perception in a single glance.

Acknowledgements

This commentary benefited from careful and detailed feedback provided by Ralph Radach.

Notes

1 See, for example, the number of cited works in this area on page 390 of Rayner (1998).

2 I distinctly remember Don Mitchell saying to me, over ten years ago, that "visual word recognition had been solved". This happened at one of those excellent workshops organized on

several occasions by Alan Kennedy and Joel Pynte, that allowed researchers like myself not using eye movement recordings, to be confronted with the results obtained with this technique.

3 The fourth era can be seen as the next step in the evolution described by Rayner (1978, 1998), and its initiation can be situated at the end of the 20th century and associated with a number of critical events such as Rayner's (1998) review, the Reichle *et al.* (1998) paper, and the conference held at Luminy, Marseille, France in November 1998 (Kennedy, Radach, Heller & Pynte, 2000).

4 The other issue discussed by Starr and Rayner (2001) concerned the amount of information that can be extracted from the right of fixation in reading.

5 I thank Ralph Radach for pointing out the early contribution of McConkie (1979).

6 Jacobs (2000) defined the notion of a "strongest falsificator" for a given model and urged cognitive modellers to define such a situation for their models. The word skipping data would appear to be a good candidate as a strongest falsificator for the E-Z Reader model.

7 Reading aloud is, of course, an interesting activity to study in its own right.

References

Balota, D. A., & Chumbley, J. I. (1984). Are lexical decisions a good measure of lexical access? The role of word frequency in the neglected decision stage. *Journal of Experimental Psychology: Human Perception and Performance*, *10*, 340–357.

Binder, K. S., Pollatsek, A., & Rayner, K. (1999). Extraction of information to the left of the fixated word in reading. *Journal of Experimental Psychology: Human Perception and Performance*, *25*, 1162–1172.

Carreiras, M., Perea, M., & Grainger, J. (1997). Orthographic neighborhood effects on visual word recognition in Spanish: Cross-task comparisons. *Journal of Experimental Psychology: Learning, Memory, and Cognition*, *23*, 857–871.

Clark, J. J., & O'Regan, J. K. (1999). Word ambiguity and the optimal viewing position in reading. *Vision Research*, *39*, 843–857.

Coltheart, M., Rastle, K., Perry, C., Langdon, R., & Ziegler, J. (2001). DRC: A dual route cascaded model of visual word recognition and reading aloud. *Psychological Review*, *108*, 204–256.

Deubel, H., O'Regan, J. K., & Radach, R. (2000). Attention, information processing and eye movement control. In: A. Kenney, R. Radach, D. Heller and J. Pynte (eds), *Reading as a Perceptual Process* (pp. 355–374). Oxford: Elsevier.

Estes, W. K. (1975). Some targets for mathematical psychology. *Journal of Mathematical Psychology*, *12*, 263–282.

Findlay, J. M., & Walker, R. (1999). A model of saccade generation based on parallel processing and competitive inhibition. *Behavioral and Brain Sciences*, *22*, 661–721.

Folk, J. R., & Morris, R. (1995). Multiple lexical codes in reading: Evidence from eye movements, naming time, and oral reading. *Journal of Experimental Psychology: Learning, Memory, and Cognition*, *21*, 1412–1429.

Forster, K. I. (1994). Computational modelling and elementary process analysis in visual word recognition. *Journal of Experimental Psychology: Human Perception and Performance*, *20*, 1292–1310.

Grainger, J. (2000). From print to meaning via words? In: A. Kennedy, R. Radach, D. Heller and J. Pynte (eds), *Reading as a Perceptual Process* (pp. 147–161). Oxford: Elsevier.

Grainger, J., & Jacobs, A. M. (1996). Orthographic processing in visual word recognition: A multiple read-out model. *Psychological Review*, *103*, 518–565.

Grainger, J., O'Regan, K., Jacobs, A., & Segui, J. (1992). Neighborhood frequency effects and letter visibility in visual word recognition. *Perception and Psychophysics*, *51*, 49–56.

Grossberg, S. (1980). How does a brain build a cognitive code? *Psychological Review*, *87*, 151.

Grossberg, S. (1999). How does the visual cortex work? Learning, attention, and grouping by the laminar circuits of visual cortex. *Spatial Vision*, *12*, 163–185.

Grossberg, S., & Stone, G. O. (1986). Neural dynamics of word recognition and recall: Attentional priming, learning, and resonance. *Psychological Review*, *93*, 46–74.

Inhoff, A. W., Briihl, D., & Schwarz, J. (1996). Compound words effects differ in reading, on-line naming, and delayed naming tasks. *Memory & Cognition*, *24*, 466–476.

Inhoff, A. W., & Topolski, R. (1994). Use of phonological codes during eye fixations in reading and in on-line and delayed naming tasks. *Journal of Memory and Language*, *33*, 689–713.

Jacobs, A. M. (2000). Five questions about cognitive models and some answers from three models of reading. In: A. Kenney, R. Radach, D. Heller and J. Pynte (eds), *Reading as a Perceptual Process* (pp. 721–732). Oxford: Elsevier.

Jacobs, A. M., & Grainger, J. (1994). Models of visual word recognition: Sampling the state of the art. *Journal of Experimental Psychology: Human Perception and Performance*, *20*, 1311–1334.

Jacobs, A. M., Rey, A., Ziegler, J. C., & Grainger, J. (1998). MROM-p: An interactive activation, multiple read-out model of orthographic and phonological processes in visual word recognition. In: J. Grainger and A. M. Jacobs (eds), *Localist Connectionist Approaches to Human Cognition*. Mahwah, NJ: Erlbaum.

Lévy-Schoen, A. (1981). Flexible and/or rigid control of oculomotor scanning behavior. In: D. Fisher, R. A. Monty and J. W. Senders (eds), *Eye Movements: Cognition and Visual Perception* (pp. 299–314). Hillsdale, NJ: Erlbaum.

Luce P. A., & Pisoni D. B. (1998). Recognizing spoken words: The neighborhood activation model. *Ear Hear*, *19*, 1–36.

Luce, R. D. (1959). *Individual Choice Behavior*. New York: Wiley.

McClelland, J. L., & Rumelhart, D. E. (1981). An interactive activation model of context effects in letter perception: Part 1. An account of basic findings. *Psychological Review* *88*(5): 375–407.

McConkie, G. W. (1979). On the role and control of eye movements in reading. In: P. A. Kolers, M. E. Wrolstad and H. Bouma (eds), *Processing of Visible Language. Vol. I* (pp. 37–48). New York: Plenum Press.

McConkie, G. W., Kerr, P. W., & Dyre, B. P. (1994). What are "normal" eye movements during reading: Toward a mathematical description. In: J. Ygge and G. Lennestrand (eds), *Eye Movements in Reading* (pp. 315–327). Oxford: Elsevier.

Morrison, R. E. (1984). Manipulation of stimulus onset delay in reading: Evidence for parallel programming of saccades. *Journal of Experimental Psychology: Human Perception and Performance*, *10*, 667–682.

Nazir, T. A., Heller, D., & Sussmann, C. (1992). Letter visibility and word recognition: The optimal viewing position in printed words. *Perception & Psychophysics*, *52*, 315–328.

O'Regan, J. K. (1990). Eye movements in reading. In: E. Kowler (ed.), *Eye Movements and their Role in Visual and Cognitive Processes* (pp. 395–453). Amsterdam: Elsevier.

O'Regan, J. K., & Jacobs, A. M. (1992). The optimal viewing position effect in word recognition: A challenge to current theory. *Journal of Experimental Psychology: Human Perception and Performance*, *18*, 185–197.

Paap, K. R., Newsome, S. L., McDonald, J. E., & Schvaneveldt, R. W. (1982). An activation-verification model for letter and word recognition: The word superiority effect. *Psychological Review*, *89*, 573–594.

Page, M. (2000). Connectionist modelling in psychology: A localist manifesto. *Behavioral and Brain Sciences, 23*, 443–512.

Perea, M., & Pollatsek, A. (1998). The effects of neighborhood frequency in reading and lexical decision. *Journal of Experimental Psychology: Human Perception and Performance, 24*, 767–779.

Radach, R. & Heller, D. (2000). Relations between spatial and temporal aspects of eye movement control. In A. Kennedy, R. Radach, D. Heller & J. Pynte (eds). *Reading as a Perceptual Process*. Oxford: Elsevier (pp. 165–192).

Rayner, K. (1978). Eye movements in reading and information processing. *Psychological Bulletin, 85,* 618–660.

Rayner, K. (1998). Eye movements in reading and information processing: 20 years of research. *Psychological Bulletin, 124*, 372–422.

Reichle, E. D., Pollatsek, A. Fisher, D. L., & Rayner, K. (1998). Toward a model of eye movement control in reading. *Psychological Review, 105*, 125–157.

Schilling, H. E. H., Rayner, K., & Chumbley, J. I. (1998). Comparing naming, lexical decision, and eye fixation times: Word frequency effects and individual differences. *Memory & Cognition, 26*, 1270–1281.

Seidenberg, M. S., & McClelland, J. L. (1989). A distributed developmental model of word recognition and naming. *Psychological Review, 96*, 523–568.

Starr, M. S., & Rayner, K. (2001). Eye movements during reading: Some current controversies. *Trends in Cognitive Sciences, 5*, 156–163.

Stevens, M., & Grainger, J. (2003). Letter visibility and the viewing position effect in visual word recognition. *Perception & Psychophysics*, in press.

Vitu, F., O'Regan, J. K., Inhoff, A. W., & Topolski, R. (1995). Mindless reading: Eye movement characteristics are similar in scanning letter strings and reading text. *Perception & Psychophysics, 57*, 352–364.

Yang, S. N., & McConkie, G. W. (2001). Eye movements during reading: A theory of saccade initiation times. *Vision Research, 41*, 3567–3585.

Section 4

Eye Movements in Human–Computer Interaction

Section Editor: Ralph Radach

Chapter 22

Voluntary Eye Movements in Human–Computer Interaction

Veikko Surakka, Marko Illi, and Poika Isokoski

The present aim is to give an overview on human–computer interaction (HCI) techniques that utilise especially voluntary eye movements for interaction with computers. Following the introductory section the chapter gives a brief outline on technical and methodological aspects that need to be considered in these techniques. As the number of communicative channels (i.e. unimodal vs. multimodal) between human and computer is one of the central issues in HCI, the main part of the chapter is organised according to the amount and nature of used modalities.

Introduction

Graphical user interfaces were introduced to public use in the early 1980s and they still are the most commonly used user interface types. A typical interaction event consists of pointing and selecting objects on a computer screen and the most widely used interaction device is a mouse. Trackballs, joysticks and touchpads are examples of other commonly used interaction devices. All these devices require the use of hands for interaction and there have been many attempts to develop alternatives for hand-based interaction techniques.

The development of alternative interaction techniques that utilise modalities other than hands has been important for two reasons. First, there are people who cannot use traditional hand-based interaction devices. However, it is important that they can use the computers as well as functionally intact people. The second reason for the development of alternative interaction techniques has been the demand for more natural and efficient interaction between the user and the computer. In the field of human–computer interaction research this would mean that the creation of new types of user interfaces should be based on the modelling of human behaviour including for example, cognitive psychology, human emotions, and also human interaction (e.g. Cassell, 2000).

The Mind's Eye: Cognitive and Applied Aspects of Eye Movement Research

ISBN: 0–444–51020–6

Basically, interaction between humans as well as interaction between humans and computers can be classified as purely unimodal or multimodal. In unimodal interaction, communicative signals are transmitted through one modality only. An example of human unimodal interaction could be a discussion over the telephone in which case the sender uses speech only and the receiver uses hearing only. Also interaction with computers can be constructed in this way. In multimodal interaction, communicative signals are transmitted through several different modalities. For example, in human face-to-face communication we frequently utilize both verbal and non-verbal cues for sending and receiving interactive signals (Hietanen, Surakka & Linnankoski, 1998; Sams, Manninen, Surakka, Helin & Kättö, 1998; Surakka & Hietanen, 1998). In normal circumstances people interact with each other through several different modalities. Speech, gaze direction, touch, facial expressions and gestures, for example, provide different interaction modalities for communicative purposes. Although human–human interaction is possible, to some degree, even through one modality only (e.g. interaction based purely on speech or touch), more natural and efficient interaction between humans is achieved when several modalities are used simultaneously for communicative purposes. People are good at integrating the use of several modalities at the same time. Multimodal integration of different modalities usually occurs without effort and quite unconsciously (Sams *et al.*, 1998; Hietanen, Manninen, Sams & Surakka, 2001). Thus, human–human interaction is essentially multimodal. This holds for both conveying and receiving communicative signals (Surakka & Hietanen, 1998).

In comparison to human interaction, HCI is more limited in both input and output methods. Most systems allow the inputs for a computer to be transmitted only through one modality and usually this modality has been the use of hands. This means that other possible input modalities that people use naturally in human–human interaction remain unused in HCI. However, it is possible to utilize these other modalities in HCI, too. Several different modalities, for example, speech, eye movements and facial muscle activity offer promising and interesting alternative methods to be used as an input for a computer. If the information from user to computer and vice versa could be transmitted through several modalities this would conceivably result in more efficient, versatile, flexible, and eventually also in more natural interaction between the user and the computer, similar to human–human interaction.

The search for alternative HCI techniques has found, among other possibilities, that tracking users' eye movements could offer a feasible option for more natural interaction techniques. Although earlier studies on eye movement tracking explored human perceptual and cognitive processes (Rayner, 1998), it is possible to use the monitoring of eye-movements with modern eye-trackers for several different purposes. During the past two decades, research has been done to study how functional the tracking of eye movements is at controlling computers. Interaction techniques that are based on the use of eye movements have been considered promising alternatives for hand-based techniques. The reason for this is that while interacting with computers, people usually look at the objects of interest and, thus, these techniques utilise the modality that people use naturally in information seeking (Sibert & Jacob, 2000).

Normal behaviour of the eyes includes both involuntary and voluntary movements. Involuntary eye movements occur without a person's intention and quite unconsciously. Voluntary eye movements are more conscious and controlled. Voluntary eye movements have been utilized more often in computer user interfaces, although involuntary eye movements can also be used. In both types of interfaces the system tracks the user's eye movements and performs predetermined actions if certain eye movements occur. The difference is that when involuntary eye movements are used, the system tries to interpret the natural eye-movements of the user whereas an interface using voluntary eye movements requires the user to consciously use his or her eyes in a certain way. Techniques that utilize voluntary eye movements have been typically used to replace the traditional mouse actions. This includes pointing and selecting objects on-screen. Many systems using eye movements have been built to aid functionally impaired people (for a review see Majaranta & Räihä, 2002). However, the focus of the current chapter is in more general user interfaces.

Research has shown that eye movements provide a promising alternative interaction technique for HCI and that some advantages can be gained over hand-based techniques. In hand-based techniques, people have to first find the object of interest by moving their eyes so that the gaze falls over an object of interest, and only then can the cursor be manually adjusted over the object. In gaze pointing techniques, the gaze is already on the correct object the moment a person looks at it and people can give the inputs for a computer without using their hands. In comparison to hand based techniques this can also be faster. The hands-free advantage is especially important for disabled people who cannot use their hands. In terms of user experiences, there are some observations that interfaces utilizing eye movements can be more enjoyable than traditional user interfaces (Sibert & Jacob, 2000; Salvucci & Anderson, 2000; Tanriverdi & Jacob, 2000).

Although the use of eye movements in HCI seems to be a promising alternative interaction technique, there are several considerations that need to be taken into account in understanding the possibilities and the limitations of using eye movements in HCI. The outline of this chapter is as follows. In the second section we introduce briefly the most important factors that need to be realised and considered when using these alternative techniques. Following this we introduce HCI techniques that have utilised voluntarily controlled eye movements. These techniques can be classified in various ways (e.g. gaze based, gaze contingent, and gaze added techniques) but as the nature and the number of communicative channels (i.e. unimodal vs. multimodal) between human and computer is one of the central issues in HCI we classify them into three categories, unimodal, multimodal, and inherently multimodal techniques (see Table 22.1). In the third and fourth sections the chapter gives examples of all these three different types of interaction techniques. The third section introduces techniques that utilise eye movements as the sole input method. These can be classified as unimodal techniques. In the fourth section we introduce multimodal techniques. Some of them can be classified primarily as multimodal interaction techniques, although they can also be used unimodally. Some of the multimodal techniques can be classified as inherently multimodal. These techniques utilize eye movements in conjunction with some other technique and they are fully functional only when two or more modalities are used together. The chapter ends by discussing the advantages and disadvantages that are related to the utilisation of eye movements in HCI as a unimodal and as a part of multimodal interaction techniques.

Table 22.1: Interaction techniques classified in respect to the required modalities for functional HCI. The abbreviations are as follows: U = unimodal, M = multimodal, and IM = inherently multimodal.

Experiment		**Modality**			**Fitts' Law**
Reference	**Selection Method**	**U**	**M**	**IM**	
Sibert *et al.*, 2001, Experiment 1	dwell time	X			Does not fit
Sibert *et al.*, 2001, Experiment 2	dwell time	X			Not tested
Miniotas 2000	dwell time	X			Fits well
Salvucci and Anderson, 2000	hardware button		X		Not tested
Zhai *et al.*, 1999, liberal MAGIC	mouse			X	Not tested
Zhai *et al.*, 1999, conservative MAGIC	mouse			X	Not tested
Partala *et al.*, 2001	electromyography			X	Not tested
Surakka *et al.*, submitted	electromyography			X	Fits well
Ware and Mikaelian, 1987, Experiment 1	dwell time	X			Fits well
Ware and Mikaelian, 1987, Experiment 1	on-screen button	X			Fits well
Ware and Mikaelian, 1987, Experiment 1	hardware button			X	Fits well
Ware and Mikaelian, 1987, Experiment 2	dwell time	X			Not tested
Ware and Mikaelian, 1987, Experiment 2	hardware button			X	Not tested

Technical and Methodological Considerations

Eye Movements and Gaze Direction

According to Sibert and Jacob (2000), the two most important elements in the behaviour of the eye in HCI are fixations and saccades. They are most suitable for the analysis of visual search tasks and exploration of the visual environment. Saccades are rapid ballistic movements that move the gaze from one fixation point to another. It is impossible to change the direction of an eye movement during a saccade. The duration of a saccade depends roughly linearly on the distance from one visual object to another (Abrams, Meyer & Kornblum, 1989). A saccade can last from 30 to 120 ms and it can cover a range from 1 to 40 degrees of visual angle (Sibert & Jacob, 2000). The velocity of a saccade is highest near the midpoint of movement and can be as high as 500 degrees per second (Rayner, 1998). Saccadic eye movements are inherently superior in speed when compared to any other modality for pointing at objects (Sibert & Jacob, 2000). Fixations are the periods during which the information from objects is received and they typically last between 200–300 ms (Rayner, 1998). Several algorithms for detecting fixations and saccades from eye tracker data have been developed. For a more detailed discussion on these algorithms we refer the reader to the taxonomy by Salvucci and Goldberg (2000).

When a subject directs her or his gaze onto an object, the eyes move so that the image of the target object appears on the fovea of the retina. This is the high acuity area of vision and it covers approximately one degree of visual arc. Images falling on any part of the fovea are seen clearly. Thus, a person has to direct her/his eyes only in such accuracy that the gaze direction corresponds to the focus of visual attention within one degree of the fovea. Because of this, it has been suggested that it is not possible to discriminate the exact target of visual attention from person's gaze direction with accuracy better than one degree (Ware & Mikaelian, 1987; Zhai, Morimoto & Ihde, 1999).

Whether to aim for user interfaces that utilise voluntary eye movements rather than involuntary eye movements has been an issue of some debate. Both types of eye-movements have been utilised in the development of alternative HCI techniques. For example, iDict is a tool that utilises involuntary eye movements to help people to read documents written in a foreign language. Users' eye movements are monitored during reading and if the normal reading process gets stuck in some part of the text, iDict automatically offers them translation assistance (Hyrskykari, Majaranta, Aaltonen & Räihä, 2000). Other examples of the use of involuntary eye movements have been proposed for reading remediation (Sibert, Gokturk & Lavine, 2000), user interfaces for virtual reality (Tanriverdi & Jacob, 2000), story-telling systems (Starker & Bolt, 1990), and to control the level of detail in rendering computer graphics (O'Sullivan, Dingliana, Bradshawe & McNamara, 2001). However, in practice it is difficult to program the system so that it really interprets people's intentions or attention correctly from eye movements. If the system makes false interpretations or if the system's actions are inconsistent with users' intentions, the users will most likely find the system disturbing. For these reasons, the use of voluntary eye movements can be considered a more promising alternative interaction technique for HCI.

Eye Tracking Technology

When using eye movements in HCI it is essential to be able to measure the eye movements as reliably and comfortably as possible. Most current eye-trackers use pupil-centre corneal-reflection method to determine the user's gaze direction from a video image. Both remote and head-mounted eye-trackers exist. A video camera located near the computer screen or attached to a headband monitors one or both eyes. An infrared light emitter located near the camera lens illuminates the eye generating a corneal reflection. Image-processing software then identifies and locates both the centre of the pupil and the centre of the corneal reflection. Gaze direction is calculated on the basis of the position of the pupil centre and the corneal reflection within the video image. The tracking accuracy of modern video image tracking systems, at its best, lies between 0.5 and 1.0 degrees. When a tracker without head motion compensation is used, the user has to keep her or his head reasonably motionless. Otherwise, the system may fail to track the eye movements properly. Systems with head tracking and motion compensation allow the user to move more freely, but they tend to feel uncomfortable because they usually need to be attached to the user in order to enable accurate tracking. Headband systems also tend to suffer from the problem that they cannot be attached to the skull tightly enough. So, even after a short time period, the headband starts to move away from its initial position, thus creating increasing inaccuracy in the tracking.

Calibration and Drift Correction

Before the eye tracking system can track users' eye movements properly, it needs to be calibrated for each user and situation individually. This means that the system is adjusted to the user's eye attributes so that it can calculate gaze vectors and on-screen intersections according to eye movements. Typically during the calibration the user is required to look at fixed points on different parts of the screen and confirm that she or he is looking at a certain point by pressing a hardware button or by fixating the point long enough.

After a successful calibration, accuracy problems usually start to appear because the discrepancy between the user's actual gaze direction and the direction measured by the computer starts to increase soon after calibration. This error, known as drift error, can be larger in some parts of the screen than in other parts of the screen (Jacob, 1991). Drift error may result from, for example, head movements and changes in the position of eyeglasses and contact lenses.

There are alternative methods to correct the drift error. It is possible to manually recalibrate the eye tracking system whenever necessary. There are some problems associated with this method. Some systems cannot be recalibrated without the help of another person. It is also time consuming and inconvenient for the user to recalibrate the system during the tasks. More advanced methods for recalibration have been developed. For example, in dynamic recalibration the system knows when the user is looking at a certain object, and it can use this object as a calibration point to automatically correct the drift (Stampe & Reingold, 1995).

Feedback

Typically computer users get visual feedback while using the mouse. In eye movement based techniques there are alternative methods that can be used to give feedback from eye movements. One possibility is to use a gaze cursor, which indicates the position of the gaze on a computer screen. It is not completely clear how the gaze cursor affects the performance of the user. It may help the user to adjust small calibration errors. However, the gaze cursor may also capture the user's attention and distract them seriously from the pointing operation. Furthermore, the cursor can be quite distracting when it trembles following unintentional eye movements. These small jittery eye movements and random inaccuracies in eye-trackers cause the cursor to be in constant trembling motion even when the users think that their eyes are stable.

Smoothing the computed position of the cursor can reduce the trembling of the cursor. A common method is to calculate the average position of the cursor with several measured data points. Averaging over a number of points makes the cursor act more stably and smoothly. If too many data points are used when calculating the average cursor position however, the cursor may lag behind the user's actual gaze location. Therefore, various algorithms exist to calculate the position of the cursor. More weight can be given to some data points than others. For example, the velocity of the eye movements can be used to alter the number of data points used in the averaging. When the eye velocity is high, fewer data points can be used, and when the velocity is low, more data points can be used for smoothing. These kinds of methods decrease the cursor trembling during fixations and allow the cursor to rapidly follow the saccadic eye movements.

Another method of visual feedback is to change the visual appearance of the selected object. This can be done, for example, by changing the colour, shape, and size of the selected object. It is possible to use also other modalities in giving feedback. A beep sound, for example, is a commonly used alternative to visual feedback and it may be convenient for certain tasks and purposes.

It is clear that the form of feedback can significantly affect all the usability aspects (i.e. effectiveness, efficiency, and satisfaction) of HCI techniques that utilise eye movements. It is equally clear that more research is needed in finding out the best forms of feedback for eye movements in HCI.

Testing New Interaction Techniques

When studying the usefulness and the effectiveness of gaze-based interaction techniques, it is important to empirically test different features and attributes of the technique used. Frøkjær, Hertzum and Hornbaek (2000) have suggested that usability comprises the aspects of effectiveness, efficiency and satisfaction. All these aspects should be considered independently of each other. For example, an interaction technique that is enjoyable may be inefficient. New interaction techniques can be compared and evaluated against existing interaction techniques. In most studies the use of eye movements as an input device in HCI has been compared to more conventional interaction devices (e.g. the mouse). This may not always be the best way to analyse, for example, the speed of a new interaction technique. Because the mouse has been used already for quite a long time in contrast to any new technique, there always exists an unbalanced background of learning and experience with the use of the two techniques to be compared. Another method that has been used for evaluating different pointing devices is to investigate their performance using the index of performance (IP) metric based on Fitts' law. At present, Fitts' law has been widely used for testing the functionality of different pointing devices (Douglas, Kirkpatrick & MacKenzie, 1999; ISO, 2000).

Fitts' law is a model of human movement stating that targets that are smaller or farther away are more difficult to acquire and the movement takes more time. There are many variations of this model but the following formula is prevalently used in HCI research. The time to move to a target is linearly proportional to $\log_2(A/W + 1)$ where A is the distance moved and W is the width of the target (Fitts, 1954; Gillan, Holden, Adam, Rudisill & Magee, 1990; MacKenzie, 1992). This means that the smaller and/or the farther the pointing targets are the more time is required for pointing accurately at them, that is, the more difficult they are to point at. For this reason, the logarithm term given above has been used as an index of difficulty (ID) in quantifying the difficulty of a task. The average pointing times for different IDs can be measured and using the movement time (MT) formula a quantification of user performance with the tested pointing device can be acquired. In short the relationship between MT and a pointing task according to Fitts' law is:

$$MT = a + b \log_2(A/W + 1)$$

In this formula MT is the movement time, a and b are empirically determined constants and the logarithm term is the ID given earlier. Constant a is the intercept coefficient and b is the slope of the least-squares regression line. The reciprocal of b (i.e. $1/b$), also known as an index of performance (IP), is often used to compare two input devices. The device with higher IP is considered better because a high IP indicates that making the task more difficult has a smaller effect on MT than with a device with lower IP. Unfortunately, when different values of a are measured for two devices, IP does not allow fair comparison because a device with high a and high IP can be slower than a device with low a and low IP. To overcome this effect, the currently preferred practice to produce IP values is to first compute ID/MT for each different task and then using the average of these as IP (ISO 2000). IP values should be compared only with knowledge of the experimental and computational methods used for acquiring them (MacKenzie, 1992; Douglas *et al.*, 1999).

It is not totally clear whether all interaction techniques that utilize eye-movements follow Fitts' law. However, as presented in Table 22.1 almost every study that has tested this support the conclusion that Fitts' law applies to these techniques (e.g. Ware & Mikaelian, 1987; Miniotas, 2000; Surakka, Illi & Isokoski, submitted).

Eye Movements as a Unimodal Interaction Technique

It would be ideal if a computer could be used and controlled simply by looking at it. For this reason there have been attempts to build unimodal interaction techniques that are based solely on eye movements. When voluntary eye movements are used for pointing, the user points at the objects simply by looking at them. Eye movements, in terms of voluntarily directed gaze, have been found to be functional for pointing to objects. However, when voluntarily directed gaze has been used unimodally as the only method for computer input, there have been difficulties in finding a suitable method for selecting objects. In general, all these methods suffer from the problem that objects become selected every time the subject looks at them. This drawback has been called the Midas touch problem (Jacob, 1991). The Midas touch problem has been partially solved with the use of a dwell time protocol. The dwell time is a predetermined time that the subject's gaze has to stay on an object before the system issues a selection command (Jacob, 1991). The length of the optimal dwell time depends on the subject and the task. There is evidence that for some tasks the dwell times can be cut down to 150 ms or even to the level of single samples (Isokoski, 2000; Jacob, 1991). It has been noted that the more difficult the task the longer the dwell time should be (Stampe & Reingold, 1995). The use of graphical selection buttons on the screen has been another method to try to avoid the Midas touch problem. After the subject has pointed the object of interest it can be selected by looking at an on-screen selection button. As soon as the subject looks at the on-screen selection button the last object that has been looked at becomes selected.

The use of the dwell time protocol has been found to be problematic because of increased selection time and because the optimal dwell time is difficult to define. The use of graphical on-screen selection buttons is problematic because there cannot be any objects between the selection button and the object that the subject would like to select. Otherwise the object of interest may not be the last object looked at, when the subject looks at the selection button.

Sibert and Jacob (2000) studied recently the use of voluntarily directed gaze as a unimodal input method (i.e. pointing and selecting objects) and compared their technique to the use of a mouse (see also Sibert, Jacob & Templeman, 2001). The subjects' task was to select a highlighted circle from a grid of 12 circles as fast as possible. A dwell time of 150 ms was used for selecting the objects. When the circle was selected, it became de-highlighted and the next target became immediately highlighted. Only the target object was selectable. The results showed that the eye gaze interaction technique was significantly faster than the mouse.

In their second experiment the task was the same except that there was a letter inside the circles. The target letter was told to the subject through a speaker. When the circle with the correct letter inside it was selected, it became highlighted and the next letter was presented. Again, the eye-movement technique was significantly faster than the mouse. Their results supported their notion that the farther away you need to point, the greater the advantage of the eye is. Fitts' law applied to the use of the mouse, but it did not fit with their eye-movement technique data. Rather the data suggested a flat or very slowly increasing linear relationship between distance and pointing time such as described for saccades by Abrams *et al.* (1989).

Another study compared the use of the mouse and unimodal eye-movement techniques with a dwell time of 250 ms. First, the subject had to move the cursor onto a square home box. After 250 ms dwell time a beep sound indicated that task time measurement was initiated. Then the subject had to move the cursor onto a target ribbon as soon as possible. Four different target distances and two target ribbon widths were used. A beep sound indicated the completion of a trial after the subject's gaze had stayed on the target for 250 ms. The pointing task times were measured as the time between the two beeps minus the 250 ms dwell time. The results showed that the average pointing time for the mouse was 2.7 times faster than for the gaze based technique. Fitts' law applied for both techniques (Miniotas, 2000).

Eye movements as a Part of Multimodal Interaction Techniques

Because of the Midas touch problem, there have been attempts to develop interaction techniques that combine the use of eye tracking and some other input devices. The use of a hardware button has been a commonly used method for object selection in these techniques. In this way, the Midas touch problem and the use of the dwell time protocol are avoided. Interaction techniques that combine the use of several input modalities have been implemented with two different methods. First, eye movement input is an additional function and the subjects can choose it if they want to employ it for computer input. These interaction techniques allow the subjects to perform all actions unimodally either with eye movements, or with some other input device (e.g., mouse), or multimodally in conjunction with each other. Second, eye movement input has been used in conjunction with other input methods so that the subject has to use a combination of different input modalities in order to give all the necessary inputs for the computer. These input techniques are inherently multimodal (see Table 22.1).

Intelligent Gaze-added Operating System

Salvucci and Anderson (2000) have argued that using eye movements as an additional input modality in mouse-operated software may be better than building user interfaces that are based on gaze only. They developed an intelligent gaze-added interface that incorporated probabilistic context sensitive modelling of user actions. The main idea was to increase the area of the more probable targets at the expense of the not so probable targets. The model affected only object selection based on the eye tracker coordinates without changing the on-screen representation of the objects. These ideas were implemented in the software called Intelligent Gaze-added Operating system (IGO). IGO allows the user to perform simple file manipulation tasks like moving, copying, and renaming files in an environment that resembles the Mac-OS desktop. IGO can be used with a mouse or with a combination of gaze for pointing and keyboard for selecting.

A usability test aimed at testing user performance of gaze-added versus mouse-only use of IGO suggested that while gaze-added use of IGO was less accurate than the use of mouse only, it was convenient enough. When the users were given a free choice in using the mouse, gaze-added technique, or the combination of both for accomplishing the given tasks, they used gaze-added technique in 67% of the tested actions on average. The task execution times did not differ significantly, but object selection with gaze-added technique was more error prone than object selection with a mouse.

Although the relatively frequent use of the gaze-added technique may have been partly motivated by curiosity rather than real preference over the mouse, Salvucci and Anderson claimed that IGO was an example of a successful gaze-added user interface. Especially the probabilistic facilitation of object selection seemed a functional idea. Unfortunately, Salvucci and Anderson did not give statistical data that would have directly supported this conclusion. Although a probabilistic task context sensitive model like the one used in IGO undoubtedly is helpful, we do not know how big an improvement it is. The techniques used in IGO seemed to work well in the tested file management operations.

Manual and Gaze Input Cascaded Pointing Technique

Zhai *et al.* (1999) designed a technique in which pointing and selections of objects remained primarily under manual control, but were aided by the tracking of gaze direction. They designed two different Manual And Gaze Input Cascaded (MAGIC) pointing techniques, liberal and conservative. In both techniques, the gaze was used to dynamically redefine the "home" position of the pointing cursor. Because subjects usually look at the objects they aim to select, the cursor can be automatically moved close to the target object using the gaze direction information. After the cursor has been moved near the target, the subject has to make only a small movement towards the target with a regular manual input device.

In the liberal approach, the cursor is moved near every new object the subject looks at. Then the subject can control the cursor with the mouse or ignore it. The cursor is moved near a new object if the gaze moves far enough from the current cursor position (e.g. 120 pixels). When the subject is controlling the cursor with the mouse, automatic cursor moving is blocked. In the more conservative MAGIC pointing technique, the cursor is moved close to the target only after the manual input device has been actuated.

Zhai *et al.* (1999) tested three techniques: (1) Manual pointing with no gaze tracking, (2) Liberal MAGIC pointing, and (3) Conservative MAGIC pointing. The results showed, that the task times significantly varied with the used techniques. The liberal MAGIC pointing technique was the fastest and the conservative technique the slowest. The size of the targets significantly affected the pointing task times. Importantly, there were no more misses with MAGIC pointing techniques than with the manual pointing technique. The distance to the target affected the manual pointing times more than the MAGIC pointing task times. However, task completion times increased in both of the MAGIC pointing techniques as well as in the manual technique as the target distance increased. Zhai *et al.* did not report how Fitts' law applied to their results.

Biosignals and Eye Movements as a Multimodal Interaction Technique

As already seen, attempts to solve the Midas touch problem tend to produce multimodal solutions. Our own research in this area has led us to the use of biosignals in conjunction with eye movements. Human physiological signals have been used, for example, in psychophysiological research for quite a long time, but the idea of using human biological signals in HCI is more recent (e.g. Kübler, Kotchoubey, Hinterberger, Ghanayim, Perelmouter, Schauer, Fritsch, Taub & Birbaumer, 1999; Partala, Jokiniemi & Surakka, 2000; Partala & Surakka, in press). Recently, there have been some interesting and promising experimentations using the monitoring of voluntarily produced changes in the electrical activity of the human body (Doherty, Bloor & Cockton, 1999; Laakso, Juhola, Surakka, Aula & Partala, 2001). For example, the thought translation device for brain–computer communication uses slow cortical potentials in order to help disabled people to communicate with computers and other people (Kübler *et al.*, 1999).

We have recently studied voluntarily directed eye movements and voluntarily produced changes in facial muscle activity as a new multimodal HCI technique. There have been two central motivations for the development of this new technique. First, in the area of HCI it is important to model human communication in order to create interaction techniques that would be natural and versatile. New interaction techniques should be somehow based on the psychology of human communication. It is known that part of human communication relies heavily on nonverbal communication, especially on facial expressions (Ekman & Friesen, 1978; Surakka, 1996, 1998; Surakka & Hietanen, 1998). Because changes in electrical activity of facial muscles are activated during communication, can be activated at will, and can be registered and monitored in real time with electromyography (EMG), they offer tempting possibilities for HCI techniques.

The other motive for the development of our new technique was that, as already pointed out, the use of eye-movements as a unimodal user interface suffers from certain problems (e.g. the Midas Touch in selecting objects). Also multimodal techniques presented thus far require the use of, for example, hands for selecting objects. Thus, our motive has been to test the suitability of voluntary facial activity to act as a replacement for the mouse button press (i.e. object selection) in association with voluntary eye movements.

In our first study we compared the use of two interaction techniques (Partala, Aula & Surakka, 2001). The use of mouse was compared to the use of a new interaction technique. The new technique combined the use of voluntary gaze direction and voluntarily produced facial muscle activity. Voluntarily directed gaze was used for object pointing and voluntary activation of the *corrugator supercilii* facial muscle (the muscle activated in frowning) was used as a replacement for the mouse button press. This muscle site was chosen because it is located near the eyes, and was considered to be appropriate to be used in conjunction with the gaze direction.

In the new technique condition, an Applied Science Laboratories series 4000 eye-tracking system measured the gaze coordinates from the user's right eye at a sampling rate of 50 Hz. For a bipolar EMG recording In Vivo Metric Ag/AgCl (E220X) surface electrodes filled with electrode paste were used. After carefully cleaning the skin the electrodes were placed on the region of *corrucator supercilii* on the left side of the face (Figure 22.1) according to the guidelines given by Fridlund and Cacioppo (1986). The electrical activity of the user's *corrugator supercilii* was measured using a Grass® Instruments, Modell 15 Rxi™ polygraph. The sampling rate of the system was 1000 Hz. In this study the data from the eye tracker and EMG were combined and analysed offline.

Seven subjects were tested in a within-subject design with two pointing techniques and three pointing distances. The pointing tasks were similar to those of Douglas and Mithal (1994) with minor adjustments. In both conditions, the subject was presented simultaneously a home square and a target circle and they were to see and click them as fast as possible in respective order. There were three different target distances: 50, 100, and 150 pixels, measured from the centre of the square to the centre of the circle. The targets appeared in one of eight different angles (four orthogonal and four diagonal directions) around the home square. As there were three different distances, there were in sum 24 different target circle positions. Each position was used two times.

In the new technique condition, the time between the clicks was measured from the voluntarily produced first peak voltage change (negative or positive) to the second peak voltage change of the facial muscle (see Figure 22.2). In the mouse condition, the time between the first and the second click was measured.

The results from three pointing and selecting distances showed that the new technique was significantly faster to use than the mouse at medium and long distances. In the mouse condition the task times increased significantly as the target distance

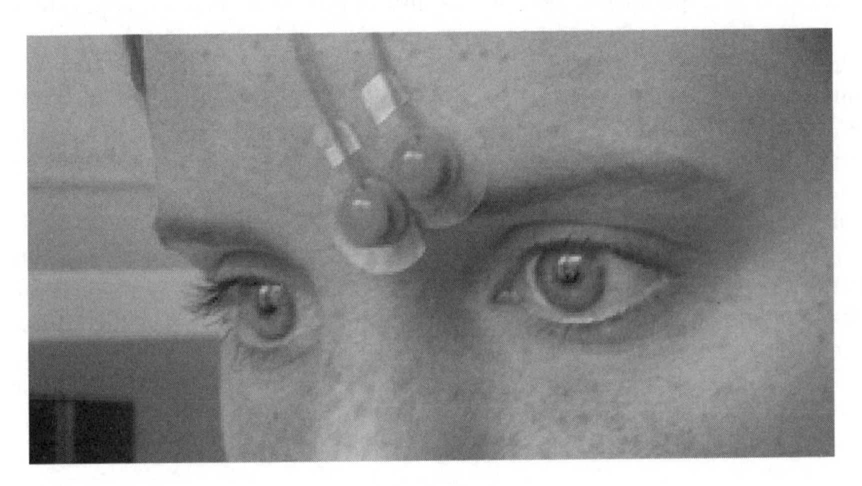

Figure 22.1: EMG electrodes were placed on the region of *corrugator supercilii,* on the left side of the face.

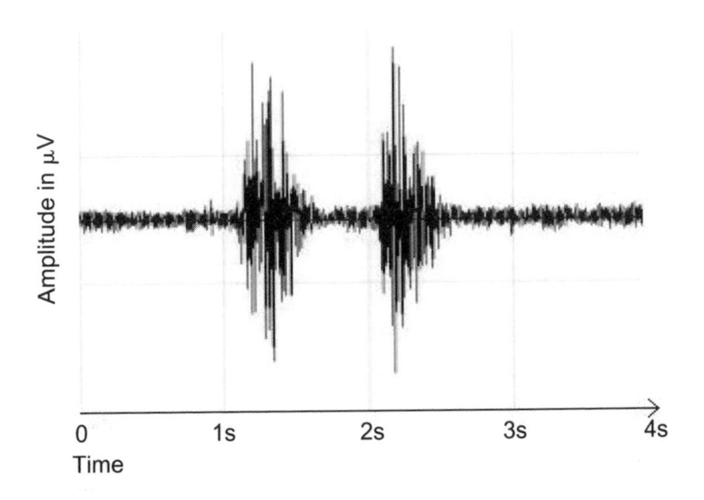

Figure 22.2: A sample from an EMG recording that displays two consecutive voluntary facial muscle activations during one task. Adapted from Partala *et al.* (2001).

increased. In the new technique condition the effect of distance was not significant and the task times were similar at all target distances (Partala *et al.*, 2001). The applicability of Fitts' law was not analysed.

Recently we have explored further the idea of combining the use of voluntarily directed gaze and voluntarily produced changes in facial muscle activity for real time object pointing in a graphical user interface (Surakka *et al.*, submitted). A computer program was written and used to combine both the eye tracker and EMG data in real time. Fourteen subjects performed pointing tasks with both the new technique and with the mouse. The experimental design was also improved so that the applicability of Fitts' law to the new technique and to the mouse could be investigated.

The experiment was a within-subject design with two pointing techniques, three pointing distances, and three target sizes. Again as in the first experiment two objects, a home square and a target circle were presented to the subjects simultaneously and subjects' task was to select the home square and then the target circle as fast and as accurately as possible.

Our results showed that the new technique worked well in practice. Even though the mouse was significantly faster to use than the new technique at short target distances, there were no statistically significant differences at medium and long distances. Pointing task times increased significantly for both techniques the farther and the smaller the targets were. This was evident from the Fitts' law regression slopes that were clearly ascending for both techniques and from the fact that the variation not explained by the regression equations was small (i.e. less than 3%). Thus, Fitts' law applied to both the mouse and the new technique (see Figure 22.3).

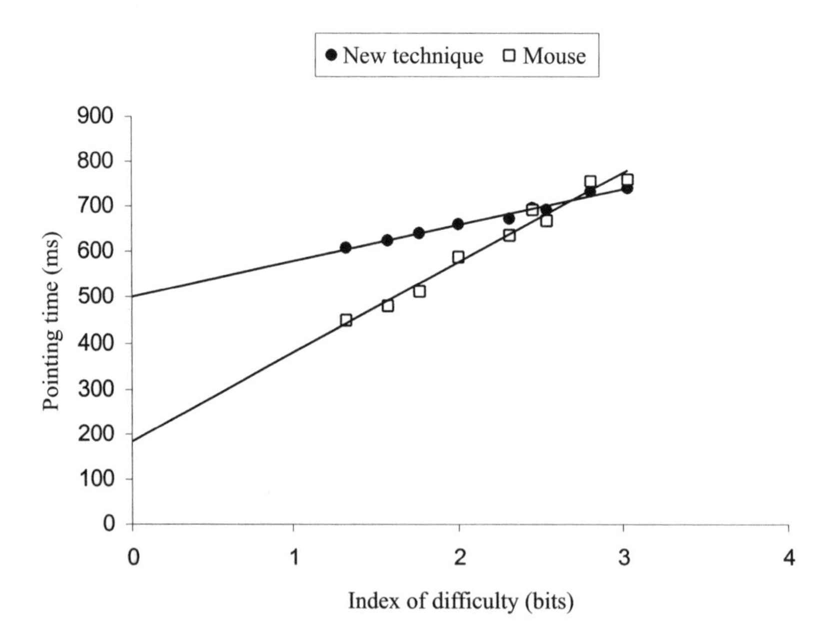

Figure 22.3: Fitts' law regression slopes for both the new technique and for the mouse.

Computed with the reciprocal of the slope of the regression line, the Fitts' law Index of Performance (IP) was 2.5 times higher for the new technique than for the mouse. This means that the new technique was potentially more efficient as compared to the mouse. Pointing error analysis showed that more pointing errors were made with the new technique than with the mouse in all target distances and sizes. However, in considering the error percentage even for the smallest targets that was 26.3, it is still not very high when considering that the subjects were instructed to perform as fast and as accurately as possible. This instruction may have resulted in a trade-off between speed and accuracy too much in favour of speed.

Comparison of the Selection Methods

Studies that have compared the use of eye movements as a unimodal input method to the use of eye-movements in the context of multimodal input techniques are rare. However, Ware and Mikaelian (1987) compared (experiment 1) three different methods to point at and select the objects in a graphical user interface. The subjects were told to point at the objects by looking at them. The object selection was to be made either with a button press, dwell time, or on-screen selection button. The task was to select a highlighted square from a menu of seven squares. Two forms of feedback were given to the subject. The first was a gaze cursor that tracked the subject's current fixation position. The second was a square, which appeared in the upper right hand corner of the last fixated menu item. The results showed that the use of the hardware button press and the use of dwell time (400 ms) were equally fast. Both were faster than the on-screen selection button. More errors were made with the on-screen selection button (22%) than with the hardware button (12%) and the dwell time (8.5%). However, the differences between methods were not statistically significant. Fitts' law was argued to apply to these techniques.

In their second experiment the subject's task was to select the highlighted square from a 4 by 4 square matrix. The results showed that the hardware button press was faster than the dwell time (400 ms) for all target sizes. More errors were made in using the hardware button than in using the dwell time. Selection task times decreased as the targets became larger. The task times, however, decreased only slightly after the target size was above 24 mm. Error percentages were below 10% in all conditions in which the target size was larger than one degree of visual angle. Fitts' law computations were not performed for the data from experiment 2.

Discussion

Eye movements have been utilized in unimodal and multimodal interaction techniques. In unimodal interaction techniques the Midas touch problem is the most severe problem. This has been partially solved with the dwell time method and on-screen button. The results have shown that the dwell time is faster and more accurate than the on-screen button (Ware & Mikaelian, 1987). Because the dwell time protocol and the on-screen button press evidently increase object selection times, different methods for object selection have been developed. Hardware button press has been a widely used method for object selection. When the dwell time protocol and the hardware button press have been compared, the hardware button press has been faster for object selection than the dwell time (Ware & Mikaelian, 1987). This finding suggests that at least in some cases multimodal systems are better than unimodal solutions.

As presented in this chapter several alternative multimode input methods exist but they require the use of hands as an additional input. Future HCI would benefit from using different input and output channels in a more imaginative manner. Our own research is one way of trying to follow that direction. Our technique utilizes signals that are active in human–human communication. These signals are used frequently and in most cases fully automatically in human interaction. In this sense this new technique can be argued to approach more natural human communication. The Midas touch problem is avoided and the subject's hands stay free for other purposes.

The objective evaluation of different techniques is somewhat problematic. In many studies, only one usability aspect has been thoroughly evaluated. Also, the lack of appropriate statistical testing in some studies makes it difficult to evaluate and compare the different user interfaces. The use of Fitts' law is one way to test the efficiency of various techniques. Although the suitability of the use of Fitts' law in some situations is under debate, ideally the IP is independent of the technique used. Furthermore, if the testing procedure described in ISO-9241–9 (Annex B) is followed, inter-study comparisons of the results are more valid than without standard testing procedures. Even some of the most recent techniques presented in this chapter, as shown in Table 22.1, did not test the applicability of Fitts' law. However, those studies (in Table 22.1) that did include Fitts' law analysis showed with only one exception that Fitts' law applies to these techniques.

In summary, previous research has demonstrated that at least the use of voluntary eye movements offer a promising interaction technique for HCI. The results from different studies show that people have voluntary control over their eye movements in the context of HCI. The results also show that people can direct their gaze with reasonable accuracy to objects of interest for controlling computers. Further, voluntarily directed gaze provides a fast and convenient pointing method. At least in some studies, eye movement interaction techniques have been equally fast or even faster than those involving more traditional interaction devices. The previous research has also suggested that the use of eye movements can be integrated with the use of other interaction modalities, thus, making it usable also for multimode applications.

Acknowledgments

This study was supported by the Finnish Academy (project number 167491), Tampere Graduate School in Information Science and Engineering, and the University of Tampere.

References

Abrams, R. A., Meyer, D. E., & Kornblum, S. (1989) Speed and accuracy of saccadic eye movements: Characteristics of impulse variability in the oculomotor system. *Journal of Experimental Psychology: Human Perception and Performance*, *15*, 529–543.

Cassell, J. (2000) Nudge nudge wink wink: Elements of face-to-face conversation for embodied conversational agents. In: J. Cassell, J. Sullivan, S. Prevost and E. Churchill (eds), *Embodied Conversational Agents*. Cambridge: The MIT Press.

Doherty, E., Bloor, C., & Cockton, G. (1999). The "cyberlink" brain-body interface as an assistive technology for persons with traumatic brain injury: Longitudinal results from a group of case studies. *Cyberpsychology & Behavior*, *2*, 249–259.

Douglas, S. A. Kirkpatrick A. E., & MacKenzie I. S. (1999). Testing pointing device performance and user assessment with the ISO9241 part 9 standard. In: *Proceedings of CHI'99 Human Factors in Computing Systems* (pp. 215–222). ACM Press.

Douglas, S. A., & Mithal, A. K. (1994). The effect of reducing homing time on the speed of a finger-controlled isometric pointing device. In: *Proceedings of CHI'94 Human Factors in Computing Systems* (pp. 411–416). ACM Press.

Ekman, P., & Friesen, W. V. (1978). *Facial Action Coding System (FACS): A Technique for the Measurement of Facial Action*. Palo Alto. CA: Consulting Psychologists Press.

Fitts, P. M. (1954). The information capacity of the human motor system in controlling the amplitude of movement. *Journal of Experimental Psychology*, *47*, 381–391.

Fridlund, A. J., & Cacioppo, J. T. (1986). Guidelines for human electromyographic research. *Psychophysiology*, *23*, 567–589.

Frøkjær, E., Hertzum, M., & Hornbæk, K. (2000). Measuring usability: Are effectiveness, efficiency and satisfaction really correlated? In: *Proceedings of CHI'2000 Human Factors in Computing Systems* (pp. 345–352). ACM Press.

Gillan, D. J., Holden, K., Adam, S., Rudisill, M., & Magee, L. (1990). How does Fitts' law fit pointing and dragging? In: *Proceedings of CHI'90 Human Factors in Computing Systems* (pp. 227–234). ACM Press.

Hietanen, J. K., Manninen, P., Sams, M., & Surakka, V. (2001) Does audiovisual speech perception use information about facial configuration. *European Journal of Cognitive Psychology*, *13*, 395–407.

Hietanen, J. K., Surakka, V., & Linnankoski, I. (1998). Facial electromyographic responses to vocal affect expressions. *Psychophysiology*, *35*, 530–536.

Hyrskykari, A., Majaranta, P., Aaltonen, A., & Räihä, K.-J. (2000). Design issues of iDict: A gaze assisted translation aid. In: *Proceedings of Eye Tracking Research and Applications Symposium 2000 (ETRA'00)* (pp. 9–14). ACM Press.

ISO (2000). *ISO-9241–9:2000 Ergonomic Requirements for Office Work with Visual Display Terminals (VDTs) – Part 9 Requirements for Non-Keyboard Input Devices*. International Standard. International Organization for Standardization.

Isokoski, P. (2000). Text input methods for eye trackers using off-screen targets. In: *Proceedings of Eye Tracking Research and Applications Symposium 2000 (ETRA'00)* (pp. 15–21). ACM Press.

Jacob, R. J. K. (1991). The use of eye movements in human computer interaction techniques: What you look at is what you get. *ACM Transactions on Information Systems*, *9*, 152–169.

Kübler, A., Kotchoubey, B., Hinterberger, T., Ghanayim, N., Perelmouter, J., Schauer, M., Fritsch, C., Taub, E., & Birbaumer, N. (1999). The thought translation device: A neurophysiological approach to communication in total motor paralysis. *Experimental Brain Research*, *124*, 223–232.

Laakso, J., Juhola, M. Surakka, V., Aula, A., & Partala, T. (2001). Neural network and wavelet recognition of facial electromyographic signals. In: *Proceedings of 10th World Congress on Health and Medical Informatics (medinfo2001)* (pp. 489–492). Amsterdam, IOS Press.

MacKenzie, I. S. (1992). Movement time prediction in human–computer interfaces. In: *Proceedings of Graphics Interface '92* (pp. 140–150). Canadian Information Processing Society.

Majaranta, P., & Räihä, K.-J. (2002). Twenty years of eye typing: Systems and design issues. In: *Proceeding of Eye Tracking Research and Applications Symposium 2002 (ETRA'02)* (pp. 15–22). ACM Press.

Miniotas, D. (2000). Application of Fitts' law to eye gaze interaction. In: *Extended Abstracts of CHI'2000 Human Factors in Computing Systems* (pp. 339–340). ACM Press.

O'Sullivan, C., Dingliana, J., Bradshaw, G., McNamara, A. (2001). Eye-tracking for interactive computer graphics. In: *Proceedings of the 11th European Conference on Eye Movements (ECEM11).*

Partala, T., Aula, A., & Surakka, V. (2001). Combined voluntary gaze direction and facial muscle activity as a new pointing technique. In: *Proceedings of INTERACT'2001* (pp. 100–107). IOS Press.

Partala, T., Jokiniemi, M., & Surakka, V. (2000) Pupillary responses to emotionally provocative stimuli. In: *Proceedings of Eye Tracking Research and Applications Symposium 2000 (ETRA'00)* (pp. 123–129). ACM Press.

Partala, T., & Surakka, V. (in press). Pupil size variation as an indication of affective processing. *International Journal of Human Computer Studies.*

Rayner, K. (1998). Eye movements in reading and information processing: 20 years of research, *Psychological Bulletin, 124*, 372–422.

Salvucci, D. D., & Anderson, J. R. (2000). Intelligent gaze-added interfaces. In: *Proceedings of CHI'2000 Human Factors in Computing Systems* (pp. 273–280). ACM Press.

Salvucci, D. D., & Goldberg, J. H. (2000). Identifying fixations and saccades in eye-tracking protocols. In: *Proceedings of the Eye Tracking Research and Applications Symposium 2000 (ETRA'00)* (pp. 71–78). ACM Press.

Sams, M., Manninen, P., Surakka, V., Helin, P., & Kättö, R. (1998). Effects of word meaning and sentence context on the integration of audiovisual speech. *Speech Communication, 26*, 75–87.

Sibert J. L., Gokturk M., & Lavine R. E. (2000). The reading assistant: Eye gaze triggered auditory prompting for reading remediation. In: *Proceedings of the 13th Annual ACM Symposium on User Interface Software and Technology* (pp. 101–107). ACM Press.

Sibert, L. E., & Jacob, R. J. K. (2000). Evaluation of eye gaze interaction. In: *Proceedings of CHI'2000 Human Factors in Computing Systems* (pp. 281–288). ACM Press.

Sibert L. E., Jacob R. J. K, & James N. Templeman (2001). Evaluation and analysis of eye gaze interaction. *NRL Report NRL/FR/5513–01–9990.* Washington, DC: Naval Research Laboratory.

Stampe, D. M., & Reingold, E. M. (1995). Selection by looking: A novel computer interface and its application to psychological research. In: J. M. Findlay, R. Walker and R. W. Kentridge (eds), *Eye Movement Research* (pp. 467–478). Amsterdam: Elsevier Science.

Starker, I., & Bolt, R. A. (1990). A gaze-responsive self-disclosing display. In: *Proceedings of CHI'90 Human Factors in Computing Systems* (pp. 9–3). ACM Press.

Surakka, V. (1996) Kasvonilmeet ja emootioiden tutkimus (Engl. Facial expression and the study of human emotions). *Psykologia, 31*, 412–420.

Surakka, V. (1998) *Contagion and Modulation of Human Emotions*. Doctoral dissertation. Acta Universitatis Tamperensis, 627. Vammala: Vammalan Kirjapaino OY.

Surakka, V., & Hietanen, J. K. (1998). Facial and emotional reactions to duchenne and non-duchenne smiles. *International Journal of Psychophysiology*, *29*, 23–33.

Surakka, V., Illi, M., & Isokoski, P. (submitted). Voluntarily directed gaze and voluntarily produced changes in facial muscle activity as a new multimodal HCI technique. *Journal of Applied Psychology*.

Tanriverdi, V., & Jacob, R. J. K. (2000). Interacting with eye movements in virtual environments. In: *Proceedings of CHI'2000 Human Factors in Computing Systems* (pp. 265–272). ACM Press.

Ware, C., & Mikaelian, H. H. (1987). An evaluation of an eye tracker as a device for computer input. In: *Proceedings of CHI+GI'87 Human Factors in Computing Systems* (pp. 183–188). ACM Press.

Zhai, S., Morimoto, C., & Ihde, S. (1999). Manual and gaze input cascaded (MAGIC) pointing. In: *Proceedings of CHI'99 Human Factors in Computing Systems* (pp. 246–253). ACM Press.

Chapter 23

Eye Tracking in Usability Evaluation: A Practitioner's Guide

Joseph H. Goldberg and Anna M. Wichansky

This chapter provides a practical guide for either the software usability engineer who is considering the benefits of eye tracking, or for the eye tracking specialist who is considering software usability evaluation as an application. The basics of industrial software usability evaluation are summarized, followed by a presentation of prior usability results and recommendations that have been derived from eye tracking. A detailed discussion of the methodology of eye tracking is then provided, focusing on practical issues of interest to the usability engineer. Finally, a call for research is provided, to help stimulate the growth of eye tracking for usability evaluation.

Usability Evaluation

Usability evaluation is defined rather loosely by industry as any of several applied techniques where users interact with a product, system, or service, and some behavioral data are collected (Dumas & Redish, 1993; Wichansky, 2000). Usability goals are often stipulated as criteria, and an attempt is made to use test participants similar to the target market users (Rubin, 1994). It is conducted before, during and after development and sales of products and services to customers. This is a common practice in computer hardware, software, medical devices, consumer products, entertainment, and on-line services such as electronic commerce or sales support.

The present chapter covers basic methodological issues first in usability evaluation, then in the eye tracking realm. An integrated knowledge of both of these areas is beneficial for the experimenter who is conducting eye tracking as part of a usability evaluation. Within each of these areas, major issues are presented by a rhetorical questioning style. By initially presenting the current section on usability evaluation, it is hoped that the practical use of an eye tracking methodology will be placed into a proper and realistic perspective.

The Mind's Eye: Cognitive and Applied Aspects of Eye Movement Research

ISBN: 0–444–51020–6

What Techniques Are Used for Usability Evaluation?

There is wide variation among academic and industry implementations of usability evaluation techniques. Some techniques are very informal, providing little more than nominal or qualitative data, and others are methodologically very rigorous, involving hundreds of participants and highly controlled administration protocols. The most frequently used techniques are various forms of usability testing, which have in common several key characteristics:

- Users are selected from target market groups or customer organizations.
- Users interact systematically with the product or service.
- They use the product under controlled conditions.
- They perform a task to achieve a goal.
- There is an applied scenario.
- Quantitative behavioral data are collected.

Other techniques such as cognitive walkthroughs or expert evaluations may partially fulfill the above requirements. Often these techniques lack the participation of "real users", they are not controlled, they may not provide a task, and there may be no systematic collection of quantitative data. The results of these *formative usability tests* may not answer the question, "Is this product usable?," but can be of value in making design decisions about a product under development.

The Industry Usability Reporting Project, an industry consortium led by the US National Institute of Standards & Technology (NIST), has recently developed a standard for reporting usability testing data (Scholtz & Morse, 2002). This standard is particularly useful when customers need to make procurement decisions about a supplier's software. This Common Industry Format (CIF) is available as ANSI/INCITS 354–2001. The requirements necessary for CIF compliance were based upon the human factors literature, existing ISO standards on usability (e.g. ISO 9241–11 and ISO 13407), and industry best practice. They include requirements for goal-oriented tasks, minimum of eight participants per user type, and quantitative metrics representative of efficiency, effectiveness and satisfaction. *Summative testing* techniques, such as those prescribed by NIST, provide a better answer than the formative techniques to the question of whether a product is usable.

What Software Is Tested?

In large software companies, there can be so much software developed and manufactured, that insufficient resources are available to test all of it for usability. Typically, new products, those undergoing drastic changes, and those for which customers have provided negative feedback are prioritized higher for usability evaluation. Successive versions of the same product may contain new features to be tested in the context of the original product in development. Where the user interfaces are the same, there is typically less emphasis on testing. In the realm of Web software, it is difficult to conduct controlled tests because websites change frequently, not only in content, but in basic architecture. Therefore, any test is really a snapshot in time of the product's usability. However, websites have the advantage that once a problem is recognized, it can be corrected quickly.

When Is Software Tested for Usability?

Software should be tested early and often throughout the product development cycle. In this way, problems with the product's conceptual model can be identified early, before too much coding has been completed and it is too late to change the product. This is done by developing prototypes that represent the product's functionality and interface design early in the product development cycle. These prototypes may range in fidelity from low (paper drawings) to medium (partially functional screen representations with little or no behavior) to high (interactive screen renditions of the product's front-end, with partial or simulated functionality). As the product is developed, Alpha code, representing feature functionality, and Beta code, representing complete product functionality, can be tested. At or near product release, summative testing as per the NIST CIF can be conducted. At every successive stage, users can be presented with more challenging tasks, and more definitive feedback and recommendations can be provided to the developers for changes to make the product more usable.

Who Conducts Usability Tests?

Usability practitioners come from a wide variety of backgrounds. Typically they have some expertise or education in experimental psychology and statistics. They often have degrees in industrial engineering, psychology, ergonomics, or human factors engineering, and a familiarity with topics such as laboratory and field data collection, and the overall scientific method. Many companies employ user interface designers and usability engineers with some or all of the above background. In other organizations, individuals with little or no formal training conduct usability tests. These team members may have developed a strong interest in usability through experiences in product management, technical writing, customer support, or other related roles.

What Metrics Are Associated with Usability Testing?

Most industry tests include some subset of behavioral metrics that can be identified with efficiency, effectiveness and satisfaction. Complete descriptions of these aspects of usability and related metrics are available in Macleod, Bowden, Bevan and Corson (1997). *Efficiency* is similar to the concept of productivity, as a function of work accomplished per unit time. Common measures are time on task (often weighted for errors or assists from the test administrator) and mean completion rate/mean task time, known as the efficiency ratio. *Effectiveness* is how well the user is able to perform the task, in terms of the extent to which the goal is accomplished. A common measure is percent task completion, again weighted for errors, assists, or references to support hotlines or documentation. *Satisfaction* is determined from subjective measures that are administered at task completion, or at the end of the usability test. These include rating scales addressing concepts such as usefulness, usability, comprehensibility, and aesthetics. Psychometrically developed batteries of subjective questions are often used, such as the Software Usability Measurement Inventory (Kirakowski, 1996) and the Questionnaire on User Interface Satisfaction (Chin, Diehl & Norman, 1988).

What Is a Typical Test Protocol?

Typically test participants are brought individually into a usability testing lab, which consists of a control room with a one-way mirror, and a user room that is often designed as a simulated usage environment appropriate to the product (Figure 23.1).

Figure 23.1: Tester instructs participant in a usability lab at Oracle.

After receiving material on the rights of human test participants (Sales & Folkman, 2000) and signing consent forms, the user is presented with a set of tasks, which may be read aloud or silently. The tester sits behind the one-way mirror, unseen by the user, and may call out instructions to begin or end tasks as appropriate. A think-aloud technique, in which the user describes out loud his mental process as he performs the steps of the task, is often used (Dumas, 2002). Think-aloud protocols are particularly useful in the formative stages of product development, when there is a lot of latitude for making changes. In summative testing, the user may perform the task silently, and merely indicate when he is done. All of this is usually captured on videotape, for possible later review. The tester typically logs various aspects of the user's behavior, including task start and end times, times to reach critical points in the task, errors, comments, and requests for assistance. Questionnaires are usually administered at the end of the test, and the user may be interviewed about his experience.

How Is the Data from a Usability Test Exploited?

The data from formative tests are used to determine where the product's usability is satisfactory and where it needs improvement. If certain tasks are performed particularly slowly, generate many errors and requests for assistance, or produce negative comments, the user interfaces supporting those features and functions are analyzed to determine the cause of problems. Then they are corrected, usually in conjunction with the user interface designers and developers.

The data from summative tests are used to assess overall usability with reference to goals that have been established by the development team. For example, a particular task may need to be performed by the average user in a set period of time with no errors, or with no outside assistance. The data also make it possible to compare the current product to previous versions, or to other products.

What Aspects of Usability Cannot be Addressed by Routine Usability Testing?

Current usability techniques are derived from industrial engineering, which concerns itself with the measurement of work in the workplace, and the behaviorist school of psychology, which emphasizes operational definitions of user performance rather than what may be inferred about the cognitive process. Thus, most observable, behavioral aspects of work and task performance (e.g. screen navigation, menu selection) can be captured. Cognitive processes are more difficult to infer.

User intent is difficult to assess using current techniques. If users spend too long looking at a specific window or application page, for example, traditional usability analysis does not provide sufficient information for whether a specific screen element was found, or whether its meaning was unclear (Karn, Ellis & Juliano, 2000). Micro-level behaviors, such as the focus of attention during a task, distractions, or the visibility of an icon, usually have little awareness to an individual, and thus are not reported in the think-aloud protocol. Reading, mental computations, problem solving, and thinking about the content of an application are also difficult to quantify using this protocol.

There are techniques aimed at assessing mental models and user intentions, such as cognitive walkthroughs (e.g., Polson, Lewis, Rieman & Olson, 1992) and protocol analyses (e.g., Card, Pirolli, Van Der Wege, Morrison, Reeder, Schraedley & Boshart, 2001). These dwell upon thought processes, but may be very time and labor-intensive and are often practical only in research settings.

Usability engineers would like better tools to understand when users are reading versus searching on a display, and to determine how learning a screen's layout impacts its usability. Current evaluation approaches can capture keystrokes and cursor movements, allowing inferences to be made about complex cognitive processes such as reading, but cursor motion does not necessarily track where the user is looking (e.g., Byrne, Anderson, Douglass & Matessa, 1999). Thus, better tools are required to assess temporal and sequential order of viewing visual screen targets, and to assess when users are reading as opposed to searching on a display. Eye tracking can provide at least some of this data.

What Types of Design Recommendations Have Been Made from Eye Tracking Results?

User interface architecture and screen design The outcome of using eye tracking as a secondary technique in usability studies should be improvements in the breadth and specificity of design recommendations. Most studies have collected data relevant to navigation, search, and other interaction with on-line applications. These findings influence overall user interface architecture of software applications, design and content of screens and menus, location and type of visual elements, and choice of user interface visualization style.

Recommendations for icon size for normal and vision-limited users were provided by Jacko, Barreto, Chu, Bautsch, Marmet, Scott and Rosa (2000), by requiring users to search for icons of varying sizes. Based upon fixation and scanpath time durations, they recommended minimum icon sizes that varied by background brightness on displays.

Menus contain both reading and visual search subtasks, and provide a well-learned interaction model for eye tracking purposes. Crosby and Peterson (1991) investigated scanning style as users searched multi-column lists for specific target items. Four styles were uncovered from eye tracking: Comparative (scanning between columns and/or rows to compare specific items); Down and Up (starting from the top of a column and continuing from its bottom to the bottom of its neighboring column, then back up to its top); Scan from Top (always scanning downward from the top of each column); and Exhaustive (scanning all areas and columns, even after finding the target item). These strategies were related to an independently measured index of cognitive style.

Because menu search can vary from highly directed to an informal browsing activity, Aaltonen, Hyrskykari and Raiha (1998) manipulated the directness of the search target through instructions. From eye tracking results, either top-down scanning of each item or a combination of top-down and bottom-up scanning strategies were evident. Analysis was not conducted, however, to predict the stimuli that drive these strategies. In similar work, Byrne, Anderson, Douglass and Matessa (1999) had users search for target characters in pulldown menus while menu length, menu target location, and target type were manipulated. Menus of six items produced a similar number of fixations, regardless of item location within the menu. Initial fixations were made to one of the first three menu items, with subsequent search occurring in serial order.

Pirolli, Card and Van der Wege (2001) used eye tracking as an aid to interpret users' behaviors in navigating hyperbolic and other network browsers. Based upon the number and duration of fixations, they recommended a hyperbolic browser in directed search tasks, as visual link traversal rate was nearly twice as fast as a traditional tree browser.

Redline and Lankford (2001) provided a very detailed analysis of users' behaviors while using an on-line survey. They studied scanpaths from eye tracking data in order to categorize users' omission or commission errors. Example behaviors included failure to observe branching instructions, premature branching from one question to another, or non-systematic fixations leading to a non-answer. Overall, the authors noted that the order in which information was read influenced whether the survey was understood as intended.

Goldberg, Stimson, Lewenstein, Scott and Wichansky (2002) evaluated a portal-style website with eye tracking techniques, allowing users to freely navigate within and among web pages. They recommended that portlets requiring the most visibility be placed in the left column of the portal page; "Customize" and "More" links found in many portlet headers may benefit from being added to the bottom of the portlet body. In a much earlier evaluation of a Prodigy web browser, Benel, Ottens, and Horst (1991) found that users spent much more time than designers anticipated looking at special display regions.

Reading Eye movements during reading are different from those exhibited during navigation and search of an interface. The eyes follow a typical scan path across the text, in the appropriate direction depending upon the language of the text (e.g. left to right, top to bottom for Western cultures). This is a highly learned behavior. When there are difficulties with legibility or comprehensibility of the text, the eyes may fixate on specific words or samples of text, or even reverse direction to reread text, slowing down reading progress and disturbing the normal pattern.

Reading research has presented a rich eye tracking environment for many years. As a task, reading is much more constrained and context-sensitive than spatial/graphical visual search. Reading pattern differences are clear between novices and experts, and complexity of written material can be discerned from eye tracking data. Rayner (1998) provided several facts about eye movements in reading. Reading consists of a series of short (saccadic) eye movements, each spanning about 7–9 letter spaces for average readers. Regressions, marked by right-to-left eye movements, occur in about 10–15% of saccades. Skilled readers quickly return to the area of the text that was poorly understood, whereas weaker readers engage in more backtracking through the text. A very distinctive return sweep, which is distinguishable from a regression, is made as one reads from one line to the next. Textual information in the periphery of the eye is also considered to be important for gaining context and preparing the next several saccades. Text that is perceived as difficult by a reader results in longer fixation durations and saccade lengths, and increasing frequency of regressions. Furthermore, fixations on words that are semantically related or close to one another may be shorter or longer than fixations on words that are unrelated, possibly indicating a cognitive "priming" effect.

Eye tracking studies for reading tasks have provided a great deal of information concerning the lag between text perception and cognitive processing (see Inhoff & Radach, 1998). For the usability engineer, eye tracking can provide information about when a participant is searching for visual targets versus reading text. For example, mean fixation duration in visual search and scene perception is 275–330 msec, but these are only 225–275 msec in reading (Rayner, 1998). Other differences exist in mean saccadic amplitude and velocity. However, there are strong individual differences in reading, so studies should preferably present within-participant manipulations.

Studies involving reading of on-line documents coupled with eye-tracking methods are indeed quite relevant to usability evaluation. Zellweger, Regli, Mackinlay and Chang (2001) designed a variety of "fluid document" strategies to enable supplementary text (similar to footnotes) to be presented in on-line documents. They evaluated four hypertext designs using eye-tracking: fluid interline (lines part, hypertext appears in between lines); fluid margin (hypertext in the margin); fluid overlay (hypertext over text); and pop-up (hypertext in a separate box). From eye tracking data, participants often did not look at footnotes located at the bottom of the document. Application of these "glosses" did not cause wildly shifting point of regard, as anticipated by the authors. Eye tracking data was confirmed by subjective results and by amount of time glosses were opened: shorter for glosses farther from the anchor text and longer for glosses local to the anchor text.

In 1998, researchers at Stanford University Center for the Study of Language and Information began to collaborate with the Poynter Institute (2000), dedicated to journalism research, in a study of on-line newspaper reading behaviors. Eye-tracking measures were used in addition to videotaped observation and survey measures with 67 participants in two cities. Among the results, readers initially looked at text briefs and captions, not photographs or graphics, upon navigation to a news-oriented website. Eye tracking revealed that text is sought out and either skimmed or read. Banner ads were read 45% of the time and fixated an average of 1 sec, long enough to be clearly perceived. Readers scanned 75% of the length of most articles. The authors found the eye-tracking data to be an important addition to surveys and recall data.

Cognitive workload Although still under active investigation, eye tracking can possibly provide significant information about an observer's cognitive workload. Several studies have presented tasks in which participants must actively control and make time-limited decisions about a system. Various eye tracking-derived metrics are generated, and possibly correlated with subjective scales of perceived workload. As cognitive workload increases, the saccadic extent, or breadth of coverage on a display, decreases (e.g., May, Kennedy, Williams, Dunlap & Brannan, 1990), consistent with a narrower attentional focus. The blink rate of the eye also decreases (Brookings,

Wilson & Swain, 1996), emphasizing the need to gather as much visual information as possible while under high workload. Although influenced by many factors, some have found that the pupil dilates in response to high workload tasks (e.g., Hoeks & Levelt, 1993). In response to difficult cognitive processing requirements, fixation durations may also increase (Goldberg & Kotval, 1999).

The usability engineer is advised to use caution before interpreting eye tracking-derived measures as indicators of workload. Large individual differences and the presence of many contaminating factors could lead to false conclusions. Within-participant manipulations are advised here, noting relative differences between conditions.

How Do Eye Tracking Results Correlate with Other Measures of Usability?

There are few published studies where authors have tried to correlate eye tracking measures with performance times, completion rates, or other global measures of commercial software usability. Available studies of man–machine interaction incorporating eye-tracking tend to be used in settings such as radar and air traffic control (e.g. David, Mollard, Cabon & Farbos, 2000). A very promising study attempting to correlate paper map reading accuracy and time with eye movement characteristics has also been reported, in which traditonal think-aloud techniques were used (Brodersen, Andersen & Weber, 2001).

In general, industry experts in human–computer interaction use eye tracking data to support recommendations for how a user interface should be changed, rather than for global assessment of a product's usability. If participants are looking directly at an icon but not selecting it, this might indicate the icon is noticeable but not interpretable as relevant to a particular goal or function. Conversely, if users do not look at an item at all, it may be necessary to change its location on the screen to make it more noticeable.

In studies where relationships have been reported, eye tracking data typically support other measures of usability. For example, Kotval and Goldberg (1998) investigated the sensitivity of eye tracking parameters to more subjective usability evaluations by interface designers. They developed several versions of a simulated drawing package, by manipulating items' grouping and labeling on the application's tool bars. Screen snapshots were then sent to a broad set of interface designers in the software industry for subjective rating and ranking. Participants in the eye tracking laboratory completed several tasks using the software, and metrics from eye tracking were then correlated with the designers' subjective ratings.

Several of their eye tracking-derived parameters were indeed related to usability ratings when aggregated across tasks. Scanpath length, number of fixations, and number of saccades were generally the most sensitive measures of rated usability differences. These parameters increased from 46% to 67% from the most highly rated interfaces to the most poorly rated interfaces. Scanpath duration, a composite eye tracking-derived measure, is the sum of the duration of saccades and fixations within a scanpath (i.e., the total time spent within a scanpath). Goldberg and Kotval's (1998) scanpath duration also increased by up to 80% from the most highly rated to the most poorly rated interfaces. Fixation duration, saccadic amplitude, and fixation/saccade ratio were not as successful in predicting rated usability differences between the interface versions. While some of these eye tracking measures may indeed predict interface usability, it must be emphasized that this work had designers rate static hard copy images of interfaces, rather than conduct true usability evaluations.

Cowen (2001) extended Kotval and Goldberg's (1998) work to live web pages, using eye tracking as an aid to usability evaluation of two tasks on each of four web pages. The pages were English language versions of a mobile phone company's home page in four different countries. Page layouts were evaluated in several objective ways, including the percentage of total page area devoted to navigation, images, logos, and other constructs; visual clutter was also computed. Each task required a participant to locate and click a specific link. Task completion success and time were measured. Four eye tracking-derived measures were computed: total fixation duration, number of fixations, average fixation duration, and spatial density of fixations. While the eye movement-derived measures were somewhat sensitive to differences in the page layout variables, completion time and success predicted these differences better. Like Kotval and Goldberg's (1998) study, true usability evaluations were still not conducted. However, the ocular measures' sensitivity to page layout variables is a promising confirmation of the potential linkage with usability.

As discussed by Goldberg (2000), eye tracking can provide an effective indication of certain aspects of usability. Visual clarity, or the recognizability of desired targets presented on a (possibly cluttered) background, should be quite accessible via eye tracking. Greater visual clarity should result in shorter scanpaths, when searching for specified targets. Usability criteria that should be difficult to ascertain via eye tracking include interface compatibility and locus of control. Other usability criteria, such as quality of feedback and error handling may or may not be ascertainable by eye tracking, depending on the specific task and domain area. Further research should be conducted to elucidate general design rules for usability improvement, using eye tracking methods.

Eye Tracking Basics

The usability engineer who seeks to add eye tracking to usability evaluations will require some basic knowledge about eye movements and eye trackers. The following information should help, but this information is by no means exhaustive. For example, types of eye trackers other than those mentioned do exist, but are not common in the domain of user interface evaluation.

Which Types of Eye Movements Are Measured in Usability Evaluations?

Eye movements exist in humans to either maintain or shift an observer's gaze between visual areas, while keeping the projected visual image on a specialized, high acuity area of the retina. There are several types of eye movements, based upon their specific function and qualities. Young and Sheena (1975) and Carpenter (1988) provided introductory reviews.

Saccades are commonly observed when watching an observer's eyes while conducting search tasks. These jerky movements occur in both eyes at once, range from about 2–10 degrees of visual angle, and are completed in about 25–100 msec. They have rotational velocities of 500–900 degrees/second, and so have very high acceleration (Carpenter, 1988). Saccades are typically exhibited with a latency of about 150 to 250 msec following the onset of a visual target. Because of their rapid velocity, there is a suppression of most vision during a saccade to prevent blurring of the perceived visual scene.

Each saccade is followed by a fixation, where the eye has a 250–500 msec dwell time to process visual information. These saccade-fixation sequences form scanpaths, providing a rich set of data for tracking visual attention on a display (Noton & Stark, 1970). Applied eye tracking methods for usability evaluation generally capture saccadic eye movements and fixations while observers search displays.

Smooth Pursuit eye movements are much slower (1–30 degrees/second) movements used to track slowly and regularly moving visual targets. These cannot be voluntarily generated, and may be independently superimposed upon saccades to stabilize moving targets on the retina.

Compensatory eye movements stabilize the image on the retina when the head or trunk is actively or passively moved. These slow eye movements occur in the opposite rotational direction from the head or trunk movement. They do not usually occur in usability evaluation studies, because of little head or body movement.

Vergence eye movements are rotational movements of the two eyes in opposite directions, in response to focusing error from visual targets moving closer or further away. The direction of rotation depends on whether a visual target is moving closer or further away from the eyes. These slow, 10 degree/second eye movements are necessary to fuse the two images from the eyes. They are usually not observed in computer-oriented studies, as the viewing plane does not typically change.

Miniature eye movements refer to those eye movements that are under 1 degree of visual angle, and include small corrective motions, as well as tremor. These are usually not of interest in usability evaluations, as eye trackers typically used in the field of HCI have accuracy error of this magnitude (see Engbert & Kliegl, this volume, for a discussion of small eye movements).

Finally, *nystagmus* eye movements refer to a unique sawtooth-pattern of eye movements caused by a rapid back-and-forth movement of the eyes. These movements occur for several reasons, such as fatigue, hot or cold water in the ears, rapid rotation of head, or special optokinetic stimuli. They are not typically observed in usability evaluations.

How Is Point-of-regard Measured on a Display?

Eye tracking methods attempt to capture a user's focus of visual attention on a display or scene through special hardware/software. The eye tracking system gathers *x*/*y* location and pupil size information at typical rates from 30 Hz — 250 Hz, with 60/50 Hz typical for usability evaluation needs. Slower sampling rates do not provide sufficient resolution, especially for scrolling or other tasks involving moving visual targets.

Faster sampling rates can create a massive data reduction problem for typical usability evaluations. The *x/y* sample locations are then further processed into fixations, which may be assigned to experimenter-defined areas-of-interest on the viewed display or scene. Fixations and saccades can be combined into scanpaths, providing further information about one's cognitive processing during a task (Just & Carpenter, 1976b). Scanpaths can be quantitatively or subjectively analyzed to provide information about the extent or complexity of visual search on a display (Goldberg & Kotval, 1999, 1998).

While commercially developed eye tracking systems have used many different properties of the eye to infer an observer's point-of-regard, the following will focus on commonly used infrared corneal reflection-pupil center systems that dominate the applied eye tracking market. The University of Derby has maintained a comprehensive website for several years, listing manufacturers of eye tracking hardware and software (Eye Tracking Equipment Database, 2002).

Infrared-type, corneal reflection eye tracking systems, in common use for applied research, rely upon the location of observers' pupils, relative to a small reflected light glint on the surface of the cornea (Young & Sheena, 1975; Mulligan, 1997). A camera lens (the "eye" camera) is focused upon the observer's eye; a second lens (the "scene" camera) may also optionally be pointed towards the current visual display or scene being viewed. A scan converter is frequently used in place of the scene camera when tasks are conducted on a computer. The pupil is located by the tracking system (and possibly modeled as an ellipse) from its relative contrast with the surrounding iris. "Bright pupil" systems illuminate the pupil with infrared light, whereas "dark pupil" systems do not illuminate the pupil; some systems may be switched between these modes, to find the most robust pupil imaging for a testing environment. Within a usability laboratory, diffuse lighting will generally provide fewer problems than pinpoint lighting sources, that could cause additional light glints on the cornea. The ability to switch between bright and dark pupil methods can also be helpful in usability testing situations.

Following calibration to the display or scene of interest, the eye tracking system returns *x/y* coordinates of gaze samples, an indication of whether the eye was tracked, and pupil size. These data may be streamed and used in real-time, or may be saved to a file for later analysis. Note that modern eye tracking systems come with software/hardware that automatically filters out blinks and other anomalies. Embedded algorithms within commercial eye tracking software can detect and filter out blinks by searching for unique light-dark-light signatures of reflected light that are associated with blinking. Special blink detection hardware is available for blink rate and fatigue research (e.g., Stern, Boyer & Schroeder, 1994).

Should a Head-mounted or Remote System Be Used?

Eye tracking systems may be mounted to the head of an observer, or may be remote from the observer. Head mounted systems (e.g., Figure 23.2) are most useful for tasks in which a great deal of free head/trunk motion is expected. Eye tracking during sports, walking, or driving are common examples. While head mounted systems can continue

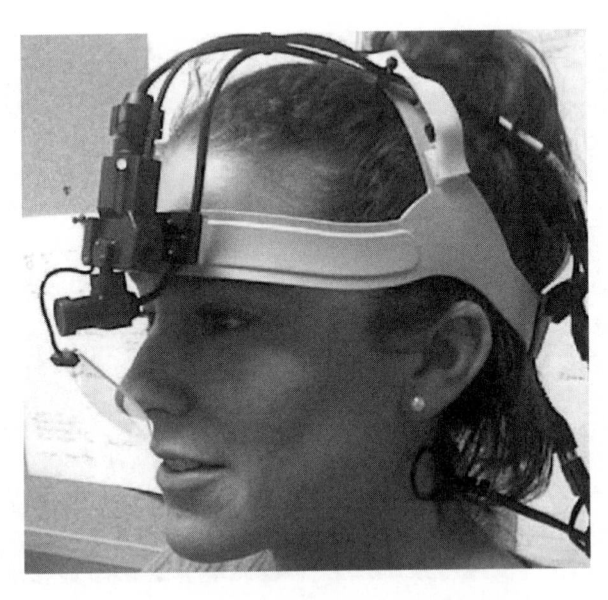

Figure 23.2: Head-mounted eye tracking system, including both scene (lower) and eye (upper) cameras.

to track the eyes even with significant head movement, they are somewhat instrusive, costly, and delicate. They typically obstruct a small portion of the observer's visual field, and the observer cannot easily forget that a system is recording his eye movements.

Remote systems, on the other hand, consist of a camera lens mounted under or beside the computer display being used by the participant in a usability evaluation (Figure 23.3). An infrared source is typically located in-line with the camera lens. The presence of the camera is quickly forgotten by the participant, making this a very non-intrusive procedure. In fact, special systems are available that can even hide the camera lens entirely from view. These systems can be a bit less expensive and are less delicate than the head mounted systems, but require the participant to maintain relatively stable head position. Chin rests could be required if head motion is not easily controlled. Eye cameras that have an autofocus capability allow some forward-backward movement of the participant's head, an additional advantage for usability testing.

Modern, remote eye tracking systems allow head motion within approximately one cubic foot, by coupling a motor/servo driven platform and magnetic head tracking with the eye camera, enabling robust location of the eye as the participant sits naturally in front of a computer. The addition of these head tracking technologies can make the cost of a remote system similar to that of a head mounted system. Costs over the past few years have ranged from about $20,000 – $50,000 for a complete eye tracking system; however, new systems in the sub-$20,000 range are now starting to appear.

Whether head-mounted or remote, the eye tracker generates data that is fed to the port of a host computer, possibly through a control box. This computer is usually dedicated to the eye tracker, containing a large amount of disk space for data storage.

Figure 23.3: Remote eye tracking system, with camera optics to the left of the computer display.

Many investigators use a two-computer setup, where software under evaluation is run on an independent, application computer. The experimenter can control the eye tracker data collection through the host computer. Optionally, the host computer can also be programmed to send signals to the eye tracker to start or stop tracking. These signals could also be generated from the application computer, in response to user-generated events such as mouse movements or button presses.

What Steps Are Required for Analyzing Eye Tracking Data?

Analysis of eye tracking data typically requires several steps, moving from raw data to fixations to computation of specific metrics. Reduction of eye tracking data into meaningful information can be a tedious process. Summaries of this process have discussed in detail the steps that are required (e.g., Goldberg, Probart & Zak, 1999; Jacob, 1995).

Raw data samples are usually aggregated off-line by the investigator into meaningful behavioral units of fixations and saccades. Although the temporal and spatial requirements that define these vary among investigators, the intent is to quantitatively describe the behavioral tendency of an observer's attentional flow on an interface. The fixations are written into a new file, listing start/end times, and locations. Additionally, a code may be included, corresponding to the quality of the original signal (i.e., tracking status) received by the eye tracker.

Because fixations are made at a rate of approximately 3 Hz, the datafile length is reduced by a factor of about 20. While reduction to fixations is typically automated by software, the investigator is strongly encouraged to review the created fixations against images of viewed displays to ensure that the fixations are valid. Further processing of the fixation file can identify which defined screen objects were captured by each fixation, and thus compute the instantaneous and cumulative viewing times by screen objects. Sequences and transitions among these objects can then be computed. Goldberg and Kotval (1999, 1998) provided a detailed discussion and classification scheme for these higher level metrics. Inhoff and Radach (1998) provided a detailed discussion of metrics in eye tracking and cognitive research. Commercially available software has also been developed to aid the data reduction process (e.g., Lankford, 2000).

Methodological Issues in Eye Tracking

There are both advantages and disadvantages to eye tracking methods for usability evaluation. The ability to record one's micro-flow of visual attention in a non-intrusive way is certainly an advantage, but this can create a huge, tedious data reduction problem. Individual eye movements are quite randomly distributed, often requiring inferential statistics for discovering scanning trends. Strong individual differences are evident, ranging from individuals who scan broadly to those who barely make observable eye movements. Hardware must be frequently calibrated, and subtle differences in eye color or eye kinematics can cause an eye tracking failure. The usability engineer must certainly weigh these issues before deciding to implement an eye tracking methodology within a usability study. The present section considers many of these methodological issues that should be understood by the usability engineer. While ignoring one or more of these may not invalidate a study, the tester might want to alter an experimental procedure to better accommodate some of these pitfalls.

If eye tracking is desired, but it is not possible to purchase or rent a system, consulting firms exist which provide evaluation services. These companies can provide on-site eye tracking services, reduce data, and provide a report. Before hiring these services, however, it would be wise to understand the scope of evaluation that will be desired, and to provide a list of questions/hypotheses that the eye tracking service needs to answer.

Does the Eye's Current Location Indicate What One Is Currently Thinking About?

Eye tracking methods generally rely upon the *eye-mind hypothesis*, which states that, when looking at a visual display and completing a task, the location of one's gazepoint corresponds to the thought that is "on top of the stack" of mental operations (Just & Carpenter, 1976a). Visual attention may, however, lead or lag the current gazepoint, under certain conditions (Rayner, 1998). In reading text, for example, rapid eye movements followed by short fixations cause one's semantic processing of text to lag behind perceptual stimulus input. Reading difficult text passages very quickly leads to longer fixation durations and re-reading of passages. Processing of linguistic targets in reading tasks is often distributed over a spatial area, making it difficult to know exactly when and where the target is processed (Inhoff & Radach, 1998). Certain task factors can also decouple the eyes from the mind, reducing the sensitivity of eye movements as indicators of mental processing. Examples of these factors include spatial and temporal uncertainty about where important information is located, low visual target salience, task interruptions, and a high peripheral visual load (Potter, 1983).

Although there are no guarantees that the eye-mind hypothesis will be valid, several steps can be taken to support this hypothesis insofar as possible (Just & Carpenter, 1976a):

- *Tasks should be designed to require the encoding and processing of visual information to achieve well-specified task goals.* Tasks in which observers are only requested to "look at" a page or scene don't provide these necessary task goals. Rather, goal-directed instructions to search for (and possibly read and process) a particular item within a specified time limit should be specified.
- *Extraneous peripheral information should be controlled within tasks.* Screens could be masked off, or highly distracting blinking, moving, or colorful peripheral items could be eliminated.
- *Scanning uncertainty should be minimized within tasks.* Observers should have a general notion of the location of visual targets, and should not be surprised by the location of target items.
- *The "behavioral unit" of experimental tasks should be large enough to be open to conscious introspection.* Eye fixations generally correlate well with verbal protocols when decision makers are solving problems or choosing from among several alternatives. For example, selection of a menu item represents a larger behavioral unit than reading specific words within each menu item.
- *Cross-modality tasks may be candidates for eye tracking methodologies.* When listening to spoken linguistic passages, for example, the eyes tend to fixate upon spatial or pictorial referents of the words in the text.

How Does the Screen Cursor Influence One's Visual Attention?

Because a moving cursor can provide a highly salient and attention-drawing visual target on a display, it is important to understand the impact of cursor movements on scanpaths. In their study of menu item selection, Byrne *et al.* (1999) noted that the eye initially fixates a visual target, and is then trailed by cursor movement. Smith, Ho, Ark and Shai (2000) had participants use a mouse, touchpad, or pointing stick to perform either a reciprocal or random pointing task on a display. Most participants' visual attention either led or lagged the cursor, but some exhibited frequent target–cursor–target switching. The eyes tended to lead the cursor for known target locations, but lead vs. lag strategies were difficult to predict, based upon task factors.

Does Task Contrast and Luminance Influence Eye Tracking?

In the case of infrared eye tracking systems, very small pupil sizes can make it difficult for the eye tracking system to model the pupil center, causing poor eye tracking. Also, very small pupils are more easily hidden by the lower eyelid, especially if the eye tracking camera is poorly positioned. The pupil contracts and expands in response to the overall luminance and contrast of an observed display. Tasks presenting screens with large bright areas (e.g., lots of white space) cause the pupil to be smaller than those that contain screens with darker backgrounds. Tasks that contain animated images, scrolling, extreme foreground-background contrast, or extreme variance in luminance cause the pupil size to frequently vary. There are several recommendations to ensure the best possible eye tracking conditions for a task:

- *Minimize the use of pinpoint task lighting.* This can cause secondary glints on the observer's corneal surface, confusing the eyetracking system. There should also be no bright, reflective surfaces in the room, which could cause inadvertent light glints.
- *Avoid bright lighting and/or bright task screens.* To ensure an observer's pupils are sufficiently large for the system to locate and model, the testing room should ideally be dimly lit and screens should be uniform in background luminance. Because usability testing labs often use bright lighting to accommodate video cameras, some compromise may be necessary. Both room lights and the infrared source on the eye tracker should be controlled by variable rheostats. Note that, if luminance is highly variable on test screens, increasing room luminance may somewhat negate the screen influences to pupil size changes, enabling a steadier pupil size.
- *Ensure proper geometry between the eye tracking camera and the observer's eye.* Whether remote or head-mounted, the eye tracking camera is generally set up to view the eye from a position that is lower than the observer's horizontal horizon. If mounted too low, the lower eyelid can obstruct proper pupil imaging, especially for small pupils.

Are There Large Individual Differences in Eye Tracking Results?

Informal estimates suggest that as high as 20% of recruited participants will have problems with loss of tracking or calibration on an eye tracking system. For instance, Crowe and Naryayanan (2000) found that observers' eyes were not tracked 18% of the time for a keyboard/mouse interface; this increased to 35% of the time for a speech interface, when more head movement was noted. Excessive head movement during and between calibrations is quite problematic for remote eye tracking systems. Those with extremely dark iris colors may not have sufficient contrast between pupil and iris for the system to find the pupil's center. Some individuals have lower eyelids that cover substantial portions of the pupil and/or iris. Nguyen, Wagner, Koons and Flickner (2002) provided a detailed discussion of these and other factors that influence the eye's ability to retro-reflect infrared light to achieve reliable eye tracking.

Glasses and contact lenses (soft or hard lenses) are frequently problematic to effective eye tracking. Lenses refract both the incident source and the reflected infrared glint from the cornea, possibly causing high-order non-linearities in calibration equations, especially for extreme right or left eye rotations. Both types of lenses may also cause secondary light glints on the front or back sides of the lenses, making it difficult for the system to determine which light glint is associated with the corneal surface. Contacts have the possible additional problem of slippage as the eye rotates. The following suggestions may alleviate some of these tracking problems:

- *Minimize head motion.* Ensure that observers' heads remain as stationary as reasonably possible during calibration procedures, and if required, during testing. Chin rests or bite bars may be necessary in extreme cases.
- *When confronted with an individual with extremely dark irises, try both bright and dark pupil tracking methods, if possible.*
- *Pay careful attention to camera setup.* When setting up the eye tracking camera relative to the observer's position, be careful that the lower eyelid does not overly obstruct the view of the iris and pupil.
- *Consider limiting glasses and contact lens wearers during recruitment.* Participants with weaker prescriptions, for example, may calibrate more successfully than those with stronger prescriptions. Contact lenses can cause additional edge reflections, which can mask the accurate capturing of a corneal light glint.
- *Over-recruit for participants.* Recruit up to 20% more participants than required in case some fail to calibrate to the eye tracker.

Which Eye Should be Tracked?

Most eye tracking systems for behavioral analysis only track a single eye. Though binocular tracking systems are available, these are only required when vergence (or any other non-conjugate) movements must be measured, as in frequent, rapid changes in distance between observed targets and the eyes. Tracking of a single eye is usually adequate for most tasks, as the eyes are yoked when tracking objects moving across one's visual field. Because the light source is typically infrared, the participant has little awareness of the eye tracking apparatus, and can see stimuli clearly from both eyes.

Most head-mounted and remote eye tracking systems allow tracking of either eye. Some investigators prefer to test only the dominant eye of the participant, because the majority of "useful" vision is obtained through this eye. Approximately 97% of the population has a visual sighting eye dominance, in which observers consistently use the same eye for primary vision. Approximately 65% of all observers are right-eye dominant, and 32% are left-eye dominant (Porac & Coren, 1976). Of those exhibiting eye dominance, most (approximately 75%) are same-side dominant as handedness. Eye dominance may be determined by a simple test:

- The participant is asked to point to a far object with an outstretched arm, using both eyes.

- While still pointing, the participant is asked to close one eye at a time.
- The eye that sees the finger pointing directly at the target is dominant.

Other simple tests for eye dominance are described by Porac and Coren (1976).

The issue of whether to use the dominant eye (as opposed to always using the right or left eye) is under current debate. Recent evidence has demonstrated that eye dominance can switch as a function of horizontal gaze angle (Khan & Crawford, 2001). These investigators showed that, for their psychomotor task, this shift occurred at an average eccentricity of about 15 degrees towards the non-dominant eye. Note, however, that at typical (50 cm) eye-display distances, these eccentricities are near the edge of typical (35 cm) display widths. In fact, most usability evaluations are conducted within smaller windows presented on these displays. Inhoff and Radach (1998) provide in-depth discussion of this and related issues.

How Is an Eye Tracker Calibrated?

Calibration procedures are a very necessary and critical part of eye tracking studies, in order to relate an observer's gaze angle, or point-of-regard, to locations in the environment (e.g., computer display). Poor calibration can invalidate an entire eye tracking study, because there will be a mismatch between the participant's point-of-regard, and the corresponding location on a display. Hardware drift and head movement are two big problems that can invalidate a set of data.

During a testing session, calibration should be conducted at multiple times, as well as between each participant. Commercial eye tracking systems generally come with calibration software that may be run independently or inserted into one's experimental test code. Building a calibration routine within experimental stimulus presentation code is a good idea, especially if the calibration can easily be conducted on demand (e.g., between trials or blocks of trials). Recalibration is recommended every few minutes, such as between trial blocks.

The calibration procedure generally requires an observer to successively fixate a series of known locations on a display, or within the environment. On a display, these locations are usually near the corners, and at the display's center in a 5-point calibration. A 9- or 13-point calibration locates these at the vertices of successively larger squares, starting from the display's center.

Calibration routines may be experimenter-, system-, or participant-controlled. An experimenter-controlled routine starts by announcing or showing the point to be observed. The experimenter then controls the length of sampling time at that location (e.g., by mouse click), to ensure a valid and reliable sample of *x*/*y*-observation locations. It is helpful, in this case, if the experimenter has an indication of sample-to-sample variance at each point, to know which, if any, points must be re-sampled. System-controlled calibration uses an automated procedure to randomly select a location to be viewed, then will sample the observer's point-of-regard until a minimum-variance criterion is achieved. The experimenter should, however, still be able to take control of the calibration in case of difficulties. Participant-controlled calibration requires the participant to look at a screen location, then press a button when ready to move on to another calibration location. Regardless of which routine is used, calibration should be easy enough to conduct on a frequent basis.

How Many Participants Are Required for Eye Tracking Studies?

The required number of participants for a research or usability study is determined by expected effect sizes and statistical power, as well as expectations within a domain. With regards to statistical power, an eye tracking study generates large amounts of data, much like a reaction time study. However, significant aggregation and averaging of data prior to statistical analyses may increase the required number of participants. For example, a datafile containing 1000 fixations could be reduced to a file containing average gaze durations within 10 areas of interest on a display. Nonparametric or parametric statistical comparisons might then be conducted among these reduced data.

Domain-dependent expectations for the required number of participants can be addressed by considering the number of participants tested in a representative sample of prior studies. Table 23.1 summarizes the number of participants from several eye tracking studies, noting whether the experimental design was within or between participants, and providing a very short task description. Overall, designs have used about 6–30 participants, which coincides with the NIST CIF requirement of at least eight participants per user type in usability tests (Wichansky, 2000). These numbers should

Table 23.1: Number of participants from recent eye tracking studies.

Number of participants	Design*	Task	Reference
30	W	Searching variably-ordered lists	Crosby and Peterson (1991)
20	W	Searching menus for stated target items	Aaltonen *et al.* (1998)
17	W	Search of four different web pages for specific information tagets	Cowan (2001)
11	W	Searching menu lists for target characters	Byrne *et al.* (1999)
10	W	Matching icons of various sizes	Jacko *et al.* (2000)
8	W	Searching alternate file tree representations	Pirolli *et al.* (2001)
7	W	Using screens from Prodigy	Benel *et al.* (1991)
25	B	Evaluation of a computer-delivered questionnaire	Redline and Lankford (2001)

Note: *W: Within participant; B: Between group design.

only be used as a domain-dependent guide for designing an eye tracking-based usability evaluation; the statistical power and expected effect sizes mentioned previously should also be considered.

Research Needs

The expanding field of applied eye tracking can provide a useful contribution to the field of usability evaluation. This chapter has provided information to aid in the cross-education of usability engineers on the basics of eye tracking, and to inform eye tracking specialists on the methods and requirements of usability evaluation. The graceful melding of these areas still, however, requires several advances in knowledge and practical methodology.

Eye tracking must not slow down usability evaluations. Usability engineers are under great time constraints during testing sessions, and more easily used tools are required to allow them to incorporate eye tracking into usability tests. Although available now, automated calibration procedures could be more broadly distributed for use in usability evaluation. In addition, the eye tracker should be extremely quick to set up for a participant, requiring a minimum of adjustment between and within participant test sessions.

Eye tracking must work in a relatively unconstrained participant testing environment. Continued development of algorithms to compensate for head movement is recommended. Participants in a usability evaluation should not feel as if they are motion-constrained while using products. Ideally, participants should be able to leave the testing area for a short time, then resume testing following a short calibration. Also, better compensation for head motion during calibration might result in fewer participants who fail to calibrate to an eye tracker.

Analysis tools for eye tracking data should be easy to use by those with little knowledge of eye tracking. Tools are currently available that will automatically parse web pages into areas of interest, then batch process those pages to provide summaries of viewing time by areas. Further automated tools are desired to determine the order of these visited areas, and to compare these orders across experimental conditions and participants. More tools that provide quantitative indications of transitions among areas are desired. Recent methodological articles by Inhoff and Radach (1998) and Goldberg and Kotval (1999) summarize some of the types of quantifiable information that is available.

Standards for eye tracking in usability evaluations should be developed and published. The usability engineer is interested in whether software is above or below a stated usability threshold, and if below, how it may be improved. While there are well-accepted, standard usability metrics, little has been done to correlate eye tracking-derived measures to those. There is a strong need for research relating these two sets of measures. Convergence into a standard eye tracking protocol is also desirable for usability testing, much like the NIST CIF standard for usability. Minimum standards, such as frequency of calibration, data sampling rates, and equipment specifications would make it easier to cross-interpret data from multiple studies. Certainly, more published reports of eye tracking in usability evaluations will be required before such a standard could be considered. Absolute, rather than relative benchmarks for eye tracking metrics would also aid cross-interpretation, and eventual convergence to a standard (Cowen, 2001).

More knowledge is required on the contribution of task factors to eye tracking-derived metrics. Little is known, for example, about how the density of a display or visibility of icons influences eye tracking results, and perceived usability. This is not a trivial issue, and has been the subject of visual search studies for many years.

Ultimately, eye tracking may become commonplace as a secondary methodology for usability evaluation. This could help to drive down the price of eye tracking systems, and support an easier process of reducing eye tracking data from gazepoint samples to meaningful behavioral inferences.

References

Aaltonen, A., Hyrskykari, A., & Raiha, K. (1998). 101 spots, or how do users read menus? In: *Proceedings of ACM CHI 1998 Conference on Human Factors in Computing Systems 1998* (pp. 132–139). New York: ACM Press.

Benel, D., Ottens, D., & Horst, R. (1991). Use of an eyetracking system in the usability laboratory. In: *Proceedings of the Human Factors Society 35th Annual Meeting* (pp. 461–465). Santa Monica, CA: HFES.

Brodersen, L., Andersen, H. H. K., & Weber, S. (2001). *Applying Eye-movement Tracking for the Study of Map Perception and Map Design.* Report to the National Survey and Cadastre Denmark, October, 2001.

Brookings, J. B., Wilson, G. F., & Swain, C. R. (1996). Psychophysiological responses to changes in workload during simulated air traffic control. *Biological Psychology, 42*: 361–377.

Byrne, M. D., Anderson, J. R., Douglass, S., & Matessa, M. (1999). Eye tracking the visual search of click-down menus. In: *Proceedings of ACM CHI 1999 Conference on Human Factors in Computing Systems 1999* (pp. 402–409). New York: ACM.

Card, S. K., Pirolli, P.,Van Der Wege, M., Morrison, J. B., Reeder, R. W., Schraedley, P. K., & Boshart, J . (2001). Information scent as a driver of web behavior graphs: Results of a protocol analysis method for web usability information scent. In: *Proceedings of ACM CHI 2001 Conference on Human Factors in Computing Systems 2001* (pp. 498–505). New York: ACM.

Carpenter, R. H. S. (1988). *Movements of the Eyes.* (2nd edn). London: Pion.

Chin, J. P., Diehl, V. A., & Norman, K. (1988). Development of an instrument measuring user satisfaction of the human–computer interface. In: *Proceedings of ACM SIGCHI '88* (pp. 213–218). New York: ACM.

Cowen, L. (2001). An eye movement analysis of web-page usability. Unpublished MSc Thesis, Lancaster University, UK.

Crosby, M. E., & Peterson, W. W. (1991). Using eye movements to classify search strategies, In: *Proceedings of the 35th Annual Meeting of the Human Factors Society* (pp. 1476–1480). Santa Monica, CA: HFES.

Crowe, E. C., & Narayanan, N. H. (2000). Comparing interfaces based on what users watch and do. In: *Proceedings of ACM/SIGCHI Eye Tracking Research & Applications Symposium* (pp. 29–36). New York: ACM.

David, H., Mollard,, R., Cabon, P., & Farbos, B. (2000). Point-of-gaze, EEG and ECG measures of graphical/keyboard interfaces. In: *Measuring Behavior 2000, 3rd International Conference on Methods and Techniques in Behavioral Research* (pp. 15–18). Nijmegen, The Netherlands.

Dumas, J. S. (2002). User-based evaluations. In: J. A. Jacko and A. Sears (eds), The *Human–computer Interaction Handbook: Fundamentals, Evolving Technologies, and Emerging Applications*, Mahwah, NJ: Erlbaum & Associates.

Dumas, J. S., & Redish, J. C. (1993). *A Practical Guide to Usability Testing.* Norwood: Ablex.

Eye Tracking Equipment Database (2002). Institute of Behavioural Sciences, University of Derby. http://ibs.derby.ac.uk/emed/.

Goldberg, J. H. (2000). Eye movement-based interface evaluation: What can and cannot be assessed? In: *Proceedings of the IEA 2000/HFES 2000 Congress (44th Annual Meeting of the Human Factors and Ergonomics Society)* (pp. 6/625–6/628). Santa Monica: HFES.

Goldberg, J. H., & Kotval, X. P. (1998). Eye movement-based evaluation of the computer interface. In: S. K. Kumar (ed.), *Advances in Occupational Ergonomics and Safety* (pp. 529–532). Amsterdam: IOS Press.

Goldberg, J. H., & Kotval, X. P. (1999). Computer interface evaluation using eye movements: Methods and constructs. *International Journal of Industrial Ergonomics*, *24*: 631–645.

Goldberg, J. H., Stimson, M. J., Lewenstein, M., Scott, N., & Wichansky, A. M. (2002). Eye tracking in web search tasks: Design implications. In: *Proceedings of ACM/SIGCHI Eye Tracking Research & Applications Symposium 2002* (pp. 51–58). New York: ACM.

Goldberg, J. H., Probart, C. K., & Zak, R. E. (1999). Visual search of food nutrition labels. *Human Factors 41*(3): 425–437.

Hoeks, B., & Levelt, W. J. M. (1993). Pupillary dilation as a measure of attention: A quantitative system analysis. *Behavior Research Methods, Instruments & Computers 25*(1): 16–26.

Inhoff, A. W., & Radach, R. (1998), Definition and computation of oculomotor measures in the study of cognitive processes. In: G. Underwood (ed.), *Eye Guidance in Reading and Scene Perception* (pp. 29–53). Oxford: Elsevier.

Jacko, J. A., Barreto, A. B., Chu, J. Y. M., Bautsch, H. S., Marmet, G. J., Scott, I. U., & Rosa, R. H. (2000). In: *Proceedings of ACM/SIGGRAPH Eye Tracking Research & Applications Symposium* (p. 112). New York: ACM.

Jacob, R. J. K. (1995). Eye tracking in advanced interface design. In: W. Barfeld and T. Furness (eds), *Advanced Interface Design and Virtual Environments* (pp. 258–288). Oxford: Oxford University Press.

Just, M. A., & Carpenter, P. A. (1976a). Eye fixations and cognitive processes. *Cognitive Psychology*, *8*: 441–480.

Just, M. A., & Carpenter, P. A. (1976b). The role of eye-fixation research in cognitive psychology. *Behavior Research Methods & Instrumentation 8*(2): 139–143.

Karn, K., Ellis, S., & Juliano, C. (2000). The hunt for usability: Tracking eye movements. *SIGCHI Bulletin*, November/December 2000 (p. 11). New York: ACM.

Khan, A., & Crawford, J. (2001). Ocular dominance reverses as a function of horizontal gaze angle. *Vision Research 41*(14): 1743–1748.

Kirakowski, J. (1996). The software usability measurement inventory: Background and usage. In: P. Jordan, B. Thomas and B. Weerdmeester (eds), *Usability Evaluation in Industry.* London: Taylor & Francis.

Kotval, X. P., & Goldberg, J. H. (1998). Eye movements and interface components grouping: An evaluation method. In: *Proceedings of the 42nd Annual Meeting of the Human Factors and Ergonomics Society* (pp. 486–490). Santa Monica: HFES.

Lankford, C. (2000). GazeTracker: Software designed to facilitate eye movement analysis. In: *Proceedings of the ACM Symposium on Eye Tracking Research & Applications* (pp. 51–55). New York: ACM Press.

Macleod, M., Bowden, R., Bevan, N., & Corson, I. (1997). The MUSIC performance measurement method. *Behaviour and Information Technology 16*(4–5): 279–293.

May, J. G., Kennedy, R. S., Williams, M. C., Dunlap, W. P., & Brannan, J. R. (1990). Eye Movement Indeces of Mental Workload. *Acta Psychologica, 75*: 75–89.

Mulligan, J. B. (1997). Image Processing for Improved Eye-Tracking Accuracy. *Behavior Research Methods, Instruments, & Computers, 29*: 54–65.

Nguyen, K., Wagner, C., Koons, D., & Flickner, M. (2002), Differences in the infrared bright pupil response of human eyes. In: *Proceedings of ACM/SIGCHI Eye Tracking Research & Applications Symposium 2002* (pp. 133–138). ACM: New York.

Noton, D., & Stark, L. (1970). Scanpaths in saccadic eye movements while viewing and recognizing patterns. *Vision Research, 11,* 929–942.

Pirolli, P., Card, S. K., & Van der Wege, M. M. (2001). Visual information foraging in a focus + context visualization. *CHI Letters 3*(1): 506–513.

Polson, P., Lewis, C., Rieman, J., & Olson, J. (1992). Cognitive walkthroughs: A method for theory-based evaluation of user interfaces. *International Journal of Man-Machine Studies, 36*: 741–773.

Porac, C., & Coren, S. (1976) The dominant eye. *Psychological Bulletin 83*(5): 880–897.

Potter, M. C. (1983). Representational Buffers: The eye-mind hypothesis in picture perception, reading, and visual search. In: K. Rayner (ed.), *Eye Movements in Reading: Perceptual and Language Processes* (pp. 413–437). New York: Academic Press.

Poynter Institute (2000). Stanford University Poynter Project. http://www.poynter.org/eyetrack2000/.

Rayner, K. (1998). Eye movements in reading and information processing: 20 years of research. *Psychological Bulletin 124*(3): 372–422.

Redline, C. D., & Lankford, C. P. (2001). Eye-movement analysis: A new tool for evaluating the design of visually administered instruments (Paper and Web). *Proceedings of American Association of Public Opinion Research*, Montreal, Canada.

Rubin, J. (1994). *Handbook of Usability Testing*. New York: Wiley Press.

Sales, B. D., & Folkman, S. (2000). *Ethics in Research with Human Participants*. Washington, DC: American Psychological Association.

Scholtz, J., & Morse, E. (2002). A new usability standard and what it means to you. *SIGCHI Bulletin*, May/June, 10–11.

Smith, B. A., Ho, J., Ark, W., & Zhai, S. (2000). Hand eye coordination patterns in target selection. In: *Proceedings of the ACM Symposium on Eye Tracking Research & Applications* (pp. 117–122), 6–8 November 2000, Palm Beach Gardens, FL.

Stern, J. A., Boyer, D., & Schroeder, D. (1994). Blink rate: A possible measure of fatigue. *Human Factors 36*(2): 285–297.

Wichansky, A. M. (2000). Usability testing in 2000 and beyond. *Ergonomics 43*(7): 998–1006.

Young, L. R., & Sheena, D. (1975). Survey of eye movement recording methods, *Behavior Research Methods & Instrumentation, 7*: 397–429.

Zellweger, P. T., Regli, S. H., Mackinlay, J. D., & Chang, B. W. (2001). The impact of fluid documents on reading and browsing: An observational study. In: *Proceedings of ACM SIGCHI 2000 Conference on Human Factors in Computing Systems* (pp. 249–256). New York: ACM Press.

Chapter 24

Processing Spatial Configurations in Computer Interfaces

Martha E. Crosby and Catherine Sophian

Complex computer interfaces often provide quantitative information that people need to integrate. In addition, people have a great deal of difficulty reasoning effectively about ratios and proportions when their properties are critical. Therefore, on many tasks the format in which ratio information is conveyed may have an enormous impact on performance. In essence, the task of proportional reasoning is to compare the relation between one pair of terms (or quantities) with the relation between another pair. One of the simplest examples of proportional reasoning occurs in the identification of shapes. The task of identifying two shapes that are geometrically similar, though different in size, is a proportional one, but also one in which spatial processing may be important. This chapter describes several studies that investigate ways in which spatial presentations of the data can facilitate the process of integrating quantitative information.

Introduction

Very often user interfaces present quantitative information, critical to task performance, in a visual form. The most familiar of such displays are perhaps the gauges and dials that allow drivers to monitor speed and other performance characteristics of their vehicles without taking their eyes from the road for any significant length of time. Characteristically, these displays present magnitude information in essentially linear form — the further to the right the pointer on a speedometer moves, for example, the greater the speed of the vehicle.

Similar linear representations are used in many computer interfaces. The scroll bar is an interesting example. The further the marker moves down the scroll bar, the further one's position is from the beginning of the document. Note that what is represented

The Mind's Eye: Cognitive and Applied Aspects of Eye Movement Research

ISBN: 0-444-51020-6

is not an absolute magnitude but a ratio — the ratio between the portion of a document that is above one's current position and the portion that is below it (or, equivalently, the ratio between either of those two portions and the total document size). Because users are not generally concerned with either the ratio properties of their document or those of the scroll bar, the fact that the information represented is fundamentally proportional has little impact on the effectiveness of the scroll bar display. However, it is well known that people have a great deal of difficulty reasoning effectively about ratios and proportions when their properties are critical. Therefore, on many tasks the format in which ratio information is conveyed may have an enormous impact on performance. Several researchers have investigated different aspects of how people make proportion judgments.

Schwartz and Moore (1998) examined the assumption that rather than people first understanding a situation and then choosing computations to apply to the situation, mathematics can be a tool to help people construct an understanding of a situation. Since early proportional reasoning is a good example of when mathematics may provide new understanding, they conducted three studies with 6th grade children reasoning about proportions. The children were ask to predict when two glasses of juice of different sizes would taste the same when they were filled from a single carton of juice made from concentrate and water. When the problem was presented as a diagram with complex numbers, or "realistically" with easy numbers, the children predicted the glasses would taste different. However, when the problem was presented realistically with complex numbers and when the problem was presented as a diagram with easy numbers, the children predicted the glasses would taste the same on the basis of proportional relations. They suggested that mathematics made a complex empirical situation cognitively tractable and facilitated the cases of successful proportional reasoning by helping the children construct mental models of the situation. Johnson-Laird *et al.* (1999) outline a theory of naive probability that relies on the concept of mental models. Their explanation clarifies some common misconceptions of probabilistic reasoning. People can infer the probabilities of events by contrasting mental models of what is true in different situations. The probability of an event depends on the proportion of models in which it occurs. Each case is equally likely unless the people already have other beliefs. This theory predicts several phenomena of reasoning about absolute probabilities, including typical biases. It correctly predicts certain cognitive illusions in inferences about relative probabilities.

Lawton (1993), noting that proportional reasoning is a difficult for adults as well as children described several factors that may contribute to the complexity. These include the type of problem, whether or not the quantity is discrete or continuous, and the person's familiarity with the type of problem. She states that most people's understanding of proportion concepts is relatively fragile and easily influenced by structural variations in the problem. Thus, another factor that may be important in how people understand proportional relationships is the degree of physical similarity between the objects in a problem. She performed two studies that tested the hypothesis that people were more likely to use proportional reasoning if the distinction between the items in a proportion was emphasized. Lawton's first experiment investigated the effects of varying the similarity between items in classic cylinders problem and the second experiment examined the effects of similarity between items in a mixture problem. The results from her two studies suggest that people more readily solve proportion problems if the contents of the items in the problem are seen as being substantially different from each other.

Hollands and Dyre (2000) developed a cyclical power model to account for the various observed systematic asymmetries in bias patterns that people have when they make part-whole proportion judgments. People tend to overestimate small proportions and underestimate large proportions, overestimate large proportions and underestimate small proportions or have a bias pattern that repeats cyclically (e.g., overestimation of proportions less than 0.25 and between 0.5 and 0.75 underestimation otherwise).

A fundamental characteristic of ratios is that they posses two dimensions. Since a ratio is a relation between two quantities, it is affected by alterations in either of those quantities. Plotting curves relating the two variables is one way of expressing this relation spatially. The changing height of the curve at different points along the horizontal axis captures the changing relation between the two variables. Each individual point on the curve, correspondingly, represents the relation between the two variables at a particular point. One way to give spatial expression to that relation would be to draw a rectangle connecting the selected point to the two axes of the graph; the shape of the resulting rectangle (how wide it is relative to how tall) then reflects how great one variable is relative to the other at that particular point.

Psychological tests of proportional reasoning typically control for perceptual cues to proportionality such as shape by using different kinds of units for comparison and test stimuli (e.g., Spinillo & Bryant, 1991). The goal of this type of design is to ensure that respondents cannot solve the task on the basis of a global perceptual judgment without understanding the mathematical relation of proportionality. However even a perceptual judgment is a kind of proportionality judgment, provided it is based on the relation between the two dimensions and not on just one of them. Mathematically a proportion is an equivalence between two ratios: A is to B as C is to D. This relation does not cease to be a proportional one when it takes a spatial form and indeed our hypothesis is that it may be first, and most easily, grasped in relation to spatial proportions.

Although psychological tests of proportional reasoning typically control for perception, there is nothing in principle that makes a perceptually based judgment of proportionality mathematically unlike other proportionality judgments. What is important is that non-relational bases for task performance be controlled. Thus, matching the fatness of two shapes that are the same height might not involve a consideration of ratios, since comparisons based on the horizontal dimension alone could determine whether the shapes were the same or different. Differences in overall magnitude are thus essential in assessments of proportions.

Proportional reasoning begins early in life. Even young children can tell whether two shapes are equally proportioned, in terms of the ratios of their heights to their widths for instance — even if they are very different in overall size (Sophian & Crosby, 1998, 1999). To the child, the proportionally matching shapes just "look the same", but the task is nevertheless one of establishing proportional equivalence. Precisely because ratio comparison seems so effortless in this context, we hypothesize that the presentation of ratio information in spatial form may facilitate the effective utilization of that information in the performance of complex tasks.

We report studies that examine this hypothesis. The first two experiments consider the role of spatial relations in proportional matching, not by eliminating them but by varying the form they take across two otherwise identical sets of problems. Both sets of problems entail making judgments about the proportional relations among two line segments. However, the configuration of the line segments is varied so as to provide different amounts of spatial support for the ratio comparison. In our third experiment, we record eye fixations during the performance of a ratio-based shape comparison task in order to learn more about the perceptual and cognitive processes underlying the interpretation of spatial representations of ratio information.

An important methodological point is that, because ratios fundamentally concern the relation between two quantities, it is crucial that non-relational bases for responding be controlled in any study of ratio comparisons. Thus, for instance, judging the relative fatness (i.e., height–width ratio) of two shapes is ostensibly a ratio comparison task, but if the two shapes are the same height then correct judgments might be based on consideration of the horizontal dimension alone and so could not be accepted unequivocally as evidence of ratio comparison. Differences in overall magnitude are thus an essential element in any ratio comparison task, and a basic element of our research designs. We investigated the hypothesis that the perceptual processes by which we identify the shapes of things enable even young children to make proportionality judgments. The degree, to which an object is wide or narrow in shape, fat or thin, is a function not simply of its horizontal extent but of the ratio between its horizontal and vertical dimensions. The focus of this chapter is on explicating the cognitive processes that underlie this kind of judgment.

Experiment 1

The first experiment evaluated young children's ability to identify proportional relationships among visual stimuli. Two conditions were compared, varying in the type of stimuli that were used and so in the degree of spatial-configurational information available. In one condition, the stimuli were solid rectangles that varied in the proportional relationship between their vertical and horizontal dimensions. Rectangular stimuli were chosen so that no extraneous cues, such as changes in the angles at which edges met, would differentiate the proportionally matching stimulus from the incorrect alternative. In the second condition, the stimuli were pairs of vertical lines differing in length, with the shorter one always positioned directly above the longer one. These line pairs embodied the same lengths as the heights and widths of the rectangles. It was hypothesized, however, that the absence of the spatial cue of shape would make this condition much more difficult than the rectangle condition.

Method

Participants Twenty-nine children participated in the experiment: Seventeen 5-year-olds (9 boys and 8 girls; age range: 5 years, 0 month to 5 years, 11 months; $M = 5$

years, 5 months) and twelve 4-year-olds (8 boys and 4 girls; age range: 4 years, 0 months to 4 years, 11 months; $M = 4$ years, 5 months). The sample was ethnically diverse, with Caucasian and Asian ethnicities predominating.

Materials Each problem consisted of a relatively large sample stimulus, centered in the upper part of the page, and two smaller choice stimuli, positioned on the left and right sides of the bottom part of the page. An example of a rectangle problem and a line problem can be seen in Figure 24.1.

Within stimulus type, problems differed in the magnitude of the ratios they embodied (the ratios of width: height for rectangles, and of length of shorter line: length of longer line for pairs of lines) and in the degree of contrast between the ratios comprising a choice pair. For "close contrast" problems, ratios were chosen such that one alternative embodied a ratio 2/3 as great as the other; these ratios were 1:2 vs. 1:3 for the large-ratio problems, and 1:1.2 vs. 1:1.8 for the small-ratio problems. For "far contrast" problems, ratios were chosen such that one alternative embodied a ratio only 1/2 as great as the other; these ratios were either 1:2 vs. 1:4 (large ratios) or 1:1.2 vs. 1:2.4 (small ratios). In addition, they varied in whether the alternative embodying the greater or the lesser ratio corresponded to the sample, and in whether that alternative was positioned on the left or the right side of the page. Each of the resulting 32 problems was printed twice, in two different colors (orange, green, dark blue, or magenta), to form the complete stimulus set. Colors were chosen so that all four colors appeared equally often within each stimulus type and also within each of the other factors in the design.

Additional materials for the study consisted of several three-dimensional toys that were used to introduce children to the matching task. There were two toy trolls, similar in appearance but contrasting in height (one standing about 14 cm tall and the other 7 cm tall), two plastic bears (one 4 cm tall and the other 2.7 cm tall), two flat yellow blocks (one $5 \times 7.5 \times 0.4$ cm and the other $3.4 \times 5 \times 0.4$ cm), two pairs of black plastic sticks (one pair consisting of sticks 5 and 7.5 cm in length; the other of sticks 3.4 and 5 cm in length), and three cone-shaped paper hats, two made from circles of radius 2.2 cm and the third from a circle of radius 3 cm. One of the smaller circles was wrapped into a tight cone, so that the resulting hat was taller than it was wide; the other two were formed into much wider cones, creating hats that were about twice as wide as they were tall.

Procedure Children were tested individually in a quiet room at their school, in a single session that lasted about 15 minutes. Each child completed four blocks of 16 problems, alternating between blocks of rectangle problems and blocks of line problems (with order counterbalanced). The session began with a warm-up activity in which the trolls, bears, hats, and either the sticks or the blocks were used to establish a rationale for the matching task. The experimenter pointed out that the trolls were very much alike except for size and said that they liked to play with the same toys too, except the small troll's toys needed to be smaller than the big troll's. This idea was illustrated with the bears and the hats, and then with materials that corresponded to whichever stimulus type would be tested first (sticks for the line stimuli; blocks for the rectangle stimuli). The experimenter asked the child to look at pictures of other

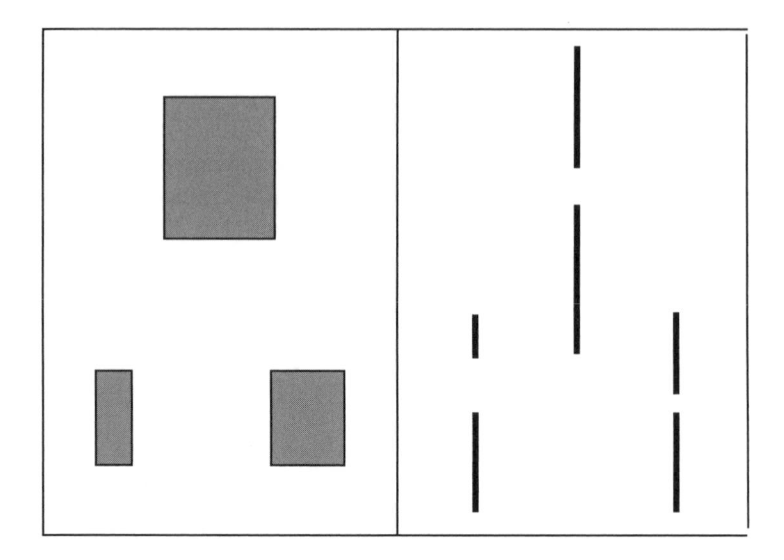

Figure 24.1: Example problems from Experiment 1. Both of these problems involve a contrast between a test choice that matches the larger sample stimulus (the alternative on the right in each example) and one that differs in that the width of the rectangle, or the length of the shorter line, is twice as long as it is in the correct alternative. Thus, both are far-contrast problems.

blocks or pairs of sticks and to help the little troll find toys in his size that were just like the toys the big troll got. She emphasized that the little troll's toys needed to match the big troll's toys in how wide the blocks were compared to how tall they were, or in how long one stick of a pair was compared to the other. In the same manner, following the first block of trials, the experimenter introduced the stimuli to be used for the second (and fourth) blocks using either the blocks or the sticks as a three-dimensional counterpart to the stimuli that would appear in those problems.

Results

Figure 24.2 displays the mean proportions of children's responses that were correct, as a function of age group, stimulus type (rectangles vs. line pairs), and degree of contrast between the pairs of ratios that constituted the choice stimuli. The data were analyzed in a 2 (age) × 2 (gender) × 2 (order: alternating blocks beginning with rectangle stimuli vs. alternating blocks beginning with line stimuli) × 2 (stimulus type: rectangles vs. line pairs) × 2 (trial set: early = first half of experiment, late = second half) × 2 (ratio contrast: close or far) ANOVA, with the last three factors as repeated measures. There were significant main effects of stimulus type, $F(1,21) = 103.96$, $MSe = 0.048$, $p < 0.001$, and contrast, $F(1,21) = 34.09$, $MSe = 0.010$, $p < 0.001$. Consistent with the hypothesized role of spatial-configurational cues in children's proportional

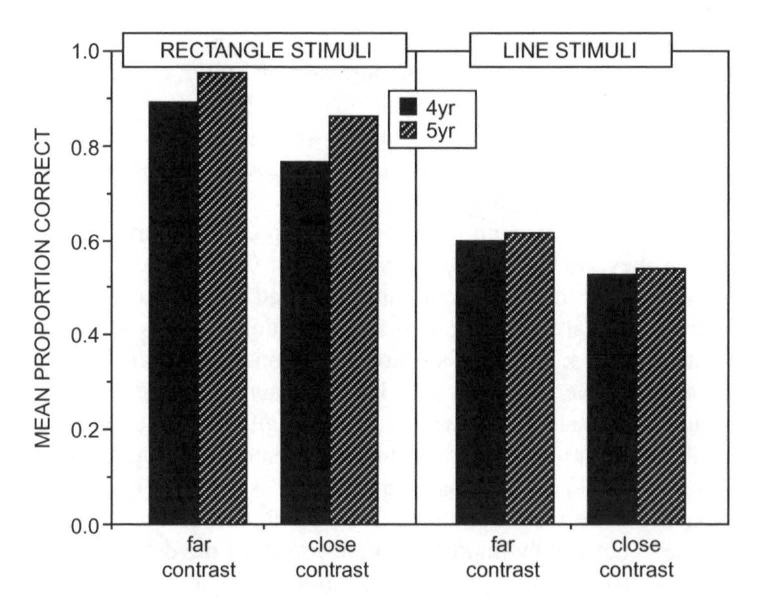

Figure 24.2: Mean proportions of correct responses by 4- and 5-year-old children to each problem type.

matching, children's performance was considerably better on the rectangle stimuli than on the pairs of lines, overall *M*s = 0.89 vs. 0.57. In addition, children did better when there was a relatively large contrast between the choice alternatives (far contrast) than when they were more similar to each other (close contrast), *M*s = 0.77 vs. 0.69. There were no significant differences between the two age groups (*M*s for 4- vs. 5-year-olds = 0.83 vs. 0.91 for rectangle stimuli, and 0.56 vs. 0.58 for line stimuli) or between children's performance on the earlier vs. the later trials in the experiment (*M*s for early vs. late trials = 0.87 vs. 0.89 for rectangle stimuli, and 0.57 vs. 0.57 for line stimuli).

One-sample t-tests were carried out on the mean scores for each stimulus type × contrast combination to determine whether children were reliably above chance on each kind of problem. Performance on the rectangle stimuli was significantly above chance (0.50) for both far-contrast, $M = 0.93$, $t(28) = 14.93$, $p < 0.001$, and close-contrast problems, $M = 0.82$, $t(28) = 11.87$, $p < 0.001$. Performance on the line stimuli, however, was reliably above chance only for the far-contrast problems, $M = 0.61$, $t(28) = 4.13$, $p < 0.001$ (for close-contrast problems: $M = 0.53$, $t(28) = 1.51$, $p > 0.10$).

Discussion

The 4- and 5-year-old children in this study were able to identify which of two smaller rectangles was the same in shape as a larger sample rectangle. Performance dropped markedly, however, when the proportional relation they were asked to judge was that between two vertical lines, rather than between the height and width of a rectangle. This result supports the hypothesized importance of spatial-configurational cues in children's early processing of proportional relations. It suggests that spatial proportions may be foundational in a developmental sense, that they are among the first proportional relations children notice and may provide an important avenue for learning about other kinds of proportional relations.

Experiment 2 asks whether adults, like children, process proportional relations more effectively when they are presented in spatial form. On the one hand, it is possible that the effects observed in Experiment 1 are restricted to young children, who are not yet facile with non-spatial forms of proportional relations. Adults, who have had more experience with diverse kinds of proportional relations, may perform just as well on stimuli like the pairs of vertical lines used in Experiment 1, which do not afford strong spatial-configurational cues, as on the rectangle stimuli. On the other hand, however, if the accessibility of proportional relations presented in spatial form for young children reflects a basic aspect of visual/spatial processing, it might be expected that this form of presentation would continue to be advantageous even for adults. Such a finding would have potentially important implications for the design of user interfaces for computer systems, because proportional information is critical for many visual processing tasks.

Experiment 2

Experiment 2 therefore replicated Experiment 1, using adult rather than child participants. Black-and-white, rather than colored, stimuli were used in order to make photocopies of the stimuli that could be prepared for each participant on which s/he could directly mark his or her responses. Again, performance on rectangles varying in the relation of height:width was compared with performance on pairs of vertical lines that embodied the same lengths as the heights and widths of the rectangles but were positioned directly above one another to minimize spatial-configurational cues to proportionality.

Method

Participants Thirty-one participants, 17 male and 14 female, from the College of Business Administration (CBA) at the University of Hawaii, took part in this experiment. These ethnically diverse students were enrolled in an upper division course in the Department of Decision Sciences.

Materials Each problem consisted of a relatively large sample stimulus, centered in the upper part of the page, and two smaller choice stimuli, positioned on the left and right sides of the bottom part of the page. Other than the fact that the rectangle problems used unfilled rectangles, and that all the stimuli were printed in black ink, the materials were the same as those used in Experiment 1.

Procedure The single session experiment was conducted during a regular meeting of the CBA class. At the beginning of the class period, the participants were given a booklet that contained four blocks of eight problems, one to a page, alternating between blocks of rectangle problems and blocks of line problems (with order counterbalanced). Each block of problems was preceded by a page of instructions, which asked participants to respond as quickly as possible without making errors and to refrain from turning back to a prior problem once they had gone on to a subsequent one. Participants were given as much time as they needed to complete the task; all finished within 15 minutes.

Results

The adult data was near ceiling. The average percent correct was 97.4% for the rectangles and 94.6% for the line pairs. Since many of the participants performed the task perfectly, the data was analyzed by Friedman's non-parametric test. In addition to the stimulus type (rectangles or line pairs), two other variables, ratio contrast (close or far) and trial set (early or late) were analyzed.

In one analysis, the effects of stimulus type and contrast were considered. The four contrast variables were: lines/close, lines/far, rectangles/close and rectangles/far. Friedman's test showed significantly different ($p = 0.004$) rankings of the ratio contrasts for each stimulus type. The mean ranks are shown in Figure 24.3. Further analysis of the contrasts showed a significant difference between the close contrast and far contrast line pairs ($p = 0.02$), as well as between the close contrast and far contrast rectangles ($p = 0.05$). Within the close contrast stimuli, there was a marginally

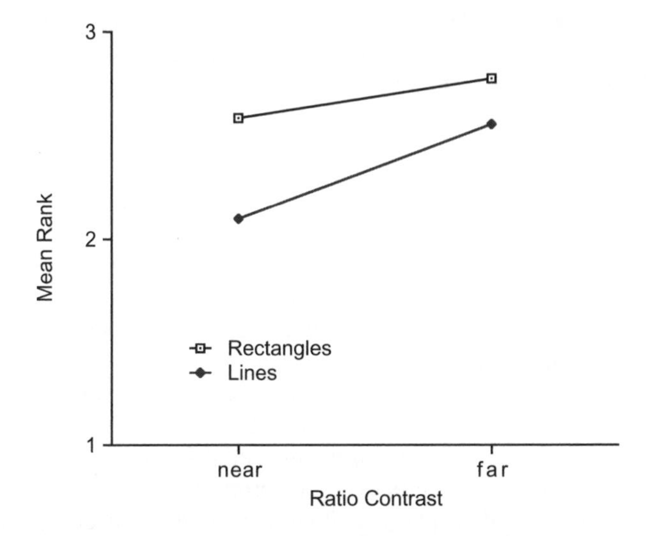

Figure 24.3: Mean rank contrasts for rectangle vs. line stimuli presented to adults in the early vs. the later part of the testing sequence.

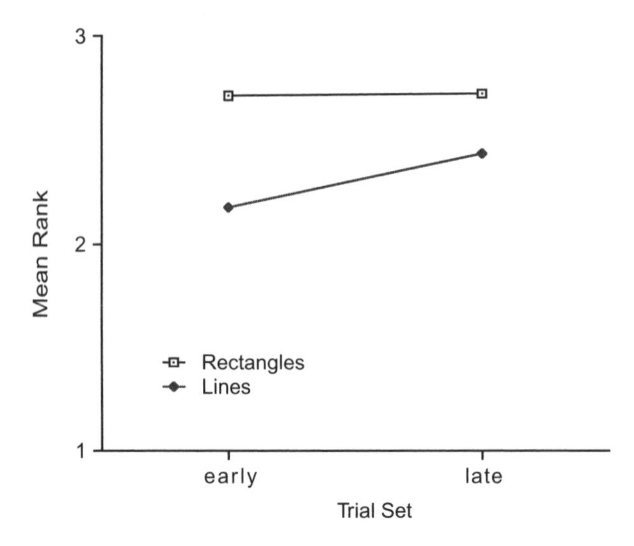

Figure 24.4: Mean rank contrasts for rectangle vs. line stimuli presented to adults in the early vs. the late part of the testing sequence.

significant difference between the line pairs and the rectangles ($p = 0.06$). There was no significant difference, however, between the far contrast line pairs vs. rectangles.

In a second analysis, the variables lines/early, lines/late, rectangles/early, and rectangles/late were compared. Friedman's test again showed the rankings for these four variables were significantly different ($p = 0.006$). These mean rankings are shown in Figure 24.4. Further analysis found a significant difference between the early and late sets of line stimuli ($p = 0.05$), but not between the two sets of rectangle stimuli. Similarly, within the early trial set there was a marginally significant difference between the line pairs and rectangles ($p = 0.07$), but the late sets of line pairs vs. rectangles did not differ significantly.

Discussion

Notwithstanding their near-ceiling performance, the adults in this experiment were still better at identifying proportional matches between rectangular stimuli than between pairs of lines that embodied the same quantitative relations. Their performance on the line stimuli was much better than that of the young children in Experiment 1, supporting the expectation that they would have considerably more proficiency with reasoning about proportional relations; yet they nevertheless continued to benefit from the presentation of the quantitative relations in spatial form. This is the first evidence we know about that shows adult processing of proportional information can be facilitated by the use of spatial displays, a finding with both theoretical and practical implications. Theoretically, this result points to important inter-relations between visual/spatial processing and higher-level, symbolic, forms of reasoning. Both the developmental primacy of reasoning about spatial proportions, and the continued advantage of spatial stimuli for adults, suggest that the basic visual processes that allow us to identify shape and other spatial configurations may be an important substrate for higher-level functions such as the identification of proportional equivalencies. Insofar as this theoretical insight provides a basis for designing information displays that make quantitative information more accessible to users, it may be of considerable practical value in optimizing computer interfaces.

Clearly, this research is only a first step toward understanding the links between visual/spatial processing and the effective interpretation of quantitative information. The adult data from Experiment 2 are limited by the presence of near-ceiling performance, a consequence of testing adults on a task initially developed for young children. In future research, a richer picture of how spatial presentations affect proportional reasoning could be obtained by using a more difficult task that would make it possible to evaluate the effects of other variables on performance with both spatial and non-spatial displays. Likewise, it would be useful to examine eye movement patterns in order to learn more about how individuals extract proportionality information from both spatial and non-spatial displays.

Experiment 3

In Experiment 3, we monitored the eye movements of a sample of adults as they performed a shape comparison task which entailed deciding which of two shapes that differed in overall size was "fatter" in the ratio of its width to its height. Two questions were of particular interest. First, would the fineness of the discrimination a problem required affect viewers' patterns of fixation? In visual search tasks the discriminability of targets from distractors affects the extent to which visual attention is demanded. Correspondingly, we hypothesized that the more similar the two ratios were that viewers had to compare, the more looking they would need to do to choose between them. Second, we were interested in how fixations would be distributed across the larger vs. the smaller of the two stimuli. This again was expected to shed light on the role of attentional processing in task performance. The stimuli were clearly too large to be encompassed in a single fixation. Two alternative possibilities then, are, (a) that the requisite proportional information is extracted from a series of fixations and integrated, in which case the larger stimulus should receive more fixations because less of it fits within the foveal region on any single fixation, or (b) that the proportional relations can be apprehended non-foveally, in which case the larger stimulus should not receive any more fixations than the smaller one and might even receive fewer (because it requires less fine-grained analysis).

Method

Thirty-one students from the University of Hawaii viewed 24 scenes depicting pairs of geometric shapes in outline form. The shape on the left was always greater in both height and width than the one on the right. The two shapes also differed in that one was fatter, that is, greater in the ratio of its width to its height, than the other. The scenes varied in whether the fatter shape was the larger one on the left or the smaller one on the right, in the general shape of both figures (hexagonal, oval, or rectangular), in the relative elongation of those shapes (narrow vs. medium or medium vs. squat), and in the degree of contrast between the ratios embodied in them (close vs. far).

Eye movement data was collected using an ASL eye movement monitor. A reflection from an infrared beam was used to automatically compute the location coordinates on the array where the eye fixated. Fixations were defined by the occurrence of at least three consecutive point locations within a 10 by 18 pixel area. Participants sat 5 feet from the 20-inch monitor on which the scenes were displayed. They responded verbally, "top" or "bottom."

Results

Accuracy was greater for widely separated ratios, $M = 94\%$, than for closer ones, $M = 80\%$. It was also greater when the correct choice was the alternative on the left ($M = 95\%$) rather than the one on the right ($M = 78\%$).

Fixations were tallied for all problems to which viewers responded correctly. The screen was partitioned into left and right regions (a 2/3 vs. 1/3 split, as the figure on the left was substantially larger than the one on the right). Numbers of fixations to each region were then compared across problems varying in the degree of contrast between the two ratios and in whether the correct stimulus was on the left or the right. Means are presented in Figure 24.5.

Effects of contrast were consistent with the discriminability hypothesis, although differences between low- and high-contrast problems varied for left vs. right fixations (significantly more right fixations on low- than on high-contrast problems, *M*s = 3.10 vs. 2.69, $F(1,30) = 18.87$, $MSe = 0.617$, $p < 0.001$; left fixation *M*s = 3.60 vs. 3.53, n. s.) and across problems involving more vs. less elongated shapes (more fixations on low- than on high-contrast problems only when the shapes were relatively elongated; *M*s = 3.72 vs. 3.24, $F(1,30) = 8.43$, $MSe = 1.91$, $p < 0.01$; for less elongated shapes *M*s = 2.99 vs. 2.95).

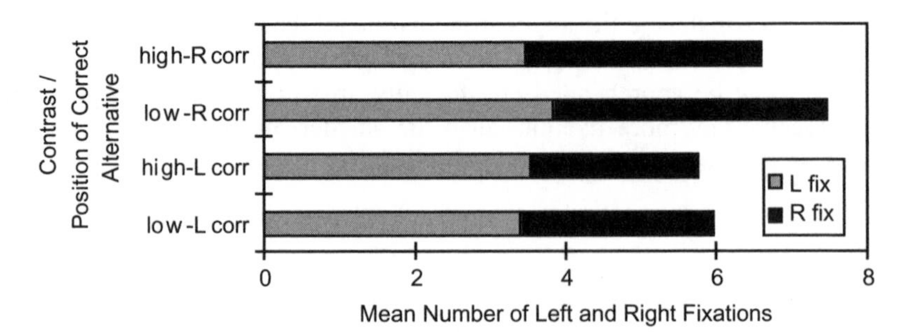

Figure 24.5: Patterns of eye fixations in ratio comparison task (Experiment 3).

Not surprisingly, viewers fixated the larger left region substantially more than the smaller right region overall, *M*s = 3.56 vs. 2.86, $F(1,30) = 14.19$, $MSe = 4.27$, $p < 0.001$. However, almost all viewers fixated both the left and right regions prior to making correct responses. One individual responded correctly 75% of the time yet fixated only one region (always the left) on a majority of those trials (55%). All of the other participants, however, fixated both regions on virtually every trial; across these 29 individuals there were only a few instances in which no fixations were recorded for one of the two regions of the scenes.

The distribution of fixations across the left vs. right regions varied with contrast and with the position of the correct alternative. There was a greater difference between the two regions on high- than on low-contrast problems, though left fixations predominated in both cases; for high-contrast problems *M*s = 3.50 vs. 2.69, $F(1,30) = 15.33$, $MSe = 3.14$, $p < 0.001$; for low-contrast problems *M*s = 3.61 vs. 3.10, $F(1,30) = 9.80$, $MSe = 1.69$, $p < 0.01$. Significantly more fixations fell on the left vs. right regions of the scenes when the correct alternative was on the left, *M*s = 3.47 vs. 2.40, simple effect $F(1,30) = 25.68$, $MSe = 3.003$, $p < 0.001$; but fixations to the two regions did not differ significantly when the correct alternative was on the right, *M*s = 3.64 vs. 3.40.

Discussion

The most striking finding from this study is the relatively small number of fixations needed to compare two spatial ratios (overall *M*, combining fixations to the left and right regions, = 6.45). Given the size of the figures, this observation suggests that much of the visual processing needed to compare the proportions of the stimuli was done non-foveally. At the same time, the finding that viewers almost always fixated both regions of the scenes at least once implies that some foveal viewing of each figure was needed to make a judgment. An interesting puzzle is how the one individual who most often fixated only the left region managed to respond correctly as often as he did.

Consistent with findings from studies of visual search, the number of fixations viewers made increased with the fineness of the discrimination a problem required. The finding that this pattern held only for right fixations suggests that viewers tended to focus on the left stimulus (perhaps because of its size, and/or because of a left-right scanning pattern) and only look at the right as much as they felt necessary to make their judgment. The attenuation of the left-right difference when the correct alternative was on the right may reflect a tendency to fixate the selected alternative as a choice was made. When this alternative was the one on the right, the resulting increase in fixations to that region would offset the usual predominance of left fixations.

Conclusions and Practical Implications

The comparative findings of Experiments 1 and 2 along with the fixation data of Experiment 3 support the conjecture with which we began, that two dimensional representations may facilitate the extraction of ratio information from computer displays.

Effects of contrast, however, suggest that there may be limits on the precision of judgments based on these representations.

Acknowledgments

This research was supported in part by Office of Naval Research grant no. N00014970578 and DARPA grant NBCH1020004 awarded to M. E. Crosby

References

Johnson-Laird, P. N., Legrenzi, P., Girotto, V., Legrenzi, M. S., & Caverni, J. P. (1999). Naive probability: A mental model theory of extensional reasoning. *Psychological Review*, *106*, 62–88.

Hollands, J. G., & Dyre, Brian P. 2000. Bias in proportion judgements: The cyclical power model. *Psychological Review*, *107*, 500–524.

Lawton, C. A. (1993). Contextual factors affecting errors in proportional reasoning. *Journal for Research in Mathematics Education*, *24*, 460–466.

Schwartz, D. L., & Moore, J. L. (1998). On the role of mathematics in explaining the material world: Mental models for proportional reasoning. *Cognitive Science*, *22*, 471–516.

Sophian, C., & Crosby, M. (1998) *Ratios that Even Young Children Understand: The Case of Spatial Proportions*. Proceedings of the Cognitive Science Society of Ireland.

Sophian, C., & Crosby, M., (1999*). Young Children Match Spatial Proportions*. Conference of the Society for Research in Child Development.

Spinillo, A. G., & Bryant, P. (1991). Children's proportional judgments: The importance of "half". *Child Development*, *62*, 427–440.

Chapter 25

Eye Tracking for Evaluating Industrial Human–Computer Interfaces

Gert Zülch and Sascha Stowasser

This chapter presents two eye movement studies in the field of evaluation of industrial manufacturing software. The first research objective is the exploration of different problem solving strategies in the context of scheduling shop-floor orders in the industrial manufacturing area. The second main research aspect concerns the analysis of different user strategies for data searches, which is structured according to object-orientation. The aim of the second investigation was to detect relationships between different types of data representation and the correct interpretation of the visualized data. Both studies are independent from each other and will hence be presented in separate sections of the chapter.

Importance of Ergonomic Software Design

The human–computer interface enables the interaction between a human user and a computer or, more specifically, the installed software-systems. In the past, ergonomic aspects were often neglected in the phase of software development and interface design, especially in production environments. Many software users complain about the difficulties in learning to use a software system or its insufficient functionality and the complexity of necessary interactions. Once the users get dissatisfied with a software, the functionality and the productivity of the system are no longer focused. The high learning effort, the insufficient usability and the maintenance effort are even more significant in this case.

Studies in enterprises show that the economic viability of a software system depends on its ergonomic design (cf. Klotz, 1994). Only 17% of all expenses for information technology in the first five years are investments in hard- and software. Twenty-six percent of all expenses are costs for technical support and administration. Nearly 57%

The Mind's Eye: Cognitive and Applied Aspects of Eye Movement Research

ISBN: 0–444–51020–6

of all expenses are incurred due to activities like searching errors, repairing data files, searching in handbooks or discussing usability problems with colleagues. These facts show the importance of an ergonomic design of interfaces. If ergonomic aspects are not considered thoroughly enough, expenses may increase enormously. In this case, the expected expenses for the first two years possibly double (cf. Klotz, 1994).

The design of industrial manufacturing software represents a main research area in the research of communication ergonomics. The aim of these research activities is to reduce complexity by using user-adequate software tools in order to support employees in their planning and control of operating tasks. The main empirical methods used for software evaluation are (cf. Grießer, 1995; Stowasser, 2002):

- eye movement registration,
- key-stroke recording,
- video-recording, and
- structured interview.

The appendix will give a description of these evaluation methods and mainly of the eye movement registration equipment and measurement which is used in the following investigations.

Investigation of Strategies Used for Solving Scheduling Problems

Detailed Planning of Shop-floor Orders

The trends in the shop-floor area are currently characterized by the fact that many organizational functions that are upstream to production areas (e.g. in operations planning) are increasingly being integrated into the manufacturing area. This shifting of functions has far-reaching consequences for process and departmental organization. The workers are confronted with new organizational methods and tasks. Thus, they are confronted with an increased number of holistic, comprehensive tasks, which can generally only be solved with computer aided procedures.

Thus, it makes sense that all procedures which are used by the workers in such a manufacturing system should be integrated into one technical system support workplace, the so-called "Leitstand system" (shop-floor control system). Using such an organizational workplace, tasks like the manufacturing of a specific parts spectrum can be controlled and monitored. In such a system the detailed planning of the shop-floor orders plays a central role, as illustrated in Figure 25.1.

Within the frame of an experimental examination, the detailed scheduling of shop-floor orders was examined as an aspect of system support (for details see Grießer, 1995). In the first examination series, in particular the influence of the task complexity upon the scheduling strategy and the results achieved was inspected. The focus of the

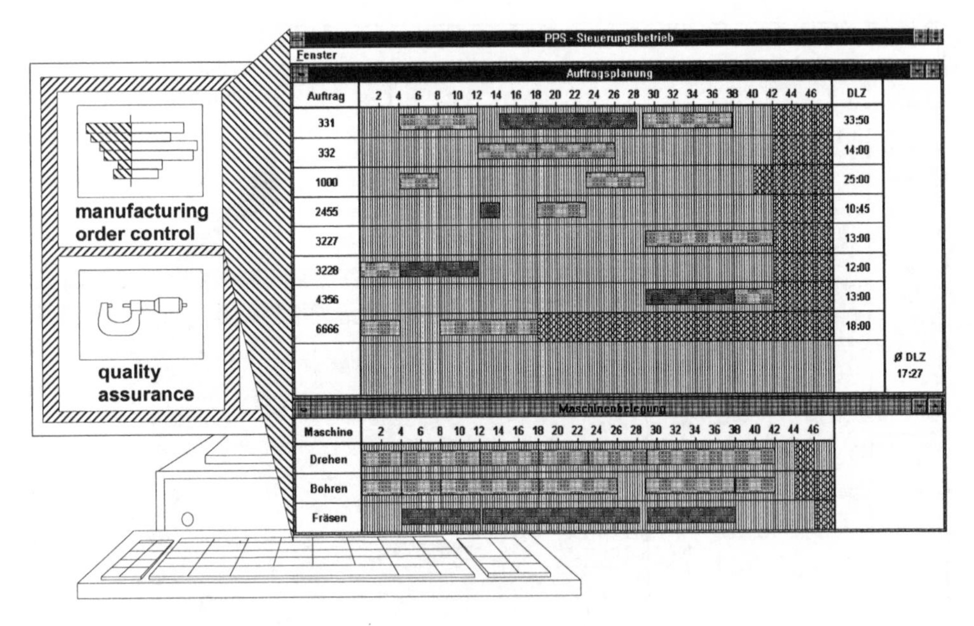

Figure 25.1: "Leitstand system" for shop-floor control.

second examination series was aimed at the impact of task complexity and colour composition on the scheduling strategies and learning effects.

Design of the Investigation of Strategies for Solving Scheduling Problems

Investigation tasks A special "Leitstand system", whose functionality was greatly reduced in comparison with industrial control work places, was conceived and developed specifically for the analysis of people's behaviour during the control of shop-floor orders. As in most standard graphic control stations, a time-related bar representation (Gantt-chart) was chosen for the overview of machines and orders. Through direct manipulation the individual work processes could be dispatched to the machines. The task is to find a valid order-to-machine assignment with an optimal throughput time solution. Hereby the intended finish date for the individual order must be reached.

Figure 25.2 shows an example of an optimal task processing. The initial situation is characterized by a multiple assignment to individual machines with various operations. In the case at hand a targeted maximum of 13 hours for the average throughput time for all orders has been set. The optimal solution lies at 11:33 hours. In the scenario shown here, the solution of the tasks has been complicated by the attempt to take further restrictions during the operating time, in the form of unforeseen machine maintenance, into account.

The individual work tasks can be ranked according to their complexity. The complexity increased, for example, with an increase in the number of orders or the

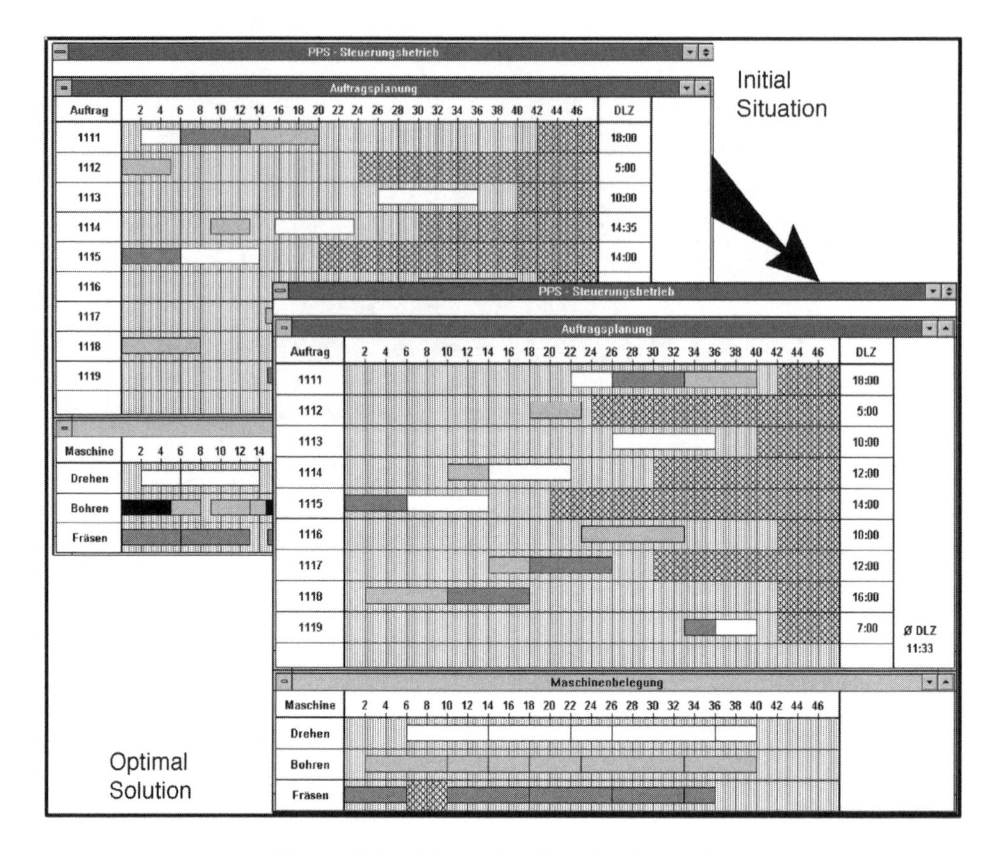

Figure 25.2: Example of a planning task.

number of interactions necessary to attain the optimal solution. Furthermore, particular events such as rush orders or machine breakdowns were simulated in order to increase the complexity of the work task. An additional criterion which was drawn upon for the determination of the complexity ranking was the standard time required to carry out the optional solution, calculated with the keystroke level method (Card *et al.*, 1983). The complexity ranking can be seen in Figure 25.3.

Evaluation methods and experimental design In order to record eye movements, the NAC Eye Mark Recorder Model V was used (see Appendix). Furthermore, video recordings and structured interviews were carried out in order to examine the above-mentioned questions.

In total two test series were carried out. Fifteen participants took part in the first test series. Upon completion of a practice run, in which the participants familiarized themselves with the procedure, five typical (practice-oriented) planning tasks with increasing complexity (task 1 to task 5) had to be completed. The processing order was pre-determined and identical for each participant. Thus, this first test series concerns a

	Low		Complexity			High
	Task 1	Task 2	Task 3	Task 4	Task 5	Task 6
Orders (number)	7	7	8	9	10	10
Operations (number)	15	15	16	16	27	27
Interactions (minimum) (number)	10	12	12	13	26	28
Rush orders			yes			yes
Machine breakdowns				yes		yes
Standard time according to Key-stroke-level-method (s)	50	55	55	58	118	126

Reasons for increased complexity of a task

Figure 25.3: Experimental design of the investigation.

single factor experiment in which the dependent variable (task complexity) is varied over five levels.

A further investigation task, which was the most complex (task 5), was conceived for the second test series. The content of tasks 1 to 5 remained identical to those in the first test series. The sequential order of the tasks differed from test person to test person so that an order effect can be excluded. However, the colour representation on the screen (which was either monochrome or polychrome) was varied in an attempt to determine the effects of coloured user-interfaces. Furthermore, the participants were confronted with the same tasks at the beginning and at the end of each test in order to examine the learning effect more closely.

Results of the Investigation of Strategies for Solving Scheduling Problems

Identification of various processing strategies When assessing the results it could be determined that the participants used (roughly-classified) either a "structured approach" or a "trial and error" approach in order to solve the presented task. The participants changed individually their approach with increasing task complexity. The "trial and error" approach was predominant in the very simple tasks (87% in task 1), whereas this shifted with increasing task difficulty to a structured approach (up to 80% in task 4). Figure 25.4 shows the key data of eye mark registration for the investigation of the strategies applied for solving the given scheduling problems. For detailed information about the measurement see the Appendix.

If one considers the average total problem solving time and the average number of interactions of the participants, separated by task and chosen strategy, one notices that those persons who chose a structured approach solved the presented task much faster and with fewer interactions. The Kruskal-Wallis test showed significant differences

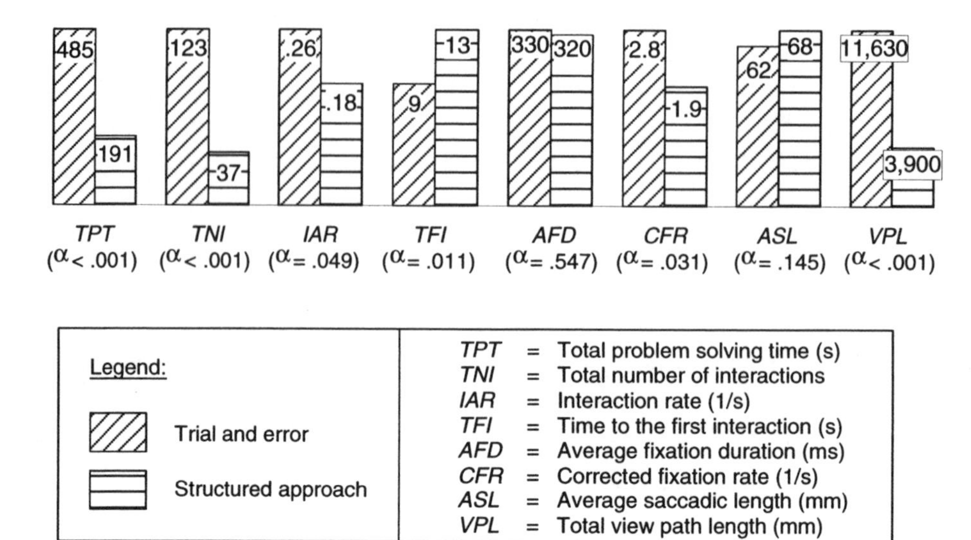

Figure 25.4: Key data of eye movement registration for identifying various problem solving strategies.

between the two strategies, not only in the total problem solving time ($\chi^2 = 4.29$; $\alpha < 0.001$), but also in the time until the first interaction ($\chi^2 = 6.51$; $\alpha < 0.01$), the total number of interactions ($\chi^2 = 3.85$; $\alpha < 0.05$), and the total mouse path length ($\chi^2 = 36.38$; $\alpha < 0.001$).

Furthermore, the Kruskal-Wallis test also showed that significant differences in visuomotor behaviour. Substantial differences between the two problem solving strategies were ascertained in the number of fixations ($\chi^2 = 16.02$; $\alpha < 0.001$), the fixation rate ($\chi^2 = 4.66$; $\alpha < 0.05$) as well as the total view path length ($\chi^2 = 14.11$; $\alpha < 0.001$).

Colour configuration and technical results The results of the examination of colour configuration showed that no significant differences between monochrome and polychrome screens were found with respect to the main variables of work behaviour. (cf. Grießer, 1995). No major differences between the two representation forms could be ascertained, either with eye mark registration or with keystroke recording (Figure 25.5).

The technical results, that is the deviation from the optimal solution as a percentage, also showed no significant differences between the monochrome (11%) and the polychrome (12%) representations.

It can thus be concluded that a colour user interface possesses no advantages. The subsequent survey of the participants showed however that the subjective evaluations from each participant favoured the colour screen. Therefore, for acceptance reasons, an equally functional colour configuration should be preferred.

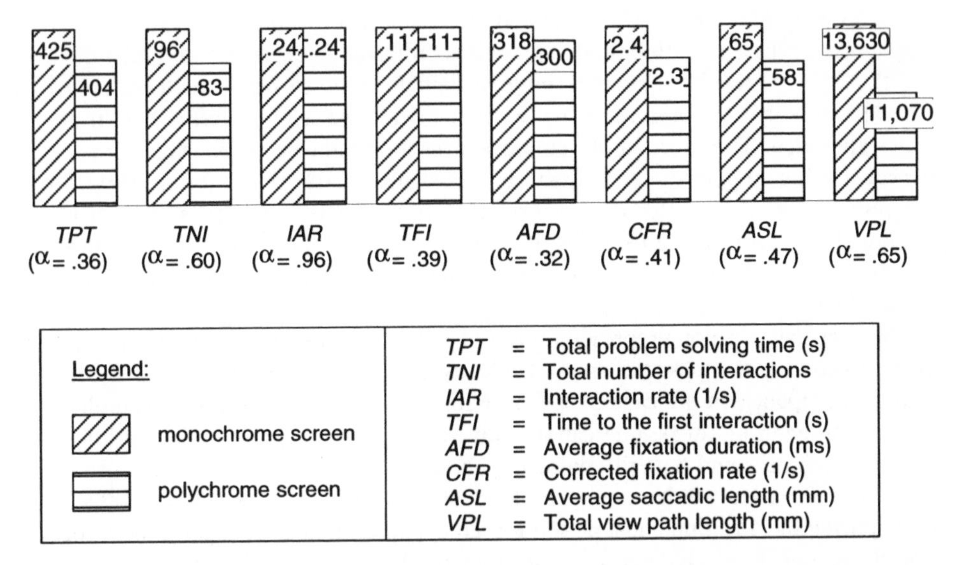

Figure 25.5: Key data of eye movement registration for examination of coloured screens.

Impact of learning effects In addition to the colour configuration, learning effects were also examined in the second test series. In accordance with the test plan each participant was asked to carry out with the same tasks at the beginning and the end of the test series.

As all participants took part in the first test series, and were therefore familiar with the use of the control station, a significant reduction in the average total processing time could be measured with repeated task processing in almost all cases. The total problem solving time was reduced by an average of 27%.

In the subsequent interview 80% of the participants stated that they had not recognized the repeated tasks as such. The remaining 20% recognized the test, but could no longer remember (according to their own accounts) the details of their solution, due to the complexity of the task. Therefore, the significant reduction in the total problem solving time in the second test series can be ascribed to a learning effect.

This fact is of particular importance if one considers the number of computer supported procedures with which the workers will be confronted in the future within the manufacturing area. At technical workplaces, for example, in which a variety of computer system procedures are integrated, the learning effect must be taken into account with every necessary procedure change. The more frequently the change takes place the heavier losses to performance can be expected.

Implications for Designing Control Systems

Various problem solving strategies were chosen intuitively by the participants for the solution of the presented tasks. The choice of a strategy is above all influenced by the complexity of the task. The "trial and error" approach is predominant in relatively simple tasks, whereas a structured approach is preferred in tasks with a moderate level of difficulty. The results show that those participants who chose a structured approach were able to solve the presented task quicker than those who preferred the "trial and error" approach. Therefore, a structured approach should be emphasized in training measures using an appropriate help system. Possible measures are, for example, adaptive procedures which offer the user suitable support when needed and react to the individual approach of the user. Furthermore, support in the form of online tutorials, in which the advantages of a structured approach are more closely considered, are conceivable.

The fact that a colour representation did not bring any improvement in the efficiency of solving the tasks was remarkable, even though it was preferred subjectively by the participants. In the case of unfavourable lighting conditions, for example as in many manufacturing areas, one can abstain from a colour representation as long as a suitable brightness and pattern contrast is chosen.

Despite good knowledge about the kind of presented tasks, a learning effect was found during the 45-minute test period for almost all participants. The problem solving time decreased by an average of 27% from the beginning of the test period, to the second processing at the end. The impact of the learning effect must be considered in particular when procedures occur in which, under certain circumstances, quick reactions and decision are necessary. Procedural or energy control tasks are examples for this. Particular attention must be paid to this effect when users are only occasionally confronted with such procedures. From a work organization's point of view, it should thus be attempted to keep the number of procedure switches to a minimum. The transfer of holistic tasks, associated with frequent changes of media and computer supported procedures, can be useful from a human factors engineering point of view, but should be weighed against the disadvantages of recurring learning effects.

Investigation of Object-orientated Data Presentation

Visualization of Object-orientated Data

Data, states, events, information, experience and knowledge are present in all production enterprises in a vast array of forms. There is a common trend for storage, administration and processing of these in a distributed and connected information system. The systematic collection and maintenance of information is a significant strategic factor for the competitiveness of a company. An important challenge for the application of information technologies does not only lie in the technical infrastructure for the storage, but also in providing suitable modelling and visualization methods for the representation and processing of the information.

Fundamental for the model- and data-based integration, an integrated, distributed, redundance-free database for the areas of design, operations planning and production of mechanical parts was developed with an object-orientated product/production model

(PPM) within the framework of the Collaborative Research Centre 346 "Computer-Integrated Design and Manufacturing of Parts" at the University of Karlsruhe. The object-orientated approach promises new perspectives (Corradi *et al.*, 1997) such as, for example, the combination of database functionality with the expressive power of object-orientated programming languages (Jonsson, 1998).

Aside from the well known strengths, object-orientated databases also posses weaknesses. According to Lang and Lockemann (1995), the object-orientated approach results in very complex databases in comparison with other data models, which can often lead to time-consuming and laborious searches for information. In addition, it must be assumed that with an increasing number of users co-operating within a distributed information system, the scale of the databases grows considerably. It must also be expected that an employee working with such complex information technologies is exposed to an even larger flood of information than in a less complex system.

A particular focus within the Collaborative Research Centre 346 lies in the conception, realization and evaluation of ergonomically favourable user-interfaces with respect to the handling, navigation and processing within object-orientated databases. Through ergonomically compatible forms of visualization, it should be possible to provide the user with a simple overview of the available data and easy access to the individually required data. The representation of thematic, abstract, condensed and detailed views of the object-orientated databases is connected with this.

Design of the Investigation of Object-orientated Data Representation

For the investigation of object-orientated data representation, two series of investigations were performed. The investigation was divided into subject dependent and subject independent examinations. In the subject independent examination with 20 students, some of the major possibilities for the representation of object sets and associations were analysed (Figure 25.6). The subject dependent examination was built upon this first examination phase and examined the various forms of representation of different views of the database as well as the visualization of object versions and their histories and archives. Twenty participants from several industry enterprises took part in this examination. In this way it could be verified to what degree the developed representation variations fulfilled the demand for industrial relevance and for a fast and intuitive access.

The central point in the assessment was the analysis of the view positions, which were recorded with the SMI-Headmounted Eyetracking Device (see Appendix; SMI 1999). In addition, the interactions of the users were recorded in the form of a log file using keystroke recording. Thus, this method was suitable for the analysis of the tactile actions of a user on the keyboard or mouse. In this case both the actual entries as well as the temporal interval between the entries were logged.

The following section will illustrate two aspects of the investigation of object-orientated visualization types:

- representation of object sets and associations,
- visualization of object archives and versions.

	First series of experiments		Second series of experiments	
Research task	Representation of object sets	Representation of associations	Representation of data structures	Visualization of archives and versions of objects
Number of test persons	20		20	
Education, profession	Students of different disciplines, all familiar with a PC		Experts from industry (production-, operations- and process planning), all familiar with a PC	
Average age	24.3 years		36.2 years	
Sex	3 female, 17 male		20 male	
Duration of eye mark registration	average 1 hour (without preparation, interviews...)		Average 1 hour (without preparation, interviews...)	

Figure 25.6: Participants in the investigation of object-orientated visualization types.

Representation of Object Sets

Developed variants for the representation of object sets A three-dimensional representation was chosen as the basis for the examination of the visualization of object sets, since this could benefit from the user's spatial perception ability. This promises advantages compared with a two-dimensional representation, in particular with the orientation within large object sets with a hypermedia structure (see e.g. Wandmacher, 1993; Carriere & Kazman, 1995; Snowdon & Jää-Aro, 1996). Cognitive psychological examinations corroborate the advantages of a graphical-spatial information configuration (see Dutke, 1994; Stowasser, 2002). One of the most frequently used arguments for the implementation of close to reality visualization is that of memory relief: The more exactly the internal relations within the computer system corresponds to the mental model, the better the computer system is understood and the easier the characteristics of the real system can be remembered or reconstructed (see Stowasser, 2002).

In the three-dimensional representation, the objects positions were constant in order to avoid stressful search processes after changes to the user interface (e.g. change of window). As a rule, it can be avoided that the visual system must adapt to a new graphic structure when the window layer is changed. Thus, the visual, cognitive effort which is required to identify the target information in the newly displayed window is reduced (see Fleischer & Becker, 1997).

Examination results for the representation of object sets In the first subject independent examination series the question concerning the maximum number of objects which can be represented simultaneously was posed. For this purpose, nine search and identification tasks of increasing complexity were given, in which the participants were supposed to identify certain objects. When only few objects were represented (< 10), all participants were able to locate the sought object upon first glance. From eye mark registration, this is proved by the relatively small number of fixations required for the discovery of the target. The searching process becomes more complex with an increasing number of simultaneously represented objects: the search time increases remarkably (see Zülch *et al.*, 2000). Wandmacher (1993) also assumes that the perception of a scene with up to seven objects can often be mastered with a single fixation. However, the examination described by Wandmacher focuses only on the perception of the scene and not, as analysed in the investigation here, on the search for a given object within this scene.

The assessment of the eye mark registration and the further data analysis showed that different approaches are used by the participants, dependent upon the complexity of the task (see Zülch *et al.*, 2000). A structured approach and an unstructured one could be roughly differentiated. In a structured approach the object quantities were searched in a structured manner (e.g. row by row or column by column), while participants with an unstructured approach allows their gaze to wander "chaotically" over the object set. The investigation showed that the unstructured approach led to the goal more slowly than the structured approach, so that the search for an object in a set of 64 simultaneously represented (in a context like the interface shown in Figure 25.7) objects led to an increase in search time due to the changeover to an unstructured approach often associated with a larger body of objects. Hence, the conclusion can be drawn, that not more than 64 objects should be represented simultaneously in a three-dimensional form of visualization in order to ensure that a certain object can be found quickly and with the smallest possible amount of mental strain.

Aside from the stipulation of a user-friendly number of represented objects, the dependence of the duration of the search process on the labelling of the object was examined. For this investigation, both symbols and alphanumeric characters were used to label the objects. In the search for an object out of a set of up to 49 objects, the search time was shorter with symbolic coding than with coding using alphanumerical names. If the number of objects rises above this level, the use of alphanumerical coding leads to a shorter search time than the use of symbolic coding (Figure 25.7).

The application of graphical symbols can however only be seen as useful when the symbols are self-explanatory, independent from the level of training or knowledge of the user. According to Fleischer and Becker (1997) a combination of several symbols for the labelling of an object is not useful since the transformation of the individual symbols into a metaphor for the object requires a greater mental processing effort than the semantic interpretation of the name.

Representation of Associations

The representation of a special object does not necessarily imply the complexity of the object. The complexity of an object can only be determined based upon associations, attributes and methods connected with it. In the second test series of the subject

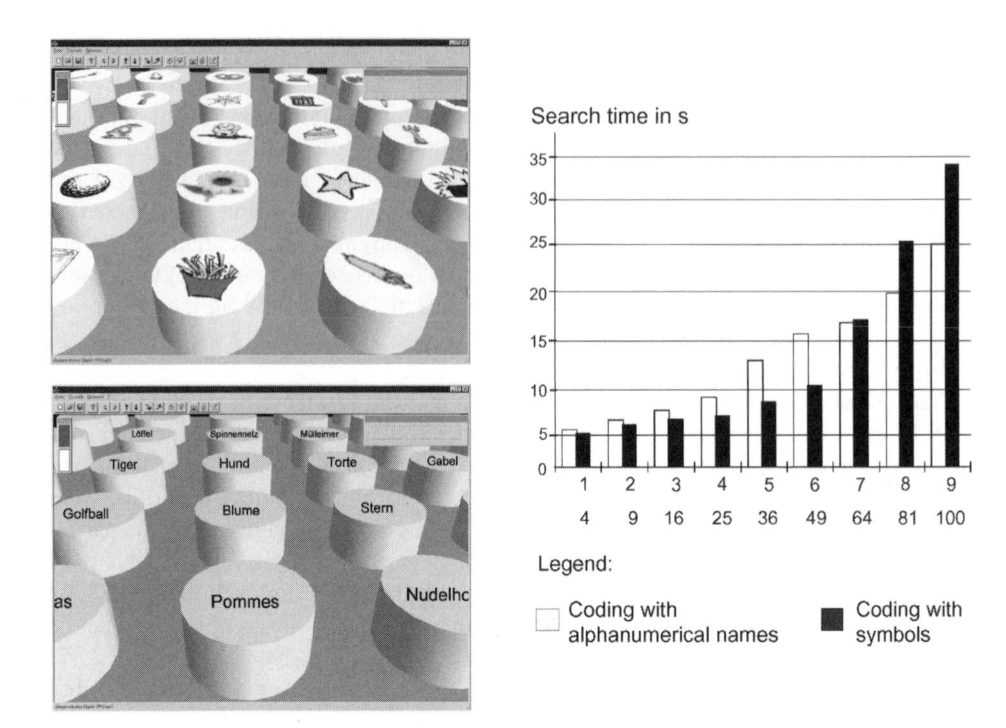

Figure 25.7: Visualization of data objects.

independent examination, various methods for depicting relationships were examined. The participants were given the task of picking out relationships of various complexity between one object and others. The relationships could be visualized with lists in a window or graphical line representations (Figure 25.8).

In a search for associations, 95% of the participants preferred, by their own account, the representation of associations with lists as opposed to with lines. Furthermore, all participants indicated that the use of lists sped up the search for associations. However, these subjective statements were only verified with eye mark registration when the number of associations represented was greater than 10 (Zülch *et al.*, 2000). When the number of associations was less than 10 the discovery of the sought association was better supported by representation with graphical lines since the special orientation, as well as the distinctive semantic characteristic of the goal stimuli (the lines) reduced the number of necessary fixations. In contrast, the flood of information connected with a line representation increases significantly with an increase in the number of presented associations, which could be confirmed with eye mark registration. The average fixation rate, a measurement of the frequency of fixations and thus the temporal variability in visuomotor behaviour, was 4.8 per second with the line representation and at 3.3 per second for the representation with lists. Representation with lists is clearer for the depiction of a great number of relationships (more than 10) of one object to another than the graphical representation of connections.

Objective analysis with eye mark registration:

Number of objects ≥ 10, then representation with window lists better.

Number of objects < 10, then representation with graphical lines better.

Subjective assessment of the test persons:

95% prefer textual lists.

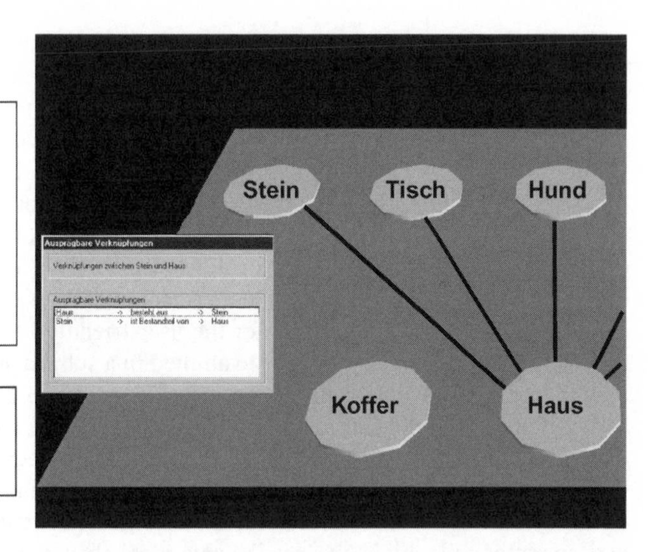

Figure 25.8: Visualization of associations.

Visualization of Object Versions and Archives

Object versions and their historical evolutions stored in archives are quite meaningful for engineering applications. An important example is the design of parts or components of a product. Within the investigation of object-orientated data visualization, the development, analysis and evaluation of ergonomically reasonable representations for the targeted identification of various object versions as well as historical data of an object has been carried out.

Design of the investigation of archives and versions of objects The identification of components, their state of validity and time of transactions on them can be called upon for the representation and assessment of temporal aspects in object modelling (see Hughes, 1992; Schreiber, 1995, p. 126; Saake *et al.*, 1997).

Identification of object versions Normally, as soon as an object is created for the first time, only one valid version of this object should be displayed to the user. If modifications to the object are carried out, the user must be made aware of the version sequence in such a way that the versions can be distinguished from one another and can be arranged according to their historical context.

State of validity of a version In the representation of data, the user must know whether the object version is valid, outdated or invalid. The state of validity of a version generally changes with time.

Transaction date The transaction date (also called "time of registration") indicates when a specific version of the object was saved in the database. Fundamentally, the transaction date can neither be changed retroactively, nor can it lie in the future.

Fundamentally different forms of visualization can be chosen for presenting object versions, the state of validity and the transaction date. In the choice of a representation variant, it is important that the user can locate the sought data quickly and securely as well as perceive and process it without error. Various characteristics such as form (e.g. fonts, pictograms, graphics), colour (e.g. red, yellow, green) and location (e.g. object proximity) can be used for the optical coding of object variants.

Various coding characteristics for the user-friendly visualization of object versions and their historical inventory were examined in a subject specific examination. Twenty people took part in this examination (see Figure 25.9). Each participant was given a series of tasks including 21 tasks of identifying versions, validity states as well as times of transaction for various objects. In this test, the objects were instanced using condensed key figures from the areas of personnel, technology and organization (key figure system defined by Groth, 1992). In different scenarios from order scheduling and operations planning, the participants were then required to process various data archives aspects (e.g. identification of the oldest and still valid version of an object, identification of the instance with transaction data 13.11.97). Initially, the different variants were presented to the user separately. In a final scenario all variants were then integrated into the user interface in order to sort out the user's preferred variant.

Generally, the variants differ with respect to the coding of their position on the screen (Figure 25.9). The object attributes are usually visualized either in direct proximity to the object (e.g. version number directly beside the object name) or at a greater distance from it (e.g. version number in an information bar). The coding of version identification and transaction date were implemented numerically. According to this, a version is described clearly by a version number as well as a transaction date (date and time). Other forms of graphical visualization of these quantitative specifications, e.g. with symbols or diagrams, are, from an ergonomics point of view, not suitable (see DIN 66 234, part 5, 1988; Wandmacher, 1993; Shneiderman, 1998).

In comparison, qualitative statements about the state of validity can be represented by means of alphanumeric symbols, colours or pictographical symbols (see DIN 66 234, part 5, 1998).

The state of validity of a version could assume the characteristic "current" (current version with the most recent date of transaction), "valid" (not the current version, some attributes however are still valid) or "invalid" (outdated version, attributes are no longer valid). An object can, according to this definition, possess one or more actual versions, including one or more valid versions as well as one or more invalid versions. Three variants with alphanumeric symbols (current, valid, invalid), one colour coded "traffic light"-representation (green denotes current, yellow denotes valid, red denotes invalid) and one abstract bar graph representation (full denotes current, half full denotes valid, empty denotes invalid) were examined for the representation of the state of validity. These three variants were, in one case, positioned directly beside the object, and in the other case, positioned at a distance (in the information bar).

Component	Investigated variants	Examples
Identification of versions	letters, in direct or distant position to the object	version number: V1, V2, ...
State of validity	letters, traffic light visualization, filling beams visualization, each in direct or distant position to the object	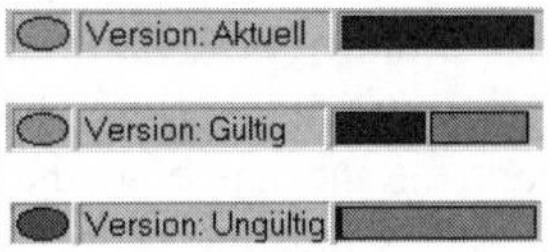 Version: Aktuell Version: Gültig Version: Ungültig
Transaction date	letters, in direct or distant position to the object	data entry time: 12.04.1998 / 13.31 h

Figure 25.9: Visualization of object versions and archives.

Examination results for the representation of object versions and archives The following results were attained in a combination of the analysis methods described in the first section. The questions posed were:

Which spatial position is preferred for the representation of version number, transaction date or state of validity with respect to the representation position of the object?
Which representation of the state of validity is preferred?
Which strategy is used preferentially for searching historical data?

The analysis of the interviews and the eye movement registration showed that all participants preferred the coding of the version number, transaction date and state of validity in direct proximity to the object. A representation of these attributes in the information bar (e.g. after the mouse pointer was moved over the object) was proven to be unsuitable; this form of coding led to a 25% longer search time. Furthermore, it was observed that an indirect positioning of the temporal attributes increased the strain on the visual system due to more frequent glance changes, which is proven by the average eye path of 20,800 mm with an indirect positioning, compared with 16,700 mm with positioning in close proximity. Thus, the attributes for the representation of versions or archives should be placed relatively close to the object. For example, in a line representation of all objects, the version number should be aligned directly adjacent to the represented object.

The examination showed clear preferences for the representation of the state of validity. The use of colour coding (the traffic light representation in this examination) substantially improves the performance with respect to orientating, searching and discovering the state of validity. A colour coding can be detected more quickly in a set of objects than other forms of coding, such as alphanumerical characters or abstract graphic representations (e.g. bar graphs). This advantage of colour coding is in part a result of the fact that colour signals can be discovered and differentiated well, even in the periphery of the fixation point.

The comparison of the identification durations of the sought objects (averaged for all participants and tasks) resulted in 17 seconds for colour coding, 22 seconds for the bar graph representation and 23 seconds for representation with written characters. Thus, invalid objects should be coded, for example, in red, either with red characters for the name in the object list or with an abstract colour labelling (e.g. a red circle beside the object name). However, it should be mentioned that the sole use of colour as a coding characteristic cannot be recommended, for several reasons (e.g. colour weakness of user, monochrome screens). It should rather be supplemented with a further form of coding (e.g. alphanumerical signs).

During the survey of the participants and the assessment of the eye mark registration it was discovered that various different strategies were used to solve the given tasks. A difference could thereby be identified between a structured approach to identifying the object's historical data and an unstructured one. In a structured approach, the presented set of objects is searched systematically for the desired attribute. Participants with unstructured approaches shift their gaze to randomly chosen objects in the hope of finding the right one in this manner.

The rate of fixation can be called upon in the description of fixation behaviour during task processing. It has been shown that a much smaller fixation rate can be ascertained with those individuals who use a structured strategy (Grießer, 1995). Fixation transfers, so-called saccades, occur between two fixations. The average saccadic length was substantially longer when the structured approach was used than when an unstructured approach was used. This behaviour can be explained by the fact that the participants with a structured strategy let their eyes flow over the rows or columns. A visual information uptake can obviously not always be assumed with this "wandered" gaze so that the saccades between two fixations are longer. The eye path

Characteristic number	All test persons	Test persons with structured approach	Test persons with unstructured approach
Total processing time (in s)	383	356	409
Fixation rate (1/s)	4.4	3.3	5.5
Average saccadic length (in mm)	27	35	18
Total view path (in mm)	40,191	37,904	42,477

Figure 25.10: Key data of the eye mark registration of object versions and histories.

behaves differently: the more difficult the pattern of stimuli is to identify, the longer is the length of eye paths. Individuals with an unstructured approach should be directed by a structured arrangement of the objects on the screen and a correspondingly clear representation of the associated attributes.

Conclusion

This chapter presented a survey of two main research fields within the field of evaluating industrial manufacturing software. The first research objective was the exploration of different problem solving strategies by means of eye movement registration. In particular the differences in education, knowledge, experience, etc. of the people using the same software system was the motivation for investigations of the design of Human–computer interfaces which support different users.

The goal of the experimental studies was to gain a more in-depth knowledge about the dependencies between different kinds of information representation and individual problem solving strategies. The study was carried out on several typical shop floor scheduling problems: a list of manufacturing tasks were given and had to be assigned to machines. It was observed that two main strategies were used to solve this kind of problems, namely trial and error vs. a structured approach.

The second main research concerned the observation of different user strategies in searching data, which is structured in an object-orientated manner. Based on existing results and ongoing research activities for visualizing data, e.g. in hypermedia and network-orientated databases, several investigations were performed to verify several presumptions regarding the connection between object-orientation and cognitive information processing. The aim of these investigations was to detect relationships between different types of data representation, correct interpretation of data and personal preferences. From this research, hints for the user-friendly design of object-orientated databases (object attributes, associations, archives and variants of objects) could be derived. For example, it has been shown that an extensive graphical visualization of all information is not favourable. The presented object set has to be limited. In order to describe relationships and associations between objects, graphical and textual elements should be used in an appropriate, context-sensitive way.

What are the aspects of future research in this field? One aspect is derived from the growing importance of information networks. Existing knowledge about a given problem should be processed to and for the user in an ergonomical manner. Navigation through the knowledge items and the handling of the knowledge items should be designed in a user-friendly way.

A second aspect concerns the dynamics of information and data. For industrial applications, process-orientated data are significant for decision making. In particular, the representation of continually ongoing process data should be focused in order to support decisions in a dynamic changing process (cf. Stowasser, 2002). Furthermore, the user-friendly processing of knowledge and experience data is still an open field. Here, the dynamic evolution of data, including their temporary validity, leads to further questions regarding the final goal of establishing some sort of management of knowledge.

Acknowledgement

This contribution is part of the Special Research Centre 346 "Computer Integrated Design and Manufacturing of Parts" founded by Deutsche Forschungsgemeinschaft (German Research Foundation).

References

Anders, G. (2000). *Messungen von Blickbewegungen im A330 Full Flight Simulator*. Berlin: Technical University of Berlin, Institute of Aeronautics and Astronautics.

Card, S. K., Moran, T. P., & Newell, A. (1983). *The psychology of human–computer interaction*. Hillsdale NJ: Erlbaum.

Carriere, J., & Kazman, R. (1995). Interacting with huge hierarchies: Beyond cone trees. In: *Proceedings of IEEE Information Visualization '95* (pp. 74–81). Los Alamitos CA: IEEE Computer Press.

Corradi, E., Bartolotta, A., & Garetti, M. (1997). Manufacturing Systems Engineering for the Extended Enterprise. In: P. Schönsleben and A. Büchel (eds), *Organizing the Extended Enterprise, Vol. 1.* (pp. 137–147). Zürich: IFIP Working Group 5.7; BWI Institute for Industrial Engineering and Management at the Swiss Federal Institute of Technology ETH.

DIN 66 234, Teil 8 (1988). *Bildschirmarbeitsplätze. Grundsätze ergonomischer Dialoggestaltung*. Berlin: Beuth-Verlag.

Dutke, S. (1994). *Mentale Modelle: Konstrukte des Wissens und Verstehens*. Göttingen: Verlag für Angewandte Psychologie.

Enderle, E., Korn, A., & Tropf, H. (1982). Echtzeit-Registrierung von Blickregistrierungen bei frei beweglichem Kopf. *FhG-Berichte*, *3*, 12–15.

Fleischer, A. G., & Becker, G. W. (1997). Such- und Merkprozesse bei der Anwendung von Fenstertechniken. *Zeitschrift für Arbeitswissenschaft*, *51(23NF)*, 65–78.

Grießer, K. (1995). *Einsatz der Blickregistrierung bei der Analyse rechnerunterstützter Steuerungsaufgaben*. Doctoral dissertation, University of Karlsruhe. (ifab-Forschungsberichte aus dem Institut für Arbeitswissenschaft und Betriebsorganisation der Universität Karlsruhe, Band 10) (ISSN 0940–0559).

Groth, U. (1992). *Kennzahlensystem zur Beurteilung und Analyse der Leistungsfähigkeit einer Fertigung*. Düsseldorf: VDI-Verlag. (Technik und Wirtschaft; Reihe 16).

Gullberg, M., & Holmqvist, K. (1999). Keeping an eye on gesture. *Pragmatics and Cognition*, *7*, 35–65.

Hughes, J. G. (1992). *Objektorientierte Datenbanken*. München, Wien: Carl Hanser Verlag.

Jonsson, U. (1998). Object-oriented modeling of products and production systems. In: G. Zülch (ed.), *Design of Organisational Structures, Work Systems and Man-Machine-Interaction* (pp. 33–45). Aachen: Shaker Verlag. (ifab-Forschungsberichte aus dem Institut für Arbeitswissenschaft und Betriebsorganisation der Universität Karlsruhe, Band 16).

Klotz, U. (1994). Objektorientierung — ein facettenreiches Leitbild verbindet Flexibilität mit humaner Arbeitsgestaltung. *Zeitschrift für Arbeitswissenschaft*, *48*, 99–112.

Lang, S. M., & Lockemann, P. C. (1995). *Datenbankeinsatz*. Berlin: Springer Verlag.

NAC (1986). *Operation Manual — Eye Mark Recorder Model V*. Tokio: NAC.

Saake, G., Türker, C., & Schmitt, I. (1997). *Objektdatenbanken*. Bonn: International Thomson Publishing.

Schreiber, D. (1995). *Objektorientierte Entwicklung betrieblicher Informationssysteme*. Heidelberg: Physica Verlag.

Shneiderman, B. *et al.* (1998). *Designing the User Interface*. Reading MA: Addison-Wesley.
SMI (1999). *iView*. Teltow: SensoMotoric Instruments.
Snowdon, D., & Jää-Aro, K.-M. (1996). Body-centred configuration in collaborative virtual environments. In: M. Bergamasco (ed.), *Proceedings of the 2nd FIVE International Conference* (pp. 48–54). Pisa (Italien): FIVE.
Stowasser, S. (2002). *Vergleichende Evaluation von Visualisierungsformen zur operativen Werkstattsteuerung*. Aachen: Shaker. (ifab-Forschungsberichte aus dem Institut für Arbeitswissenschaft und Betriebsorganisation der Universität Karlsruhe, Band 26).
Wandmacher, J. (1993). *Software-Ergonomie*. Berlin, New York: Verlag Walter de Guyter.
Young, L. R., & Sheena, D. (1975). Survey of eye movement recording methods. *Behavior Research Methods and Instrumentation*, *7*, 397–429.
Zülch, G., Fischer, A. E., & Jonsson, U. (2000). Objektorientierte Modellierung und Visualisierung von Planungs- und Methodenwissen. In: H. Krallmann (ed.), *Wettbewerbsvorteile durch Wissensmanagement* (pp. 151–202). Stuttgart: Schäffer-Poeschel. (HAB-Forschungsberichte, Band 11).
Zülch, G., & Stowasser, S. (1999). Einsatz der Blickregistrierung zur Gestaltung von Prüfarbeitsplätzen in der Bekleidungsindustrie. *Zeitschrift für Arbeitswissenschaft*, *53(25NF)*, 2–9.
Zülch, G., & Stowasser, S. (2001). VISOR — Towards a three-dimensional shop-floor visualization. In: M. J. Smith and G. Salvendy (eds), *Systems, Social and Internationalization Design Aspects of Human–Computer Interaction* (pp. 217–221). Mahwah NJ, London: Lawrence Erlbaum Associates.
Zülch, G., Stowasser, S., & Keller, V. (2001). Sichtenkonzept zur kommunikationsergonomischen Darstellung von objektorientierten Unternehmensdaten. *Zeitschrift für Arbeitswissenschaft*, *55(27NF)*, 113–123.
Zwerina, H. (1992). Erkennung von Sehzeichen in unterschiedlichen Strukturen auf dem Bildschirm. Doctoral dissertation, University of Karlsruhe.

Appendix: Equipment and Evaluation Methods

Eye Movement Registration Equipment at the ifab-Institute

Precise knowledge of eye positions is useful for many types of investigations, not exclusively in basic research on the ocular system but also in many application-orientated research fields. In order to record eye movements for the purpose of evaluating industrial manufacturing software, two head-mounted registration systems can be found in the Laboratory for Human-Machine Interaction (Figure 25.11). The older system, a NAC Eye Mark Recorder Model V (NAC, 1986), serves as a stationary system for the analysis of informational aspects of human–computer interfaces. The second system, a SMI Headmounted Eyetracking Device (SMI, 1999), is used both in stationary and in field experiments.

Eye movement registration with NAC Eye Mark Recorder An NAC Eye Mark Recorder model V (NAC, 1986), which operates according to the corneal-reflection method (cf. Young & Sheena, 1975), was used to record eye movements. Initially, the

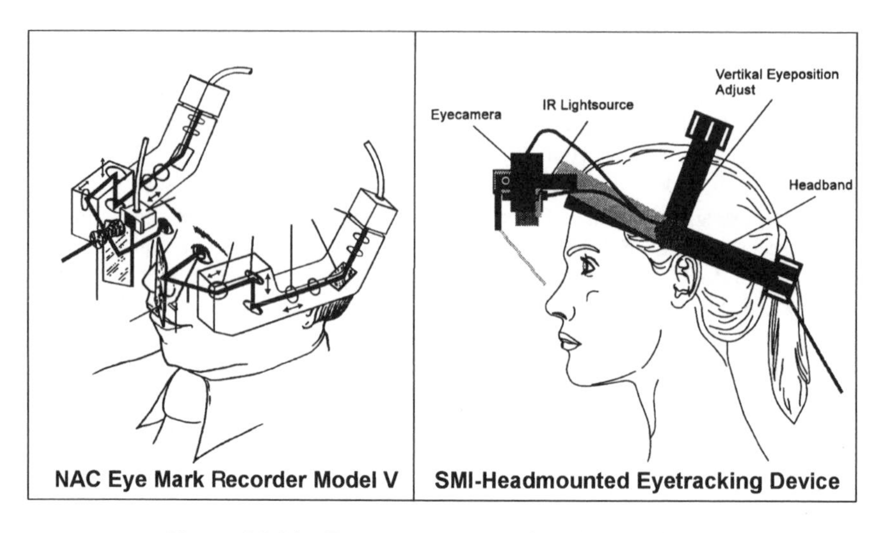

Figure 25.11: Eye movement registration devices.

basic equipment merely allowed the view direction to be determined, with respect to the head position. An assignment of individual view positions to the fixed objects in the visual field can only be achieved through a time-consuming, manual assessment of the still images recorded on video. In order to avoid a restricted head mobility and yet be able to provide an automated assessment, the standard model of NAC Eye Mark Recorder model V has been extended. The technical equipment which is required, in addition to the actual eye mark registration device, for controlling and assessment of the experiment is described in detail in this section.

The eye movements are recorded, together with the image from the visual field camera, by a eye mark registration device and forwarded to the NAC-Camera-Controller. This information is combined in a mixed image and transformed into a monochrome video signal. In addition, various reference points within the visual field, which light actively in the infrared spectrum, are recorded by the visual field camera. These reference points are used to separate head and eye movements from another in the subsequent image processing. This is necessary in order to be able to automatically associate individual view points to specific objects within the visual field.

The NAC-Camera-Controller video signal is digitalized and analysed completely for the image analysis with help from a process computer (SAM — Sensorsystem für Automation und Messtechnik) (Enderle *et al.*, 1982). A maximum of 25 individual images per second can be assessed with this configuration. The co-ordinates of the corneal reflection and two of four reference points lit in the infrared spectrum are recorded for each individual image. Head and body movements can be separated mathematically from the eye movements using the position of the reference points so that the complete test structure allows for free head movements and, in addition, makes an online calculation of the co-ordinates in the object space possible.

Eye movement registration with SMI Headmounted Eyetracking Device A Headmounted Eyetracking Device (HED-II) with Headtracker from SensoMotoric Instruments (SMI, 1999) was used in experimental investigations for the analysis of visualization forms of user interfaces. This computer supported eye mark registration system had already proven itself in various examinations (e.g. Gullberg & Holmqvist, 1999; Zülch & Stowasser, 1999; Anders, 2000; Zülch *et al.*, 2001; Zülch & Stowasser, 2001; Stowasser, 2002).

The SMI-System uses the light reflecting qualities of the cornea and operates on the basis of point of regard measurement (SMI, 1999; Young & Sheena, 1975). For this the eye is illuminated with an infra-red light source with a wave-length of 880 nm. At the same time, the system records a video image of the eye with an eye camera using a half mirror which reflects only infra-red light. This is analysed with computer support and a sampling rate of 50 Hz by locating the nearly circular pupil as well as the reflection of the light beam on the cornea (corneal reflection). By using two threshold values the corneal reflection point and the pupil are filtered from the recorded video image. The view position is calculated online from the position of the corneal reflection point in relation to the pupil centre.

At the same time the participant's visual field is recorded with a scene camera. Since eye and scene camera are "rigidly" mounted on the headband with a common framework, a relationship between the view position and the viewed object can be created. A superimposition of the view position on the visual field, which is carried out directly with help from a computer system, produces a colour scene image with integrated identification of the participant's current view position.

Measurement and Evaluation Methods

In order to record numerical data of view positions on an object online, some definitions are necessary. First of all, this concerns the definition of a fixation (Figure 25.12). A fixation may consist of a number of view positions. If there are more than 5 view positions in a circle with a radius of 7 mm we count them as 1 fixation. All distances between fixation points are measured from the centre of this circle. In order to exclude view positions without any mental information processing, a fixation must have a minimum duration of 200 ms. This definition is in accordance with other authors, e.g. Zwerina (1992).

In order to evaluate user interfaces it is necessary to define areas of interest. This refers to those parts of a screen in which a defined class of information is placed. Thus, several fixation points can be found in one area of interest. Fixations in one area of interest following one after the other is called a gaze. For certain applications, in particular when monitoring technical systems, sequences of screens or screens with dynamically changing information items are to be observed by the user. For this purpose, scenes have to be defined, including the transition from one scene to another. One specific measure of this is the initial fixation in a scene (Stowasser, 2002). It numerically describes the extent to which orientation in the scene is required from the user.

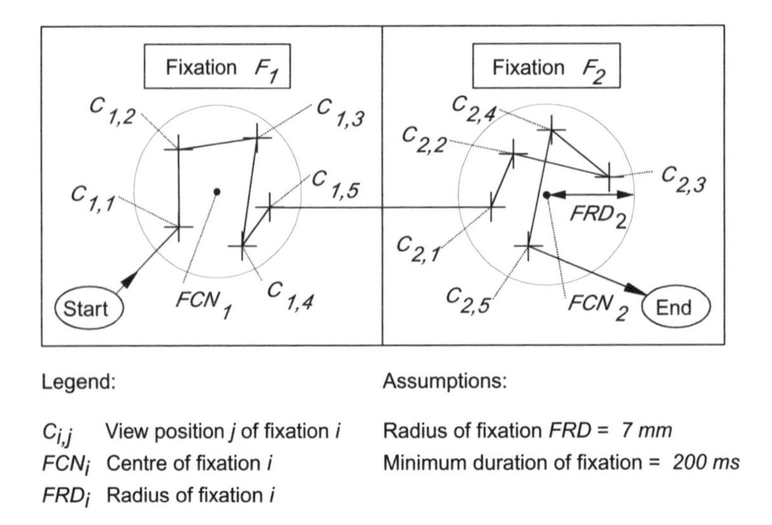

Figure 25.12: Definitions of fixations for eye movement registration.
Source: Grießer (1995) and Stowasser (2002).

From the eye tracking device itself, only the view points of a user are recorded. From this raw data, the fixations and gazes are calculated using special evaluation software. This data can be further processed (Figure 25.13). For example, view paths for a certain time period of the observation can be traced. A so-called Sankey-diagram may show the strength of gazes between the areas of interest. This kind of diagram is well known from material flow investigations.

The following are among the most important eye movement registration variables which can be recorded with the described procedure (see Grießer, 1995; Stowasser, 2002):

- the number of view positions,
- the number of fixations,
- the number of fixations per area of interest,
- the average duration of fixations,
- total fixation duration,
- average length of saccades,
- total view path,
- the average number of fixations per gaze,
- the average gaze length,
- the average gaze duration,
- the frequency of transitions between two areas of interest,
- the frequency of specific view sequences, and
- the fixation rate.

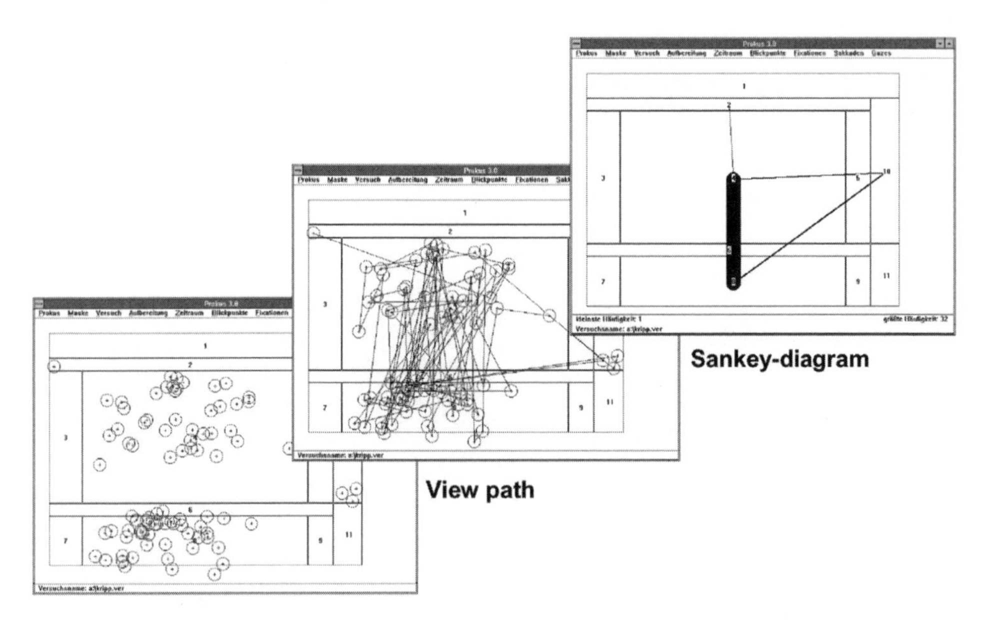

Figure 25.13: Evaluation methods for eye movement registration.

Additional Examination Methods Besides Eye Movement Registration

In addition to eye movement registration, the keystroke-recording method has been implemented as a further quantitative evaluation method. With this method, all user actions are recorded in a log file. Total processing time, length of time before first interaction, total number of interactions or rate of interactions, meaning the number of interactions per unit of time, are among the most important registration variables which can be recorded with help from keystroke-recording. Additionally, entire processing sequences can be analysed through a log file analysis. The keystroke data and the eye movement data are synchronized, so the combination of the user's visual and manual processes can be analysed.

As a source of supplemental data, structured interviews are suitable in particular for the recording of demographic data and for detailed analysis of the processing strategies used by the participants. In addition, video recordings were made in order to document the entire test procedure.

Chapter 26

Eye-movements and Interactive Graphics

Carol O'Sullivan, John Dingliana and Sarah Howlett

In this chapter, we will discuss the usefulness of eye-tracking for computer applications that attempt to render simulations at interactive rates. Some recent research that uses this technology is presented. While eyetracking has been relatively well-explored for the purposes of accelerating real-time rendering, it has not been very much considered with respect to the actual simulation process. In fact, perceptual issues relating to physically-based simulation have been largely neglected in the field of computer graphics. To evaluate simulations, we explore the relationship between eye-movements and dynamic events, such as collisions. We will describe some gaze-contingent systems that we and others have developed that exploit eye-tracking to enhance the perceived realism of images and simulations. Some experiments that were carried out to evaluate the feasibility of this approach are also presented.

Introduction

Researchers in the field of computer graphics are concerned with producing images and animations that are as realistic as possible, mainly to human viewers. Creating images consists of two phases: *modelling* a scene containing objects and environmental effects and storing an appropriate representation in digital format; and *rendering*, or displaying, these scenes using a variety of platforms and technologies. The two problems are not independent — the quality of the final image depends greatly on the accuracy of the models, as does the speed at which the scene may be rendered. The more complex and detailed the model, the longer it will take to produce a graphical representation of it. If we now consider the problem of producing an animation, we can see that both the above phases will form an integral part of this process, as each frame of an animation consists of a computer generated image of a digitally-represented model. At each time-step of the animation, the positions and orientations of the models must be updated by some means, known as *simulation*, and an image rendered of the updated

The Mind's Eye: Cognitive and Applied Aspects of Eye Movement Research

ISBN: 0–444–51020–6

scene. When this animation is shown to a viewer, at the very least 10 frames per second must be displayed, or else the animation will appear to be very slow and jerky. In fact, most movie animations are generated at around 25 frames per second, while with interactive applications such as games the ideal frame rate is often much higher than this. In the case of a movie, the frames of the animation can be created in advance and then played back in real-time to the viewer. Hence, the speed at which the frames are rendered is not critical. However, for interactive animations each frame must be generated while the viewer watches, thus providing a formidable challenge for the graphics engine. Strategies that reduce this computational burden often produce poor-quality images and motions as a trade-off, but by taking perceptual factors into account and integrating eye-movement analysis, adaptive systems can be developed which improve the perceived quality of the degraded graphics.

Perception, Eye-movements and Graphics

In this section we will consider in more detail the phases of modelling objects with a limited amount of data, rendering these objects quickly and simulating their motions realistically, in each case examining how eye-movement analysis can be used to understand and improve performance and quality.

Modelling

In time-critical computer graphics applications, such as Virtual Reality (VR), three-dimensional objects are often represented as meshes of polygons (see Figure 26.1). The requirement in interactive systems for real-time frame rates means that a limited

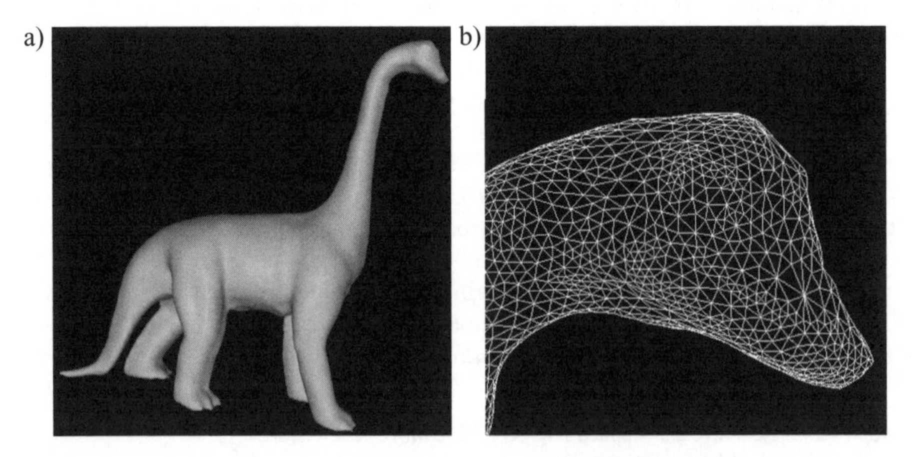

Figure 26.1: (a) A dinosaur model, consisting of 47,904 triangles and 23,984 vertices. (b) A detailed view of the triangles forming the head.

number of polygons can be displayed by the graphics engine in each frame of a simulation. Therefore, meshes with a high polygon count often have to be simplified in order to achieve acceptable display rates. The number of polygons and hence the *Level Of Detail (LOD)* of the model needs to be reduced (see Figure 26.2).

This can be achieved in two ways: a representation of the object at several levels of detail with a fixed polygon count can be generated, although switching between such levels of detail can cause a perceivable "pop" in an animation; or special meshes, called multi-resolution or progressive meshes, can be built that can be refined at run-time, i.e. parts of an object may be refined or simplified based on the current view, thus allowing a smooth transition from lower levels to higher levels of detail. Many techniques exist for producing both types of simplification, but they usually choose the areas of the object to simplify based on properties of the surface, such as curvature, colour or texture. For example, the popular QSlim simplification software, described in Garland (1999), is based on such principles. Several techniques exist which use perceptual principles to choose which objects or parts of objects to simplify at run-time, as discussed on p. 561, but perceptual factors are not considered while actually building the levels of detail. Simplifying objects based on surface properties alone does improve the visual quality of the resulting mesh significantly, with fewer jagged edges and facets visible. However, if the *semantics* of the object are not considered when simplifying, at lower levels of detail the object may become unrecognisable sooner than is necessary. For example, the ears of a bunny are the most important features that make it recognisable. But how can such semantics of be captured in some objective and measurable way?

When building level of detail meshes, we need to determine which triangles should be retained at highest detail for longest. Eye-tracking can be used to provide

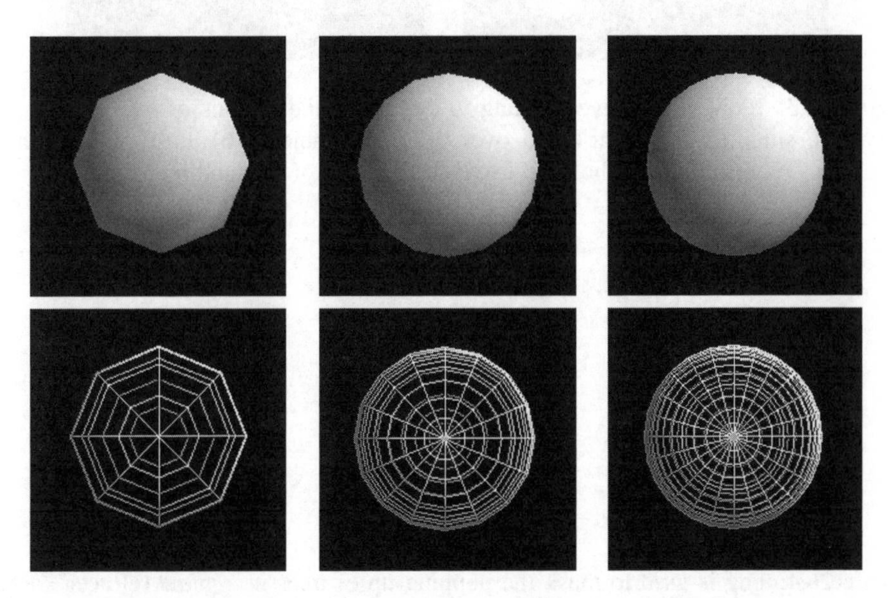

Figure 26.2: A sphere modelled at different levels of detail.

an objective measure of a viewer's interest in a specific region. The IPoMM software (Interactive Perception of Multiresolution Meshes), described in Janott (1999) and Janott and O'Sullivan (1999), uses an interest-dependent strategy for decreasing the LOD of a given object. We display models at the highest level of detail to a number of human viewers, rotating it in all directions to eliminate view-dependency, while simultaneously tracking their eye-movements. A counter associated with each triangle is incremented each time it is fixated by the viewer. The number of fixations is then interpreted as a measure of the salience of that triangle, and the order with which regions are coarsened is updated accordingly. Using this strategy, regions of the object with higher saliency are retained at high resolution for as long as possible, while less interesting regions are simplified earlier. In this way, the object remains recognisable for longer (see Figure 26.3). Although this system has been used to generate multi-resolution mesh structures to date, it is equally applicable to the generation of fixed LOD meshes.

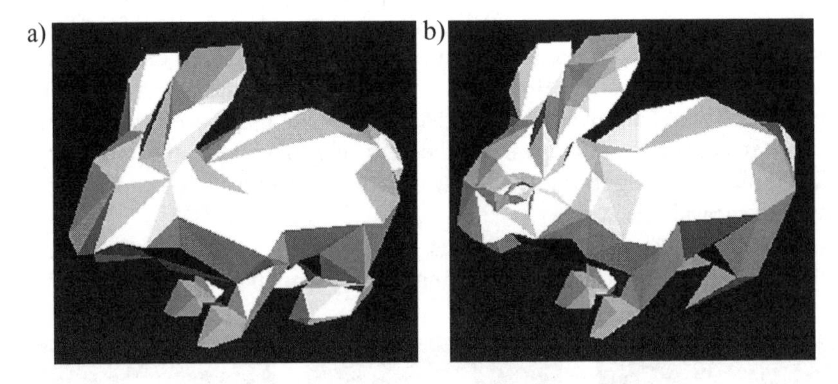

Figure 26.3: IPoMM uses eye-tracking to determine the regions of interest. (a) 358 polygons, simplification distributed over the whole object. (b) 358 polygons, high resolution at the head leaves less detail for the hind legs.

Figure 26.4: Fog is used to mask the popping up of trees in a game (eRacer ©Rage Software).

Rendering

We have discussed the importance of being able to reduce the Level of Detail (LOD) of models for interactive graphical systems, thus reducing the potential load on the graphics engine. This reduction in accuracy will also cause the visual quality of the graphics to be degraded in a number of ways. We have seen that preprocessing the models using eye-movement data can help to ameliorate this situation, but the situation can be further improved when the graphics engine is actually rendering the frames of an interactive animation. Adaptive systems use heuristics to determine which objects, or parts of objects, to refine at runtime. The system described in Funkhouser and Séquin (1993) was the first such system, which used factors such as the distance of an object from the viewer and its velocity to choose from a number of fixed LOD models to render.

The problems that arise with such an approach include popping effects that occur when the graphics engine suddenly switches from a lower to a higher level of detail. Nevertheless, this may be the method of choice for extremely time-critical systems, for example when the processing power is particularly limited and a small number of LOD models are stored for each simple object. Several visual tricks can be used to help mask the popping and we will discuss how real-time eye-movement analysis can help in the next subsection. However, for more complex models, such as mountainous terrains, the costs of storing several versions of such a model would be prohibitive. If smooth, progressive switching between LODs is required, only parts of the object should be refined at runtime, based on heuristics as in the above-mentioned system. Gaze-contingent techniques, which incorporate eye-tracking or models of visual attention to direct this process, are discussed on pp. 561–562.

Level of detail popping and change-blindness When processing power is limited, the graphics engine of an interactive system such as a game cannot render all objects that should be visible, nor can it render those objects at a consistently high level of detail. This means that whole objects must suddenly appear and disappear, as in Figure 26.4, or a lower or higher LOD for the object will suddenly be displayed, as discussed above. Tricks such as masking the changes using fog are possible, but these may not always be effective or appropriate. We propose that the phenomenon of *change blindness*, i.e. the inability of the brain to detect certain changes following saccades, could be a more effective means of hiding such events. For example, when a house suddenly pops up in a computer game, if this had happened during an eye-movement, the change should be less noticeable.

Many studies have shown that people are very slow to notice quite large changes made to an image during a saccade. (See Henderson & Hollingworth, 1999; Rensink *et al.*, 2000; Simons, 2000; Grimes, 1996; O'Regan, 2001 and many others for a full discussion of this phenomenon.) Saccades can either be detected using eye-tracking hardware — see Triesch *et al.* (2002) for a system designed specifically for interactive graphics systems — or induced using the flicker paradigm (i.e. blanking the screen) or mud-splashes.

We wish to exploit change-blindness to mask LOD popping in an interactive graphics program such as a game. Therefore, we decided to first examine eye-movements while people play such games, our main aim being to establish the durations and frequency of saccades. This information is important if we want to develop a real-time system where changes occur during saccades. We did not find any previous study of eye-movements while playing games, but in Andrews and Coppola (1999), eye-movements were recorded while viewers performed several tasks, including viewing a scene, watching repetitively moving patterns and performing an active search task. They recorded average fixation durations of 0.3, 0.6 and 0.22 seconds respectively for each of these tasks. To establish what patterns occur when playing an interactive graphical game, we conducted some preliminary experiments using the SMI EyeLink eye-tracker and recorded fixation and saccade durations. Eight subjects passively viewed video clips of a rollercoaster, and played two different types of game: a racing game, where the participant had to navigate a car around a racing track, and another where they had to shoot asteroids which could appear at any location on the screen.

The video clip provided the highest fixation duration average of 0.4 seconds, while the average fixation duration for the games was 0.22 and 0.25 seconds for the racing game and the asteroids respectively. Saccade duration averages were 0.038 and 0.045 seconds for the racing and asteroids games respectively. These results are consistent with those reported in Andrews and Coppola (1999), in that our passive task, i.e. viewing a video clip, had higher fixation durations than our active tasks, i.e. playing a game. Three of the participants were experienced games players and it was found that the average fixation duration for these players was higher for both games (by 0.126 and 0.08 seconds respectively). This is probably because they have less need to look around in order to effectively navigate or aim. The results found for the fixation duration when looking at the video clip was roughly the same as results found for other subjects. We also found that the saccade duration for experienced players is on average 0.09 seconds less than that for non-experienced players.

Using the results of these experiments, we are building an adaptive framework that allows objects to be added to a scene, or the fixed level of detail to be raised for objects that have increased in salience, during a saccade. Rather than allowing the popping

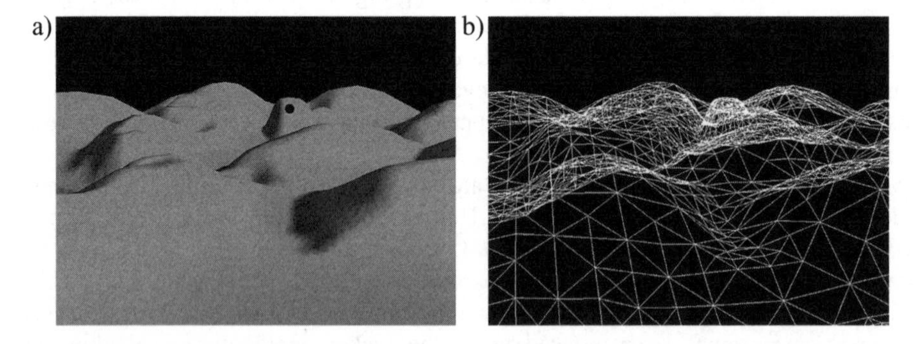

Figure 26.5: A multi-resolution mesh. (a) The terrain as it looks to the viewer – the black dot shows the fixation position. (b) The terrain in wireframe mode, with increased detail where the viewer is looking.

to occur immediately when the viewer is at a fixed distance from the location of the object, once a particular threshold has been passed the object will then be popped during the next saccade. A list of such objects is maintained between subsequent saccades.

Gaze-directed rendering On pp. 556–558 we saw how an object can be modelled to allow for simplification at run-time, and now we will discuss how the graphics engine will choose when and where to simplify such objects. Funkhouser and Séquin (1993) established the principle of exploiting the limitations of the human visual system to choose LODS for objects, by taking factors such as the velocity and size of the objects into account. Reddy (1997) was the first to examine in detail how a model of human visual acuity could guide simplification, determining when a change from one LOD to the next will be imperceptible to the viewer.

It is a well-known fact that people's ability to perform many types of visual processing tasks diminishes the more peripherally stimuli are presented (see Aubert & Foerster, 1857; Weymouth, 1958). A complete discussion of the physiological reasons for this eccentricity effect and the cortical magnification factor (M) which can in some circumstances be used to quantify it can be found elsewhere. (See DeValois & DeValois, 1988; Rovamo & Virsu, 1979; Carrasco & Frieder, 1997.) This effect has been investigated in the field of reading and scene perception also. (See Rayner, 1986; Shioiri & Ikeda, 1989; van Diepen *et al.*, 1998 and others.) Bearing this in mind, it is clear that simplifications to objects or regions of objects in the periphery of a viewer's visual field should be less perceptible than those close to their fixation point. Exploiting this effect while rendering objects at multiple LODS is known as *gaze-directed rendering*. Examples of systems that use this technique may be found in Ohshima *et al.* (1996) and Murphy and Duchowski (2001). However, eccentricity alone is often not sufficient to predict visual acuity. For example, in Strasburger *et al.* (1994) both size and contrast had to be increased to neutralise the eccentricity effect. Considering eccentricity together with other perceptual factors when simplifying is therefore a more favourable approach. A method which considers eccentricity, contrast, spatial frequency and silhouette-preservation is described in Luebke *et al.* (2000).

As discussed above, there are circumstances when choosing a fixed LOD for an object at runtime will not be desirable or feasible. Instead, a special mesh structure, called a multi-resolution or progressive mesh, can be used to avoid having to display a whole object at the highest level of detail even though only part of it may be currently important based on the user's view. The system described in Luebke *et al.* (2000) used just such a view-dependent simplification technique. When an eye-tracker is used with such meshes, detail can be added wherever the user is looking. This will give the impression that the mesh is much more detailed than it actually is. Figure 26.5 shows an implementation of a method called ROAM (Duchaineau *et al.*, 1997), which has been adapted to incorporate foveation. The focus of the viewer's attention is shown in red, which causes the mesh to be retained at a higher level of detail close to that point — see Watson (2000) for further details.

An important issue is whether the performance gains won by using a tracking device offset the expense and inconvenience involved. Currently, systems with high spatial and temporal resolution are quite intrusive and expensive, while those that are less expensive and/or intrusive simply cannot produce results at the rate and accuracy required for real-time graphics. However, if the benefits of this technology become well-established and techniques which exploit it are developed, better non-intrusive and low-cost solutions are sure to follow. Alternatively, systems could use models of visual attention to predict where the eye is likely to be directed while watching an animation, and simply use eye-tracking to verify that the model is correct. In Yee *et al.* (2001), a saliency map is used to predict the location of the eye while viewing dynamic scenes, and they claim that this compensates for the lack of an eye-tracker. However, they found that when they implemented parafoveal degradation, although significant speedups were achieved when producing the animation, this factor was only useful for the first few viewings of the animation, as people then tended to look away from the predicted regions to explore the field of view more fully. Therefore, any model of this type, even if top-down processes are taken into account, is unlikely to be robust enough for highly interactive situations, such as a game where the player is exploring a virtual environment fully in order to carry out some task.

Simulation

When objects are being animated, their motions need to be updated by some means for each time-step of the animation. In traditional animation, artists drew objects and characters in a number of key positions, known as key-frames, and human animators filled in the individual images to produce the effect of these objects moving, called in-betweening. In computer-based animation, similar techniques can be used, in that scripts and key-frames are provided for the objects in a scene and the computer simply takes over the role of in-betweening. Such key-framed animations are often used in interactive graphical systems also, for example to add cheap animation effects to characters in a game. However, the fixed nature of such scripted motions restricts interactivity, so other techniques which actually generate the motions automatically are often desirable.

The process of updating the positions and orientations of objects based on the laws of physics is referred to as physically-based animation or simulation. The process of evaluating the current physical state of the objects in the scene and their interactions with each other, and thus determining the appropriate physical response, provides a further significant burden on the graphics engine. In fact, this may be the most significant overhead for many systems, as modern systems relieve much of the rendering burden by delegating many tasks to specialized graphics acceleration hardware. Because of the time constraints imposed by systems that need interactive frame rates, physically accurate movements are not always possible to generate in the time allowed. Therefore, as in the case of rendering, simulation accuracy must sometimes be sacrificed for speed.

Several researchers have proposed strategies for simplifying simulations while maintaining some degree of physical accuracy. In Chenney and Forsyth (1997), objects outside the current view are no longer updated based on physical laws. In Carlson and Hodgins (1997), the motions of legged creatures not in view are updated based on simplified rules, while Barzel *et al.* (1996) and Chenney and Forsyth (2000) maintain that plausibility as opposed to accuracy is acceptable in many situations, and examine ways in which physically plausible simulations may be generated. However, perceptual principles have to date been largely neglected when considering the issue of level of detail (LOD) for simulation.

One of the most important behaviours of objects that are simulated in a physically accurate or plausible way, is the way that they react when they touch or collide with each other. Without a mechanism to handle such collisions, objects would simply float through each other. Collision handling is, unfortunately, extremely expensive in terms of computational power, and can often be the major bottleneck in physically-based animations consisting of many interacting objects. The avalanche simulation described in Mirtich (2000) is an example of such a situation where the collision processing slows down the simulation to a totally non-interactive rate — in this case 97 seconds on average to compute the simulation for just one frame of the animation. In previous studies, we have investigated techniques for graceful degradation of collision handling and the perceptual impact of such degradation, in particular with respect to eccentricity. (See O'Sullivan & Dingliana, 2001; Dingliana & O'Sullivan, 2000; O'Sullivan *et al*., 1999.) To our knowledge, these are the only studies to date that explore the perceptual aspects of level of detail for physically-based simulation in interactive graphics. The issues of collision perception and gaze-contingent collision handling will be explored in more detail in the next section.

Collisions and Eye-movements

While a viewer is watching a simulated collision between two objects, several factors will determine whether they perceive that collision to be "correct", i.e. consistent with their expectations of how those objects should behave. In O'Sullivan & Dingliana (2001), we described studies that investigated some of these factors, in particular eccentricity, separation, distractors, causality and accuracy of physical response. These are presented in the next subsection. Eye-movement data could be extremely useful in determining which collisions to resolve at a higher LOD and in the following subsection we will describe a new system for gaze-contingent collision handling using an eye-tracker and the results from an initial evaluation experiment will also be presented in the final subsection.

Perception of Collisions

Newtonian mechanics can be used to describe the behaviour of objects in the physical world, using dynamic concepts such as force and mass. However, most people have intuitive preconceptions concerning mechanical events that, although incorrect according to Newtonian mechanics, are highly stable and widespread (Clement, 1982). It has also been shown that people use only one dimension of information when making dynamical judgements (Profitt & Gilden, 1989). Therefore, when a dynamic event involves more than one dimension of information, such as velocity and rotation (i.e. an extended body motion as opposed to a particle that has only one dimension of information), humans are less able to correctly identify anomalous physical behaviour. The same authors also discovered that judgements about collisions were made based on heuristics, and that people were influenced by kinematic data, such as velocity after impact and the way that the colliding objects ricochet (Gilden & Profitt, 1989). We maintain that we can exploit this inaccuracy in human perception to produce more visually plausible physical simulations. We wish to determine the circumstances under which this degradation in accuracy will be imperceptible. Robust factors that can significantly affect a viewer's perception of a collision may then be used to prioritise collision processing in a perceptually-adaptive system. Some experiments that investigated these issues are described in O'Sullivan & Dingliana (2001).

Causality refers to the ability to detect whether one event causes another. For example, a collision of a moving object with a stationary one will cause the second object to move, whereas a stationary object that starts to move by itself is perceived to be autonomous (Michotte, 1963; Scholl & Tremoulet, 2000). We ran an experiment similar to Michotte's famous causality tests and found that adding a time delay between object contact and collision response reduced the perception of causality and thereby the plausibility of the collision event itself. Therefore, we can conclude that constant frame rates are imperative in any real-time collision handling system and hence interruptible collision detection is the only feasible solution for large numbers of complex objects.

Interrupting collision detection before it is complete either leads to interpenetrations, which are usually unacceptable, or more frequently to objects which bounce off each other at a distance. We found that the separation of objects when they collide provides a strong visual impression of an erroneous collision, but that this effect may be ameliorated by factors such as occlusion of the collision points, eccentricity (i.e. peripheral events) and the presence, number and type of distractors (e.g. visually similar distractors have a stronger masking effect).

We also found that, despite reduced collision detection resolution, it is possible to produce a random collision response that is as believable as the more accurate ones, thus further masking collision anomalies. As in Profitt and Gilden (1989), we conclude that people seem to be capable of correctly perceiving errors in collision response only when there is one salient feature (such as gap size), whereas when the simulation becomes more complex, they rely on their own naïve or common-sense judgements of dynamics, which are more often than not inaccurate.

In the following sections, we will discuss strategies for building these factors into a gaze-contingent collision handling system. Further experiments involving the measurement of eye-movements are being conducted to identify the effect of these factors in more complex scenarios with larger numbers of colliding entities and we also present some initial results.

Gaze-contingent Collision Handling

Collision Handling is an important part of dynamic simulation, as contacts and collisions are the primary source for interaction within the simulation world. Collision handling incorporates three computationally expensive processes, which are key candidates for optimisation using an adaptive level of detail approach. The first stage of collision handling involves *detecting* when virtual objects in a simulation scene are in contact or indeed interpenetrating with one another. Should such a condition be found, it is the role of the *contact modelling* process to identify the points or areas of contact. Then the *collision response* module has the twofold duty of resolving any interpenetrations by moving the objects back to a safe position and then applying the correct forces or impulses on the colliding objects to simulate how the objects would behave in the real world.

Adaptive level of detail modulation in collision handling is made possible by using multi-resolution volume models to represent objects in the simulation world. Aside from the display models used in rendering, each object is associated with a functional model to represent its volume. This volume data can then be used as a parameter in all parts of the collision handling phase of simulation. Numerous approaches exist for modelling systems based on multiresolution volume models. (See Hubbard, 1996; Gottschalk *et al.*, 1996; Bergen, 1997 and others.) Hierarchies built with spheres are particularly useful for collision detection (Hubbard, 1995). Figure 26.6 shows an example of a sphere tree representation of an object at different levels of detail.

An interruptible collision handler works by progressive refinement of collision data. Potentially colliding objects are first dealt with at the lowest level of detail e.g. the bounding sphere (Figure 26.7(a)). At this level alone the collision handler is able to make a coarse judgement of whether or not two objects are colliding and how they should react to the collision. The circles show the volume model currently processed while the arrow shows the direction of forces calculated at this level. Once this coarse judgement is made and preliminary output is generated, consisting of the new states of the objects (e.g. new velocities and positions), the system has the option of refining the data by inspecting the objects at the next higher resolutions of volume data (Figure 26.7(b), (c), (d)). We can see in (d) that in the particular case shown, there is in fact no collision. It should be noted that each higher level of volume data requires progressively more computation time.

A *scheduler* keeps track of the processing time spent on computing the current simulation frame and interrupts this refinement process when the allocated time for collision handling has been exceeded. The simulation process proceeds by using the highest resolution data that the collision handling system was able to compute in the allocated time.

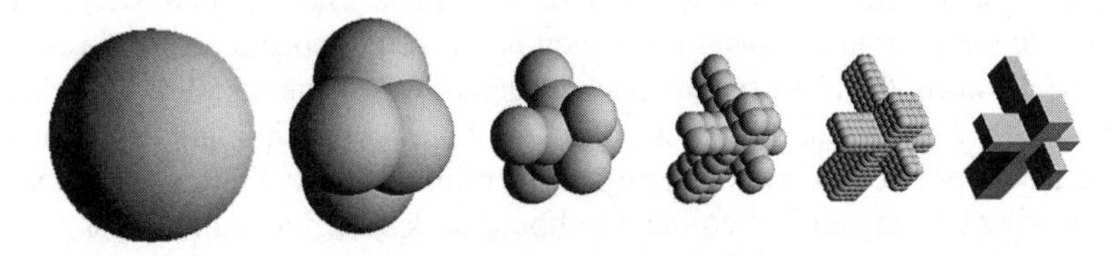

Figure 26.6: Multiple levels of resolution of a sphere tree model of an object and the original object on the far right.

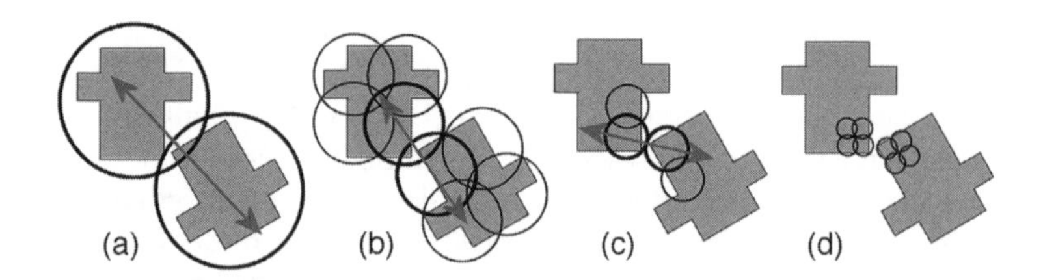

Figure 26.7: Collision handling data gathered at four different resolutions.

This time-critical approach alone can be used to guarantee real-time frame rates but can be strategically optimised by refining objects to higher levels in regions of the scene that are more important. Different strategies can be implemented for determining the importance of scene regions, for example by taking account of the factors described on pp. 563–564. In a gaze-contingent prioritisation scheme we can use an eye-tracker to determine the user's gaze location at any instant during the simulation. Eccentricity can then be used as the primary measure of importance, or combined with other factors as part of a more complete perceptual model.

Given any prioritisation strategy, we would ideally wish to sort all objects in the scene based on their priority and apply simulation refinement to objects in order of their priority values. However, in practice, the computational cost of performing a complete sort can become unjustifiably prohibitive so a more practical approach is to use a small number of priority groups into which colliding objects are bin sorted. Each priority group is then allocated its share of processing time by the scheduler, with more processing being spent on higher priority groups. This method, whilst still preserving some level of prioritisation, bears considerably less overhead expense than a full continuous sort and in practice delivers improvements even with only two priority groups.

Evaluation

To evaluate the effectiveness of a gaze-dependent prioritisation scheme for interactive simulation, an experiment was performed. Ten participants (computer science staff and students) were presented with 36 short simulations of rigid bodies colliding and bouncing off each other inside a closed cube. The simulation was run on a desktop PC with graphics acceleration and a 22-inch screen. Participants were instructed to react to the quality of the simulations in two different ways. The first task was to respond, by clicking the mouse button, whenever they perceived the occurrence of a frame containing one or more "bad" collisions during the course of the simulation. A bad collision here refers to one resulting from a coarse level approximation of a collision as described in the previous section. In an initial training phase, they were shown examples of what both good and bad collisions should look like. The second task was to rate the overall quality on a scale of one to five, at the end of each simulation. During the training phase, examples of the best and worst quality simulations were shown, and participants were told that the two extremes should receive a rating of five and one respectively. They were also told that simulations with quality ranging between both of these limits were also possible. They then practised on a further number of simulations and were observed to ensure that they had understood the instructions.

Four distinct types of simulations were presented to the participants in random order. The first type of simulation (denoted as *all good*) resolved all collisions at the highest resolution of the volume model. It was possible to deal with objects at this high a resolution in the experiment as the maximum number of objects dealt with was relatively small. A second type of simulation (*all bad*) dealt with all collisions at the very lowest level of resolution i.e. object collisions were dealt with at the bounding sphere level, resulting in objects repulsing each other at a distance in almost all cases. It should be noted that this distant repulsion is not always obvious to viewers as inter-object occlusion sometimes prevents the gap from being visible in the projected display.

Two further types of collisions had combinations of good and bad collisions occurring in the scene at the same time throughout the simulation. In both of these, a *high-priority region* was chosen in the scene where collisions were dealt with at the *all-good* level while outside of this region all objects were dealt with at the coarse level. In one of these, the *tracked* simulations, the user's gaze position was tracked and used as the centre of the high-priority region. In the other case, the *random* simulations, a random position was chosen every 5 frames to serve as the centre of the high priority region. Having a randomly located priority region of the same size in the scene ensures that roughly the same proportion of good and bad collisions is maintained as in the *tracked* case.

Each simulation type was shown with 5, 10 and 15 objects and three repetitions each were shown for these twelve cases, see Figure 26.8. In varying the number of simulated objects, the size of the cube, within which the objects were contained, was correspondingly resized to maintain a constant density of objects at all times within the container. This was in order to ensure consistency in the number of collisions occurring in the simulation. The size of the boxes displayed were not scaled to fill the screen, as then the size of the objects would vary between conditions. Necessarily, this reduced the active field of view for the smaller number of objects.

After a short training phase, in which participants were shown isolated cases of good and bad collisions, eye-movements were recorded at all times with an SMI EyeLink eye-tracker and the 36 simulation runs were shown in random order.

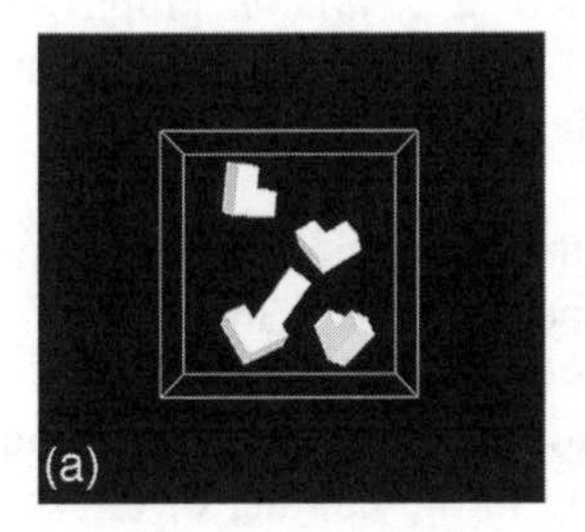

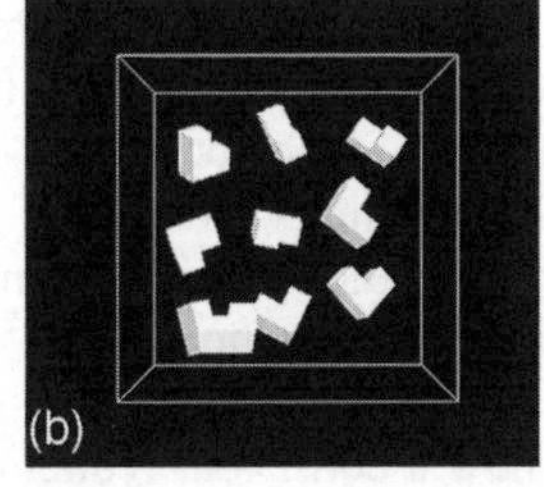

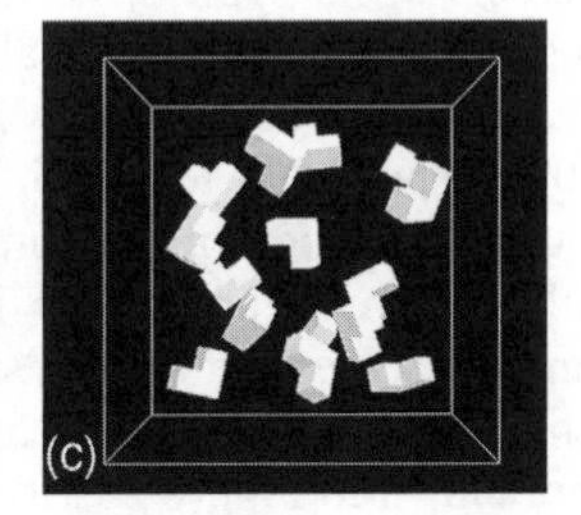

Figure 26.8: Screenshots of the experiment with 5, 10 and 15 objects contained in proportionately-sized boxes.

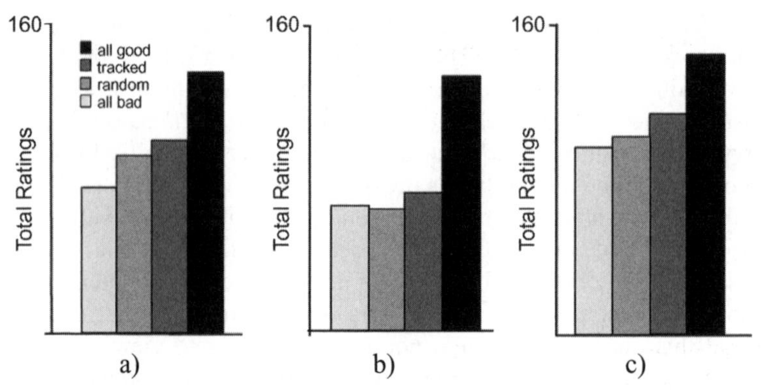

Figure 26.9: Results from rating task: (a) Five objects. (b) Ten objects. (c) Fifteen objects.

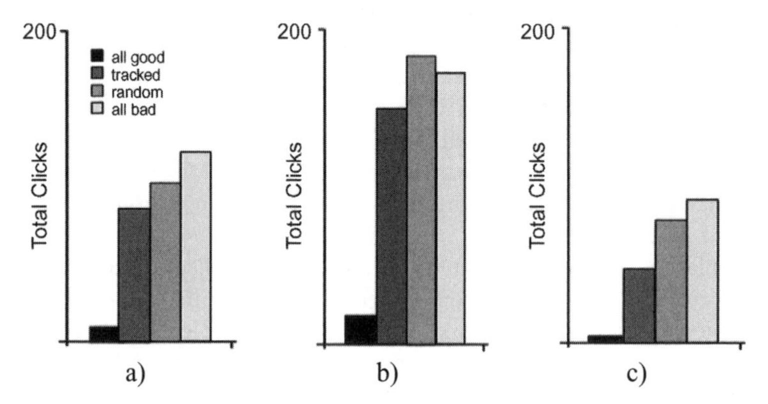

Figure 26.10: Results from clicks task: (a) Five objects. (b) Ten objects. (c) Fifteen objects.

Results

Figure 26.9 shows the participants' ratings for the different sets of simulations organised by number of objects in the simulation. Of most interest here, in the context of a gaze-contingent system, is the comparison of the *tracked* and *random* graphs. The results clearly show an overall improvement in the perception of the *tracked* simulation. A single-factor ANOVA showed a weak significance of $> 70\%$, $> 55\%$ and $> 80\%$ respectively for the 5, 10 and 15 object results.

An examination of the number of clicks for each of the simulations shows similar results. There is an overall improvement in the number of clicks in each of the simulation cases with weak statistical significance of $> 60\%$, $> 75\%$ and $> 80\%$ respectively for the 5, 10 and 15 cases. The graphs of the total clicks during the simulations, shown in Figure 26.10, show a consistent reduction in the number of clicks for the 15 object simulations. It is reasonable to assume that this is due to an increase in the number occluded objects as well as in the number of similar distracters as discussed on p. 564.

Why there isn't a stronger statistical significance has to do with the difficulty in setting up fair experimental simulations. The complexity and multi-dimensionality of the simulation process makes it difficult to design an experimental task that is both unambiguous and fairly representative of the variables being evaluated. This is particularly so in the case of gaze-contingent simulation where a random variable (i.e. the gaze position) is an active factor which affects the outcome of the simulation that is given to the participants for evaluation. Modulating simulation detail levels at random or gaze-dependent locations in the scene introduces a significant level of non-determinism into the simulation. This makes it impossible to show all the participants an identical set of simulations. A time of ten seconds was chosen as the duration for each simulation to give the participants a representative sample of collisions.

Some participants reported that they used peripheral vision in certain cases to decide on their rating for the simulation. As discussed in the previous section, a potentially more effective solution would be to have simulation levels degrade progressively with the projected distance from the gaze position. The experiment, however, used a simpler, two-level scheme for prioritisation (i.e. fine resolution within the high-priority region and coarse resolution everywhere else). While there are some studies that examine the ideal size of high-resolution regions for scene viewing, as in Loschky and McConkie (2000), there is no documented study that suggests an ideal radius for a high priority region for simulation purposes. Future work is planned to examine the effects of modulating the radius of the high priority region in different simulation cases and the full field of view will also be used, thus allowing more extensive exploitation of the eccentricity effect.

References

Andrews, T., & Coppola, D. (1999). Idiosyncratic characteristics of saccadic eye movements when viewing different visual environments. *Vision Research*, *39*, 2947–2953.

Aubert, H., & Foerster, O. (1857). Beiträage zur Kentniss des indirekten Sehens (i): Untersuchungen über den Raumsinn der Retina. *Albert von Graefes Archiv für Ophthamologie*, *3*, 1–37.

Barzel, R., Hughes, J., & Wood, D. (1996). Plausible motion simulation for computer graphics animation. In: *Computer Animation and Simulation'96* (pp. 183–197). Vienna: Springer.

Bergen, G. V. D. (1997). Efficient collision detection of complex deformable models using AABB trees. *Journal of Graphics Tools*, *2*(4), 1–13.

Carlson, D., & Hodgins, J. (1997). Simulation levels of detail for real-time animation. In: *Proceedings Graphics Interface* (pp. 1–8). Toronto: Canadian Information Processing Society

Carrasco, M., & Frieder, K. (1997). Cortical magnification neutralizes the eccentricity effect in visual search. *Vision Research*, *37*(1), 63–82.

Chenney, S., & Forsyth, D. (1997). View-dependent culling of dynamic systems in virtual environments. In: *ACM Symposium on Interactive 3D Graphics* (pp. 55–58).

Chenney, S., & Forsyth, D. (2000). Sampling plausible solutions to multi-body constraint problems. In: *Proceedings Siggraph 2000* (pp. 219–228). New York: ACM Press.

Clement, J. (1982). Students' preconceptions in introductory mechanics. *American Journal of Physics*, *50*(1), 66–71.

DeValois, R., & DeValois, K. (1988). *Spatial Vision*. New York: Oxford University.
Dingliana, J., & O'Sullivan, C. (2000). Graceful degradation of collision handling in physically based animation. *Computer Graphics Forum (Eurographics 2000 Proceedings)*, *19*(3), 239–247.
Duchaineau, M., Wolinsky, M., Sigeti, D., Miller, M., Aldrich, C., & Mineev-Weinstein, M. (1997). Roaming terrain: Real-time optimally adapting meshes. In: *Proceedings of IEEE Visualization '97* (pp. 81–88). New York: ACM Press/Addison-Wesley Publishing Co.
Funkhouser, T., & Séquin, C. (1993). Adaptive display algorithm for interactive frame rates during visualization of complex virtual environments. In: *Proceedings SIGGRAPH '93* (pp. 247–254). New York: ACM Press.
Garland, M. (1999). *Quadric-Based Polygonal Surface Simplification*. PhD thesis, Carnegie Mellon University, Tech. Rept. CMU-CS-99–105.
Gilden, D., & Profitt, D. (1989). Understanding collision dynamics. *Journal of Experimental Psychology: Human Perception and Performance*, *15*(2), 372–383.
Gottschalk, S., Lin, M., & Manocha, D. (1996). OBB-tree: A hierarchical structure for rapid interference detection. In: *Proceedings of ACM SIG-GRAPH '96* (pp. 171–180).
Grimes, J. (1996). On the failure to detect changes in scenes across saccades. In: K. Akins (ed.), *Perception (Vancouver Studies in Cognitive Science)*, (Vol. 2, pp. 89–110). Oxford: Oxford University Press.
Henderson, J., & Hollingworth, A. (1999). The role of fixation position in detecting scene changes across saccades. *Psychological Science*, *5*, 438–443.
Hubbard, P. (1995). Collision detection for interactive graphics applications. *IEEE Transactions on Visualization and Computer Graphics*, *1*(3), 218–230.
Hubbard, P. (1996). Approximating polyhedra with spheres for time-critical collision detection. *ACM Transactions on Graphics*, *15*(3), 179–210.
Janott, M. (1999). Interactive perception of multiresolution meshes. Final year thesis, Computer Science Dept., Trinity College, Dublin.
Janott, M., & O'Sullivan, C. (1999). Using interactive perception to optimise the visual quality of approximated objects in computer graphics. In: *10th European Conference on Eye Movements, Book of Abstracts* (pp. 110–111). The Netherlands: Utrecht University.
Loschky, L., & McConkie, G. (2000). User performance with gaze contingent multiresolutional displays. In: *Proceedings of the Eye Tracking Research and Applications Symposium* (pp. 97–103). New York: ACM Press
Luebke, D., Hallen, B., Newfield, D., & Watson, B. (2000). Perceptually driven simplification using gaze-directed rendering. *Technical Report, CS–2000–04*. University of Virginia.
Michotte, A. (1963). *The Perception of Causality*. New York: Basic Books.
Mirtich, B. (2000). Timewarp rigid body simulation. In: *Proceedings SIGGRAPH 2000* (pp. 193–200). New York: ACM Press/Addison-Wesley Publishing Co.
Murphy, H., & Duchowski, A. (2001). Gaze-contingent level of detail rendering. In: *Proceedings of EuroGraphics 2001 (Short Presentations)* (pp. 219–228). UK: Eurographics Association.
Ohshima, T., Yamamoto, H., & Tamura, H. (1996). Gaze-directed adaptive rendering for interacting with virtual space. In: *Proceedings IEEE VRAIS'96*. (pp. 103–110). New York: Springer.
O'Regan, J. (2001). Thoughts on change blindness. In: L. Harris and M. Jenkin (eds), *Vision and Attention* (pp. 281–302). New York: Springer.
O'Sullivan, C., & Dingliana, J. (2001). Collisions and perception. *ACM Transactions on Graphics*, *20*(3).
O'Sullivan, C., Radach, R., & Collins, S. (1999). A model of collision perception for real-time animation. In: N. Magnenat-Thalmann and D. Thalmann (eds), *Computer Animation and Simulation'99* (pp. 67–76). Vienna: Springer.

Profitt, D., & Gilden, D. (1989). Understanding natural dynamics. *Journal of Experimental Psychology: Human Perception and Performance*, *15*(2), 384–393.

Rayner, K. (1986). Eye movements and the perceptual span in beginning and skilled readers. *Journal of Experimental Child Psychology*, *41*, 211–236.

Reddy, M. (1997). Perceptually modulated level of detail for virtual environments. PhD thesis, University of Edinburgh.

Rensink, R., O'Regan, J., & Clark, J. (2000). On the failure to detect changes in scenes across brief interruptions. *Visual Cognition*, *7*(1), 127–146.

Rovamo, J., & Virsu, V. (1979). An estimation and application of the human cortical magnification factor. *Experimental Brain Research*, *37*, 495–510.

Scholl, B., & Tremoulet, P. (2000). Perceptual causality and animacy. *Trends in Cognitive Sciences*, *4*(8), 299–309.

Shioiri, S., & Ikeda, M. (1989). Useful resolution for picture perception as a function of eccentricity. *Perception*, *18*, 347–361.

Simons, D. (2000). Current approaches to change blindness. *Visual Cognition*, *7*(1/2/3), 1–15.

Strasburger, H., Rentschler, I., & Harvey Jr., L. O. (1994). Cortical magnification theory fails to predict visual recognition. *European Journal of Neuroscience*, *6*, 1583–1588.

Triesch, J., Sullivan, B., Hayhoe, M., & Ballard, D. (2002). Saccade contingent updating in virtual reality. In: *ETRA'02, Symposium on Eye-tracking Research and Applications* pp. 95–120. New York: ACM Press.

van Diepen, P., Wampers, M., & d'Ydewalle, G. (1998). Functional division of the visual field: Moving masks and moving windows. In: G. Underwood (ed.), *Eye Guidance in Reading and Scene Perception* (pp. 337–355). Oxford: Elsevier.

Watson, M. (2000). Applications of gaze tracking. Final year thesis, Computer Science Dept., Trinity College, Dublin.

Weymouth, R. (1958). Visual sensory units and the minimal angle of resolution. *American Journal of Ophthamology*, *46*, 102–113.

Yee, H., Pattanaik, S., & Greenberg, D. (2001). Spatiotemporal sensitivity and visual attention for efficient rendering of dynamic environments. *ACM Transactions on Graphics*, *20*(2), 39–65.

Commentary on Section 4

Eye Tracking in Human–Computer Interaction and Usability Research: Ready to Deliver the Promises

Robert J. K. Jacob and Keith S. Karn

Introduction

This section considers the application of eye movements to user interfaces, both for analyzing interfaces (measuring usability) and as an actual control medium within a human–computer dialogue. The two areas have generally been reported separately; but this book seeks to tie them together. For usability analysis, the user's eye movements are recorded during system use and later analyzed retrospectively; but the eye movements do not affect the interface in real time. As a direct control medium, the eye movements are obtained and used in real time as an input to the user–computer dialogue. They might be the sole input, typically for disabled users or hands-busy applications, or they might be used as one of several inputs, combining with mouse, keyboard, sensors, or other devices.

Interestingly, the principal challenges for both retrospective and real time eye tracking in human–computer interaction (HCI) turn out to be analogous. For retrospective analysis, the problem is to find appropriate ways to use and interpret the data; it is not nearly as straightforward as it is with more typical task performance, speed, or error data. For real time use, the problem is to find appropriate ways to respond judiciously to eye movement input, and avoid over-responding; it is not nearly as straightforward as responding to well-defined, intentional mouse or keyboard input. We will see in this chapter how these two problems are closely related.

These uses of eye tracking in HCI have been highly promising for many years, but progress in making good use of eye movements in HCI has been slow to date. We see promising research work, but we have not yet seen wide use of these approaches in practice or in the marketplace. We will describe the promises of this technology, its limitations, and the obstacles that must still be overcome. Work presented in this book and elsewhere shows that the field is indeed beginning to flourish.

The Mind's Eye: Cognitive and Applied Aspects of Eye Movement Research

ISBN: 0–444–51020–6

History of Eye Tracking in HCI

The study of eye movements pre-dates the widespread use of computers by almost 100 years (for example, Javal, 1878/1879). Beyond mere visual observation, initial methods for tracking the location of eye fixations were quite invasive — involving direct mechanical contact with the cornea. Dodge and Cline (1901) developed the first precise, non-invasive eye tracking technique, using light reflected from the cornea. Their system recorded only horizontal eye position onto a falling photographic plate and required the participant's head to be motionless. Shortly after this, Judd, McAllister and Steel (1905) applied motion picture photography in order to record the temporal aspects of eye movements in two dimensions. They recorded the movement of a small white speck of material placed on the participants' corneas rather than light reflected directly from the cornea. These and other researchers interested in eye movements made additional advances in eye tracking systems during the first half of the twentieth century by combining the corneal reflection and motion picture techniques in various ways (see Mackworth & Mackworth, 1958 for a review).

In the 1930s, Miles Tinker and his colleagues began to apply photographic techniques to study eye movements in reading (see Tinker, 1963 for a thorough review of this work). They varied typeface, print size, page layout, etc. and studied the resulting effects on reading speed and patterns of eye movements. In 1947 Paul Fitts and his colleagues (Fitts, Jones & Milton, 1950) began using motion picture cameras to study the movements of pilots' eyes as they used cockpit controls and instruments to land an airplane. The Fitts *et al.* study represents the earliest application of eye tracking to what is now known as *usability engineering* — the systematic study of users interacting with products to improve product design.

Around that time Hartridge and Thompson (1948) invented the first head-mounted eye tracker. Crude by current standards, this innovation served as a start to freeing eye tracking study participants from tight constraints on head movement. In the 1960s, Shackel (1960) and Mackworth and Thomas (1962) advanced the concept of head-mounted eye tracking systems making them somewhat less obtrusive and further reducing restrictions on participant head movement. In another significant advance relevant to the application of eye tracking to human–computer interaction, Mackworth and Mackworth (1958) devised a system to record eye movements superimposed on the changing visual scene viewed by the participant.

Eye movement research and eye tracking flourished in the 1970s, with great advances in both eye tracking technology and psychological theory to link eye tracking data to cognitive processes. See for example books resulting from eye movement conferences during this period (i.e., Monty & Senders, 1976; Senders, Fisher & Monty, 1978; Fisher, Monty & Senders, 1981). Much of the work focused on research in psychology and physiology and explored how the human eye operates and what it can reveal about perceptual and cognitive processes. Publication records from the 1970s, however, indicate a lull in activity relating eye tracking to usability engineering. We presume this occurred largely due to the effort involved not only with data collection, but even more so with data analysis. As Monty (1975) puts it: "It is not uncommon to spend days processing data that took only minutes to collect" (pp. 331–332). Work in several human factors/usability laboratories (particularly those linked to military aviation) focused on solving the shortfalls with eye tracking technology and data analysis during this timeframe. Researchers in these laboratories recorded much of their work in US military technical reports (see Simmons, 1979 for a review).

Much of the relevant work in the 1970s focused on technical improvements to increase accuracy and precision and reduce the impact of the trackers on those whose eyes were tracked. The discovery that multiple reflections from the eye could be used to dissociate eye rotations from head movement (Cornsweet & Crane, 1973) increased tracking precision and also prepared the ground for developments resulting in greater freedom of participant movement. Using this discovery, two joint military/industry teams (US Airforce/Honeywell Corporation and US Army/EG&G Corporation) each developed a remote eye tracking system that dramatically reduced tracker obtrusiveness and its constraints on the participant (see Lambert, Monty & Hall, 1974; Monty, 1975; Merchant *et al.*, 1974 for descriptions). These joint military/industry development teams and others made even more important contributions with the automation of eye tracking data analysis. The advent of the minicomputer in that general timeframe provided the necessary resources for high-speed data processing. This innovation was an essential precursor to the use of eye tracking data in real-time as a means of human–computer interaction (Anliker, 1976). Nearly all eye tracking work prior to this used the data only retrospectively, rather than in real time (in early work, analysis could only proceed after film was developed). The technological advances in eye tracking during the 1960s and 1970s are still seen reflected in most commercially available eye tracking systems today (see Collewijn, 1999 for a recent review).

Psychologists who studied eye movements and fixations prior to the 1970s generally attempted to avoid cognitive factors such as learning, memory, workload, and deployment of attention. Instead, their focus was on relationships between eye movements and simple visual stimulus properties such as target movement, contrast, and location. Their *solution* to the *problem* of higher-level cognitive factors had been "to ignore, minimize or postpone their consideration in an attempt to develop models of the supposedly simpler lower-level processes, namely, sensorimotor relationships and their underlying physiology" (Kowler, 1990, p. 1). This attitude began to change gradually in the 1970s. While engineers improved eye tracking technology, psychologists began to study the relationships between fixations and cognitive activity. This work resulted in some rudimentary, theoretical models for relating fixations to specific cognitive processes (see for example work by Just & Carpenter, 1976a, 1976b). Of course scientific, educational, and engineering laboratories provided the only home for computers during most of this period. So eye tracking was not yet applied to the study of human–computer interaction at this point. Teletypes for command line entry, punched paper cards and tapes, and printed lines of alphanumeric output served as the primary form of human–computer interaction.

As Senders (2000) pointed out, the use of eye tracking has persistently come back to solve new problems in each decade since the 1950s. Senders likens eye tracking to a Phoenix raising from the ashes again and again with each new generation of engineers designing new eye tracking systems and each new generation of cognitive psychologists tackling new problems. The 1980s were no exception. As personal computers proliferated, researchers began to investigate how the field of eye tracking could be applied to issues of human–computer interaction. The technology seemed particularly handy for answering questions about how users search for commands in computer menus (see, for example, Card, 1984; Hendrickson, 1989; Altonen, Hyrskykari & Räihä, 1998; Byrne *et al.,* 1999). The 1980s also ushered in the start of eye tracking in real time as a means of human–computer interaction. Early work in this area initially focused primarily on disabled users (e.g., Hutchinson *et al.*, 1989; Levine, 1981, Levine, 1984). In addition, work in flight simulators attempted to simulate a large, ultra-high resolution display by providing high resolution wherever the observer was fixating and lower resolution in the periphery (Tong & Fisher, 1984). The combination of real-time eye movement data with other, more conventional modes of user–computer communication was also pioneered during the 1980s (Bolt, 1981, 1982; Levine, 1984; Glenn *et al.*, 1986; Ware & Mikaelian, 1987).

In more recent times, eye tracking in human–computer interaction has shown modest growth both as a means of studying the usability of computer interfaces and as a means of interacting with the computer. As technological advances such as the Internet, e-mail, and videoconferencing evolved into viable means of information sharing during the 1990s and beyond, researchers again turned to eye tracking to answer questions about usability (e.g., Benel, Ottens & Horst, 1991; Ellis *et al.*, 1998; Cowen, 2001) and to serve as a computer input device (e.g., Starker & Bolt, 1990; Vertegaal, 1999; Jacob, 1991; Zhai, Morimoto & Ihde, 1999). We will address these two topics and cover their recent advances in more detail with the separate sections that follow.

Eye Movements in Usability Research

Why "Rising From the Ashes" Rather Than "Taking Off Like Wildfire?"

As mentioned above, the concept of using eye tracking to shed light on usability issues has been around since before computer interfaces, as we know them. The pioneering work of Fitts, Jones and Milton (1950) required heroic effort to capture eye movements (with cockpit-mounted mirrors and movie camera) and to analyze eye movement data with painstaking frame-by-frame analysis of the pilot's face. Despite large individual differences, Fitts and his colleagues made some conclusions that are still useful today. For example, they proposed that fixation frequency is a measure of a display's importance; fixation duration, a measure of difficulty of information extraction and interpretation; and the pattern of fixation transitions between displays; a measure of efficiency of the arrangement of individual display elements.

Note that it was also Paul Fitts whose study of the relationships between the duration, amplitude, and precision of human movements published four years later (Fitts, 1954) is still so widely cited as "Fitts' Law." A look at the ISI Citation Index[1] reveals that in the past 29 years Fitts *et al.*'s 1950 cockpit eye movement study was only cited 16 times[2] while Fitts' Law (Fitts, 1954) has been cited 855 times. So we ask, why has

Fitts' work on predicting movement time been applied so extensively while his work in the application of eye tracking been so slow to catch on? Is it simply a useless concept? We think not. The technique has continually been classified as promising over the years since Fitts' work. Consider the following quotes:

> For a long time now there has been a *great need* for a means of recording where people are looking while they work at particular tasks. A whole series of unsolved problems awaits such a technique (Mackworth & Thomas, 1962, p. 713; emphasis added).
>
> [T]he eyetracking system has a *promising future* in usability engineering (Benel, Ottens & Horst, 1991, p. 465; emphasis added).
>
> [A]ggregating, analyzing, and visualizing eye tracking data in conjunction with other interaction data holds *considerable promise* as a powerful tool for designers and experimenters in evaluating interfaces (Crowe & Narayanan, 2000, p. 35; emphasis added).
>
> Eye-movement analysis does appear to be a *promising* new tool for evaluating visually administered questionnaires (Redline & Lankford, 2001; emphasis added).
>
> Another *promising* area is the use of eye-tracking techniques to support interface and product design. Continual improvements in . . . eye-tracking systems . . . have increased the usefulness of this technique for studying a variety of interface issues (Merwin, 2002, p. 39; emphasis added).

Why has this technique of applying eye tracking to usability engineering been classified as simply "promising" over the past 50 years? For a technology to be labeled "promising" for so long is both good news and bad. The good news is that the technique must really be promising; otherwise it would have been discarded by now. The bad news is that something has held it up in this merely promising stage. There are a number of probable reasons for this slow start, including technical problems with eye tracking in usability studies, labor-intensive data extraction, and difficulties in data interpretation. We will consider each of these three issues in the following sections.

Technical Problems with Eye Tracking in Usability Studies

Technical issues that have plagued eye tracking in the past, making it unreliable and time consuming are resolving slowly (see Collewijn, 1999; Goldberg & Wichansky, this volume). By comparison to techniques used by Fitts and his team, modern eye tracking systems are incredibly easy to operate. Today, commercially available eye tracking systems suitable for usability laboratories are based on video images of the eye. These trackers are mounted either on the participant's head or remotely, in front of the participant (e.g., on a desktop). They capture reflections of infrared light from both the cornea and the retina and are based on the fundamental principles developed in the pioneering work of the 1960s and 1970s reviewed earlier. Vendors typically provide software to make setup and calibration relatively quick and easy. Together these properties make modern eye tracking systems fairly reliable and easy to use. The ability to track participants' eyes is much better than with systems of the recent past. There are still problems with eye tracking a considerable minority of participants (typically 10 to 20% cannot be tracked reliably). Goldberg and Wichansky (this volume) present some techniques to maximize the percentage of participants whose eyes can be tracked. For additional practical guidance in eye tracking techniques see Duchowski (2003).

The need to constrain the physical relationship between the eye tracking system and the participant remains one of the most significant barriers to incorporation of eye tracking in more usability studies. Developers of eye tracking systems have made great progress in reducing this barrier, but existing solutions are far from optimal. Currently the experimenter has the choice of a remotely mounted eye tracking system that puts some restrictions on the participant's movement or a system that must be firmly (and uncomfortably) mounted to the participant's head. Of course the experimenter has the option of using the remote tracking system and not constraining the user's range and speed of head motion, but must then deal with frequent track losses and manual reacquiring of the eye track. In typical WIMP (i.e., windows, icons, menus, and pointer) human–computer interfaces, constraining the user's head to about a cubic foot of space may seem only mildly annoying. If, however, we consider human–computer interaction in a broader sense and include other instances of "ubiquitous computing" (Weiser, 1993), then constraining a participant in a usability study can be quite a limiting factor. For example, it would be difficult to study the usability of portable systems such as a personal digital assistant or cell phone, or distributed computer peripherals such as a printer or scanner while constraining the user's movement to that typically required by commercially available remote eye trackers. Recent advances in eye tracker portability (Land, 1992; Land, Mennie & Rusted, 1999; Pelz & Canosa, 2001; Babcock, Lipps & Pelz, 2002) may largely eliminate such constraints. These new systems can fit un-tethered into a small backpack and allow the eye tracking participant almost complete freedom of eye, head, and whole body movement while interacting with a product or moving through an environment. Of course such systems still have the discomfort of the head-mounted systems and add the burden of the backpack. Another solution to the problem of eye tracking while allowing free head movement integrates a magnetic head tracking system with a head mounted eye tracking system (e.g., Iida, Tomono & Kobayashi, 1989). These systems work best in an environment free of ferrous metals and they add complexity to the eye tracking procedure. Including head tracking also results in an inevitable decrease in precision due to the integrating of the two signals (eye-in-head and head-in-world).

We see that currently available eye trackers have progressed considerably from systems used in early usability studies; but they are far from optimized for usability research. For a list, and thorough discussion, of desired properties of eye tracking systems see Collewijn (1999). We can probably safely ignore Collewijn's call for a 500 Hz sampling rate, as 250 Hz is sufficient for those interested in fixations rather than basic research on saccadic eye movements. See the comparison of "saccade pickers" and "fixation pickers" in Karn *et al.* (2000). For wish lists of desired properties of eye tracking systems specifically tailored for usability research see Karn, Ellis and Juliano (1999, 2000) and Goldberg and Wichansky (this volume).

Labor-intensive Data Extraction

Most eye trackers produce signals that represent the orientation of the eye within the head or the position of the point of regard on a display at a specified distance. In either case, the eye tracking system typically provides a horizontal and vertical coordinate for each sample. Depending on the sampling rate (typically 50 to 250 Hz), and the duration of the session, this can quickly add up to a lot of data. One of the first steps in data analysis is usually to distinguish between fixations (times when the eye is essentially stationary) and saccades (rapid re-orienting eye movements). Several eye tracker manufacturers, related commercial companies, and academic research labs now provide analysis software that allows experimenters to extract quickly the fixations and saccades from the data stream (see for example Lankford, 2000; Salvucci, 2000). These software tools typically use either eye position (computing dispersion of a string of eye position data points known as *proximity analysis*), or eye velocity (change in position over time). Using such software tools the experimenter can quickly and easily know when the eyes moved, when they stopped to fixate, and where in the visual field these fixations occurred. Be forewarned, however, that there is no standard technique for identifying fixations (see Salvucci & Goldberg, 2000 for a good overview). Even minor changes in the parameters that define a fixation can result in dramatically different results (Karsh & Breitenbach, 1983). For example, a measure of the number of fixations during a given time period would not be comparable across two studies that used slightly different parameters in an automated fixation detection algorithm. Goldberg and Wichansky (this volume) call for more standardization in this regard. At a minimum, researchers in this field need to be aware of the effects of the choices of these parameters and to report them fully in their publications.[3]

The automated software systems described above might appear to eliminate completely the tedious task of data extraction mentioned earlier. While this may be true if the visual stimulus is always known as in the case of a static observer viewing a static scene, even the most conventional human–computer interfaces can hardly be considered static. The dynamic nature of modern computer interfaces (e.g., scrolling windows, pop-up messages, animated graphics, and user-initiated object movement and navigation) provides a technical challenge for studying eye fixations. For example, knowing that a person was fixating 10 degrees above and 5 degrees to the left of the display's center does not allow us to know what object the person was looking at in the computer interface unless we keep track of the changes in the computer display. Note that if Fitts were alive today to repeat the study of eye tracking of military pilots he would run into this problem with dynamic electronic displays in modern cockpits. These displays allow pilots to call up different flight information on the same display depending on the pilot's changing needs throughout a flight. Certainly less conventional human–computer interaction with ubiquitous computing devices provide similar challenges. Recent advances integrating eye tracking and computer interface navigation logging enable the mapping of fixation points to visual stimuli in some typical dynamic human–computer interfaces (Crowe & Narayanan, 2000; Reeder, Pirolli & Card, 2001). These systems account for user and system initiated display changes such as window scrolling and pop-up messages. Such systems are just beginning to become commercially available, and should soon further reduce the burden of eye tracking data analysis.

A dynamically changing scene caused by head or body movement of the participant in, or through, an environment provides another challenge to automating eye tracking data extraction (Sheena & Flagg, 1978). Head-tracking systems are now often integrated with eye tracking systems and can help resolve this problem (Iida, Tomono & Kobayashi, 1989), but only in well defined visual environments (for a further description of these issues see Sodhi, Reimer, Cohen, Vastenburg, Kaars & Kirschenbaum, 2002). Another approach is image processing of the video signal captured by a head-mounted scene camera to detect known landmarks (Mulligan, 2002).

Despite the advances reviewed above, researchers are often left with no alternative to the labor-intensive manual, frame-by-frame coding of videotape depicting the scene with a cursor representing the fixation point. This daunting task remains a hindrance to more widespread inclusion of eye tracking in usability studies.

Difficulties in Data Interpretation

Assuming a researcher, interested in studying the usability of a human–computer interface, is not scared off by the technical and data extraction problems discussed above, there is still the issue of making sense out of eye tracking data. How does the usability researcher relate fixation patterns to task-related cognitive activity?

Eye tracking data analysis can proceed either *top-down* — based on cognitive theory or design hypotheses, or *bottom-up* — based entirely on observation of the data without predefined theories relating eye movements to cognitive activity (see Goldberg, Stimson, Lewenstein, Scott & Wichansky, 2002). Here are examples of each of these processes driving data interpretation:

Top-down based on a cognitive theory. Longer fixations on a control element in the interface reflect a participant's difficulty interpreting the proper use of that control.

Top-down based on a design hypothesis. People will look at a banner advertisement on a web page more frequently if we place it lower on the page.

Bottom-up. Participants are taking much longer than anticipated making selections on this screen. We wonder where they are looking.

Reviewing the published reports of eye tracking applied to usability evaluation, we see that all three of these techniques are commonly used. While a top-down approach may seem most attractive (perhaps even necessary to infer cognitive processes from eye fixation data), usability researchers do not always have a strong theory or hypothesis to drive the analysis. In such cases, the researchers must, at least initially, apply a data-driven search for fixation patterns. In an attempt to study stages of consumer choice, for example, Russo and Leclerc (1994) simply looked at video tapes of participants' eye movements, coded the sequence of items fixated, and then looked for and found common patterns in these sequences. Land, Mennie and Rusted (1999) performed a similar type of analysis as participants performed the apparently simple act of making a cup of tea. Even when theory is available to drive the investigation, researchers usually reap rewards from a bottom-up approach when they take the time to replay and carefully examine scan paths superimposed on a representation of the stimulus.

To interpret eye tracking data, the usability researcher must choose some aspects (dependent variables or metrics) to analyze in the data stream. A review of the literature on this topic reveals that usability researchers use a wide variety of eye tracking metrics. In fact the number of different metrics is fewer than it may appear at first due to the lack of standard terminology and definitions for even the most fundamental concepts used in eye tracking data interpretation. Readers may feel bogged down in a swamp of imprecise definitions and conflicting uses of the same terms. If we look closely at this mire we see that differences in eye tracking data collection and analysis techniques often account for these differences in terminology and their underlying concepts. For example, in studies done with simple video or motion picture imaging of the participants' face (e.g., Fitts, *et al.*, 1950; Card, 1984; Svensson *et al.*, 1997) a "fixation" by its typical definition cannot be isolated. Researchers usually realize this, but nevertheless, some misuse the term "fixation" to refer to a series of consecutive fixations within an area of interest. In fact, the definition of the term "fixation" is entirely dependent on the size of the intervening saccades that can be detected and which the researcher wants to recognize. With a high-precision eye tracker, even small micro-saccades might be counted as interruptions to fixation (see Engbert & Kliegl, this volume, for a discussion).

Eye Tracking Metrics Most Commonly Reported in Usability Studies The usability researcher must choose eye tracking metrics that are relevant to the tasks and their inherent cognitive activities for each usability study individually. To provide some idea of these choices, Table 1 summarizes 21 different usability studies that have incorporated eye tracking.[4] The table includes a brief description of the users, the task and the eye tracking related metrics used by the authors. Note that rather than referring to the same concept by the differing terms used by the original authors, we have attempted to use a common set of definitions as follows:

Fixation: A relatively stable eye-in-head position within some threshold of dispersion (typically ~2°) over some minimum duration (typically 100–200 ms), and with a velocity below some threshold (typically 15–100 degrees per second).

Gaze Duration: cumulative duration and average spatial location of a series of consecutive fixations within an area of interest. Gaze duration typically includes several fixations and may include the relatively small amount of time for the short saccades between these fixations. A fixation occurring outside the area of interest marks the end of the gaze.[5] Authors cited in Table 1 have used "dwell."[6] "glance," or "fixation cycle," in place of "gaze duration."

Table 1: Summary of 21 usability studies incorporating eye tracking.

Authors/Date	Users and Tasks	Eye Tracking Related Metrics
Fitts *et al.* (1950)	40 military pilots. Aircraft landing approach.	• Gaze rate (# of gazes/minute) on each area of interest • Gaze duration mean, on each area of interest • Gaze % (proportion of time) on each area of interest • Transition probability between areas of interest
Harris and Christhilf (1980)	4 instrument-rated pilots. Flying maneuvers in a simulator	• Gaze % (proportion of time) on each area of interest • Gaze duration mean, on each area of interest
Kolers, Duchnicky and Ferguson (1981)	20 university students. Reading text on a CRT in various formats and with various scroll rates.	• Number of fixations, overall • Number of fixations on each area of interest (line of text) • Number of words per fixation • Fixation rate overall (fixations/s) • Fixation duration mean, overall
Card (1984)	3 PC users. Searching for and selecting specified item from computer pull-down menu.	• Scan path direction (up/down) • Number of fixations, overall
Hendrickson (1989)	36 PC users. Selecting 1 to 3 items in various styles of computer menus.	• Number of fixations, overall • Fixation rate overall (fixations/s) • Fixation duration mean, overall • Number of fixations on each area of interest • Fixation rate on each area of interest • Fixation duration mean, on each area of interest • Gaze duration mean, on each area of interest • Gaze % (proportion of time) on each area of interest • Transition probability between areas of interest
Graf and Kruger (1989)	6 participants. Search for information to answer questions on screens of varying organization.	• Number of *voluntary* (>320 ms) fixations, overall • Number of *involuntary* (<240 ms) fixations, overall • Number of fixations on target
Benel, Ottens and Horst (1991)	7 PC users. Viewing web pages.	• Gaze % (proportion of time) on each area of interest • Scan path
Backs and Walrath (1992)	8 engineers. Symbol search and counting tasks on color or monochrome displays.	• Number of fixations, overall • Fixation duration mean, overall • Fixation rate overall (fixations/s)
Yamamoto and Kuto (1992)	7 young adults. Confirm sales receipts (unit price, quantity, etc.) on various screen layouts.	• Scan path direction • Number of instances of backtracking

Table 1: continued.

Authors/Date	**Users and Tasks**	**Eye Tracking Related Metrics**
Svensson *et al.* (1997)	18 military pilots. Fly and monitor threat display containing varying number of symbols.	• Gaze duration mean, on each area of interest • Frequencies of gaze durations dwells on area of interest
Altonen *et al.* (1998)	20 PC users. Select menu item specified directly or by concept definition.	• Scan path direction • *Sweep* — scan path progressing in the same direction • Number of fixations per sweep
Ellis *et al.* (1998)	16 PC users with web experience. Directed web search and judgment.	• Number of fixations, overall • Fixation duration mean, overall • Number of fixations on each area of interest • Time to 1st fixation on target area of interest • Gaze % (proportion of time) on each area of interest
Kotval and Goldberg (1998)	12 university students. Select command button specified directly from buttons grouped with various strategies.	• Scan path duration • Scan path length • Scan path area (convex hull) • Fixation spatial density • Transition density • Number of fixations, overall • Fixation duration mean, overall • Fixation/saccade time ratio • Saccade length
Byrne *et al.* (1999)	11 university students. Choosing menu items specified directly from computer pull-down menus of varying length.	• Number of fixations, overall • First area of interest fixated • Number of fixations on each area of interest
Flemisch and Onken (2000)	6 military pilots. Low-level flight and navigation in a flight simulator using different display formats.	• Gaze % (proportion of time) on each area of interest
Redline and Lankford (2001)	25 adults. Fill out a 4-page questionnaire (of various forms) about lifestyle.	• Scan path
Cowen (2001)	17 PC users with web experience. Search/ extract information from web pages.	• Fixation duration total • Number of fixations, overall • Fixation duration mean, overall • Fixation spatial density
Josephson and Holmes (2002)	8 university students with web experience. Passively view web pages.	• Scan path

Table 1: continued.

Authors/Date	Users and Tasks	Eye Tracking Related Metrics
Goldberg, Stimson, Lewenstein, Scott and Wichansky (2002)	7 adult PC users with web experience. Search/extract information from web pages.	• Number of fixations on each area of interest • Fixation duration mean, on each area of interest • Saccade length • Fixation duration total, on each area of interest • Number of areas of interest fixated • Scan path length • Scan path direction • Transition probability between areas of interest
Albert (2002)	24 intermediate to advanced web users. Web search for purchase and travel arrangements on sites with varying banner ad placement.	• Number of fixations on area of interest (banner ad) • Gaze % (proportion of time) on each area of interest • Participant % fixating on each area of interest
Albert and Liu (in press)	12 licensed drivers. Simultaneous driving and navigation using electronic map in simulator.	• Number of dwells, overall • Gaze duration mean, on area of interest (map) • Number of dwells on each area of interest

Area of interest: Area of a display or visual environment that is of interest to the research or design team and thus defined by them (not by the participant).
Scan path: Spatial arrangement of a sequence of fixations.

When we count up the number of times each metric is used (both from the 21 studies included in Table 1 and the three studies reported by Zülch & Stowasser, this volume), we find the most commonly used metrics listed below. The number in parentheses after each metric is the number of studies in which it is used out of the total of the 24 studies reviewed.

Number of fixations, overall (11)
Gaze % (proportion of time) on each area of interest (7)
Fixation duration mean, overall (6)
Number of fixations on each area of interest (6)
Gaze duration mean, on each area of interest (5)
Fixation rate overall (fixations/s) (5)

Each of these six most frequently used metrics is discussed briefly below. For more detailed discussion of these and other metrics see Goldberg and Kotval (1998); Kotval and Goldberg (1998) and Zülch and Stowasser (this volume).

Number of fixations, overall. The number of fixations overall is thought to be negatively correlated with search efficiency (Goldberg & Kotval, 1998; Kotval & Goldberg, 1998). A larger number of fixations indicates less efficient search possibly resulting from a poor arrangement of display elements. The experimenter should consider the relationship of the number of fixations to task time (i.e., longer tasks will usually require more fixations).

Gaze % (proportion of time) on each area of interest. The proportion of time looking at a particular display element (of interest to the design team) could reflect the importance of that element. Researchers using this metric should be careful to note that it confounds frequency of gazing on a display element with the duration of those gazes. According to Fitts *et al.* (1950) these should be treated as separate metrics, with duration reflecting difficulty of information extraction and frequency reflecting the importance of that area of the display.

Fixation duration mean, overall. Longer fixations (and perhaps even more so, longer gazes) are generally believed to be an indication of a participant's difficulty extracting information from a display (Fitts *et al*, 1950; Goldberg & Kotval, 1998).

Number of fixations on each area of interest. This metric is closely related to gaze rate, which is used to study the number of fixations across tasks of differing overall duration. The number of fixations on a particular display element (of interest to the design team) should reflect the importance of that element. More important display elements will be fixated more frequently (Fitts *et al*, 1950).

Gaze duration mean, on each area of interest. This is one of the original metrics in Fitts *et al.* (1950). They predicted that gazes on a specific display element would be longer if the participant experiences difficulty extracting or interpreting information from that display element.

Fixation rate overall (fixations/s). This metric is closely related to fixation duration. Since the time between fixations (typically short duration saccadic eye movements) is relatively small compared with the time spent fixating, fixation rate should be approximately the inverse of the mean fixation duration.

Other Promising Eye Tracking Metrics Although the metrics presented above are the most popular, they are not necessarily always the best metrics to apply. Other important metrics to consider include:

Scan path (sequence of fixations) and derived measures such as the transition probability between areas of interest — can indicate the efficiency of the arrangement of elements in the user interface.

Number of gazes on each area of interest — is a simple, but often forgotten, measure. Gazes (the concatenation of successive fixations within the same area of interest) are often more meaningful than counting the number of individual fixations.

Number of involuntary and number of voluntary fixations — Graf & Kruger (1989) have proposed that short fixations (<240 ms) and long fixations (>320 ms) be classified as involuntary and voluntary fixations respectively. Further research is needed to validate this method of classifying fixations.

Percentage of participants fixating an area of interest — can serve as a simple indicator of the attention-getting properties of an interface element.
Time to 1st fixation on target area of interest — is a useful measure when a specific search target exists.

Other aspects of ocular-motor performance such as blinks (e.g., Stern, Boyer & Schroeder, 1994), pupil changes (e.g., Hoeks & Levelt, 1993; Marshall, 1998; Backs & Walrath, 1992), vergence, and accommodation can be exploited. These have been considered annoying problems by most eye movement researchers in the past, but they may be a rich source of data. For example, Brookings, Wilson and Swain (1996) report that blink rate is more sensitive to workload (related to task difficulty) than many other more conventionally used eye tracking measures including saccade rate and amplitude in a demanding visual task (air traffic control).

Other researchers have come up with innovative techniques for analyzing and presenting existing eye tracking metrics. Wooding (2002), for example, has introduced the "Fixation Map" for conveying the most frequently fixated areas in an image. Land *et al.* (1999) refer to "Object-related actions" as a neat way to combine eye tracking data with other participant behaviors such as reaching and manipulation movements. Harris and Christhilf (1980) used an innovative plot of gaze percentage versus average gaze duration to classify types of displays and the ways they are used by pilots. Josephson and Holmes (2002) apply optimal matching (or string-edit) analysis for comparing fixation sequences. To study more truly a participant's fixation of an object, Pelz, Canosa and Babcock (2000) combine times when the eye is moving with respect to the head, but fixed relative to the fixated visual object. Such situations occur when the participant visually pursues a moving object or compensates for head movements via the vestibuloocular reflex (VOR). Salvucci (2000) used an automated data analysis system to test predictions made by various models of cognitive processes.

Difficulties relating eye tracking data to cognitive activity is probably the single most significant barrier to the greater inclusion of eye tracking in usability studies. The most important question to ask when incorporating eye tracking into a usability study is "what aspects of eye position will help explain usability issues?" As discussed above, the most relevant metrics related to eye position vary from task to task and study to study. Sometimes the experimenter has to risk going on a bit of a "fishing expedition" (i.e., collect some eye tracking records and examine them closely in various ways before deciding on the most relevant analyses).

Current and Future Directions for Applying Eye Tracking in Usability Engineering

From the literature reviewed in the preceding sections, we see that the application of eye tracking in usability engineering is indeed beginning to flourish. In this volume, we have additional contributions to this growing field. Goldberg and Wichansky (this volume) provide a thorough introduction to two groups of readers: eye tracking scientists who wish to apply their work to product usability evaluations; and usability engineers who wish to incorporate eye tracking into their studies. Goldberg and Wichansky also provide some practical tips that will be helpful to anyone who is interested in integrating eye tracking in usability engineering. A great deal of more basic research using eye tracking continues to produce results that are applicable to the design of human–computer interaction. The studies by Crosby and Sophian, reported in the current volume, are examples of such work. Here they show that shape comparison can be an effective way to compare ratio data visually. Surprisingly few fixations are needed to compare two shapes. Zülch and Stowasser (current volume) report a series of usability studies using eye tracking. Their laboratory is one of the few of which we are aware that uses these techniques fairly regularly. In their chapter, Zülch and Stowasser report results from a series of studies of industrial manufacturing software where users must solve scheduling problems, search for object or relationship information in list or graphic representation, or search for/extract information from a visual database. Comparing structured and unstructured approaches in problem solving and search tasks, they find differences in eye tracking data.

These studies and the historical work reviewed in previous paragraphs, point to some springboards for further study. We list what we believe to be the greatest opportunities for future work here:

Many usability studies that have incorporated eye tracking have indicated a difference between novice and more experienced participants (Fitts *et al*, 1950; Crosby & Peterson, 1991; Card, 1984; Aaltonen *et al.*, 1998) and individual differences (Yarbus, 1967 [1965]; Card, 1984; Andrews & Coppola, 1999). Eye tracking seems like an especially useful tool to study repetitive or well-practiced tasks and "power usability" (Karn, Krolczyk & Perry, 1997) and the process by which people evolve from novice users to expert users.

When users search for a tool, menu item, icon, etc. in a typical human–computer interface, they often do not have a good representation of the target. Most of the literature in visual search starts with the participant knowing the specific target. We need more basic research in visual search when the target is not known completely. A more realistic search task is looking for the tool that will help me do a specific task, having not yet seen the tool.

More work is needed to resolve the technical issues with eye trackers and the analysis of the data they produce. These issues include constraints on participant movement; tracker accuracy, precision, ease of setup; dealing with dynamic stimuli; and labor-intensive data extraction.

While there is a wealth of literature dealing with fixation patterns both in reading and in picture perception, little data exists on the viewing of pictures and text in combination as they often occur in instruction materials, news media, advertising, multimedia content, etc. (Stolk *et al.*, 1993; Hegarty, 1992; Hegarty & Just, 1993). This seems like fertile ground for the application of eye tracking in usability evaluation.

While there has been considerable use of eye tracking in usability engineering over the 50+ years since Fitts' pioneering work, the concept has not caught on with anywhere near the popularity of Fitts' Law for human limb movement. We see however, that just in the past ten years, significant technological advances have made the incorporation of eye tracking in usability research much more feasible. As a result, we are already seeing a rapid increase in the adoption of eye tracking in usability labs. We anticipate that future application of these techniques will allow the human–computer interaction design community to learn more about users' deployment of visual attention and to design product interfaces that more closely fit human needs.

Finally we remind the reader always to explore multiple facets of usability. Various measures of usability are necessary to gather the whole picture (see for example Frokjær, Hertzum & Hornbæk, 2000). Eye tracking alone is not a complete usability engineering approach, but it can make a significant contribution to the assessment of usability.

Input from the Eye

Background

We turn now to eye movements as a real time input medium. First, why would one want to use eye movements interactively in a user interface? We can view the basic task of human–computer interaction as moving information between the brain of the user and the computer. Our goal is to increase the useful bandwidth across that interface with faster, more natural, and more convenient communication mechanisms. Most current user interfaces provide much more bandwidth from the computer to the user than in the opposite direction. Graphics, animations, audio, and other media can output large amounts of information rapidly, but there are hardly any means of inputting comparably large amounts of information from the user. Today's user–computer dialogues are thus typically rather one-sided. New input devices and media that use "lightweight," passive measurements can help redress this imbalance by obtaining data from the user conveniently and rapidly. The movements of a user's eyes can thus provide a convenient high-bandwidth source of additional user input.

Eye trackers have existed for a number of years, but their use has largely been confined to laboratory experiments. The equipment is gradually becoming sufficiently robust and inexpensive to consider use in real user–computer interfaces. What is now needed is research in appropriate interaction techniques that incorporate eye movements into the user–computer dialogue in a convenient and natural way.

The simplest solution would be to substitute an eye tracker directly for a mouse — install an eye tracker and use its *x, y* output stream in place of that of the mouse. Changes in the user's line of gaze would directly cause the mouse cursor to move. But the eye moves very differently from the intentional way the hand moves a mouse; this would work poorly and be quite annoying.

There are significant differences between a manual input source like the mouse and eye position to be considered in designing eye movement-based interaction techniques:

Eye movement input is distinctly faster than other current input media (Ware & Mikaelian, 1987, Sibert & Jacob, 2000). Before the user operates any mechanical pointing device, he or she usually looks at the destination to which he or she wishes to move. Thus the eye movement is available as an indication of the user's goal before he or she could actuate any other input device.

"Operating" the eye requires no training or particular coordination for normal users; they simply look at an object. The control-to-display relationship for this device is already established in the brain.

The eye is, of course, much more than a high-speed cursor positioning tool. Unlike any other input device, an eye tracker also tells where the user's interest is focused. By the very act of pointing with this device, the user changes his or her focus of attention; and every change of focus is available as a pointing command to the computer. A mouse input tells the system simply that the user intentionally picked up the mouse and pointed it at something. An eye tracker input could be interpreted in the same way (the user intentionally pointed the eye at something, because he or she was trained to operate this system that way). But it can also be interpreted as an indication of what the user is currently paying attention to, without any explicit input action on his or her part.

This same quality is also a problem for using the eye as a computer input device. Moving one's eyes is often an almost subconscious act. Unlike a mouse, it is relatively difficult to control eye position consciously and precisely at all times. The eyes continually dart from spot to spot, even when its owner thinks he or she is looking steadily at a single object, and it is not desirable for each such move to initiate a computer command.[7]

Similarly, unlike a mouse, eye movements are always "on." There is no natural way to indicate when to engage the input device, as there is with grasping or releasing the mouse. Closing the eyes is rejected for obvious reasons — even with eye tracking as input, the principal function of the eyes in the user–computer dialogue is for communication *to* the user. Eye movements are an example of a more general problem with many new passive or non-command input media, requiring either careful interface design to avoid this problem or some form of explicit "clutch" to engage and disengage the monitoring.

Also, in comparison to a mouse, eye tracking lacks an analogue of the integral buttons most mice have. Using blinks as a signal is a less than ideal solution because it detracts from the naturalness possible with an eye movement-based dialogue by requiring the user to think about when to blink.

Finally, eye tracking equipment is still far less stable and accurate than most manual input devices.

The problem with a simple implementation of an eye tracker interface is that people are not accustomed to operating devices simply by moving their eyes. They expect to be able to look at an item without having the look *mean* something. At first, it is empowering to be able simply to look at what you want and have it happen, rather than having to look at it (as you would anyway) and then point and click it with the mouse. Before long, though, it becomes like the "Midas Touch." Everywhere you look, another command is activated; you cannot look anywhere without issuing a command. The challenge in building a useful eye tracker interface is to avoid this Midas Touch problem. Ideally, the interface should act on the user's eye input when he or she wants it to and let the user just look around when that's what he wants, but the two cases are impossible to distinguish in general. Instead, researchers develop interaction techniques that address this problem in specific cases.

The key is to make wise and effective use of eye movements. Like other passive, lightweight, non-command inputs (e.g., gesture, conversational speech), eye movements are often non-intentional or not conscious, so they must be interpreted carefully to avoid annoying the user with unwanted responses to his actions. A user does not say or mean much by a movement of the eyes — far less than by a keyboard command or mouse click. The computer ought to respond with correspondingly small, subtle responses. Rearranging the screen or opening a new window would typically be too strong a response. More appropriate actions might be highlighting an object for future action, showing amplifying information on a second screen, or merely downloading extra information in case it is requested.

We have also found in informal evaluation of eye movement interfaces that, when all is performing well, eye gaze interaction can give a subjective feeling of a highly responsive system, almost as though the system is executing the user's intentions before he or she expresses them (Jacob, 1991). This, more than raw speed, is the real benefit we seek from eye movement-based interaction. Work presented in this book shows some of the ways to achieve this goal. For example, Illi, Isokoski, and Surakka present a sophisticated user interface that subtly incorporates the eye movement data with other lightweight inputs, and fuses them to decide more accurately just when the system should react to the eye.

Survey of Past Work in Eye Tracking for HCI Input

Using eye movements for human–computer interaction in real time has been studied most often for disabled (typically quadriplegic) users, who can use only their eyes for input (e.g., Hutchinson, 1989; Levine, 1981, 1984), report work for which the primary focus was disabled users). Because all other user–computer communication modes are unavailable, the resulting interfaces are rather slow and tricky to use for non-disabled people, but, of course, a tremendous boon to their intended users.

One other case in which real-time eye movement data has been used in an interface is to create the illusion of a better graphic display. The chapter by O'Sullivan, Dingliana, and Howlett covers a variety of new ways to do this for both graphics and behavior. Earlier work in flight simulators attempted to simulate a large, ultra-high resolution display (Tong & Fisher, 1984). With this approach, the portion of the display that is currently being viewed is depicted with high resolution, while the larger surrounding area (visible only in peripheral vision) is depicted in lower resolution. Here, however, the eye movements are used essentially to simulate a better display device, but do not alter the basic user–computer dialogue.

A relatively small amount of work has focused on the more general use of real-time eye movement data in HCI in more conventional user–computer dialogues, alone or in combination with other input modalities. Richard Bolt did some of the earliest work and demonstrated several innovative uses of eye movements (Bolt, 1981; Bolt, 1982; Starker & Bolt, 1990). Floyd Glenn (Glenn *et al.*, 1986) used eye movements for several tracking tasks involving moving targets. Ware and Mikaelian (1987) reported an experiment in which simple target selection and cursor positioning operations were performed approximately twice as fast with an eye tracker than with any of the more conventional cursor positioning devices. More recently, the area has flourished, with much more research and even its own separate conference series, discussed below in this section.

In surveying research in eye movement-based human–computer interaction we can draw two distinctions, one in the nature of the user's eye movements and the other, in the nature of the responses. Each of these could be viewed as *natural* (that is, based on a corresponding real-world analogy) or *unnatural* (no real world counterpart). In eye movements as with other areas of user interface design, it is helpful to draw analogies that use people's already-existing skills for operating in the natural environment and then apply them to communicating with a computer. One of the reasons for the success of direct manipulation interfaces is that they draw on analogies to existing human skills (pointing, grabbing, moving objects in physical space), rather than trained behaviors; virtual reality interfaces similarly exploit people's existing physical navigation and manipulation abilities. These notions are more difficult to extend to eye movement-based interaction, since few objects in the real world respond to people's eye movements. The principal exception is, of course, other people: they detect and respond to being looked at directly and, to a lesser and much less precise degree, to what else one may be looking at. We can view the user eye movements and the system responses separately as natural or unnatural:

User's eye movements: Within the world created by an eye movement-based interface, users could move their eyes to scan the scene, just as they would a real world scene, unaffected by the presence of eye tracking equipment (i.e., natural eye movement). The alternative is to instruct users of the eye movement-based interface to move their eyes in particular ways, not necessarily those they would have employed if left to their own devices, in order to actuate the system (i.e., unnatural or learned eye movements).

Nature of the response: Objects could respond to a user's eye movements in a natural way, that is, the object responds to the user's looking in the same way real objects do. As noted, there is a limited domain from which to draw such analogies in the real world. The alternative is unnatural response, where objects respond in ways not experienced in the real world.

This suggests a taxonomy of four possible styles of eye movement-based interaction:

Natural eye movement/Natural response: This area is a difficult one, because it draws on a limited and subtle domain, principally how people respond to other people's gaze. Starker and Bolt provide an excellent example of this mode, drawing on the analogy of a tour guide or host who estimates the visitor's interests by his or her gazes (Starker & Bolt, 1990). Another example related to this category is the use of eye movements in videoconferencing systems (Vertegaal, 1999). Here the goal is to transmit the correct eye position from one user to another (by manipulating camera angle or processing the video image) so that the recipient (rather than a computer) can respond naturally to the gaze. The work described above, in which eye movement input is used to simulate a better display, is also related to this category.

Natural eye movement/Unnatural response: In our work (Jacob, 1991), we have used natural (not trained) eye movements as input, but we provide responses unlike those in the real world. This is a compromise between full analogy to the real world and an entirely artificial interface. We present a display and allow the user to observe it with his or her normal scanning mechanisms, but such scans then induce responses from the computer not normally exhibited by real world objects.

Unnatural eye movement/Unnatural response: Most previous eye movement-based systems have used learned ("unnatural") eye movements for operation and thus, of necessity, unnatural responses. Much of that work has been aimed at disabled or hands-busy applications, where the cost of learning the required eye movements ("stare at this icon to activate the device") is repaid by the acquisition of an otherwise impossible new ability. However, we believe that the real benefits of eye movement interaction for the majority of users will be in its naturalness, fluidity, low cognitive load, and almost unconscious operation. These benefits are attenuated if unnatural, and thus quite conscious, eye movements are required.

Unnatural eye movement/Natural response: The remaining category created by this taxonomy is anomalous and not seen in practice.

Current Research in Eye Tracking for HCI Input

There is a variety of ways that eye movements can be used in user interfaces and the chapter in this section by O'Sullivan, Dingliana, and Howlett describes a range of them, including some using retrospective analysis for generating better interactive displays. The work they describe uses eye movement input in order to simulate a better interactive graphic display — where better might mean higher resolution, faster update rate, or more accurate simulation. We could say that, if techniques like this work perfectly, the user would not realize that eye tracking is in use, but would simply believe he or she were viewing a better graphic display than could otherwise be built; the eye movement input ought to be invisible to the user. This is in contrast to other work in eye movement-based interaction, such as that described in the chapter by Illi, Isokoski and Surakka, where the eye specifically provides input to the dialogue and may actuate commands directly.

Gaze-based rendering can be applied in several ways: One would be to make a higher resolution display at the point where the user's fovea is pointed. We know that the eye can see with much higher resolution near the fovea than in the periphery, so a uniform, high-resolution display "wastes" many pixels that the user can't see. However, this requires a hardware device that can modify its pixel density in real time.

Some work has been done along these lines for flight simulators, to create the illusion of a large, ultra-high resolution display in a projected dome display described above (Tong & Fisher, 1984). Two overlapping projectors are used, one covering the whole dome screen, and one covering only the high resolution inset; the second one is attached to a mirror on a servomechanism, so it can be rapidly moved around on the screen, to follow the user's eye movements.

Another way to use eye movements is to concentrate graphics rendering power in the foveal area. Here, a slower, but higher quality ray tracing algorithm generates the pixels near the fovea, and a faster one generates the rest of the screen, but the hardware pixel density of the screen is not altered (Levoy and Whitaker, 1990). However, as graphics processors improve, this technique may become less valuable, and it continues to be limited by the hardware pixel density of the screen. A similar approach can also be used, but applied to the simulation rather than the rendering. Even with fast graphics processors and fixed pixel density, the calculations required to simulate three-dimensional worlds can be arbitrarily complex and time consuming. This approach focuses higher quality simulation calculations, rather than rendering, at the foveal region. Another approach described in the chapter by O'Sullivan, Dingliana, and Howlett uses known performance characteristics of the eye, rather than real time eye tracking, to inform the design of the interface. By knowing when a saccade is likely to occur and that the visual system nearly shuts down during saccades, the system can use that time period to change the level of detail of a displayed object so that the change will not be noticed. Finally, the chapter presents ways to use retrospective analysis of eye movements to build a better interface. The authors analyze a user's eye movements viewing an object and use that information to determine the key perceptual points of that object. They can then develop a graphical model that captures higher resolution mesh detail at those points and less detail at others, so that it can be rendered rapidly but still preserve the most important details.

One of the most interesting points raised here is the scheduler component, which keeps track of the processing time spent on computing the current simulation frame and interrupts the process when the allocated time has been exceeded. The authors state "The simulation process proceeds by using the highest resolution data that the collision handling system was able to compute in the allocated time." This is a novel aspect of algorithms for interactive graphics systems — the need for producing the best computation possible in a fixed time (typically, in time for the next video refresh), rather than producing a fixed computation as fast as possible. We too have found this particularly important in virtual reality interfaces and developed a constraint-based language for specifying such interaction at a high level, which allows an underlying runtime system to perform optimization, tradeoffs, and conversion into discrete steps as needed. This allows us to tailor the response speeds of different elements of the user interface within the available computing resources for each video frame (Jacob, Deligiannidis & Morrison, 1999; Deligiannidis & Jacob, 2002).

We turn now to eye movement-based interaction in which the eye input is a first-class partner in the dialogue, rather than being used to improve the display quality but not affect the dialogue. The chapter by Illi, Isokoski and Surakka describes a variety of real time, interactive uses of eye movements in user interfaces. They provide a survey of much work in this area and lay out the problems and issues in using eye movements as an input medium.

As we have seen, the Midas Touch or clutch problem plagues the design of eye movement-based interfaces — how can we tell which eye movements the system should respond to and which it should ignore? Illi, Isokoski, and Surakka describe an approach that uses additional physiological measurements from the user to help solve this problem. By measuring other lightweight inputs from the user and combining the information from the eye with the information from these additional sensors, the system can form a better picture of the user's intentions. Here, facial muscle activity is measured in real time with EMG sensors. Eye movements, like EMG and other such measurements, share the advantage that they are easy for the user to generate — they happen nearly unconsciously. But they share the drawback that they are difficult to measure and to interpret. By combining several such imperfect inputs, we may be able to draw a better conclusion about the user's intentions and make a more appropriate response, just as one combines several imperfect sensor observations in the physical world to obtain better information than any of the individual sensors could have yielded alone.

Other recent researchers have also found ways to advance the use of eye movements in user interfaces. The MAGIC approach of Zhai, Morimoto and Ihde (1999) is carefully tailored to exploit the characteristics of the eye and the hand and combine their use to provide good performance. Salvucci and Anderson show a way to improve performance and ease of use by adding intelligence to the underlying software. Their system favors more likely targets to compensate for some of the inaccuracy of the eye tracker (Salvucci, 2000). The chapter by Illi, Isokoski, and Surakka describes both of these techniques in further detail. Other researchers are beginning to study the eye movements of automobile drivers, ultimately leading to a lightweight style interface that might combine eye movements with other sensors to anticipate a driver's actions and needs. Selker, Lockerd and Martinez (2001) have developed a new device that detects only the *amount* of eye motion. While this gives less information than a conventional eye tracker, the device is extremely compact and inexpensive and could be used in many situations where a conventional eye tracker would not be feasible.

Most eye movement-based interfaces respond to the user's instantaneous eye position. We can extend this to more subtle or lightweight interaction techniques by replacing "interaction by staring at" with a "interaction by looking around." An additional benefit of this approach is that, because the interaction is based on recent history rather than each instantaneous eye movement, it is more tolerant of the very brief failures of the eye tracker that we often observe. Our approach here is for the computer to respond to the user's glances with continuous, gradual changes. Imagine a histogram that represents the accumulation of eye fixations on each possible target object in an environment. As the user keeps looking at an object, histogram value of the object increases steadily, while histogram values of all other objects slowly decrease. At any moment we thus have a profile of the user's "recent interest" in the various displayed objects. We respond to those histogram values by allowing the user to select and examine the objects of interest. When the user shows interest in an object by looking at it, the system responds by enlarging the object, fading its surface color out to expose its internals, and hence selecting it. When the user looks away from the object, the program gradually zooms the object out, restores its initial color, and hence deselects it. The program uses the histogram values to calculate factors for zooming and fading continuously (Tanriverdi & Jacob, 2000).

We also see a natural marriage between eye tracking and virtual reality (VR), both for practical hardware reasons and because of the larger distances to be traversed in VR compared to a desktop display. The user is already wearing a head-mounted display; adding a tiny head-mounted eye tracking camera and illuminator adds little to the bulk or weight of the device. Moreover, in VR, the head, eye tracker, and display itself all move together, so the head orientation information is not needed to determine line of gaze (it is of course used to drive the VR display). Objects displayed in a virtual world are often beyond the reach of the user's arm or the range of a short walk. We have observed that eye movement interaction is typically faster than interaction with a mouse. More significantly for this purpose, we have also found that the time required to move the eye is hardly related to the distance to be moved, unlike most other input devices (Sibert & Jacob, 2000). This suggests that eye gaze interaction is most beneficial when users need to interact with distant objects, as is often the case in a virtual environment. We have indeed observed this benefit in a comparison between eye and hand interaction in VR (Tanriverdi & Jacob, 2000), and this may be a promising direction for further use of eye movement interaction.

Future Directions of Eye Tracking for HCI Input

We have seen that eye movements in HCI have been studied for many years. They continue to appear to be a promising approach, but we do not yet see widespread use of eye movement interfaces or widespread adoption of eye trackers in the marketplace. We should remember that there has historically been a long time lag between invention and widespread use of new input or output technologies. Consider the mouse, one of the more successful innovations in input devices, first developed around 1968 (Engelbart & English, 1968). It took approximately ten years before it was found even in many other research labs; and perhaps twenty before it was widely used in applications outside the research world. And the mouse was based on simple, mechanical principles that were well understood from the start, rather than experimental computer vision algorithms and exotic video processing hardware.

When Jacob (first author of this chapter) and his colleagues started work on this issue in 1988 at the Naval Research Laboratories, they saw two main problems. One was better and less expensive eye tracking hardware, and the other was new interaction techniques and ways to use eye movements in interface design. They began working on solving the second problem, intending that the eye tracker industry would advance the first. They used existing large, clumsy, and expensive eye trackers as a test bed to study interfaces that might someday run on convenient, inexpensive, and ubiquitous new eye trackers. The goal was to develop the techniques that people would use by the time this new equipment appeared. Unfortunately, we are still waiting for the new equipment. Eye trackers are continuing to improve, but rather slowly.

This is partly due to the nature of the eye tracker industry. It consists of small companies without large capital investment. They might sell only tens or hundreds of eye trackers in a year. This makes it difficult to invest in the large engineering effort that would be required to develop a really good, inexpensive unit. But without such a unit, the market will continue to be limited to tens or hundreds per year — a "chicken and egg" problem. The cycle may at last be breaking. The basic hardware components of an eye tracker are a video camera, frame grabber, and a processor capable of analyzing the video in real time. The first two components are not only becoming quite inexpensive but are often incorporated into ordinary computer workstations, mainly for use in teleconferencing. Small video cameras and moderate quality frame grabbers are beginning to appear as standard components of desktop workstations. While they are intended for teleconferencing, they could also be useful for eye tracking. Further, current CPU chips can perform the needed processing directly, unlike earlier systems that required dedicated analogue electronics for the necessary speed.

There are still a few wrinkles. We need a frame buffer with fairly high resolution and pixel depth (though it need not be color). A more difficult problem is that the camera must be focused tightly on the eye; ideally the eye should fill the video frame, in order to get enough resolution to distinguish just where the pupil is pointing. This is usually solved with a servo mechanism to point the camera or else a chinrest to hold the user steady. A camera with wide field of view and very high resolution over the entire field might also solve this problem someday. The final requirement is a small infrared light aimed at the eye to provide the corneal reflection. This is not difficult to provide, but is clearly not a default component of current camera-equipped desktop workstations.

The necessary accuracy of an eye tracker that is useful in a real-time interface (as opposed to the more stringent requirements for basic eye movement research) is limited, since a user generally need not position his or her eye more accurately than about one degree to see an object sharply. The eye's normal jittering (microsaccades) and slow drift movements further limits the practical accuracy of eye tracking. It is possible to improve accuracy by averaging over a fixation, but not in a real-time interface. The accuracy of the best current eye trackers that can be used for these applications approaches one-degree useful limit. However, *stability* and *repeatability* of the measurements leave much to be desired. In a research study it is acceptable if the eye tracker fails very briefly from time to time; it may require that an experimental trial be discarded, but the user need not be aware of the problem. In an interactive interface, though, as soon as it begins to fail, the user can no longer rely on the fact that the computer dialogue is influenced by where his or her eye is pointing and will thus soon be tempted to retreat permanently to whatever backup input modes are available. While eye trackers have dropped somewhat in price, their performance in this regard has not improved significantly. Performance appears to be constrained less by fundamental limits, than simply by lack of effort in this narrow commercial market.

One of the most promising current hardware technologies is the IBM Blue Eyes device that combines bright pupil and dark pupil eye tracking in a single compact unit (Zhai *et al.*, 1999). One method will sometimes work better than the other for certain subjects or at a moment when there is an extra reflection or artifact in the image. By rapidly toggling between these two modes, the device can use whichever gives better results from moment to moment. The Eye-R device of Selker *et al.* (2001) represents another possible direction in eye tracking hardware, toward an inexpensive, widely deployable device.

We can also step back and observe a progression in user interface devices that begins with experimental devices used to measure some physical attribute of a person in laboratory studies. As such devices become more robust, they may be used as practical medical instruments outside the laboratory. As they become convenient, non-invasive, and inexpensive, they may find use as future computer input devices. The eye tracker is such an example; other physiological monitoring devices may also follow this progression.

Conclusions

We have reviewed the progress of using eye tracking in human–computer interaction both retrospectively, for usability engineering and in real time, as a control medium within a human–computer dialogue. We primarily discussed these two areas separately, but we also showed that they share the same principal challenges with eye tracking technology and interpretation of the resulting data. These two areas intersect in software applications where fixations are both recorded for analysis and the display is changed contingent on the locus of a user's fixation. This sort of technology was pioneered for the study of reading (for a thorough review and current methodological developments in this area see Rayner, 1998; Kennedy, Radach, Heller & Pynte, 2000) and has more recently been applied to more complex visual displays (e.g., McConkie, 1991; McConkie & Currie, 1996; Karn & Hayhoe, 2000) and to virtual environments (Triesch, Sullivan, Hayhoe & Ballard, 2002). Hyrskykari, Majaranta, Aaltonen and Räihä (2000) report a practical example of such an application of gaze dependent displays in human–computer interaction. While a user reads text in a non-native language, the system detects reader difficulties (by monitoring for long fixations and regressions) and then pops up translation help automatically.

Although progress has been slow, the concept of using eye tracking in human–computer interaction is clearly beginning to blossom. New work, described in this section and elsewhere in this volume, provides examples of this growing field. We can see this growth in the establishment of a new conference series covering this area, the Eye Tracking Research and Applications Symposium (ETRA) sponsored by the Association for Computing Machinery (ACM). The new field of perceptual or perceptive user interfaces (PUI) also brings together a variety of work on lightweight or sensing interfaces that, like eye tracking, observe the user's actions and respond. It, too, has established a conference series (PUI), beginning in 1997. Finally, we can observe clear growth within the ACM Human Factors in Computing Systems Conference (CHI), which is the premier conference in the general field of human–computer interaction. The first paper at CHI on eye tracking in HCI appeared in 1987 (Ware & Mikaelian, 1987), two more in 1990 (Starker & Bolt, 1990; Jacob, 1990); followed by one in 1996, three in 1998, four in 1999, and six or more in each of 2000, 2001 and 2002.

From the perspective of mainstream eye movement research, HCI, together with related work in the broader field of communications and media research, appears as a new and very promising area of applied work. Hence, at the last European Conference on Eye Movements in Turku, Finland in September 2001 a significant group of researchers from these areas participated and more were invited to contribute to the current volume. It is obvious that both basic and applied work can profit from integration within a unified field of eye movement research.

In this commentary chapter we have reviewed a variety of research on eye movements in human–computer interaction. The promise of this field was clear from the start, but progress has been slow to date. Does this mean the field is no longer worthwhile? We think not. Application of eye tracking in human–computer interaction remains a very promising approach. Its technological and market barriers are finally being reduced. Reports in the popular press (e.g., Kerber, 1999; Gramza, 2001) of research in this area indicate that the concepts are truly catching on. Work, in these chapters and elsewhere, demonstrates that this field is now beginning to flourish. Eye tracking technology has been used to make many promises in the past 50 years. However, we do believe the technology is maturing and has already delivered promising results. We are quite confident that the current generation of workers in this field will finally break the circle of boom and decline that we have seen in the past and make applications of eye movements in HCI an integrated part of modern information technology.

Acknowledgements

We thank Noshir E. Dalal for help gathering reference material; Jill Kress Karn for moral support and meticulous proofreading; John W. Senders for suggestions relating to the history section; collaborators in eye movement research Linda Sibert and James Templeman at the Naval Research Laboratory annd Vildan Tanriverdi and Sal Soraci at Tufts; the Naval Research Laboratory, Office of Naval Research, and the National Science Foundation for supporting portions of this research; and Ralph Radach for his gracious invitation to contribute this chapter and encouragement and help along the way.

Notes

1 This ISI Citation search includes three indices (*Science Citation Index Expanded, Social Sciences Citation Index, and the Arts & Humanities Citation Index*) for the years 1973 to the present.

2 There have certainly been more than 16 studies incorporating eye tracking in usability research, but we use this citation index as a means of judging the relative popularity of these two techniques that Paul Fitts left as his legacy.

3 The problem of defining and computing eye movement parameters has recently been the subject of intensive methodological debate in the neigbouring area of reading research (see Inhoff & Radach, 1998 and Inhoff & Weger, this volume, for detailed discussions). The

chapters by Vonk and Cozijn and by Hyönä, Lorch and Rinck in the section on empirial research on reading deal with specific problems of data aggregation that to a certain degree also generalize to the area of usability research.

4 The list provided in Table 1 is not a complete list of all applications of eye tracking in usability studies, but it provides a good sense of how these types of studies have evolved over these past 50 years.

5 Some other authors use "gaze duration" differently, to refer to the *total* time fixating an area of interest during an entire experimental trial (i.e., the sum of all individual gaze durations).

6 "Dwell" is still arguably a more convenient word and time will tell whether "dwell" or "gaze" becomes the more common term.

7 In their comprehensive theoretical discussion on saccadic eye movments, Findlay and Walker (1999) distinguish between three levels of saccade generation: automatic, automated, and voluntary. The least well known of these levels is the automated one, representing the very frequent class of saccades made on the basis of learned oculomotor routines. On all three levels the generation of eye movements is mediated by bottom up and top down processing, but only a rather small a minority of eye movmements appears to be made on on a "voluntary" basis.

References

Albert, W. (2002). Do web users actually look at ads? A case study of banner ads and eye-tracking technology. In: *Proceedings of the 11th Annual Conference of the Usability Professionals' Association*, Florida.

Albert, W. S., & Liu, A. (in press). The effects of map orientation and landmarks on visual attention while using an in-vehicle navigation system. To appear in: A. G. Gale (ed.), *Vision in Vehicles 8*. London: Oxford Press.

Aaltonen, A. Hyrskykari, A., & Räihä, K. (1998). 101 Spots, or how do users read menus? In: *Proceedings of CHI 98 Human Factors in Computing Systems* (pp. 132–139). ACM Press.

Andrews, T., & Coppola, D. (1999). Idiosyncratic characteristics of saccadic eye movements when viewing different visual environments. *Vision Research*, *39*, 2947–2953.

Anliker, J. (1976). Eye movements: On-line measurement, analysis, and control. In: R. S. Monty, and J. W. Senders (eds), *Eye Movements and Psychological Processes* (pp. 185–199). Hillsdale, NJ: Lawrence Erlbaum Associates.

Babcock, J., Lipps, M., & Pelz, J. B. (2002). How people look at pictures before, during, and after image capture: Buswell revisited. In: *Proceedings of SPIE, Human Vision and Electronic Imaging*, *4662* (pp. 34–47).

Backs, R. W., & Walrath, L. C. (1992). Eye movement and pupillary response indices of mental workload during visual search of symbolic displays. *Applied Ergonomics*, *23*, 243–254.

Bolt, R. A. (1981). Gaze-orchestrated dynamic windows. *Computer Graphics*, *15*, 109–119.

Bolt, R. A. (1982). Eyes at the interface. In: *Proceedings of the ACM Human Factors in Computer Systems Conference* (pp. 360–362).

Benel, D. C. R., Ottens, D., & Horst, R. (1991). Use of an eye tracking system in the usability laboratory. In: *Proceedings of the Human Factors Society 35th Annual Meeting* (pp. 461–465). Santa Monica: Human Factors and Ergonomics Society.

Brookings, J. B., Wilson, G. F., & Swain, C. R. (1996). Psychophysiological responses to changes in workload during simulated air traffic control. *Biological Psychology*, 42, 361–377.

Byrne, M. D., Anderson, J. R., Douglas, S., & Matessa, M. (1999). Eye tracking the visual search of click-down menus. In: *Proceedings of CHI 99* (pp. 402–409). NY: ACM Press.

Card, S. K. (1984). Visual search of computer command menus. In: H. Bouma and D. G. Bouwhuis (eds), *Attention and Performance X, Control of Language Processes*. Hillsdale, NJ: Lawrence Erlbaum Associates.

Collewijn, H. (1999). Eye movement recording. In: R. H. S. Carpenter and J. G. Robson (eds.), *Vision Research: A Practical Guide to Laboratory Methods* (pp. 245–285). Oxford: Oxford University Press.

Cornsweet & Crane (1973). Accurate two-dimensional eye tracker using first and fourth Purkinje images. *Journal of the Optical Society of America, 63*, 921–928.

Cowen, L. (2001). An eye movement analysis of web-page usability. Unpublished Masters' thesis, Lancaster University, UK.

Crosby, M. E., & Peterson, W. W. (1991). Using eye movements to classify search strategies. In: *Proceedings of the Human Factors Society 35th Annual Meeting* (pp. 1476–1480). Santa Monica: Human Factors and Ergonomics Society.

Crowe, E. C., & Narayanan, N. H. (2000). Comparing interfaces based on what users watch and do. In: *Proceedings Eye Tracking Research and Applications Symposium 2000* (pp. 29–36). New York: Association for Computing Machinery.

Deligiannidis, L., & Jacob, R. J. K. (2002). DLoVe: Using Constraints to Allow Parallel Processing in Multi-User Virtual Reality. In: *Proceedings of the IEEE Virtual Reality 2002 Conference*, IEEE Computer Society Press. (Available at http://www.cs.tufts.edu/~jacob/papers/vr02.deligiannidis.pdf).

Dodge, R., & Cline, T. S. (1901). The angle velocity of eye movements. *Psychological Review, 8*, 145–157.

Duchowski, A. T. (2003). *Eye Tracking Methodology: Theory and Practice*. London: Springer-Verlag.

Ellis, S., Candrea, R., Misner, J., Craig, C. S., Lankford, C. P., & Hutshinson, T. E. (1998). Windows to the soul? What eye movements tell us about software usability. In: *Proceedings of the Usability Professionals' Association Conference 1998* (pp. 151–178).

Engelbart, D. C., & English, W. K. (1968). A research center for augmenting human intellect. In: *Proceedings of the 1968 Fall Joint Computer Conference* (pp. 395–410). AFIPS.

Findlay, J. M., & Walker, R. (1999). A model of saccade generation based on parallel processing and competitive inhibition. *Behavioral & Brain Sciences 22*(4), 661–721.

Fisher, D. F., Monty, R. A., & Senders, J. W. (eds) (1981). *Eye Movements: Cognition and Visual Perception.* Hillsdale, NJ: Lawrence Erlbaum.

Fitts, P. M. (1954). The information capacity of the human motor system in controlling the amplitude of movement. *Journal of Experimental Psychology, 47*, 381–391.

Fitts, P. M., Jones, R. E., & Milton, J. L. (1950). Eye movements of aircraft pilots during instrument-landing approaches. *Aeronautical Engineering Review 9*(2), 24–29.

Flemisch F. O., & Onken, R. (2000). Detecting usability problems with eye tracking in airborne battle management support. In: *Proceedings of the NATO RTO HFM Symposium on Usability of information in Battle Management Operations* (pp. 1–13). Oslo.

Frøkjær, E., Hertzum, M., & Hornbæk, K. (2000). Measuring usability: Are effectiveness, efficiency and satisfaction really correlated? In: *Proceedings of CHI 2000 Human Factors in Computing Systems* (pp. 345–352). ACM Press.

Glenn, F., Iavecchia, H., Ross, L., Stokes, J., Weiland, W., Weiss, D., & Zaklad, A. (1986). Eye-voice-controlled interface. In: *Proceedings of the 30th Annual Meeting of the Human Factors Society* (pp. 322–326). Santa Monica: Human Factors Society.

Goldberg, J. H., & Kotval, X. P. (1998). Eye movement-based evaluation of the computer interface. In: S. K. Kumar (ed.), *Advances in Occupational Ergonomics and Safety* (pp. 529–532). Amsterdam: ISO Press.

Goldberg, J. H., Stimson, M. J., Lewenstein, M. Scott, N., & Wichansky, A. M. (2002). Eye tracking in web search tasks: Design implications. In: *Proceedings of the Eye Tracking Research & Applications Symposium 2002* (pp. 51–58). New York ACM.

Graf, W., & Krueger, H. (1989). Ergonomic evaluation of user-interfaces by means of eye-movement data. In: M. J. Smith and G. Salvendy (eds), *Work with Computers: Organizational, Management, Stress and Health Aspects* (pp. 659–665). Amsterdam: Elsevier Science.

Gramza, J. (2001). What are you looking at? *Popular Science*, March, 54–56.

Harris, R. L., & Christhilf, D. M. (1980). What do pilots see in displays? In: *Proceedings of the Human Factors Society — 24th Annual Meeting* (pp. 22–26). Los Angeles: Human Factors Society.

Hartridge, H., & Thompson, L. C. (1948). Methods of investigating eye movements, *British Journal of Ophthalmology*, *32*, 581–591.

Hegarty, M. (1992). The mechanics of comprehension and comprehension of mechanics. In: K. Rayner (ed.), *Eye Movements and Visual Cognition: Scene Perception and Reading.* New York: Springer Verlag.

Hegarty, M., & Just, M.A. (1993). Constructing mental models of machines from text and diagrams. *Journal of Memory and Language*, *32*, 717–742.

Hendrickson, J. J. (1989). Performance, preference, and visual scan patterns on a menu-based system: Implications for interface design. In: *Proceedings of the ACM CHI'89 Human Factors in Computing Systems Conference* (pp. 217–222). ACM Press.

Hoeks, B., & Levelt, W. J. M. (1993). Pupillary dilation as a measure of attention: A quantitative system analysis. *Behavior Research Methods, Instruments & Computers*, *25*, 16–26.

Hutchinson, T. E., White, K. P., Martin, W. N., Reichert, K. C., & Frey, L. A. (1989). human–computer interaction using eye-gaze input. *IEEE Transactions on Systems, Man, and Cybernetics*, *19*, 1527–1534.

Hyrskykari, A., Majaranta, P., Aaltonen, A., & Räihä, K. (2000). Design issues of iDict: A gaze-assisted translation aid. In: *Proceedings of the Eye Tracking Research and Applications Symposium 2000* (pp. 9–14). NY: ACM Press.

Iida, M. Tomono, A., & Kobayashi, Y. (1989). A study of human interface using and eye-movement detection system. In: M. J. Smith and G. Salvendy (eds), *Work with Computers: Organizational, Management, Stress and Health Aspects* (pp. 666–673). Amsterdam: Elsevier Science.

Inhoff, A. W., & Radach, R. (1998). Definition and computation of oculomotor measures in the study of cognitive processes. In: G. Underwood (ed.), *Eye Guidance in Reading and Scene Perception*, (pp. 29–53). Oxford: Elsevier

Jacob, R. J. K. (1990). What you look at is what you get: Eye movement-based interaction techniques. In: *Proceedings of the ACM CHI'90 Human Factors in Computing Systems Conference* (pp. 11–18). Addison-Wesley/ACM Press.

Jacob, R. J. K. (1991). The use of eye movements in human–computer interaction techniques: What you look at is what you get. *ACM Transactions on Information Systems*, *9*, 152–169.

Jacob, R. J. K., Deligiannidis, L., & Morrison, S. (1999). A software model and specification language for non-wimp user interfaces. *ACM Transactions on Computer–Human Interaction*, *6*(1), 1–46. (Available at http://www.cs.tufts.edu/~jacob/papers/tochi.pmiw.html [HTML] or http://www.cs.tufts.edu/~jacob/papers/tochi.pmiw.pdf [PDF]).

Javal, E. (1878). Essai sur la physiologie de la lecture. *Annales d'Oculistique*, *79*, 97–117, 155–167, 240–274; *80* (1879), 61–73, 72–81, 157–162, 159–170, 242–253.

Josephson, S., & Holmes, M. E. (2002). Visual attention to repeated Internet images: Testing the scanpath theory on the world wide web. In: *Proceedings of the Eye Tracking Research & Applications Symposium 2002* (pp. 43–49). New York: ACM.

Judd, C. H., McAllister, C. N., & Steel, W. M. (1905). General introduction to a series of studies of eye movements by means of kinetoscopic photographs. In: J. M. Baldwin, H. C. Warren and C. H. Judd (eds), *Psychological Review, Monograph Supplements*, *7*, 1–16. Baltimore: The Review Publishing Company.

Just, M. A., & Carpenter, P. A. (1976a). Eye Fixations and Cognitive Processes. *Cognitive Psychology*, *8*, 441–480.

Just, M. A., & Carpenter, P. A. (1976b). The role of eye-fixation research in cognitive psychology. *Behavior Research Methods & Instrumentation*, *8*, 139–143.

Karn, K., Ellis, S., & Juliano, C. (1999). The hunt for usability. Workshop conducted at CHI 99 Human Factors in Computing Systems, Conference of the Computer–Human Interaction Special Interest Group of the Association of Computing Machinery. Pittsburgh. In: *CHI 99 Extended Abstracts*. (p. 173). NY: ACM Press.

Karn, K., Ellis, S., & Juliano, C. (2000). The hunt for usability: Tracking eye movements. *SIGCHI Bulletin*, November/December 2000 (p. 11). New York: Association for Computing Machinery. (Available at http://www.acm.org/sigchi/bulletin/2000.5/eye.html).

Karn, K., Goldberg, J., McConkie, G., Rojna, W., Salvucci, D., Senders, J., Vertegaal, R., & Wooding, D. (2000). "Saccade Pickers" vs. "Fixation Pickers": The effect of eye tracking instrumentation on research. (Panel presentation). Abstract in: *Proceedings of the Eye Tracking Research and Applications Symposium 2000* (p. 87). NY: ACM Press.

Karn, K., & Hayhoe, M. (2000). Memory representations guide targeting eye movements in a natural task. *Visual Cognition*, *7*, 673–703.

Karn, K., Krolczyk, M., & Perry, T. (1997). Testing for power usability. Workshop conducted at CHI 97 Human Factors in Computing Systems. Conference of the Computer–Human Interaction Special Interest Group of the Association of Computing Machinery. Atlanta. In: *CHI 97 Extended Abstracts*. (p. 235). NY: ACM Press.

Karsh, R., & Breitenbach, F. W. (1983). Looking at looking: The amorphous fixation measure. In: R. Groner, C. Menz, D. Fisher and R. A. Monty (eds), *Eye Movements and Psychological Functions: International Views* (pp. 53–64). Hillsdale, NJ: Lawrence Erlbaum Associates.

Kennedy, A., Radach, R., Heller, D., & Pynte, J. (eds, 2000). *Reading as a Perceptual Process*. Oxford: Elsevier.

Kerber, R. (1999). Cleanup crew has eye on web sites. *The Boston Globe*, August 22, 1999. A1, A16–17.

Kolers, P. A., Duchnicky, R. L., & Ferguson, D. C. (1981). Eye movement measurement of readability of CRT displays. *Human Factors*, *23*, 517–527.

Kotval, X. P., & Goldberg, J. H. (1998). Eye movements and interface components grouping: An evaluation method. In: *Proceedings of the 42nd Annual Meeting of the Human Factors and Ergonomics Society* (pp. 486–490). Santa Monica: Human Factors and Ergonomics Society.

Kowler, E. (1990). The role of visual and cognitive processes in the control of eye movement. In: E. Kowler (ed.), *Eye Movements and their Role in Visual and Cognitive Processes*. Amsterdam: Elsevier Science.

Lambert, R. H., Monty, R. A., & Hall, R. J. (1974). High-speed data processing and unobtrusive monitoring of eye movements. *Behavioral Research Methods & Instrumentation*, *6*, 525–530.

Lankford, C. (2000) Gazetracker™: Software designed to facilitate eye movement analysis. In: *Proceedings of the Eye Tracking Research and Applications Symposium 2000* (pp. 51–55). NY: ACM Press.

Land, M.F. (1992). Predictable eye-head coordination during driving. *Nature, 359*, 318–320.

Land, M. F., Mennie, N., & Rusted, J. (1999). The role of vision and eye movements in the control of activities of daily living. *Perception, 28*, 1311–1328.

Levine, J. L. (1981). *An Eye-Controlled Computer*. Research Report RC-8857. New York: IBM Thomas J. Watson Research Center, Yorktown Heights.

Levine, J. L. (1984). Performance of an eyetracker for office use. *Computational Biology and Medicine*, 14, 77–89.

Levoy, M., & Whitaker, R. (1990). Gaze-directed volume rendering. In: *Proceedings of the 1990 Symposium on Interactive 3D Graphics* (pp. 217–223). Utah: Snowbird.

Mackworth, J. F., & Mackworth, N. H. (1958). Eye fixations recorded on changing visual scenes by the television eye-marker. *Journal of the Optical Society of America, 48*, 439–445.

Mackworth, N. H., & Thomas, E. L. (1962). Head-mounted eye-marker camera. *Journal of the Optical Society of America, 52*, 713–716.

Marshall, S. P. (1998). Cognitive workload and point of gaze: A re-analysis of the DSS directed-question data. *Technical Report CERF 98–03*. San Diego, CA: Cognitive Ergonomics Research Facility, San Diego State University.

McConkie, G. W. (1991). Perceiving a stable visual world. In: J. Van Rensbergen, M. Devijver and G. d'Ydewalle (eds), *Proceedings of the Sixth European Conference on Eye Movements*. (pp. 5–7). Leuven, Belgium: Laboratorium voor Experimental Psychologie, Katholieke Universiteit Leuven.

McConkie, G. W., & Currie, C. B. (1996). Visual stability across saccades while viewing complex pictures. *Journal of Experimental Psychology: Human Perception and Performance, 22*, 563–581.

Merchant, J. Morrissette, R., & Porterfield, J. L. (1974). Remote measurement of eye direction allowing subject motion over one cubic foot of space. *IEEE Transactions on Biomedical Engineering*, BME-21, 309–317.

Merwin, D. (2002). Bridging the gap between research and practice. *User Experience,* Winter, 38–40.

Monty, R. A. (1975). An advanced eye-movement measuring and recording system. *American Psychologist, 30*, 331–335.

Monty, R. A., & Senders, J. W. (eds) (1976). *Eye Movements and Psychological Processes.* Hillsdale, NJ: Lawrence Erlbaum.

Mulligan, J. (2002). A software-based eye tracking system for the study of air-traffic displays. In: *Proceedings of the Eye Tracking Research and Applications Symposium 2002* (pp. 69–76). New York: Association for Computing Machinery.

Pelz, J. B., & Canosa, R., (2001). Oculomotor behavior and perceptual strategies in complex tasks. *Vision Research, 41*, 3587–3596.

Pelz, J. B., Canosa, R., & Babcock, J. (2000). Extended tasks elicit complex eye movement patterns. In: *Proceedings of the Eye Tracking Research and Applications Symposium 2000.* (pp. 37–43) New York: ACM Press.

Rayner, K. (1998). Eye movements in reading and information processing: 20 years of research. *Psychological Bulletin, 124*, 372–422.

Redline, C. D., & Lankford, C. P. (2001). Eye-movement analysis: A new tool for evaluating the design of visually administered instruments (paper and web). Paper presented at 2001

AAPOR Annual Conference, Montreal, Quebec, Canada, May 2001. In: *Proceedings of the Section on Survey Research Methods*, American Statistical Association.

Reeder, R. W., Pirolli, P., & Card, S. K. (2001). WebEyeMapper and WebLogger: Tools for analyzing eye tracking data collected in web-use studies. In: *Extended Abstracts of the Conference on Human Factors in Computing Systems, CHI 2001* (pp. 19–20). New York: ACM Press.

Russo, J. E., & Leclerc, F. (1994). An eye-fixation analysis of choice process for consumer nondurables. *Journal of Consumer Research*, *21*, 274–290.

Salvucci, D. D. (2000). An interactive model-based environment for eye-movement protocol analysis and visualization. In: *Proceedings of the Eye Tracking Research and Applications Symposium 2000* (pp. 57–63). NY: ACM Press.

Salvucci, D. D., & Anderson, J. R. (2000). Intelligent gaze-added interfaces. In: *Proceedings of the ACM CHI 2000 Human Factors in Computing Systems Conference* (pp. 273–280). Addison-Wesley/ACM Press.

Salvucci, D. D., & Goldberg, J. H. (2000). Identifying fixations and saccades in eye-tracking protocols. In: *Proceedings of the Eye Tracking Research and Applications Symposium 2000* (pp. 71–78). NY: ACM Press.

Selker, T., Lockerd, A., & Martinez, J. (2001). Eye-R, a glasses-mounted eye motion detection interface. In: *Proceedings of the ACM CHI 2001 Human Factors in Computing Systems Conference Extended Abstracts* (pp. 179–180). NY: ACM Press.

Senders, J. W. (2000). Four theoretical and practical questions. Keynote address presented at the Eye Tracking Research and Applications Symposium 2000. Abstract in: *Proceedings of the Eye Tracking Research and Applications Symposium 2000* (p. 8). New York: Association for Computing Machinery.

Senders, J. W., Fisher, D. F., & Monty, R. A. (eds) (1978). *Eye Movements and the Higher Psychological Functions.* Hillsdale, NJ: Lawrence Erlbaum.

Shackel, B. (1960). Note on mobile eye viewpoint recording. *Journal of the Optical Society of America*, *59*, 763–768.

Sheena, D., & Flagg, B. N. (1978). Semiautomatic eye movement data analysis techniques for experiments with varying scenes. In: J. W. Senders, D. F. Fisher and R. A. Monty (eds), *Eye Movements and the Higher Psychological Functions* (pp. 65–75). Hillsdale, NJ: Lawrence Erlbaum.

Sibert, L. E., & Jacob, R. J. K. (2000). Evaluation of eye gaze interaction. In: *Proceedings of the ACM CHI 2000 Human Factors in Computing Systems Conference* (pp. 281–288). Addison-Wesley/ACM Press. (Available at http://www.cs.tufts.edu/~jacob/papers/chi00.sibert.pdf [PDF]).

Simmons, R. R. (1979). Methodological considerations of visual workloads of helicopter pilots. *Human Factors*, *21*, 353–367.

Sodhi, M., Reimer, B., Cohen, J. L., Vastenburg, E., Kaars, R., & Kirschenbaum, S. (2002). On-road driver eye movement tracking using head-mounted devices. In: *Proceedings of the Eye Tracking Research & Applications Symposium 2002* (pp. 61–68). New York: ACM.

Starker, I., & Bolt, R. A. (1990). A gaze-responsive self-disclosing display. In: *Proceedings of the ACM CHI'90 Human Factors in Computing Systems Conference* (pp. 3–9). Addison-Wesley/ACM Press.

Stern, J. A., Boyer, D., & Schroeder, D. (1994). Blink rate: A possible measure of fatigue. *Human Factors*, *36*, 285–297.

Stolk, H. Boon, K., & Smulders, M. (1993). Visual information processing in a study task using text and pictures. In: G. d'Ydewalle and J. Van Rensbergen (eds), *Perception and Cognition* (pp. 285–296). Amsterdam: Elsevier Science.

Svensson, E., Angelborg-Thanderz, M. Sjöeberg, L., & Olsson, S. (1997). Information complexity: Mental workload and performance in combat aircraft. *Ergonomics*, *40*, 362–380.
Tanriverdi, V., & Jacob, R. J. K. (2000). Interacting with eye movements in virtual environments. In: *Proceedings of the ACM CHI 2000 Human Factors in Computing Systems Conference* (pp. 265–272). Addison-Wesley/ACM Press. (Available at http://www.cs.tufts.edu/~jacob/papers/chi00.tanriverdi.pdf [PDF]).
Tinker, M. A. (1963). *Legibility of Print*. Ames, Iowa: Iowa State University Press.
Tong, H. M., & Fisher, R. A. (1984). *Progress Report on an Eye-Slaved Area-of-interest Visual Display*. Report No. AFHRL-TR-84–36, Air Force Human Resources Laboratory, Brooks Air Force Base, Texas. Proceedings of IMAGE III Conference.
Triesch, J., Sullivan, B. T., Hayhoe, M. M., & Ballard, D. H. (2002). Saccade contingent updating in virtual reality. In: *Proceedings of the Eye Tracking Research & Applications Symposium 2002* (pp. 95–102). New York: ACM.
Vertegaal, R. (1999). The GAZE groupware system: Mediating joint attention in multiparty communication and collaboration. In: *Proceedings of the ACM CHI'99 Human Factors in Computing Systems Conference* (pp. 294–301). Addison-Wesley/ACM Press.
Ware, C., & Mikaelian, H. T. (1987). An evaluation of an eye tracker as a device for computer input. In: *Proceedings of the ACM CHI+GI'87 Human Factors in Computing Systems Conference* (pp. 183–188). New York: ACM Press.
Weiser, M. (1993). Some computer science issues in ubiquitous computing. *Communications of the ACM 36*(7), 75–84.
Wooding, D. S. (2002). Fixation Maps: Quantifying eye-movement traces. In: *Proceedings of the Eye Tracking Research & Applications Symposium 2002* (pp. 31–36). New York: ACM.
Yarbus, A. L. (1967) *Eye Movements and Vision* (Trans. B. Haigh). New York: Plenum Press. (Original work published 1965).
Yamamoto, S., & Kuto, Y. (1992). A method of evaluating VDT screen layout by eye movement analysis. *Ergonomics*, *35*, 591–606.
Zhai, S., Morimoto, C., & Ihde, S. (1999). Manual and gaze input cascaded (MAGIC) pointing. In: *Proceedings of the ACM CHI'99 Human Factors in Computing Systems Conference* (pp. 246–253). Addison-Wesley/ACM Press.

Section 5

Eye Movements in Media Applications and Communication

Section Editor: Jukka Hyönä

Chapter 27

Eye Movements in the Processing of Print Advertisements

Ralph Radach, Stefanie Lemmer, Christian Vorstius, Dieter Heller and Karina Radach

Research on the visual processing of advertisements has so far focused on formal aspects such as the relative size of print or the composition of graphical elements. Little is known, however, about effects of specific content-related design factors. One such factor is the complexity of the pragmatic relation between image and text. We refer to an advertisement as "explicit" when it depicts the target product (e.g. a stereo or a car) together with a related headline in a semantically straightforward way. In contrast, an "implicit" advertisement includes pictures and text neither of which are directly related to the product.

In two experiments, eye movements were measured during the viewing of advertisements containing a large pictorial element, a headline and a fictitious product name. Items targeting on identical products were designed in an explicit and an implicit version. Participants were asked to view the stimuli item-by-item in preparation for an evaluation on scales of valence and interestingness or for a short paraphrasing of ad content.

In support of our hypothesis, viewers spent significantly more time on implicit advertisements and also rated them as more positive and interesting. While mean fixation durations and saccade amplitudes did not differ, there was a substantial difference in the number of fixations. The variation of task had profound consequences both for viewing behavior on regions of interest (headline, picture and brand name) and recall, but did not affect the main effect of ad complexity.

The Mind's Eye: Cognitive and Applied Aspects of Eye Movement Research

ISBN: 0-444-51020-6

Introduction

Advertisements are an important part of our visual world. From the perspective of clients who order advertisements, they are supposed to be effective. What does this mean? In a short term perspective, potential customers should spend a lot of time looking at an advertisement, retain a vivid memory and find the ad pleasing and interesting. On the long run, it is expected that positive evaluation and stable memory transform into a favorable attitude towards products or brands and many purchase decisions. A central goal of advertisement research is to understand the mechanisms that make ads effective and to contribute to increasing their success.

Potential customers who look at advertisements can respond in different ways: They may consider some ads boring or even annoying, they may like them as a source of entertainment or they may use them to search for useful information. Over time, people will develop attitudes and preferences in response to the many ads they have seen. In addition to a particular response to the appearance and specific content of an ad, more general design and content factors are likely to play a role in determining what viewers like and dislike. From an economical perspective knowledge about these factors is a critical condition for effectiveness. From the perspective of viewers, this type of research may well contribute to increase the proportion of "good" advertisements and therefore make our visual world more comfortable.

The research reported in this chapter was triggered by a practical problem. Professionals who produce advertisements are often interested in work that they consider original and creative. This includes ads where the relations between picture, headline and other elements are rather complex and where the message cannot immediately be understood. The idea is that advertisements that pose an intellectual challenge will take more effort to understand (often including an element of humor), but may be liked more and remembered better. This view can lead to disagreement with clients who often tend to believe that an effective advertisement should be clear and straightforward. This includes depicting the product and presenting text that also directly relates to the product or brand, e.g. describing its properties or advantages. From the perspective of cognitive psychology, this controversy deals with the complexity of the pragmatic relation between ad elements. More specifically, the question is whether more complex ads are indeed more effective. To deal with this problem, complexity needs to be clearly defined and it must be determined how precisely ad effectiveness should be measured.

Some Theoretical Background

In a discussion of the history of systematic research on advertisement, Hansen (1995) draws a distinction between three periods. He describes these as "the age of the recognition vs. recall debate" (1930–1970), the "persuasion age" (starting around 1960) and the "age of model based research" (starting around 1990). In fact, the bulk of the literature on advertisement research is still on memory for ads and on issues around "persuasion" like preference or emotional appraisal of ad stimuli or the development of attitudes towards brands or products. "Model-based" research most often includes a general theoretical framework of the advertisement process, for example in the tradition of the famous AIDA (attention → interest → desire → action) theory (Lewis, 1898; Strong, 1925). Lewis is believed to have been the first to suggest that effective salesmanship requires salesmen to attract customers' attention, maintain their interest, and create a desire. In modern approaches, attempts are made to provide more detailed accounts of ad-related consumer behavior, including various proposed sequences of processing stages and behavioral choices (see e.g. Rosbergen, 1998, for a detailed discussion).

In this chapter we will look at advertisement research primarily from the somewhat reduced angle of visual information processing. We will also provide some background on the evaluation of ads and on memory in terms of recall and recognition to the extent to which they are relevant for the experiments that will be reported.

Visual Processing of Advertisements

Looking at the time course of visual processing of print ads, it is intuitively clear that two or three distinct time periods need to be distinguished. This distinction is in line with research on different aspects of picture processing, for example in the inspection of medical x-ray images (Nodine, Kundel, Polikoff & Toto, 1987) or in recent studies on effects of change blindness (see the section on this subject in Hyönä, Munoz, Heide & Radach, 2002). One common observation is that there is a quick phase of orientation and global processing, followed by a more detailed scanning of selected areas. Perhaps the best theoretical conceptualization of the routes of perceptual and cognitive processing involved in this process is the model of complex scene perception by de Graef (1992).

In Table 27.1, a typical newspaper reading situation is assumed. There are several likely ways for a reader to send her eyes to and across a print advertisement. The most likely is that the eyes land on the ad as part of a general pattern of scanning

Table 27.1: Simple sketch of a time line of processing a print advertisement.

	Processing mechanism	Time period
Initial selection	• peripheral saliency • automized scan path • visual search	• pre fixation: extrafoveal • processing concurrently with other information
Early processing	• skimming over text • fast picture scanning	• first few fixations 150 ms to 600 ms
Late processing	• reading of text and scanning of picture details • semantic processing	• 600 ms up to several seconds

the page that is also used to identify interesting non-ad text or pictures. In other cases a reader may actively look for advertisement information on specific products. It is also possible that a reader is using a scanning strategy that explicitly includes avoidance of ads (see the chapter by Stenfors *et al.*, this volume for a discussion of ad avoidance for the case of internet pages). In this situation, the only chance for the ad to be selected as the goal for a saccade would be that it contains some visual features that make it extremely salient in the context of the page. In any case there will be a decision to send the eye to the advertisement made on the basis of extrafoveal information in relation to some top down processing. Is this decision taken and the eyes have landed on the advertisement, there will most likely be a period of time during which some quick acquisition of pictorial and text information will take place. The major goal in this phase is to verify that it is indeed desirable to spend time with the ad, e.g. in terms of identifying its topic or even checking an initial hypothesis about its content. There is a high chance that early processing leads to a negative decision and the viewing of the ad is terminated. If the decision is positive, however, and more detailed processing takes place, the viewer will stop when the information presented is exhausted. In the ideal case, processing will last until all critical information is encoded and the message is understood.

The first studies using eye movements to study the processing of ads appeared in the 1960s (Robinson, 1963; Starch, 1966). Since then, oculomotor measures have become a technically feasible and potentially powerful option in the arsenal of available methods. However, this type of work still represents a relatively small segment in the substantial literature on ad effectiveness. In general, the goal of most studies is to gather information about towards which part of an advertisement respondents direct their attention and how deeply people process the information offered by an advertisement (von Keitz, 1988). An excellent overview of the literature on eye tracking in advertisement research has been provided by Rosbergen (1998). Some of the most important studies are listed in Table 27.2, based on a more detailed table in Rosbergen's dissertation.

Looking at this table, it is apparent that "aspects of eye movements" are conceptualized in terms of "amount of attention". Looking for an explicit definition of attention in Rosbergen, Pieters and Wedel (1997) we found statements like the following: "Capacity theories of attention (see e.g. Broadbent 1971; Kahneman 1973) as well as information-processing models (see e.g., Greenwald & Leavitt, 1984; MacInnis & Jaworski, 1989) assume that the attention allocated to an ad is a function of consumers' motivation, opportunity, and ability, which are affected by, for instance, physical properties of the advertisement and consumers' characteristics" (p. 305). Being in sympathy with the enthusiasm present in Rosbergen *et al.* (1997) and other, similar publications, two remarks seem to be in place. One is that it is not at all evident how *precisely* claims about the role of motivation, opportunity, ability, etc. for visual processing of advertisements may follow from the classic work by Broadbent, Kahneman and others. And, do we really know much about *whether* and *how* these are, in turn, affected by physical properties of advertisements? Careful empirical work and extensive theoretical discussions (to which Rosbergen *et al.* make valuable contributions) will be necessary to carry such claims from face plausibility to solid theoretical grounding.

Table 27.2: A compilation of eye tracking studies in advertising research (modified after Rosbergen, 1998). The term "attention" refers to various spatial and temporal eye movement measures.

Reference	**Factors examined**	**Aspects of eye movements that were analyzed**
Robinson (1963)	Ad size	Amount of attention per ad & number of saccades between ad quadrants
Starch (1966)	None	Amount of attention per ad
Krugman (1968)	Repetition	Number of 1 × 1 inch ad elements that are fixated
Treistman and Gregg (1979)	None	Amount of attention per ad element
Witt (1977, in Kroeber-Riel 1979)	Erotic illustration	Number of eye fixations per ad element
Edell and Staelin (1983)	Ad structure	Amount of attention per ad element
Young (1984)	Size & clutter	Looked at outdoor board, product advertised and copy or not
Bogart and Tolley (1988)	None	Number of fixations per ad
Janiszewski (1993)	Length of copy	Amount of attention per ad element
Janiszewski and Warlop (1993)	Classical conditioning	Order of fixation on soda brands
Krugman *et al.* (1994)	Type of warning	Amount of attention to the warning and time to first fixation on the warning
Lohse (1997)	Color, size & position	Amount and order of attention to ads

Our second remark is that the above statement appears typical for the widespread belief that attention is such a basic concept that its meaning should be more or less clear to anyone reading the publications listed in Table 27.2. Throughout the advertisement literature, the term "attention" is used interchangeably in terms of selection (either referring to selection itself and/or something that is being selected), preference, processing time, depth of processing, mental effort, etc. without reflecting the epistemic and logical problems related to the concept (see e.g. Allport, 1992 and Radach, Inhoff & Heller, 2002, for discussions).

In this field of research there are many studies that looked at eye movements on ads taken from newspapers or journals, but in a few studies ad variants were also specifically designed to allow for experimental variations. For example, Lohse (1997) examined eye movement patterns of consumers on yellow-page advertisements. Thirty-two fictitious yellow pages containing 348 ads, assembled in four books, were presented to 32 subjects. Participants were asked to imagine that they had just moved to the Boston area. Their task was to locate different products or services and to choose three favorite businesses under each heading. The stimuli were controlled for different combinations of layout and design features. Ad type, locations of display ads on the page, size of ad, color, use of graphics, whether or not a listing had a bold typeface, serial position of the ad (alphabetic order), and number of types of information in the ad (hours, years in business, slogan, brand names, specialties) were implemented as dependent variables. Eye tracking data such as fixation duration, fixation number, and total viewing time were computed as independent variables.

The data gathered in this experiment support earlier findings, indicating that ad size, graphics and color all have impact on the visual processing of advertisements (Hornik, 1980; Rossiter, 1981, 1988; Schindler, 1986; Valiente, 1973). The results of this study demonstrate e.g. that larger ads were more likely to be fixated, participants noticed 93% of the large display ads but only 26 % of the plain listings. Subjects noticed more color ads than ads without color (92% vs. 84%) and looked at color ads before ads without color. They viewed color ads 21% longer than equivalent ads in black and white. Subjects also fixated 96% of ads with graphics. Quite interesting are effects of position of an advertisement on the page. Since people scan ads on a page in alphabetic order, advertisements at the end of a page are nearly never looked at. Lohse (1997) stresses that his data have implications for pricing decisions in telephone directories, the graphic layout and the marketing for paper and electronic telephone directories.

An interesting new research perspective was introduced by Rosbergen *et al.* (1997). They used local gaze duration patterns analyzed with a latent class regression model to identify groups of consumers that respond differently to visual properties of ads. For this purpose, four different versions of an existing advertisement were professionally designed and inserted into a custom developed issue of a popular Dutch weekly women's magazine. The design variables of interest included size and position of elements like pictorial, pack shot (i.e. a picture of a shampoo bottle), and headline. Analyses of gaze duration data revealed three different populations of consumers with qualitatively different "patterns of attention" labeled as scanning, initial attention, and sustained attention. Interestingly, this classification is quite similar to the time line suggested in Table 27.1. The principle differences is that Rosbergen *et al.* see their types of responses as characteristics of potential customers, whereas we see our phases of processing as means of response that are at the disposal of every viewer.

Recently, Rayner, Rotello, Stewart, Keir and Duffy (2001) raised the question of how different goals may influence the visual processing devoted to different aspects of an advertisement. Another interesting purpose of their experiment was to investigate how viewers process text compared to pictorial elements of an advertisement, and how both types of information are integrated. Specific goals were given through instruction. In their "car condition" (US American) participants were told to imagine that they had just moved to England and needed to buy a car from the number of available makes and models. In the "skin care condition" subjects were asked to imagine that they had moved to England and needed to decide which skin lotion and body wash to buy. Participants were shown 24 color advertisements that included both groups of critical items and filler items. The ads were taken from British magazines unknown in the US market. After viewing the ads, subjects were asked to take part in a recall and a recognition test, and to recall the ads they had liked and disliked. The eye movement data support earlier findings: subjects spent more time on text elements than on pictorial elements, they tended not to go back and forth between text and the picture (see also Carroll, Young & Guertin, 1992; Hegarty, 1992). As part of the experimental setting, the initial fixation was always located in the center of the ad. However, the first saccade usually went to the large print, regardless of its spatial position within the ad. In general, viewers tended to first read the text. Also, following the instruction, subjects spent more time on the type of ads they were asked to pay attention to. Rayner *et al.* (2001) concluded that the viewing time on the stimuli was modulated by strategies associated with the instructions viewers were given. As a major conclusion they suggest that participants' goals should be considered in future research.

As the above discussion has shown, the bulk of the work that has so far been done on the topic of eye movements and ad processing deals with these two phases of initial selection and early processing. Most of the studies examined how variations in design features like size, color or arrangement of elements influenced eye movements. Most often this was phrased in terms of "attracting a customer's attention". In contrast to this, in the experiments reported in this chapter it is presupposed that a decision in favor of prolonged processing has been made and that the viewer has the intention to acquire and understand the information presented.

Evaluation of Advertisements

An important way to assess the effectiveness of an advertisement is the application of rating scales. Haley (1990) compared six different "copy testing methods" for TV commercials (e.g. recall, communication, persuasion) which included the most common types of measures. The objective was to determine the accuracy of each copy-testing method to predict "sales winners". For this purpose, pairs of commercials of five different products were included in the test. Each pair contained two different commercials that had shown significant differences in the levels of sales response in 1-year split cable sales tests. The results demonstrated that all types of measures had some predictive power, but likeability turned out to be the single best predictor of sales effectiveness. This scale predicted sales winners 87% of the time (Haley, 1990).

Biel (1998) presents results of a study by the Ogilvy Center for Research & Development, San Francisco. The objective of their work was to investigate the relationship between an advertising's likeability and its potential to motivate costumers. First, measures of brand persuasion and of commercial liking were taken among 895 consumers for a sample of 73 prime-time TV commercials. Sales effectiveness was determined pre- and post exposure via phone, following a procedure developed by Mapes and Ross. Participants were also asked to rate likeability on a 5-point scaled question ("liked it a lot", "liked it somewhat", "neutral", "disliked it somewhat", "disliked it a lot"). Based on these data, the sample was divided into three subgroups: people who are enthusiastic about the commercial they have seen, those who were mildly positive and people who were neutral. In a second phase of the study, changes in brand preferences were measured for the three subsamples. The highest change in brand preference appeared with 16.2% in the group of enthusiasts compared to changes in the other two groups (mildly positive 9.5%, neutral 8.2%). Biel (1998) concluded "that it pays to produce advertising that people like, and the more they like it, the more they are persuaded by it" (p. 114).

Although the studies discussed so far all deal with TV commercials, it is likely that their results can be generalized to print advertisements as well. Stapel (1998) reports that he collected data about likeability and interest in post testing of print advertisements over the last 20 years. On the basis of a large data set, he examined the relationship between performance in memory measures (recall and recognition) and scoring of likeability and interestingness by cross tabulation. The results indicate that both high likeability and interestingness tend to massively increase the recall and recognition scores. In conclusion, scales of likeability and interest can be seen as established and effective sources of information about advertising effectiveness.

Advertisement-Related Issues in Research on Memory

As noted above, quantitative advertising research started with the development of memory measures. In the late 1920s Gallup and Robinson developed the "proven recall technique" followed by Starch in the late 1930s who introduced the recognition technique (specifically by going through a newspaper or magazine again after some time has passed). Today both recognition and recall are important and frequently used methods and the majority of investigations in the field of quantitative advertising research are based on these techniques.

There is a longstanding debate on the specific advantages and limitations of using recall vs. recognition in the context of advertisement. Thorson and Rothschild (1984), reviewing the substantial literature on this issue, note that both involve different cognitive operations, imply different levels of informational organization and initiate different learning procedures. An interesting conclusion drawn by Thorson and Rothschild (1984) is that the usefulness of recognition and recall appears to depend on the length of delay. With short delays, recall is the more useful indicator of memory for commercials, with longer delays recognition becomes a more effective indicator. It is further suggested that "particulars of execution and product characteristics would affect recognition" whereas "global characteristics of commercials would determine recall" (p. 302).

One of the most important variables moderating memory measures is exposure duration. Several studies have shown that exposure duration influences recall and recognition performance positively. Up to a duration of about 2 to 4 seconds recognition and recall accuracy increase linearly, depending on the complexity of the pictorial stimulus (Potter & Levy, 1969; Shepard, 1967; Fleming & Sheikhian, 1972; Intraub, 1979). Based on these observations, Rossiter and Percy (1983) conclude that advertisers seeking high recognition scores should try to employ not just "attention-getting" but rather "attention-holding" pictures, which "hold the consumer's attention" for at least 2 seconds, as this should guarantee high memorability. So far we have not found literature that draws a direct line from exposure duration to subject-controlled processing time. However, it seems straightforward to assume that both ways to increase the time spent processing an ad should have similar positive consequences on memory. In this context it may be worth mentioning that it appears inadvisable to control exposure duration in experimental research. Rosbergen *et al.* (1997) note that average exposure durations to advertisements tend to be significantly shorter when viewers control exposure (e.g. Kiss & Wettig, 1972) than when fixed presentation durations controlled by experimenters are used (Janiszewski, 1993; Kroeber-Riel, 1984). Moreover, control over the exposure duration may affect not only the duration itself, but also influence eye movement patterns.

A framework of memory research that is highly relevant for the present study is the level of processing approach by Craik and Lockhart (1972). This theoretical idea claims that human cognition involves a hierarchy of processing levels referred to as the "depth of processing", where greater "depth" implies a greater degree of semantic or cognitive analysis. More specifically, Craik and Lockhart (1972) suggest that the persistence of a memory trace is a function of depth of analysis, with deeper levels of analysis associated with more elaborate, longer-lasting and stronger traces. It follows that retention is a function of depth and not primarily speed of analysis. According to this view, memory can be seen as a continuum from the transient products of sensory analyses to the highly durable products of semantic-associative operations with deeper analysis involving longer processing time.

The level of processing approach was criticized to be circular, as deep processing is measured by better retention while better memory is supposed to be based on deeper processing (e.g. Baddeley, 1978). To avoid this problem, Saegert and Young (1984) used the terms "semantic" versus "non-semantic processing" instead of "deep" versus "shallow processing". Saegert (1978) examined the effect of semantic vs. nonsemantic processing on retention with advertising stimuli. He presented 40 magazine ads at a rate of five seconds per ad. For each stimulus a question about the respective brand name was asked. This question directed processing either to semantic features (e.g. "Have you used this brand before?") or to nonsemantic features (e.g. "Is the brand name in blue letters?"). The results of surprise recall and recognition tasks demonstrated superior memory for ads as a function of the processing given by the viewer. Subjects showed a better performance in both recall and recognition for those brand names associated with a semantic question. Further experiments corroborated these results. Reid and Soley (1980) found again the two-to-one superiority of recall for a semantic-processing condition with television ads. Saegert and Young (1984) were able to replicate these results in a further study looking at twenty-four-hour recall as a function of semantic vs. nonsemantic — processing manipulations.

One further merit of the level of processing (or: semantic vs. non-semantic) approach is that it provides a theoretical framework focusing on the viewer as an active information processor. This is in contrast to more traditional conditioning theories of learning that appear to be centered more on variables such as the characteristics of the to-be-remembered stimulus and the number of stimulus presentations (Saegert &Young, 1984).

Experiment I

As discussed above, the major purpose of the present study was to test the hypothesis that pragmatically more complex advertisements are more effective. To make this question feasible for an experimental design, we defined two types of advertisements: In *explicit advertisements* pictorial and textual elements are in direct semantic relation to each other and the message is presented in a straightforward way. In contrast, *implicit advertisements* include complex relations between image and text elements that point to the product only indirectly. To understand the message, elaborated cognitive processing is required including operations like drawing inferences and discerning metaphors and analogies. Following these definitions, advertisements were created by professional designers such that for a given product, an explicit and an implicit ad version were available.

Following our discussion above, we intended to measure advertisement effectiveness on three different levels. First, in terms of characteristics of visual processing as revealed by eye movement measures. Second, an evaluation of ad content was induced by asking participants to rate each stimulus on scales of likeability and interestingness, as it is standard practice in advertisement research. This also solved the problem of giving viewers a reasonable task. It appears safe to assume that people (after entering the *late processing* stage proposed in Table 27.1) often look at an advertisement poster or bill board with the intention to form an opinion about whether they like what they see. The instruction to evaluate an ad can be seen as a standardized version of this frequent type of dealing with ad content. The third level of measuring effectiveness, the examination of memory turned out to be difficult using a within subject design with a limited set of stimuli. Since two versions of an ad for the same product were presented, it was impossible to apply a recall task, as elements of both versions were likely to be mixed up during recall. However, by assigning a different product name to each ad version, an attempt was made to implement at least a partial recognition task.

Methodology

Task and procedure Materials consisted of advertisements targeting at 10 different products (car, stereo, lottery, hearing aid, diapers, etc.). For each product, an explicit and an implicit ad version was prepared by professional designers. All these stimuli contained a headline, a large picture and a product logo and name. The size and spatial arrangement of these elements was held constant between explicit and implicit versions. As an example, Figure 27.1 shows a pair of car ads including an explicit and implicit version with scan paths superimposed on the figures.

During the experimental block, the ads were presented in a fixed random order such that all motives appeared in the first half of the block for one time and in the second half for the second time. The order of presentation was counterbalanced for the two versions of each advertisement. In addition, each ad version was assigned one of two fictitious product names such that a participant would see one name in the explicit version and the other name in the implicit version or vice versa. On the basis of these

a)

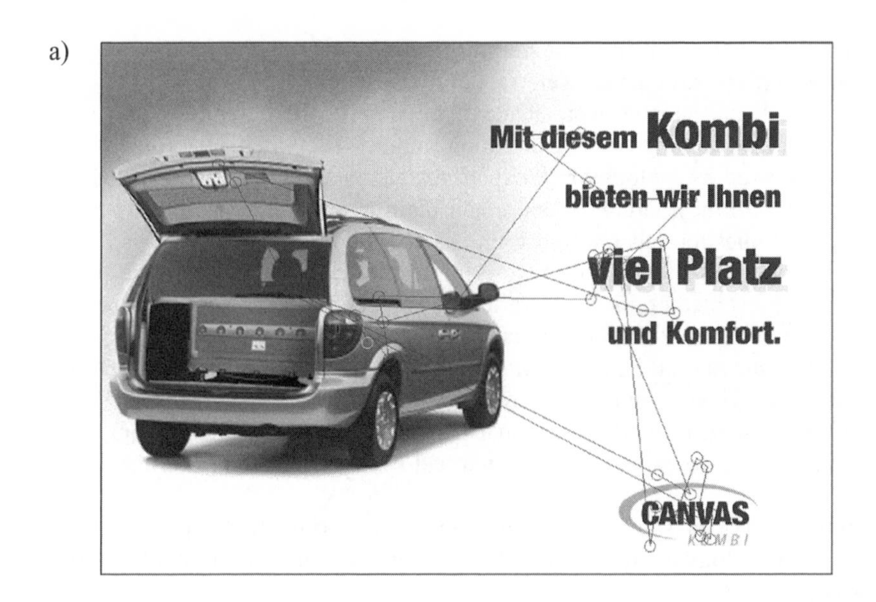

b)

Figure 27.1 Explicit (a) and implicit (b) versions of an advertisement for a van with superimposed scan paths. The headline for the explicit version translates as "With this van we offer you a lot of space and comfort. The headline for the implicit version is "With this van we can do this five times again". In the experiment, both advertisements were presented in 265 colours at a resolution of 1024 × 786 pixels on a 21 inch monitor.

variations, there were four different stimulus lists (2 order of presentation × 2 product name), to which 16 participants were randomly assigned.

Upon arrival, participants were made familiar with the equipment and task and completed a quick training of the calibration routine. They then performed four practice trials, followed by a block of 20 experimental trials. Subjects were asked to view the ad stimuli until they had understood their main message and were ready to answer two evaluative questions that would immediately follow. These questions ("How do you like the advertisement?" and "How interesting do you consider the advertisement?") were presented on the screen after the subject had terminated the presentation of the respective item. Responses were given on a seven point scale using a virtual keyboard also shown on the screen. After participants had completed the experimental block, they were given a short distractor task (to count backwards from 100 in steps of 7) and then asked to mark all product names they would recognize on a list that contained a total of 40 names, 20 of which had been used in the experiment.

Participants Sixteen subjects took part in the experiment. All participants were students of psychology or students of other fields with a minor in psychology and took part for course credit.

Apparatus Eye movements were recorded with an SR Research Ltd EyeLink infrared eye tracking system at a sampling rate of 250 Hz (4 ms temporal resolution). The relative accuracy of the system is in the order of a few minutes of arc. Absolute accuracy as expressed of short term repeatability of fixation position mapping (McConkie, 1981) was estimated in independent test sessions to be about 0.25 deg. The on-line saccade detector of the eye tracker was set to detect saccades using an acceleration threshold of 9500°/sec^2 and a velocity threshold of 30°/sec. Stimuli were displayed on a 21 inch EyeQ monitor subtending a visual angle of 34° horizontally and 25° vertically at a viewing distance of 67 centimeters. The display was generated using a matrox millennium video card running at a refresh rate of 100 Hz.

Results and Discussion

Table 27.3 shows the effects of the implicit vs. explicit variation on several eye movement parameters. There is a substantial difference in mean viewing duration per item, $F\ (1,15) = 16.999$, $p < 0.01$, amounting to a mean increase of 1334 ms for implicit advertisements. This effect is exclusively based on a larger number of fixations (5.9 per item, $F\ (1,15) = 13.779, p < 0.01$), while at the same time, mean fixation duration, $F < 1$, and mean saccade amplitude, $F < 1$, remained virtually identical for both conditions. The difference in viewing time and number of fixations is accompanied by a major discrepancy in standard errors, suggesting that some subjects responded especially sensitively to the variation of pragmatic complexity. In general, there were substantial differences in mean viewing time and number of fixations between participants. For example, individual mean viewing times for implicit ads ranged from 3933 ms to 11,912 ms.

Table 27.3: Means and standard errors for general eye movement parameters as a function of advertisement type in experiment 1. Saccade length is given in pixels and 100 pixels correspond to approximately 3.32 deg. of visual angle.

	Explicit		**Implicit**	
	M	**SE**	**M**	**SE**
Viewing time per item	6039	1428	7373	2122
Number of fixations per item	25.4	4.6	31.3	9.1
Fixation duration (ms)	186	30	184	28
Saccade amplitude (pixels)	194	26.5	196	13

Table 27.4: Means and standard errors of ratings for valence and interest on a seven point scale and mean values for recognition of brand names in Experiment 1. Please note that smaller numbers index better ratings (see text for further details).

	Explicit		**Implicit**	
	M	**SE**	**M**	**SE**
Valence (mean rating)	4.37	0.70	3.25	0.54
Interest (mean rating)	4.77	0.88	3.61	0.84
Recognition (mean percentage)	0.47	0.19	0.51	0.23

Table 27.4 presents mean values for ad evaluation and the recognition of brand names. The valence or likeability rating (How did you like the advertisement?) was significantly better for implicit ad versions, $F(1,15) = 24.780$, $p < 0.01$. The same was also true for the rating on interestingness, $F(1,15) = 20.599$, $p < 0.01$. The results from the two rating scales show a correlation of $r = 0.62$ ($p < 0.01$), indicating that there is substantial overlap in the information conveyed by both scales. Looking at correlations between mean rating values and eye movement measures, it turned out that there was only a weak relation between valence and viewing duration ($r = 0.28$, n.s.) and number of fixations ($r = 0.19$, n.s.). Correlations between interestingness and eye movement parameters were much larger and significant (viewing time: $r = 0.56$, $p < 0.001$; number of fixations: $r = 0.58$, $p < 0.001$).

In recognition performance, also depicted in Table 27.4, there was no difference between explicit and implicit ads, $F < 1$. This can have two reasons: First, it is possible that longer viewing durations and more positive ratings are not associated with better memory in the type of task and stimuli used in Experiment 1. However, since for both conditions recognition performance was at chance level it appears likely that our methodology led to a floor effect such that possible differences could not be picked up at all. We had expected that the recognition task would be quite easy to do and therefore used alternative product names that closely resembled the correct ones. Contrary to our expectation, recognition performance was quite poor, perhaps because the brand names were irrelevant to the task that the participants were asked to perform.

Taken together, the results of Experiment 1 support our main hypothesis that implicit ads are more effective in two respects. First, participants take significantly more time to process implicit ads, which may reflect a higher mental effort to encode and process the information presented and/or a deeper (more semantic) mode of processing. In addition, the implicit ads were rated to be liked better and also found to be more interesting, which in itself can be seen as an important indicator of ad effectiveness.

Despite these clear-cut results, Experiment 1 also had a number of weaknesses. The effect of the explicit/implicit variation on memory could not be appropriately tested, as in both conditions recognition was at chance level. In addition, there were problems with some of the stimuli, which contained overlap between regions belonging to textual vs. pictorial elements.[1] Also, it could be claimed that the instruction to view an advertisement in preparation for a judgment on "how nice and interesting this ad is" is artificial and may have had some influence on the viewing behavior. In particular, given the significant correlation between interestingness and viewing time, one might argue that when a participant is asked to determine whether an advertisement is interesting, the viewing duration may reflect the mental effort to answer this question rather than "understanding" the ad.

Experiment II

To address the problems with Experiment 1, a second experiment was designed. The main difference is that, instead of 10 products, there were now 20 products with each an explicit and an implicit ad version. This allowed for the implementation of a design where each participant would be confronted only with either the explicit or implicit variant of each advertisement. Also, in this experiment the task was varied as a between-subject variable. All participants were asked to view each ad in a way that would allow understanding of its main message. However, only 16 of the 32 subjects then completed the evaluation tasks as in Experiment 1. The instruction for the remaining 16 subjects was to answer the question "What has been advertised?" as clearly as possible in one sentence. The idea was that this somewhat neutral task would correspond to a more cursory way to deal with ads that is quite typical when reading a magazine or looking at a billboard. A further improvement was that we now used a short block of eight training trials with a fixed presentation time of 8 seconds. The purpose of this manipulation was to induce a reference for a typical response time that would help avoiding the extreme differences in viewing durations that were present in Experiment 1.

Methodology

Task and procedure In this experiment, materials included explicit and implicit ad variants for a total of 20 products. The general layout was identical to the stimuli used in Experiment 1, but great care was taken to avoid overlap between picture and text regions. Also, the number of letters and words in the headline text and brand name were matched as closely as possible on an item by item basis. The order of product presentation was identical for all participants, following a fixed random order. Subjects were assigned to one of two task conditions, one including a rating task and the other including a quick paraphrasing of the ad content (see above). Within each task condition, there were two stimulus lists, with either the explicit or implicit version of each ad appearing at the respective list position. Prior to the experiment, participants performed a block of eight practice trials where sample ads were presented for a fixed duration of 8 seconds.

Subjects were asked to view each ad stimulus until they had understood its main message and were ready to answer two evaluative questions that would immediately follow. For 16 of the 32 participants the task was identical to the one used in Experiment 1. The remaining subjects were asked to answer the question "What has been advertised?" as clearly as possible in one sentence after viewing the item. After completing the main experiment (and a short distractor task) all subjects were asked to recall as many ads as they could and to note as many details as possible. Finally, they were shown cards with 20 advertisements, including 10 that were present in the experimental block that a given participant had just completed and 10 that were part of the alternative stimulus list. For each instruction and stimulus list, two sublists were constructed such that for each cell of the design known and unknown items for recognition would be counterbalanced.

Results

Table 27.5 presents the effects of both factors, type of instruction and ad complexity on several eye movement parameters. Overall, the viewing times for the paraphrasing task were substantially shorter in comparison to the evaluation condition. There were large differences in mean viewing duration per item, amounting to 823 ms for the evaluation task and 750 ms for the paraphrasing task. Both the effects of ad complexity, F $(1,30) = 16.55$, $p < 0.000$, and task, $F(1,30) = 13.731$, $p < 0.001$, on viewing time were significant, but their interaction was not, $F < 1$. As in Experiment 1, the results for number of fixations are very similar. The effects of ad complexity, $F(1,30) = 23.786$, $p < 0.01$, and task, $F(1,30) = 12.866$, $p < 0.01$, are significant, but there is no significant interaction, $F < 1$. In contrast, the effects of both factors on mean fixation durations and mean saccades length are all non-significant.

Mean values for ad ratings in the evaluation task and memory measures are presented in Table 27.6. Replicating the results of experiment 1, mean valence, $F(1,30) = 24.015$, $p < 0.01$, and interestingness, $F(1,30) = 18.584$, $p < 0.01$, ratings were significantly better for implicit advertisements. Mean recognition rates, also shown in Table 27.6 indicate that participants had no problems to differentiate the ads that they had previously seen from their counterparts in the alternative list. Apparently, this led to a ceiling effect and no significant differences emerged. The mean rate of recall refers to the percentage of stimuli that were recalled, no matter what the specific content of

Table 27.5: Means and standard errors for general eye movement parameters as a function of advertisement type in Experiment 2. Saccade length is given in pixels and 100 pixels correspond to approximately 3.32 deg. of visual angle.

	Instruction I (evaluation)				**Instruction II (paraphrase)**			
	Explicit		**Implicit**		**Explicit**		**Implicit**	
	M	**SE**	**M**	**SE**	**M**	**SE**	**M**	**SE**
Viewing time	5582	1435	6405	1166	3976	1265	4726	1570
Number of fixations	22.7	4.5	26.2	4.2	16.8	5.6	19.9	6.1
Fixation duration	196	18.6	195	16.4	191	27.2	188	27.9
Saccade length	178	12.5	176	18.6	173	19.7	170	16.6

Table 27.6: Means and standard errors of ratings for valence and interest on a seven point scale and mean values for recognition of brand names in Experiment 2. Please note that smaller numbers index better ratings (see text for further details).

	Instruction I (evaluation)				**Instruction II (paraphrase)**			
	Explicit		**Implicit**		**Explicit**		**Implicit**	
	M	**SE**	**M**	**SE**	**M**	**SE**	**M**	**SE**
Valence rating	3.5	1.0	4.6	0.6	–	–	–	–
Interest rating	3.2	0.9	4.3	0.5	–	–	–	–
Recognition	96.3	6.2	92.5	8.6	92.5	11.8	89.4	12.9
Recall	49.4	21.8	44.4	16.7	44.4	16.7	51.9	15.2

the report was. About half of the stimuli were recalled, but, again this global measure did not differentiate between tasks and ad complexity.

However, when looking at *what* was being recalled, a more promising picture emerged. First of all, the task had a major influence on the content of recall: Participants who were asked to view the advertisement stimuli in preparation for ratings of valence and interestingness recalled pictorial information in 10.3% of the trials but brand names in only 0.3% of the trials. The opposite pattern was present in participants who viewed the ad stimuli in preparation for paraphrasing its content (more specifically: reporting what was advertised). Here there was almost no recall of pictorial information (1.5%) but brand names were recalled for 11.6% of the items. Our next step of analysis was to look at differences in the recall rates between implicit and explicit ads in both tasks. There was a significant difference of 6.9% for pictorial information in the evaluation condition (13.8 vs. 6.9%; $F(1,30) = 9.648$, $p < 0{,}01$) and a difference of 4.4% for brand names in the paraphrasing condition, that did not reach significance (13.8 vs. 9.4; $F(1,30) = 0.968$, $p < 0.33$).

To get a more detailed picture of viewing behavior, all stimuli were divided into three regions of interest, referred to as headline, image and brand name. Figure 27.2 shows a typical example for this classification. Importantly, the borders of the respective regions of interest were identical for both the explicit and implicit ad versions for a given product such that a direct comparison is possible. In the top left corner of Figure 27.2, a square is depicted that represents an area of 128 × 128 pixels. We divided each stimulus image, presented at a resolution of 1024 × 768 pixels, into 8 × 6 such areas, referred to as "one tile". This scaling can be used to compute a relative viewing measure in addition to the absolute viewing time per region. Results for both analyses are shown in Figure 27.3.

As is apparent from Figure 27.3a, the type of task had a major impact on the distribution of absolute viewing times over the regions present in the advertisement stimuli. In the evaluation condition, a large proportion of the absolute viewing time was directed towards the picture region and only a small proportion towards the brand name area. In the paraphrasing condition however, viewing time was dramatically shorter for the picture region and substantially increased for the brand name area. In fact, viewing time was even longer for the much smaller name region. Looking at Figure 27.3b, presenting mean (relative) viewing times per tile, different proportions between

Figure 27.2: Example for a regions of interest (headline, image and brand name) classification using an advertisement for carrot juice. The text in the headline area translates as: "Tested by experienced experts: 100 percent natural ingredients".

a)

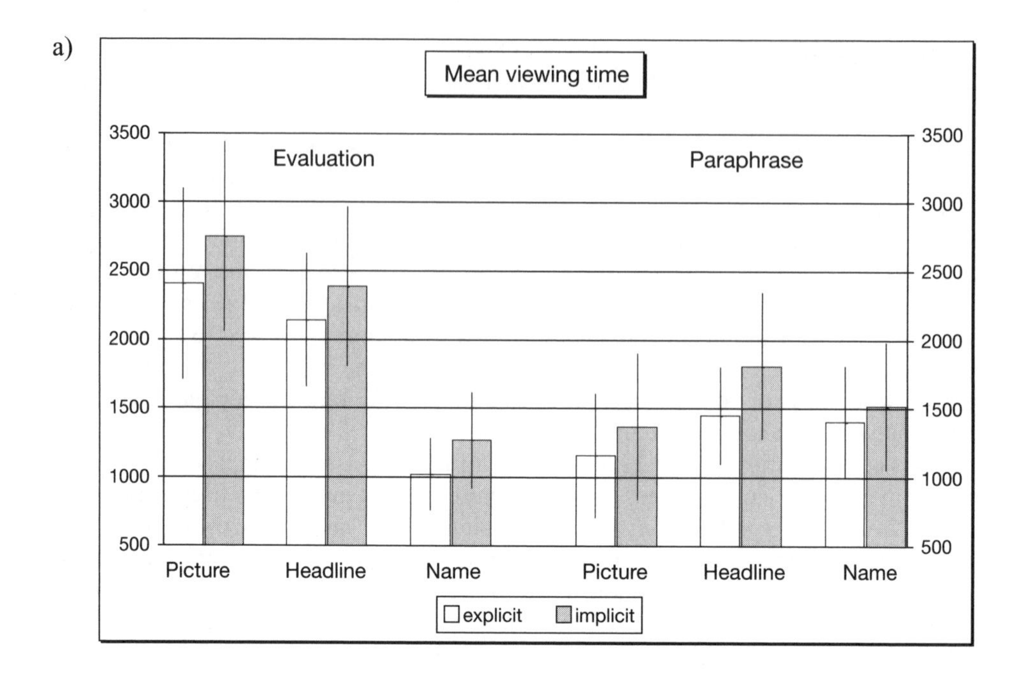

b)

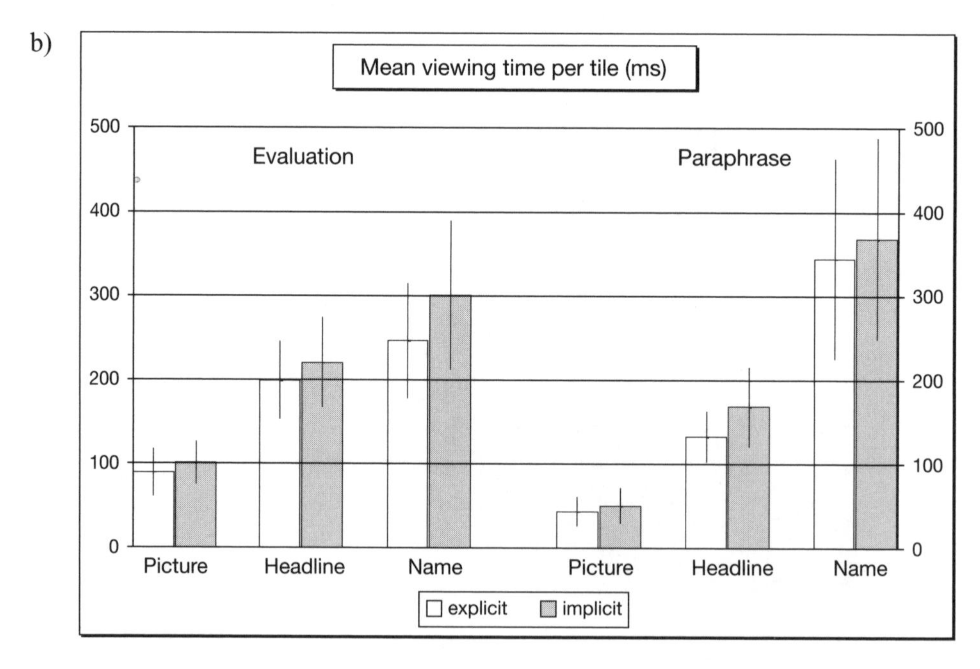

Figure 27.3 (a) Mean viewing time (in ms) for regions of interest as a function of task and ad complexity. In (b), viewing time is scaled by a unit of area corresponding to 128 × 128 pixels, referred to as "one tile".

region of interest emerge. It becomes clear that relative viewing time is always largest for the brand name areas and smallest for the relatively large pictures. Most importantly, despite all variation over tasks and regions, a substantial explicit–implicit difference is present in all cases.

General Discussion

Taken together, our experiments provide convincing evidence in favor of our main hypothesis. Advertisements with a higher degree of pragmatic complexity are looked at substantially longer and more fixations are made when viewers attempt to understand their message. This difference in visual processing time is entirely due to an increased number of fixations, both mean fixation duration and saccade amplitude remain unaffected. This is in line with results in the domain of reading research, where it has been shown that increases in gaze duration for words that are more difficult to process are also largely based on making more fixations rather than increasing their duration (Blanchard, 1985; O'Regan & Levy-Schoen, 1987). Our task variation confirmed the effect of ad complexity for a condition that may reflect a more cursory, informal processing of advertisement content. The paraphrasing task was done in less time but produced virtually the identical effects. We conclude that the advantage of implicit ads with respect to processing time is valid for a broad range of goals and strategies when dealing with advertisements.

In Experiment 1 as well as in the evaluation condition of Experiment 2, there were significant differences in ratings of valence or likeability ("How do you like the advertisement?") and interestingness ("How interesting do you consider the advertisement?"). As we have noted in the introductory section of this chapter, scales of likeability have proven to be a very powerful predictor of sales behavior. It can therefore be concluded that implicit advertisements are likely to be more effective from an economic point of view. To verify the memory component of ad effectiveness turned out to be more difficult. In Experiment 1, recognition of brand names was at chance level, probably because these names were irrelevant for the evaluation task. Also, in Experiment 2, global rates of recognition and recall were insensitive to the variation of ad complexity. However, in a more detailed analysis of what participants had recalled, some evidence for an advantage of implicit ads emerged.

To our knowledge, our study is the first to combine a variation in task with a region of interest analysis. The main result from this analysis is that the task has profound consequences for the distribution of viewing time and fixation positions.[2] A greater proportion of viewing time was devoted to the elements that were more relevant to the task. In the case of the evaluation task, more fixations were directed at the pictures, which, as participants reported, was often a major base of judgment in the rating task. In contrast, the brand name was less relevant for this task and received only few fixations. In the paraphrasing task, that was carried out in much less time, the proportion of fixations on headline, picture and brand name was more balanced, reflecting an attempt to arrive at a concise characterization of the topic or message of the ad. It appears that this accentuation of relevant elements during the viewing of the ads translated into what was retained in memory. In the evaluation condition, some pictorial information was recalled, but participants had almost no idea about names, in the paraphrasing condition the picture was completely reversed. Seen from a slightly different angle, it is also evident that a quite substantial *relative* viewing time, as present for names in the evaluation condition, does not guarantee good memory. Most important from a theoretical perspective is the fact that the processing time advantage for implicit ads was present for all regions of interest in both tasks.

Although these results appear straightforward, it may be pointed out that there is a major weakness in our design. As reported above, care has been taken to control for properties of the text present in the headlines and to keep the design and wording of brand names identical between conditions. However, one major difference between items within a pair of explicit and implicit ads are the different pictures. One may argue that implicit ads are looked at for longer durations and are rated more positive because their pictures are more pleasant and interesting (or more complex in terms of visual processing). Although we have tried to avoid extremes when selecting the pictures for our study, we cannot exclude the possibility that some of the observed effects are due to uncontrolled variance in pictorial information. However, this concern was addressed directly in a follow-up experiment. Here explicit and implicit ad variants were designed that used the same picture but advertised different products. Preliminary results indicate that, although the pictorial information is absolutely identical, marked advantages for implicit ads are again present.

In addition to the quantitative analyses reported above, we have also made a number of interesting, more informal observations. First, concerning the order with which the stimulus elements were scanned, it turned out that in about half the trials the scan path started with either picture or headline, with a tendency to some consistency within subjects. In most cases, a relatively larger number of fixations were made on this initial element (e.g. when reading the entire headline text) before going to the next element. This is in harmony with observations reported by Rayner *et al.* (2001). However, contrary to their findings, we found that participants looked back and forth between different elements, especially in terms of scanning several times over picture and headline. This may be caused by the relatively demanding tasks in our experiments in comparison to the more basic search instruction used by Rayner *et al.* A quantitative analysis of this sequential scanning behavior is beyond the scope of this chapter. An appropriate strategy for analyzing such data could be developed in analogy to what is customary when dealing with word fixation patterns in reading research (e.g. Inhoff & Radach, 1998). Successive fixations on a specific region may be aggregated to "passes", representing a gaze on the respective area before moving to the next area. This measure not only represents a reasonable aggregation of the fixation data but it also well-suited to pick up the issue of switching between elements within an advertisement.

The properties of viewing behavior just described can all be observed in Figure 27.1. In addition, an interesting, rather unexpected observation can be made. After looking at picture, headline and brand name, often in more than one pass, participants sometimes direct their eyes to regions that appear to contain no useful information. This leads to a chain of additional fixations before the viewer either terminates the trial or moves on to additional passes over the relevant ad elements. Our interpretation of this behavior is that the visual information necessary for processing the present ad has been successfully encoded and the eyes are "parked" temporarily while higher level cognitive processing operations are being carried out. This pattern is quite frequent, especially in the data for implicit ads. It cannot be based on measurement artefacts, as fixations on relevant pictorial details during the same trial are often extremely accurate. It has some similarity with the distinction between "scanning" and "processing" fixations introduced by Groner (1978) and with the slowing of eye movements when sentence and clause "wrap up" operations are performed during the reading of continuous text (Just & Carpenter, 1980; Rayner, Kambe & Duffy, 2000). In any case, this phenomenon represents an interesting puzzle for further research on the visual processing of "intelligent" advertisements and other cognitively demanding compositions of texts and pictures.

Notes

1 It is for this reason that analyses of viewing behavior for specific regions of interest are reported only for Experiment 2. However, an informal analysis using approximated areas for headline, picture and product name suggested that the main effect holds for all three types of regions.
2 Detailed analyses not reported in this chapter confirmed that the pattern of results was identical for viewing durations and number of fixations per region.

References

Allport, A. D. (1992). Attention and control: Have we been asking the wrong questions? A critical review of twenty-five years. In: D. E. Meyer and S. M. Kornblum (eds), *Attention and Performance XIV: Synergies in Experimental Psychology, Artificial Intelligence, and Cognitive Neuroscience.* Cambridge, MA: MIT Press.

Baddeley, A. D. (1978). The trouble with levels: A reexamination of Craik and Lockhart's framework for memory research. *Psychological Review, 85*(3), 139–152.

Biel, A. L. (1998). Likeability — Why advertising that is well liked sells well. In: J. P. Jones (ed.), *How Advertising Works The Role of Research.* Thousand Oaks, London, New Dehli: Sage Publications.

Blanchard, H. E. (1985). A comparison of some processing time measures based on eye movements. *Acta Psychologica*, 58, 1–15.

Bogart, L., & Tolley, B. S. (1983). The search for information in newspaper advertising. *Journal of Advertising Research*, 9–19.

Broadbent, D. E. (1971). *Decision and stress.* London: Academic Press.

Carroll, P. J., Young, J. R., & Guertin, M. S. (1992). Visual analysis of cartoons: A view from the far side. In: K. Rayner (ed.), *Eye Movements and Visual Cognition: Scene Perception and Reading* (pp. 444–461). New York: Springer-Verlag.

Craik, F. I., & Lockhart R. S. (1972). Levels of processing: A framework for memory research. *Journal of Verbal Learning and Verbal Behaviour*, 11, 671–684.

De Graef, P. (1992). Scene-context effects and models of real-world perception. In: K. Rayner (ed.), *Eye Movements and Visual Cognition: Scene Perception and Reading* (pp. 243–259). New York: Springer Verlag.

Edell, J. E., & Staelin, R. (1983). The information processing of pictures. *The Journal of Consumer Research*, *10*, 45–61.

Fleming, M. L., & Sheikhian, M. (1972). Influence of pictorial attributes on recognition memory. *AV Communication Review*, *20*, 423–441.

Greenwald, A. G., & Leavitt, C. (1984). Audience involvement in advertising: Four levels. *Journal of Consumer Research*, *11*, 581–592.

Groner, R. (1978). *Hypothesen im Denkprozess*. Bern: Huber.

Haley R. J. (1990). *Final report of the ARF copy research validity project*. New York: Advertising Research Foundation Copy Research Workshop, July.

Hansen, F. (1995) Recent developments in the measurement of advertising effectiveness: The third generation. *Marketing and Research Today*, 23 (November), 259–269.

Hegarty, M. (1992). The mechanics of comprehension and comprehension of mechanics. In: K. Rayner (ed.), *Eye Movements and Visual Cognition: Scene Perception and Reading* (pp. 428–443). New York: Springer-Verlag.

Hornik, J. (1980). Quantitative analysis of visual perception of printed advertisements. *Journal of Advertising Research*, *20*(6), 41–48.

Hyönä, J., Munoz, D., Heide, W., & Radach, R. (2002). *The Brain's Eye: Neurobiological and Clinical Aspects of Oculomotor Research*. Amsterdam: Elsevier Science.

Inhoff, A. W., & Radach, R. (1998). Definition and computation of oculomotor measures in the study of cognitive processes. In: G. Underwood (ed.), *Eye Guidance in Reading and Scene Perception* (pp. 29–53). Oxford: Elsevier.

Intraub, H. (1979) The role of implicit naming in pictorial encoding. *Journal of Experimental Psychology: Human Learning and Memory*, *5*, 78–87.

Janiszewski, C. (1993). Preattentive Mere Exposure Effects. *Journal of Consumer Research*, *20* (December), 376–392.

Janiszewski, C., & Warlop, L. (1993). The influence of classical conditioning procedures on subsequent attention to the conditioned brand. *Journal of Consumer Research*, 20, 171–189.

Just, M. A., & Carpenter, P. A. (1980). A theory of reading: From eye fixations to comprehension. *Psychological Review*, *87*, 329–354.

Kahneman, D. (1973). *Attention and Effort*. Englewood Cliffs, NJ: Prentice Hall.

Kiss, T., & Wettig, H. (1972). Die Anzeigenwirkung in Abhängigkeit von Wirkungsfaktoren der Zeitschriften. *European Society for Opinion and Marketing Research*, 101–139.

Kroeber-Riel, W. (1979). Activation research: Psychobiological approaches in consumer research. *Journal of Consumer Research*, *5*, 240–250.

Kroeber-Riel, W. (1984). Effect of emotional pictorial elements in ads analyzed by means of eye movement monitoring. In: T. C. Kinnear (ed.), *Advances in Consumer Research,* Vol. 11 (pp. 591–596). Provo, UT: Association for Consumer Research.

Krugman, H. E. (1968). Processes underlying exposure to advertising. *American Psychologist*, *23*, 245–253.

Krugman, D. M., Fox, R. J., Fletcher, J. E., Fischer, P. M. & Rojas, T. H. (1994). Do adolescents attend to warnings in cigarette advertising? An eye-tracking approach. *Journal of Advertising Research*, November/December: 39–52.

Lewis, St. E. (1898) in Strong, E. K. J. (1925). *The Psychology of Selling and Advertising*. London: McGraw-Hill.

Lohse, G. L. (1997). Consumer eye movement patterns on yellow pages advertising. *Journal of Advertising*, *XXVI*(1), 61–73.

MacInnis D. J., & Jaworski, B. J. (1989). Information processing from advertisements: Toward an integrative framework. *Journal of Marketing*, *53* (October), 1–23.

MacInnis D. J., Jaworski, B. J., & Moorman, C. (1991). Enhancing and measuring consumers' motivation, opportunity, and ability to process brand information from ads. *Journal of Marketing*, *55*, 32–53.

McConkie, G. W. (1981). Evaluating and reporting data quality in eye movement research. *Behaviour Research Methods & Instrumentation*, *13*, 97–106.

Nodine, C. F., Kundel, H. L., Polikoff, J. B., & Toto, L. C. (1987). In: G. Lüer, U. Lass and J. Shallo-Hoffmann (eds), *Eye Movement Research. Physiological and Psychological Aspects* (pp. 349–363). Göttingen: Hogrefe.

O'Regan, J. K., & Levy-Schoen, A. (1987). Eye movement strategy and tactics in word recognition and reading. In: M. Coltheart (ed.), *Attention and Performance XII: The Psychology of Reading* (pp. 363–383). Hillsdale, NJ: Erlbaum.

Potter, M .C., & Levy, E. I. (1969). Recognition memory for a rapid sequence of pictures. *Journal of Experimental Psychology*, *81*, 10–15.

Radach, R., Inhoff, A. W., & Heller, D. (2002). The role of attention in normal reading. In: E. Witruk, A. Friederici, and Th. Lachmann (eds), *Basic Mechanisms of Language and Language Disorders*. Dordrecht: Kluwer.

Rayner, K., Kambe, G., & Duffy, S. A. (2000). The effect of clause wrap-up on eye movements during reading. *Quarterly Journal of Experimental Psychology*, *53*, 1061–1080.

Rayner, K., Rotello, C. M., Stewart, A. J., Keir, J., & Duffy, S. A. (2001). Integrating text and pictorial information: Eye movements when looking at print advertisements. *Journal of Experimental Psychology: Applied*, *7*(3), 219–226.

Reid, L. N., & Soley L. C. (1983). Decorative models and the readership of magazine ads. *Journal of Advertising Research*, *23* (April), 27–32.

Robinson, E. J. (1963). How an advertisement's size affects responses to it. *Journal of Advertising Research*, *3* (December), 16–24.

Rosbergen, E. (1998). *Assessing visual attention to print advertising through statistical analysis of eye-movement data*. Capelle a/d IJssel: Labyrint Publication.

Rosbergen, E., Pieters, R., & Wedel, M. (1997). Visual attention to advertising: A segment-level analysis. *Journal of Consumer Research*, *24* (December), 305–314.

Rossiter, J. (1981). Predicting Starch Scores. *Journal of Advertising Research 21*(5), 63–68.

Rossiter, J. R. (1988). The increase in magazine ad readership. *Journal of Advertising Research*, *28*, 35–39.

Rossiter J. R., & Percy L. (1983). Visual communication in advertising. In: R. J. Harris (ed.), *Information Processing Research in Advertising*. Hillsdale, NJ: Erlbaum.

Saegert, J. (1978). A demonstration of levels of processing theory in memory for advertisements. In: W. L. Wilkie (ed.), *Advances in Consumer Research* (Vol. 6). Ann Arbor, MI: Association for Consumer Research.

Saegert J., & Young R. K. (1984) Levels of processing and memory for advertisements. In: L. Percy and A. G. Woodside (eds), *Advertising and Consumer Psychology*. Lexington MA, Toronto: Lexington Books.

Schindler, P. S. (1986). Color and contrast in magazine advertising. *Psychology and Marketing*, *3* (Summer), 69–78.

Shepard, R. N. (1967). Recognition memory for words, sentences, and pictures. *Journal of Verbal Learning and Verbal Behavior*, *6*, 156–163.

Stapel, J. (1998). Recall and recognition: A very close relationship. *Journal of Advertising Research*, July/August, 41–45.

Starch, D. (1966). *Measuring advertising readership and results*. New York: McGraw-Hill.

Strong, E. K. Jr (1925). *The Psychology of Selling and Advertising*. London: McGraw Hill.

Thorson E., & Rothschild M. L. (1984). Recognition and recall of commercials: Prediction from a text-comprehension analysis of commercial scripts. In: L. Percy and A. G. Woodside (eds), *Advertising and Consumer Psychology*. Lexington, MA, Toronto: Lexington Books.

Treistman, Joan & John P. Gregg (1979). Visual, Verbal, and Sales Responses to Print Ads. *Journal of Advertising Research*, *19*(4), 41–47.

Valiente, R. (1973). Mechanical correlates of ad recognition. *Journal of Advertising Research*, *13*, 13–18.

Von Keitz, B. (1988). Eye movement research: Do consumers use the information they are offered? *European Research*, *16* (November), 217–224.

Chapter 28

Behavioural Strategies in Web Interaction: A View from Eye-movement Research

Iréne Stenfors, Jan Morén and Christian Balkenius

The use of eye-tracking has revealed some interesting facts about how users interact with web pages. This interaction is dependent on what information is desired and presented, and on the spatial arrangement of the information. The interaction between the user and the web page is highly user-driven. Since they are aware of where Internet ads (banners) are located they know how to avoid them, using active avoidance strategies which seem to be a result of conditioning. The conditioning not to look at Internet ads while surfing may also be transferred to a new context. The learned inhibition against looking at web advertising is thus both more general and less domain specific than previously assumed.

Introduction

To design functional commercial web sites, it is necessary to know how Internet users are affected by the design of the site (Landauer, 1996). For a long time, Internet usability testing has been seen mainly as a question of aesthetics. As such, it has been entrusted to people concerned with layout and design rather than usability factors and man-machine interaction (Moeller, 2001). However, the interactivity of web pages creates a communication event unique to the Internet medium and it is essential that the design of a web site functions well (Nielsen, 2000; Spool, Scanlon, Schroeder, Snyder & De Angelo, 1998). This is not only a matter of making a good impression on the visitors, but of communicating the content in an efficient way (Platt, 1999).

Both the communicative value and the immediate behavioural responses of Internet users can be measured by eye-tracking methods. Eye-tracking has proved to be a powerful tool in evaluating the interaction between a user and a web site since it allows simultaneous recording of attention and behaviour.

The Mind's Eye: Cognitive and Applied Aspects of Eye Movement Research

ISBN: 0–444–51020–6

The tests traditionally used for web design are developed in the realm of experimental psychology and include for example tests like questionnaires, interviews, protocol simulations and field trials (Rubin, 1994). A lot of usability tests today also use a verbal protocol in conjunction with some sort of performance measurement (van Waes, 2000). One benefit of thinking-aloud protocols is that they do not have to be analysed before the collected information can be used.

A problem with the more traditional Internet usability testing is that it pays more attention to what the subjects included in the test report and what actual behavioural steps they take than to what they actually visually attend to (Wichansky, 2000). This is of course a consequence of the methods and procedures used. It is usually the case that a user is tested while being videotaped or — worse — carefully being watched by the test team standing next to the test subject.

In the studies outlined below we have used the IriSense® eye-tracking system that has been developed specifically to be used in web-interaction studies. Unlike many other eye-tracking systems (Sibert & Jacob, 2000), it can be used on subjects that wear glasses or contact lenses. The system makes it possible to almost unobtrusively record the attention of the subject. The quantitative eye-tracking data in usability testing is combined with qualitative data from interviews. In that way, a comparison between the subject's visual behaviour (i.e. what she did) and the subject's report (i.e. her explanations of her behaviour) can be made. This technology is described in the next section.

The following sections summarise some of our findings relating to the interaction with the web. They describe a number of areas where the use of eye-tracking has revealed interesting facts about how users interact with web pages. When designing a web page it is important to understand what captures the initial interest of the user and how the browsing patterns on the site depends on previous expectations of how web pages work as well as on new learning about the structure of the page or site. In the third section, we report on experimental studies that show what strategies users use to search for information on a web site. It is important to realise that the different search strategies reflect goal-directed behaviour directed towards finding the required information. In this context, it is not surprising that users exploit various avoidance strategies to evade looking at information that they do not want or need. Some of these strategies are described in the fourth section.

In the fifth section we investigate what kind of external cognitive structures that are used to decrease the memory load of the user. These strategies often depend on some form of anchoring, i.e. an external cue support, in the visual elements on the screen. This is contrasted to results that indicate that users tend to buffer spatial locations on the screen in memory and use them later for actions without refocusing on the target elements. Finally in the sixth section, we discuss implication of our results for the understanding of web-interaction.

The IriSense® Technology[1]

Traditionally, eye-tracking technology has been based on corneal reflection detection techniques: An infrared (IR) light projected on the eye (the cornea) makes it possible to calculate the point in the scene at which the subject is looking. This technology can work quite well in many cases and is relatively inexpensive and feasible to realise in hardware. The method is not without its drawbacks, however (Sibert & Jacob, 2000): The lengthy calibration, the sensitivity to ambient light (e.g. sunlight) during recording and the demand for a clear and unobstructed eye, e.g. glasses and especially contact lenses pose a problem.

Instead of using IR reflections, we have developed a system based on passive monitoring of the eye-position. The advantages of this approach are clear. As it does not depend on finding salient features with IR, the IriSense® system is not nearly as sensitive to glasses, contacts, or cosmetics as IR based systems. In the studies reported below, calculations were made to a precision of less than two degrees, which is sufficient for an accurate estimation of the visual element fixated on a web page. The temporal resolution is 20 ms, which again is reasonable for the situations we are interested in.

Search Strategies

In his studies, Yarbus (1967) presented findings that showed that an observer's intentions, expectations, and strategies for scanning a visual field are moderated by the particular task and stimulus materials with which one is confronted. This moderating internal power is traditionally referred to as a plan or a schema of visual activity. The plan of visual activity reflects a cognitive strategy, which is modifiable as a function of available stimulus information and task demands.

Our eye-tracking studies on Internet interaction have confirmed Yarbus' findings showing that during task relevant scanning or searching web ads are not fixated by the subjects (Stenfors, 1998). Users do not passively react to contrast or intensity of the elements on the page. Instead, the scanning of a web page is strongly influenced by the task at hand.

To minimise search time and maximise the control of visual activity, subjects use the centre of the web page as a starting point and begin orienting from that location (Stenfors, 1998). The orienting phase is defined by the initial fixations made by a subject to establish the identity of a web-page before any interaction with the page. A possible reason for this could be that in this way there is an equal distance to all other positions on the web page. This finding was supported by the behaviour of the subjects in another test in which they were asked to predict the location of a yellow dot on the screen (Stenfors, Morén & Balkenius, 2001). When the subjects did not know at which of two locations the dot would appear, they often chose to fixate at a position in the middle between the two expected locations.

Once the content on the web page appears, the subjects begin an orienting phase. This phase does not usually start until most of the page content is visible (Stenfors, 1998). It appears that subjects avoid looking at the page before enough information is available (see the next section). The initial orienting phase varies considerably in time depending on the familiarity with the web page, the site and the Internet in general. For example, in one of our studies of user behaviour on Internet help pages the time for the initial orientation phase ranges from a few seconds to several minutes (Area 17 AB, unpublished report, 2001).

When the items on the web page create a figural pattern (e.g. a scroll list) that is correlated in some meaningful way with the content, the user will learn to exploit the pattern to decrease search time (Nygren, Allard & Lind, 1995). This has also been confirmed in later eye-tracking studies where it was found that subjects select areas on the web pages that are well grouped and structured (Stenfors, 1998; Area 17 AB, unpublished report, 2001): These areas are first scanned before moving on to the more unstructured areas. Also, Goldberg and Kotval (1999) have shown that the grouping of interface items raise expectations of a relation to something that they have in common. This was manifested in shorter scan-paths covering smaller areas.

The recordings of the subjects' eye movements show that the subjects agree at an early stage about what locations on the web page that are most task relevant and informative: Regular eye movement patterns have been found in all web pages tested (Stenfors, 1998; Area 17 AB, unpublished report, 2001). The areas of task relevant information present a pattern of dense fixations whereas areas of little or no task relevant information (e.g. the ads) are fixated very little, if at all. Thus, the areas judged to be highly informative attract the fixations of the eyes (Goldberg & Kotval, 1999).

Only the initial fixations on the web pages seem to be stimulus driven, and only to a limited extent (Stenfors, 1998). This was also confirmed by a scene perception study by Henderson and Hollingworth (1999) who showed that the initial fixations an observer performs when looking at a visual scene are the result of a stimulus-driven response. Thereafter a semantic frame for the visual scene is established and the successive fixations are located at semantically interesting areas. Experienced Internet users seem to have adopted certain schemas of what visual objects to avoid on the web page and what actions to take in order to keep the search time short. When a page is visited repeatedly, the time for the orienting phase decreases rapidly (Figure 28.1).

For a first time visitor, it is important that the design of the web page is clear and well structured. This makes it easy for the user to understand what different items on the page represent and to find relevant information (Moeller, 2001). For a more frequent user of a web page, however, there are other needs: A user accustomed to the web page will learn strategies for scanning the page. One apparent reason for this is that systematic search yields better results than random search. This was showed in a study made by Arani, Drury and Karawan (1984). The design of the web page can be detrimental to the efficiency of these strategies. When the user has learned a more efficient scanning, the search time becomes shorter.

Even small differences in the web page design can have an effect on the scanning. For example, Nygren *et al.* (1995) found that scanning horizontally aligned items was slower than scanning vertically aligned items. The scanning rate for horizontal lists was found to be 1.2 times the scanning of vertical lists. It appears as if it is more efficient for the user if a vertical alignment rather than horizontal alignment is used since the varying widths in a column make up pattern that can be learned by the user (Nygren & Allard, 1996). Since the peripheral visual system detects features like width, size and orientation these figural patterns are quickly captured.

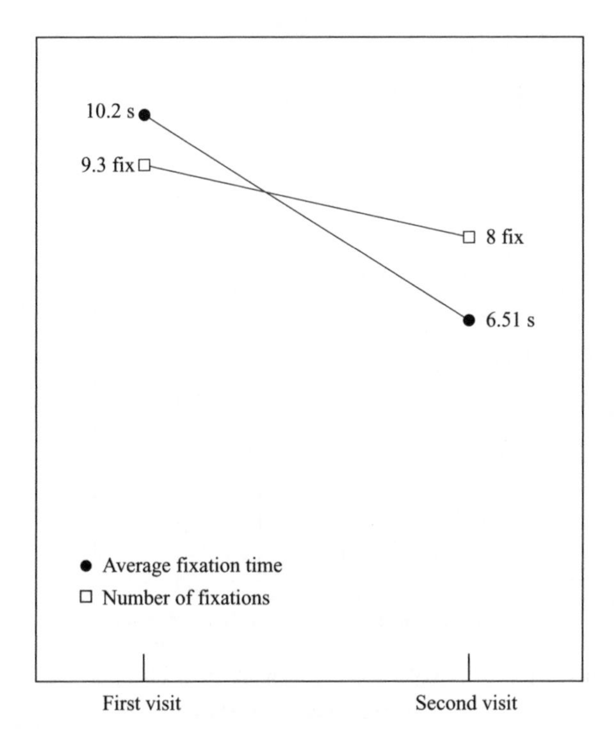

Figure 28.1: The orienting time decreases when a page is visited a second time.
Source: Stenfors (1998).

Another factor that influences the search behaviour is that subjects have a tendency to confirm the information they have found by searching the page for additional information (Stenfors & Holmqvist, 1999). This can result in hesitant behaviour when the subjects are not able to judge the quality of the information. This is found especially when task relevant information may be gained from both pictorial and textual objects on the same page.

The subjects' behaviour becomes more hesitant when there is a mismatch between pictures and text (Stenfors & Holmqvist, 1999). Such a mismatch occurs if the picture does not convey the same meaning as the test. This is very common on web-pages with elaborate graphics. Regardless of how salient a picture may initially be, the subjects tend to choose the text in these cases since it is the less ambiguous alternative. On the other hand, a picture well representing the contents of its link is experienced as a better guide in the search process (Goldberg & Kotval, 1999). This may be explained by the fact that informative pictures have a strong impact on the reduction of semantic uncertainty. The relative choice between text and picture links is shown in Figure 28.2. When pictures are very lucid, subjects prefer to use pictures rather than text links. When the pictures are unclear, subjects prefer to use text links.

It is interesting to note that much of the learning about where to find information on a web page may be unconscious. That visual learning can be unconscious was

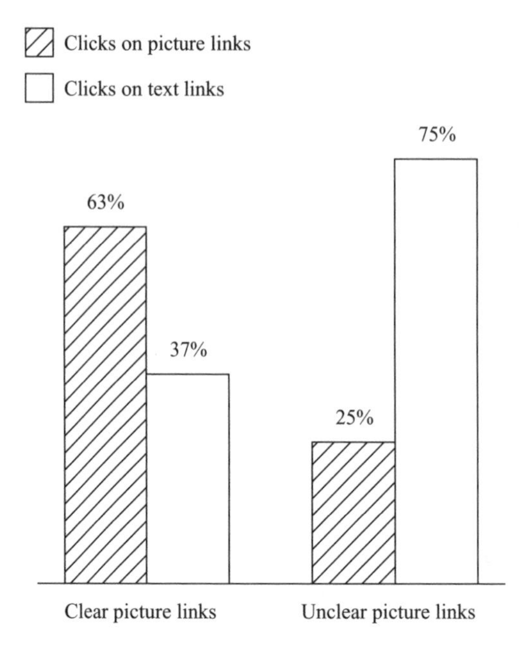

Figure 28.2: Relative choice between text and picture links.
Source: Stenfors (1998).

illustrated in an eye-tracking study where subjects where asked to predict the location of a visual target (Balkenius, Holsanova & Holmqvist, 2001). The target location was indicated by the colour of a visual cue, but the subjects were not told about this relation. As could be expected, the subjects became better at predicting the target location with practice. However, they did not in general report that they had seen or understood that the cues predicted the location of the target, although their behaviour clearly showed that they had learned this relation between the cues and the target.

Avoidance Strategies

It seems that the users use a plan to avoid web ads and even employ different kinds of avoidance strategies against web ads to achieve top-down control over their visual attention. These avoidance strategies can be either active or passive.

When the subjects have established which areas of the web page that are task relevant they do not depart from these. The subjects seem to be able to sort out the relevant animations from the irrelevant ones. One may expect that any visual object in motion that appears on the screen during the search process would force the subjects to react and look at the element in motion. But this is not the case. The subjects adopt a form of passive avoidance where they ignore the stimuli and avoid reacting to them in any way. Highly salient visual cues such as high colour contrast and motion that usually directs attention appears to have no effect on the experienced Internet user.

That such stimuli are not easy to avoid, however, is shown by the use of active avoidance strategies. In her studies, Stenfors (1998; Stenfors & Holmqvist, 1999) presented several examples of strong avoidance strategies. The most frequently used strategy is to move the mouse pointer while waiting for the web page to download. This usually takes the form of making the mouse pointer dance in small circles in the middle of the screen while following it with the eyes. Since the movement of the pointer attracts attention, the subjects avoid looking at ads (and express their boredom of waiting). Playing with the pointer in the middle of the screen also has the additional function that the pointer is at the average position where the next element to click will appear. This parallels the role of looking in the middle between two target locations as described above.

To investigate whether such learned avoidance behaviour was transferred to other domains we let subjects look at a screen where a central visual cue predicted where a target stimulus would appear on the screen (Stenfors *et al.*, 2001). The design was similar to the one in the study by Balkenius *et al.* (2001) described above. As in that experiment, the subjects were not told that the colour of the central fixation point indicated the location of the target that was subsequently presented.

Five subjects, which were all frequent Internet users were placed in front of a computer screen and instructed to look at a target (a yellow dot) when present. Otherwise they were free to look wherever they wanted. Forty stimulus sequences were shown, each consisting of a central target (a red or blue dot) shown in the middle of the screen for 500 ms followed by the target shown either to the right or to the left of the central target. The lateral target was visible for 500 ms followed by a 1000 ms pause before the start of the next sequence.

The experiment was designed so that when the central target was red the lateral target appeared to the left, and when blue, the lateral target appeared to the right. The two lateral targets were presented in pseudo-random order. For the last twenty stimulus sequences, distractors were presented at the top of the screen. The distractors were banners of the types used for ads on the Internet. They all had the same shape, and were changed every second sequence.

Despite the simple relation between the cue and target location, none of the subjects discovered the pattern. Instead they tested various simple hypotheses. Interestingly, very few saccades were made to the banners. When saccades were made, they occurred when the distractor appeared the first time, or (rarely) when it changed. Two subjects did not make any saccades towards the banners. Three subjects made saccades towards the banners followed by short fixations. There were ten such saccades in total for all subjects. The subjects timed these fixations in such a way that they did not interfere with the task.

The subjects preferred to look at the central yellow dot (95% of the sequences) rather than to look at the banners (5%). Although the central target had no meaning to the subjects, they all used it as a fixation point even when the only other visible thing was the banner. This shows that the avoidance behaviour against web ads is very strong and that after the web ad is identified as such by the subject, no overt attention is paid to it.

The study shows that not even a new element dynamically appearing on the screen may interrupt the subjects' search behaviour. Subjects behave as if the new information is not task related. This may be a result of a learning process in which a learned inhibition towards Internet ads is so strong that it generalises to other domains.

Anchoring and Buffering

Several types of external cues support the search through a web page. The most prominent are the visual elements on the page that serve to anchor the search behaviour in different ways. Anchoring can be either global or local. In addition anchoring can be categorised as being either static or dynamic.

An element such as a menu bar is used as a global anchor (Stenfors, 1998; Area 17 AB, unpublished report, 2001). By returning with the eye to such elements, the orientation on the page or site is quickly recovered. To function successfully as an anchor, global anchors should be static. If these elements change their appearance, they can no longer be used as anchors and will instead cause additional orienting of the page.

A more interesting type of anchoring is dynamic and acts locally on the web page. It seems that the subjects' primary attention focus is often located at the position of the mouse pointer (Stenfors, 1998). The mouse pointer functions both as a guide for the eyes when searching among links, but also as an anchor of where interesting information was found. The subjects anchor the mouse pointer on a location when they want to confirm its information by looking at other locations on the page. Placing the mouse pointer on a certain spot allows the eyes, searching the web page yet another time, to be safely guided back to the same spot again. In this way the mouse pointer is used as an external representation (Zhang & Norman, 1994). Approximately one third of the subjects appear to use this type of dynamic anchoring at times (Area 17 AB, unpublished report, 2001).

Also, the mouse pointer has been found to be used as an indicator of where to fixate next; it seems as if the subjects may start moving the mouse pointer in one direction while continuing to search the page in the other direction. When finishing the search, the eyes are directed to where the mouse pointer may have been located. An apparent example of this is when a subject, without initially looking at the scrollbar to the right, starts scrolling while searching among the links to the left on a web page.

Apart from using the mouse pointer as an anchor, subjects can also potentially use the cursor to mark a text field on a page or select text to mark the corresponding area on the page. We have not yet investigated these types of anchoring, however.

The use of the external stimuli as an aid to avoid memory burden is well documented in many situations (Zhang & Norman, 1994). For example, in an experiment where subjects were asked to copy a tiled pattern, they often returned to look at the template although they could in principle remember it (Ballard, Hayhoe, Pook & Rao, 1997). Before the subject picked up a coloured tile, they would first move their gaze to its location to guide the action with their eyes.

It is therefore surprising to see in our studies (Stenfors 1998; Stenfors *et al.*, 1999; Area 17 AB, unpublished report, 2001) that some of our subjects frequently do not return their gaze to a link or button before clicking on it. In these cases the subjects have fixated the link but continue to scan the web page before they decide to click on the link. At this time, the pointer moves to the link and the subject clicks without returning their gaze. Apparently he or she has buffered the location of the link for later use. As Wang, Lin and Drury (1997: 102) states in their study (1997): "Any memory for previous fixation locations improves search performance, in that more targets are detected per unit time."

Discussion

The subject's behaviour on a web page is not only dependent on what information is desired and presented but also on the format and spatial arrangement of the information (Spool *et al.*, 1998). A basic rule for presentation of information on a web page is that less is better: The information should be arranged in such a way that the shortest search time is obtained (Nielsen, 2000).

There is of course a conflict between wanting to present a lot of information on the web page and wanting to make it as searchable as possible and keep the search time short for the user. It is preferable if the user has available information in one single screen rather than to force scrolling, for example.

The user is not a passive receiver of information. The various cues are not only guides for a passive reading, but is also used for an active, ongoing process of evaluation and discrimination of the elements of the site. This process is used for an active, efficient search for the information that the user presently is interested in. This process is also dynamic, and so the user can readily learn to exploit new cues regarding the information layout on the page. The designer can thus not spoon-feed information to the user; it is the user that is in control.

The Internet is a user-driven experience (Platt, 1999) and as the study made by Stenfors *et al.* (2001) shows, the subjects are able to decide not to look at Internet ads. The fixation data shows that the subjects are well aware of how traditional Internet ads look and where they are located (Stenfors & Holmqvist 1999; Stenfors *et al.*, 2001). Elements that are in flashy colours and are moving are not to be seen. This is also confirmed by the comments made by some of the subjects during the interviews, for example: "I don't look up there [referring to the upper area of the screen where banners usually are located] since I know that there's nothing but advertisements" (Stenfors, 1998). This behaviour is also strengthened by the fact that the user's attention is lowest in the upper part of the screen, while attention is highest in the lower part of the screen (Danska Teknologirådet, 1987).

But the fact that a visual object is not fixated does not necessarily implicate that the object has not been attended to. An observer may be looking at a certain visual object while shifting attention from one peripheral area to another without changing eye position. Can this explain the lack of fixations on the Internet ads? As the eye recording data in the study by Stenfors *et al.* (2001) show, all fixations near the areas where ads were located are shorter than 80 ms; the length of these fixations are not in the suggested interval of 100–150 ms to be cognitively processed (i.e. allow recognition). This means that although the subjects may have shifted their peripheral attention to the ads, these areas have not been submitted to cognitive processing. This is also indicated in the result of the recall tests: The subjects just do not remember any ads on the web pages included in the tests.

There can be a number of explanations for this. First, it is possible that the subjects are so involved with the experimental task that they do not have the time or the attention to spare to look at the top of the screen. This does not seem implausible. A closer look at the data does cast a good deal of doubt on this explanation, however. The task was run at such a leisurely pace that the subjects frequently made exploratory fixations on various parts of the screen. When unsure of where the next fixation target would appear, they would place their gaze somewhere in between the two target locations, but make occasional fixations on other parts of the screen as well.

When they were wrong (frequently), the subjects still had ample time to move the gaze to the correct part of the screen. Interviews with the subjects confirmed this; they spent a good deal time speculating about the nature of the task, and none of them considered the task as stressful or high-paced (Stenfors *et al.*, 2001).

Also, the web page ads, i.e. the banners, did not show up until after a number of iterations of the task. This ought to induce a surprise reaction due to the stimulus' unexpectedness; more so, as no mention of anything like it happening during the experiment was mentioned beforehand. Many of the subjects were aware that something was happening on the top of the screen, but none of them considered it (whatever it was) important, or worth looking at, even though the graphics of the ads where not static, but changed frequently.

The other explanation is that the subjects had been conditioned not to look at banners while surfing, that this conditioning results in active avoidance strategies rather than passive disinterest, and that they transferred that conditioning into this new context (Balkenius & Morén, 2000). All of the subjects were university students, and well versed with Internet surfing, and previous studies have shown that surfers rapidly learn to ignore web advertising (Stenfors, 1998). The context in this case is of course somewhat similar to the domain of web surfing — the subjects are watching a computer screen — but both the social setting and the task is very different.

One indication for this explanation is the fact that some of the subjects did not look even once at the banners. This indicates that they are making use of previous knowledge, as they would otherwise have needed to look at least once to be able to determine that the banner graphic really did not contribute to the task in any way.

If the second explanation is correct, this means that the learned inhibition towards looking at web advertising is both more general and less domain specific than previously assumed. It would mean that anything that sits on the top centre of the screen, is graphical in nature (colourful, moving) and with a vaguely horizontal layout is ignored, no matter in what context it is presented. It would be interesting to find out just how far from the original context this inhibitory behaviour will work.

The commercial value of Internet ads depends primarily on the number of Internet users observing them, i.e. how visible the ads are. How deep an impact a specific ad makes is therefore determined by the number of observers, its visibility, and of course the effectiveness of the ad itself. Eye-tracking studies show that the ads' impact on the subjects search behaviour seems to be very small (Stenfors, 1998). When searching for task relevant information, the disturbance of irrelevant visual elements is almost non-existing: The unattended stimuli have nearly no influence on the search behaviour (Stenfors & Holmqvist, 1999).

In summary, it is clear that the use of visual attention and behaviour on a web page is different from that in other situations. Subjects do not react to novelty that is unrelated to their own actions such as the sudden appearance of an ad or animation. On the other hand, visual motion is used to find the pointer on the screen by moving the mouse. Another discrepancy lies in the use of buffering to remember where to click and later move the pointer to that location without visual guidance. Although these behaviours are not unique to web interaction, they are seen more frequently there.

Note

1 IriSense® is a registered trademark of Area 17 AB.

References

Arani, T., Drury, C. G., & Karawan, M. H. (1984). A variable-memory model of visual search. *Human Factors 26*(6), 631–639.

Balkenius, C. (2000). Attention, habituation and conditioning: Toward a computational model. *Cognitive Science Quarterly 1*(2), 171–214.

Balkenius, C., Holsanova, J., & Holmqvist, K. (2001). Implicit learning of anticipatory saccades. In: *Eleventh European Conference on Eye-Movements* (p. P76), August 22–25, Åbo.

Balkenius, C., & Morén, J. (2000). A computational model of context processing, In: J-A. Meyer, A. Berthoz, D. Floreano, H. L. Roitblat and S. W. Wilson (eds), *From Animals to Animats 6: Proceedings of the 6th International Conference on the Simulation of Adaptive Behaviour*, (pp. 256–265). Cambridge, MA: The MIT Press.

Ballard, D. H., Hayhoe, M. M., Pook, P. K., & Rao, R. P. N. (1997). Deictic codes for the embodiment of cognition. *Behavioral and Brain Sciences 20*(4), 723–767.

Danska Teknologirådet (1987). *Menneske-Maskin-Samspil*, ECR-199.

Goldberg, J. H., & Kotval, X. P. (1999). Computer interface evaluation using eye movements: Methods and constructs — its psychophysical foundation and relevance to display design. *International Journal of Industrial Ergonomics 24*(6), 631–645.

Henderson, J. M., & Hollingworth, A. (1999). High-level scene perception. *Annual Review of Psychology*, *50*, 243–271.

Landauer, T. K. (1996). *The Trouble with Computers: Usefulness, Usability, and Productivity*. Cambridge MA: The MIT Press.

Moeller, E. W. (2001). The Latest Web Trend: Usability? In: *Professional Communication Conference, IPCC, Proceedings*, IEEE International (pp. 151–158).

Nielsen, J. (2000). *Designing Web Usability: The Practice of Simplicity*. Indianapolis: New Riders Publishing.

Nygren, E., & Allard, A., (1996). *Display Design Principles Based on a Model of Visual Search*. Report no. 64/96. Uppsala Sweden: Uppsala University Center for Human–Computer Studies.

Nygren, E., Allard, A., & Lind, M., (1995). *Experiments in Visual Search I. Effects of Patterns of Highlight Items on List Search*. Report no. 55/95. Uppsala, Sweden: Uppsala University Center for Human–Computer Studies.

Platt, A.-B. (1999). The usability risk. In: *Proceedings of the 18th IEEE Symposium on Reliable Distributed Systems* (Vol. 18, pp. 396–400). 18–21 October 1999, Lausanne, Switzerland.

Rubin, J. (1994). *Handbook of Usability Testing*. New York, NY: John Wiley & Sons.

Sibert, L. E., & Jacob, R. J. K. (2000). Evaluation of eye gaze interaction. In: *CHI 2000 Conference Proceedings* (pp. 281–288). New York: ACM.

Spool, J. M., Scanlon, T., Schroeder, W., Snyder, C., & De Angelo, T. (1998). *Web Site Usability: A Designer's Guide*. San Francisco: Morgan Kaufmann.

Stenfors, I. (1998). Visual behaviour of internet users relative to web page design and internet advertisements. Masters thesis, Lund University Cognitive Science, Lund.

Stenfors, I., & Holmqvist, K. (1999). The strategic control of gaze direction when avoiding Internet ads. In: *10th European Conference on Eye Movements* September 23–25 (p. 142). Utrecht.

Stenfors, I., Morén, J., & Balkenius, C. (2001). Avoidance of internet ads in a visual search task. In: *Eleventh European Conference on Eye-Movements.* August 22–25 (p. S47). Åbo.

van Waes, L. (2000). Thinking aloud as a method for testing the usability of websites: The influence of task variation on the evaluation of hypertext. In: *IEEE Transactions on Professional Communication*, *43*(3), 279–291.

Wang, M.-J.J, Lin, S.-C., & Drury, C. G. (1997). Training for strategy in visual search. *International Journal of Industrial Ergonomics 20*(2), 101–108.

Wichansky, A. M. (2000). Usability testing in 2000 and beyond. *Ergonomics 43*(7), 998–1007.

Yarbus, A. L. (1967). *Eye Movements and Vision*. New York: Plenum Press.

Zhang, J., & Norman, D. (1994). Representations in distributed cognitive tasks. *Cognitive Science*, *18*, 87–122.

Chapter 29

Determining the Parameters for the Scrolling Text Display Technique

Y. Kitamura, K. Horii, O.Takeuchi, K. Kotani
and G. d'Ydewalle

This paper reports on empirical efforts to determine the threshold values and the feasible ranges of the parameters for scrolling text display technique (i.e., Times Square scrolling). Parameters associated with the technique, such as speed of scrolling and maximum possible length of a sentence for one scrolling, were examined through the use of an Eye Link System, and answers to the following research questions were experimentally provided: (1) how fast should the text scrolling be?; (2) how well do experimental subjects retain the text at different speeds?; (3) at what speed do the subjects fail to comprehend the scrolling text?; and (4) how many Japanese characters can subjects read at a time?

Introduction

With the recent rapid advances in multimedia, a variety of visual presentation techniques have been developed. Horizontal right-to-left scrolling text display (i.e., Times Square scrolling) is one major technique, in which linguistic information (i.e., characters) is presented, as in the case of an electric bulletin board, scrolling from right to left at a certain speed, in a limited visual display area on a monitor.

Previous studies, such as Elson (1982), Granaas, McKay, Laham, Hurt, and Juola (1984), Chen, and Tsoi (1988), Kang, and Muter (1989) and Juola, Tiritoglu, and Pleunis (1995), have dealt with scrolling text either by RSVP or the Times Square method, but none of these have focused on eye movement, recorded by computer. Kolers, Duchnicky, and Ferguson (1981) have indeed studied eye movement when scrolling text moving upward was shown. In our experiment, we studied text scrolling smoothly from right to left, pixel by pixel.

The Mind's Eye: Cognitive and Applied Aspects of Eye Movement Research

ISBN: 0-444-51020-6

The parameters associated with scrolling, such as speed of scrolling, and maximum possible length of a sentence for one scrolling have, however, so far been determined, working mainly from the mere experience and intuition of programmers, and thus without empirical underpinnings.[1]

In this experiment, therefore, efforts were made to determine basically and empirically the optimal scrolling speed and the maximum possible sentence-length for one scrolling.

In reading text under ordinary conditions, the alternation of fixations and saccades is repeated from the beginning to the end of the text. Information is extracted during fixations, and the saccades occur to move on to the next source of information.

When reading the text scrolling from right to left, the locus of the eye movement is as shown in Figure 29.1, in which the *X*-axis shows the time, and the *Y*-axis indicates location of the eye. Ideally, the eyes follow the scrolling text and recognize it during smooth pursuit eye movement, which is then followed by a saccadic eye movement to a new position — a process that is repeated to the end of the text.

Purposes

The purposes of this paper are to answer the following four research questions concerning the parameters of the text scrolling:

1. How fast (or slow) should the text scrolling be?
2. How well do the subjects retain the text at different speeds?
3. At what speed do the subjects fail to comprehend the scrolling text?
4. How many Japanese characters can the subjects read at a time?

Experiment

Apparatus

An EyeLink System (sampling rate: 250 MHz) produced by the SMI Corp. was used for this experiment. The system is made up of a subject PC (Compaq Deskpro 4000) for presenting the stimuli and collecting the data from the eye camera mounted on each subject, and an operator PC (Compaq Deskpro 4000) for regulating the stimulus presentation, and accumulating the data collected from the subject PC.

Materials and Display Method

The twelve texts used for this experiment were selected from the *Vox Populi, Vox Dei* column in the *Asahi* Newspaper, a major Japanese newspaper. They were written in standard Japanese with *hiragana* (i.e., the Japanese cursive syllabary) with a moderate amount of Chinese characters scattered through the texts. They were of approximately 200 words in length.

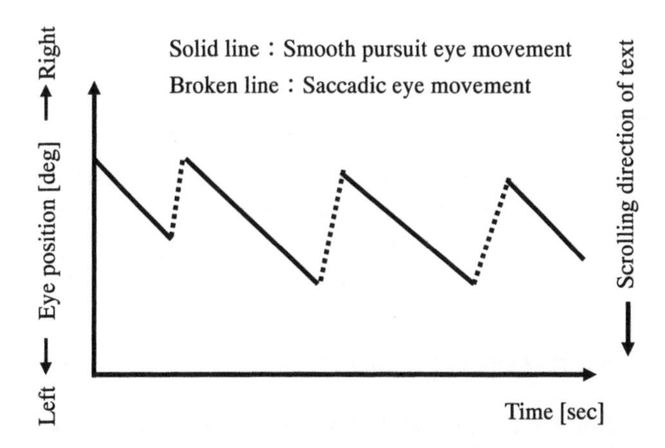

Figure 29.1: The schematic locus of eye in reading the scrolling text.

The texts were displayed in a window on the monitor. The window was 24° in width. This width was derived experimentally, with the object of not hindering natural reading in scrolling texts (Kamei & Kikuchi, 1997). The distance between the monitor and the subject was 41 cm, and the size of one character displayed was 1° horizontally and vertically measured as a visual angle. The scrolling speeds chosen for the experiment were 3, 5, 7, 10, 13, 16, 19, 22, and 25°/s. The scrolling speeds and the texts were randomly combined and displayed to the subjects, so that any influence from differences among the texts could be cancelled out.

Subjects

Five subjects (20–26 years old) with normal vision participated in this experiment. They were college students at Kansai University, whose native language is Japanese.

Procedure

Prior to the experiment, it was explained to the subjects how the texts were to be presented on the monitor, and they were instructed to read and retain their informational content. The subjects were then asked to write down what they had retained on paper provided. What they had written was then checked by a member of our research team, using a 10-point scale based on idea-units predetermined for each text. An example of the idea-units can be found in the Appendix.

The subjects were also asked to supply, by using a one-to-five Likert scale, their own rating of the relative easiness of reading at different scrolling speeds. One on this scale meant "the least ease of reading", and three "the average ease", while five indicated "the most ease".

Results

Scrolling Speed and Relative Easiness of Reading

Figure 29.2 shows the relationship between the scrolling speed and the subject-reported relative ease of reading. Given a scrolling speed below 10°/sec, the subjects found the texts to be relatively easy to read, while they found them difficult or uncomfortable to read when given a scrolling speed above 10°/sec.

Scrolling Speed and Retention of Content

A minimum score of five points (out of a maximum of ten) on the retention of predetermined idea-units was adopted as the threshold level of retention of content. For example, as Figure 29.3 shows, subject B's score for retention dropped below five at a scrolling speed of 10°/sec.

Scrolling Speed and Loci

Compare Figures 29.4–1, 29.4–2, and 29.4–3 for the relation between scrolling speed and loci. These are examples of a subject that behaved typical to the other subjects. The loci in Figure 29.4–1, which shows the eye-movement curve at a scrolling speed of 3°/sec, are mainly pairings of fixation and saccade. This indicates that the speed for scrolling is too slow. The loci in Figure 29.4–2 (at 7°/sec) are mainly pairings of smooth pursuit eye movement and saccadic eye movement. On the other hand, the loci in Figure 29.4–3 (at 22°/sec) are combinations of not only smooth pursuit eye movement and saccadic eye movement, but also fixation, which suggests that the subject gave up reading at this scrolling speed.

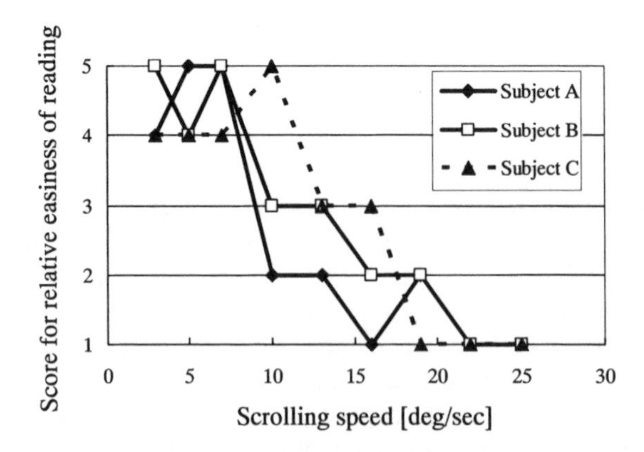

Figure 29.2: Relation of scrolling speed and relative easiness of reading.

Note: 1° is equivalent to one character.

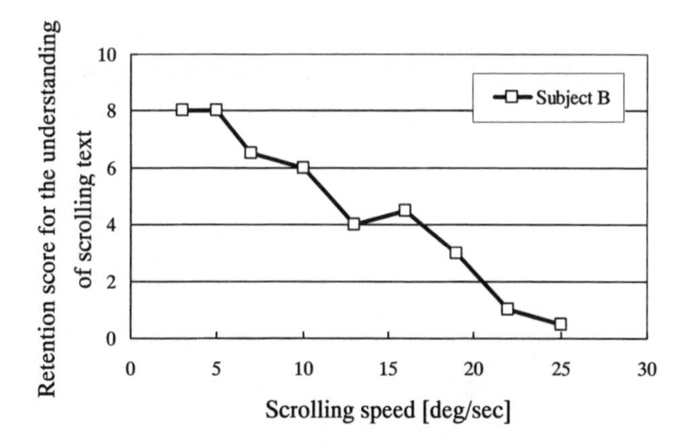

Figure 29.3: Scrolling speed and retention score.

Note: 1° is equivalent to one character.

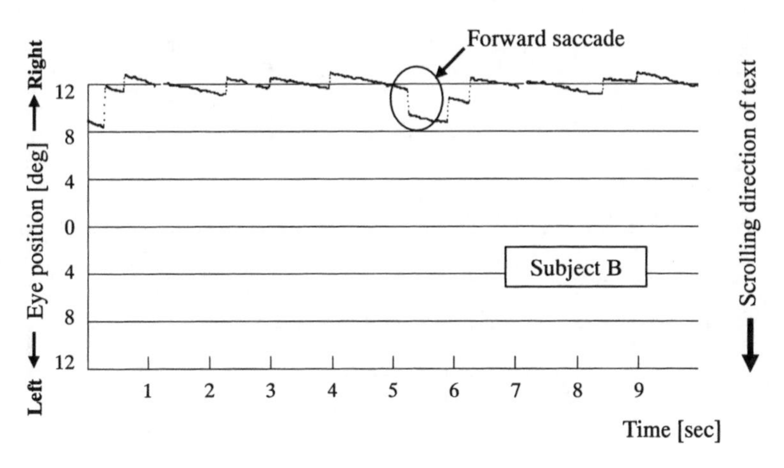

Figure 29.4–1: Locus at scrolling speed 3°/sec.

Note: 1° is equivalent to one character.

A similar behavior (Merrill & Stark, 1963), termed OKN (optokinetic nystagmus), must be mentioned here. From a moving train, a man is taking a look outside, absent-mindedly, and suddenly notices a shrub in the desert outside the train. His eyes follow the shrub for a short moment. This eye movement, OKN, is totally involuntary and is not related to the eye movements examined in this article. Some researchers, such as Honrubia, Downey, Mitchell, and Ward (1968) and Jürgens, Becker, Reiger and Widderich (1981), indicated that the range and the speed of the smooth pursuit eye movements of OKN are far too larger than those of our experiment result. This is because the voluntary eye movement in our experiment is concerned with cognition of reading.

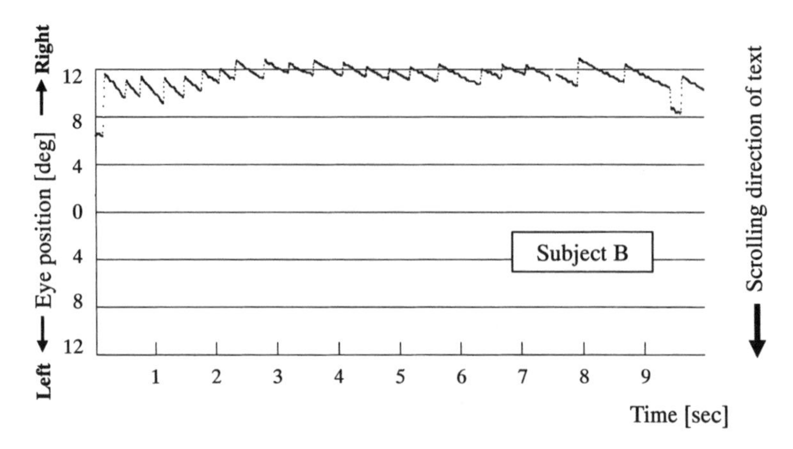

Figure 29.4–2: Locus at scrolling speed 7°/sec.

Note: 1° is equivalent to one character.

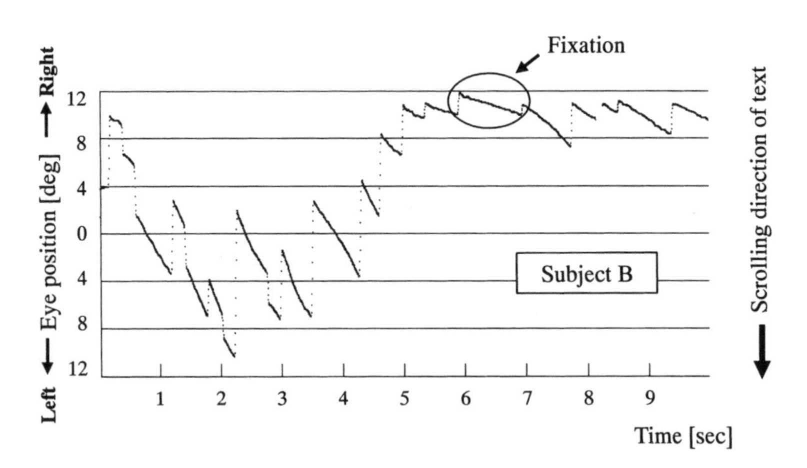

Figure 29.4–3: Locus at scrolling speed 22°/sec.

Note: 1° is equivalent to one character.

Table 29.1 is a summary of the relationships between the scrolling speeds and eye-movement properties. This table shows that between the speeds of 7 and 10°/sec, all our subjects repeated smooth pursuit eye movements and saccadic eye movements in the direction opposite to that of the scrolling display, which is a sign of ease of reading.

Average Velocity of Smooth Pursuit Eye Movements

Figures 29.5–1 (3°/sec), 29.5–2 (7°/sec), and 29.5–3 (22°/sec) respectively show the relation between the scrolling speed and the smooth pursuit eye movement velocity.

Table 29.1: A summary of the relationships between the scrolling speeds and eye movement properties (1° is equivalent to one character.)

	Eye-movement properties			
			Saccadic eye movement	
Scrolling speed [deg/sec]	**Fixation**	**Smooth pursuit eye movement**	**Opposite direction of scrolling display**	**Direction of scrolling display**
3	F	R	L	L
5	L	L	F	L
7	R	F	F	R
10	R	F	F	R
13	R	F	F	L
16	R	F	F	F
19	R	F	F	F
22	L	F	F	L
25	L	F	F	L

Note: F: Frequent, L: A few, R: Rare.

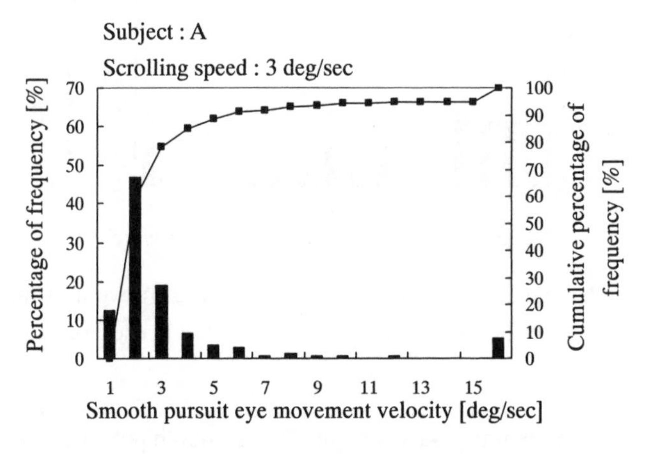

Figure 29.5–1: Percentage of frequency of smooth pursuit eye movement velocity.
Note: Scrolling speed: 3°/sec. 1° is equivalent to one character.

As is typical, the *X*-axis indicates the velocities of smooth pursuit eye movements, with the percentage of frequency shown on the *Y*-axis. As can be seen in these figures, regardless of scrolling speed, the velocity of smooth pursuit eye movements is around 2 to 4°/sec. In Figures 29.5–1 and 29.5–2, the velocity over 15°/sec seemingly appears,

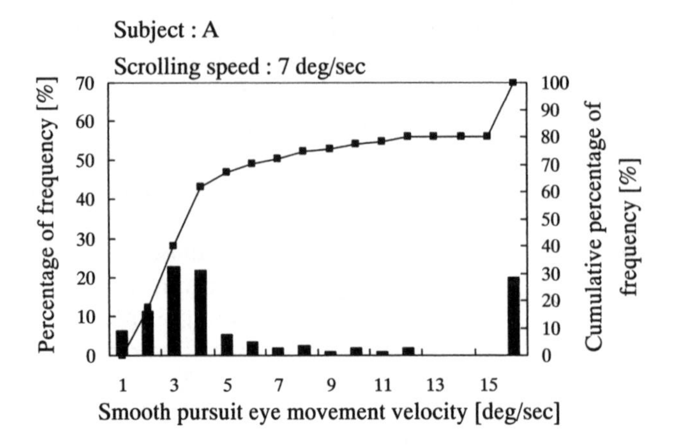

Figure 29.5–2: Percentage of frequency of smooth pursuit eye movement velocity.
Note: Scrolling speed : 7°/sec. 1° is equivalent to one character.

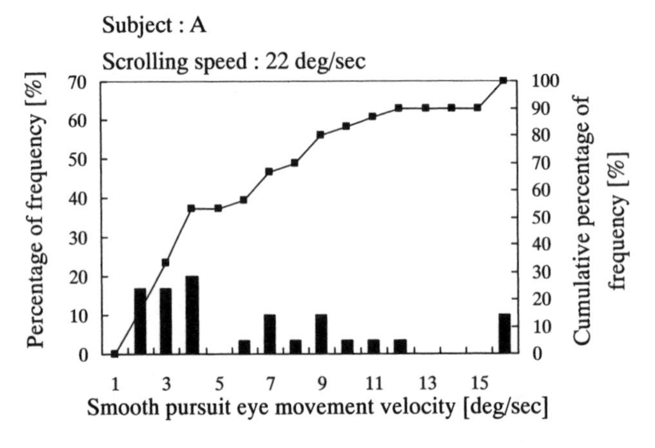

Figure 29.5–3: Percentage of frequency of smooth pursuit eye movement velocity.
Note: Scrolling speed : 22°/sec. 1° is equivalent to one character.

but it is the accumulation of tiny saccades and small smooth pursuits, whereas in Figure 29.5–3, it is the accumulation of OKN.

In other words, smooth pursuit eye movements invariably center around the velocity of 2 to 4°/sec in reading the scrolling text. This generalization led us to compute the average velocity of the smooth pursuit eye movements made by our subjects, as shown in Figure 29.6.

In this figure, the solid line is the hypothetical line projected for a situation in which the smooth pursuit eye movement velocity and the scrolling speed are identical in speed. All subjects showed smooth pursuit eye movement velocities that fell lower

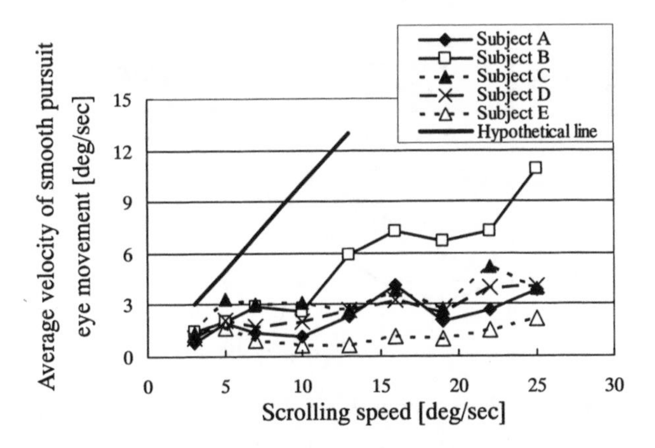

Figure 29.6: Average velocity of smooth pursuit eye movement.

Note: 1° is equivalent to one character.

than this hypothetical line. Excepting subject B, who showed different smooth pursuit eye movement velocities, the remaining subjects clustered around 1 to 5°/sec.

Relative Smooth Pursuit Eye Movement Velocity

The concept of relative smooth pursuit eye movement velocity, original to the authors, was adopted, and this relative velocity was calculated according to the formula shown below.

Formula 1

(Relative smooth pursuit eye movement velocity) = (scrolling speed) – (smooth pursuit eye movement velocity)

Figure 29.7 shows the plotted results of the calculations, indicating that the relative smooth pursuit eye movement velocity at the circled area was below 10°/sec. From the information shown in Figures 29.2 and 29.3, it was evident that the subjects in this circled area comprehended the text by 50% or more. In other words, when the relative smooth pursuit eye movement velocity is over 10°/sec, all the subjects failed to comprehend the text. This is because such relative smooth pursuit eye movement velocity produces only blurred images on the retina, resulting in incomprehension. When the relative smooth pursuit eye movement velocity is 0°/sec, there is no blur of the image. The greater the relative smooth pursuit eye movement velocity, the more blurred the image becomes. So the image of the text on the retina can no longer be recognized once the relative smooth pursuit eye movement velocity rises above 10°/sec.

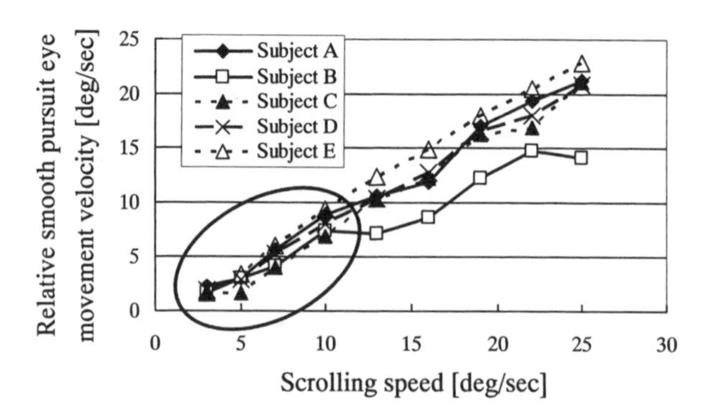

Figure 29.7: Relative smooth pursuit eye movement velocity.
Note: 1° is equivalent to one character.

Number of the Characters Read During a Single Smooth Pursuit Eye Movement

The number of Japanese characters to be covered in one smooth pursuit eye movement can be calculated using the following formula:

Formula 2

(Number of characters) = (relative smooth pursuit eye movement velocity) × (smooth pursuit eye movement time in length)

Note that 1°/sec is equal to 1 Japanese character/sec.

The average number of the characters read during one smooth pursuit eye movement made by each subject is shown in Figure 29.8. The circled area in this figure shows scrolling speeds of 10°/sec or less, within which range all our subjects read a maximum of five characters. We may, therefore, safely conclude that the scrolling text allows readers to read at most five Japanese characters at one smooth pursuit eye movement.[2]

Conclusions

The results of our experiment showed that (1) the maximum possible scrolling speed (in which reader's comprehension is ensured) was 10°/sec; (2) smooth pursuit eye movements in the direction of scrolling and saccadic eye movements in the opposite direction were repeatedly and constantly alternated during the reading of the scrolling sentences presented at speeds of between 7 and 10°/sec; and (3) when the relative velocity between the smooth pursuit eye movement and the scrolling (interpreted as the velocity of scrolled images on the retina) was more than 10°/sec, our subjects failed to comprehend the sentences. Lastly, (4) the number of Japanese characters (the mixture of *hiragana* and Chinese characters) that could be read in one smooth pursuit eye movement, was five at maximum.

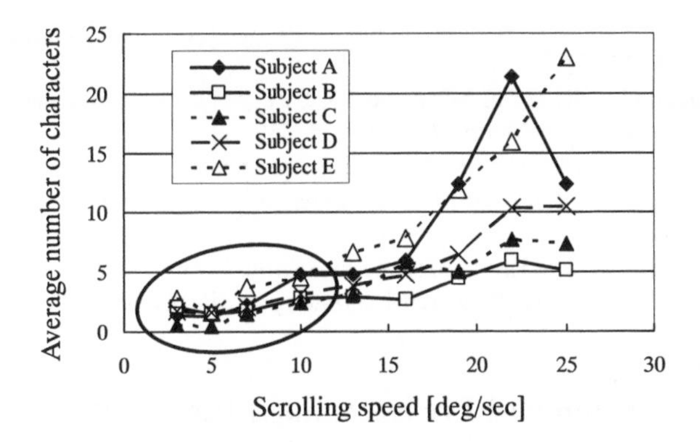

Figure 29.8: Average number of characters to be read during a smooth pursuit eye movement.

Note: 1° is equivalent to one character.

Notes

1 Dyson and Haselgrove's study in 2001 and Komine and Isono's study in 1996 are two rare attempts at empirical validation. Also, at our own laboratory, Kamei and Kikuchi (1997) conducted research in the same vein through the use of a lower precision/resolution EMR-7 produced by the NAC Corp.

2 Ikeda and Saida (1978) reports that the average number is around 8 or 9 English characters at one fixation and 3 or 4 Japanese characters in reading texts presented statically.

References

Chen, H. C., & Tsoi, K. C. (1988). Factors affecting the readability of moving text on a computer display. *Human Factors*, *30*, 25–33.

Dyson, M. C., & Haselgrove, M. (2001). The influence of reading speed and line length on the effectiveness of reading from screen. *International Journal of Human Computing Studies*, *54*, 585–612.

Elson, I. J. (1982). Designing readable scrolling displays. *Displays*, 155–157.

Granaas, M. M., McKay, T. D., Laham, R. D., Hurt, L. D., & Juola, J. F. (1984). Reading moving text on a CRT screen. *Human Factors*, *26*, 97–104.

Honrubia, V., Downey, W. L., Mitchell, D. P., & Ward, P. H. (1968). Experimental studies on optokinetic nystagmus. II. Normal humans. *Acta Otolaryngologica*, *65*, 441–448.

Ikeda, M., & Saida, S. (1978). Span of recognition in reading. *Vision Research*, *18*, 83–88.

Juola, J. F., Tiritoglu, A., & Pleunis, J. (1995). Reading text presented on a small display. *Applied Ergonomics*, *26*, 227–229.

Jürgens, R., Becker, W., Reiger, P., & Widderich, A. (1981). Interaction between goal-directed saccades and the vestibulo-ocular reflex (VOR) is different from interaction between quick phase and VOR. In: A. Fuchs and W. Becker (eds), *Progress in Oculomoter Research*. New York: Elsevier/North-Holland.

Kamei, N., & Kikuchi, H. (1997). Perceptual and cognitive characteristics of eye movements in reading scrolling texts. Unpublished Bachelor's thesis, Faculty of Engineering, Kansai University, Osaka, Japan.

Kang, T. J., & Muter, P. (1989). Reading dynamically displayed text. *Behaviour and Information Technology*, *8*, 33–42.
Kolers, P. A., Duchnicky, R. L., & Ferguson, D. C. (1981). Eye movement measurement of readability of CRT displays. *Human Factors*, *23*, 517–527.
Komine, K., & Isono, H. (1996). A study of desirable speed for scrolling superimposed characters. In: *Proceedings of the Institute of Electronics, Information and Communication Engineers* (p. 390).
Merrill, E. C., & Stark, L. (1963). Optokinetic nystagmus: Double strip experiment. *Quarterly Progress Report*, 70, 357–359. Research Laboratory of Electronics/MIT.

Appendix: An Example of Predetermined Idea Units

A sample sentence (Translation ours. Original sentence written in standard Japanese with a mixture of *Hiragana* and Chinese characters.)

> When Kyoko found Kazuo on the opposite side of the road and tried to call him, a white dog jumped onto the road, and the driver of a green trailer truck that was running at top speed on the road, noticed it and hit his brake hard.

Idea Units Derived from the Sentence

- Kyoko found Kazuo.
- Kazuo was on the opposite side of the road.
- Kyoko tried to call him.
- A dog jumped onto the road.
- The dog was white.
- A trailer truck came up.
- The trailer truck was running at full speed.
- The driver noticed it.
- The driver braked hard.

Chapter 30

Reading or Scanning? A Study of Newspaper and Net Paper Reading

Kenneth Holmqvist, Jana Holsanova, Maria Barthelson and Daniel Lundqvist

Net paper readers have been shown to read deeper into articles than newspaper readers. It has also been claimed that newspaper readers rather scan than read newspapers. Do these findings mean that net paper readers read proportionally more than newspaper readers? This paper presents results showing that in fact net paper readers scan more and read less than newspaper readers. We furthermore investigate whether this result can be explained by the differences in layout, navigation structure and purpose of reading between the two media.

Introduction

Today, we use the Internet for several purposes: to get information, entertainment and to do errands. During February 2002, 5,080,000 Swedes were connected to the Internet, which is 58.2% of the whole Swedish population above 2 years of age (80% of the age group 18–49 years).[1] Sixty-two percent of the Swedish Internet users state that they read news and net papers online. This activity is second to e-mailing (84%).[2] In 1994, Aftonbladet.se opened the first net paper site in Sweden. In February 2002, this site had 1,836,000 unique visitors.

News providers try to adjust to the trend and are present on the Internet. Due to the competition to catch the readers' attention, news providers are anxious to understand the behaviour of net paper users. Some years ago, they could only get feedback from logs and online surveys. Nowadays, they conduct eye-tracking studies.

Traditional printed media exist in parallel to the new media. Reading traditional newspapers implies looking for headlines, briefs, photos and drop quotes. Does the new medium influence our way of searching for news? Does net paper reading differ from newspaper reading?

The Mind's Eye: Cognitive and Applied Aspects of Eye Movement Research

ISBN: 0–444–51020–6

On the one hand, the new media differ considerably from the old: Online readers read from a computer screen and move around by clicking on links and menu buttons. They navigate through a virtual space and may run into problems orienting themselves in the complex net structure (see De Léon & Holsanova, 1997).

On the other hand, readers themselves report on important differences in the use of the traditional and new media. In Barthelson (2002), experienced online news readers describe the medium differences as follows: Reading a newspaper is something they do with pleasure and if possible in a situation that allows few distraction (along with breakfast, in a coffee break after lunch, on the train or in the subway). It is a relaxing activity to traverse through the folds and it usually takes quite a long time. In contrast, reading a net paper is something you do in much shorter breaks, perhaps between two e-mails, usually in your office in the early morning, or during lunch. The purpose is to become updated on a few particular questions. The users make brief visits to the news sites several times a day with the expectation of obtaining a quick overview over the latest events.

A net paper consists of a front page (also called the first page), topic pages (sometimes called section pages) and article pages. A page is a two-dimensional surface containing text and images that can be accessed by scrolling up and down in a web browser window. The information for each page is stored at a different URL.

All these two-dimensional surfaces are then linked together in a complex multi-dimensional structure. By hyper-linking, hitting the back button, choosing a different

Back Forward Stop Refresh Home
Address: http://www.dn.se/

bildirekt.se
Köp och sälj begagnade bilar!

Prova DN i 4 månader, betala för 3.

SÖK PÅ DN.SE
Sök
▸ Gör DN till startsida

Onsdag 7 november 2001

E-post
info@dn.se

DAGENS NYHETER.

▸Prenumerera
▸Annonsera

Börsen: 0,53 %
Senaste nytt

SVERIGE
▸ Bankrån i Bålsta [10:37]
▸ Misstänkt vätska även i Gambros filter [09:31]
▸ Björnungar i Lycksele avlivas [06:54]

UTRIKES
▸ Domare ihjälskjuten i Baskien [09:10]
▸ Israel drog tillbaka trupper från Ramallah [07:49]
▸ Afghanska oppositionen närmare Mazar-i-Sharif [07:43]

EKONOMI

Uppror mot betygsstopp
Stockholms stad startar nu ett uppror mot Skolverkets beslut att slopa betygsgränserna i de nationella proven. Skolborgarrådet Jan Björklund uppmanar resten av Sverige att följa efter och använda Stockholms betygsgränser.
▸ Sverige

SAS minskar i betydelse
WEBB-TV: SAS går sämre än någonsin tidigare. I går meddelade bolaget att ytterligare 2.500 anställda måste bort. Men hela det europeiska flyget är i djup kris, och står inför en framtid där allt färre aktörer får plats i luften.
▸ Ekonomi
SAS AB +/-0.00% ▸Följ utvecklingen
▸ Sabena går i konkurs

Massiva hjälpinsatser i Afghanistan
TYCK TILL: FN:s barnfond Unicef hoppas kunna vaccinera över fem miljoner

WEBBFRÅGAN
Har du skänkt några pengar till de nödlidande afghanerna?
Ja
Nej
SKICKA
▸ Ställningen
▸ Läs artikeln

Nyheter
Sverige
Utrikes
Ekonomi
Sport
/DN Fotboll
/Målservice
Kultur

Tjänster
Börs
Bostad
Resor

<Annonser>
Köp DN-cykeln
Svensktillverkad
Shimano bromsnav
Komplett rostskydd
Belysning
Reflexer
Ett års garanti
Tre växlar (tillägg 695 kr)
DN.se/butiken

Figure 30.1: The front page of the DN net paper (www.dn.se).

bookmark, or typing in a new web location, the readers can move around in this virtual space. Net papers usually organise their pages in a hierarchical tree structure, with the front page as the root and the topics and stories branching out from it.

Previous Research

So far, only a small number of explorative and experimental eye tracking studies that concern newspaper and net paper reading have been conducted. One of the main questions has been what item on a page first catches the readers' eye. This issue of the so-called priority order on a page has often been studied (see Barthelsson, 2002; Garcia & Stark, 1991; Hansen, 1994; Lewenstein, Edwards, Tatar and Devigal, 2000). The distribution of fixations in different regions has been focused on in several papers (see Barthelsson, 2002; Hansen, 1994, Widman & Polansky, 1990). Sometimes the transitions between regions and between pages have also been analysed (see Lewenstein *et al.*, 2000) to find out how readers navigate in the virtual space. Last but not least, the processing and reading of newspaper advertisements has been investigated (see Lundqvist & Holmqvist, forthcoming, Widman & Polansky, 1990). These studies are all concerned with layout questions, such as the position of various graphical elements on a page, the role of photographs versus text, or the role of colour in the layout. Apart from the explorative studies by Garcia and Stark (1991) and Lewenstein *et al.* (2000), the overall reading behaviour has often been investigated as a by-product of these results. The focus was on results that had an immediate relevance for designers and journalists.

Widman and Polansky (1990) tested 129 readers of the Stockholm newspaper *Dagens Nyheter* for advertisement reading. The unpublished internal report shows that 39% of all ads are seen, and the bigger the ad, the more likely that it is seen and remembered. Widman and Polansky also investigates the importance of different positions and contents of newspaper ads, noting that a position slightly to the left of the middle is optimal for the smaller ads. Photographs, pictures and colour in ads are reported to increase fixation frequency on the ad.

Lundqvist and Holmqvist (forthcoming) tested 14 readers of *Dagens Nyheter* in a follow-up study to Widman and Polansky, but focusing entirely on the effect of size on perception, memory and attitude towards ads. It is concluded that these variables strongly correlate, and that size and attitude data explain around 50% of the variance in the perception and memory data. Ad content and design were not taken into account as a variable.

Garcia and Stark (1991) tested 90 readers of three newspapers at three different sites in the US. The editions (prototype A and B) given to the readers were manipulated with respect to colour so as to show whether colour is an attractor that can direct the reader. Participants could read as long as they wished and their eye movements were tracked during the reading session. A videotape with newspaper pages and the tracks of the eye movements across the pages were used to find out whether elements on the page were processed, read or read in depth. The material was considered "read" if the reader's eyes moved across one or more lines of print from left to right. When at least one half of any text was read, it was considered "read in depth." "Processing" was rather vaguely defined: the reader's attention stops long enough at an individual element for information to be acquired. These results give an immediate feedback to journalists and news designers, but the lack of precise definitions and measures make them difficult for researchers to interpret.

A major finding in Garcia and Stark's study was that readers do not really read but rather *scan* newspapers. At certain *entry points* they stop scanning and start reading the story that the entry point belongs to. Pictures and graphics were identified as the main entry points, followed by front-page promotion boxes.[3] Readers usually enter the page through the dominant photo and then move to a prominent headline or another dominant photo. But virtually all elements, anywhere, even editorial text, can serve as entry points into reading. This, the authors argue, is strong evidence against the classical inverted information pyramid. The inverted information pyramid tells us that important information should be given most space, and placed at the top, while gradually less important information should be given less space further down. Instead they speak in favour of a creative design using more graphical elements on the folds.

Garcia and Stark also report about processing text and photos. Headlines, cut lines and briefs are processed often and in depth. Only 25% of the articles are processed. Text processing is highest in the news section, lowest in the sports. Readers devote more time to photo groupings when they are in colour. Size increases the attraction to a photo. Garcia and Stark's discussion ends by defining newspaper design as the task "to give readers material that is worthy of their scan, that makes them stop scanning and start reading" (ibid., 1991: 67).

Hansen (1994) studied 12 readers of the Copenhagen newspaper *Det Fri Aktuelt*. He investigated the order in which objects on folds were scanned and the readers' priorities with respect to text. He found that pictures are first seen, then icons and graphics followed by headlines of different sizes and text, with form items at the bottom of order hierarchy. Hansen's hierarchical order reflects that of Garcia and Stark's entry points.

When investigating readers' priorities, Hansen considered the length of articles, their placement and their genre (news, features, debate, sport, etc.). He measured how many centimetres of an article were read by a subject and calculated a depth index (the amount of text that has been actually read as a percentage of the whole text length) and a total response index (depth index as a percentage times the number of readers). His results show that only short articles are fully read. The longer the article, the smaller the proportion of it will be read. In this respect, Hansen's results resemble the result in Garcia and Stark (1991) showing that a mere 25% of all articles are seen, and only 12% are read deeper than half of their length. Concerning article placement, subjects were most engaged in reading on pages 2–6 and after that, their interest in reading decreased with the exception of the last page of the newspaper. Stories on the left-hand side of the fold were seen significantly earlier than stories on the right hand side. Articles with cut lines had significantly better total response values than neutral articles. News, features and debates had approximately the same total response whereas sport articles had significantly less. Hansen stresses the importance of designing the newspaper layout so that it quickly leads the reader to the information that the designer wants to emphasise.

Lewenstein *et al.* (2000) conducted an explorative eye-tracking study investigating naturally occurring reading behaviour of 67 users at different US news sites. Like its newspaper predecessor (see Garcia & Stark, 1991) study, the net paper follow-up has a broad scope and merely descriptive statistics.

The subjects (experienced users of online newspaper sites) were asked to read in the manner they usually read, for as long a session as was typical for them. Their eye movements were recorded by the SMI EyeLink eye tracker. After the reading session, subjects were interviewed about their media habits. The screens viewed were coded by format (headlines, articles, briefs, photos, etc.), by topics (science, medicine, national, international, etc.), by action (keyboard events) and by visit and time spent. Eye fixations were overlaid on screen dumps of the pages. Using this kind of data, they studied front page entries, fixation order on the pages, returns to front page and story reading order.

The study of Lewenstein *et al.* has been subjected to some criticism due to the lack of experimental control (e.g. Jacobson, 2000). From a scientific point of view, their findings do not qualify as being predictive in other areas, merely to provide a descriptive base for further research. However, the study is strong in ecological validity.

A major conclusion drawn by Lewenstein *et al.* (2000) is that text is the preferred entry point among online newsreaders. First fixations on the front page of a net paper go to text (78%) rather than to photos or graphics. The online newsreaders fixate on briefs and captions first and not on the dominant pictorial elements in the page. This result very much contrasts with the previous findings from Garcia & Stark and Hansen that pictures and images are the foremost entry points in newspapers (see Jacobson, 2000 for criticism of this result).

Scanning Versus Reading

Several researchers focused on the question whether newspapers and net papers are read or scanned. Results in Hansen (1994) show that only short articles in a newspaper are fully read. As pointed out by Garcia and Stark (1991), readers do not really read newspapers. They search until they find an interesting piece of news to read. Only 25% of all articles are seen, and only 12% are read deeper than half of their length. In contrast, net paper stories are read to a depth of 75% on average (see Lewenstein *et al.*, 2000). Net paper readers, as Lewenstein *et al.* conclude, go deeper into articles than newspaper readers.

Does the deeper reading found in net papers also mean that net paper readers read proportionally more than newspaper readers? Do newspaper readers have to scan more to find the stories they want to read because of the differences in how information is laid out in the two media? In the first experiment, we want to test the first hypothesis that readers of newspapers scan more than do readers of net papers.

Hypothesis 1: More reading and less scanning in net papers.

Experiment 1

We made two recordings of eye movement data from readers of two net papers and two newspapers.

Design of Net Paper Reading Recording

Twelve subjects read the net papers *Dagens Industri* (www.di.se) and *Aftonbladet* (www.aftonbladet.se), which are net versions of a financial and an evening paper. The readers were experienced with these net papers. Recordings were made during one day in November 2000, using the SMI iView remote set at 50 Hz. Subjects read from a regular 120 MHz Pentium computer with a 100 Mbit Ethernet connection and a 17 inch screen. Each subject read both net papers in a regular browser (Explorer). Eye movement data were recorded in the stimulus screen co-ordinate system (1024*768).

Subjects were instructed to read each net paper for as long as they liked. They were asked to read in the same way as they normally read these net papers. They had to stay within the pages belonging to the paper, but could read these pages in any order they liked. We planned to let our subjects switch to the other net paper after 15 minutes, but virtually all were bored and stopped reading after 5–8 minutes. In post questionnaires, subjects stated that the presence of the eye tracker did not influence their reading.

Design of Newspaper Reading Recording

Fifteen subjects read the European newspaper *Metro* (Stockholm edition) and 14 readers the Stockholm newspaper *Svenska Dagbladet*, which are both morning papers. Each subject read one newspaper. Recordings were made during two days in March 2001. *Metro* is distributed free. *Svenska Dagbladet* is a regular morning paper. Both newspapers are in the tabloid format. The newspapers were placed on a board for support, which allowed recording co-ordinate data using the SMI iView remote set. Subjects turned pages themselves. Data were recorded as 1024*768 coordinates in the reference system of the newspaper.

Subjects were instructed to read the newspaper for as long as they liked. They were asked to read the newspaper as they would normally read it. They could turn pages back and forth as they pleased. Each subject read for 15 minutes or more. After one hour, if they had not stopped, we stopped them. In post questionnaires, subjects stated that the presence of the eye tracker did not influence their reading.

Data Treatment

From the raw data of both the net and the newspaper recording, we calculated fixations using SMI's algorithm with a minimum fixation time of 100 ms. These fixation data were then run through a custom-made reading filter.

The reading filter identifies fixations that appear before, between or after at least two successive forward-going saccades. Also, fixations that appear before and after return sweeps are identified. Correction saccades following a return sweep are not recognised, nor backward reading saccades.

The sizes of reading saccades and return sweeps were calibrated from samples of data in each recording. The filter therefore only identifies forward going saccades of a size characteristic to reading, excluding the longer scanning saccades in the same direction. Also, the text and columns in newspapers are much smaller than in net papers, yielding shorter reading saccades in the newspaper data. We used the criterion that 95% of all the reading saccades that we saw in the scan path plots of four selected pages per data set should have the start and stop fixation recognised by our filter.

The fixations that are not part of reading patterns are assumed to be part of scanning processes.

Results

In newspaper reading, 55.1% of all fixations by all subjects were recognised by our filter as being part of a reading pattern. For the net paper reading, 44.4% of all fixations are part of a reading pattern, significantly less (independent, two-tailed *t*-test, $p < 0.001$).

For newspapers, reading proportion strongly correlates with the time spent on a fold (0.75, $p < 0.0001$).

Intermediate Discussion

Hypothesis 1, inspired by Lewenstein *et al.* (2000), that there is less scanning and more reading activity in net paper reading than in newspaper reading, must be rejected.

In order to explain the difference, we need to study how scanning versus reading is distributed across the different layouts in net papers and newspapers. In our newspaper data, reading proportion varies across the different folds.

Figure 30.2 shows the average proportion of reading through all the folds of the *Metro* newspaper during the average 25 minutes of reading. The first ten folds with more editorial material have a reading percentage of about 60%, and then reading proportion drops sharply for folds 11–19, which are dominated by ads.

How does reading proportions vary in net papers? Specifically, does it increase on article pages and decrease on front pages? Exactly which layout elements in net papers are scanned and which are read?

Hypothesis 2: More reading on article areas than on other areas.

Hypothesis 3: Scanning mainly takes place over link lists.

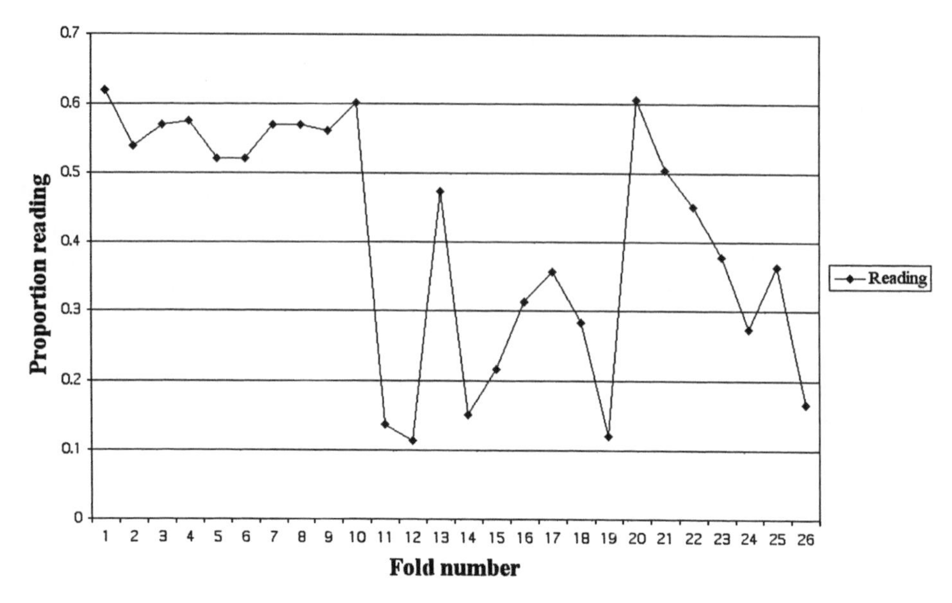

Figure 30.2: Average proportion of reading over folds in the *Metro* newspaper.

Experiment 2

In the second experiment, we wanted to confirm that reading percentages increase during article reading and decrease between articles. If this were the case, it would yield a zig-zag pattern in the plot of reading percentage over pages, with a low reading value on the pages (presumably front pages) being scanned for entry points and high reading values on the articles.

Also, we wanted to test whether link lists in net papers have taken over the role of entry points in newspapers. If this were the case, people who scan a lot would spend more fixation time in the link lists, while people who read more would spend more time in the reading area.

Data Recording

In November 2001, we recorded 12 readers of the two net papers *Dagens Nyheter* (www.dn.se, with a Stockholm newspaper owner) and *Sydsvenska Dagbladet Snällposten* (www.sydsvenskan.se, with a Malmö newspaper owner). Recordings were made during one day in November 2001, using the SMI iView remote set at 50 Hz. Subjects read from a regular 120 MHz Pentium computer with a 100 Mbit Ethernet connection and a 17 inch (43 cm) screen. Data was recorded as screen co-ordinates in a custom-made browser. The browser compensates eye-tracking data for scrolling, saves screen dumps of the pages visited, and allows all normal browser activities. Also, importantly, the browser signalises page changes (link clicks, back buttons etc.) to the eye-tracker data file. We also recorded data on video.

Each reader read both net papers. They were asked to read the net paper as they would normally read it. We imposed a reading time restriction of 5 minutes for each net paper.

Directly after the reading session, each subject was shown the video of his/her own reading session, showing the computer screen and the gaze marker (cf. Hansen, 1991). In this retrospective phase, subjects were asked to comment on what they read, how they found their way through the net paper structure etc. in a semi-structured interview.

Data Treatment

From the raw data, we calculated fixations using SMI's algorithm with a minimum fixation time of 100 ms. These fixation data were then run through the same custom-made reading filter as in Experiment 1.

Retrospective comments were recorded and thematically sorted. Full results from the retrospective phase can be read in Barthelsson (2002).

All pages of both net papers were divided into areas of interest, following the design of the page (see Figure 30.3 and Table 30.1). The proportion of fixation time spent in each area and transitions between them were calculated.

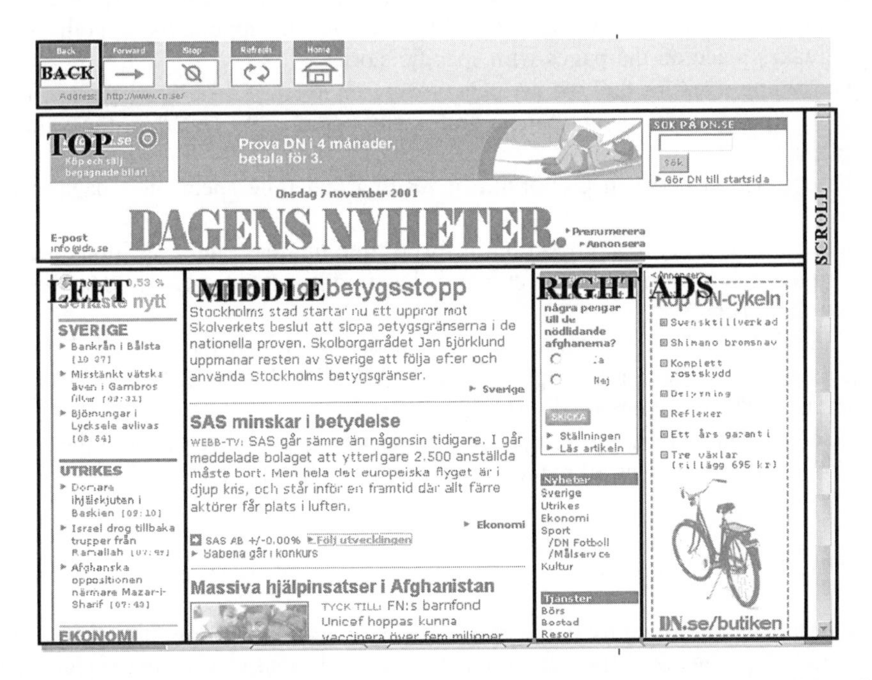

Figure 30.3: Areas on the DN front page coded according to their content and function.

Table 30.1: The different areas on the DN page coded according to their content and function.

Area name	Content and function
Back	Browser back button in the top navigation area
Top	Promotion box, name plate, advertisements and search box
Left	Headline links ordered by topics (national, international, economy, sports, culture) including the exact hour, linked to full articles
Middle	Lead editorials: Top stories with dominant visual elements such as headlines, photos and graphics, briefs and links to topic pages and to full articles
Right	Links to topic pages and different services available (travels, accommodation, stock market, etc.)
Ads	Advertisements
Scrollbar	Scrollbar

Results

A zig-zag pattern emerges for all readers (example in Figure 30.4),[4] showing an alternation in the amount of scanning and reading. This result supports the hypothesis that reading takes place on the pages with specific stories.

The reading rates for the two net papers were on average 44.8%, very close to the value in Experiment 1 for net papers and again significantly less than the value for newspapers.

The correlation between proportion of reading and time spent on a page is only 0.25.

Figure 30.5 exemplifies proportion of dwell time in areas, and transitions between them on the first page of the DN net paper. The middle reading area attracts the most fixation time, here 59%.

Of the different links on the front page, readers preferred the left link list. They looked significantly less at the right link lists (ANOVA, Tukey HSD, $p < 0.01$), even if both lists linked to thematically similar material.

The correlation between reading pattern frequency against dwell time in the reading area is 0.55, $p < 0.01$ over all readers and the entire reading session.

Intermediate Discussion

Hypothesis 2 is supported by the general zig-zag pattern for reading proportion found in the data. Reading takes place on article pages, and decreases when the reader returns to the front page.

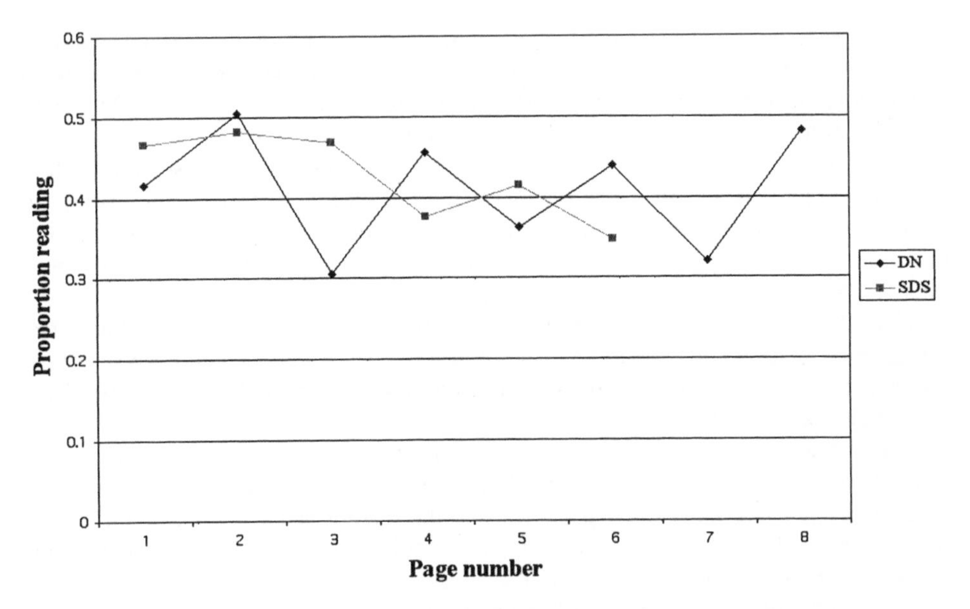

Figure 30.4: Proportion reading over pages in net papers.

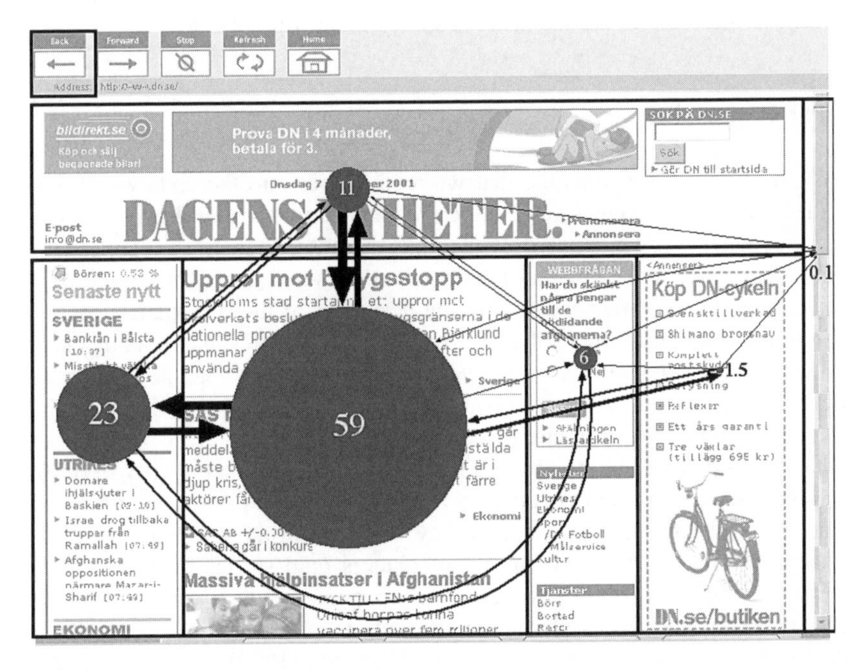

Figure 30.5: Proportional dwell time and transitions between areas on the DN front page.

The zig-zag pattern is particularly clear with the DN net paper. DN has a design that allows for alternating between scanning for links on the front page (odd page numbers) and reading the text that appears when clicking the link (even page numbers). The more complex structure of the SDS net paper does not give such a clear alternation.

The correlation in Experiment 1 between reading activity and time spent on a fold reflects the fact that scanning a newspaper is made in search of entry points. When no interesting entry points are found, the reader does not continue to scan the fold but turns the page. For the folds with the lowest reading rates (below 15%), this happens after 3–5 seconds.

The non-existent correlation with net papers shows that if you do not find an interesting entry point in a net paper, you cannot turn the page. You have to keep scanning. Net paper readers choose their own path through the paper, and the majority of net paper pages are never seen.

Hypothesis 3 is supported by the detected correlation between reading pattern frequency against dwell time in the reading area. The correlation indicates that scanning in net papers mainly takes place outside the reading area; that is, in and across the link lists. This result can be interpreted as indicating that link lists in net papers takes over the role of entry points in newspapers.

The lengths of our recordings differ greatly between net and newspaper data. But the scanning and reading proportions largely remain constant through the recordings (see Figures 30.2 and 30.4). The variation we can see in the data reflects the content on the folds. For instance, reading proportion drops after fold 10 in the newspaper data, because the section starting there displays many full size ads.

General Discussion

Lewenstein *et al.* (2000) conclude that we read deeper into the texts we find in a net paper than we do in the newspaper. Since readers can be very selective in what they read in a net paper — by clicking only the links that interest them — it appears to mean that net papers offer more efficient reading than do newspapers. In this view, turning page after page in the order decided by the designers of the newspaper forces you to read, or at least scan, many texts that you would not have chosen to read in a net paper.

Still, as we have shown here, net paper readers scan more than newspaper readers. If efficiency in a media means that you read more and spend less time searching for what to read, then newspapers are more efficient.

In the semi-structured interviews after the reading session, a majority of our readers claimed that they scan more in net papers than in newspapers. They claim to scan more in order to find the two or three stories they will read in the net paper.

Concerning article topics, the net paper readers read stories thematically close to their own specific profession or interests. All net paper readers were reading the news category that could be termed "scandals and catastrophes".

The newspaper readers in our study, like Hansen's (1994) and Garcia and Stark's (1991) readers, were much less selective. They read (parts) of text on all the different pages containing editorial material, including a wide variety of genres and topics.

Not only do net paper readers go deeper into articles than newspaper readers, as noted by Lewenstein *et al.* (2000). They also visit much fewer and more specialised texts. And in looking for these texts, they have to scan more than newspaper readers.

The linear structure of the newspaper invites linear browsing. News on each page therefore has at least the possibility of being seen. In net papers, most texts are never open for the reader to look at them. Instead of being opened in a browsing process, the net paper texts must compete for reader attention by means of links from the front page. Using a link to catch a reader to a story means feeding the reader with much poorer information on the story content than when the story presents itself spread out on the page of a newspaper.

The poorer chances of links catching reader interest may be the explanation why net paper readers have to scan more, why they read only certain types of texts and why they are bored so soon.

Experiment 2 showed a general zig-zag pattern in the reading proportion with higher scanning rates on the front page. Nine out of 12 subjects in Experiment 2 preferred to return to the front page after exploring the details of an article. In the retrospective phase we asked them about this behaviour, and our subjects stated that they wanted to return to the front page to ensure that they did not miss anything. When they probed deeper into an article they felt that they had left the main trail. In other words, only the front page is conceived of as a reliable provider of entry points.

The preference of the left link list detected in Experiment 2 could be a result of reader expectations of the design of net papers and web pages in general. But there are other studies of media perception that also show a preference for the left hand side of media layouts. For instance, Widman and Polansky (1990) show that of all small ads, those placed slightly to the left of the middle on a fold get the most fixations. Hansen (1994) showed that articles on the left page on a fold are seen significantly earlier than articles on the right page.

In a study of the use of Danish yellow pages, Hansen (1999) gave strong evidence for the dominance of the left hand side. Not only did Hansen's subjects look more at the ads slightly to the left, they also phoned these companies more frequently, and they never once looked at or phoned a company advertising in the rightmost column.

Garcia and Stark (1991: 33) claim to have found the opposite, however. According to them, entry points on the right hand side of a fold are seen first. Newspaper designers that we have talked to often say they think the right hand side to be more looked at in their products.

Also, in perceptual psychology (see Arnheim, 1974; Gombrich, 1977), and within the so-called sociosemiotic approach (see Kress & van Leeuwen, 1990, 1996), the right hand side is regarded as more informative: "any pictorial object looks heavier at the right side of the picture" (Arnheim, 1974: 34). Kress and van Leeuwen assume that visual layouts form a grammar where the dimensions left-right, up-down and centre-margins are associated with different information values. The left-hand side is associated with known information, whereas new information is on the right.

Today, the links as a substitute for entry points are taken from net papers into newspaper design. In November 2001, the SDS *newspaper* presented a new design with a list of links in the right column of the front page, motivating it by writing that they want to make it easier for readers to quickly find the story they are interested in.

Notes

1 Jupiter MMXI.
2 Intelligence.se.
3 Compare areas on the front page, Figure 30.3.
4 In netpaper reading, all subjects look at different pages in different orders. We therefore only show one typical subject reading the two netpapers SDS (6 pages) and DN (8 pages).

References

Arnheim, R. (1974). *Art and Visual Perception. A Psychology of the Creative Eye*. Berkeley, Los Angeles & London: University of California Press.

Barthelson, M. (2002). Reading behaviour in online news reading. Graduation project. Lund University: Department of Cognitive Science.

De Léon, D., & Holsanova, J. (1997). Revealing user behaviour on the world-wide web. *Lund University Cognitive Studies*, 60.

Garcia, M. R., & Stark, P. (1991). *Eyes on the News*. St. Petersburg, Florida: The Poynter Institute.

Gombrich, E. H. (1977). *Art and Illusion. A Study of the Psychology of Pictorial Representation* (5th edn). London: Phaidon Press.

Hansen, J. P. (1991). The use of eye mark recordings to support verbal retrospection in software testing. *Acta Psychologica*, *76*, 31–49.

Hansen, J. P. (1994). Analyse af læsernes informationsprioritering, Unpublished report. Kognitiv Systemgruppen, Forskningscenter Risø, Roskilde.

Hansen, J. P. (1999). Reading of Yellow Pages, presentation at the Lund-Risø eyetracking seminar, Lund.

Jacobson, A. (2000). *Projects Too Much From Too Little*, http://www.poynter.org/centerpiece/071200alan.htm.

Kress, G., & van Leuwen, T. J. (1990). *Reading Images*. Deakin University: Deakin University Press.

Kress, G., & van Leuwen, T. J. (1996). *Reading Image: The Grammar of Visual Design*. London & New York: Routledge.

Lewenstein, M., Edwards, G., Tatar, D., & DeVigal, A. (2000). *Poynter Eyetrack Study*. http://www.poynter.org/eyetrack2000.

Lundqvist, D., & Holmqvist, K. (forthcoming). Bigger is better: How size of newspaper advertisement and reader attitude relate to attention and memory. Manuscript under review.

Widman, L., & Polansky, S. H. (1990). Annonsläsning: En ögonrörelseundersökning av DN-läsare. Unpublished report. Stockholm: Dagens Nyheter.

Chapter 31

Reading Native and Foreign Language Television Subtitles in Children and Adults

Wim De Bruycker and Géry d'Ydewalle

In the first part of the chapter, studies on attention allocation to subtitled television programmes are reviewed. Subtitles are processed automatically, with no major age differences. With remote foreign languages, switching the languages in soundtrack and subtitle (reversed subtitling) leads to less time being spent in the subtitles. In the second part of the chapter a new study is presented, focusing on word-by-word reading behaviour in subtitles. Less word-by-word reading occurs with reversed than with standard subtitling. With standard subtitling, more reading occurs with two-line than with one-line subtitles: Longer sentences are less redundant with the pictoral information.

Introduction

Watching television is a common and important source of information for most people nowadays. Moreover, since communication and media become more and more international, in many countries a considerable amount of television programmes is imported from abroad. This implies that viewers have to deal with many different languages. The foreign languages are translated to the local language either by dubbing or by subtitling. Smaller language communities typically apply subtitling, because of its lower cost. Subtitling provides the viewer with three sources of information: the visual image, the foreign language soundtrack, and the native language subtitles. The three sources of information are partially redundant. The information in the subtitles should ideally be a completely overlapping translation of the information in the soundtrack. Moreover, the visual image and the sequence of events in the

The Mind's Eye: Cognitive and Applied Aspects of Eye Movement Research

ISBN: 0–444–51020–6

movie typically provide abundant information, which occasionally makes processing either the soundtrack or the subtitles superfluous. Finally, people unconsciously lip-read to a certain extent (Campbell, 1999).

Surprisingly, not much research has been done so far as to how information is processed in such a complex situation. Older theoretical work on attention, such as Broadbent (1958), Neisser (1967), Atkinson and Shiffrin (1968), and the unitary-resource theories (e.g., Kahneman, 1973), emphasized the serial nature of the human information processing system: At any given moment, only one among the sensory inputs is fully analysed, and switching to another sensory input takes time and effort. The emphasis on sequential processing has gradually been abandoned in favour of more parallel and flexible processing of multiple inputs. Typical examples are the multiple-resource theories (e.g., Gopher *et al.*, 1982; Navon & Gopher, 1979, 1980) where various sets of processing resources, each having its own capacity, are used simultaneously and in combination, allowing observers to flexibly divide and tune attention to multiple information sources.

Nevertheless, the research tradition on attention has mainly focused on simple stimulus presentations with no redundancy between the various input sources. As pointed out above however, in a subtitled television programme there is considerable overlap between the information channels. In the first part of this chapter we present some former studies, in which we looked at the dynamics of attention in the complex situation of television subtitling. How is an observer able to divide and shift attention between visual image, soundtrack, and subtitle? Are there costs involved? Do observers listen to the foreign language in the soundtrack? The second part of the chapter focuses on a new study, in which we investigate word-by word reading behaviour of the subtitles.

These research questions also serve some practical goals. If subtitles are effectively processed (and read), subtitling may be considered a teaching tool for children who are in the process of learning to read, or for people experiencing reading difficulties. Television subtitling embeds printed text in a familiar, meaningful, and stimulating context. It also provides opportunities to see the connections between the text on the one hand, and events, characters, and other background information on the other hand, thus aiding in the ability to comprehend the information presented (see also Linebarger, 2001). Similarly, if both the foreign language in the soundtrack and the native language in the subtitles are processed, this may result in viewers acquiring (components of) the foreign language in an informal and motivating way. In contrast to foreign language learning, which is a conscious process and occurs after explicit and formal teaching, foreign language acquisition is an unconscious, implicit process taking place through informal contact with the foreign language (Krashen, 1981). To allow for such language acquisition, at least the foreign language should be effectively processed. In a subtitled television programme the presence of the native language, and to a minor extent also the pictoral information, should facilitate acquisition of the foreign language, by allowing to infer connections between the foreign language soundtrack and the other information channels (i.e., the native language subtitles and the visual image).

Attention Allocation in Standard Subtitling

One possibility to study how attention is allocated to the various information sources in a subtitled television programme, is to use a double task procedure. While watching a movie, observers have to react as quickly as possible whenever a particular stimulus (light flash + tone pulse) appears. Stimuli are presented when soundtrack and/or subtitle are either available or not. Reaction times to the stimuli are taken as a measurement for the amount of processing resources allocated to the first task (i.e., watching television). The presence of subtitles only (without soundtrack), and soundtrack only (without subtitles) independently slow down the reaction times, relative to when neither soundtrack nor subtitles are available. The slowest reaction times are obtained whenever both a speaker and subtitles are present, which suggests that the observers process the subtitle and also make an effort to follow the soundtrack. Apparently, the information sources from the subtitles and soundtrack are being processed almost simultaneously (d'Ydewalle & Van de Poel, 2003; Sohl, 1989; Vanachter, De Bruycker & d'Ydewalle, 2003; for an overview, see d'Ydewalle, 2001).

An alternative approach to study how attention is divided between visual image and subtitle, is to register eye movements of people watching a subtitled television programme. Strong evidence is obtained that shifting attention to a subtitle at its presentation onset is almost automatic and effortless. D'Ydewalle *et al.* (1987) report an average latency time of 318 ms to switch to one-line subtitles and 350 ms to two-line subtitles. Moreover, time spent in reading the subtitles does not change as a function of the availability of the soundtrack and as a function of expert knowledge of the spoken language. However, participants in this study all had longstanding experience with television subtitling, and the automaticity of paying attention to subtitles could also be explained as an acquired habit.

D'Ydewalle and Van Rensbergen (1989) showed that automatically paying attention to subtitles is already apparent with children of Grade 4. These children take on average 227 ms to shift attention to the subtitle, and spend about half of the presentation time of the subtitles reading them. With children of Grade 2 the results are more ambiguous. When watching "Garfield", a cartoon in which verbal interactions are critically important to understand the story, Grade 2 children produce the same eye movement pattern as older children. However, when Grade 2 children watch "Popeye", a cartoon with numerous actions and in which the verbal message is less important, processing of the subtitles is less evident, and the average latency time is extremely long (827 ms). Moreover, when Grade 2 children are presented the two cartoons one after the other, a transfer of the eye movement pattern from the first to the second cartoon occurs. It appears that shifting attention to the subtitles is not yet as compulsory for young children as it is for older children and adults, who have established a habit of automatically switching attention to subtitles at their presentation onset.

Nevertheless, long-term experience with subtitled television programs does not suffice as an explanation for the automatic nature of reading subtitles. D'Ydewalle *et al.* (1991) carried out an experiment in which American participants, who are not familiar with television subtitling, watched an American movie with English subtitles. Participants spent considerable time in the subtitle area, showing that reading subtitles is independent of long-term familiarity with television subtitling, and is not solely due to habit formation. A more extensive overview of our former studies can be found in d'Ydewalle and Gielen (1992).

One of the most consistent results in our former eye movement studies on television subtitles, is that relatively more time is spent in looking at two-line than at one-line subtitles. Again, this pattern is found in both adults and children from Grade 4 on, but is absent with children of Grade 2. Praet *et al.* (1990) offered three possible explanations for this somewhat surprising result.

First, the presence of two lines of text may produce more lateral (vertical) interference than one line of text. A line close to the one being read interferes with parafoveal vision by reducing the size of the visual reading field (Bouma, 1980). Accordingly, less information can be extracted during fixations, resulting in more saccades and more time needed to process the subtitle. This explanation was dismissed as manipulating the interline distance in two-line sentences in a regular reading situation (i.e., without pictures and soundtrack) does not affect the number of fixations, the mean fixation duration, and the total fixation time in the sentence. According to the second explanation, more time is spent with two-line subtitles because of the greater syntactic and semantic complexity as compared to one-line subtitles. However, in a regular reading situation reading a two-line sentence seems to be easier. One-line sentences show relatively more and longer fixations (possibly reflecting more processing demands). This result led Praet *et al.* to abandon the second explanation.

Instead, the third explanation points at the informational value of the subtitle relative to the information from the visual image as an important determinant of the processing time. A one-line subtitle (often a short exclamation or outcry) typically does not add much information to what can already be extracted from the picture and the auditory message. Two-line subtitles on the other hand convey more content information that is less redundant with the available pictorial and auditory information. For example, a short one-line sentence like "Get out of here!" is relatively easy to infer, given the auditory and pictoral cues (angry intonation, angry face, pointing at the door, etc.). With longer sentences such as "Get out of this town, before I call the police!" participants can rely less on pictoral and auditory information, and are relatively more dependent on the subtitle to grasp the content of the sentence.

The explanation is evidenced by a comparison of reading behaviour in a subtitling condition (i.e., with pictures and sound) and a regular reading condition (i.e., without pictures and sound). In the subtitling condition, the percentage of time spent reading is larger with two lines than with one line, while in the reading condition the results are in the opposite direction: Participants proportionally spend more time in a one-line sentence than in a two-line sentence. Indirect evidence also comes from an unpublished experiment (Gielen, 1988) in which the type of information in the television program was manipulated. Participants either watched fragments of a news broadcast or excerpts of an action movie. With the news fragment, which consists of a substantial amount of (verbal) information over a short period of time, partially presented by a newsreader, participants have shorter latency times and spend proportionally more time in the subtitles than with the movie fragment, in which pictorial and auditory information are relatively more important.

Attention Allocation in Reversed Subtitling

Next to standard subtitling (soundtrack in foreign language, subtitles in native language), some studies have also taken interest in how attention is allocated to the various information sources in reversed subtitling. With reversed subtitling the languages in soundtrack and subtitles are switched: The foreign language is in the subtitles, and the native language is in the soundtrack. Interest in reversed subtitling stems from studies on foreign language acquisition through watching subtitlted television programmes. Several studies proved that, at least with adults, the most promising results for foreign language acquisition are found when using reversed subtitling instead of standard subtitling (Holobow *et al.*, 1984; Lambert *et al.*, 1981; Lambert & Holobow, 1984). This is rather surprising, since participants need at least process the foreign language in order to allow for any acquisition effects. With reversed subtitling the foreign language is in the subtitles. Why would people bother reading a language they do not know, instead of paying attention to the simultaneously presented pictorial information?

In order to investigate how native and foreign language subtitles are processed, Pavakanun (1992) registered eye movements of adults watching movies with either standard or reversed subtitling. The study was done with nine different foreign languages, in order to study the influence of similarity between the native language and the foreign language. Here we will focus only on the languages written in Western alphabet: South African (Afrikaans), German, Spanish, and Italian. Participants were native Dutch speakers and were not proficient in the foreign language they were presented. German and South African are very similar to Dutch. In fact, South African is merely a variant of the Dutch spoken in Belgium and the Netherlands. Spanish and Italian are more remote languages. Figure 31.1a averages the percentage of time spent in the subtitle for the close languages (South African and German), while Figure 31.1b gives the same averages for the remote languages (Spanish and Italian). For close languages, the results with both standard and reversed subtitling follow the same

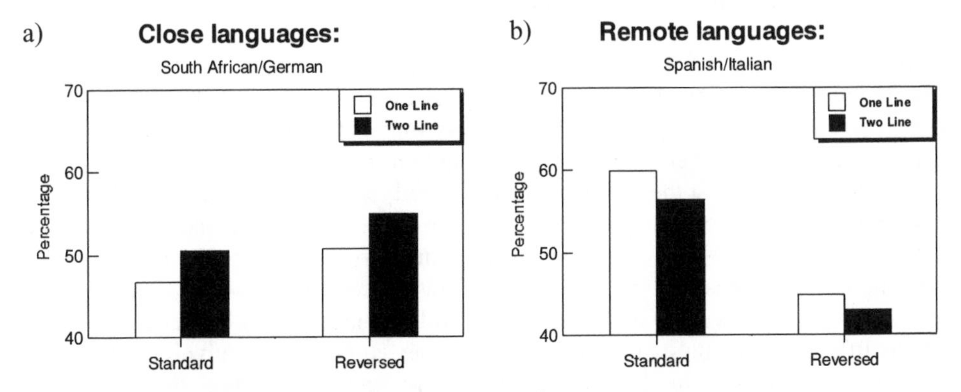

Figure 31.1: Percentage time spent in subtitles as a function of standard vs. reversed subtitling and the foreign language (close vs. remote).

pattern as in the previous studies: Relatively more processing time is allocated to two-line than to one-line subtitles; latency times are also longer with two-line than with one-line subtitles.

For the percentage of time spent in the subtitle, there is an interaction between subtitling mode and the foreign language. With languages most similar to the participants' native language, more time is spent in processing the foreign language subtitles (reversed subtitling) than in processing the native language subtitles (standard subtitling). With more remote languages, this is the opposite: More time is spent in the subtitles with standard than with reversed subtitling. Finally, latency times are always longer with reversed than with standard subtitling. Still, in all conditions and with all the foreign languages a substantial amount of time is spent in the subtitles: The smallest percentage is found with two-line reversed Spanish subtitles, and even then 40.29% of the presentation time of the subtitles is spent looking at them. Pavakanun concluded therefore that in both subtitling modes participants do read the subtitles.

As to whether the soundtrack of the movie is processed, no direct evidence is offered by this experiment. However, indirect support for the processing of the soundtrack comes from a set of language and comprehension tests that were administered after the participants had watched the subtitled television programme. In both the standard and the reversed subtitling conditions, and for each of the foreign languages, participants perform well above chance level on tests of foreign language word identification and sentence construction. This means that with standard subtitling, participants must have paid attention, at least to some extent, to the foreign language soundtrack. On the other hand, the comprehension test shows that in both subtitling conditions participants had captured the story line. Since none of the participants in this study was proficient in the foreign language they were presented, participants with reversed subtitling must have listened to the native language soundtrack in order to follow the story line.

Reading in Standard and Reversed Subtitling

From the studies reported in the previous sections, it seems safe to conclude that subtitles are indeed processed to a certain extent. However, our former studies only looked at the percentage of time spent in the subtitle as well as the latency time to shift attention to the subtitle. Yet, the fact that attention is paid to a subtitle does not automatically imply that there is true word-by-word reading in that subtitle. Moreover, if such reading occurs, are subtitles read in the same way in standard and reversed subtitling, especially since reversed subtitles are written in the foreign language? Looking at subtitles is a more or less automatic process. Pavakanun (1992) showed that this is also true with reversed subtitling. Still, the fact that with foreign languages rather different from the native language (remote languages) relatively less time is spent in reversed than in standard subtitles, may lead us to expect that systematic reading will be less apparent with reversed subtitling. Finally, the question still needs to be answered to what extent children show the same pattern of reading in subtitles as adults. D'Ydewalle and Van Rensbergen (1989) suggest that with standard subtitling no major differences are found between age groups. No previous research has investigated possible differences between children and adults with reversed subtitling.

In order to answer these questions, we conducted a new experiment in which we measured eye movements of both adults and 10- to 12-year-old children while they were watching either a standard or a reversed subtitled movie. This time, we extended the list of dependent variables used in the previous experiments, in order to investigate in detail word-by-word reading in subtitles.

University students (age 19–26) and Grade 5–6 children (age 10–12) volunteered to participate in the experiment. In both age groups there was an equal number of male and female participants, who were randomly assigned to one of the two subtitling conditions (standard or reversed). All participants were native Dutch speakers from monolingual families, and did not have any active knowledge of the foreign language they were presented. The movie fragment consisted of a 15-minute excerpt of a Swedish cartoon movie. In both versions (i.e., with standard and with reversed subtitling), presentation of the subtitles followed as closely as possible the 6-second rule. The rule states that a full-length two-line subtitle of 64 characters (including spaces and punctuation marks) is displayed for 6 s; shorter subtitles are timed proportionally. The 6-s rule is used by most television stations throughout the world, and has been proven to offer the optimal presentation time of subtitles (Warlop *et al.*, 1986). Participants were asked to watch the movie in the same way as they would do when watching television at home. While they were watching, their eye movements were registered with the EyeLink eye tracking system, which has a sampling rate of 250 Hz. When the movie fragment had ended, participants were asked to identify the foreign language. None of them could tell with certainty, but when insisting to guess, most adults correctly stated that it was probably one of the Scandinavian languages.

For each participant, we calculated the percentage of subtitles that was skipped (i.e., not fixated at all), the mean latency time (i.e., the time between the presentation onset of the subtitle and the first fixation in the subtitle), and the mean percentage of time spent in the subtitles (i.e., the total duration of all fixations and saccades in the subtitle, divided by the total presentation time of the subtitle). Since this latter variable is a function of the frequency and the length of fixations in the subtitles, we also measured the number of fixations per subtitle, and the mean fixation duration.

Other dependent variables included the mean amplitude of the saccades in visual degrees (1° covered approximately 1.5 characters or spaces), the mean percentage of regressive eye movements in the subtitle (relative to the total number of saccades in the subtitle, and excluding return sweeps from one line to another), and the number of shifts from the visual image to the subtitle (i.e., the number of saccades from a fixation in the visual image to a fixation in the subtitle area, not including the first image-to-subtitle shift following the presentation onset of the subtitle).

The average length of the subtitles included for analysis was not the same for each participant. For some participants, an off-line screening of the data showed too large inaccuracies in eye movement registration for some parts of the movie; subtitles in these movie parts were excluded from analysis. Accordingly, analyses were done on a different set of subtitles for each participant. For this reason we divided the number of fixations by the number of words in the subtitles. The value of this variable is obviously influenced by the number of words that were either skipped or (re)fixated.

Furthermore, the first 3 min of the movie fragment were considered an adaptation period for the participants, and were also excluded from analysis. After the adaptation period, 114 subtitles remained in the standard subtitling condition and 138 in the reversed condition. Finally, for the analyses of latency time, fixation duration, and number of image-to-subtitle shifts, only subtitles with at least one fixation were included (88% of the subtitles). Similarly, for the analyses of saccade amplitude and percentage of regressive eye movements, only subtitles with at least two fixations were taken into account (80% of the remaining subtitles).

In the following sections we will present the results of the experiment. Because we wanted to give an extensive inventory of reading behaviour in subtitles, all analyses included the subtitling mode (standard vs. reversed) and the age group of the participants (children vs. adults) as between-subjects independent variables, and the number of text lines in the subtitle (one line vs. two lines) as a within-subjects independent variable. For the analysis of saccade amplitude, the direction of the saccade (forward vs. backward) was added as an additional within-subjects independent variable. All reported differences are significant at α-level 0.05, except where stated otherwise.

Differences between Standard and Reversed Subtitling

In line with the results for remote languages in Pavakanun (1992), more evidence for word-by-word reading is obtained with standard than with reversed subtitling. With standard subtitling, only 4% of the subtitles are skipped (i.e., not fixated), while this is the case with 21% of the reversed subtitles. Also, latency times are shorter with standard than with reversed subtitling (362 vs. 513 ms). Participants with standard subtitling spend 41% of the presentation time in the subtitle area; with reversed subtitling this is only 26%. The probability of fixating a word is also higher with standard than with reversed subtitling (0.91 vs. 0.59). Finally, participants with standard subtitling show more image-to-subtitle shifts than participants with reversed subtitling (0.49 vs. 0.24). No significant differences between the two subtitling modes are found for the duration of the fixations, the amplitude of the saccades, and the percentage of regressive eye movements.

In short, participants with standard subtitling shift attention to the subtitles faster, stay longer in the subtitle area, and return to it more frequently after having focused on it a first time. Of course, this was to be expected, since the participants need the native language subtitles to follow the story line. With reversed subtitling, the native language is in the soundtrack, so participants need not shift to the subtitle immediately. Still, even with reversed subtitling most subtitles are processed for a substantial amount of time, and about 60% of the words are fixated. This means we cannot completely exclude reading behaviour when the subtitle is in the foreign language.

With standard subtitling, several further findings confirm the results of previous studies (see e.g., Praet *et al.*, 1990) that more regular word-by-word reading occurs in two-line than in one-line subtitles. There is slightly (but significantly) less skipping of two-line than of one-line standard subtitles (98% vs. 94%). Two-line standard subtitles show relatively more fixations than one-line standard subtitles (0.97 vs. 0.85 fixations per word), resulting in more time being spent in two-line than in one-line standard subtitles (45% vs. 37%). Finally, there are also less regressive eye movements and more image-to-subtitle shifts in two-line than in one-line standard subtitles (32% vs. 44%, and 0.73 vs. 0.25 respectively). Again, no differences are found between one-line and two-line standard subtitles for fixation duration and saccade amplitude, while there are slightly (albeit not significant) longer latency times for two-line than for one-line subtitles (391 vs. 333 ms).

With reversed subtitling, there is an opposite pattern of results. In two-line reversed subtitles there is a much longer mean latency time than in one-line reversed subtitles (610 vs. 416 ms). Also, slightly less time is spent in two-line than in one-line reversed subtitles (24% vs. 28%), due to the smaller number of fixations in two-line than in one-line subtitles (0.48 vs. 0.69 fixations per word). This means that less than half of the words in two-line reversed subtitles is fixated. Moreover, the long mean latency time also suggests that in these subtitles the participants only occasionally grasp some keywords, without really reading the sentences as a whole. However, this conclusion needs to be qualified by a few exceptions. Although there are fewer fixations in two-line than in one-line reversed subtitles, they are somewhat longer (242 vs. 212 ms). Also more one-line than two-line reversed subtitles are skipped (24% vs. 18%), and there are more regressive eye movements and less image-to-subtitle shifts with one line than with two lines of text (47% vs. 35%, and 0.14 vs. 0.35 respectively). Finally, saccade amplitude does not differ as a function of the number of lines in reversed subtitles.

Typically, there are no significant differences between one-line subtitles with standard subtitling and one-line subtitles with reversed subtitling. Accordingly, we conclude that there is more word-by-word reading in two-line than in one-line standard subtitles, and less word-by-word reading in two-line than in one-line reversed subtitles, while there are no differences between standard and reversed one-line subtitles. This means that attention is tuned to the information available. With standard subtitling, word-by-word reading is observed, especially in subtitles with two-lines of text. Two-line standard subtitles typically convey verbal information that cannot easily be inferred from the pictorial information on the screen. One-line subtitles are more redundant to the information in the picture, and therefore systematic reading is less apparent (Praet *et al.*, 1990). Reversed subtitling shows even less reading, as the subtitles are written in the foreign language. Still, participants may incidentally focus on one or more keywords in the subtitle.

Information redundancy may also explain the very short fixation durations in the subtitles. With standard subtitling the mean fixation duration is 178 ms for adults and 248 ms for children. Similarly, d'Ydewalle *et al.* (1985) estimated an average fixation time of 124 ms per word in the subtitle by adults. They did not measure the fixation time directly, but inferred it from regressing the number of words to the total viewing time in the subtitle. In a regular reading situation, individual words are fixated by adults for about 200 to 250 ms (Rayner, 1998; Rayner & Pollatsek, 1987) and by 10 to 12 year old children for 270 to 300 ms (Spragins *et al.*, 1976; Taylor, 1965). In a subtitling situation, additional information from the other input sources (picture and sound) may facilitate the processing of the subtitle. Indeed, more time is spent in sentences when they simply have to be read (without sound and pictorial background) than in a subtitling situation (Praet *et al.*, 1990).

Another possible explanation for the short fixation durations is that in the present experiment, fixation duration is calculated as a function of the total fixation time in the subtitle, divided by the number of fixations in the subtitle. This averaged measure also includes the refixations on some words. The number of refixations, which typically have shorter durations than first fixations, must have been high, as the percentage regressive eye movements was around 40%.

Differences between Children and Adults

The present experiment largely confirms the results of d'Ydewalle and Van Rensbergen (1989): Grade 5–6 children do not show a radically different reading pattern in subtitles than adults. Overall, the children in the present study do not skip the subtitles more frequently, do not make more fixations, do not show more regressive eye movements, and do not make more image-to-subtitle shifts than the adults. Exceptions are the longer fixation duration of the children (resulting in more time spent in the subtitle), a larger difference in fixation duration between one-line and two-line subtitles, a smaller saccade amplitude, and longer latency to shift attention to the subtitle. However, none of these results are specific for the subtitling situation. Older studies already convincingly demonstrated that children make longer fixations while reading than adults (see, e.g., Lefton *et al.*, 1979). Furthermore, studies on general saccade characteristics made clear that the latency of a saccade to a visual target is longer for children than for adults (e.g., Miller, 1969). Finally, it is also typical for children to make shorter saccades when reading text than adults (Rayner, 1998).

General Conclusions

This chapter shows that in standard subtitling, subtitles are fluently processed, and that true reading is apparent. This is certainly the case with two-line subtitles, since they provide a considerable amount of content information that cannot be inferred directly from the pictorial information. With reversed subtitling, systematic reading is less clear. While there are no major differences between one-line reversed and one-line standard subtitles, the findings with two-line reversed subtitles suggest that only occasionally some words in the subtitle are processed.

The eye movement studies were not designed to infer to what extent the soundtrack of the movie was processed. However, Pavakanun (1992) provides some indirect evidence that when watching subtitled television, people do indeed attend to the soundtrack, in both standard and reversed subtitling. Moreover, a recent series of experiments using a double task methodology (d'Ydewalle & Van de Poel, 2003; Sohl, 1989; Vanachter *et al.*, 2003; for an overview, see d'Ydewalle, 2001) demonstrate that with standard subtitling, children (Grade 4 and 6) and adults have sufficient resources available to process the native language subtitles and to simultaneously attend to the foreign language soundtrack. Younger children discarded the foreign language soundtrack when native language subtitles were available. With reversed subtitling a similar pattern emerged. Adults are able to process both the native language soundtrack and the foreign language subtitles. Children on the other hand do not have enough information processing capacities, and therefore concentrate almost exclusively on the information source that is easiest to process, the native language soundtrack, while paying less attention to the foreign language subtitles.

Since older children and adults are able to pay attention to both subtitles and soundtrack, the native language as well as the foreign language is processed. This offers the possibility for incidental foreign language acquisition to occur. And indeed, several studies (d'Ydewalle & Pavakanun, 1995, 1996, 1997; d'Ydewalle & Van de Poel, 1999, Lambert *et al.*, 1981; Pavakanun & d'Ydewalle, 1992) obtained substantial language acquisition effects on vocabulary tests, with both children and adults. However, performance in tests on syntax and grammar acquisition remained relatively poor. Of course, mastering vocabulary is only a first step in language acquisition. Acquiring syntax and grammar may require repeated exposure to the foreign language, or may involve some formal learning. Adults seem to benefit more from reversed subtitling, while children show more foreign language acquisition with standard subtitling.

According to Lambert *et al.* (1981), the superiority of reversed subtitling among adults stems from the processing precedence of the native language. With reversed subtitling, the dominant native language is processed relatively effortlessly in the transient auditory channel, leaving the observers time to match this message with the foreign language subtitle, thus giving them the opportunity to grasp the foreign language. With standard subtitling, processing of the foreign language soundtrack is offset by the more automatic tendency to process the native language subtitles.

D'Ydewalle and Pavakanun (1995) offered a different explanation. As reading subtitles is automatic and obligatory, observers with reversed subtitling automatically read the foreign language subtitles, while at the same time listening to the native language soundtrack to capture the story line. With standard subtitling on the contrary, the native language subtitles suffice to understand the movie, and therefore less attention is paid to the foreign auditory message. The longer latency times and the lack of systematic word-by-word reading in reversed subtitles (particularly two-line subtitles) favour more the native language processing precedence hypothesis of Lambert *et al.* than the subtitle processing precedence.

Why children do not benefit more from reversed than from standard subtitling remains unclear. According to d'Ydewalle and Van de Poel (1999), children are not provided with sufficient reading capacities to allow reading foreign language subtitles. However, the explanation is ruled out by the results presented in this chapter, as the children do not show a different reading pattern in subtitles than adults.

Acknowledgements

This work was made possible by a grant from the Flemish Government of Belgium (CAW 96/06) and by a I.A.P.-grant from the Federal Government of Belgium, Convention N° P4/19.

References

Atkinson, R. C., & Shiffrin, R. M. (1968). Human memory: A proposed system and its control processes. In: K. W. Spence and J. T. Spence (eds), *The Psychology of Learning and Motivation: Advances in Research and Theory* (Vol. 2, pp. 89–195). New York: Academic Press.

Bouma, H. (1980). Visual reading processes and the quality of text displays. *IPO Annual Progress Report, 15*, 83–90.

Broadbent, D. E. (1958). *Perception and Communication*. London: Pergamon.

Campbell, R. (1999). Language from faces: Uses of the face in speech and in sign. In: L. S. Messing and R. Campbell (eds), *Gesture, Speech, and Sign* (pp. 57–73). New York: Oxford University Press.

d'Ydewalle, G. (2001). The mind at the crossroad of multiple ongoing activities: A challenge to cognitive psychology. In: L. Bäckman and C. von Hofsten (eds), *Psychology at the Turn of the Millennium, Vol. 1: Cognitive, Biological, and Health Perspectives* (pp. 153–178). Hove, UK: Psychology Press.

d'Ydewalle, G., & Gielen, I. (1992). Attention allocation with overlapping sound, image, and text. In: K. Rayner (ed.), *Eye Movements and Visual Cognition: Scene Perception and Reading* (pp. 415–27). New York: Springer-Verlag.

d'Ydewalle, G., Muylle, P., & Van Rensbergen, J. (1985). Attention shifts in partially redundant information situations. In: R. Groner, G. W. McConkie and C. Menz (eds), *Eye Movements and Human Information Processing* (pp. 375–84). Amsterdam: North-Holland.

d'Ydewalle, G., & Pavakanun, U. (1995). Acquisition of a second/foreign language by viewing a television program. In: P. Winterhoff-Spurk (ed.), *Psychology of Media in Europe: The State of the Art — Perspectives for the Future* (pp. 51–64). Opladen: Westdeutscher Verlag.

d'Ydewalle, G., & Pavakanun, U. (1996). Le sous-titrage à la télévision facilite-t-il l'apprentissage des langues? (Does television subtitling facilitate language acquisition?). In: Y. Gambier (ed.), *Les Transferts Linguistiques dans les Médias Audiovisuels* (pp. 217–23). Lille: Presses Universitaires du Septentrion.

d'Ydewalle, G., & Pavakanun, U. (1997). Could enjoying a movie lead to language acquisition? In: P. Winterhoff and T. Van der Voort (eds), *New Horizons in Media Psychology* (pp. 145–55). Opladen: Westdeutscher Verlag.

d'Ydewalle, G., Praet, C., Verfaillie, K., & Van Rensbergen, J. (1991). Watching subtitled television: Automatic reading behaviour. *Communication Research, 18*, 650–666.

d'Ydewalle, G., & Van de Poel, M. (1999). Incidental foreign-language acquisition by children watching subtitled television programs. *Journal of Psycholinguistic Research, 28*, 227–244.

d'Ydewalle, G., & Van de Poel, M. (2003). Do children listen to the spoken foreign language while watching subtitled television programs? Manuscript in preparation.

d'Ydewalle, G., & Van Rensbergen, J. (1989). Developmental studies of text-picture interactions in the perception of animated cartoons with text. In: H. Mandl and J. R. Levin (eds), *Knowledge Acquisition from Text and Pictures* (pp. 223–248). Amsterdam: Elsevier.

d'Ydewalle, G., Van Rensbergen, J., & Pollet, J. (1987). Reading a message when the same message is available auditorily in another language: The case of subtitling. In: J. K. O'Regan and A. Lévy-Schoen (eds), *Eye Movements: From Physiology to Cognition* (pp. 313–321). Amsterdam: Elsevier.

Gielen, I. (1988). Het verwerken van ondertitels in functie van de taalkennis van de klankband en typen van informatie in het beeld (Processing of subtitles as a function of knowledge of the foreign language in the soundtrack and type of information in the visual image). Unpublished Licentiate Thesis, University of Leuven, Belgium.

Gopher, D., Brickner, M., & Navon, D. (1982). Different difficulty manipulations interact differently with task emphasis: Evidence for multiple resources. *Journal of Experimental Psychology: Human Perception and Performance*, *8*, 146–157.

Holobow, N. E., Lambert, W. E., & Sayegh, L. (1984). Pairing script and dialogue: Combinations that show promise for second or foreign language learning. *Language Learning*, *34*, 59–76.

Kahneman, D. (1973). *Attention and Effort*. Englewood Cliffs, NJ: Prentice Hall.

Krashen, S. (1981) *Second Language Acquisition and Second Language Learning*. Oxford, UK: Pergamon.

Lambert, W. E., Boehler, I., & Sidoti, N. (1981). Choosing the languages of subtitles and spoken dialogues for media presentations: Implications for second language acquisition. *Applied Psycholinguistics*, *2*, 133–148.

Lambert, W. E., & Holobow, N. E. (1984). Combinations of printed script and spoken dialogue that show promise for students of a foreign language. *Canadian Journal of Behavioural Science*, *16*, 1–11.

Lefton, L. A., Nagle, R. J., Johnson, G., & Fisher, D. F. (1979). Eye movement dynamics of good and poor readers: Then and now. *Journal of Reading Behaviour*, *11*, 319–328.

Linebarger, D. L. (2001). Learning to read from television: The effects of using captions and narration. *Journal of Educational Psychology*, *93*, 288–298.

Miller, L. R. (1969). Eye movement latency as a function of age, stimulus uncertainty and position in the visual field. *Perceptual and Motor Skills*, *28*, 631–636.

Navon, D., & Gopher, D. (1979). On the economy of the human-processing system. *Psychological Review*, *86*, 214–255.

Navon, D., & Gopher, D. (1980). Task difficulty, resources, and dual task performance. In: R. S. Nickerson (ed.), *Attention and Performance* (Vol. 8, pp. 297–315). Hillsdale, NJ: Erlbaum.

Neisser, U. (1967). *Cognitive Psychology*. Englewood Cliffs, NJ: Prentice-Hall.

Pavakanun, U. (1992). Incidental acquisition of foreign language through subtitled television programmes as a function of similarity with native language and as a function of presentation mode. Unpublished Doctoral Dissertation, University of Leuven, Belgium.

Pavakanun, G., & d'Ydewalle, G. (1992). Watching foreign television programs and language learning. In: F. L. Engel, D. G. Bouwhuis and G. d'Ydewalle (eds), *Cognitive Modelling and Interactive Environments in Language Learning* (pp. 193–198). Berlin: Springer-Verlag.

Praet, C., Verfaillie, K., De Graef, P., Van Rensbergen, J., & d'Ydewalle, G. (1990). A one line text is not half a two line text. In: R. Groner, G. d'Ydewalle and R. Parham (eds), *From Eye to Mind: Information Acquisition in Perception, Search and Reading* (pp. 205–213). Amsterdam: North-Holland.

Rayner, K. (1998). Eye movements in reading and information processing: 20 years of research. *Psychological Bulletin*, *124*, 372–422

Rayner, K., & Pollatsek, A. (1987). Eye movements in reading: A tutorial review. In: M. Coltheart (ed.), *Attention and Performance: Vol. 12. The Psychology of Reading* (pp. 327–362). Hove, UK: Erlbaum.

Sohl, G. (1989). Het verwerken van de vreemdtalige gesproken tekst in een ondertiteld TV-programma (Processing the foreign language soundtrack in a subtitled television program). Unpublished Licentiate Thesis, University of Leuven, Belgium.

Spragins, A. B., Lefton, L. A., & Fisher, D. F. (1976). Eye movements while reading and searching spatially transformed text: A developmental examination. *Memory and Cognition*, *4*, 36–42.

Taylor, S. E. (1965). Eye movements in reading: Facts and fallacies. *American Educational Research Journal*, *2*, 187–202.

Vanachter, I., De Bruycker, W., & d'Ydewalle, G. (2003). Attention allocation while watching subtitled television programs. Manuscript in preparation.

Warlop, L., Van Rensbergen, J., & d'Ydewalle, G. (1986). *Ondertiteling op de B.R.T.* (Subtitling at the Belgian Radio and Television). (Psychological Reports No. 55). Leuven, Belgium: University of Leuven, Laboratory of Experimental Psychology.

Chapter 32

Eye Movements and Gestures in Human Face-to-face Interaction

Marianne Gullberg

Gestures are visuospatial events, meaning carriers, and social interactional phenomena. As such they constitute a particularly favourable area for investigating visual attention in a complex everyday situation under conditions of competitive processing. This chapter discusses visual attention to spontaneous gestures in human face-to-face interaction as explored with eye-tracking. Some basic fixation patterns are described, live and video-based settings are compared, and preliminary results on the relationship between fixations and information processing are outlined.

Introduction

This paper is concerned with visual attention to gestures in face-to-face interaction. It addresses a few seemingly simple questions: do addressees look at speakers' gestures in interaction? If they do, at which ones do they look, and why? Despite the widespread interest in the visual perception of hands in neurology (e.g. Decety & Grèzes, 1999; Goldenberg, 2001; Grèzes *et al.*, 1999; Hermsdörfer *et al.*, 2001; Neville *et al.*, 1997; Peigneux *et al.*, 2000; Perani *et al.*, 2001; Rizzolatti, *et al.*, 2001), in studies of Sign Language perception (e.g. Bavelier *et al.*, 2001; Corina *et al.*, 1996; Neville *et al.*, 1997; Rettenbach *et al.*, 1999; Swisher, 1993), and in gesture recognition in man-machine interfaces (e.g. Braffort, *et al.*, 1999; Wachsmuth & Sowa, 2002), we know surprisingly little about the attention afforded to gestures in human interaction. In gesture research the interest in these questions is motivated by an ongoing debate regarding whether or not gestural information is communicatively relevant to participants in interaction. It will be suggested in this chapter that these questions are also of relevance for eye movement research. In particular, the answers to these question may inform research concerning task-specific behaviour and visual attention in complex, natural situations.

The Mind's Eye: Cognitive and Applied Aspects of Eye Movement Research

ISBN: 0–444–51020–6

The gestures we are concerned with here are the (mainly manual) movements speakers perform unwittingly while they speak as part of the expressive effort (Kendon, 1993; McNeill, 1992). This definition excludes functional actions, manual object manipulations and movements such as scratching or playing with hair, since these movements are not part of the speaker's intended message. The narrow definition still leaves a broad spectrum of movements to consider. It covers conventionalised gestures like the OK-sign. It also includes a wide range of spontaneous, non-conventional movements that imitate real actions, iconically represent, or indicate entities talked about (Kendon, 1986; McNeill, 1998). Gestures thus defined are symbolic movements, and they are closely temporally and semantically related to language and speech. Simply put, gestures and speech tend to express the same meaning at the same time. However, even if gestural information often is redundant with regard to speech, gestures can also express additional information not present in concomitant speech.

It is partly this latter property that motivates an ongoing debate in gesture research regarding their communicative relevance. Because gestures can encode additional information to that expressed in speech, the need to attend to this information may not be essential to addressees in spoken face-to-face interaction. Whilst a number of studies have been concerned with gaze towards gestures (Goodwin, 1986; Kendon, 1990; Streeck, 1993, 1994; Streeck & Knapp, 1992; Tuite, 1993), there is a conspicuous lack of precise perceptual data concerning attention to gestures. Furthermore, there has been no systematic investigation of the relationship between visual attention to gestures and the processing and integration of the gestural information. This empirical gap has largely motivated the studies outlined in this chapter.

From the point of view of vision research, gestures in interaction also offer a challenging opportunity to study what catches attention in complex natural settings, and the influence of specific tasks and activities on fixation patterns. Gestures constitute a potential locus for visual attention by virtue of being visuospatial phenomena that represent movement in the visual field. Gestures could also be the locus of visual attention because they are symbolic movements that encode meaning closely related to but not necessarily identical to that expressed in language and speech. Gestures could thus be visually attended to for low-level perceptual reasons or for reasons related to higher cognitive processes such as information extraction. Finally, gestures are (mostly) interactional, social phenomena. As such, their occurrence in situations that are governed by socially and culturally determined norms for behaviour is likely to modulate visual behaviour towards them. There is thus potential tension between different mechanisms governing visual attention: the tendency to attend to movement, the need to look at what you are seeking information about, and the social conventions that govern gaze.

This paper will outline the results from a number of recent and forthcoming studies that exploit eye tracking to investigate the attention addressees allocate to gestures in interaction. Since some of these studies are presently unpublished, and the methodology is novel, I will describe the general procedure in some detail and present some key results. This chapter will cover three main foci: (1) the basic fixation patterns in face-to-face interaction; (2) the effect of the medium of presentation, comparing live vs. video settings; and (3) the relationship between fixation and information uptake in a complex setting. In a final section, I will discuss the findings and relate them to some issues raised in eye movement research, particularly with respect to the generalisability of findings from lab settings to naturalistic contexts. Specifically, I will discuss the effect of the task and context on gaze patterns.

Eye Movements and Gestures in Live Interaction

Social, dyadic face-to-face interaction is one of the most primary and common types of human activity. Gaze is of fundamental importance to such interaction and has previously received considerable attention in various disciplines. Numerous studies in social psychology, e.g., have investigated the influence of personality, gender, culture, interactional style and setting, mental health, etc., on gaze patterns in interaction (for comprehensive overviews, see Argyle & Cook, 1976; Fehr & Exline, 1987; Kendon, 1990). However, none of these studies have been based on precise measurements of eye movements. We therefore know very little about the exact patterns that arise from this particular activity, and virtually nothing about how behaviour in this setting relates to findings for other natural activities such as driving, or tea- or sandwich-making (Hayhoe, 2000; Land & Hayhoe, 2001; Land *et al.*, 1999; Shinoda *et al.*, 2001).

Despite the lack of precise measurements, a number of claims have been made in the literature regarding participants' gaze patterns in interaction. For instance, a frequently reported observation is that the speaker's face dominates as a target for addressees (Argyle & Cook, 1976; Fehr & Exline, 1987; Kendon, 1990). It is generally assumed that this is a reflection of a (culture-specific) politeness norm for maintained mutual gaze signalling sustained interest and attention.

With respect to gestures, three candidates for direct fixation can be derived from the literature. First, gestures performed in peripheral gesture space are often presumed to attract overt visual attention. Gesture space can be divided into central and peripheral gesture space (cf. McNeill, 1992). Central space refers to a shallow disc of space in front of the speaker's body, delimited by the elbows, the shoulders, and the lower abdomen. This area is outlined by a rectangle in Figure 32.1. Peripheral gesture space is everything outside this area. The majority of a speaker's gestures are performed in central gesture space. If an addressee is fixating the speaker's face, then all gestures in principle occur in the addressee's peripheral visual space. Nonetheless, a gesture performed in the speaker's peripheral gesture space occurs in the addressee's extreme peripheral visual field. It is therefore assumed that it could attract fixation, as it would otherwise be too challenging for peripheral vision. Second, pointing gestures are suggested as potential targets for fixation. Pointing gestures direct attention to the target they are indicating. However, it is often presumed that it is necessary to look at the gesture first in order to compute the trajectory towards the target of the pointing gesture. Finally, gestures that speakers themselves look at have also been suggested as candidates for fixation. When speakers look at their own gestures, they are assumed to intentionally direct the addressee's attention to the target of their gaze (Goodwin, 1986; Streeck, 1993, 1994; Streeck & Knapp, 1992; Tuite, 1993). Speakers' gaze shifts are thus claimed to have the same deictic function as pointing.

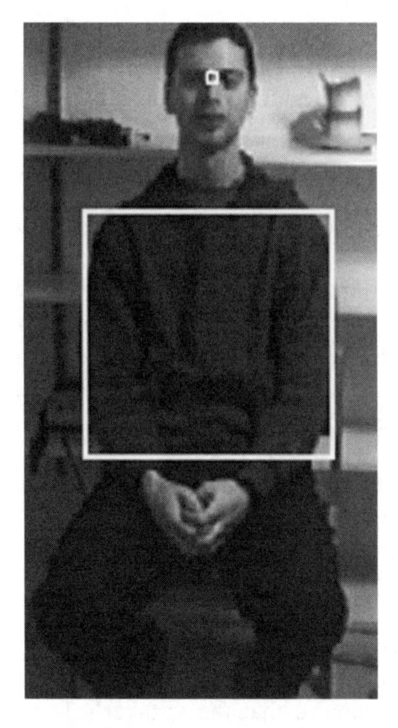

Figure 32.1: The speaker's central gesture space as a rectangle. Everything outside the rectangle represents the speaker's peripheral gesture space. The addressee's fixation as a small white circle at its default location in interaction.

By exploring the possibilities of eye-tracking in face-to-face interaction, we have attempted to investigate these proposals and also to charter the basic fixation patterns in face-to-face interaction (Gullberg & Holmqvist, 1999, 2002). Fifteen pairs of Swedish participants, unacquainted prior to the experiment, were randomly assigned the role as "speaker" or "addressee" (i.e. the wearer of the eye-tracker). In order to allow for spontaneous gesturing while maintaining control over the gestural content, a story-retelling task was used. Speakers memorised a printed cartoon and were instructed to convey the story as well as possible to the addressees who would have to answer questions about it. Addressees were instructed to make sure they understood the story and were encouraged to ask questions and engage in the interaction. Addressees were fitted with a head-mounted (HED) SMI iView© eye-tracker, a monocular 50 Hz pupil and corneal reflex video imaging system. This eye-tracker is well suited to interactional studies as both participants have an unobstructed face view of each other. The device samples data to an average spatial accuracy of 1 degree. Each addressee was calibrated using a nine-point matrix on the wall. The location and size of the matrix was equivalent to the area that the speaker would later occupy. After calibration of the addressees, speakers were introduced into the room and seated 180 cm away from the addres-sees (measured back to back) facing them. The speakers' stories generated natural narratives and a range of spontaneous gestures, all of which were analysed and considered as potential targets for addressees' fixations.

During the story-retelling interaction, the addressees' eye movements were recorded with the corneal reflex camera. The eye-tracker also has a scene-camera on the head-band. The iView software creates a circle overlay indicating the gaze position that is then merged with the video of the scene image. Since the scene-camera moves with the head, the eye-in-head signal indicates the gaze point with respect to the world. Head movements therefore appear on the video as full-field image motion. Given that both the target stimulus (the speaker) and the field of vision itself moved, the merged video data of the subject's gaze position on the scene image were analysed frame-by-frame.

Fixations were defined as instances where the gaze marker remained for at least 120 ms directly on a fixated object.

Post-test questionnaires showed that subjects did not identify gestures as the target of the study and were not disturbed by the equipment. In fact, the speech and gestural behaviour of speakers did not differ quantitatively or qualitatively from data collected in an identical situation without eye-trackers (Gullberg, 1998). Addressees' gaze data include fixations of socially unacceptable areas that they might have avoided had they been concerned about the equipment. We interpret this as meaning that the apparatus did not interfere with the addressees' natural behaviour. The ecological validity of the data is therefore not compromised.

Basic Fixation Patterns in Interaction

In both studies the default location of the addressee's fixations is the speaker's face, viz. the nose bridge or eye area as seen in Figure 32.2 (Gullberg & Holmqvist, 1999, 2002).[1] In a context where the background does not represent any particular interest and there are no objects present relevant to the ongoing talk (cf. Argyle & Graham, 1976), addressees on average devote as much as 96% of their total viewing time to the face. This finding is consistent with earlier reports on addressees' gaze in interaction. The remainder of the time is spent fixating objects in the room behind the speaker or the speaker's immobile body parts. Only 0.5% of the total viewing time is devoted to gestures. In terms of number of fixations, only a minority of gestures are overtly fixated by addressees, or on average 7% of all gestures.

There is very little continuous scanning of the scene as a whole in these data and almost no cases of smooth pursuit of gestures. Saccades to fixation locations outside the face, including gestures, are direct and accurate despite the distances involved (cf. Land *et al.*, 1999). Typically, the eye moves from the face directly to a gesture in progress, stays on this target for an average fixation duration of 458 ms (SD = 436 ms), then returns directly back to the default location, i.e. the face. Note that fixations on gestures are spatially unambiguous. In all cases of gesture fixation the entire fixation marker is clearly located directly on the hand, not tangential to it, and not in the vicinity of a gesture in progress. This is also true for landing sites such as objects in the room and immobile body parts.

Figure 32.2: Example of the data. The addressee is fixating the default location, the speaker's face, despite the gesture in progress.

Gestures thus receive remarkably little overt visual attention. If addressees attend to gestures, they appear to do so in peripheral vision. Gestures that do attract direct fixations tend to display one or both of two features. They are either gestures with so-called post-stroke holds or gestures that speakers themselves have first looked at, speaker-fixated gestures. A hold is the momentary cessation of movement in a gesture while the hand is maintained immobile in gesture space (Kendon, 1972, 1980). Gestures that stop moving thus attract direct fixations to a greater extent than moving gestures. This finding was unexpected (but see Nobe *et al.,* 1998), as the standard assumption is that movement — rather than offset of movement — attracts attention. However, it is possible that the movement of inalienable body parts is not "salient" enough to draw overt fixations in this particular context (cf. Raymond, 2000). Since gestures are pervasive in interaction during speech, they represent almost constant movement in the addressees' visual field. It is conceivable that the cessation or offset of movement in gesture space instead represents a sudden change in the visual field and that this then evokes fixations. It may also be relevant that the cessation of movement takes place in gesture space, i.e. in symbolic space, and not anywhere in the visual field. A gesture that stops moving because the speaker drops her hands into the lap, i.e. effectively stops gesturing, does not attract fixation.

Figure 32.3: Example of a speaker-fixated gesture that is also fixated by the addressee (= white circle).

Speaker-fixated gestures also tend to be fixated by addressees in interaction as exemplified in Figure 32.3. This result is consistent with the claims in the literature. It is also in accordance with what is known about the powerful effect of speakers' gaze and head orientation on joint attention (Deák *et al.*, 2000; Doherty & Anders, 1999; Driver *et al.*, 1999; Gibson & Pick, 1963; Langton, 2000; Langton & Bruce, 1999; Langton *et al.*, 1996; Langton *et al.*, 2000; Moore & Dunham, 1995). The novelty is that gestures themselves can be the target of such joint attention, not just serve as indicators. Also, it has been claimed that speakers' gaze shifts induce automatic or reflexive shift of attention in addressees (cf. Langton *et al.*, 2000). An important observation in these data, however, is that speakers' gaze at their own gestures does not lead to automatic *overt* attention shift in addressees, i.e. to rapid saccades to and fixations of the target. The data do not, of course, tell us anything about the *covert* attention shifts (Hoffman, 1998; Johnson, 1995; Posner, 1980). Nevertheless, it is noteworthy that not all but only 23% of all speaker gaze shifts to gestures lead to overt fixation by addressees.

The results for the two other predicted gesture types, pointing gestures and gestures performed in peripheral gesture space, vary between the studies, and are inconclusive. This variation is presumably due to the fact that the relevant features tend to cluster in spontaneous gesture data. The individual attraction force of these features is therefore difficult to assess. Gullberg and Kita (forthcoming) consequently attempted to establish the individual effect of location of gesture performance (in central/peripheral gesture space), hold (presence/absence), and speaker-fixation (presence/absence) (see also pp. 695–697). Addressees were shown video clips of speakers retelling short stories. Each video clip contained only one gesture with the relevant feature. This gesture was designated as the target gesture. It was embedded in sequences of other gestures so as not to draw attention as a singleton. The target gesture had not been manipulated, but natural examples had been carefully selected from a bigger database such that each target gesture displayed the desired feature. The videos were projected life-sized on a wall and the eye movements of the addressees were recorded using the same set-up and task as described above.

The results showed that the location of the gesture in central or peripheral gesture space had no impact on addressees' fixations. In contrast, holds and speaker-fixation both individually attracted fixations significantly more often than gestures without these features. 12.5% of holds and 12.5% of speaker-fixated gestures were fixated as opposed to 0% of the gestures without these features. We also found a significant difference in saccade onset latencies to gestures with different features. Holds were fixated on average 108 ms after the onset of the gestural hold. In contrast, speaker-fixated gestures were fixated on average 808 ms after the onset of the speaker-fixation, i.e. after the speakers had directed their gaze towards their own gesture. Note that addressees were fixating the speaker's eye region at the moment of the onset both of holds and of speaker's fixations of their own gestures. This means that, in the case of holds, addressees responded very quickly to the cessation of movement perceived peripherally. In contrast, even though they were fixating the speaker's eyes and therefore should have detected the gaze shift immediately on the onset, saccades to the target gestures were only initiated after a considerable time delay. These findings suggest that gestures may be fixated for different reasons. We propose that fixations of holds are stimulus-driven or bottom-up guided whilst fixations of speaker-fixated gestures appear to be goal-driven (cf. Yantis, 1998). At this point, we cannot exclude the possibility that the gestural movement immediately preceding the holds does not initiate some form of saccadic planning. However, it seems unlikely given that gestural movements not followed by holds do not lead to saccades to gestures.[2]

More on Context: Live vs. Video, Social and Size Related Effects

The basic gaze pattern outlined above thus displays a number of properties that appear to be specific to the context of human social interaction. These findings have been challenged, however, in two studies of addressees' attention to the manual gestures of an anthropomorphic agent presented on a computer screen (Nobe *et al.*, 1998, 2000). In these studies, addressees fixated the vast majority of gestures (70–75%). This is in stark contrast to the mere 7% fixated in live interaction. The reduced number of gesture fixations and possibly the dominance of the face in the live setting could therefore be motivated by a social norm for maintained mutual gaze in face-to-face interaction. In the absence of any social pressure for eye contact, as in a video setting for instance, fixation behaviour towards speakers and their gestures may therefore look different. However, the studies by Nobe *et al.* differed from ours on a range of parameters other than the presence/absence of a live interlocutor. Most importantly, they used a different type of agent (anthropomorphic vs. human), a different size of the visual scene (computer screen vs. life size), and different types of gestures (conventionalised vs. spontaneous). To enable an evaluation of these differences, Gullberg & Holmqvist (2002) therefore undertook to specifically study the social effect of the presence/absence of a live interlocutor and the effect of the size of the display on gaze behaviour *ceteris paribus*. We compared fixation behaviour towards the same speakers and their gestures in three conditions: face-to-face live, on a 28-inch video screen, and on life-sized video projected on a wall. The task and the equipment were the same as outlined above, and the procedure for the live condition was identical. In addition to being filmed by the scene-camera on the eye-tracker, speakers in the live condition were filmed with a separate video camera. The resulting video served as the stimulus in the video conditions, where it was projected to two new sets of addressees. The design thus yielded fixation data for exactly the same gestures in three conditions.

Based on the assumption that the pattern for live interaction is motivated by social norms for sustained eye contact, we hypothesised that the absence of a live interlocutor might lead to less time spent on the face, more fixations on gestures and other targets, as well as more scene scanning behaviour. However, we also hypothesised that the smaller display size might result in fewer gestures being fixated. The logic here is that with the reduced angles on a small screen, gesture detection would be possible even if the fixation marker remained on the speaker's face.

The socially motivated hypothesis was only minimally borne out. The results showed that the face dominated as a target overwhelmingly in all three conditions. The average viewing time on the face did drop on video, and more so in the small screen condition than the life-sized condition (from 96% live to 94% on life-sized and 91% on small screen video), but not significantly so. The number of gestures fixated decreased in both video conditions, although only significantly so on small screen video (from 7% of all gestures fixated live to 4.5% on life-sized and 3% on small screen video). Notice that this finding is in accordance (only) with the size-related hypothesis. With reduced viewing times on the face and fewer gesture fixations, the video conditions instead showed more fixations on body parts and empty space. This was partially predicted by the social hypothesis. The number of fixations of immobile body parts increased significantly in the small screen video condition (from 35% live and 32% on life-sized video to 57% on small screen video). Fixations on empty space increased significantly in the life-sized video condition (from 5% live and on small screen video to 21% on life-sized video). While there was a somewhat increased tendency for scene scanning in the small screen video condition, in principle, the typical saccade pattern outlined above for the live condition held in both video conditions as well. In sum, gaze behaviour in the live and life-sized video conditions was very similar overall. The small screen video condition showed the greatest number of differences from the live condition, even if most did not reach significance in this study. We interpret this as meaning that the reduced display size had a greater impact on general gaze behaviour than the absence of a live interlocutor.

Despite the overall reduction in fixation rate of gestures on video, by and large, the same gestures were fixated across the conditions. Specifically, the (proportional) attraction effect of holds and speaker-fixated gestures was maintained on video. In other words, holds and speaker-fixated gestures were fixated significantly more often than gestures without these features in all conditions. Importantly, speaker-fixated gestures were clearly affected by the absence of a live interlocutor. The number of fixations on speaker-fixated gestures was significantly reduced in both video conditions (from 23% live to 8% in both video conditions). This is consistent with the view that the speaker's gaze is essentially a social cue to joint attention, and that it is more powerful in a fully social setting than on video. There was also an overall if non-significant decrease in fixation of holds across the conditions (from 33% live to 20% on life-sized and 15% on small screen video). This decrease appears to be motivated by both social and size related factors, but these findings are more difficult to interpret. If holds attract fixations largely as a response to a sudden change in the visual field, then this reaction should be maintained even on video. These questions must be studied in more detail.

To summarise, in all conditions the face dominates as a fixation target and there is very little scanning behaviour of the rest of the scene. Gestures are only given a minimal amount of overt visual attention, and the same gestural features attract fixations across conditions. The social effects appear to be less powerful overall than size effects with the exception of the impact on speaker-fixation. The pattern outlined for the interactional situation is thus not exclusively socially conditioned. Gaze behaviour towards a human interlocutor on video is clearly more similar to behaviour towards an interlocutor live than to other types of video stimuli. This seems to reflect the particular status of the human face as an inherent focus of attention. This well-documented bias seems to have a biological basis in neural circuitry dedicated to face processing (for an overview, see Farah, 2000), and is manifest very early in infants' preference for faces as targets of attention (Valenza *et al.*, 1996).

Gestures, Fixations, and Information Uptake

Up until this point, gestures have mainly been considered as visuospatial phenomena. However, as shown in the introduction, gestures are also symbolic movements that carry meaning. The issue of whether addressees attend to and process gestural information is controversial in the field of gesture studies. Simplifying matters somewhat, it is sometimes argued that gestures have little real communicative value since addressees cannot reliably assign meaning to gestures in the absence of speech (e.g. Krauss *et al.*, 1991; Krauss *et al.*, 1996). However, there is a growing body of evidence showing that addressees do process gestural information (cf. Kendon, 1994). For instance, information expressed only in gestures re-surfaces in retellings, either as speech, as gesture, or both (Cassell *et al.*, 1999; McNeill *et al.*, 1994). Questions about the size and relative position of objects are better answered when gestures are part of the description than when gestures are absent (Beattie & Shovelton, 1999a; 1999b). Stroop-test designs also show cross-modal interference effects in the processing of gestural and spoken information (Langton & Bruce, 2000; Langton *et al.*, 1996). When subjects are shown static pictures of a person pointing either up, down, left, or right, and hear or see an incongruent corresponding word, their responses to words are affected by gestures, and responses to gestures by words. These data suggest that gesture processing is automatic and occurs in parallel to processing of speech.

Given that addressees direct very little overt visual attention to gestures in interaction, gestures must generally be attended to covertly. However, in contrast to Sign Language, we cannot automatically assume that non-fixated gestures are nonetheless attended to peripherally in the sense that the information encoded is processed and integrated into a representation of meaning. In order to investigate the division of labour between foveal and peripheral processing of gestural information, we need to study the uptake of information and the allocation of visual attention simultaneously. The question then arises whether there is any evidence that fixations or overt visual attention lead to better uptake of gestural information. Put differently, are any of the gesture fixations that do occur driven by the need to extract information, as often assumed in the non-technical literature?

The relationship between attention in a broad sense, information processing, and fixations is not a trivial one in complex naturalistic settings. Recent studies on change blindness (see e.g. the section on this topic in Hyönä, *et al.*, 2002) show that subjects occasionally fail to detect startling changes in a visual scene. Provided that their conscious attention is directed at a specific task, subjects fail to notice the switch of the main protagonist in a story or men in gorilla suits walking across the scene (e.g. Mack & Rock, 1998; Simons, 2000; Simons *et al.*, 2002; Simons & Levin, 1998). Complex scene perception is particularly challenging as the fixation marker is no straightforward indicator of what aspect of a scene is being attended to. As remarked by O'Regan *et al.*: "what an observer 'sees' at any moment in a scene is not the *location* he or she is directly fixating with the eyes, but the *aspect* of the scene he or she is currently attending to, that is, presumably, what he or she is processing with a view to encoding for storage into memory." (O'Regan *et al.*, 2000: 209). What aspects are relevant and what information is extracted (and retained) is still an empirical question (e.g. Aginsky & Tarr, 2000; Tatler, 2002).

Gullberg and Kita (forthcoming) investigated the relationship between addressees' information uptake from gestures and their fixations of gestures by exposing subjects to video clips of speakers retelling short stories and performing spontaneous gestures in face-to-face interaction in Dutch (Kita, 1996). Each video clip contained one relevant gesture (the target gesture) that was embedded among other spontaneous gestures. The target gestures represented motion of a protagonist left or right. This information (left or right direction) was only present in the target gesture and not in concurrent speech. The information could not be inferred from surrounding gestures. Each target gesture also displayed one of the features assumed to attract fixations: articulation in peripheral gesture space, hold, and speaker-fixation (see p. 692). Subjects watched four video clips of four different speakers retelling stories and gesturing in sequence. The videos were projected life-sized against a wall. Subjects' eye movements were recorded with the SMI iView head-mounted eye-tracker as outlined above. After watching the four videos, subjects answered questions about the target events by drawing pictures of the protagonists of the story. The data were coded for fixation on target gesture and for matched reply. A target gesture was coded as fixated if the fixation marker was immobile on the gesture for a minimum of 120 ms. As explained above, fixations on gestures were spatially unambiguous. Either a gesture was clearly fixated, or the fixation marker stayed on the speaker's face. A drawing was coded as a matched reply if the direction in the drawing matched the direction of the gesture as seen on video from the addressee perspective (see Figures 32.4 and 32.5).[3]

Figure 32.4: Example of a speaker performing a directional gesture that is also fixated by the addressee (= white circle).

Figure 32.5: Example of the addressee's drawing matching the direction of the target gesture in Figure 32.4 as seen from the addressee-perspective.

As seen in the second section, gestures with holds and speaker-fixated gestures attracted fixations in 12.5% of the cases, respectively. However, addressees picked up the gestural information reliably above chance only from speaker-fixated gestures (82.5%), i.e. when speakers themselves had first looked at the gestures. In other words, addressees fixated some gestures (with hold) whose information they did not process. Conversely, addressees processed information from many gestures that they did not fixate (speaker-fixated), provided that speakers themselves had first look at the gestures. There was no significant difference in uptake for fixated vs. non-fixated gestures, and no effect of fixation duration.

The lack of information uptake above chance from the hold fixations could simply be due to the fact that the directional information is not present once the gesture has stopped moving. In contrast, the reliable uptake effect from speaker-fixated gestures even when these were *not* overtly fixated by the addressees is striking. This finding supports the claims that speakers' gaze shifts direct addressees' covert attention semi-automatically to the target of the gaze (cf. Langton & Bruce, 1999; Langton *et al.*, 2000). It also confirms that peripheral vision is sufficient to process gestural information. It does not, however, tell us whether gestures are fixated for the purpose of information extraction. In fact, the observation that uptake is not determined by direct fixations rather seems to suggest that overt fixations on speaker-fixated gestures are essentially socially motivated. The overt following of a speaker's gaze shift to the target of that gaze may be determined by social norms for joint attention. This finding is not per se in contradiction with statements to the effect that foveating gives an advantage to information extraction (e.g. Tatler, 2002). However, it does show that fine-grained information extraction is possible in complex situations even without direct fixation provided that covert attention has been directed to a specific target by speaker's gaze.

General Discussion and Conclusions

The studies reviewed in this chapter show that the human face overwhelmingly dominates as a target for overt visual attention in live interaction as well as on video. Gestures, in contrast, attract very few direct fixations both live and on video. They are fixated mainly if they momentarily cease to move as in gestural holds, or if speakers themselves have looked at them first. Holds are fixated quickly after their onset, suggesting a bottom-up response, whereas speaker-fixated gestures are fixated after a considerable delay, suggesting a top-down mechanism. Gestures thus appear to be fixated in their capacities as visuospatial entities, due to change in the visual field (holds), and possibly also for social reasons related to joint attention (speaker-fixated gestures). We know less about whether gestures are directly fixated for the purpose of extracting information. Most information processing appears to be covert or to be done in peripheral vision. At the very least, there is no simple relationship between fixations on a gesture and uptake of the gestural information in this complex setting.

Some of these results seem challenging in view of what is known about eye movements from studies in lab settings. However, they are not incompatible with recent findings from studies of eye movements in "conditions of competitive, parallel processing" (Tatler, 2002) as present in the real world. A number of studies have investigated subjects' eye movements as they perform natural tasks such as driving, making tea or sandwiches (Hayhoe, 2000; Land & Hayhoe, 2001; Land *et al.*, 1999; Shinoda *et al.*, 2001, Tatler, 2001), copying blocks (Pelz *et al.*, 2001), or drawing portraits (Tchalenko, this volume). In these activities the eye is typically directed to different areas according to the requirements of the current task, and many of the studies are concerned with eye-hand co-ordination. The results generally point in the direction of task and context sensitivity of fixation patterns. The studies reviewed above confirm the strong influence of the task, the activity type, and the context on gaze behaviour. They also highlight the contextual constraints on overt responses to attentional processes. Subjects do not automatically overtly respond to anything that catches their attention in interaction. Their responses are constrained by factors such as social norms for interaction, the status of the human face, and even kinaesthetic knowledge about body movements. Specifically, these factors interact in complex ways to influence behaviour. Finally, the results also suggest that the temporal resolution of overt behaviour is affected by the context. For instance, addressees' slow overt responses to speakers' gaze shifts are in stark contrast to the claims in the literature regarding the automatic effect on attention of such gaze shifts. Taken together, these issues form an important cluster to consider in discussions of the generalisability of findings from lab contexts to more complex, naturalistic settings.

Many issues need further investigation. Most pressing is perhaps the need to take the dynamic aspects of gesture performance into account when considering what triggers addressees' fixations on gestures. Despite the interactional perspective, we have treated gestures as individual events, isolated from each other in time and space, that can be fixated or not. However, since gestures are de facto linked to each other in gesture units that unfold over time (Kendon, 1972), a given gesture fixation is equally likely to be influenced by the number and the nature of preceding gestures as by the properties of any individual gesture. The influence of properties of speech should also be considered. For instance, noise in the speech signal or lowered comprehensibility of speech may also lead to increased attention to gestures. The assumption is that when the speech channel is compromised, addressees will rely more on gestures for decoding the message (Rimé *et al.*, 1988; Rogers, 1978). Other potentially influential factors include deictic expressions directing attention explicitly to gestures (e.g. "it was this big"), as well as interruptions and dysfluencies (Seyfeddinipur & Kita, 2001).

Many questions and unsolved puzzles thus remain. Hopefully the findings outlined here nonetheless show the importance of studying complex, dynamic and interactive contexts where inherent foci of interest (the human face), knowledge of the world (how bodies move), and social factors (principles of joint attention) conspire and influence behaviours at spatial and temporal levels alike. By considering such contexts, we hope to contribute to a multi-faceted picture of how visual attention works in the real world.

Acknowledgements

I thank Jukka Hyönä and an anonymous reviewer for helpful and constructive comments on an earlier version of this chapter. I also gratefully acknowledge the financial support of Birgit and Gad Rausing's Foundation for Research in the Humanities.

Notes

1 A similar pattern has been suggested in American Sign Language (ASL) interaction. Signers' default visual focus in face-to-face signing is reported to be the face. Interestingly, however, the default locus of attention appears to be a region about the *lower* half or the chin region of the signer's face rather than the eye region (Corina *et al.*, 1996). This shift to the lower region of the signer's face could be a modification to enable foveal perception of oral grammatical components while simultaneously allowing for good sign perception in peripheral vision. For linguistic reasons, signers need to ensure that both motion detection and hand configuration details are adequately processed. Little is known, however, about actual fixation patterns during sign interaction.

2 We are currently investigating the effect of the preceding movement vs. the hold itself by comparing fixation behaviour to the same gestures without and with artificially introduced holds.

3 There is no evidence that addressees reversed the directions in the drawings in order to represent the direction as expressed from the speaker's viewpoint. Had addressees been reversing the viewpoints, we would have expected within-subject consistency of such reversals. There is no such consistency in the data, however.

References

Aginsky, V., & Tarr, M. J. (2000). How are different properties of a scene encoded in visual memory? *Visual Cognition*, *7*, 147–162.

Argyle, M., & Cook, M. (1976). *Gaze and Mutual Gaze*. Cambridge: Cambridge University Press.

Argyle, M., & Graham, J. A. (1976). The Central Europe experiment: Looking at persons and looking at things. *Journal of Environmental Psychology and Nonverbal Behavior*, *1*, 6–16.

Bavelier, D., Brozinsky, C., Tomann, A., Mitchell, T., Neville, H., & Liu, G. (2001). Impact of early deafness and early exposure to Sign Language on the cerebral organization for motion processing. *Journal of Neuroscience*, *21*, 8931–8942.

Beattie, G., & Shovelton, H. (1999a). Do iconic hand gestures really contribute anything to the semantic information conveyed by speech? *Semiotica*, *123*, 1–30.

Beattie, G., & Shovelton, H. (1999b). Mapping the range of information contained in the iconic hand gestures that accompany spontaneous speech. *Journal of Language and Social Psychology*, *18*, 438–462.

Braffort, A., Gherbi, R., Gibet, S., Richardson, J., & Teil, D. (eds) (1999). *Gesture-based Communication in Human–Computer Interaction*. (Vol. 1739). Berlin: Springer Verlag.

Cassell, J., McNeill, D., & McCullough, K.-E. (1999). Speech-gesture mismatches: Evidence for one underlying representation of linguistic and nonlinguistic information. *Pragmatics & Cognition*, *7*, 1–33.

Corina, D., Kritchevsky, M., & Bellugi, U. (1996). Visual language processing and unilateral neglect: Evidence from American Sign Language. *Cognitive Neuropsychology, 13*, 321–356.
Deák, G. O., Flom, R. A., & Pick, A. D. (2000). Effects of gesture and target on 12- and 18-month-olds' joint visual attention to objects in front of or behind them. *Developmental Psychology, 36*, 511–523.
Decety, J., & Grèzes, J. (1999). Neural mechanisms subserving the perception of human actions. *Trends in Cognitive Sciences, 3*, 172–178.
Doherty, M. J., & Anders, J. R. (1999). A new look at gaze: Pre-school children's understanding of eye-direction. *Cognitive Development, 14*, 549–571.
Driver, J., Davis, G., Ricciardelli, P., Kidd, P., Maxwell, E., & Baron-Cohen, S. (1999). Gaze perception triggers reflexive visuospatial orienting. *Visual Cognition, 6*, 509–540.
Farah, M. J. (2000). *The Cognitive Neuroscience of Vision*. Oxford: Blackwells.
Fehr, B. J., & Exline, R. V. (1987). Social visual interaction: A conceptual and literature review. In: A. W. Siegman and S. Feldstein (eds), *Nonverbal Behavior and Communication* (pp. 225–326). Hillsdale, NJ: Erlbaum.
Gibson, J. J., & Pick, A. D. (1963). Perception of another person's looking behavior. *American Journal of Psychology, 76*, 386–394.
Goldenberg, G. (2001). Imitation and matching of hand and finger postures. *NeuroImage, 14*, S132-S136.
Goodwin, C. (1986). Gestures as a resource for the organization of mutual orientation. *Semiotica, 62*, 29–49.
Grèzes, J., Costes, N., & Decety, J. (1999). The effects of learning and intention on the neural network involved in the perception of meaningless actions. *Brain, 122*, 1875–1887.
Gullberg, M. (1998). *Gesture as a Communication Strategy in Second Language Discourse. A Study of Learners of French and Swedish*. Lund: Lund University Press.
Gullberg, M., & Holmqvist, K. (1999). Keeping an eye on gestures: Visual perception of gestures in face-to-face communication. *Pragmatics & Cognition, 7*, 35–63.
Gullberg, M., & Holmqvist, K. (forthcoming). What speakers do and what listeners look at. Visual attention to gestures in face-to-face interaction and on video.
Gullberg, M., & Holmqvist, K. (2002). Visual attention towards gestures in face-to-face interaction vs. on screen. In: I. Wachsmuth and T. Sowa (eds), *Gesture and Sign Language based Human–Computer Interaction* (pp. 206–214). Berlin: Springer Verlag.
Gullberg, M., & Kita, S. (forthcoming). Attention to gestures. Information processing and fixations.
Hayhoe, M. (2000). Vision using routines: A functional account of vision. *Visual Cognition, 7*, 43–64.
Hermsdörfer, J., Goldenberg, G., Wachsmuth, C., Conrad, B., Ceballos-Baumann, O., Bartenstein, P., Schwaiger, M., & Boecker, H. (2001). Cortical correlates of gesture processing: Clues to the cerebral mechanisms underlying apraxia during the imitation of meaningless gestures. *NeuroImage, 14*, 149–161.
Hoffman, J. E. (1998). Visual attention and eye movements. In: H. Pashler (ed.), *Attention* (pp. 119–153). Hove: Psychology Press Ltd.
Hyönä, J., Muñoz, D., Heide, W., & Radach, R. (eds) (2002). *The Brain's Eye: Neurobiological and Clinical Aspects of Oculomotor Research*. Amsterdam: Elsevier.
Johnson, M. H. (1995). The development of visual attention: A cognitive neuroscience perspective. In: M. S. Gazzaniga (ed.), *The Cognitive Neurosciences* (pp. 735–747). Cambridge, MA: MIT Press.
Kendon, A. (1972). Some relationships between body motion and speech: An analysis of an example. In: A. W. Siegman and B. Pope (eds), *Studies in Dyadic Communication* (pp. 177–210). New York: Pergamon.

Kendon, A. (1980). Gesticulation and speech: Two aspects of the process of utterance. In: M. R. Key (ed.), *The Relationship of Verbal and Nonverbal Communication* (pp. 207–227). The Hague: Mouton.

Kendon, A. (1986). Some reasons for studying gesture. *Semiotica*, *62*, 3–28.

Kendon, A. (1990). *Conducting Interaction*. Cambridge: Cambridge University Press.

Kendon, A. (1993). Human gesture. In: K. R. Gibson and T. Ingold (eds), *Tools, Language and Cognition in Human Evolution* (pp. 43–62). Cambridge: Cambridge University Press.

Kendon, A. (1994). Do gestures communicate?: A review. *Research on Language and Social Interaction*, *27*, 175–200.

Kita, S. (1996). Listeners' up-take of gestural information. *MPI Annual Report, 1996*, 78.

Krauss, R. M., Chen, Y., & Chawla, P. (1996). Nonverbal behavior and nonverbal communication: What do conversational hand gestures tell us? *Advances in Experimental Social Psychology*, *28*, 389–450.

Krauss, R. M., Morrel-Samuels, P., & Colasante, C. (1991). Do conversational hand gestures communicate? *Journal of Personality and Social Psychology*, *61*, 743–754.

Land, M. F., & Hayhoe, M. (2001). In what ways do eye movements contribute to everyday activities. *Vision Research*, *41*, 3559–3565.

Land, M. F., Mennie, N., & Rusted, J. (1999). Eye movements and the roles of vision in activities of daily living: Making a cup of tea. *Perception*, *28*, 1311–1328.

Langton, S. R. H. (2000). The mutual influence of gaze and head orientation in the analysis of social attention direction. *Quarterly Journal of Experimental Psychology*, *53*, 825–845.

Langton, S. R. H., & Bruce, V. (1999). Reflexive visual orienting in response to the social attention of others. *Visual Cognition*, *6*, 541–567.

Langton, S. R. H., & Bruce, V. (2000). You must see the point: Automatic processing of cues to the direction of social attention. *Journal of Experimental Psychology: Human Perception and Performance*, *26*, 747–757.

Langton, S. R. H., O'Malley, C., & Bruce, V. (1996). Actions speak no louder than words: Symmetrical cross-modal interference effects in the processing of verbal and gestural information. *Journal of Experimental Psychology: Human Perception and Performance*, *22*, 1357–1375.

Langton, S. R. H., Watt, R. J., & Bruce, V. (2000). Do the eyes have it? Cues to the direction of social attention. *Trends in Cognitive Sciences*, *4*, 50–59.

Mack, A., & Rock, I. (1998). *Inattentional Blindness*. Cambridge, MA: MIT Press.

McNeill, D. (1992). *Hand and Mind. What the Hands Reveal about Thought*. Chicago: Chicago University Press.

McNeill, D. (1998). Speech and gesture integration. In: J. Iverson and S. Goldin-Meadows (eds), *The Nature and Functions of Gesture in Children's Communication* (pp. 11–27). San Francisco: Jossey-Bass.

McNeill, D., Cassell, J., & McCullough, K.-E. (1994). Communicative effects of speech mismatched gestures. *Research on Language and Social Interaction*, *27*, 223–237.

Moore, C., & Dunham, P. J. (eds) (1995). *Joint Attention*. Hillsdale, NJ: Erlbaum.

Neville, H. J., Coffe, S. A., Lawson, D. S., Fischer, A., Emmorey, K., & Bellugi, U. (1997). Neural systems mediating American Sign Language: Effects of sensory experience and age of acquisition. *Brain and Language*, *57*, 285–308.

Nobe, S., Hayamizu, S., Hasegawa, O., & Takahashi, H. (1998). Are listeners paying attention to the hand gestures of an anthropomorphic agent? An evaluation using a gaze tracking method. In: I. Wachsmuth and M. Fröhlich (eds), *Gesture and Sign Language in Human–Computer Interaction* (pp. 49–59). Berlin: Springer.

Nobe, S., Hayamizu, S., Hasegawa, O., & Takahashi, H. (2000). Hand gestures of an anthropomorphic agent: Listeners' eye fixation and comprehension. *Cognitive Studies. Bulletin of the Japanese Cognitive Science Society*, *7*, 86–92.
O'Regan, J. K., Deubel, H., Clark, J. J., & Rensink, R. A. (2000). Picture changes during blinks: Looking without seeing and seeing without looking. *Visual Cognition*, *7*, 191–211.
Peigneux, P., Salmon, E., van der Linden, M., Garraux, G., Aerts, J., Delfiore, G., Degueldre, C., Luxen, A., Orban, G., & Franck, G. (2000). The role of lateral occipitotemporal junction and area MT/V5 in the visual analysis of upper-limb postures. *NeuroImage*, *11*, 644–655.
Pelz, J., Hayhoe, M., & Loeber, R. (2001). The coordination of eye, head, and hand movements in a natural task. *Experimental Brain Research*, *139*, 266–277.
Perani, D., Fazio, F., Borghese, N. A., Tettamanti, M., Ferrari, S., Decety, J., & Gilardi, M. C. (2001). Different brain correlates for watching real and virtual hand actions. *NeuroImage*, *14*, 749–758.
Posner, M. I. (1980). Orienting of attention. *Quarterly Journal of Experimental Psychology*, *32*, 3–25.
Raymond, J. E. (2000). Attentional modulation of visual motion perception. *Trends in Cognitive Sciences*, *4*, 42–50.
Rettenbach, R., Diller, G., & Sireteanu, R. (1999). Do deaf people see better? Texture segmentation and visual search compensate in adult but not in juvenile subjects. *Journal of Cognitive Neuroscience*, *11*, 560–583.
Rimé, B., Boulanger, B., & d'Ydewalle, G. (1988). Visual attention to the communicator's nonverbal behavior as a function of the intelligibility of the message. Paper presented at the Symposium on TV Behavior, 24th International Congress of Psychology, Sydney, Australia, 28 August–2 September 1988.
Rizzolatti, G., Fogassi, L., & Gallese, V. (2001). Neurophysiological mechanisms underlying the understanding and imitation of action. *Nature Reviews Neuroscience*, *2*, 661–670.
Rogers, W. T. (1978). The contribution of kinesic illustrators toward the comprehension of verbal behavior within utterances. *Human Communication Research*, *5*, 54–62.
Seyfeddinipur, M., & Kita, S. (2001). Gesture and dysfluencies in speech. In: C. Cavé, I. Guaïtella and S. Santi (eds), *Oralité et Gestualité: Interactions et Comportements Multimodaux Dans la Communication* (pp. 266–270). Paris: L'Harmattan.
Shinoda, H., Hayhoe, M. M., & Shrivastava, A. (2001). What controls attention in natural environments? *Vision Research*, *41*, 3535–3545.
Simons, D. J. (2000). Attentional capture and inattentional blindness. *Trends in Cognitive Sciences*, *4*, 147–155.
Simons, D. J., Chabris, C. F., Schnur, T., & Levin, D. T. (2002). Evidence for preserved representations in change blindness. *Consciousness and Cognition*, *11*, 78–97.
Simons, D. J., & Levin, D. T. (1998). Failure to detect changes to people during real-world interaction. *Psychonomic Bulletin and Review*, *5*, 644–649.
Streeck, J. (1993). Gesture as communication I: Its coordination with gaze and speech. *Communication Monographs*, *60*, 275–299.
Streeck, J. (1994). Gesture as communication II: The audience as co-author. *Research on Language and Social Interaction*, *27*, 239–267.
Streeck, J., & Knapp, M. L. (1992). The interaction of visual and verbal features in human communication. In: F. Poyatos (ed.), *Advances in Nonverbal Communication: Interdisciplinary Approaches Through the Social and Clinical Sciences, Literature and the Arts* (pp. 3–23). Amsterdam: Benjamins.
Swisher, M. V. (1993). Perceptual and cognitive aspects of recognition of signs in peripheral vision. In: M. Marschark and M. D. Clark (eds), *Psychological Perspectives on Deafness* (pp. 209–228). Hillsdale: Erlbaum.

Tatler, B. W. (2001). Characterising the visual buffer: real-world evidence for overwriting early in each fixation. *Perception*, 30, 993–1006.
Tatler, B. W. (2002). What information survives saccades in the real world? In: J. Hyönä, D. Muñoz, W. Heide and R. R. (eds), *The Brain's Eyes: Neurobiological and Clinical Aspects of Oculomotor Research*. Amsterdam: Elsevier.
Tuite, K. (1993). The production of gesture. *Semiotica*, *93*, 83–105.
Valenza, E., Simion, F., Macchi Cassia, V., & Umilta, C. (1996). Face preference at birth. *Journal of Experimental Psychology: Human Perception and Performance*, *22*, 892–903.
Wachsmuth, I., & Sowa, T. (eds) (2002). *Gesture and Sign Language in Human–Computer Interaction*. (Vol. 2298). Berlin: Springer Verlag.
Yantis, S. (1998). Control of visual attention. In: H. Pashler (ed.), *Attention* (pp. 223–256). Hove: Psychology Press Ltd.

Chapter 33

Eye Movement and Voluntary Control in Portrait Drawing

J. Tchalenko, L. Dempere-Marco, X. P. Hu and G. Z. Yang

In drawing portraits from life, the eye plays a central role as it is the means by which visual input is acquired from the external world, as well as the means by which the hand is guided as it draws and the results are evaluated. Eye movements were measured in 12 subjects ranging in skill from the professional to the novice. A fundamental paper-to-model-to-paper rhythm was investigated, as well as eye-hand coordination patterns and the spatial accuracy with which a pre-drawn line could be followed with the eye. Although more subjects drawing in different ways need to be examined, this first quantitative observation of the artist at work has opened the field for the direct study of the cognitive processes involved in drawing and artistic creativity.

Introduction

A central question of the picture production process in art is how does the artist transform a vision of the external world into the drawn or painted work. In portrait drawing from life, the question concerns the way the artist acquires visual information by looking at the model and then produces an interpretation of this information by drawing on the paper. The eye has the central role in this process, as it is the means by which visual data is entered into the brain, as well as the means by which the hand is guided and the results are evaluated. Our approach was therefore to study the painter's eye movements in order to establish the ground data required to start understanding the picture production process. The lack of any previous data on the subject led us to investigate the basic eye–hand coordination pattern, and enquire into the eye control skills shown by experienced painters. Our present study is restricted to artists who draw portraits from life in a realistic style.

The Mind's Eye: Cognitive and Applied Aspects of Eye Movement Research

ISBN: 0–444–51020–6

Most artists drawing from life construct a portrait detail by detail, and they are continually moving their gaze from the paper (P) to the model (M) and back again. This P–M–P cycle structures the entire portrait drawing work. The action during the cycle, at least with professional painters, is precise, rhythmically repetitive and uninterrupted by extraneous events, and hence apposite to the study of eye movement with eyetrackers. The context of such a study is that of a real-life situation observed with the help of instruments more commonly associated with laboratory testing.

Prior to our study, the eyetracker does not seem to have been used with painters. In fact, Konecni (1991) seems to be the only investigator who timed eye movements as portraits were being drawn. He used a video camera to time the glances of six subjects drawing sketch portraits, and found a frequency of glances to the model varying between 19 and 25/minute, with little apparent differences between artists and novices. We will see that these movement rates agree well with our own findings. The work reported in Tchalenko (1991) involved about 100 hours of video film of a painter taken from the model's point of view during the painting of a portrait, and eventually became the starting point for our eyetracker studies. It was partly inspired by Lord (1980) who sat for Alberto Giacometti and described the drawing process in considerable detail, albeit not on the level of eye and hand movements. Despite this lack of quantitative eye movement data, several references shed indirect light on our subject. Within the cognitive literature, Livingston and Hubel (1995) and Frith and Law (1995) outlined how brain imaging studies in other areas can inform us on drawing skills. The visual information of a scene is divided into individual processing components on the basis of which the brain computes the necessary movements for drawing. More pertinent to the present study, Solso (2000) and Solso (2001) described some fMRI tests we performed on HO, the principle subject of our eyetracker studies, as he was drawing while his brain was being scanned. The feasibility of this type of investigation and the preliminary results are presented and discussed. Cohen and Bennett (1997) and Snyder and Thomas (1997) examined the question of preconceived notions the drawer — novices in the first reference and autistic children in the second — may, or may not, have about the form of objects in the external world that they are drawing.

A question we are sometimes asked is whether our approach has common ground with the theories on art suggested by art historians. Gombrich (1963) is probably the most frequently quoted reference on the subject. His starting point is the analysis of the finished picture and he shows how Nature was interpreted, and how paintings and drawings were viewed, at different periods of History. His conclusions are essentially about the changing perception of works of art. The starting point for our present study is quite different. It is a direct observation of the painter at work, and in particular of the artist's eye movements. We only refer to the finished picture, which can be either by an expert or a novice, when it provides information on the movement of the drawing hand. Our analysis concentrates uniquely on the physiological processes taking place during the act of drawing. Eventually, when sufficient data will have been gathered, such observations may form the basis of future art theories.

Eyetracker Studies

The eye's different functions — input of visual data, output of the drawn line and evaluation of the results — are determined and driven by the requirements of the painter's task. The hand leads the task, while the eye supplies the appropriate information as and when required. Thus the eye when it looks at the model is not reacting to unsolicited visual stimuli, but is voluntarily moved to pinpoint a particular detail required at that particular moment in time. It is moved away from that point when the painter deems that sufficient information has been captured. This behaviour is different from our normal way of perceiving the external world, and perception and cognition studies made in areas such as face recognition or the viewing of pictures will not necessarily be relevant to drawing from life.

A consequence of the task-driven nature of portrait drawing is that, at all times, the painter's eye movements are under his, or her, control. Moving from paper to model or vice-versa, from one detail to another, staying fixed or moving together with the pencil, etc. are all volitional actions in the sense that the artist may choose to do them differently or not at all. The basic parameters of the painter's eye and hand movements are outlined in the following two sections of this chapter, "Eye Movements" and "Eye-Hand Coordination". As these eye movements are repeated many hundreds or thousands of times during the making of a portrait, and many millions of times during the lifetime of a painter, they take on for each artist a characteristic structure and rhythm. The question of whether a painter acquires in this way a greater eye control than the non-painter is discussed in the last section, "Eye Control."

Throughout this study, our priority was to provide the artist with as normal working conditions as possible, in particular allowing free movement with minimum vision restrictions. The modern eyetracker device (ISECS) of our latest tests is simply a pair of spectacles without cable attachments. Our main systematic work was however done on slightly earlier equipment (EYEPUTER and ASL 501: see Appendix I for all equipment characteristics). The experimental set up first devised with subject HO (Miall & Tchalenko 2001) was subsequently adopted for all drawing tests with other subjects (Table 33.1). It consisted in seating the painter in front of a drawing board held vertically on an easel at arm's length. The model was seated next to the board and at a same average distance from the artist's eyes. This distance was decided upon because it suited the artist and simplified subsequent eyetracker computations. For self-portraits, the mirror image was similarly made to be at the same distance as the board.

In some tests with HO we also recorded the hand's position with the help of a motion sensor. For the 5-hour *Portrait of Nick*, the eyetracker was worn for about 10 to 15 minutes at the start of each hourly session; for all other tests it was worn throughout. Eyetracker output was in the usual form of digital data file and video scene of the painter's view with superimposed gaze point as filmed by a head-mounted camera. Independently, a continuous close-up video recording of the drawing board was also made for each artist in order to verify the hand's position and follow the picture's progress.

Table 33.1: Eyetracker tests: subjects tested and equipment used. For equipment abbreviations and explanations see Appendix I.

Subject	Ability	Drawing test & equipment	Eye control test & equipment	Reference
HO	Full-time painter, mainly portraits. Work in many international galleries and collections.	Select model (EP)		Miall and Tchalenko (2001)
		Portrait of Nick (EP)		Tchalenko (2001a)
		Portrait *Luke 2* (EP)		Tchalenko (2001b)
		Self-portrait (EP)	QuickGlance	
		Memory Luke 3 (501)		
		From photo (EP)		
3 controls (for HO)	Novices, do not draw	From photo (EP)	EP	Miall and Tchalenko (2001)
ME	Painter, draws very frequently, work in galleries and collections	Self-portrait (ISECS)	ASL 504	Tchalenko *et al.* (2001) This chapter (2003)
JT	Ex-art school, draws very frequently.	Self-portrait (ISECS)	ASL 504	Tchalenko *et al.* (2001) This chapter (2003)
PL	PhD art school, draws regularly.	Self-portrait (ISECS)	ASL 504	Tchalenko *et al.* (2001) This chapter (2003)
CS	2nd year art school, draws.	Portrait of B (ASL 501)		This chapter (2003)
DL	2nd year art school, draws.	Portrait of B (ASL 501)		This chapter (2003)
IA	2nd year art school, draws.	Parallel lines (ASL 501)		This chapter (2003)
BD	1st year art school, draws occasionally.		ASL 504	Tchalenko *et al.* (2001) This chapter (2003)
MM	PhD art school, does not draw.		ASL 504	Tchalenko *et al.* (2001) This chapter (2003)

Eye Movements

We will first examine separately the two basic fixation types occurring during drawing, those located on the paper (P) and those located on the model (M), before considering their alternation in the P-M-P cycle.

Location of Fixations on the Paper

For all subjects studied we found that fixations on the paper when actually drawing were located slightly away from the pencil tip, at a distance of about 0.5 to 1.0 degree, and in the position least occluded by the drawing hand. As the subjects we studied were all right-handed and drawing on a vertical board from a seated position, when drawing a horizontal line fixations were located above the line and near, or slightly ahead, of the starting point, and when drawing a vertical line they were located to the left of it. With HO who attributes extreme importance to the precision of single lines, a fixation remained stable until the pencil had moved about 1 degree from the starting point, at which stage a saccade would occur to reposition the next fixation in the same relationship to the line as previously. In this way, the eye was continually lagging behind and catching up with the pencil in fits and starts, suggesting that the eye's function was one of evaluating the segment that had just been drawn as well as of guiding the pencil. With ME who draws more rapidly and with many more lines than HO, fixations followed the same pattern but with more flexibility: they were less stable and moved back and forth along the line as it was being drawn. In none of the cases studied did the eye follow the pencil in a smooth pursuit movement. Ballard *et al.* (1992) had already observed in their block-moving exercise that smooth pursuit was not used even in cases where it could have been. When, during the drawing of a line, the artist stopped his pencil to refer back to the model and then returned to the paper, the fixation on the paper also followed the above pattern. We shall refer to this behaviour as the "normal" eye position when drawing.

To our knowledge, these eye movements on the paper during drawing have not been documented to date. We tested subject IA who was skilled at drawing closely spaced parallel lines. In Figure 33.1a, the subject was drawing a vertical line, top to bottom, about 20 cm long and about 1 mm (about 0.1 degree) to the right of a reference line. The vertical eye position shows the catching up movement from one fixation to another, while the horizontal position remains nearly constant. The average distance between one fixation and the next was 0.9 degree.

The behaviour was the same when the line was drawn at greater distances from the reference line: fixations were centred on the reference line and followed the pencil downward in episodes. However, beyond 1 cm distance, the behaviour changed. Figure 33.1b is for a line drawn 2 cm (about 2 degrees) away. Both vertical and horizontal eye positions were now changing, indicating that the eye was moving back and forth between the reference line and the "normal" drawing position as the line was being drawn. Seeing the reference line in parafoveal vision was obviously not adequate for drawing a parallel line, so the eye alternated between the reference and the drawn line.

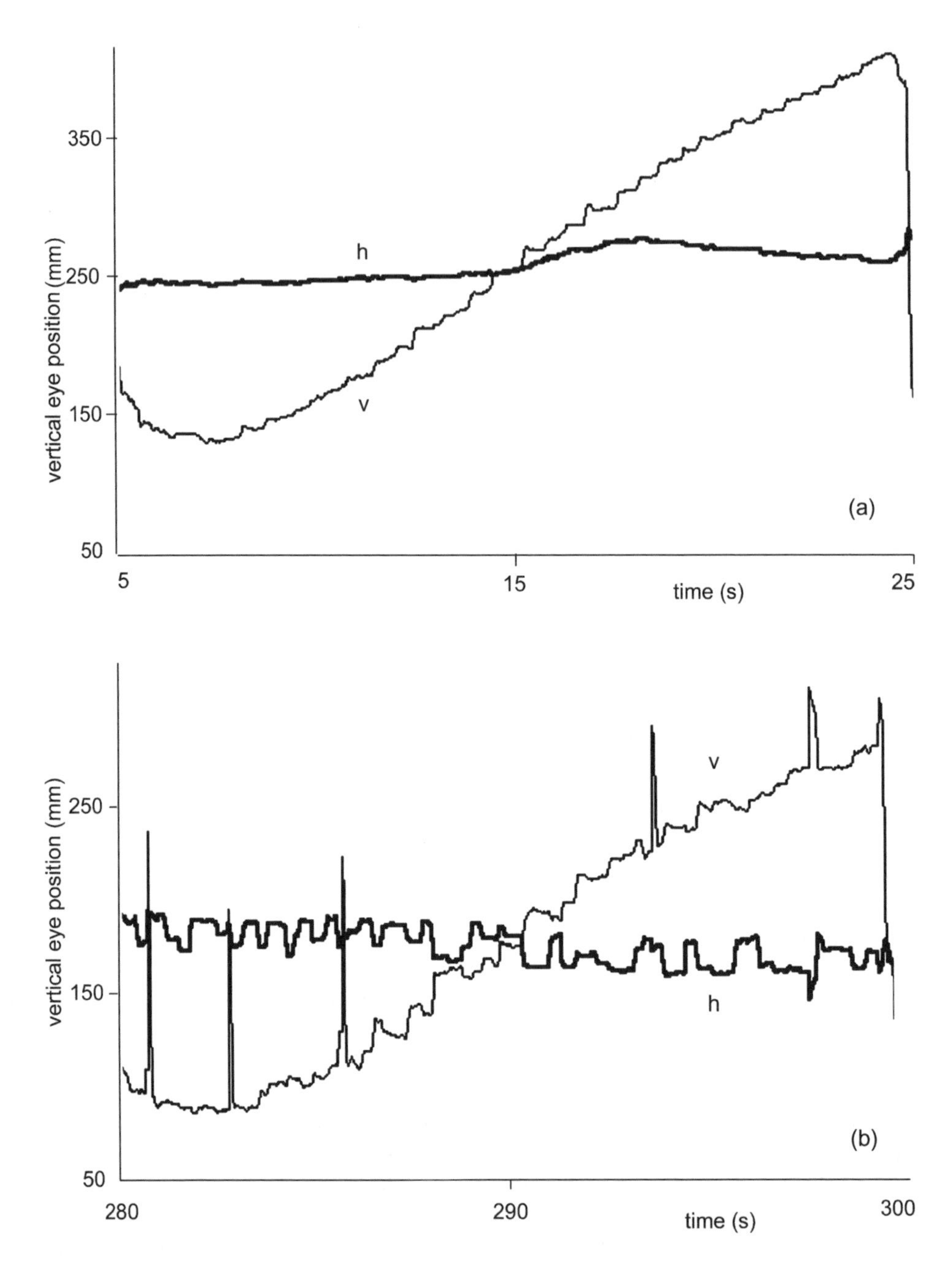

Figure 33.1: Fixation location when drawing parallel vertical lines. Subject IA. Line drawn is (a) 1 mm and (b) 2 cm away from reference line. h horizontal eye position, v vertical eye position.

Location of Fixations on the Model

All subjects with some experience in drawing (HO, ME, JT, PL) showed stable single fixations on the detail of the model that they were drawing. This feature was especially marked with HO: a saccade originating on the paper would find its target after one or two adjustment saccades and then lock onto the point for the duration of the fixation, i.e. one second or more. Fixation durations and frequencies are given in Table 33.2. For HO with whom we have done the greatest number of systematic tests, average fixation duration was 0.6 to 1.0 s and rate 12 fixations/minute. We compared HO's performance to that of three non-drawer control subjects in a separate test of drawing a quick portrait from a photograph (Miall & Tchalenko, 2001). This showed that novices did not maintain single stable fixations, producing instead several, often quite separate, short fixations, referred to here as multiple fixations.

How does this eye behaviour of the artist differ when seeing a face but not drawing it? We gave HO the task of selecting a model for drawing from four possible candidates he had never seen before. The subjects entered his field of view one at a time and sat down in front of him as he was wearing the eyetracker, and we measured his eye movements for the first 30 to 40 seconds.

With all four faces the first fixation was always on the person's left eye,[1] but after that, with one exception, the patterns were all totally unpredictable (Figure 33.2). With this type of test, we never observed a systematic scanning pattern outlining the person's contours and features, as found by Yarbus (1967), albeit for photographs rather than live faces. The exception showed a concentration of fixations on the two eyes, as when "eye contact" occurs between two persons, and the painter eventually selected that candidate to draw. Locher *et al.* (1993) working from photographs of faces found that subjects acquired essential "human contact" information in the first 100 ms flash presentation of a face, and in our case we think that something of this nature occurred during the first of HO's fixations on the candidate's left eye. After this first fixation, the lack of consistent scan path pattern suggests that the painter was not assessing any

Table 33.2: Fixations on the model: basic timings.
Data on HO from Miall and Tchalenko (2001). All other data from this chapter.

Subject and drawing type	**Portrait duration**	**Fixations/ minute**	**Range of fix. duration (s)**
HO *Nick, Luke 2, etc.*	12 min–5 hrs	12	0.6–1.0
HO *brief sketches*	2 min	22	0.6–1.0
ME *self-portrait*	30 min	24	0.4–1.8
JT *self-portrait*	30 min	22	0.8–2.0
PL *self-portrait*	30 min	28	0.3–1.2
CS *model's eye*	15 min	multiple	1.0–1.8
DL model's eye	15 min	28	0.7–1.9

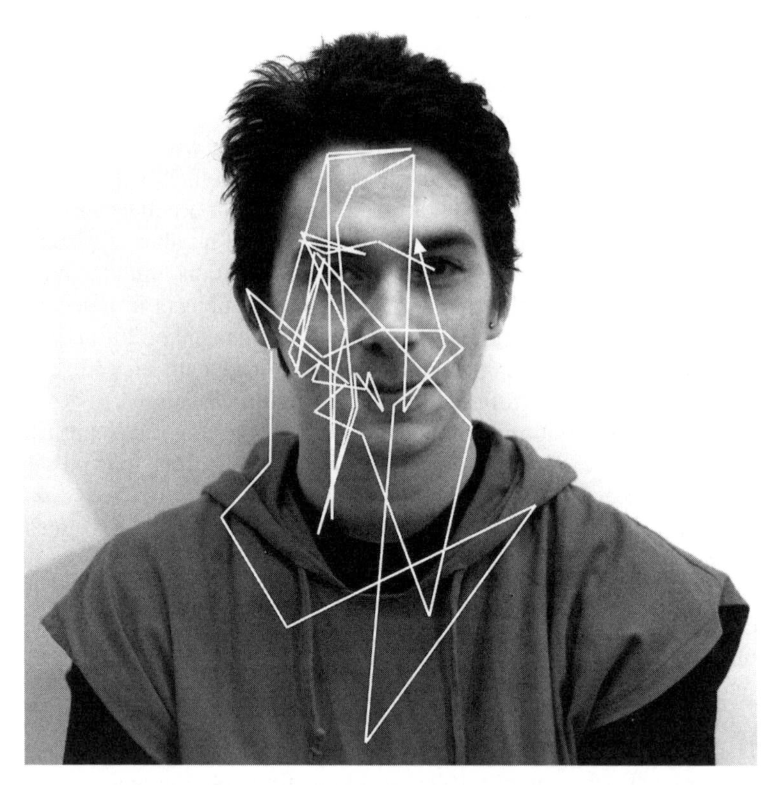

Figure 33.2: Fixations on seeing a live face for the first time. Subject HO, first 13 seconds. This is one of the three candidates that HO did not select to draw.

particular feature of the face to determine his choice of candidate, but was probably reacting spontaneously to the person in front of him, and opting for the one he felt most empathy with. This is indeed what HO confirmed to us after the test.

The clearest measurable difference between the drawing and non-drawing situation was in the fixation frequencies on the model: 12/minute when drawing, 140/minute when not drawing. The latter rate is situated at the lower end of everyday life activities for most people (Rayner & Pollatsek, 1992). Average fixation duration showed a smaller difference: 0.6 s — 1.0 s when drawing and 0.4 s when not drawing. These values apply to HO: for the drawing situation, they are taken from a large number of eyetracker studies (see Table 33.2), and for the non-drawing situation, from tests lasting 30 to 40 seconds.

Paper to Model to Paper Time Sequence (P–M–P)

Figure 33.3 and Table 33.2 show the time sequence of fixations on model and paper for painters of varying degrees of professionalism and experience. All subjects entered a regular working rhythm from the outset and maintained it throughout the session, although variations would occur for particular parts of the drawing (see Miall & Tchalenko, 2001). As found by Pelz *et al.* (2001) for simple block-moving tasks, the overall regularity of the painter's rhythmic pattern constituted the task's coordinative structure, and reflected its intrinsic dynamics and specific subtask demands.

Compared to HO, the other three subjects who drew regularly (ME, JT, PL) showed higher frequencies of looking to model (22–28 fixations/minute instead of 12 fixations/minute), but durations of the same order, indicating that they were spending less time looking at the paper. Although HO's portraits contained fewer lines, he spent more time drawing each line, reflecting his concern with precision and detail.

The last two examples shown in Figure 33.3 are from tests with art school students who had some experience in drawing but no formal training in drawing from life. They were asked to draw the model's eye. Subject CS followed a P–M–P rhythm comparable to HO's, but fixations on the model were multiple (up to about 10 per glance) with many lasting less than 0.10 s. Furthermore, fixations were not targeted on any specific detail but were located at various points of the face. It is unlikely that any useful visual input could take place under these conditions (Rayner & Pollatsek, 1992), and indeed, the resulting drawing was of a generic eye unrelated to the model's. In contrast, subject DL's fixations were single and stable, but close-up video film of the hand showed that it was continuously marking lines on the paper, even when the eye was on the model, and that these lines were not closely connected to what the eye was targeting on the model. In both these cases, the basic P–M–P eye movement cycle, although present, was not fulfilling its role of capturing specific visual information from the external world, interpreting it and laying it down on the paper. As suggested by Suppes *et al.* (1983), with unskilled operators the eye may wander while the subject is wondering what to do next.

Eye–hand Coordination

On page 703 we described the simplest case of interrelation between the eye's fixations on the paper and the hand drawing the line. Taking into account the fixations on the model and the P–M–P cycle, we can now explore the broader picture of eye–hand coordination during the drawing process.

Fixations in the Vicinity of the Hand

Figure 33.4 shows the first line drawn in HO's 5-hour *Portrait of Nick.* It depicts the back edge of the model's upper eyelid of the right eye, and is about 5 cm long and concave downwards with a changing curvature. It was drawn left-to-right with the pencil stopping twice on the paper's surface while the eye went back to look at the model — hence the three sections E1–E2–E3. The temporal movement sequence, analysed on the basis of eye tracker and video data, indicates that the eye and hand were working together, with coordination between them maintained by delaying the

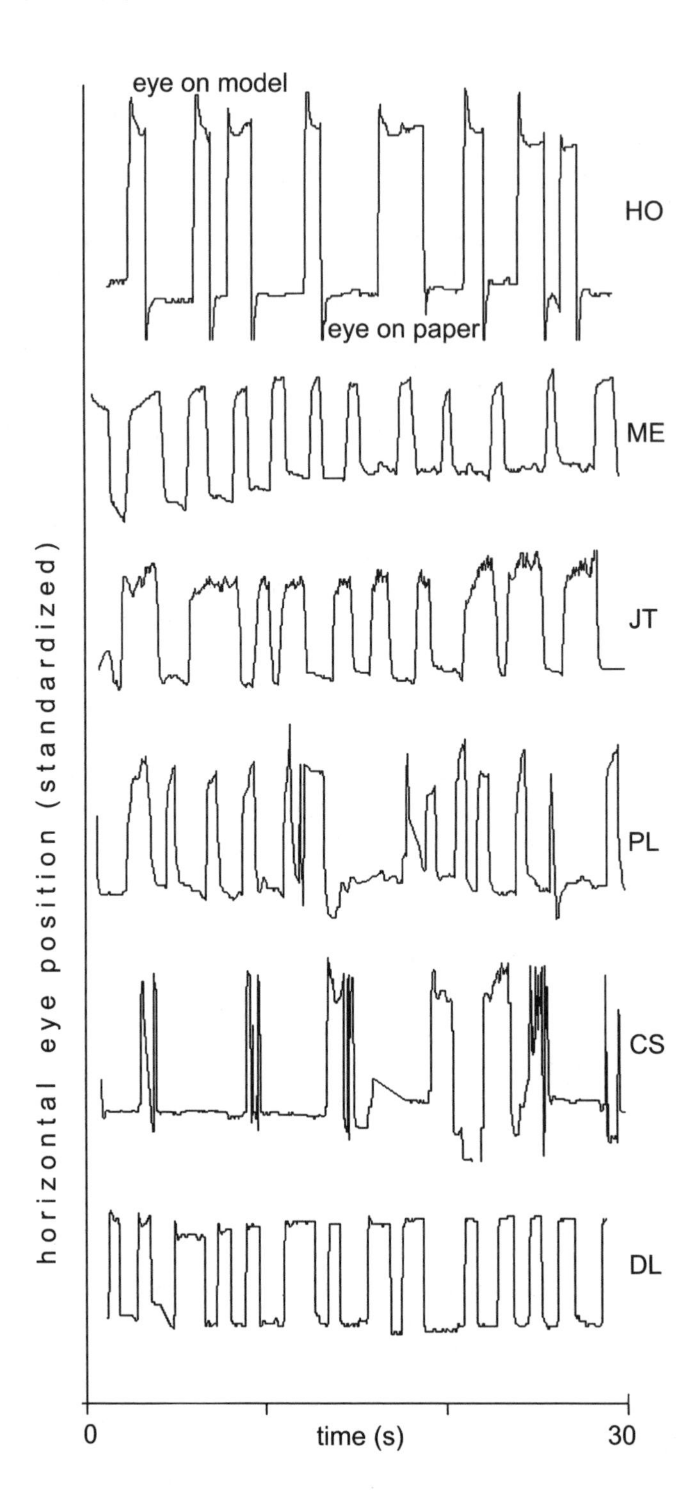
eye on model
eye on paper
HO
ME
JT
PL
CS
DL
horizontal eye position (standardized)
0
time (s)
30

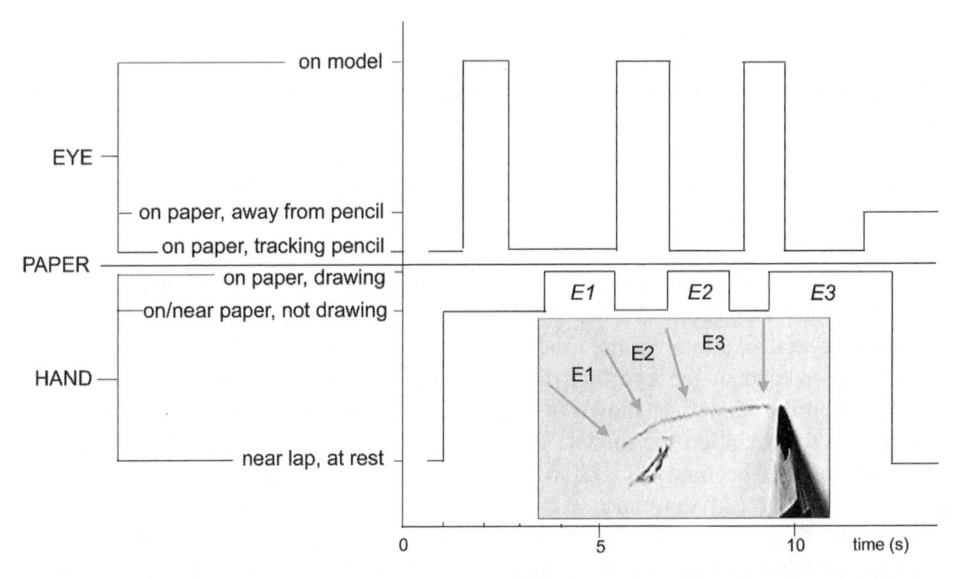

Figure 33.4: Schematic representation of horizontal eye and hand positions when drawing a simple line (Humphrey Ocean, *Portrait of Nick, 1998*).

hand movements until the eye was available for guiding the movement, as in the block-moving exercises described in Pelz *et al.* (2001). We can interpret this behaviour as being the simplest case of a fading visual memory, with the hand drawing until the memory needs to be refreshed, although the timing is exceptionally long if thought of in terms of mental image maintenance and visual buffer memory (Kosslyn, 1994). Two factors may favour such extended timings: (1) the information required to continue an existing line is essentially restricted to a single element — the line's angle, and thus forms part of the abstract schematic representation underlying our memory for scenes (Intraub, 1997); and (2), seeing the line as it is being drawn reinforces the mental image and thus aids in its memory maintenance, as for the subjects in Epelboim *et al.* (1995) who were making use of a visual display as an extension of their memory.

Figure 33.3: Comparative P-M-P time sequences during drawing. Each graph represents the horizontal eye position, with looking at the model shown as high levels, and looking at the paper as low levels. (The *Y* axis distance between the two levels represents the horizontal physical distance between model and paper, and has been standardised on the graph to facilitate temporal comparison between subjects). JT's apparent eye movement when looking at model is due to head movement and is not a fixation instability.

Fixations Away from the Hand

The before-last line drawn in the same portrait represents the outer contour of the hair on the model's right side (Figure 33.5). At 20 cm it is also the longest single line of the picture. Of gentle uniform curvature, it was drawn in two consecutive strokes, H1–H4 and H5–H6, with the hand resting at the lap in between. H5–H6 was then reinforced to become H7–H8.

Drawing started at H1 after several long fixations on the model, and the eye followed the pencil downwards with two small saccades, before leaving the trace for three fixations elsewhere on the picture while the hand continued to H4. There followed several fixations on the model, after which the eye and hand met at the start of H5, but instead of the eye then following the pencil, it fixated elsewhere on the paper and on the model, as shown in Figure 33.5. By the time the pencil had reached H6 the eye was back on the model. The line just drawn, H5–H6, was then reinforced very precisely to become H7–H8, with, as before, the eye only coinciding with the pencil at the start. In summary, a curved line about 10 cm long (H5–H6) was first drawn, and then very precisely reinforced (H7–H8), with the eye only locating the starting point and then entirely foveating elsewhere. This eye–hand behaviour is shown schematically in Figure 33.6.

The H1–H8 line denotes a more complex eye-hand coordination than E1–E2–E3. The painter now draws 10 cm before referring back to the model, suggesting that the 1.7 cm retained in E1–E2–E3 constituted a deliberate strategy, a sort of minimal memory solution as found by Ballard *et al.* (1992) for their block-moving exercises.

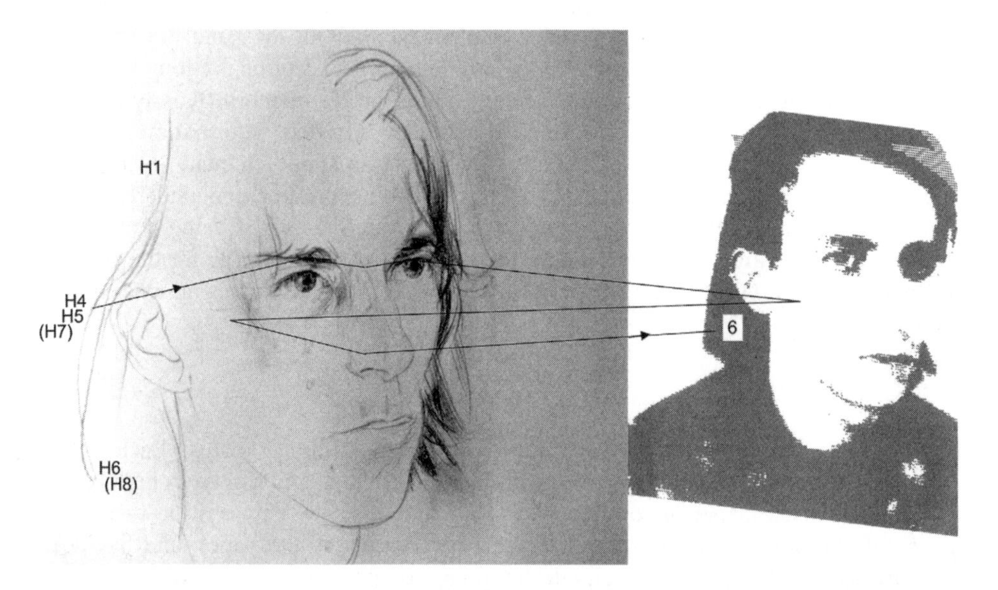

Figure 33.5: The eye's movement path between the paper (left) and the model (right) during the drawing of line H5–H6 (Humphrey Ocean, *Portrait of Nick, 1998*).

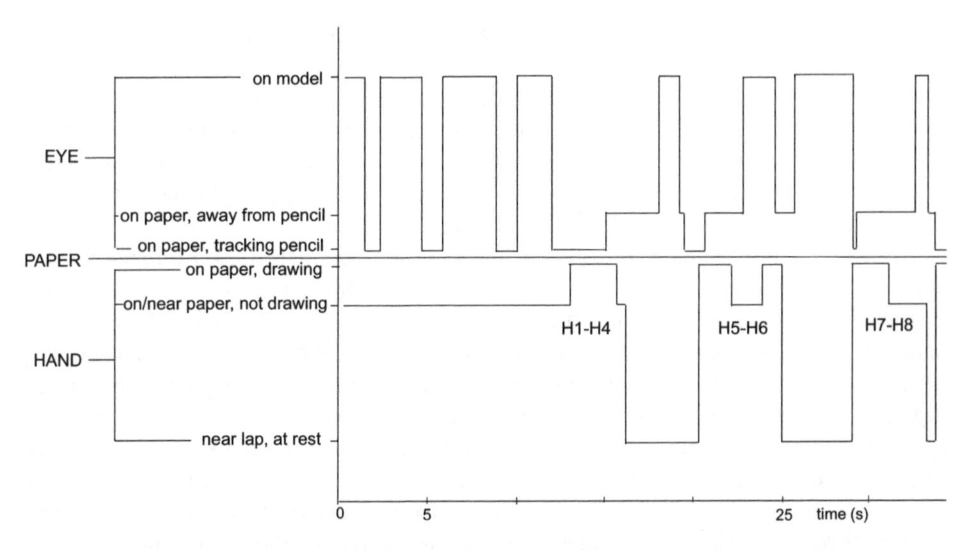

Figure 33.6: Schematic representation of horizontal eye and hand positions when drawing a complex line (Humphrey Ocean, *Portrait of Nick, 1998*).

Rather than keeping the entire E line in memory, the painter referred back to it in episodes. With the H1–H8 line, his strategy was different: during most of the drawing, his fixations were centred on other parts of the picture not directly related to that line, thus allowing for perception of the "picture so far" to become an additional input into the drawing process, for example by situating the line with respect to others previously drawn.

The reinforcing of H5–H6 into H7–H8 highlights the complexity of the cognitive process. Whereas drawing without seeing could be explained by motor commands based on a visual mental image, retracing a 10 cm line perfectly about 10 seconds after the original, and without ever having foveated the original, suggests the presence of a motor memory component to the drawing process. It is the memory of the action — not the memory of the vision — of drawing H5–H6 that allowed the painter to draw the identical H7–H8.

The two types of lines seen above have in common the fact that they are not reproducing lines which have a separate existence on the model's face. E1–E2–E3 is the painter's resolution of a zone of changing light, colour and texture above the model's right eye; to draw it, the painter relied on the visual information provided by that zone. H1–H8 is the painter's resolution of a combination of features: the shape of a single strand of hair, the general shape of the hair on that side of the face and the left limits of the picture; to draw it, the painter required at least as much visual information from the picture drawn so far as from the model's face itself. The drawing process is therefore a continually changing balance between the use of visual input from the external world and visual input from the growing picture, and the painter's eye movements are a good indication of the state reached by that balance at any given stage of the drawing.

Further Observations on Eye–hand Coordination and Motor Memory

Two sets of observations on subject HO are considered important enough to be added to the above remarks, even though future work with other painters is needed to confirm their generality. The first concerns a feedback loop between eye and hand occurring as the painter fine-tunes the line he is drawing; the second concerns the painter's acquisition of a long-term memory of the drawn picture.

Not infrequently painters rehearse a line they are about to draw by repeating the hand's action several times with the pencil tip just off the paper's surface. We studied this with HO who would at times practise a line with a dozen, or more, strokes as his pencil gradually approached the paper until it started marking the surface. During this hand movement, fixations remained stable and near the line, or were occasionally interrupted as the painter looked at other parts of the drawing or at the model.[2] Figure 33.7 is a motion sensor record of the pencil's movement during the drawing of one of the lines forming the lips in *Portrait of Nick.* The first 10 strokes were off the paper and the line was only drawn during the last four strokes. Close-up video film showed that the pencil path described an elongated ellipse encompassing the future line, gradually narrowing down onto the line's starting point. This suggests that the initial motor command to the hand was being adjusted in steps as the eye was observing the hand's movement, and that the process continued until the result corresponded to the mental image of the line the artist wanted to draw.

How long does a painter retain the memory of a portrait drawn some time ago? We tested HO by asking him to redraw the 12-minute portrait *Luke 2* described in detail in

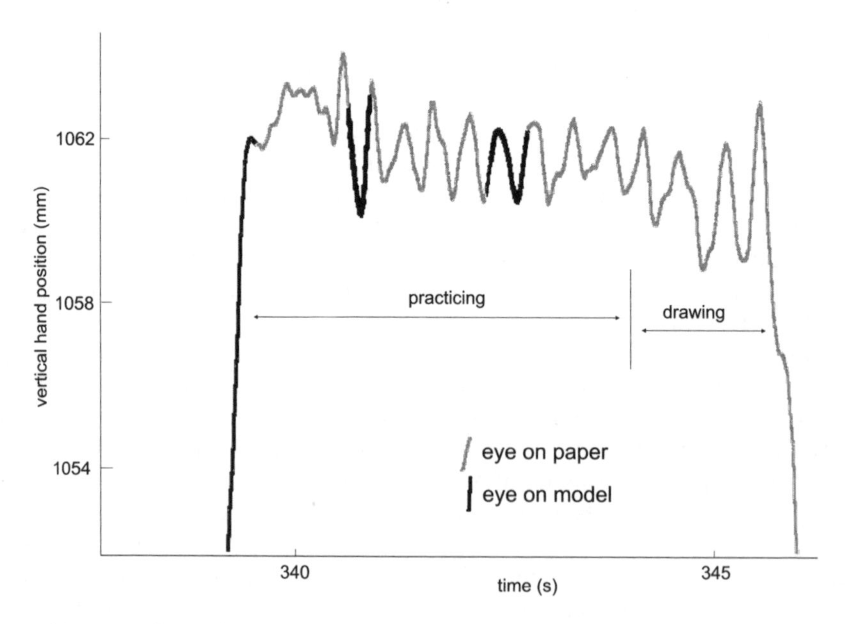

Figure 33.7: Horizontal hand position while practising a line (Humphrey Ocean, *Portrait of Nick, 1998*).

Miall and Tchalenko (2001), and which he had not seen for over one year. Fixations were now entirely located on the paper, with no observable sign of refresh or memory recall such as aversion fixations (Rayner & Pollatsek, 1992), pauses or glances away from the drawing. The overall lines of *Luke 3* were almost exactly the ones of *Luke 2*, although the new lines were individually much longer, i.e. drawn with far fewer strokes. For example, the line representing the contour of the nose was drawn in four segments in *Luke 2*, and as single segment in *Luke 3*. It suggests that the artist retained a near-perfect long term memory of the drawing he had made one year earlier, but not of the drawing act itself as had been the case with the buffer memory described for H1–H8.

Eye Control

Voluntary Control of Eye Movements

We have already noted that the experienced painter differed from the novice in his ability to repeatedly target saccades onto a small detail of the model's face, and to lock-on to that detail in a steady fixation. This suggests an eye control factor which can also be demonstrated with the "eye signature test" in which the subject writes his/her name on a blank screen with the eyes alone and without seeing either the cursor or the line (Tchalenko, 2001a). A smooth and accurate movement of this type is generally considered impossible (Kowler, 1990), but taking into account the straight-line trajectories of saccadic eye movements, we found that subjects who drew regularly from life were generally much better at producing legible results than those who did not (Figure 33.8).

A stricter test of eye control consists in moving the eye slowly from one point to another, a task first examined by Yarbus (1967). Figure 33.9a,b compares the eye's movements between the corners of a square in saccadic and in slow movement. For the latter, and unknown to the subject who has the impression of moving smoothly, the eye's trajectory is made up of smaller segments in an alternation of *constituent saccades* and inter-saccadic intervals referred to here as *fixations*. Preliminary tests on 15 subjects with a simplified eye tracker (Quick Glance) suggested a strong correspondence between the ability to trace a line in this way and write one's name (Tchalenko, 2001a,b), and we decided to see whether the method could be quantified and used to differentiate between subjects of differing aptitudes in drawing from life. We give below the first results of this study.

Tracing a Line by Eye: Temporal Behaviour

Tracing is defined here as the eye's movements when joining points or following a line displayed on the computer screen or on a blank paper. Tests were made with an ASL 504 eyetracking system used at the maximum sampling rate of 0.020 s (50 Hz). Subjects were seated 65 cm from the computer screen and provided with a back head-rest, but were not clamped.

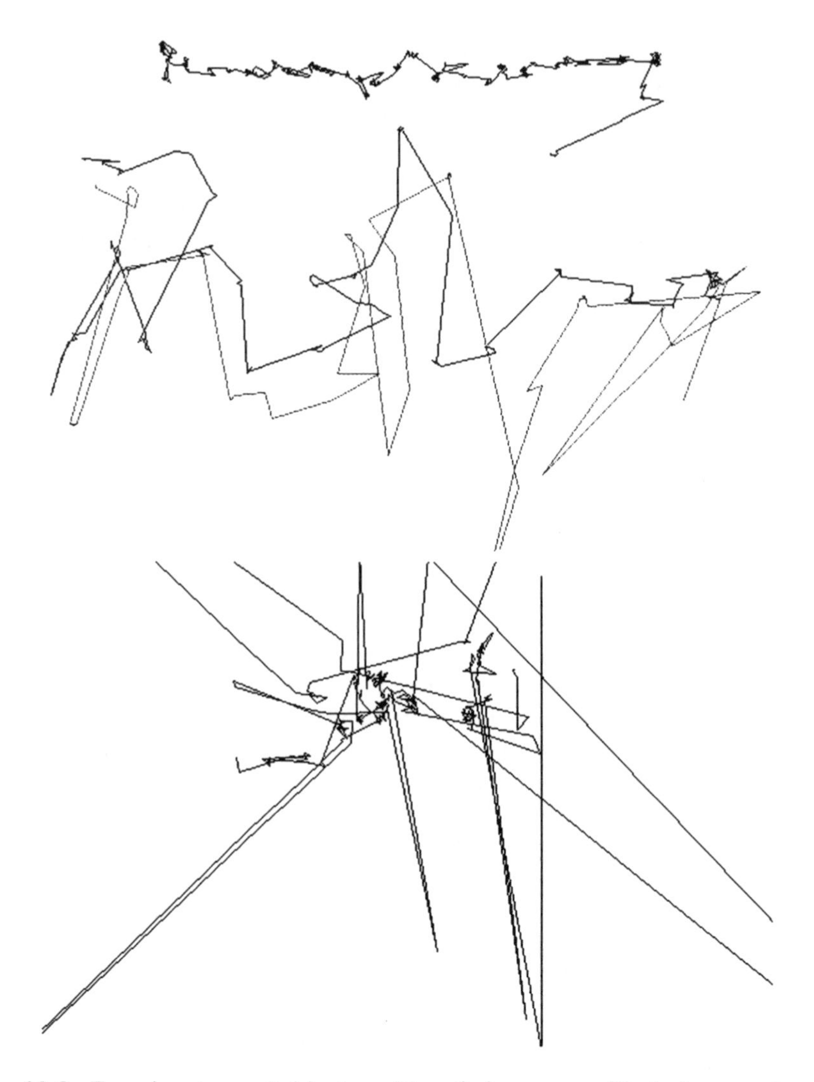

Figure 33.8: Eye signatures. Subjects writing their names with their eyes alone. All signatures are at the same scale: top "Humphrey", middle "John" two consecutive tests illustrating consistency, bottom "Carol".

Tracing a spiral was found to be a good test for tracing ability as it combined in one exercise a variety of movement directions and curvatures (Figure 33.9c). Subjects followed the original reference line from the centre outward and tests were repeated three times at different speeds. Furthermore, each subject was tested at three different times of a same day in order to ascertain reproducibility of results. The questions examined were, firstly, how regular was the overall movement in time, and, secondly, how accurate were the fixations with respect to the reference spiral.

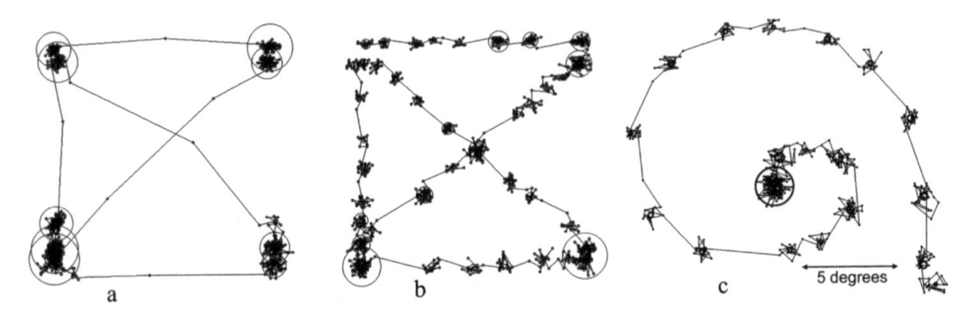

Figure 33.9: Saccades (a) and slow movement (b) between the corners of a square, and slow movement following a spiral (c). The reference spiral is not shown. Subject JT was not seeing the cursor or the line being drawn. Circles represent fixations calculated from 50Hz sampling points, as explained in Appendix I.

Defining *average saccade amplitude* (a) as the mean amplitude of the constituent saccades, and *tracing speed* (V) as the average speed at which the eye traces the drawn line from beginning to end, we found that saccade amplitude for the five subjects tested systematically increased in linear proportion to tracing speed (Figure 33.10):

$$a = 0.21\text{V} + 0.83 \qquad (1)$$

The theoretical equation for this relationship based on the known dynamics of saccades is also shown on Figure 33.10 as Equation 2 (see Appendix II for computation and discussion) and is seen to be in reasonable correspondence with our empirical data for tracing speeds above 10 degrees/s. However, despite the fact that the five subjects tested were of very different drawing skills, Equation 1 applied equally well to all of them, and it was not possible to discriminate between drawing abilities on this basis alone. We think that this may be partly due to the fact that at very low tracing speeds, where the difference between subjects is greatest, very small saccades become confused with system noise, and our method is no longer appropriate for this type of study.

Tracing a Line by Eye: Spatial Accuracy

To assess how accurately the eye was targeting the line being traced, we measured the angular distance between each 50 Hz sampling point and its orthogonal projection onto the line. We called this distance the *Spatial Accuracy* (SA). Figure 33.11 shows an example of results for subject JT at an average tracing speed of 12 degrees/s. Standard deviations calculated with respect to the mean SA value were remarkably consistent for each subject, unaffected by tracing speed and identical for straight line and spiral tests (Table 33.3). The ranking obtained corresponds well with subjects' drawing experience, suggesting that the standard deviation of Spatial Accuracy is a valuable measure of a subject's eye control capabilities.

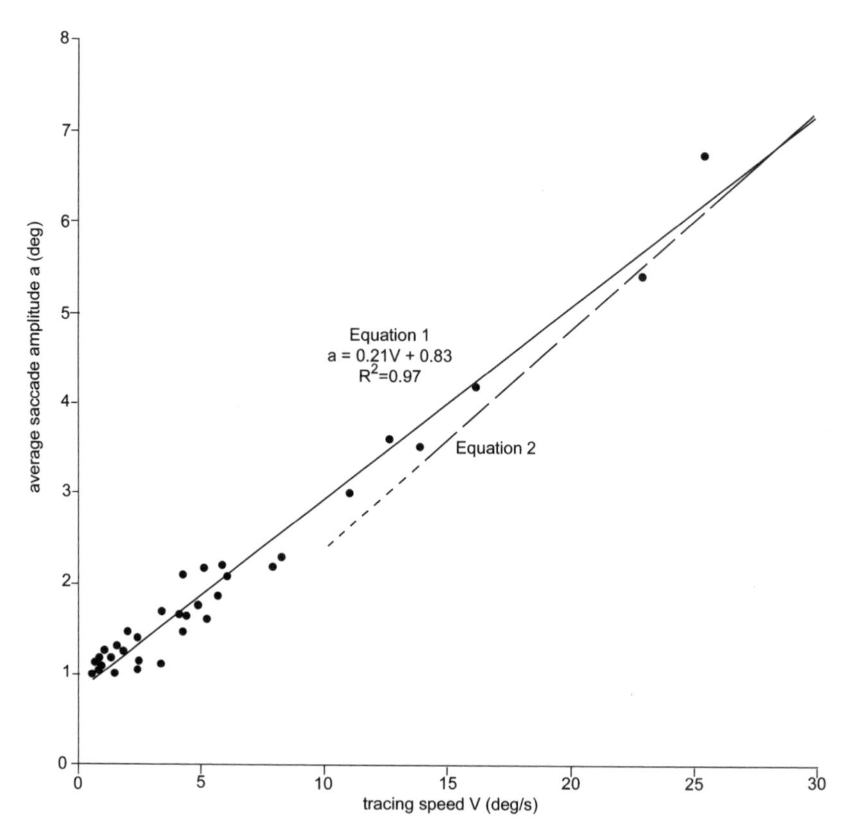

Figure 33.10: Tracing a spiral by eye. Relationship between constituent saccade amplitude (a) and tracing speed (V). Results are for five subjects: JT, BD, MM, ME, PL.

Concluding Remarks

In this first study of eye movements in portrait drawing from life we investigated the picture production process and established some of the basic physiological parameters involved. If we were to generalize on the basis of the 12 subjects of different abilities studied so far (Table 33.1) we could postulate the following description of the drawing process.

Drawing from life entails eye movements from paper to model and back — the P–M–P cycle — at a rhythm of between 12 and 28 fixations on the model per minute, and fixation durations on the model of 1s or over, precise values depending on the artist and the type of drawing — quick sketch or fully drawn portrait. The function of the glance to model being to acquire detailed visual information, this is best achieved with single and stable fixations. The information is then available in the painter's visual memory for about 2 seconds before needing to be refreshed, although in the advanced stages of the drawing, work may proceed for longer periods on the basis of the lines

Table 33.3: Spatial accuracy tests: mean and standard deviation values in angular degrees.

Subject	JT	ME	MM	BD	PL
Spiral at average speed (3 tests/subject)	0.38 ± 0.07	0.47 ± 0.08	0.54 ± 0.16	0.78 ± 0.20	1.13 ± 0.27
Spirals fast, average, slow (9 tests/subject)	0.40 ± 0.08	0.46 ± 0.10	0.52 ± 0.14	0.74 ± 0.21	1.25 ± 0.28
Straight lines fast, average, slow (9 tests/subject)	0.37 ± 0.09		0.57 ± 0.19	0.88 ± 0.34	
Spatial accuracy (SA) (average of all above tests)	0.38 ± 0.08	0.47 ± 0.09	0.54 ± 0.16	0.80 ± 0.25	1.29 ± 0.28

already existing on the paper. There is also some evidence of a motor component to the painter's memory. On the paper, the artist's fixations do not coincide with the pencil point but are located at a distance of 0.5 to 1 degree, and as the pencil moves, small saccades keep up with the line as it is being drawn. Closed-loop type situations may arise between eye and hand when the hand practises a line to be drawn without actually marking the paper, gradually honing in to its final position and direction.

As eye movements during drawing are essentially volitional and controlled by the subject, the question arises whether experienced painters have better control over these movements than beginners. One way of assessing this is by observing the eye movements when following — or tracing — slowly a pre-drawn line. Although the subject is under the impression of moving smoothly, eye movements are actually decomposed into an alternation of constituent saccades and fixations, with a linearly proportional law between average saccade size and overall tracing speed. The accuracy with which the tracing takes place, i.e. the subject's eye control, can be quantified by measuring the standard deviation of the distance between gaze point and the line being traced. Our results show that the practice of drawing from life is clearly associated with a higher degree of eye control measured in this way.

The P–M–P cycle, which we have observed with all subjects studied, forms the universal principal of life drawing, and future studies with different subjects should in time refine the results reported here. Detailed observation of the way individual painters work — such as Humphrey Ocean's practice movements — will gradually enrich our rather schematic present knowledge. The subject of eye control, which has only been touched upon so far, may prove to be the most rewarding aspect of this study as it unites drawing with another fine-controlled eye–hand skill, i.e. surgery.

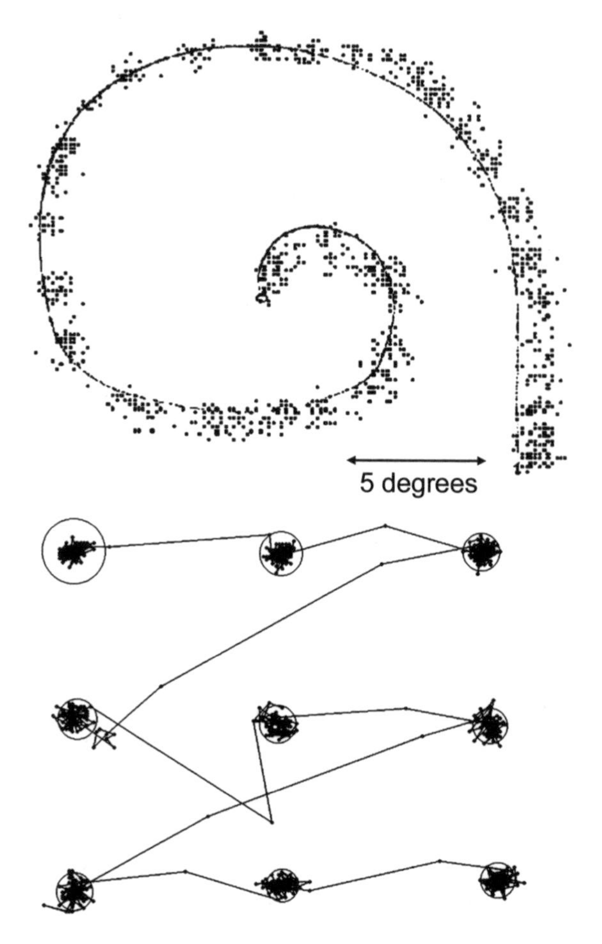

Figure 33.11: Spatial accuracy during tracing by eye. The continuous spiral is the reference line followed by the eye. The nine-point calibration test preceded the spiral tests. Subject JT.

Acknowledgements

We wish to thank Chris Miall for his constant support and advice, and Stephen Oliver and Don Jarrett for generously making available their technical eyetracking expertise and equipment. The projects described were supported by the Welcome Trust and UK Engineering and Physical Sciences Research Council (EPSRC).

Notes

1 HO is right-eyed, although we don't know if this is significant.

2 We wish to correct an initial conclusion reached in Miall and Tchalenko (2001) where it was mentioned that the eye followed the practicing hand in a tracking or smooth motion. Further analysis of the material revealed that the head oscillation mentioned in endnote 15 was more important than originally appreciated and that the painter was moving his body synchronously with the rehearsing stroke. Corrected fixation positions were, however, stable and not different from normal drawing.

References

Ballard, D. H., Hayhoe, M. M., Li, F., & Whitehead, S. D. (1992). Hand–eye coordination during sequential tasks. *Philosophical Transactions of the Royal Society of London B, Biological Sciences*, *337*, 331–339.

Becker, W. (1991). Saccades. In: R. H. S. Carpenter (ed.), *Eye Movements* (pp. 95–137). London: Macmillan Press.

Cohen, D. J., & Bennett, S. (1997). Why can't most people draw what they see? *Journal of Experimental Psychology: Human Perception and Performance*, *23*(3), 609–621.

Epelboim, J., Steinman, R. M., Kowler, E., Edwards, M., Pizlo, Z., Erkelens, C. J., & Collewijn, H. (1995). The function of visual search and memory in sequential looking tasks. *Vision Research*, *35*(23/24), 3401–3422.

Frith, C., & Law, J. (1995). Cognitive and physiological processes underlying drawing skills. *Leonardo*, *28*(3), 203–205.

Gombrich, E. H. (1963). *Art and Illusion*. NJ: Princeton University Press.

Intraub, H. (1997). The representation of visual scenes. *Trends in Cognitive Science*, *1*(6), 217–221.

Konecni, V. J. (1991). Portraiture: An experimental study of the creative process. *Leonardo*, *24*(3), 325–328.

Kosslyn, S. M. (1994). *Image and Brain*. Cambridge, MA: MIT, Press.

Kowler, E. (1990). The role of visual and cognitive processes in the control of eye movement. In: E. Kowler (ed.), *Eye Movements and Their Role in Visual and Cognitive Processes* (pp. 1–63). Amsterdam: Elsevier.

Livingston, M., & Hubel, D. (1995). Through the eyes of monkeys and men. In: R. Gregory, J. Harris, P. Heard and D. Rose (eds), *The Artful Eye* (pp. 52–65). Oxford: Oxford University Press.

Locher, P., Unger, R., Sociedade, P., & Wahl, J. (1993). At first glance: Accessibility of the physical attractiveness stereotype. *Sex Roles*, *28*(11/12), 729–743.

Lord, J. (1980). *A Giacometti Portrait*. New York: Farrar Strauss Giroux.

Miall, R. C., & Tchalenko, J. (2001). A painter'eye movements: A study of eye and hand movement during portrait drawing. *Leonardo*, *34*(1), 35–40.

Pelz, J., Hayhoe, M., & Loeber, R. (2001). The coordination of eye, head, and hand movements in a natural task. *Experimental Brain Research*, *139*(3), 266–277.

Rayner, K., & Pollatsek, A. (1992). Eye movements and scene perception. *Canadian Journal of Psychology*, *46*(3), 342–376.

Snyder, A. W., & Thomas, M. (1997). Autistic artists give clues to cognition. *Perception*, 26, 93–96.

Solso, R. (2000). The cognitive neuroscience of art. *Journal of Consciousness Studies*, *7*(8/9), 75–85.

Solso, R. (2001). Brain activity in a skilled versus a novice artist: An fMRI study. *Leonardo*, *34*(1), 31–34.

Suppes, P., Cohen, M., Laddaga, R., & Floyd, H. (1983). A procedural theory of eye movements in doing arithmetic. *Journal of Mathematical Psychology*, 27, 341–369.

Tchalenko, J. (1991). *The Making of Double-Portrait*. London: Dulwich Picture Gallery catalogue publication.

Tchalenko, J. (2001a). Free-eye drawing. *Point*, 11, 36–41.

Tchalenko, J. (2001b). Eye–hand coordination in portrait drawing. *ECEM 11, Final Program and Abstracts* (p. 42). Turku, Finland.

Tchalenko, J., Dempere-Marco, L., Hu, X., & Yang, G. (2001). Quantitative analysis of eye control in surgical skill assessment. *ECEM 11, Final Program and Abstracts* (p. 43). Turku, Finland.

Yarbus, A. (1967). *Eye Movements and Vision*. New York: Plenum Press.

Zingale, C. M., & Kowler, E. (1987). Planning sequence of saccades. *Vision Research*, *27*(8), 1327–1341.

Appendix I — Eyetracker Equipment

The Painter's Eye Movements was a Wellcome Trust 1998 sciart project undertaken with the University Laboratory of Physiology, Oxford, on their AlphaBio Eyeputer (EP) eyetracker. This is a head-mounted 30 Hz system providing fixation accuracies better than 2 degrees. A Polhemus Fastrack motion-analysis system provided simultaneous recording of the pencil position.

Investigations with subjects AE, CS, DL, IA and HO for the drawing of *Luke 3* were carried out in 2000 at Camberwell College of Arts, London (CCA) as part of the ongoing Drawing & Cognition project, using a 50 Hz head-mounted ASL 501 (Applied Science Laboratories) system providing an accuracy better than 1 degree. Concurrently, the Eye Control study that developed from a preliminary investigation at CCA with a simplified eyetracker (QuickGlance by EyeTech Digital Systems) was followed by a Wellcome Trust 2001 sciart project undertaken jointly by CCA and the Department of Computing, Imperial College of Science, Technology and Medicine, London. We used the ASL 504 System (50 Hz), a remote device with nothing attached to the subject who is seated with a back head-rest in front of a computer screen. Maximum accuracy is 0.5 degree. We adopted a standard ASL fixation algorithm using an improved visual angle calculation. In this dispersion algorithm, a fixation is initiated when the standard deviation of the x and y screen coordinates of 5 consecutive points (0.100 s duration) are below 0.5 degrees. New threshold restrictions are then applied as the fixation grows in duration before the next saccade. As part of the same project, three artist subjects ME, JT and PL were investigated with an ISECS eyetracker by QinetiQ, U.K. This is a 25 Hz device which looks, and feels, like a pair of spectacles, and provides an accuracy of about 1 deg. All these systems provide measurements in data file form as well as a video film with a superposed cursor marking the eye's gaze position.

Appendix II — Tracing a Line by Eye: Theoretical Relationship

For single saccades greater than 5 degrees in amplitude, Becker (1991) suggested the following correlation between the duration (D) of a saccade and its amplitude (a):

$$D = D_0 + d.a \tag{1a}$$

where $D_0 = 0.025$ s and $d = 0.0025$ s/deg.

For a tracing path consisting of a succession of saccades alternating with fixations, the overall duration (Δ) can be written as:

$$\Delta = N{\cdot}D + N{\cdot}F \tag{1b}$$

where N is the number of constituent saccades or fixations, and F the average duration of a fixation. Our spiral tests showed that values of F are about 0.300 s for saccades of 3 degrees and decrease asymptotically to 0.200 s for saccades greater than 5 degrees. This dependence of inter-saccadic fixation duration on tracing speed may reflect the fact that, in saccade sequences, individual saccades are controlled by an organized plan for the entire sequence, as suggested by Zingale and Kowler (1987). Taking $F = 0.200$ s and solving for Equations 1a and 1b leads to:

$$a = 0.225.V \,/\, 1–0.0025V \tag{2}$$

Equation 2 was found to provide a good approximation of our data for tracing speeds above about 10 degrees/s. At lower tracing speeds, as saccade amplitudes approach 1 degree, the algorithm adopted to differentiate between saccades and fixations may no longer be ideal for describing the observed movement. Higher sampling rates and greater system accuracies will be required to ascertain whether very slow voluntary eye movements can still be adequately described in terms of a succession of saccades and fixations.

Commentary on Section 5

Eye Movements in Communication and Media Applications

Alastair G. Gale

In recent years there has been a proliferation of the application of eye movement techniques in applied areas. This is partly due to the increasing availability of high quality eye movement recording systems at a more affordable price. This has drawn more people into using eye movement techniques, which is to be encouraged, however this must be tempered with the need for appropriate understanding and usage of such equipment.

In such applied situations it is often saccadic movements that are of main interest, and eye movement recording is typically used in conjunction with other behavioural response techniques such as post-experimental subjective rating scales, questionnaires or interviews. This combination of quantitative and qualitative data provides a very fruitful approach to examining real world questions.

In the area of communication and media an observer's eye movements can be studied for many reasons, but primarily the relationship between descriptive text and visual images is of interest. This is a theoretically intriguing area which crosses the line between typical reading research and that research concerned more with picture viewing. In practical terms, print or web based advertisers need to know how best to structure their advertisements to create maximum reader impact and potential sales leads; likewise journalists need to ensure that their carefully crafted text is actually read.

Unfortunately, communication and media applications are often investigated by means of short-term research funded by a commercial agency or a client who wants results yesterday and who may well have little interest in appropriate experimental design or scientific rigour. Objectivity of research in this area is important and it is a necessity that researchers clarify when results are simply indicative or are statistically rigorous and capable of extrapolation to a wider population. Research in this broad area is often complicated by the study design and in particular the specific instructions, or lack of them, given to the participant. There is a large difference between being instructed to simply "look at" a web site as opposed to "search for an item to purchase" on the same web site. Additionally it is important to clarify the accuracy

The Mind's Eye: Cognitive and Applied Aspects of Eye Movement Research

ISBN: 0–444–51020–6

with which the observer's gaze location is derived and consideration needs to be given to whether this actually coincides with where they are visually attending.

One area where eye movement recording is increasingly used is in marketing and advertising; whether print or web based. In the chapter by Radach, Lemmer, Vorstius, Heller and Radach the content-related design factors in print advertisements are studied. They are particularly interested in the relationship between the text and image in an advert. They consider adverts as being either explicit or implicit; the distinction refers to the complexity of the relationships between advertisement elements. In the former, both textual and pictorial elements are in direct semantic relation to one another and refer to the product; whereas in the latter neither refers to the product directly. The research interest is not in the initial advert selection or its early visual processing but in the processes beyond these stages and two experimental studies are reported. Not surprisingly inter-participant differences in eye movement parameters were found. Statistically significant differences are reported between the two advert types. Participants rated implicit adverts as more interesting and also spent significantly more time on them (exhibited more fixations) with mean fixation duration and saccade amplitude not affected. They conclude that implicit adverts are the more likely to be effective.

Another example of media composed of text and images is web pages and their design is studied by Stenfors, Morén and Balkenius. The researchers use their own eye movement technique citing the merits and drawbacks of IR based methods. The technique used has a precision of less than two degrees which they consider sufficient for an accurate estimation of the visual element fixated on a web page; although there are several remote eye movement recording systems which have a higher accuracy. This raises an interesting point. In recording eye movements there is no single technique that suits all situations. Any eye monitoring system can give an estimation of the location of the participant's point of gaze, itself typically assumed to be where the observer is visually attending (but which may not be). The useful accuracy of the system employed may vary with the particular observer as well as the experimental conditions. Additionally, it is important to consider the observer's individual useful field of view, which also may not be constant across an experimental trial.

Interesting information is presented from several studies including data concerning how participants avoid looking at banner adverts. Unfortunately participant numbers are not generally stated in these studies, which means the reader has to access previous work, some of which is not readily in the public domain. One study reported used five subjects but no statistical analysis is presented. It is thus difficult to know the veracity of some of the statements.

Scrolling textual displays are to be found as advertisements on web pages as well as in large public electronic display billboards. Kitamura, Horii, Takeuchi, Kotani and d'Ydewalle examined the parameters that affect the reading of scrolling Japanese text. The speed was varied of right to left scrolling text on a monitor and the participants had to read the text and rate its ease of reading. The resultant eye movement patterns found somewhat resembled OKN, being a combination of smooth pursuit and saccade movements. From the single participant's eye movement trace shown at different scrolling speeds this behaviour was exhibited almost at the right hand edge of the display monitor and only traversed the monitor at high scrolling speeds. Ease of reading fell off with a scrolling speed above 10°/sec. The maximum number of Japanese characters participants could read was found to be five. Although a small number of participants were used and no statistical analysis is reported the work is intriguing.

The concept of web page layout is further examined by Holmqvist, Holsanova, Barthelson and Lundqvist who are interested in the comparison between reading traditional newspapers and internet-based papers. The research topic here is not the advertisements per se but in the actual page content; the layout, navigation and structure within each media. The research investigates whether papers or net papers are the more deeply read. It is concluded that with net papers the readers scan more and read less than when they deal with newspapers — the latter media is the more efficient in that less time is spent scanning to find what to read. In the experimental design for the two experiments reported it is good to see the internet connection speed is specified as this is so clearly a variable in this domain.

A less well investigated area is the topic of television and film subtitles — a complex area where potentially the observer can shift attention between the visual scene, the soundtrack and the subtitle. De Bruycker and d'Ydewalle report a study on television subtitles using children and adults, and where the subtitle is presented either in their native language (film soundtrack in a foreign language) or where the soundtrack is in the native language and the subtitle is in the foreign language (reversed subtitling). A range of eye movement parameters was measured. Results indicated that true reading of the standard subtitles was apparent whereas evidence was less clear for systematic reading of reversed subtitles. Clear differences between children and adults were also found in their reading behaviour. Although not specifically investigated the authors additionally comment on the extent of processing of the soundtrack. A point made here is that sometimes the subtitle can be redundant depending upon the visual image pictorial cues (e.g. a pointing gesture) which relates to the next chapter.

Interpersonal communication is an area of longstanding interest to social psychologists and increasingly to other researchers. How one attends to hand gestures in face to face interaction is a fascinating topic and Gullberg details several types of interaction studies where one individual wears a head mounted eye tracker. Not too surprisingly perhaps, the face itself dominates as a target for visual attention, both in live interaction as well as where the other person is shown on a monitor. Gestures themselves appear to be attended to in either a bottom-up approach, when there is a momentary cessation in the gesture movement. In contrast, speaker-fixated gestures are only fixated by the participant after a delay and Gullberg proposes this represents a top-down attentional process. The issue of fixation locus and visual attention locus is considered. Data are mainly considered in terms of percentages and detailed statistical analysis is not presented. Overall, the work demonstrates the effects of contextual constraints on visual attention. It is argued that various processes, such as social interaction norms, mediate how someone responds to that which attracts their visual attention in interaction situations. Many interesting questions are thus raised. In this area of interpersonal interaction the real excitement will be in using dual remote systems where the scanning behaviour of both participants can be examined unencumbered.

Somewhat related to interpersonal interaction is the final chapter in this section which examines how a painter studies a model and then subsequently draws them. Tchalenko, Dempere-Marco, Hu and Yang report several studies using different eye movement equipment and painters of various art experience. The fundamental interest is in examining the painters' eye movements to understand the picture production process. How the artists looked both at the model and the paper were studied, as were the artists' eye–hand co-ordination. The data tends to be treated very descriptively with much interpretation and no statistical analysis or consideration of possible non-correspondence between eye gaze and visual attention locations. The work is fundamentally interesting, however, and clearly there is a need to extend good scientific rigour to such an area.

Overall these studies are very diverse and so demonstrate the breadth of research in this domain. Different recording techniques are employed, often monitoring different eye movement parameters coupled with varying data analysis and results are presented in different fashions. In conclusion, communication and media studies are an expanding research area and the next ECEM will clearly build upon these investigations.

Author Index

Subject Index